T0238798

Lecture Notes in Artificial Intelligence 9622

Subseries of Lecture Notes in Computer Science

More information about this series at http://www.springer.com/series/1244

Ngoc Thanh Nguyen · Bogdan Trawiński
Hamido Fujita · Tzung-Pei Hong (Eds.)

Intelligent Information and Database Systems

8th Asian Conference, ACIIDS 2016
Da Nang, Vietnam, March 14–16, 2016
Proceedings, Part II

Editors
Ngoc Thanh Nguyen
Wrocław University of Technology
Wrocław
Poland

Bogdan Trawiński
Wrocław University of Technology
Wrocław
Poland

Hamido Fujita
Iwate Prefectural University
Takizawa
Japan

Tzung-Pei Hong
National University of Kaohsiung
Kaohsiung
Taiwan

ISSN 0302-9743 ISSN 1611-3349 (electronic)
Lecture Notes in Artificial Intelligence
ISBN 978-3-662-49389-2 ISBN 978-3-662-49390-8 (eBook)
DOI 10.1007/978-3-662-49390-8

Library of Congress Control Number: 2016930675

LNCS Sublibrary: SL7 – Artificial Intelligence

Printed on acid-free paper

This Springer imprint is published by SpringerNature
The registered company is Springer-Verlag GmbH Berlin Heidelberg

Preface

ACIIDS 2016 was the eighth event of the series of international scientific conferences for research and applications in the field of intelligent information and database systems. The aim of ACIIDS 2016 was to provide an internationally respected forum for scientific research in the technologies and applications of intelligent information and database systems. ACIIDS 2016 was co-organized by the Vietnam–Korea Friendship Information Technology College (Vietnam) and Wrocław University of Technology (Poland) in co-operation with IEEE SMC Technical Committee on Computational Collective Intelligence, Bina Nusantara University (Indonesia), Ton Duc Thang University (Vietnam), and Quang Binh University (Vietnam). It took place in Da Nang (Vietnam) during March 14–16, 2016.

The ACIIDS conference series is well established. The first two events, ACIIDS 2009 and ACIIDS 2010, took place in Dong Hoi City and Hue City in Vietnam, respectively. The third event, ACIIDS 2011, took place in Daegu (Korea), while the fourth event, ACIIDS 2012, took place in Kaohsiung (Taiwan). The fifth event, ACIIDS 2013, was held in Kuala Lumpur in Malaysia, while the sixth event, ACIIDS 2014, was held in Bangkok, Thailand. The last event, ACIIDS 2015, took place in Bali (Indonesia).

We received papers from 36 countries all over the world. Each paper was peer reviewed by at least two members of the international Program Committee and international reviewer board. Only 153 papers with the highest quality were selected for oral presentation and publication in the two-volume proceedings of ACIIDS 2016.

Papers included in the proceedings cover the following topics: knowledge engineering and the Semantic Web, social networks and recommender systems, text processing and information retrieval, database systems and software engineering, intelligent information systems, decision support and control systems, machine learning and data mining, computer vision techniques, intelligent big data exploitation, cloud and network computing, multiple model approach to machine learning, advanced data mining techniques and applications, computational intelligence in data mining for complex problems, collective intelligence for service innovation, technology opportunity, e-learning and fuzzy intelligent systems, analysis of image, video, and motion data in life sciences, real-world applications in engineering and technology, ontology-based software development, intelligent and context systems, modeling and optimization techniques in information systems, database systems, and industrial systems, smart pattern processing for sports, and intelligent services for smart cities.

Accepted and presented papers highlight the new trends and challenges of intelligent information and database systems. The presenters showed how new research could lead to novel and innovative applications. We hope you will find these results useful and inspiring for your future research.

We would like to extend our heartfelt thanks to Mr. Jarosław Gowin, the Deputy Prime Minister of the Republic of Poland and Minister of Science and Higher Education for his support and honorary patronage over the conference.

We would like to express our sincere thanks to the honorary chairs, Mr. Minh Hong Nguyen (Deputy Minister of Information and Communications, Vietnam), and Prof. Tadeusz Więckowski (Rector of the Wrocław University of Technology, Poland) for their support.

Our special thanks go to the program chairs, special session chairs, organizing chairs, publicity chairs, liaison chairs, and local Organizing Committee for their work for the conference. We sincerely thank all members of the international Program Committee for their valuable efforts in the review process, which helped us to guarantee the highest quality of the selected papers for the conference. We cordially thank the organizers and chairs of special sessions, who essentially contributed to the success of the conference.

We also would like to express our thanks to the keynote speakers (Prof. Tzung-Pei Hong, Prof. Saeid Nahavandi, Prof. Jun Wang, and Prof. Piotr Wierzchoń) for their interesting and informative talks of world-class standard.

We cordially thank our main sponsors, Vietnam–Korea Friendship Information Technology College (Vietnam), Wrocław University of Technology (Poland), IEEE SMC Technical Committee on Computational Collective Intelligence, Bina Nusantara University (Indonesia), Ton Duc Thang University (Vietnam), and Quang Binh University (Vietnam). Our special thanks are due to Springer for publishing the proceedings, and all our other sponsors for their kind support.

We wish to thank the members of the Organizing Committee for their very substantial work and the members of the local Organizing Committee for their excellent work.

We cordially thank all the authors for their valuable contributions and other participants of this conference. The conference would not have been possible without their input.

Thanks are also due to many experts who contributed to making the event a success.

March 2016

Ngoc Thanh Nguyen
Bogdan Trawiński
Hamido Fujita
Tzung-Pei Hong

Conference Organization

Honorary Chairs

Minh Hong Nguyen	Deputy Minister of Information and Communications, Vietnam
Tadeusz Więckowski	Rector of Wrocław University of Technology, Poland

General Chairs

Ngoc Thanh Nguyen	Wrocław University of Technology, Poland
Bao Hung Hoang	Vietnam–Korea Friendship Information Technology College, Vietnam

Program Chairs

Bogdan Trawiński	Wrocław University of Technology, Poland
Tzung-Pei Hong	National University of Kaohsiung, Taiwan
Hamido Fujita	Iwate Prefectural University, Japan
Duc Dung Nguyen	IOIT – Vietnamese Academy of Science and Technology, Vietnam

Special Session Chairs

Dariusz Król	Wrocław University of Technology, Poland
Lech Madeyski	Wrocław University of Technology, Poland
Bay Vo	Ton Duc Thang University, Vietnam

Publicity Chairs

Khanh Van Hoang Thi	Vietnam–Korea Friendship Information Technology College, Vietnam
Adrianna Kozierkiewicz-Hetmańska	Wrocław University of Technology, Poland

Liaison Chairs

Ford Lumban Gaol	Bina Nusantara University, Indonesia
Tan Hanh	Posts and Telecommunications Institute of Technology, Vietnam

Mong-Fong Horng	National Kaohsiung University of Applied Sciences, Taiwan
Jason J. Jung	Chung-Ang University, Korea
Ali Selamat	Universiti Teknologi Malaysia, Malaysia

Organizing Chairs

The Son Tran	Vietnam–Korea Friendship Information Technology College, Vietnam
Elżbieta Kukla	Wrocław University of Technology, Poland

Local Organizing Committee

Marcin Maleszka	Wrocław University of Technology, Poland
Zbigniew Telec	Wrocław University of Technology, Poland
Bernadetta Maleszka	Wrocław University of Technology, Poland
Marcin Pietranik	Wrocław University of Technology, Poland
Nguyen Quang Vu	Vietnam–Korea Friendship Information Technology College, Vietnam
Ngo Viet Phuong	Vietnam–Korea Friendship Information Technology College, Vietnam
Le Tu Thanh	Vietnam–Korea Friendship Information Technology College, Vietnam

Webmaster

Jarosław Bernacki	Wrocław University of Technology, Poland

Steering Committee

Ngoc Thanh Nguyen (Chair)	Wrocław University of Technology, Poland
Longbing Cao	University of Technology Sydney, Australia
Suphamit Chittayasothorn	King Mongkut's Institute of Technology Ladkrabang, Thailand
Ford Lumban Gaol	Bina Nusantara University, Indonesia
Tu Bao Ho	Japan Advanced Institute of Science and Technology, Japan
Tzung-Pei Hong	National University of Kaohsiung, Taiwan
Dosam Hwang	Yeungnam University, Korea
Lakhmi C. Jain	University of South Australia, Australia
Geun-Sik Jo	Inha University, Korea
Jason J. Jung	Chung-Ang University, Korea
Hoai An Le-Thi	University of Lorraine, France

Toyoaki Nishida	Kyoto University, Japan
Leszek Rutkowski	Technical University of Czestochowa, Poland
Ali Selamat	Universiti Teknologi Malaysia, Malaysia

Keynote Speakers

Tzung-Pei Hong	National University of Kaohsiung, Taiwan
Saeid Nahavandi	Deakin University, Victoria, Australia
Jun Wang	City University of Hong Kong, Hong Kong, SAR China
Piotr Wierzchoń	Adam Mickiewicz University in Poznań, Poland

Special Sessions Organizers

1. Multiple Model Approach to Machine Learning (MMAML 20165)

Tomasz Kajdanowicz	Wrocław University of Technology, Poland
Edwin Lughofer	Johannes Kepler University Linz, Austria
Bogdan Trawiński	Wrocław University of Technology, Poland

2. Workshop on Real-World Applications in Engineering and Technology (RWAET 2016)

Pandian Vasant	Universiti Teknologi PETRONAS, Malaysia
Vish Kallimani	Universiti Teknologi PETRONAS, Malaysia
Sujan Chowdhury	Universiti Teknologi PETRONAS, Malaysia

3. Special Session on Intelligent Services for Smart Cities (IS4SC 2016)

David Camacho	Universidad Autónoma de Madrid, Spain
Jason J. Jung	Chung-Ang University, Korea
Paulo Novais	University of Minho, Portugal
Salvatore Venticinque	Seconda Università Degli Studi di Napoli, Italy

4. Special Session on Ontology-Based Software Development (OSD 2016)

Zbigniew Huzar	Wrocław University of Technology, Poland
Bogumiła Hnatkowska	Wrocław University of Technology, Poland

5. Special Session on Intelligent Big Data Exploitation (IBDE 2016)

Gottfried Vossen	University of Münster, Germany
Stuart Dillon	University of Waikato, New Zealand

6. Special Session on Intelligent and Context Systems (ICxS 2016)

Maciej Huk	Wrocław University of Technology, Poland
Jan Kwiatkowski	Wrocław University of Technology, Poland
Anita Pinheiro Sant'Anna	Halmstad University, Sweden

7. Special Session on Analysis of Image, Video, and Motion Data in Life Sciences (IVMLS 2016)

Kondrad Wojciechowski	Polish–Japanese Academy of Information Technology, Poland
Marek Kulbacki	Polish–Japanese Academy of Information Technology, Poland
Jakub Segen	Gest3D, USA
Andrzej Polański	Silesian University of Technology, Poland

8. Special Session on Collective Intelligence for Service Innovation, Technology Opportunity, E-Learning, and Fuzzy Intelligent Systems (CISTEF 2016)

Chao-Fu Hong	Aletheia University, Taiwan
Tzu-Fu Chiu	Aletheia University, Taiwan
Kuo-Sui Lin	Aletheia University, Taiwan

9. Special Session on Advanced Data Mining Techniques and Applications (ADMTA 2016)

Bay Vo	Ton Duc Thang University, Vietnam
Tzung-Pei Hong	National University of Kaohsiung, Taiwan
Bac Le	University of Science, VNU-HCM, Vietnam

10. Special Session on Modeling and Optimization Techniques in Information Systems, Database Systems, and Industrial Systems (MOT 2016)

Le Thi Hoai An	University of Lorraine, France
Pham Dinh Tao	National Institute for Applied Science – Rouen, France

11. Special Session on Computational Intelligence in Data Mining for Complex Problems (CIDMCP 2016)

Habiba Drias	University of Science and Technology USTHB Algiers, Algeria
Gabriella Pasi	University of Milano-Bicocca, Italy

12. Special Session on Smart Pattern Processing for Sports (SP2S 2016)

S.M.N. Arosha Senanayake	Universiti Brunei Darussalam, Brunei
Chu Kiong Loo	University of Malaya, Malaysia

International Program Committee

Ajith Abraham	Machine Intelligence Research Labs, USA
Muhammad Abulaish	Jamia Millia Islamia, India
El-Houssaine Aghezzaf	Ghent University, Belgium
Waseem Ahmad	International College of Auckland, New Zealand
Jesus Alcala-Fdez	University of Granada, Spain

Haider Alsabbagh	Basra University, Iraq
Ahmad Taher Azar	Benha University, Egypt
Le Hoai Bac	University of Science, VNU-HCM, Vietnam
Amelia Badica	University of Craiova, Romania
Costin Badica	University of Craiova, Romania
Emili Balaguer-Ballester	Bournemouth University, UK
Zbigniew Banaszak	Warsaw University of Technology, Poland
Dariusz Barbucha	Gdynia Maritime University, Poland
John Batubara	Bina Nusantara University, Indonesia
Ramazan Bayindir	Gazi University, Turkey
Maumita Bhattacharya	Charles Sturt University, Australia
Jacek Błażewicz	Poznań University of Technology, Poland
Veera Boonjing	King Mongkut's Institute of Technology Ladkrabang, Thailand
Mariusz Boryczka	University of Silesia, Poland
Urszula Boryczka	University of Silesia, Poland
Abdelhamid Bouchachia	Bournemouth University, UK
Zouhaier Brahmia	University of Sfax, Tunisia
Stephane Bressan	National University of Singapore, Singapore
Peter Brida	University of Zilina, Slovakia
Andrej Brodnik	University of Ljubljana, Slovenia
Grazyna Brzykcy	Poznań University of Technology, Poland
The Duy Bui	University of Engineering and Technology, VNU Hanoi, Vietnam
Robert Burduk	Wrocław University of Technology, Poland
David Camacho	Universidad Autonoma de Madrid, Spain
Frantisek Capkovic	Institute of Informatics, Slovak Academy of Sciences, Slovakia
Leopoldo Eduardo Cardenas-Barron	Tecnologico de Monterrey, Mexico
Oscar Castillo	Tijuana Institute of Technology, Mexico
Dariusz Ceglarek	Poznań School of Banking, Poland
Rituparna Chaki	University of Calcutta, India
Kit Yan Chan	Curtin University, Australia
Somchai Chatvichienchai	University of Nagasaki, Japan
Meng Chang Chen	Academia Sinica, Taiwan
Rung-Ching Chen	Chaoyang University of Technology, Taiwan
Shyi-Ming Chen	National Taiwan University of Science and Technology, Taiwan
Yi-Ping Phoebe Chen	La Trobe University, Australia
Yiu-Ming Cheung	Hong Kong Baptist University, Hong Kong, SAR China
Suphamit Chittayasothorn	King Mongkut's Institute of Technology Ladkrabang, Thailand
Tzu-Fu Chiu	Aletheia University, Taiwan
Sung-Bae Cho	Yonsei University, Korea

Kazimierz Choroś	Wrocław University of Technology, Poland
Kun-Ta Chuang	National Cheng Kung University, Taiwan
Robert Cierniak	Czestochowa University of Technology, Poland
Dorian Cojocaru	University of Craiova, Romania
Phan Cong-Vinh	NTT University, Vietnam
Jose Alfredo Ferreira Costa	UFRN – Federal University of Rio Grande do Norte, Brazil
Keeley Crockett	Manchester Metropolitan University, UK
Bogusław Cyganek	AGH University of Science and Technology, Poland
Ireneusz Czarnowski	Gdynia Maritime University, Poland
Piotr Czekalski	Silesian University of Technology, Poland
Paul Davidsson	Malmo University, Sweden
Roberto De Virgilio	Università degli Studi Roma Tre, Italy
Phuc Do	Vietnam National University, HCMC, Vietnam
Tien V. Do	Budapest University of Technology and Economics, Hungary
Grzegorz Dobrowolski	AGH University of Science and Technology, Poland
Rafał Doroz	University of Silesia, Poland
Habiba Drias	University of Science and Technology Houari Boumediene, Algeria
El-Sayed M. El-Alfy	King Fahd University of Petroleum and Minerals, Saudi Arabia
Irraivan Elamvazuthi	Universiti Teknologi PETRONAS, Malaysia
Rim Faiz	University of Carthage, Tunisia
Victor Falea	Alexandru Ioan Cuza University of Iasi, Romania
Thomas Fober	University of Marburg, Germany
Simon Fong	University of Macau, Macau
Dariusz Frejlichowski	West Pomeranian University of Technology, Poland
Hamido Fujita	Iwate Prefectural University, Japan
Mohamed Gaber	Robert Gordon University, UK
Patrick Gallinari	LIP6 – University of Paris 6, France
Junbin Gao	Charles Sturt University, Australia
Dariusz Gąsior	Wrocław University of Technology, Poland
Janusz Getta	University of Wollongong, Australia
Dejan Gjorgjevikj	Ss. Cyril and Methodius University in Skopje, Macedonia
Daniela Godoy	ISISTAN Research Institute, Argentina
Gergo Gombos	Eotvos Lorand University, Hungary
Fernando Gomide	University of Campinas, Brazil
Antonio Gonzalez-Pardo	Bilbao Center for Applied Mathematics, Spain
Janis Grundspenkis	Riga Technical University, Latvia
Claudio Gutierrez	Universidad de Chile, Chile
Sung Ho Ha	Kyungpook National University, Korea
Sajjad Haider	Institute of Business Administration, Karachi, Pakistan
Marcin Hajdul	Institute of Logistics and Warehousing, Poland
Pei-Yi Hao	National Kaohsiung University of Applied Sciences, Taiwan

Harisno	Bina Nusantara University, Indonesia
Tutut Herawan	University of Malaya, Malaysia
Marcin Hernes	Wrocław University of Economics, Poland
Bogumiła Hnatkowska	Wrocław University of Technology, Poland
Huu Hanh Hoang	Hue University, Vietnam
Jaakko Hollmen	Aalto University School of Science, Finland
Tzung-Pei Hong	National University of Kaohsiung, Taiwan
Mong-Fong Horng	National Kaohsiung University of Applied Sciences, Taiwan
Chia-Ling Hsu	Tamkang University, Taiwan
Jen-Wei Huang	National Cheng Kung University, Taiwan
Maciej Huk	Wrocław University of Technology, Poland
Zbigniew Huzar	Wrocław University of Technology, Poland
Dosam Hwang	Yeungnam University, Korea
Roliana Ibrahim	Universiti Teknologi Malaysia, Malaysia
Dmitry Ignatov	National Research University Higher School of Economics, Russia
Lazaros Iliadis	Democritus University of Thrace, Greece
Hazra Imran	Athabasca University, Canada
Agnieszka Indyka-Piasecka	Wrocław University of Technology, Poland
Mirjana Ivanovic	University of Novi Sad, Serbia
Sanjay Jain	National University of Singapore, Singapore
Chuleerat Jaruskulchai	Kasetsart University, Thailand
Khalid Jebari	LCS Rabat, Morocco
Joanna Jędrzejowicz	University of Gdańsk, Poland
Piotr Jędrzejowicz	Gdynia Maritime University, Poland
Janusz Jeżewski	Institute of Medical Technology and Equipment ITAM, Poland
Geun-Sik Jo	Inha University, Korea
Kang-Hyun Jo	University of Ulsan, Korea
Jason J. Jung	Chung-Ang University, Korea
Janusz Kacprzyk	Systems Research Institute, Polish Academy of Sciences, Poland
Tomasz Kajdanowicz	Wrocław University of Technology, Poland
Nadjet Kamel	Ferhat Abbas University of Setif, Algeria
Hung-Yu Kao	National Cheng Kung University, Taiwan
Mehmet Hakan Karaata	Kuwait University, Kuwait
Nikola Kasabov	Auckland University of Technology, New Zealand
Arkadiusz Kawa	Poznań University of Economics, Poland
Muhammad Khurram Khan	King Saud University, Saudi Arabia
Pan-Koo Kim	Chosun University, Korea
Yong Seog Kim	Utah State University, USA
Attila Kiss	Eotvos Lorand University, Hungary
Jerzy Klamka	Silesian University of Technology, Poland

Frank Klawonn	Ostfalia University of Applied Sciences, Germany
Goran Klepac	Raiffeisen Bank, Croatia
Joanna Kołodziej	Cracow University of Technology, Poland
Marek Kopel	Wrocław University of Technology, Poland
Józef Korbicz	University of Zielona Góra, Poland
Jacek Koronacki	Institute of Computer Science, Polish Academy of Sciences, Poland
Raymond Kosala	Bina Nusantara University, Indonesia
Leszek Koszałka	Wrocław University of Technology, Poland
Jan Kozak	University of Silesia, Poland
Adrianna Kozierkiewicz-Hetmańska	Wrocław University of Technology, Poland
Bartosz Krawczyk	Wrocław University of Technology, Poland
Ondrej Krejcar	University of Hradec Kralove, Czech Republic
Dalia Kriksciuniene	Vilnius University, Lithuania
Marzena Kryszkiewicz	Warsaw University of Technology, Poland
Adam Krzyzak	Concordia University, Canada
Elżbieta Kukla	Wrocław University of Technology, Poland
Marek Kulbacki	Polish–Japanese Academy of Information Technology, Poland
Marek Kurzyński	Wrocław University of Technology, Poland
Kazuhiro Kuwabara	Ritsumeikan University, Japan
Halina Kwaśnicka	Wrocław University of Technology, Poland
Mark Last	Ben-Gurion University of the Negev, Israel
Anabel Latham	Manchester Metropolitan University, UK
Hoai An Le Thi	Université de Lorraine, France
Chang-Hwan Lee	DongGuk University, Korea
Kun Chang Lee	Sungkyunkwan University, Korea
Yue-Shi Lee	Ming Chuan University, Taiwan
Philippe Lenca	Telecom Bretagne, France
Chunshien Li	National Central University, Taiwan
Jiuyong Li	University of South Australia, Australia
Rita Yi Man Li	Hong Kong Shue Yan University, Hong Kong, SAR China
Horst Lichter	RWTH Aachen University, Germany
Sebastian Link	University of Auckland, New Zealand
Igor Litvinchev	Nuevo Leon State University, Mexico
Lian Liu	University of Kentucky, USA
Rey-Long Liu	Tzu Chi University, Taiwan
Heitor Silverio Lopes	UTFPR, Federal University of Technology, Parana, Brazil
Edwin Lughofer	Johannes Kepler University Linz, Austria
Lech Madeyski	Wrocław University of Technology, Poland
Nezam Mahdavi-Amiri	Sharif University of Technology, Iran
Bernadetta Maleszka	Wrocław University of Technology, Poland
Marcin Maleszka	Wrocław University of Technology, Poland
Yannis Manolopoulos	Aristotle University of Thessaloniki, Greece

Konstantinos Margaritis	University of Macedonia, Greece
Francesco Masulli	University of Genoa, Italy
Mustafa Mat Deris	Universiti Tun Hussein Onn Malaysia, Malaysia
Tamas Matuszka	Eotvos Lorand University, Hungary
Joao Mendes-Moreira	University of Porto, Portugal
Gerardo Mendez	Instituto Tecnologico de Nuevo Leon, Mexico
Hector D. Menendez	University College London, UK
Jacek Mercik	Wrocław School of Banking, Poland
Radosław Michalski	Wrocław University of Technology, Poland
Peter Mikulecky	University of Hradec Kralove, Czech Republic
Marek Miłosz	Lublin University of Technology, Poland
Yang-Sae Moon	Kangwon National University, Korea
Leo Mrsic	IN2data Ltd Data Science Company, Croatia
Grzegorz J. Nalepa	AGH University of Science and Technology, Poland
Mahyuddin K.M. Nasution	Universitas Sumatera Utara, Indonesia
Prospero Naval	University of the Philippines, Philippines
Richi Nayak	Queensland University of Technology, Australia
Fulufhelo Nelwamondo	Council for Scientific and Industrial Research, South Africa
Dieu Ngoc Vo	Ho Chi Minh City University of Technology, Vietnam
Huu-Tuan Nguyen	Vietnam Maritime University, Vietnam
Loan T.T. Nguyen	Broadcasting College II, Vietnam
Thai-Nghe Nguyen	Can Tho University, Vietnam
Vinh Nguyen	University of Melbourne, Australia
Toyoaki Nishida	Kyoto University, Japan
Yusuke Nojima	Osaka Prefecture University, Japan
Mariusz Nowostawski	Norwegian University of Science and Technology, Norway
Alberto Nunez	Universidad Complutense de Madrid, Spain
Manuel Nunez	Universidad Complutense de Madrid, Spain
Richard Jayadi Oentaryo	Singapore Management University, Singapore
Kouzou Ohara	Aoyama Gakuin University, Japan
Tomasz Orczyk	University of Silesia, Poland
Shingo Otsuka	Kanagawa Institute of Technology, Japan
Marcin Paprzycki	Systems Research Institute, Polish Academy of Sciences, Poland
Jakub Peksiński	West Pomeranian University of Technology, Poland
Danilo Pelusi	University of Teramo, Italy
Xuan Hau Pham	Quang Binh University, Vietnam
Tao Pham Dinh	National Institute for Applied Sciences, France
Xuan-Hieu Phan	Vietnam National University, Hanoi, Vietnam
Maciej Piasecki	Wrocław University of Technology, Poland
Dariusz Pierzchała	Military University of Technology, Poland
Marcin Pietranik	Wrocław University of Technology, Poland
Elvira Popescu	University of Craiova, Romania

Piotr Porwik	University of Silesia, Poland
Bhanu Prasad	Florida A&M University, USA
Paulo Quaresma	Universidade de Evora, Portugal
R. Rajesh	Central University of Bihar, India
Mohammad Rashedur Rahman	North South University, Bangladesh
Ewa Ratajczak-Ropel	Gdynia Maritime University, Poland
Patricia Riddle	University of Auckland, New Zealand
Sebastian A. Rios	University of Chile, Chile
Miguel Angel Romero Orth	Universidad de Chile, Chile
Manuel Roveri	Politecnico di Milano, Italy
Przemysław Różewski	West Pomeranian University of Technology, Poland
Leszek Rutkowski	Czestochowa University of Technology, Poland
Alexander Ryjov	Lomonosov Moscow State University, Russia
Virgilijus Sakalauskas	Vilnius University, Lithuania
Daniel Sanchez	University of Granada, Spain
Cesar Sanin	University of Newcastle, Australia
Moamar Sayed-Mouchaweh	Ecole des Mines de Douai, France
Juergen Schmidhuber	Swiss AI Lab IDSIA, Switzerland
Bjorn Schuller	University of Passau, Germany
Jakub Segen	Gest3D, USA
Ali Selamat	Universiti Teknologi Malaysia, Malaysia
S.M.N. Arosha Senanayake	Universiti Brunei Darussalam, Brunei
Natalya Shakhovska	Lviv Polytechnic National University, Ukraine
Alexei Sharpanskykh	Delft University of Technology, The Netherlands
Andrzej Siemiński	Wrocław University of Technology, Poland
Guillermo R. Simari	Universidad Nacional del Sur, Argentina
Dragan Simic	University of Novi Sad, Serbia
Andrzej Skowron	Warsaw University, Poland
Adam Słowik	Koszalin University of Technology, Poland
Vladimir Sobeslav	University of Hradec Kralove, Czech Republic
Kulwadee Somboonviwat	University of Electro-Communications, Japan
Zenon A. Sosnowski	Białystok University of Technology, Poland
Jerzy Stefanowski	Poznań University of Technology, Poland
Serge Stinckwich	University of Caen-Lower Normandy, France
Stanimir Stoyanov	University of Plovdiv Paisii Hilendarski, Bulgaria
Suharjito	Bina Nusantara University, Indonesia
Jerzy Świątek	Wrocław University of Technology, Poland
Paweł Świątek	Wrocław University of Technology, Poland
Andrzej Świerniak	Silesian University of Technology, Poland
Edward Szczerbicki	University of Newcastle, Australia
Julian Szymański	Gdańsk University of Technology, Poland

Ryszard Tadeusiewicz	AGH University of Science and Technology, Poland
Yasufumi Takama	Tokyo Metropolitan University, Japan
Zbigniew Telec	Wrocław University of Technology, Poland
Krzysztof Tokarz	Silesian University of Technology, Poland
Behcet Ugur Toreyin	Cankaya University, Turkey
Bogdan Trawiński	Wrocław University of Technology, Poland
Krzysztof Trawiński	European Centre for Soft Computing, Spain
Hong-Linh Truong	Vienna University of Technology, Austria
Ualsher Tukeyev	Al-Farabi Kazakh National University, Kazakhstan
Olgierd Unold	Wrocław University of Technology, Poland
Pandian Vasant	Universiti Teknologi PETRONAS, Malaysia
Joost Vennekens	Katholieke Universiteit Leuven, Belgium
Jorgen Villadsen	Technical University of Denmark, Denmark
Maria Virvou	University of Piraeus, Greece
Bay Vo	Ton Duc Thang University, Vietnam
Gottfried Vossen	ERCIS Münster, Germany
M. Abdullah-Al Wadud	King Saud University, Saudi Arabia
Can Wang	CSIRO, Australia
Lipo Wang	Nanyang Technological University, Singapore
Yongkun Wang	University of Tokyo, Japan
Izabela Wierzbowska	Gdynia Maritime University, Poland
Konrad Wojciechowski	Silesian University of Technology, Poland
Michal Woźniak	Wrocław University of Technology, Poland
Krzysztof Wróbel	University of Silesia, Poland
Marian Wysocki	Rzeszow University of Technology, Poland
Guandong Xu	University of Technology Sydney, Australia
Xin-She Yang	Middlesex University, UK
Lina Yao	University of Adelaide, Australia
Shi-Jim Yen	National Dong Hwa University, Taiwan
Lean Yu	Chinese Academy of Sciences, AMSS, China
Sławomir Zadrożny	Systems Research Institute, Polish Academy of Sciences, Poland
Drago Zagar	University of Osijek, Croatia
Danuta Zakrzewska	Łódź University of Technology, Poland
Constantin-Bala Zamfirescu	Lucian Blaga University of Sibiu, Romania
Katerina Zdravkova	Ss. Cyril and Methodius University in Skopje, Macedonia
Vesna Zeljkovic	Lincoln University, USA
Aleksander Zgrzywa	Wrocław University of Technology, Poland
Zhongwei Zhang	University of Southern Queensland, Australia
Zhi-Hua Zhou	Nanjing University, China
Zhandos Zhumanov	Al-Farabi Kazakh National University, Kazakhstan
Maciej Zięba	Wrocław University of Technology, Poland

Program Committees of Special Sessions

Multiple Model Approach to Machine Learning (MMAML 2016)

Emili Balaguer-Ballester	Bournemouth University, UK
Urszula Boryczka	University of Silesia, Poland
Abdelhamid Bouchachia	Bournemouth University, UK
Robert Burduk	Wrocław University of Technology, Poland
Oscar Castillo	Tijuana Institute of Technology, Mexico
Dariusz Ceglarek	Poznań High School of Banking, Poland
Manuel Chica	European Centre for Soft Computing, Spain
Rung-Ching Chen	Chaoyang University of Technology, Taiwan
Suphamit Chittayasothorn	King Mongkut's Institute of Technology Ladkrabang, Thailand
José Alfredo F. Costa	Federal University (UFRN), Brazil
Bogusław Cyganek	AGH University of Science and Technology, Poland
Ireneusz Czarnowski	Gdynia Maritime University, Poland
Patrick Gallinari	Pierre et Marie Curie University, France
Fernando Gomide	State University of Campinas, Brazil
Tzung-Pei Hong	National University of Kaohsiung, Taiwan
Roliana Ibrahim	Universiti Teknologi Malaysia, Malaysia
Konrad Jackowski	Wrocław University of Technology, Poland
Piotr Jędrzejowicz	Gdynia Maritime University, Poland
Tomasz Kajdanowicz	Wrocław University of Technology, Poland
Yong Seog Kim	Utah State University, USA
Bartosz Krawczyk	Wrocław University of Technology, Poland
Elżbieta Kukla	Wrocław University of Technology, Poland
Kun Chang Lee	Sungkyunkwan University, Korea
Edwin Lughofer	Johannes Kepler University Linz, Austria
Hector Quintian	University of Salamanca, Spain
Andrzej Sieminski	Wrocław University of Technology, Poland
Dragan Simic	University of Novi Sad, Serbia
Adam Słowik	Koszalin University of Technology, Poland
Kulwadee Somboonviwat	University of Electro-Communications, Japan
Zbigniew Telec	Wrocław University of Technology, Poland
Bogdan Trawiński	Wrocław University of Technology, Poland
Krzysztof Trawiński	European Centre for Soft Computing, Spain
Olgierd Unold	Wrocław University of Technology, Poland
Pandian Vasant	University Technology Petronas, Malaysia
Michał Woźniak	Wrocław University of Technology, Poland
Zhongwei Zhang	University of Southern Queensland, Australia
Zhi-Hua Zhou	Nanjing University, China

Workshop on Real-World Applications in Engineering and Technology (RWAET 2016)

Vassili Kolokoltsov	University of Warwick, UK
Gerhard-Wilhelm Weber	METU, Turkey
Junzo Watada	Waseda University, Japan
Morteza Kalaji	Universiti Teknologi PETRONAS, Malaysia
Kwon-Hee Lee	Dong-A University, South Korea
Igor Litvinchev	Nuevo Leon State University, Mexico
Mohammad Abdullah-Al-Wadud	King Saud University, Saudi Arabia
Subhash Kamal	Universiti Teknologi PETRONAS, Malaysia
Vo Ngoc Dieu	HCMC University of Technology, Vietnam
Gerardo Maximiliano Mendez	Instituto Tecnologico de Nuevo Leon, Mexico
Leopoldo Eduardo Cárdenas Barrón	Tecnológico de Monterrey, Mexico
Hassan Soliemani	Universiti Teknologi PETRONAS, Malaysia
Denis Sidorov	Irkutsk State University, Russia
Weerakorn Ongsakul	Asian Institute of Technology, Thailand
Goran Klepac	Raiffeisen Bank Austria, Croatia
Jean Leveque	Universiti Teknologi PETRONAS, Malaysia
Herman Mawengkang	The University of Sumatera Utara, Indonesia
Igor Tyukhov	Moscow State University of Mechanical Engineering, Russia
Hayato Ohwada	Tokyo University of Science, Japan
Joga Setiawan	Universiti Teknologi PETRONAS, Malaysia
Ugo Fiore	Federico II University, Italy
Leo Mrsic	University College for Law and Finance Effectus Zagreb, Croatia
Nguyen Trung Thang	Ton Duc Thang University, Vietnam
Nikolai Voropai	Energy Systems Institute, Russia
Shifcraw Jufar	Univcrsiti Tcknologi PETRONAS, Malaysia
Xueguan Song	Dalian University of Technology, China
Ruhul A. Sarker	UNSW, Australia
Vipul Sharma	Lovely Professional University, India

Special Session on Intelligent Services for Smart Cities (IS4SC 2016)

Antonio Gonzalez	Basque Center for Applied Mathematics-TECNALIA, Spain
Jason J. Jung	Chung-Ang University, Korea
Pankoo Kim	Chosun University, Korea
Héctor D. Menéndez	University College London, UK
Xuan Hau Pham	Quang Binh University, Vietnam
Javier del Ser	Tecnalia, Spain

Special Session on Ontology-Based Software Development (OSD 2016)

Jose Maria Alvarez-Rodríguez	Universidad Carlos III de Madrid, Spain
Veera Boonjing	King Mongkut's Institute of Technology Ladkrabang, Bangkok Thailand
Somchai Chatvichienchai	University of Nagasaki, Japan
Rung-Ching Chen	Chaoyang University of Technology, Taiwan
Mustafa Mat Deris	Universiti Tun Hussein Onn, Malaysia
Iwona Dubielewicz	Wrocław University of Technology, Poland
Mirjana Ivanovic	University of Novi Sad, Serbia
Mark Last	Ben-Gurion University of the Negev, Israel
Adam Pease	R&D Manager at IPsoft, Hong Kong Polytechnic University, Hong Kong, SAR China
Sławomir Zadrożny	Systems Research Institute, Polish Academy of Sciences, Poland

Special Session on Intelligent Big Data Exploitation (IBDE 2016)

Alfredo Cuzzocrea	University of Trieste, Italy
Ernesto Damiani	University of Milan, Italy
Stuart Dillon	University of Waikato, New Zealand
Stefan Gruner	University of Pretoria, South Africa
Birgit Hofreiter	Technical University of Vienna, Austria
Christopher Holland	Manchester Business School, UK
Alexander Löser	Beuth University of Applied Sciences, Berlin, Germany
Ute Masermann	Decadis AG, Koblenz, Germany
Tadeusz Morzy	Technical University of Poznań, Poland
Florian Stahl	ERCIS, University of Münster, Germany
Heike Trautmann	ERCIS, University of Münster, Germany
Gottfried Vossen	ERCIS, University of Münster, Germany

Special Session on Intelligent and Context Systems (ICxS 2016)

Qiangfu Zhao	University of Aizu, Japan
Goutam Chakraborty	Iwate Prefectural University, Japan
Anita Sant'Anna	Halmstad University, Sweden
Michael Spratling	University of London, UK
Anna Fabijańska	Łódź University of Technology, Poland
Józef Korbicz	University of Zielona Góra, Poland
Jerzy Świątek	Wrocław University of Technology, Poland
Maciej Piasecki	Wrocław University of Technology, Poland
Michał Kędziora	Wrocław University of Technology, Poland
Nguyen Thanh Binh	Ho Chi Minh City University of Technology, Vietnam
Quan Thanh Tho	Ho Chi Minh City University of Technology, Vietnam
Ha Manh Tran	Ho Chi Minh City International University, Vietnam
Nguyen Khang Pham	Can Tho University, Vietnam

Nguyen Thai-Nghe	Can Tho University, Vietnam
Pedro Medeiros	University Nova of Lisbon, Portugal
Jan Kwiatkowski	Wrocław University of Technology, Poland
Maciej Huk	Wrocław University of Technology, Poland
Emilio Luque	University Autonoma of Barcelona, Spain
Dolores Rexachs	University Autonoma of Barcelona, Spain
Philip Moore	Lanzhou University, China
Norbert Jankowski	Nicholas Copernicus University, Poland
Bartlett W. Mel	University of Southern California, USA
Gregory Hager	Johns Hopkins University, USA
Shimon Ullman	Weizmann Institute of Science, Israel
Santosh S. Venkatesh	University of Pennsylvania, USA
Garrison W. Cottrell	University of California, USA
Xiao-Ping Zhang	Ryerson University, Canada
Grażyna Suchacka	Opole University, Poland
Ryszard Tadeusiewicz	AGH, Poland
William Dally	Stanford University, UK
Marek Wróblewski	University of Stuttgart, Germany
Wen Gao	Peking University, China
Elan Barenholtz	Florida Atlantic University, USA

Special Session on Analysis of Image, Video, and Motion Data in Life Sciences (IVMLS 2016)

Artur Bąk	Polish–Japanese Academy of Information Technology, Poland
Leszek Chmielewski	Warsaw University of Life Sciences, Poland
Aldona Barbara Drabik	Polish–Japanese Academy of Information Technology, Poland
Marcin Fojcik	Sogn og Fjordane University College, Norway
Adam Gudyś	Silesian University of Technology, Poland
Celina Imielińska	Vesalius Technologies LLC, USA
Henryk Josiński	Silesian University of Technology, Poland
Ryszard Klempous	Wrocław University of Technology, Poland
Ryszard Kozera	The University of Life Sciences – SGGW, Poland
Julita Kulbacka	Wrocław Medical University, Poland
Marek Kulbacki	Polish–Japanese Academy of Information Technology, Poland
Aleksander Nawrat	Silesian University of Technology, Poland
Jerzy Paweł Nowacki	Polish–Japanese Academy of Information Technology, Poland
Eric Petajan	LiveClips LLC, USA
Andrzej Polański	Silesian University of Technology, Poland
Joanna Rossowska	Polish Academy of Sciences, Institute of Immunology and Experimental Therapy, Poland
Jakub Segen	Gest3D LLC, USA

Aleksander Sieroń	Medical University of Silesia, Poland
Michał Staniszewski	Polish–Japanese Academy of Information Technology, Poland
Adam Świtoński	Silesian University of Technology, Poland
Agnieszka Szczęsna	Silesian University of Technology, Poland
Kamil Wereszczyński	Polish–Japanese Academy of Information Technology, Poland
Konrad Wojciechowski	Polish–Japanese Academy of Information Technology, Poland
Sławomir Wojciechowski	Polish–Japanese Academy of Information Technology, Poland

Special Session on Collective Intelligence for Service Innovation, Technology Opportunity, E-Learning, and Fuzzy Intelligent Systems (CISTEF 2016)

Chang, Ya-Fung	Tamkang University, Taiwan
Chen, Chi-Min	Aletheia University, Taiwan
Chiu, Chih-Chung	Aletheia University, Taiwan
Chiu, Kuan-Shiu	Aletheia University, Taiwan
Chiu, Tzu-Fu	Aletheia University, Taiwan
Chou, Chen-Huei	College of Charleston, USA
Hsu, Chia-Ling	Tamkang University, Taiwan
Hsu, Fang-Cheng	Aletheia University, Taiwan
Lin, Kuo-Sui	Aletheia University, Taiwan
Lin, Min-Huei	Aletheia University, Taiwan
Lin, Yuh-Chang	Aletheia University, Taiwan
Maeno, Yoshiharu	NEC Corporation, Japan
Sun, Pen-Choug	Aletheia University, Taiwan
Wang, Henry	Chinese Academy of Sciences, China
Wang, Leuo-Hong	Aletheia University, Taiwan
Yang, Feng-Sueng	Aletheia University, Taiwan
Yang, Hsiao-Fang	National Chengchi University, Taiwan
Yang, Ming-Chien	Aletheia University, Taiwan

Special Session on Advanced Data Mining Techniques and Applications (ADMTA 2016)

Bay Vo	Ton Duc Thang University, Vietnam
Tzung-Pei Hong	National University of Kaohsiung, Taiwan
Bac Le	University of Science, VNU-HCM, Vietnam
Chun-Hao Chen	Tamkang University, Taiwan
Chun-Wei Lin	Harbin Institute of Technology Shenzhen Graduate School, China
Wen-Yang Lin	National University of Kaohsiung, Taiwan
Yeong-Chyi Lee	Cheng Shiu University, Taiwan
Le Hoang Son	University of Science, Ha Noi, Vietnam
Le Hoang Thai	University of Science, Ho Chi Minh City, Vietnam

Vo Thi Ngoc Chau	Ho Chi Minh City University of Technology, Ho Chi Minh City, Vietnam
Van Vo	Ho Chi Minh University of Industry, Ho Chi Minh City, Vietnam

Special Session on Modeling and Optimization Techniques in Information Systems, Database Systems, and Industrial Systems (MOT 2016)

Le Thi Hoai An	University of Lorraine, France
Pham Dinh Tao	INSA-Rouen, France
Pham Duc Truong	University of Cardiff, UK
El-Houssaine Aghezzaf	University of Gent, Belgium
Azeddine Beghadi	University of Paris 13, France
Raymond Bisdorff	Université du Luxembourg
Jin-Kao Hao	University of Angers, France
Van-Dat Cung	INPG, France
Joaquim Judice	University Coimbra, Portugal
Amédéo Napoli	LORIA, France
Yann Germeur	LORIA, France
Conan-Guez Brieu	University of Lorraine, France
Gely Alain	University of Lorraine, France
Le Hoai Minh	University of Luxembourg, Luxembourg
Vo Xuan Thanh	University of Lorraine, France
Do Thanh Nghi	Can Tho University, Vietnam
Ibrahima Sakho	University of Lorraine, France

Special Session on Computational Intelligence in Data Mining for Complex Problems (CIDMCP 2016)

Sid Ahmed Benraouane	University of Minnesota, USA
Maria Gini	University of Minnesota, USA
Imed Kacem	University of Lorraine, France
Nadjet Kamel	University of Sétif, Algeria
Mehmed M. Kantardzic	University of Louisville, USA
Saroj Kaushik	Indian Institute of Technology Delhi, India
Samir Kechid	USTHB, Algeria
In-Young Ko	KAIST, Korea
Qin Lv	University of Colorado at Boulder, USA
Ana Maria Madureira	Polytechnic of Porto, Portugal
Brahim Medjahed	University of Michigan, USA
Madjid Merabti	Liverpool John Moores University, UK
Farid Meziane	University of Salford, UK
Erich J. Neuhold	Vienna University, Austria
Mourad Oussalah	Nantes University, France
Myeong Cheol Park	KAIST, Korea
Nelishia Pillay	University of KwaZulu-Natal, South Africa
Kalai Anand Ratnam	APUTI, Malaysia

Jae Jeung Rho	KAIST, Korea
Houari Sahraoui	University of Montreal, Canada
Lakhdar Sais	University of Artois, France
Thouraya Tebibel	ESI, Algeria
Farouk Yalaoui	Troyes University of Technology, France
Ning Zhong Maebashi	Institute of Technology, Japan

Special Session on Smart Pattern Processing for Sports (SP2S 2016)

Minoru Sasaki	Gifu University, Japan
Michael Yu Wang	National University of Singapore, Singapore
Eran Edirisinghe	Loughborough University, UK
Ajith Abraham	MIR Labs, USA
James F. Peters	University of Manitoba, Canada
Sergio Velastin	Kingston University, UK
Tadashi Ishihara	Fukushima University, Japan
Darwin Gouwanda	Monash University, Malaysia
Aaron Leung	The Hong Kong Polytechnic University, Hong Kong, SAR China
Toshiyo Tamura	Osaka Electro-Communication University, Japan
T. Nandha Kumaar	University of Nottingham, Malaysia
James Goh Cho Hong	National University Singapore, Singapore
Tsuyoshi Takagi	Kyushu University, Japan
William C. Rose	University of Delaware, USA

Contents – Part II

Advanced Data Mining Techniques and Applications

Real World Applications in Engineering and Technology

Ontology-Based Software Development

Contents – Part I

Text Processing and Information Retrieval

Database Systems and Software Engineering

Intelligent Information Systems

Machine Learning and Data Mining

Computer Vision Techniques

Intelligent Big Data Exploitation

Redhyte: Towards a Self-diagnosing, Self-correcting, and Helpful Analytic Platform

Wei Zhong Toh[1,2], Kwok Pui Choi[1], and Limsoon Wong[1](✉)

[1] National University of Singapore, 13 Computing Drive, Singapore 117417, Singapore
tohweizhong@u.nus.edu, stackp@nus.edu.sg, wongls@comp.nus.edu.sg
[2] NCS Pte Ltd, 5 Ang Mo Kio Street 62, Singapore 569141, Singapore

Abstract. We present a platform named Redhyte, short for an interactive platform for "Rapid exploration of data and hypothesis testing". Redhyte aims to augment the conventional statistical hypothesis testing framework with data-mining techniques in a bid for more wholesome and efficient hypothesis testing. The platform is self-diagnosing (it can detect whether the user is doing a valid statistical test), self-correcting (it can propose and make corrections to the user's statistical test), and helpful (it can search for promising or interesting hypotheses related to the initial user-specified hypothesis). In Redhyte, hypothesis mining consists of several steps: context mining, mined-hypothesis formulation, mined-hypothesis scoring on interestingness, and statistical adjustments. To capture and evaluate specific aspects of interestingness, we developed and implemented various hypothesis-mining metrics. Redhyte is an R shiny web application and can be found online at https://tohweizhong.shinyapps.io/redhyte, and the source codes are housed in a GitHub repository at https://github.com/tohweizhong/redhyte.

Keywords: Statistical hypothesis testing · Hypothesis analysis · Hypothesis mining · Data mining

1 Introduction

Much data is collected today for a variety of initial purposes. In the hands of a careful professionally-trained statistician or analyst who has a deep knowledge of the problem domain, many insights can be reliably gained from such data. However, it is often the case that an analysis project has to be carried out by someone who may lack domain knowledge or lack training, and sometimes even a professional analyst may be overwhelmed (e.g., by the volume and complexity of the data or the pressure of time) and may make mistakes [8]. A self-diagnosing, self-correcting and helpful analytic system is envisioned here to make analysis of data not only easy but also rigorous in such situations.

Self-diagnosing. All statistical tests have assumptions (e.g., observations are independent and identically distributed, observations are normally distributed)

N.T. Nguyen et al. (Eds.): ACIIDS 2016, Part II, LNAI 9622, pp. 3–12, 2016.
DOI: 10.1007/978-3-662-49390-8_1

and their conclusions are correct only when those assumptions are met. In traditional studies (e.g., a case-cohort study), subjects are carefully selected and experiments are designed so that such assumptions are met. But in today's big-data setting, we often just assemble and pull in all relevant data we could get our hands on, and there would typically be no careful selection to ensure such assumptions are met. A self-diagnosing analytic system, while making it convenient for a user to express and do a statistical test, also checks whether the test the user is doing is valid on his data. The challenging research questions include: (i) There may be no known way to check some assumptions, and thus deep statistical research is needed to figure out how to check them; (ii) the only known ways to check some assumptions are computationally costly, and thus deep algorithmic research is needed to figure out how to make these checks computationally more feasible; and (iii) how to explain to the user in a way that he can understand exactly why his requested statistical test is invalid.

Self-correcting. It is not sufficient to simply tell the user that he is performing a statistical test that is invalid on his data. The user may not know what action to take to deal with it. A self-correcting analytic system goes one step further, and tells the user how to deal with this. The challenges include: (i) How to identify alternative tests or correction steps, (ii) how to decide which alternative or correction is the most suitable one, (iii) how to explain such correction steps to the user in a way he could understand, and (iv) how to make it convenient for the user to choose and execute these corrections. Moreover, (v) for some situations, there is no known way to work around the problem, and novel idea is needed to develop the alternatives that can work in such situations.

Helpful. Beyond self-diagnosing and self-correcting, a good analytic system should also be helpful in the following sense. Initially, the user specifies a hypothesis that he wants to test. Given this initial hypothesis, the system now has some idea about what the user is interested in, and it should suggest some useful related hypotheses to the user that may give him some deeper insight into his problem. The challenges include: (i) How to identify related hypothesis, (ii) how to rank them, and (iii) how best to communicate them to the user.

In this manuscript, we describe Redhyte, which is an interactive platform for "Rapid exploration of data and hypothesis testing". Redhyte works by allowing the user to specify an initial hypothesis to be tested using one of the classical statistical tests (viz. t-test or χ^2 test), checks the validity of the test, makes corrections to the test if the initial test is detected to be invalid, as well as suggests informative related hypotheses. We believe Redhyte is a first, albeit small, step toward a self-diagnosing, self-correcting and helpful analytic platform.

The main part of this paper is organized as follows. Section 2 is a description of the Redhyte system, in particular its key functionalities. Section 3 is a case study to illustrate the features of Redhyte. The case study is based on the adult dataset from the UCI machine learning repository. Finally, Sect. 4 summarizes the work and discusses related works.

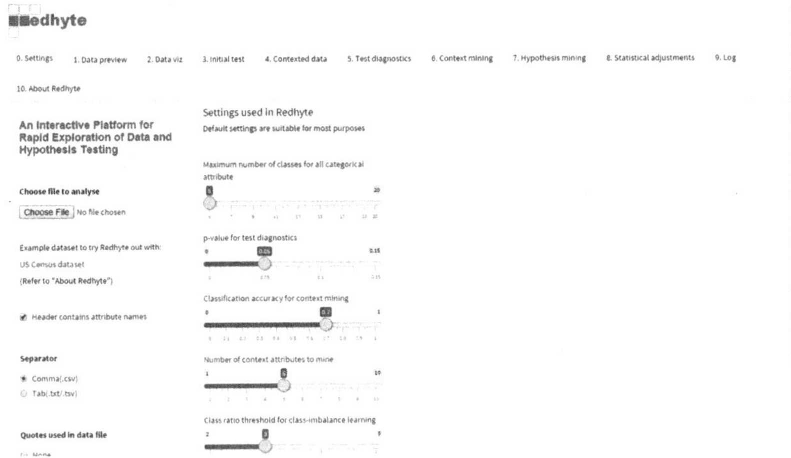

Fig. 1. A screenshot of Redhyte.

2 System Description

This section describes various modules that are more vital in Redhyte's workflow (and user-friendly interface); the less vital modules are omitted.

2.1 User Interface

Redhyte is fundamentally a web application that renders in a web browser, such as Google Chrome or Mozilla Firefox. Redhyte's user-facing interface is organised into tabs, as shown in Fig. 1, with each tab housing a specific functionality that Redhyte provides. The tabs are ordered from left to right, mirroring the expected workflow of an analysis: the user makes some brief checking and exploration of his input data (the data-preview and data-visualization modules), the user specifies his initial hypothesis (the initial-tests module), the user looks at the validity assessments of the test on his hypothesis (the test-diagnostic module), the user gets information on factors that could strengthen or contradict his hypothesis (the context-mining module), and the user looks at related hypothesis mined by Redhyte (the mined-hypothesis formulation and scoring module).

2.2 Initial Hypothesis Set-Up and Tests

After loading the input dataset into the platform, the user is prompted to set up the initial hypothesis. To establish a common lingo between the user and Redhyte regarding the initial hypothesis and all other steps in the Redhyte workflow, Redhyte defines the following terms in a hypothesis: "target attribute", "comparing attribute", and "context attribute". As an example, a hypothesis comparing resting heart rate between smokers and non-smokers amongst the males only, would have resting heart rate as the target attribute, smoking status as the comparing attribute, and gender as the context attribute. In addition, an {attribute = value} pair, e.g. {gender = male}, is called a context item. After the hypothesis is set up, depending on whether the target attribute is numerical or categorical, an initial t-test or χ^2 test is used to assess the hypothesis. Naturally, the user uses the Redhyte graphics interface to specify his hypothesis.

Following Liu et al. [11], we write $H = \langle P, A_{diff} = v_1|v_2, A_{target} = v_{target}\rangle$ to denote a hypothesis H. The set of items P is the context, which is the subset of the subjects in the dataset satisfying all items in P. The comparing attribute is A_{diff}, and $P_1 = P \cup \{A_{diff} = v_1\}$ and $P_2 = P \cup \{A_{diff} = v_2\}$ define the two subpopulations to be compared. The target attribute, on which the two subpopulations is being compared, is A_{target}.

2.3 Test Diagnostics

The test diagnostics tab aims to work on several issues:

- If the initial test is a t-test [7], the assumptions of normal distributions and equal variances are checked, using the Shapiro-Wilk test [16] and F-test [5] respectively. If either test is significant, the initial test is re-assessed using the Wilcoxon rank sum test [12], the non-parametric equivalent of the t-test.
- If the initial test was a "collapsed" χ^2 test [14], Redhyte computes the individual χ^2 contributions of each class in the comparing attribute. A collapsed χ^2 test refers to a χ^2 test where one or both groups in the initial hypothesis consist of more than one class of the comparing attribute.
- In both cases, the Cochran-Mantel-Haenszel test [3] is further used on other attributes in the dataset, to assess whether the effect of the comparing attributes on the target attribute is influenced by these co-variates.

2.4 Context Mining

In order to identify potential confounding attributes to the initial hypothesis, Redhyte uses classification models. Specifically, it constructs two random-forest [2] models to predict, using all other attributes not involved in the initial hypothesis as predictors or covariates, the target and comparing attributes. The idea is: If an attribute A is able to help classify the target or the comparing attribute, then A might possibly be related to either attribute. Thus it might be interesting to consider using A as a context attribute for the initial hypothesis.

The random-forest models confer a measure of variable importance, ranking the attributes according to how well they contribute to the classification of the target and comparing attribute. Shortlisting the top few attributes from the variable importance measure gives the mined context attributes.

2.5 Mined-Hypothesis Formulation and Scoring

After shortlisting the mined context attributes, each attribute is used as a context attribute and inserted into the initial hypothesis to form mined hypotheses, by means of stratification. For example, if occupation is a mined context attribute, then examples of mined hypotheses could be restricting the initial hypothesis to all teachers only or all engineers only.

Stratification due to a mined hypothesis can bring about one of three outcomes: The trend observed in the hypothesis could either be (i) amplified, (ii) unchanged, or (iii) reversed (Simpson's reversals). After forming the mined hypotheses, they are ranked using the various hypothesis-mining scores (difference lift, contribution, independence lift, and adjusted independence lift) to evaluate which of these three outcomes they fit. The "difference lift" and "contribution" scores are given by Liu et al. [11]. The former compares a hypothesis $H = \langle P, A_{diff} = v_1|v_2, A_{target} = v_{target}\rangle$ with a new hypothesis $H^* = \langle P \cup \{A = v\}, A_{diff} = v_1|v_2, A_{target} = v_{target}\rangle$, which has an extra item $\{A = v\}$ in its context, to see whether the trend specified in H has changed (amplification or reversal) substantially in H^*. So we also use the "independence lift" and "adjusted independence lift" scores, which are defined in the full Redhyte report [18]. These two scores always agree with difference lift in the direction of the change in trend while also take into consideration the sizes of the subpopulations being compared in H and H^*. Due to space contraint, definitions and detailed treatment [18] of these scores are omitted here.

2.6 Statistical Adjustments

Besides inserting mined context items into the initial hypothesis to detect issues like Simpson's reversals, these mined context attributes can also be held accounted for using regression models. The regression model is constructed using the target attribute as the response variable and the mined context attributes as predictors. We call this resultant model the adjustment model. For numerical target attributes, linear regression is used as the adjustment model [6]. For categorical target attributes, the logistic regression model is used instead [4].

To construct the adjustment model, Redhyte first uses the stepwise regression algorithm to further shortlist, from the mined context attributes, a subset of them to be used for adjustments in the adjustment model. Next, the construction of the adjustment model and its use are as follows:

- For numerical target attributes, the target attribute is used as the dependent variable and the shortlisted mined context attributes, with all pairwise interaction terms, are used as predictors/covariates. The constructed adjustment model gives the required numerical adjustments of the target attribute (computed as actual values found in dataset minus fitted values from model). A t-test is then done on the numerical adjustments, to compare with the initial t-test.
- For categorical target attributes, the target attribute is used as the dependent variable, while the shortlisted mined context attributes and the comparing attribute, with all pairwise interaction terms, are used as predictors/covariates. The constructed adjustment model lends itself a way to conduct "what-if" analysis (i.e., what if the entire dataset consists of samples that differ only in the target and the comparing attribute?) For instance, if

the mined context attributes are gender and occupation, we ask: What if the entire dataset consists of samples that are all males and all engineers? The logistic regression model provides a means for such an analysis, by "substituting" these covariate values into the model equation.

3 Use-Case

In this section, we use the adult dataset (from the UCI machine learning repository, http://archive.ics.uci.edu/ml) to illustrate a simple use-case for hypothesis mining by Redhyte. The adult dataset contains the demographical data of 32,561 adults. The target attribute in this dataset is the binary income attribute, taking these values: >50 K and ≤50 K. Consider the hypothesis below:

> In the context of {race = White}, is there a difference in INCOME between {>50 K} vs. {≤50 K} when comparing the samples on OCCUPATION between {Adm-clerical} and {Craft-repair}?

3.1 Initial Test

The initial test suggests that the relationship between income and occupation is significant ($p < 0.05$), with white administrative clerks earning more than white craft repairers, as shown in Fig. 2.

	Income >50K	Income ≤50K	Total
Adm-clerical	439 (14.2%)	2645 (85.8%)	3084
Craft-repair	844 (22.8%)	2850 (77.2%)	3694
Total	1283	5495	6778

Fig. 2. Contingency table of the initial hypothesis.

Using default settings, Redhyte identifies five mined context attributes after context mining, namely sex, relationship, workclass, education, and education.num. In particular, considering the context items {Sex = Male}, {Sex = Female} and {Workclass = Self-emp-not-inc} leaves us with the contingency tables in Fig. 3, which illustrate two instances of a Simpson's Paradox [17], with both genders and workclass resulting in reversals of the trend in Fig. 2.

3.2 Hypothesis-Mining Metrics

The hypothesis-mining metrics evaluated on the three items ({Sex = Male}, {Sex = Female}, and {Workclass = Self-emp-not-inc}) are given in Fig. 4.

Based on our initial hypothesis, the default settings in Redhyte is used to illustrate the above, and to generate 27 other mined hypotheses, suitably scored and ranked using the hypothesis-mining metrics, for the user to inspect.

{Sex = Male}	Income>50K	Income≤50K	Total
Adm-clerical	251 (24.2%)	787 (75.8%)	1038
Craft-repair	829 (23.5%)	2695 (76.5%)	3524
Total	1080	3482	4562

{Sex = Female}	Income>50K	Income≤50K	Total
Adm-clerical	188 (9.2%)	1858 (90.8%)	2046
Craft-repair	15 (8.8%)	155 (91.2%)	170
Total	203	2013	2216

{Workclass = Self-emp-not-inc}	Income>50K	Income≤50K	Total
Adm-clerical	16 (34.8%)	30 (65.2%)	46
Craft-repair	90 (18.0%)	409 (82.0%)	499
Total	106	439	545

Fig. 3. Contingency tables of mined hypothesis with {Sex = Female}, {Sex = Male}, and {Workclass = Self-emp-not-inc}.

Context items	Diff lift	Contrib	Indep lift	Adjusted indep lift	p-value
{Sex = Male}	−0.08	−0.31	−0.06	−0.02	0.69
{Sex = Female}	−0.04	0.31	−0.09	−0.05	0.98
{Workclass = Self-emp-not-inc}	−1.94	−0.11	−1.89	−0.05	0.01

Fig. 4. Hypothesis-mining metrics evaluated for the selected context items.

3.3 Statistical Adjustments

Following through with statistical adjustments in Redhyte, sex, relationship, workclass, and education are recommended for use in statistical adjustment. Since the target attribute, income, is a categorical one, the adjustment model is a logistic regression model, to be used for "what-if" analysis. In particular, as shown in Fig. 5, after adjusting for {sex = Male}, {relationship = Husband},

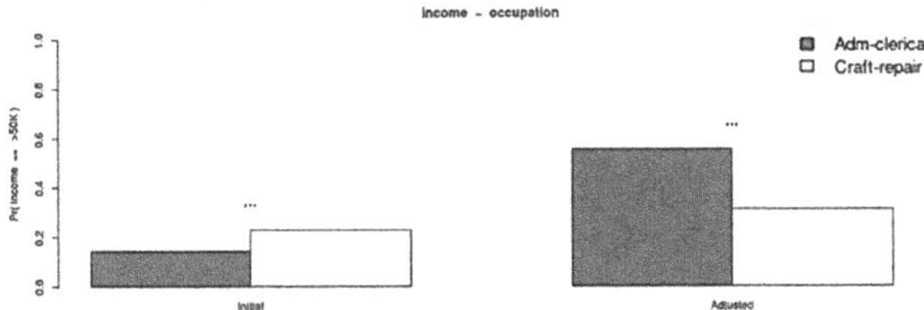

Fig. 5. A visualization show the proportions of administrative clerks and craft repairers earning more than 50 k, before (left chart) and after (right chart) adjusting for sex, relationship, worksclass, and education.

{workclass = Self-emp-not-inc} and {education = Bachelors}, a Simpson's reversal is observed in the hypothesis that administrative clerks earn more than craft repairers.

4 Conclusion

Hypothesis testing is one of the mainstay tools in data analysis, as it allows the analyst to make comparisons between different groups of samples. Conventionally, data-analysis workflows primarily consist of the following steps: (i) have a scientific question in mind, (ii) formulate an assertion or hypothesis, (iii) collect and clean relevant data, and finally (iv) test the hypothesis using statistical techniques, in order to decide whether to reject the hypothesis. Putting together a hypothesis with a statistical test allows the analyst to make justifiable conclusions from the data, and this process is often prompted by the initial question or hypothesis in mind. In other words, collection of data in conventional data analysis settings are often driven by domain requirements and scientific questions a priori.

From a statistical viewpoint, having some initial scientific questions to drive the collection of data confers an important upshot: The collected data is well specified. To be more precise, with proper sampling methods, issues such as lack of independence, dissimilar distributions, unequal variances, class imbalance etc. can be addressed and alleviated. However, the big-data setting brings about two interesting scenarios, specifically the collection of data without a scientific question a priori, and the "large p, small n" phenomenon [20].

Data collected without any initial scientific questions poses a problem: Assumptions of many statistical techniques, including hypothesis testing, are more often violated than not. Moreover, having a large number of attributes in a dataset requires adequate treatment and analysis to properly account for these attributes. Formulating a hypothesis concerning a small number of attributes and testing it in a large dataset while ignoring the other attributes is not only wasteful, but flawed (due to issues such as confounding factors). For example, given a hypothesis concerning two attributes, say A and B, for a certain class of a third categorical attribute C, the initial hypothesis could be amplified, i.e. the trend observed between A and B is strengthened when we consider the certain class of C. The trend could also be reversed; this is commonly known as Simpsons Reversal [13]. As in the use case presented earlier, while the initial observation suggests the hypothesis that craft repairers earn more than administrative clerks (cf. Fig. 2), after automatically detecting and adjusting for confounding factors by Redhyte, the completely opposite hypothesis that administrative clerks earn more than craft repairers emerges (cf. Fig. 5). A conventional, domain knowledge-driven approach of analysing data gives no simple or systematic way to reveal such phenomena, leaving discoveries of such to intuition and chance. An epitome of such a phenomenon is the UC Berkeley gender-bias case [1].

In this paper we have introduced Redhyte, a platform for statistical hypothesis testing on datasets collected without initial scientific questions. The workflow in Redhyte is as follows: (i) User first suggests an initial hypothesis, which could be rough, intuitive, and domain knowledge-driven. (ii) Redhyte first works on some initial statistical test on the initial hypothesis, and assesses the validity of the statistical test applied to the initial hypothesis. (iii) Redhyte uses data-mining techniques to search for potential confounding attributes (context mining), and uses them to form variants of the initial hypothesis, by means of stratification. These variants of the initial hypothesis are then scored and ranked, to let the user home in on the more interesting ones. (iv) Finally, Redhyte attempts to adjust for these potential confounding attributes using regression techniques.

It is easy to make mistakes involving statistics. Powerful statistical tools—R, Minitab, SPSS, etc.—certainly remove a lot of the difficulty of doing statistical calculations. However, they do not check whether the user is applying the statistical tests correctly. The special aspects of Redhyte compared to commonly-used statistical tools are that it explicitly supports checking whether what the user is doing is valid, guiding him to do his statistical test correctly, and recommending to him related hypotheses that might lead to some deeper insight. Redhyte is thus a step towards building a self-diagnosing, self-correcting, and helpful analytic system, albeit it current supports only very simple statistical tests.

On the data-mining side, the closest research related to Redhyte is perhaps the work of Liu et al. on exploratory hypothesis testing and analysis [9,11]. In these works, algorithms for mining and visualization of hypotheses from large datasets are described. More importantly, they also presented algorithms for linking together related hypotheses and measures for ranking hypotheses (and we have proposed refinements of these [18] and implemented them in Redhyte). Works from the data-mining field on the clustering and grouping of frequent itemsets and association rules [10,15,19,21] may also be useful for generating related hypotheses (we do not consider these clustering methods here because we start from a single user-specified hypothesis, and so face a much lower mining and clustering complexity than these methods). Moreover, these methods are not concerned with ensuring the validity of a user's statistical test or guiding him toward a valid statistical test.

Acknowledgements. This work was supported in part by a Singapore Ministry of Education tier-2 grant (MOE2012-T2-1-061) and by NCS Pte Ltd, Singapore.

References

1. Bickel, P., Hammel, E., O'connell, J.: Sex bias in graduate admissions: data from Berkeley. Sci. **187**, 398–404 (1975)
2. Breiman, L.: Random forests. Mach. Learn. **45**, 5–32 (2001)
3. Cochran, W.G.: Some methods for strengthening the common χ^2 tests. Biometrics **10**, 417–451 (1954)
4. Cox, D.R.: The regression analysis of binary sequences (with discussion). J. R. Stat. Soc. B **20**, 215–242 (1958)
5. Fisher, R.A.: On a distribution yielding the error functions of several well-known statistics. Proc. Int. Congr. Math. **2**, 805–813 (1924)

6. Freedman, D.A.: Statistical Models: Theory and Practice. Cambridge University Press, Cambridge (2009)
7. Gosset, W.S.: The probable error of a mean. Biometrika **6**, 1–25 (1908)
8. Ioannidis, J.P.A.: Why most published research findings are false. PLoS Med. **2**, e124 (2005)
9. Liu, G., Suchitra, A., Zhang, H., Feng, M., Ng, S.K., Wong, L.: AssocExplorer: an association rule visualization system for exploratory data analysis. In: Proceedings of 18th ACM SIGKDD Conference on Knowledge Discovery and Data Mining, pp. 1536–1539 (2012)
10. Liu, G., Zhang, H., Wong, L.: A flexible approach to finding representative pattern sets. IEEE Trans. Knowl. Data Eng. **26**, 1562–1574 (2014)
11. Liu, G., Zhang, H., Feng, M., Wong, L., Ng, S.K.: Supporting exploratory hypothesis testing and analysis. ACM Trans. Knowl. Discov. Data **9**, Article 31 (2015)
12. Mann, H.B., Whitney, D.R.: On a test of whether one of two random variables is stochastically larger than the other. Ann. Math. Stat. **18**, 50–60 (1947)
13. Pavlides, M., Perlman, M.: How likely is Simpson's paradox? Am. Stat. **63**, 226–233 (2009)
14. Pearson, K.: On the criterion that a given system of deviations from the probable in the case of a correlated system of variables is such that it can be reasonably supposed to have arisen from random sampling. Philos. Mag. Ser. **5**(50), 157–175 (1900)
15. Poernomo, A.K., Gopalkrishnan, V.: CP-summary: a concise representation for browsing frequent itemsets. In: Proceedings of 12th ACM SIGKDD International Conference on Knowlegde Discovery and Data Mining, pp. 687–696 (2009)
16. Shapiro, S.S., Wilk, M.B.: An analysis of variance test for normality (complete samples). Biometrika **52**, 591–611 (1965)
17. Simpson, E.H.: The interpretation of interaction in contingency tables. J. R. Stat. Soc. B **13**, 238–241 (1951)
18. Toh, W.Z.: Redhyte: an interactuve platform for rapid exploration of data and hypothesis testing. Project report, National University of Singapore (2015). http://www.comp.nus.edu.sg/wongls/psZ/tohweizhong-fyp2015.pdf
19. Wang, C., Parthasarathy, S.: Summarizing itemset patterns using probabilistic models. In: Proceedings of 12th ACM SIGKDD International Conference on Knowledge Discovery and Data Mining, pp. 730–735 (2006)
20. West, M.: Bayesian factor regression models in the "large p, small n" paradigm. Bayesian Stat. **7**, 723–732 (2003)
21. Yan, X., Cheng, H., Han, J., Xin, D.: Summarizing itemset patterns: a profile-based approach. In: Proceedings of 11th ACM SIGKDD International Conference on Knowledge Discovery and Data Mining, pp. 314–323 (2005)

Learning to Filter User Explicit Intents in Online Vietnamese Social Media Texts

Thai-Le Luong[1,2](✉), Thi-Hanh Tran[2], Quoc-Tuan Truong[2], Thi-Minh-Ngoc Truong[2], Thi-Thu Phi[2], and Xuan-Hieu Phan[2]

[1] University of Transport and Communications, Hanoi, Vietnam
luongthaile80@utc.edu.vn

[2] University of Engineering and Technology, Vietnam National University, Hanoi, Vietnam
{hanhtt.mi13,tuantq_57,ngocttm.mi13,thupt_570,hieupx}@vnu.edu.vn

Abstract. Today, Internet users are much more willing to express themselves on online social media channels. They commonly share their daily activities, their thoughts or feelings, and even their intention (e.g., *buy a camera*, *rent an apartment*, *borrow a loan*, etc.) about what they plan to do on blogs, forums, and especially online social networks. Understanding intents of online users, therefore, has become a crucial need for many enterprises operating in different business areas like production, banking, retail, e–commerce, and online advertising. In this paper, we will present a machine learning approach to analyze users' posts and comments on online social media to filter posts or comments containing user plans or intents. Fully understanding user intent in social media texts is a complicated process including three major stages: *user intent filtering*, *intent domain identification*, and *intent parsing and extraction*. In the scope of this study, we will propose a solution to the first one, that is, building a binary classification model to determine whether a post or comment carries an intent or not. We carefully conducted an empirical evaluation for our model on a medium–sized collection of posts in Vietnamese and achieved promising results with an average accuracy of more than 90 %.

Keywords: Intention mining · User intent identification · Social media text understanding · Content filtering · Text classification

1 Introduction

The past decade has seen an explosive growth of online social media services. In this highly interactive ecosystem, users become the key players[1] who incessantly contribute and enrich the social media channels via their online activities and behaviors. In this cyberspace, people tend to express themselves and are willing to share their daily activities, their thoughts and feelings, and even their intents about anything they would do. As a result, user posts and comments on online

[1] *Time* Person of the Year (2006): You (i.e., the Internet users).

N.T. Nguyen et al. (Eds.): ACIIDS 2016, Part II, LNAI 9622, pp. 13–24, 2016.
DOI: 10.1007/978-3-662-49390-8_2

forums and social networks can actually reflect a lot about the public opinion and people's intention. Analyzing those posts and comments, therefore, becomes an effective approach for enterprises and businesses to understand what their potential customers really care and want, helping them to have a better online marketing plan and finally penetrate the market faster and more efficiently.

Being aware of this important trend, many previous researches focused on the understanding of user intents behind their online activities like web search [1,8,10,12,13,18] or computer/mobile interactions [5,6]. Most of these studies attempted to guess or determine the user *implicit* intents behind their search queries and browsing behaviors. Understanding search intent helps improving the quality of web search significantly. *Explicit* intent, on the other hand, is a directly or explicitly written statement by a user about what he or she plans to do. According to Bratman (1987), intent or intention is a mental state that represents a commitment to carrying out an action or actions in the future [3]. As more and more users are willing to share their intents explicitly on the web, we have an opportunity to access to an invaluable source of knowledge about a huge number of online users or probably potential customers. However, there have been few previous studies really focusing on analyzing and identifying user *explicit* intents from their posts or comments on forums or social networks. This is explainable. In spite of its huge potential for application, the identification of user explicit intents is actually a natural language understanding problem which is inherently a hard research direction in natural language processing.

It, however, does not mean that this problem is unsolvable. In this paper, we will present a definition of user explicit intents in the form of a quintuple (5–tuple) and propose a three–stage process for understanding or identifying them from user posts or comments on online forums or social networks. This process consists of three major stages: (1) the filtering phase that will determine which posts/comments hold an explicit intent; (2) the domain identification phase that helps to recognize what an intent is about (e.g., *finance*, *real estate*, *tourism*, *automobile*, etc.); and (3) the intent parsing and extraction that helps to acquire all intent's information. In this process, the first and the second phases can be seen as classification problems. The last one is actually an information extraction task that extracts the intent's properties or constraints. As a user intent can be about anything in any domain, it is hard to pre–define a fixed set of domains and a fixed set of intent properties. As a result, understanding user explicit intent in open domain is extremely challenging. We, therefore, cannot solve the whole problem at once. The process should be broken down into sub–problems with feasible solutions. In this work, we will propose a machine learning approach to the first phase, that is, building a classifier to filter user posts or comments from social media to determine which ones actually carry a user explicit intent. All in all, our work has the following contributions:

- We propose a definition of user explicit intent ($\mathbf{I}_u^e$) that consists of five elements. The detailed explanation is given in Sect. 3.1.
- We also propose a three–stage process or roadmap for full understanding of user explicit intents. The description and explanation are in Sect. 3.2.

- We attempted to solve the first problem, intent filtering for user text posts or comments, with maximum entropy classification. We also built a medium-sized data set of text posts in Vietnamese collected from online forums and social networks for evaluation and achieved promising results.

The remainder of the paper is organized as follows. Section 2 reviews related work. Section 3 describes the process of user intent identification from online social media texts. Section 4 presents our main study: building a classifier to filter text posts or comments carrying a user intent. Experimental results and analysis are reported in Sect. 5. Finally, conclusions are given in Sect. 6.

2 Related Work

User intent understanding can be defined in different ways for different application domains. In this section, we will review several studies on understanding user goals or intents that are more or less related to our work.

A major number of previous studies working on the problem of identifying user goals or intents behind their web search activities. Lee et al. (2005) proposed the use of features like user–click behavior and anchor–link distribution to identify user goals in web search. They classified user goals into two classes: navigational and informational [12]. Ashkan et al. (2009) proposed a method for understanding user intents underlying their search queries [1]. Their method used ad click–through logs and query specific information to determine whether a query carries a commercial intent. Hu et al. (2009) proposed the use of Wikipedia concepts for identifying intent behind user's queries [8]. Li (2010) proposed a machine learning approach for understanding user query intent by recognizing intent heads and intent modifiers using Markov and semi–Markov conditional random fields (CRFs) [13]. Jethava et al. (2011) used tree structure distribution to determine different dimensions or facets or user intents behind their search queries [10]. Shen et al. (2011) proposed sparse hidden dynamic conditional random fields to model user intents from their search sessions. This method can model the dynamics between intent labels and user behavior variables [18]. The user intents behind their search queries can also be classified into *commercial* and *non–commercial*. Hu et al. (2009) proposed the use of skip–chain CRFs to determine a query is commercial or not [9]. Dai et al. (2006) also proposed the use of machine learning to identify online commercial intention [7].

Some other researches model the intent behind user actions on their computers or mobile devices. Chen et al. (2002) used Naive Bayes classifier to model user's action intention on a computer. This simply recognize five types of action: browse, click, query, save, and close [5]. Church and Smyth (2009) focused on studying the information need of mobile users. They studied what mobile users need when the context changes like at home, at work, or on–the–go [6].

Among the previous studies, the following are more relevant to our work. Chen (2014) [4] attempted to understand the user intent behind their questions posted on community question answering sites. They classified the question intent into five categories: subjectivity, locality, navigationality, procedurality, and causality. This helps users understand others' questions better and give more relevant answers. Kroll and Strohmaier (2009) [11] determined the user intents/goals in text documents. They constructed and enriched a taxonomy of human intentions and a knowledge base with 135 action categories. To parse intents in a document, they took each sentence as a query to the knowledge base. The intent assignment was performed based on the full–text index search (using Lucene). These studies limit the intents in a small number of categories. The latter also used search–based method to query intent from a knowledge base rather than an accurate intent identification.

3 User Intent Identification from Social Media Texts

3.1 User Explicit Intents

In a broad sense, intent or intention refers to an agent's specific purpose in performing an action or a series of actions. According to Bratman (1987) [3], intent or intention is a mental state that represents a commitment to carrying out an action or actions in the future. Intention involves mental activities such as planning and forethought. Intent can be stated explicitly or implicitly, directly or indirectly. In scope of our work, we will only focus on user *explicit* intents. Figure 1 shows several text posts by users on online forums and social networks. Some of which contain explicit intents and some do not.

In order to model and analyze user intents on online social media, we formally define a user explicit intent as a quintuple (5–tuple) as follows:

$$\mathbf{I}_u^e = \langle u, \mathbf{c}, d, w, \mathbf{p} \rangle \tag{1}$$

in which:

- u is the user identifier, e.g., user nickname or id on social media services.
- $\mathbf{c}$ is the current context or condition around this intent. For example, a user may currently be pregnant, sick, or having baby. Context $\mathbf{c}$ also includes the time at which the intent was expressed or posted on online.
- d is the domain of the intent. For example, the three sample intents shown in Fig. 1 belong to *housing*, *finance–banking*, and *education*, respectively.
- w is a key word or phrase representing the intent. It may be the name of a thing or an action of interest. The w values of the three intents listed in Fig. 1 can be *rent–house*, *borrow–loan*, and *study–english*, respectively.
- $\mathbf{p}$ is a list of properties or constraints associated with an intent. It consists of a list of property–value pairs related to the intent. For example, for the first intent in Fig. 1, $\mathbf{p}$ can be {location="*Phuong Mai, Bach Khoa or Ton That Tung*", number–people="*4*", price="*3 million vnd*"}.

Online Vietnamese social media texts	Intent?
Tình hình là mình đang cần thuê nhà quanh khu vực Phương Mai, Bách Khoa hoặc Tôn Thất Tùng cho ba người lớn và một cháu nhỏ. Tầm tiền khoảng 3 triệu. Bạn nào có thông tin gì xin liên hệ với mình theo số 0905231880. Cảm ơn nhiều. *(I am looking for a house to rent near Phuong Mai, Bach Khoa or Ton That Tung street for three adults and one child. The price is about 3 million vnd. Please contact me at 0905231880 if you have any information. Thank you a lot.)*	Explicit intent
Thực tế thì bây giờ nếu bạn vay tiền ở bất kỳ ngân hàng nào bạn cũng phải chịu lãi suất cao. *(Actually, if you borrow money from any banks at this time, you have to pay high loan interest rate)*	Non-intent
Mình đang định vay ngân hàng một khoản bằng bảng lương của mình. Không biết có mẹ nào ở đây có kinh nghiệm về việc này có thể tư vấn cho mình được không ạ. Mặc dù mình biết không thể vay được nhiều tiên theo cách này nhưng mình thấy nó đơn giản và hơn nữa cơ quan mình lại trả lương qua tài khoản ATM. *(I intend to borrow an amount of money from any bank using my payroll. If any mom here has experience about this, please give me a tip ...)*	Explicit intent
Với số tiền bạn có thì khó có thể mua được một căn hộ tại ở khu vực Cầu Giấy hoặc Thanh Xuân. *(It is impossible to buy an apartment in Cau Giay or Thanh Xuan areas with your amount of money.)*	Non-intent
Mình đang tìm một lớp luyện IELTS 6.5, học 2 ngày một tuần (trong đó một ngày là thứ 7 hoặc chủ nhật), từ 16h30 đến 18h30. Nhà mình ở Long Biên, mình đi làm ở Lò Đúc. Mẹ nào biết lớp học nào gần khu vực này thì cho mình xin thông tin với nhé. Mình cảm ơn nhiều. *(I am looking for an IELTS 6.5 class, studying 2 days a week (one is Saturday or Sunday), from 16:30 to 18:30. I live in Long Bien and work at Lo Duc. If any mom knows any class in these areas, please let me know. Thanks a lot.)*	Explicit intent

Fig. 1. Examples of texts with non–intent and explicit intents

3.2 Process of Analyzing and Understanding User Intents

The process of analyzing and understanding user intents includes three major stages as shown in Fig. 2, that are:

1. **User Intent Filtering**: This phase helps to filter text posts on online social media channels to determine which posts contain user intents and which do not. Posts carrying user intents will be forwarded to the next stage below.
2. **Intent Domain Identification**: Given a text paragraph or a text post containing a user intent, this phase will analyze and identify the domain of the intent. As explained in the previous subsection, the domain of an intent can be about *education*, *real–estate*, *finance–banking*, *tourism–vacation*, *automobile* or any other area that the intent is related to.
3. **Intent Parsing and Extraction**: Given a text post containing an intent and its domain, this stage will parse, analyze, and extract all the information about the intent. In other words, this step will extract important information from the text to fill the key word/phrase w and the list of properties/constraints $\mathbf{p}$ of the intent as defined in Formula 1 above.

Figure 3 shows a specific example of the user intent understanding process. The input is a text post on social media talking about the plan of a couple to find and book a honeymoon trip after getting married. User Intent Filtering module

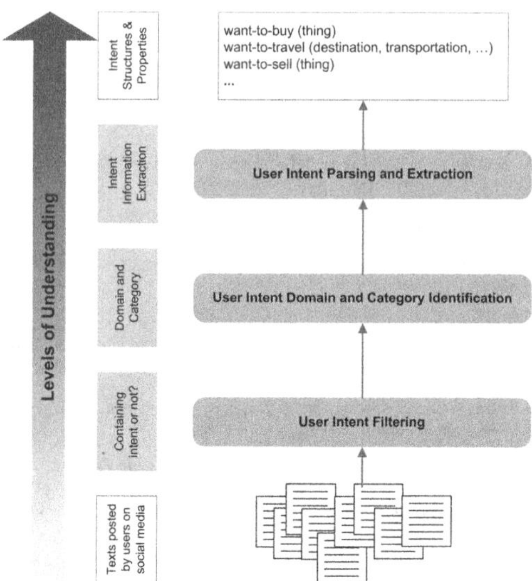

Fig. 2. Process of mining/identifying user intent from (online social media) texts

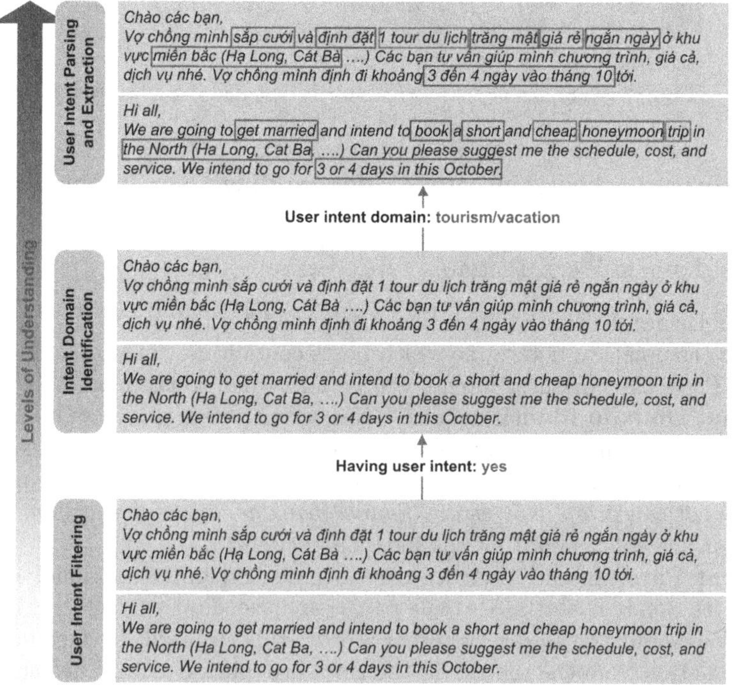

Fig. 3. Example of the user intent mining process

determined that this post holds an intent. In the next step, Intent Domain Identification module determined its domain (*tourism/vacation*). The post and its domain were then forwarded to the final phase, User Intent Parsing and Extraction. At this step, the properties/constraints of the intent were parsed and extracted: $\mathbf{p}$ = {price="*giá rẻ (cheap)*", duration="*3 đến 4 ngày (3 or 4 days)*", time="*tháng 10 (October)*", destination="*Hạ Long, Cát Bà*"}.

The process of full understanding of user intents is complex and needs a combination of different methods. The first phase, User Intent Filtering, is probably the simplest among the three phases. This is a binary classification problem. The second stage is more challenging because the number of domains is probably large. It is harder to solve this problem because we need to handle a large output space. The third stage is the most difficult. We need to parse and extract all relevant information in the texts. This is extremely hard because the list of properties or constraints $\mathbf{p}$ of an intent can vary a lot depending on its domain.

4 Filtering User Intents in Online Social Media Texts

As stated earlier, the whole process of understanding user intents in online social media texts is complicated and challenging. It needs a holistic solution combining different methods. In this section, we only focus on solving User Intent Filtering.

4.1 User Intent Filtering as a Binary Classification Problem

As described above, user intent filtering takes text posts/comments as inputs and determine which ones carry user intents. This can be seen as a binary classification problem. User intents can be diverse, they can be implicit or indirect. However, in this study, we only consider explicit intents. All text posts/comments with implicit intents will be classified into the class *no–intent*. Thus, we have two classes: *EI* (explicit intent) and *NI* (non–intent).

Basically, we can use any classification method for building a classifier. However, we decided to use maximum entropy (MaxEnt) for several reasons. First, MaxEnt is suitable for sparse data like natural language [2,16]. Second, MaxEnt can encode a variety of rich and overlapping features at different levels of granularity for better classification. Also, MaxEnt is very fast in training/inference.

4.2 Building Filtering Model with Maximum Entropy Classification

The MaxEnt principle is to build a classification model based on what have been known from data and assume nothing else about what are not known. This means MaxEnt model is the model having the highest entropy while satisfying constraints observed from empirical data. Berger et al. (1996) [2] showed that MaxEnt model has the following mathematical form:

$$p_\theta(y|x) = \frac{1}{Z_\theta(x)} \exp \sum_{i=1}^{n} \lambda_i f_i(x, y) \tag{2}$$

where x is the data object that needs to be classified, y is the output class label. $\theta = (\lambda_1, \lambda_2, \ldots, \lambda_n)$ is the vector of weights associated with the feature vector $F = (f_1, f_2, \ldots, f_n)$, and $Z_\theta(x) = \sum_{y \in \mathcal{L}} \exp \sum_i \lambda_i f_i(x, y)$ is the normalizing factor to ensure that $p_\theta(y|x)$ is a probabilistic distribution. Feature in MaxEnt is defined as a two–argument function: $f_{<cp,\ l>}(x, y) \equiv [cp(x)][y = l]$, where $[e]$ returns 1 if the logical expression e is $true$ and returns 0 otherwise. Intuitively feature $f_{<cp,\ l>}(x, y)$ indicates correlation between a useful property, called *context predicate* (cp), of the data object x and an output class label $l \in \mathcal{L}$.

Training or estimating parameters for MaxEnt model is to search the optimal weight vector $\theta^* = (\lambda_1^*, \lambda_2^*, \ldots, \lambda_n^*)$ that maximizes the conditional entropy $H(p_\theta)$ or maximizes the log-likelihood function $L(p_\theta, \mathcal{D})$ with respect to a training data set $\mathcal{D}$. Because the log-likelihood function is convex, the search for the global optimum is guaranteed. Recent studies [15] have shown that quasi-Newton methods like L–BFGS [14] are more efficient than the others. Once trained, the MaxEnt model will be used to predict class labels for new data. Given a new object x, the predicted label is $y^* = \text{argmax}_{y \in \mathcal{L}}\ p_{\theta^*}(y|x)$.

4.3 Feature Templates for Building the Filtering Model

For building the classification model with MaxEnt, we need to define our feature templates. Table 1 shows two types of features in our model. The first is n–gram. We used 1–grams (word tokens themselves), 2–grams (two consecutive word tokens), and 3–grams (three consecutive word tokens). When combining consecutive word tokens to form 2–grams and 3–grams, we did not join two consecutive word tokens if there is a punctuation mark between them.

We also used a dictionary for look–up features. Two consecutive word tokens were joined and looked up in the dictionary. This dictionary contains key phrases indicating there is an intent or not. Here are some examples: ***muốn mua*** (***want–buy***), ***cần tìm*** (***looking–for***), ***đang cần*** (***currently–need***), ***định vay*** (***intend–borrow***), ***cần bán*** (***need–sell***), ***muốn thuê*** (***want–rent***), and many more.

Table 1. Feature templates to train the MaxEnt model for user intent filtering

N–grams	Context predicate templates
1–grams	$[w_{-2}]$, $[w_{-1}]$, $[w_0]$, $[w_1]$, $[w_2]$
2–grams	$[w_{-2}w_{-1}]$, $[w_{-1}w_0]$, $[w_0w_1]$, $[w_1w_2]$
3–grams	$[w_{-2}w_{-1}w_0]$, $[w_{-1}w_0w_1]$, $[w_0w_1w_2]$
Dictionaries	Text templates for matching dictionaries
2–words	$[w_{-2}w_{-1}]$, $[w_{-1}w_0]$, $[w_0w_1]$, $[w_1w_2]$ in dictionary

5 Evaluation

5.1 Experimental Data

In order to evaluate the classification model, we collected a medium–sized collection of Vietnamese text posts and comments on online social media channels like Facebook and Webtretho (one of the most active forums in Vietnam). The collection consists of 1315 text posts/comments. A group of students were asked to label the data. They read the texts and assigned labels (either *EI* or *NI*) to the texts based on the agreement among them. The resulting collection contains 588 explicit–intent posts and 727 non–intent posts. The collection were then divided randomly into four parts. We in turn took three parts for training and the one left for test to perform 4–fold cross–validation tests. The experimental results will be reported in the next subsection.

5.2 Experimental Results and Analysis

Table 2 shows the experimental results of the 4th fold. *Human* is the number of manually annotated intents in the corresponding test set. *Model* is the number of explicit–intent posts/comments classified by the model. *Match* is the number of correctly classified posts/comments by the model, that is, the true positive. The last three columns are precision, recall, and F_1–score calculated based on *Human*, *Model*, and *Match* values. We achieved the macro–averaged F_1–measure of 91.98 and the micro–averaged F_1–measure of 92.07. This is a significantly high result because we only have n–gram and one dictionary look–up features.

Figure 4 shows the accuracy (i.e., micro–averaged F_1–score) of the four folds and the average value over the four folds. For each fold, we report to results, the first is the test result using n–gram features only while the second used both n-gram and dictionary look–up features. As we can see, classification using dictionary look–up features can give a better performance. Dictionary look–up features can improve the accuracy for more than 1.5 % on average. With the results of 4–fold cross–validation tests, we can see that the results are quite stable over the four folds. This shows that the classification model can work well on this data set.

We also calculated the average precision, recall, and F_1–measure of the two classes: non–intent and explicit–intent over the four folds. The results are shown

Table 2. Feature templates to train the MaxEnt model for user intent filtering

Class	Human	Model	Match	Precision	Recall	F_1–score
Non–intent	181	185	170	91.89	93.92	92.90
Explicit intent	147	143	132	92.31	89.80	91.03
Average_{macro}				**92.10**	**91.86**	**91.98**
Average_{micro}	328	328	302	**92.07**	**92.07**	**92.07**

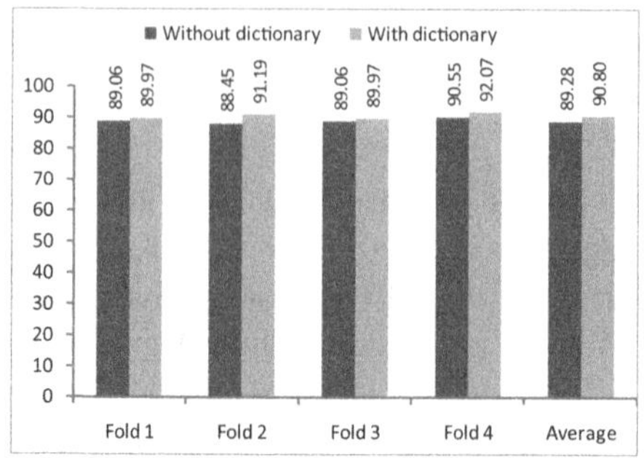

Fig. 4. The accuracy of the 4-fold cross-validation tests

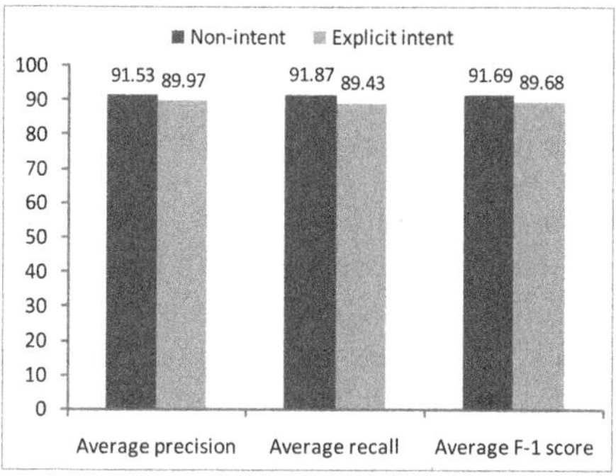

Fig. 5. The average precision, recall, and F_1-score of non-intent and explicit-intent over the 4 folds (with dictionary)

in Fig. 5. As we can see, the performance of explicit-intent class is a bit lower than that of non-intent. This is in part because the number of posts/comments carrying explicit intents is smaller (588 versus 727).

There are several hard posts/comments for classification. Some non-intent posts/comments have all keywords or phrases that commonly appear in explicit-intent texts. This is highly ambiguous and needs more sophisticated and high-level features to distinguish. For example, a post like "*cách đây vài năm mình định mua Camry nhưng sau đó ...*" (*I intended to buy a Camry couple of years ago but after that ...*) will be ambiguous. This contains an intent in the past and cannot be classified into explicit-intent. However, many of its keywords and phrases (in Vietnamese) indicate that it is an intent. Another example is that "*chị em nào muốn mua loại sữa này cho em bé thì suy nghĩ kỹ nhé*" (*think thoroughly if you want to buy this milk product*). This post/comment is actually a piece of advice or a warning message, not an explicit intent. However, it is classified into explicit-intent class. To deal with these difficult cases, we need to integrate more high-level features to capture past tense, sentence type, etc.

6 Conclusions

In this work, we have built a classification model based on the maximum entropy method to classify text posts/comments on online social media to determine which ones carry user explicit intents. This is the first stage (user intent filtering) of a complex process that aims at fully understanding user intents. We have achieved an average F_1–score of 90.80, a promising result for further work on this problem. We also realized that we need to add better and higher level features to the model in order to effectively discriminate highly ambiguous text posts/comments. This will be our focus in the future work.

Acknowledgements. This work was supported by the project QG.15.29 from Vietnam National University, Hanoi (VNU).

References

1. Ashkan, A., Clarke, C.L.A., Agichtein, E., Guo, Q.: Classifying and characterizing query intent. In: Boughanem, M., Berrut, C., Mothe, J., Soule-Dupuy, C. (eds.) ECIR 2009. LNCS, vol. 5478, pp. 578–586. Springer, Heidelberg (2009)
2. Berger, A., Pietra, S.A.D., Pietra, V.J.D.: A maximum entropy approach to natural language processing. Comput. Linguist. **22**(1), 39–71 (1996)
3. Bratman, M.: Intention, plans, and practical reason. Harvard University Press, Cambridge (1987)
4. Chen, L.: Understanding and exploiting user intent in community question answering. Ph.D. Dissertation, Birkbeck University of London (2014)
5. Chen, Z., Lin, F., Liu, H., Liu, Y., Ma, W.Y., Wenyin, L.: User intention modeling in web applications using data mining. J. WWW **5**(3), 181–191 (2002)
6. Church, K., Smyth, B.: Understanding the intent behind mobile information needs. In: The 14th IUI (2009)
7. Dai, H., Nie, Z., Wang, L., Wen, J.R., Zhao, L., Li, Y.: Detecting online commercial intention. In: The WWW (2006)
8. Hu, J., Wang, G., Lochovsky, F., Sun, J.T., Chen, Z.: Undertanding user's query intent with wikipedia. In: The WWW (2009)
9. Hu, D.H., Shen, D., Sun, J.-T., Yang, Q., Chen, Z.: Context-aware online commercial intention detection. In: Zhou, Z.-H., Washio, T. (eds.) ACML 2009. LNCS, vol. 5828, pp. 135–149. Springer, Heidelberg (2009)
10. Jethava, V., Liliana, C.B., Ricardo, B.Y.: Scalable multi-dimensional user intent identification using tree structured distributions. In: The ACM SIGIR (2011)
11. Kroll, M., Strohmaier, M.: Analyzing human intentions in natural language text. In: The K-CApP (2009)
12. Lee, U., Liu, Z., Cho, J.: Automatic identification of user goals in web search. In: The WWW (2005)
13. Li, X.: Understanding the semantic structure of noun phrase queries. In: ACL (2010)
14. Liu, D., Nocedal, J.: On the limited memory BFGS method for large-scale optimization. Math. Program. **45**, 503–528 (1989)
15. Malouf, R.: A comparison of algorithms for maximum entropy parameter estimation. In: COLING, pp. 1–7 (2002)

16. Nigam, K., Lafferty, J., McCallum, A.: Using maximum entropy for text classification. In: IJCAI Workshop on Machine Learning for Information Filtering, pp. 61–69 (1999)
17. Rose, D.E., Levinson, D.: Understanding user goals in web search. In: WWW (2004)
18. Shen, Y., Yan, J., Yan, S., Ji, L., Liu, N., Chen, Z.: Sparse hidden-dynamic conditional random fields for user intent understanding. In: The WWW (2011)

Retrieving Rising Stars in Focused Community Question-Answering

Long T. Le[1(✉)] and Chirag Shah[2]

[1] Department of Computer Science, Rutgers University, 110 Frelinghuysen Road, Piscataway, NJ 8854-8019, USA
longtle@cs.rutgers.edu

[2] School of Communication and Information, Rutgers University, 4 Huntington Street, New Brunswick, NJ 08901-1071, USA
chirags@rutgers.edu

Abstract. In Community Question Answering (CQA)' forums, there is typically a small fraction of users who provide high-quality posts and earn a very high reputation status from the community. These top contributors are critical to the community since they drive the development of the site and attract traffic from Internet users. Identifying these individuals could be highly valuable, but this is not an easy task. Unlike publication or social networks, most CQA sites lack information regarding peers, friends, or collaborators, which can be an important indicator signaling future success or performance. In this paper, we attempt to perform this analysis by extracting different sets of features to predict future contribution. The experiment covers 376,000 users who remain active in Stack Overflow for at least one year and together contribute more than 21 million posts. One of the highlights of our approach is that we can identify rising stars after short observations. Our approach achieves high accuracy, 85 %, when predicting whether a user will become a top contributor after a few weeks of observation. As a slightly different problem in which we could observe a few posts by a user, our method achieves accuracy higher than 90 %. Our approach provides higher accuracy than baselines methods including a popular time series analysis. Furthermore, our methods are robust to different classifier algorithms. Identifying the rising stars early could help CQA administrators gain an overview of the site's future and ensure that enough incentive and support is given to potential contributors.

1 Introduction

The prevalence of the Internet and related technologies has changed the way people seek and share information. Because there are now vast quantities of information on the Web, search engines such as Google, Bing, Yahoo, and Baidu were developed to support effective and efficient information seeking and retrieval. When looking for information online, one often starts by submitting simple queries to search engines. Unfortunately, search engines might not return

N.T. Nguyen et al. (Eds.): ACIIDS 2016, Part II, LNAI 9622, pp. 25–36, 2016.
DOI: 10.1007/978-3-662-49390-8_3

satisfactory answers or a more personalized answer is required [11]. Furthermore, many questions and their answers are not yet available online, and require human wisdom or experience to supply correct and specialized answers. Community Question-Answering (CQA) services such as Yahoo! Answers, WikiAnswers, and Stack Overflow, have emerged to provide this type of support to online information seekers.

"A question and answer (Q&A) Web site is purposefully designed to allow people to ask and respond to questions on a broad range of topics" (p. 866, [6]). CQA sites are community-based and designed to support both users who ask and answer questions. Participants ask questions publicly and receive answers from anyone in the community, usually from those individuals with knowledge or expertise related to the query. This model borrows from the idea that everyone can learn something from the "*wisdom of crowds*" [22]. Any user can participate in CQA and support the community, but users often have two main purposes: looking for an answer and helping the community by sharing their knowledge. Several popular CQA sites have been attracting millions of Internet users on a daily basis highlighting the popularity of these forums[1].

Because of this popularity, CQA sites have become an important source of information for Internet users. However, previous studies have shown that a majority of the content on CQA sites is contributed by a small group of users [1,12]. Furthermore, high quality content can attract traffic in CQA sites. In this work, we propose a new framework to identify the *rising star* in CQA. The rising star is the low profile user who has strong potential to contribute to the community later. This is a difficult task because unlike social networks, CQA sites do not provide information about participants' peers or collaborators. Discovering the rising star early can help CQA administrators give support or incentives to these potential users. For example, the earlier administrators recognize a rising star's contribution, the sooner they can encourage and cultivate their activity in CQA.

Below are our main contributions:

- We propose a method to detect rising stars. Our framework extracts different types of features, which are grouped into four categories: personal features, community features, temporal features, and consistent features.
- Applying our method to a popular CQA site (Stack Overflow) to show the efficacy with different classifiers. We also provide analysis about the importance of each feature with respect to prediction.

The rest of our paper is organized as follows. Section 2 discusses related works. We present the formal definition of our problem and our framework in Sect. 3. Section 4 describes the dataset and results. Finally, our conclusion is presented in Sect. 5.

2 Background and Related Work

Community Question-Answering. CQA is a popular venue where people can ask questions and share their knowledge. In traditional information seeking,

[1] http://highscalability.com/blog/2014/7/21/stackoverflow-update-560m-pageviews-a-month-25-servers-and-i.html.

users often issue queries to popular search engines such as Google, Bing, or Yahoo!. The advantage of using search engines is returning a list of related information immediately. Unfortunately, the user might not be satisfied with the search results or he might want more personalized answers [11]. Since the community's experience can provide good answers to complex questions, *wisdom of crowds* [22] users are able to ask and receive more customized questions and answers. In this way CQA provides a new approach to information seeking that is more personalized and community-based.

Adamic et al. [1] investigated the users' activities in CQA. The authors clustered questions and answers based on the content to understand users' activity on different topics. The results showed that the majority of users participated in a small number of topics. Based on these features, the amount of work and activity can predict answer quality. Shah et al. [18] reviewed major challenges and potential research problems associated with CQA. Since CQA sites are maintained and developed to help the community with question and answer exchanges, there are many interesting problems that require investigation. For example, the users' expectations, motivations, engagements were important to understand in order to design for and improve users' experiences. In [17], authors compared CQA and virtual reference. The work showed the difference between users' expectations and perception in these two virtual worlds.

Quality of the Posts and Users' Experiences in CQA. The quality of the posts in CQA is important since it affects the longevity of the community. The quality of questions and answers are investigated in [9,19] respectively. Several works also estimated how many answers a question would receive [5,24]. These works found the length of the post, the topics, and the quality of previous posts are strong indicators of the success of the new post. Shah et al. [20] investigated unanswered questions in CQA and found that many fact-based questions do not receive the answers from community.

User Behavior in Online Community. CQA sites operate primarily based on users' activities. The contents, including questions and answers, are generated by participants making user behavior in CQA especially interesting to researchers. White et al. [23] studied the effects of community size to question-answering activities and found that increasing the size of community could improve the effectiveness of the site such as the answering rates. Understanding when users will leave the community is also studied widely [13,14]. These works used the behaviour of users' *ego-nets* to make behavior and activity pattern predictions. Pal et al. [15] provided an overview of user evolution in CQA such as activities of different types of users in the site or their effect on the participation of others. Pudipeddi et al. [16] investigated when a user will leave the community in specifically in Stack Overflow and found that the time gaps between the post provides strong signal for churn. In [4,10], the authors searched rising stars in publication networks and social networks. These works used peers to estimate the quality of publication such as co-authors of papers or the publication avenue. Information of the peers is quite useful in social network [8]. In CQA, this type of information is not available.

3 Proposed Methods

3.1 Problem Definition

As prior research indicates, predicting rising stars was previously studied in publication or social network contexts where information about peers and collaborators was readily available. However, in CQA sites, information about peers, collaborators, or relationships is not apparent, making it difficult to predict a user's future performance. In this work, we want to address several questions related to identifying rising stars in a CQA community. The main questions are:

- Given a user with all of his posts in the first T weeks, will he become a top contributor to the site?
- Given a user with his first T posts, will he become a top contributor?

In order to address these questions, we need to understand the list of factors used to identify the users who have the potential to become top contributors. It might be impossible to reveal all factors affecting the performance of a user. For example, we might not know that a user changed careers and left a site. However, we try to identify as many features as possible and measure the importance of these features. We also want to see the difference between average users and high-profile users. To make predictions regarding rising star behavior, we observe their contributions for a short period or a few posts to predict whether he will become a top contributor in the long term.

3.2 Our Method

Our method includes three main steps. In the first step, we extract features that provide intuitive ideas about the future performance of users. Then, we build training and testing sets. The last step is executing an applied classification algorithm to find the rising stars.

Feature Extraction. In order to find the rising stars, we built a user's profile using several features. The features are divided into four groups:

- Personal Features: These features are based on the characteristics of users. Personal features include number of posts, the topics that the user participated in during the observation period, average length of the posts, and the ratio between questions and answers. Personal features of a user definitively affect future performance.
- Community Features: The features are based on the response of the community to users' posts. These features include the average score that the community gives to the posts and the average comments the posts receives. In general, a higher value indicates higher post quality.
- Temporal Features: These features represent the activity of users. Temporal features include the average duration between the posts and the time gap between two recent posts, the increasing/decreasing activity of users.

- Consistent Features: These features measure the consistency of user's posts such as standard deviation gap between the posts and std. scores of the posts. Standard time gap between recent posts.

Table 1 describes the features used in our experiment.

Table 1. List of features are classified into four groups of features.

Personal Features
Number of posts
Participated topics
Avg. length of post
Ratio between # questions and # answers
Community Features
Avg. score
Avg. comments the post received
Avg. Favorite marked
Temporal Features
Avg. duration between posts
Duration between the two recent posts
Trending of the posts
Consistent Features
Std. time gap between posts
Std. scores between the posts
Std. time gap between posts in 2nd half
More or less active

Building Training Set. In order to build the training data, we extracted features for each user as seen in Table 1 for different observation durations or the first few posts. A rising star user is defined as a user who will be among the top-K users who have highest score after 1 year ($K = 1\,\%$ or $10\,\%$). Normal users are the rest of the community. Since the majority of users in CQA are very inactive, we applied a sampling technique to make class balanced [3]. The details of the settings are described in Sect. 4.2.

Classification. Since our framework could use almost any classification model we compared the performance of different models in this study. In particular, we tested with the below classification algorithms [2]. Let $X = x_1, x_2, ..., x_n$ is the list of features. The list of classification algorithms are summarized as:

- Logistic regression (log-reg): Log-reg is a generalized linear model with sigmoid function: $P(Y = 1|X = \frac{1}{1+exp(-b)})$, where $b = w_0 + \sum(w_i.x_i)$, w_i are the inferred parameters from regression.
- Support vector machine (SVM) with Radial basis (RBF) kernel. The RBF kernel is defined as $K(x, x') = exp(-\frac{1}{2}||x - x'||)^2$.
- Decision trees: The Tree-based method is a nonlinear model that partitions features into smaller sets and fits a simple model into each subset. The decision tree includes two-stage processes: tree growing and tree pruning. These steps stop when a certain depth is reached or each partition has a fixed number of nodes.
- Random Forest (RF): RF is an average model approach. We use a bag of 100 decision trees. Given a sample set, the RF method randomly samples data and builds a decision tree. This step also selects a random subset of features for each tree. The final outcome is based on the average of these decisions.
- Adaptive Boosting (AdaBoost): This approach uses a list of weak learners to obtain a stronger learner by performing adaptive sampling. The general idea is putting higher weight on difficult examples.

4 Experiments

This section is organized as follows: data description, characterization of data and features, experimental setup, results, and discussion.

4.1 Data Description

Stack Overflow is a focused CQA site, which hosts programming related questions. All questions and answers must be related to programming, which is different from other general CQA sites. The data-dump of Stack Overflow was released to the public[2]. Table 2 lists the characteristics of the data set used in our experiment. Users in Stack Overflow can engage in different activities such as posting questions, giving answers, voting for the best answer, and up-voting or down-voting a post. They can also earn reputation by posting high quality questions and answers.

Table 2. Description about data.

Site	Period	# of Users	# of Posts	Topics	Reputations
StackOverflow	July '08 to Sep '14	3.4 M	21.2 M	Tags	Score

Data Pre-processing. A majority of users in CQA are inactive. We consider the life span of a user as the period between the last and the first post in our dataset. Since users join the site at different points in time, the first post is aligned at time $t = 0$. We only select sets of users who remain in the community for at least one year. After pre-processing, we identified a set of 376 K users.

[2] https://archive.org/details/stackexchange.

Defining the Rising-Star. In our definition, a rising star is a new user who will attain the top earned score in the future. It is not fair to compare a user who joined and stayed in the community for 3 years and a user who involved in the community for 1 year. Thus, we compare their scores after a fixed period (i.e., 1 year) of their association with the community. Rising stars are classified as the top-K percent of users. Here, the value K is selected depending on our purpose. For example, the top-1 % of users are very exceptional contributors in the community and the top-10 % of users are very good contributors. In our experiment, we evaluate for top-1 % and top-10 % users.

4.2 Experimental Setup

Competing Methods. We compare our method with the following baseline methods:

- RAND: Users are selected randomly.
- ARIMA (Autoregressive integrated moving average): This is the most general time series analysis technique [21].

In ARIMA, we build the time series for each user with the score they earn at time t_i. ARIMA includes three factors in its prediction. These are the (i) Autoregression: the output depends linearly on the previous value, (ii) Integration, and (iii) Moving average: a regression of current value of the series including current value and noise.

Metrics. We pick up a set of rising star users and "normal users". We then predict whether the user will become star. The rising star user is defined as part of the top-K(%) users having the highest score after belonging to the community for one year. In our experiment, we test the top-1(%) and the top-10(%). Since the number of users in top-K(%) is much smaller than the community's general population, we use a sampling method to create a balanced data set [3]. Since the class is balanced, the probability that a user becomes a rising star is 50 % in the RAND method.

4.3 Results

Overview Results. Figure 1 shows the correctness when predicting whether a user becomes a rising star. Our method is denoted as *PCTC* which stands for **P**ersonal- **C**ommunity- **T**emporal- **C**onsistency. The x-axis is the duration of our observation while the y-axis is the precision of prediction. The observation can be the first T weeks or the first T posts. It shows that our method can predict the long term performance after a short observation period. Further, the longer a user is observed, the better the prediction. Figure 1 also implies that it is more difficult to predict whether a user will become a top-10 % contributor than predicting the top-1 % contributors after observing for T weeks. The reason is the distinction between top-1 % and normal users is clearer than the top-10 %. The result in Fig. 1 is achieved by applying *Log-reg* algorithm. Next, we will examine the fluctuation of performance when we apply different classification algorithms.

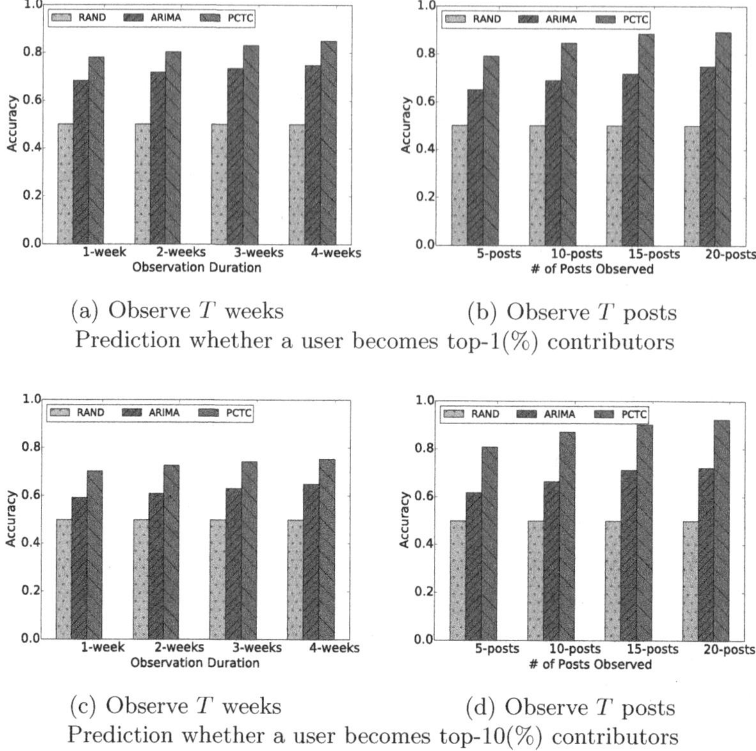

(a) Observe T weeks (b) Observe T posts

Prediction whether a user becomes top-1(%) contributors

(c) Observe T weeks (d) Observe T posts

Prediction whether a user becomes top-10(%) contributors

Fig. 1. Compare the correctness in predicting whether a user will become a top contributor after one year by observing their performance in the first T weeks ($T = 1, 2, 3, 4$) or the first T posts ($T = 5, 10, 15, 20$). The result of $PCTC$ is presented when applying *log-reg* classification algorithm.

Applying Different Classification Algorithms. We apply different classifications and compare the efficacy. Table 3 shows the efficacy of *LogReg, SVM, Decision Trees, RF* and *AdaBoost.* The result in Table 3 summarizes the correctness when detecting whether a user will become a top-1 % contributor. The decision trees have the lowest accuracy while RF and AdaBoost have the highest accuracy. In general, the accuracy is not affected greatly when applying different classification algorithms. The results show that our method is robust when testing it against different classification algorithms.

We also measure $F1$ score, which considers both precision and recall. Precision is the fraction of instances that are relevant, while recall is the fraction of relevant instances that are retrieved. The value of $F1$ is defined as $F1 = 2 * \frac{precision*recall}{precision+recall}$. Figure 2 shows that our approaches achieve higher $F1$ scores.

Table 3. Comparing different classifiers. The performance is affected slightly by choosing different classification algorithms.

Observe duration	LogReg	SVM	Decision trees	RF	AdaBoost
Observe the first T weeks					
1 week	80.6	80.2	78.5	79.1	79.2
2 weeks	83.4	83.3	80.2	83.3	83.4
3 weeks	84.8	84.9	81.1	84.9	85.9
4 weeks	85.6	85.7	82.3	86.6	86.7
Observe the first T posts					
5 posts	79.1	79.8	78.6	79.9	80.1
10 posts	84.6	84.5	82.3	84.6	84.2
15 posts	88.5	85.5	84.5	88.8	88.7
20 posts	89.6	89.9	87.3	91.1	91.2

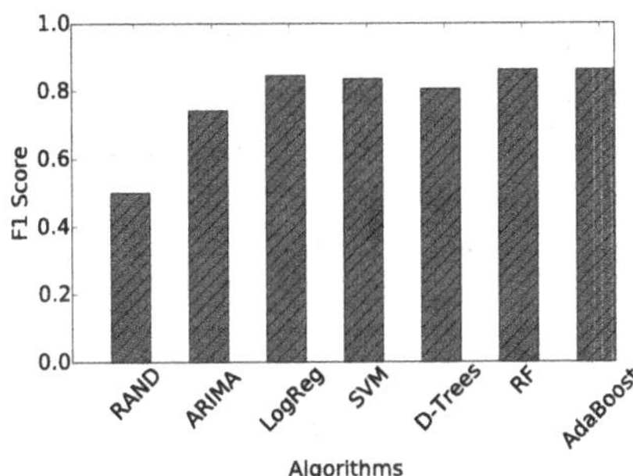

Fig. 2. Compare the $F1$ score (higher is better).

4.4 Discussion

Feature Importance. We want to quantify the feature importance in classification. A popular method to evaluate the feature importance in Random Forests is measuring the mean square error of prediction [7]. When constructing the decision trees in RF, we use out-of-bag to compute the prediction error. In order to quantify a feature f, we randomly permute f and recompute prediction error. The change is used to evaluate the importance of features in prediction. Figure 3a plots the importance of each feature in our prediction framework. The feature importance is normalized so that the sum is equal to 1. The higher value indicates that the feature is more important in prediction. We see that the number of posts, the average gap between the posts, and the average score of the posts are the most important features.

Using Different Groups of Features. Another way to quantify the feature importance is to isolate the features and perform classifications separately. There

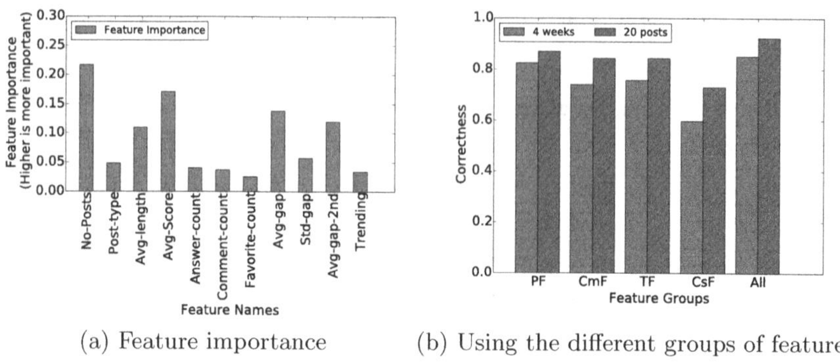

(a) Feature importance (b) Using the different groups of features

Fig. 3. Evaluate the importance of different features.

are four different groups of features in our approach. We also measure the efficacy when using each group of features separately. Figure 3a depicts the correctness when using different features separately. The higher efficacy indicates that the group of features is more important. We see that the Personal Features group achieves highest accuracy, which is close to the accuracy when using all features. The Consistency Features group performs slightly higher than RAND in the case where we observe for 4 weeks. The results match with measuring feature importance in Fig. 3a.

5 Conclusions and Future Work

In this work, we investigated the problem of identifying rising stars in CQA sites. The results show that it is possible to identify rising stars in their early stages. Those potential users contribute the majority of site content to the community and will be active participants in the community's development. A large collection of knowledge and high quality posts could help the community to retain current users and attract new users. Early detection of these users could help administrators provide incentive and support to these rising stars.

There are limitations with our work primarily due to the nature of the data. For example, it would be helpful if we could see the effects of incentives on site users. Future work will also investigate in greater detail top performers in CQA. For example, we can examine whether rising stars form a "small community" or if they are separate users within a large community. We also plan to apply our method to general CQA sites such as Yahoo! Answers to see whether there are differences between non-focused and focused CQA sites.

Acknowledgments. This work is partially funded by the US National Science Foundation (NSF) BCC-SBE award no. 1244704.

References

1. Adamic, L.A., Zhang, J., Bakshy, E., Ackerman, M.S.: Knowledge sharing and yahoo answers: everyone knows something. In: WWW, pp. 665–674 (2008)
2. Bishop, C.M.: Pattern Recognition and Machine Learning (Information Science and Statistics). Springer, New York (2006)
3. Chawla, N.V.: Data mining for imbalanced datasets: an overview. In: Maimon, O., Rokach, L. (eds.) The Data Mining and Knowledge Discovery Handbook, pp. 853–867. Springer, Heidelberg (2005)
4. Daud, A., Abbasi, R., Muhammad, F.: Finding rising stars in social networks. In: Meng, W., Feng, L., Bressan, S., Winiwarter, W., Song, W. (eds.) DASFAA 2013, Part I. LNCS, vol. 7825, pp. 13–24. Springer, Heidelberg (2013)
5. Dror, G., Maarek, Y., Szpektor, I.: Will my question be answered? predicting "question answerability" in community question-answering sites. In: Blockeel, H., Kersting, K., Nijssen, S., Železný, F. (eds.) ECML PKDD 2013, Part III. LNCS, vol. 8190, pp. 499–514. Springer, Heidelberg (2013)
6. Harper, F.M., Raban, D., Rafaeli, S., Konstan, J.A.: Predictors of answer quality in online q&a sites. In: CHI, pp. 865–874 (2008)
7. Hastie, T., Tibshirani, R., Friedman, J.: The Elements of Statistical Learning. Springer Series in Statistics. Springer, New York (2009)
8. Le, L.T., Eliassi-Rad, T., Provost, F., Moores, L.: Hyperlocal: inferring location of ip addresses in real-time bid requests for mobile ads. In: SIGSPATIAL LBSN 2013, pp. 24–33 (2013)
9. Li, B., Jin, T., Lyu, M.R., King, I., Mak, B.: Analyzing and predicting question quality in community question answering services. In: WWW, pp. 775–782 (2012)
10. Li, X.-L., Foo, C.S., Tew, K.L., Ng, S.-K.: Searching for rising stars in bibliography networks. In: Zhou, X., Yokota, H., Deng, K., Liu, Q. (eds.) DASFAA 2009. LNCS, vol. 5463, pp. 288–292. Springer, Heidelberg (2009)
11. Liu, Q., Agichtein, E., Dror, G., Maarek, Y., Szpektor, I.: When web search fails, searchers become askers: understanding the transition. In: SIGIR, pp. 801–810 (2012)
12. Movshovitz-Attias, D., Movshovitz-Attias, Y., Steenkiste, P., Faloutsos, C.: Analysis of the reputation system and user contributions on a question answering website: stackoverflow. In: ASONAM, pp. 886–893 (2013)
13. Ngonmang, B., Viennet, E., Tchuente, M.: Churn prediction in a real online social network using local community analysis. In: ASONAM, pp. 282–288 (2012)
14. Oentaryo, R.J., Lim, E.-P., Lo, D., Zhu, F., Prasetyo, P.K.: Collective churn prediction in social network. In: ASONAM, pp. 210–214 (2012)
15. Pal, A., Chang, S., Konstan, J.A.: Evolution of experts in question answering communities. In: ICWSM, pp. 274–281 (2012)
16. Pudipeddi, J.S., Akoglu, L., Tong, H.: User churn in focused question answering sites: characterizations and prediction. In: WWW Companion 2014, pp. 469–474 (2014)
17. Shah, C., Kitzie, V.: Social q&a and virtual reference - comparing apples and oranges with the help of experts and users. J. Am. Soc. Inf. Sci. Technol. **63**, 2020–2036 (2012)
18. Shah, C., Oh, S., Oh, J.S.: Research agenda for social q&a. Libr. Inf. Sci. Res. **31**(4), 205–209 (2009)
19. Shah, C., Pomerantz, J.: Evaluating and predicting answer quality in community qa. In: SIGIR, pp. 411–418 (2010)

20. Shah, C., Radford, M., Connaway, L., Choi, E., Kitzie, V.: How much change do you get from 40$? analyzing and addressing failed questions on social q&a. In: ASIST, pp. 1–10 (2012)
21. Shumway, R.H., Stoffer, D.S.: Time Series Analysis and Its Applications: With R Examples. Springer Texts in Statistics. Springer, New York (2011)
22. Surowiecki, J.: The Wisdom of Crowds. Anchor, New York (2005)
23. White, R.W., Richardson, M.: Effects of expertise differences in synchronous social q&a. In: SIGIR, pp. 1055–1056 (2012)
24. Yang, L., Bao, S., Lin, Q., Wu, X., Han, D., Su, Z., Yu, Y.: Analyzing and predicting not-answered questions in community-based question answering services. In: AAAI (2011)

Cloud and Network Computing

Fuzzy Inference System for Mobility Prediction to Control HELLO Broadcasting in MANET

Tran The Son[1(✉)], Hoang Bao Hung[1], Vo Duy Thanh[1], Nguyen Vu[1], and Hoa Le-Minh[2]

[1] Korea – Vietnam Friendship IT College, Hoa Qui, Ngu Hanh Son, Danang, Vietnam
{sontt, hunghb, thanhvd, vun}@viethanit.edu.vn
[2] Department of Physics and Electrical Engineering, Northumbria University, Newcastle, UK
hoa.le-minh@northumbria.ac.uk

Abstract. HELLO messages are used in many routing protocols in Mobile Ad Hoc Networks (MANET) for updating the neighbour information as well as the changes of network topology. However, periodically broadcasting HELLO messages in a fixed interval wastes of a lot of energy and generates the huge amount of redundant control packets across MANET if network mobility is low. In contrast, the frequency of sending HELLO messages will not quickly enough keep track the change of neighbours if network mobility is high. To solve that problem, this paper introduces a Fuzzy Inference System (FIS) for controlling HELLO frequency based on the prediction of nodes' mobility. The model helps to reduce the redundant HELLO messages while still ensuring to properly check neighbours and network topology. It is observed that energy consumption of the network saves 5 mW over 100 s (18 mW over 1 h) compared to that of the conventional HELLO broadcasting scheme.

Keywords: Fuzzy inference system · Mobile ad hoc networks · HELLO broadcasting · Average encounter rate

1 Introduction

Mobile Ad Hoc Networks (MANET) is considered as a candidate for the next generation mobile wireless networks [1] because it allows users to communicate to others in a self-organised fashion without the assistance of communication infrastructure. That nature opens many applications of MANET at those places where have no existence of network infrastructure such as battle fields, natural disaster sites. Though it has been developed since the last decade, MANET still has many technical issues that need to be resolved before deploying it to real-life applications [2].

In terms of routing, the movement of nodes and the lack of node's capability are big challenges for any routing protocols in MANET [3, 4]. Links could be broken during communication sessions; node's energy could be quickly depleted inducing a remarkable degradation of routing performance. Therefore, a routing protocol designed for MANET should adapt to the changes of the network as well as be effective to save resource and energy for network nodes.

N.T. Nguyen et al. (Eds.): ACIIDS 2016, Part II, LNAI 9622, pp. 39–50, 2016.
DOI: 10.1007/978-3-662-49390-8_4

There are many routing protocols proposed for MANET [3, 4] such as AODV [5], DSR [6], OLSR [7], most of them use HELLO messages to check link availability connecting to neighbours. Those messages are periodically sent to neighbours to inform the existence of a node to others. Also, by receiving HELLO messages from others, a node knows the existence of neighbours around. In fact, the frequency of broadcasting HELLO messages is usually fixed in most of routing protocols [5–7]. This may lead to an occupation of remarkable amount of bandwidth and a waste of lot of energy if the network mobility is low. In contrast, if the network mobility is high, HELLO messages sending in a fixed frequency may not quickly update the changes of the network topology in the routing table.

Because the dynamic changes of the network reflects into the routing table through the process of updating of HELLO messages, observing the routing table's variations will give a view to the mobility of the network. Based on the observation of the routing table, the authors in [8] have designed an adaptive routing protocol to adjust the HELLO frequency. The model is designed based on the fact that the more the new neighbours is, the more the network mobility is. As a result, their proposed model helps to reduce the number of HELLO messages while still being able to properly check link availability, thus saving node energy. However, their proposed model has not considered the network density which is an important factor that impacts to the appearance of new neighbors.

From the same perspective, the authors in [9] have proposed an adaptive HELLO protocol which has two operation modes: the Network Search Mode (NSM) and the Normal Operation Mode (NOM). The NSM is triggered whenever the number of neighbours is less than a threshold n_c, In this mode, the HELLO messages are sent in a fixed frequency to speed up the join of a new neighbour. When the number of neighbours is higher than n_c, nodes switch to NOM mode to adapt the frequency of broadcasting HELLO messages w.r.t the link change rate. If the link change rate increases, the HELLO interval will be decreased and vice versa. As a result, this model helps to reduce the number of control messages. However, the selection of n_c is a challenge of this model, because it is also dependent on the network density.

Based on the traffic demand measured at each node, the authors in [10] have introduced an HELLO messaging scheme to adapt HELLO interval. If a node has no packets to send, it will increase HELLO interval and vice versa. This could lead to an inaccurate detection of neighbours when there is no traffic to forward, but it still guarantees to check link availability when nodes have data to send. Notwithstanding, node mobility and node density are not taken into account in this model, thus, inducing a large amount of redundant HELLO messages as discussed previously.

This paper introduces an HELLO broadcasting scheme for neighbour discovery in MANET using FIS. The proposed model is able to predict the mobility of nodes (i.e., low, medium, high) and then adjust the HELLO broadcasting frequency accordingly. Therefore, it can help nodes save energy consumption especially in a low mobility condition while still ensuring to accurately discover the number of neighbours of nodes. In this model, FIS is chosen to deploy instead of a neural network (NN) because of two reasons: (i) FIS helps a node to work as a reasoner for controlling the HELLO frequency based on estimated mobility; (ii) FIS is simpler for implementing and working in network layer than a NN which requires a lot of computation and hard to deploy in network layer.

The rest of this paper is structured as follows. Section 2 introduces the mobility prediction model using FIS. Based on that model, Sect. 3 proposes a scheme for controlling HELLO broadcasting at each node. Section 4 evaluates the accuracy of the prediction model as well as simulation results when applying the proposed model for routing in MANET. Finally, Sect. 5 concludes the paper.

2 Mobility Prediction Model Using FIS

2.1 Network Model

A MANET is modelled by a random graph $G(V, L)$, where V is a set of nodes moving in an area A and L is a set of links between pairs of nodes. A link $\{u, w\}$ from node u to node w appears when node w is coming into the communication range of node u. At this instant, node w is recognized as a new encounter of node u and vice versa. Each node is equipped with a single radio with fixed transmission range R.

2.2 Definitions

Definition 1 (Encounter). Two nodes encounter each other when the distance between them becomes smaller than the communication range R [11]. In this circumstance, each node is considered as a new encounter of the others. The encounter e between node n_A and node n_B is represented by the time t at which 2 nodes meet each other within a duration Δt.

$$e = \{n_A, n_B, t, \Delta t\} \tag{1}$$

Definition 2 (Average Encounter Rate). The average encounter rate (AER) is the average number of *new* encounters experienced by each node in a duration T. Let $E_n(A)$ be the set of *new* encounters observed by node A within duration T, the AER_A can be calculated as follow [11]:

$$AER_A = \frac{|E_n(A)|}{T} \tag{2}$$

where $|E_n(A)|$ is the cardinality of set $E_n(A)$.

This paper uses the AER analysis published in the previous work [11]. Assume that node A and node B in the network move at the speed v from A_1 to A_2 and from B_1 to B_2 respectively (see Fig. 1). After a certain duration T, they both travel a distance $\Delta x = v \times T$. Let ω be the angle created by segments A_1A_2, A_2B_2; θ be the angle generated by the segment B_1B_2 and the horizontal axis. In a given density λ, assume that nodes are moving in a random direction, the AER can be estimated as

$$AER_A = \frac{|E_n(A)|}{T} = \frac{2\lambda}{\pi}\psi_A(v), \tag{3}$$

Where

$$\psi_A(v) = \frac{1}{T}\int_{r_{\min}}^{R}\int_{\omega_{\min}}^{\pi} r\theta_{\max} d\omega dr; \tag{4}$$

$$\omega_{\min} = \arccos\left(\frac{r^2 + 2R\Delta x - R^2}{2r\Delta x}\right), \tag{5}$$

$$\theta_{\max} = \arccos\left(\frac{R^2 - \Delta x^2 - k^2}{2k\Delta x}\right), \tag{6}$$

The term $\psi_A(v)$ in Eq. (3) represents the *relative* mobility of node A w.r.t its neighbours [11]; $r_{\min}$ is chosen in (0, R −2Δd) such that node B is still detected as a new encounter of node A within duration T. If $r < r_{\min}$, node B becomes an old encounter.

From Eq. (3), it can be seen that a node can predict the mobility if it knows AER and the network density λ. The first component, AER is determined by observing new neighbours coming into the communication range at each node. The second one, node density λ is unknown because nodes are distributed randomly and move autonomously as the nature of MANET. However, it can be predicted based on the degree of each node. In a given graph G (V, L), the degree k of a node, also known as the number of neighbours, distributes as a Poisson distribution [12] and represents as follows

$$p_k = \binom{n}{k} p^k (1-p)^{n-k} \approx \frac{N^k e^{-N}}{k!}, \tag{7}$$

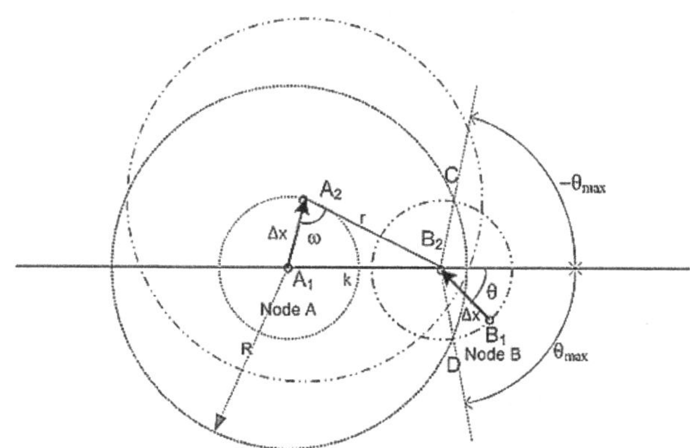

Fig. 1. AER analysis

where N is the average number of neighbours of a node; n is the total number of nodes in the network; p is the probability of existent a link between two nodes.

In Eq. (7), N can be identified based on the average density $\bar{\lambda}$ of the network as follow

$$\bar{\lambda} = \frac{n}{A} = \frac{N}{\pi R^2}, \tag{8}$$

where A is the area of the network.

Figure 2 illustrates the distributions of node degree k plotted based on Eq. (7) in different network densities. As the figure shows, the values of node degree k vary as a Poisson distribution around the average values N identified by Eq. (8).

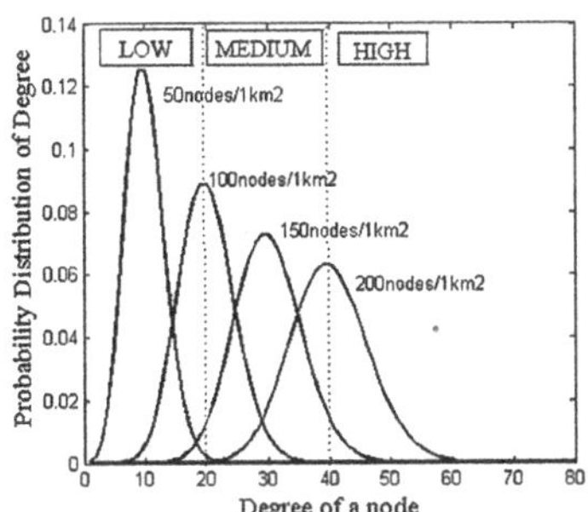

Fig. 2. Node degree distribution at different densities

2.3 FIS for Mobility Prediction

Figure 2 implies that if a node knows its degree k, it can predict whether the network density λ is *low*, *medium*, or *high*. This prediction helps a node estimate its relative mobility w.r.t the number of neighbours around based on Eq. (3). That motivates us to design a Fuzzy Inference System [13] for Mobility Prediction (FIS-MP) which is integrated into each node to predict network mobility with two input variables: AER and node degree k. As the nature of any fuzzy inference system, FIS-MP has three processes: *Fuzzification*, *Inference*, and *Defuzzification* as shown in Fig. 3.

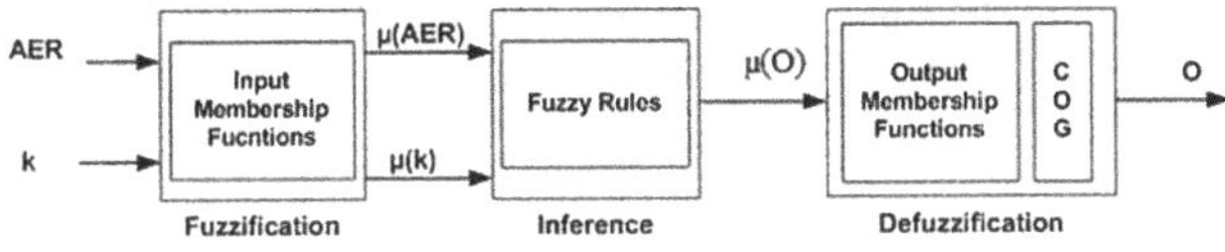

Fig. 3. FIS-MP at each node.

In the first process, i.e., fuzzification, all input variables (i.e., AER and k) need to be classified into different levels, e.g *low*, *medium*, and *high*. To do so, firstly, the velocity

v and the network density λ should be classified. Suppose that the velocity v is ranging from 0 m/s to 20 m/s and beyond (denoted by 20 +), it is classified into *low*, *medium*, *high* levels as described in Table 1.

Similarly, the density λ is considered from 0 to 2.10^{-4} nodes/m^2 (0 to 200 nodes/km^2) and beyond (denoted by $2*10^{-4}$ +). It can be categorised into *low*, *medium*, *high* levels as listed in Table 1 (see also Fig. 2). Based on the classifications of v and λ, the Eq. (3) allows us to classify the AER as shown in Table 1 with $R = 250$ m.

Table 1. Classification table

	Low	Medium	High
Velocity v (m/s)	$0 < v \leq 10$	$10 < v \leq 15$	$15 < v \leq 20$ +
Density λ (nodes/m^2)	$0 < \lambda \leq 0.5*10^{-4}$	$0.5*10^{-4} < \lambda \leq 1.5*10^{-4}$	$1.5*10^{-4} < \lambda \leq 2.0*10^{-4}$ +
AER	$0 < AER \leq 0.25$	$0.25 < AER \leq 0.56$	> 0.56
Node degree k	[0–20)	[20–40)	≥40
Output O (estimated mobility)	$0 < O \leq 0.4$	$0.4 < O \leq 0.8$	$0.8 < O \leq 1.0$

The above classification allows FIS to fuzzify the input values based on membership functions. To minimize the complexity of the proposed system, the triangular membership functions are chosen as depicted in Fig. 4.

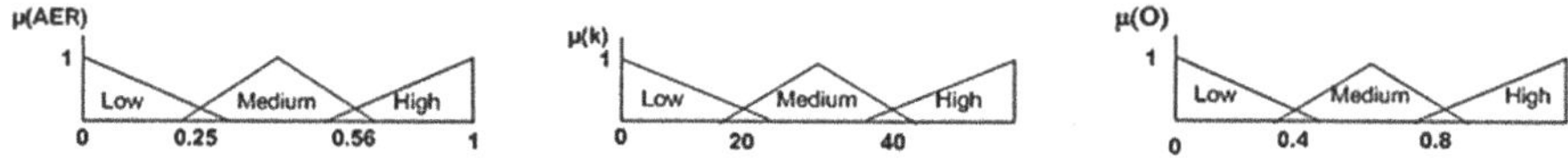

Fig. 4. Membership functions

In the second step, the FIS infers whether the mobility of a node is *low*, *medium*, or *high* based on the set of rules as described in Table 2. These rules are constructed based on the analysis and AER observation (see Fig. 7) when varying the network density and node mobility from low to medium and high levels as defined in Table 1.

Eventually, the output of the inference process, μ (O) is converted to a crisp value O by using the *centre of gravity* (COG) method [10] as follows

$$O = \frac{\int_0^1 \mu(O).xdx}{\int_0^1 \mu(O)dx}, \tag{9}$$

where x is output variable in the range of [0, 1] as shown in Table 1.

Table 2. Fuzzy rules

#	Node Degree (k)	AER	Mobility (estimate)
1	Low	Low	Low
2	Low	Medium	Medium
3	Low	High	High
4	Medium	Low	Low
5	Medium	Medium	Medium
6	Medium	High	High
7	High	Low	Medium
8	High	Medium	Medium
9	High	High	High

This outcome helps a node recognise its relative mobility w.r.t others, and then control the HELLO broadcasting process as proposed in the next section.

3 Controlling Hello Broadcasting Using FIS-MP

The HELLO frequency is controlled by adjusting the HELLO interval T_{HELLO}, and it is done by a mapping table called T_{HELLO} Mapping Table as illustrated in Fig. 5. This table is controlled by the estimated mobility O which is the outcome of FIS-MP. If O is high, HELLO interval is mapped to low value to increase the HELLO's frequency. In contrast, if O is low, HELLO interval is mapped to high level to decrease the HELLO's frequency. Otherwise, HELLO interval is kept to medium level.

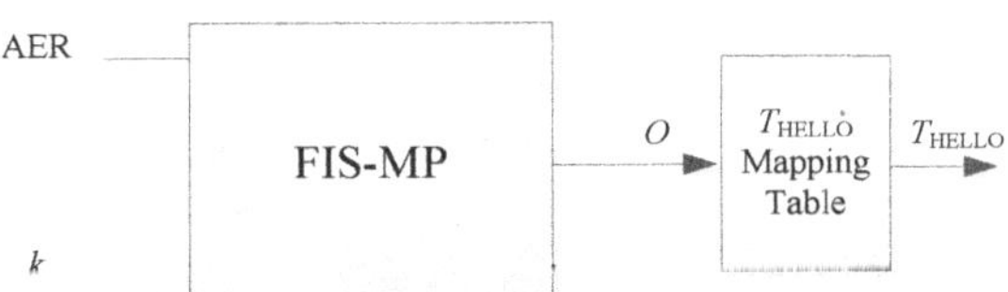

Fig. 5. Controlling HELLO broadcasting using FIS-MP

To classify T_{HELLO} into low, medium, and high levels, let's examine two nodes A and B as illustrated in Fig. 6. After T_{HELLO} node A and node B move in opposite direction. Node A needs to determine the T_{HELLO} such that it still guarantees to recognise that node B is a neighbour. If T_{HELLO} satisfies this case, all other cases (i.e., node B comes close to node A; or node B moves not in opposite direction of node A) also satisfy. From the Fig. 6, the distance d between two nodes is calculated as

$$d = d_0 + 2vT_{HELLO}, \tag{10}$$

where d_0 is the distance between two nodes before sending HELLO messages.

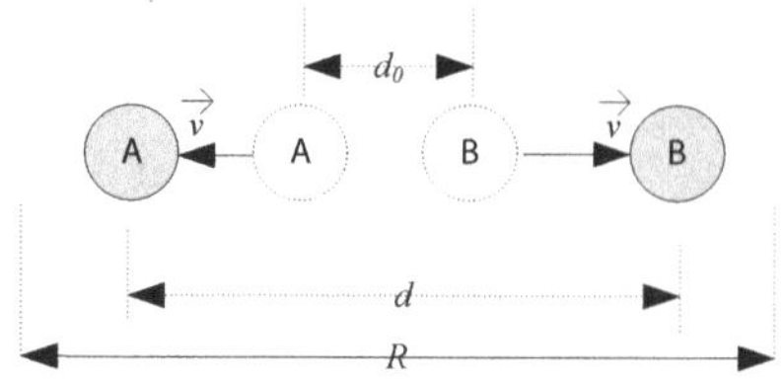

Fig. 6. Neighbour observation

It can be clearly seen that d_0 is a random value in (0, R), and can be represented by

$$d_0 = \alpha R,\ 0 < \alpha < 1 \tag{11}$$

Because node B is still to be the neighbour of node A, hence

$$d \leq R \Leftrightarrow d_0 + 2vT_{HELLO} \leq R \tag{12}$$

Or

$$\alpha R + 2vT_{HELLO} \leq R \tag{13}$$

$$T_{HELLO} \leq \frac{(1-\alpha)R}{2v} \tag{14}$$

The Eq. (14) allows classifying T_{HELLO} w.r.t mobility as shown in Table 3.

Table 3. T_{HELLO} mapping table

Mobility	Low	Medium	High
Output O^a (estimated mobility)	$0 < O \leq 0.4$	$0.4 < O \leq 0.8$	>0.8
T_{HELLO} (s)[b]	≤6.25	≤4.16	≤3.1
T_{HELLO} (s)[c]	≤3.75	≤2.5	≤1.87

[a] see also Table 1 – Sect. 2; [b] $\lambda = .5$; [c] $\lambda = .7$

4 Simulation Results and Evaluation

4.1 Simulation Setup

The proposed model has been deployed on AODV [5] which is a commonly used routing protocol in MANET. In this protocol, HELLO messages are used to detect new encounters as defined in Eq. (1). This helps to compute the AER which is an input variable of the FIS-MP as shown in Fig. 5. The second input variable, node degree k is obtained by monitoring the neighbour table that is available in AODV.

To provide an accurate AER, the encounter lifetime (defined in Eq. (1)) is set to be equal to the observation T (see Eq. (2)), i.e., 5 s. This setting implies that a new encounter which has been previously detected will no longer be considered as a new encounter because it has been met. Table 4 describes the simulation setup in details.

Table 4. Simulation Setup

+ Simulator	*ns-3 ver. 3.20 (installed on Ubuntu 12)*
+ Area	*1000 m × 1000 m.*
+ No. of nodes	*50, 100, 150, 200*
+ Simulation time	*100 s*
+ Mobility model	*Random Waypoint (pause time 0 s)*
+ Velocity (maximum)	*[0–30] m/s*
+ Communication range	*250 m*
+ Physical and MAC Layer	*Wi-Fi IEEE 802.11b, 2.4 GHz*
+ Encounter lifetime Δt	*5 s*
+ Observation time T	*5 s*
+ Transmitted power	*18 dBm*
+ Power supply	*50 Joules*

The proposed model was evaluated by following metrics:

- *Prediction Error Ratio (PER)*: It is calculated by counting number of prediction errors caused by FIS-MP over the total number of observations at a certain velocity level (i.e., low, medium, or high).
- *Neighbour Table Accuracy (NTA)*: this metric monitors the neighbour table and compare to the neighbour table of AODV with $T_{\text{HELLO}} = 1$ s as a reference point.
- *Residual Energy (RE):* this metric measures the residual energy at each node. This metric helps to evaluate the efficiency of the proposed model.

4.2 Results and Evaluation

To evaluate the proposed system, we firstly examine the relation among AER, the velocity v and the node density λ based on Eq. (3). The theoretical results in Fig. 7(a) agree with simulation results in Fig. 7(b). That confirms the prediction ability of the FIS-MP.

The PER of FIS-MP at different levels of mobility is shown in Fig. 8. As the figure shows, the PER increases when the mobility of the network increases, i.e., 15, 20, 25 and 30 m/s. This is because under a high mobility condition, a new encounter could come and leave before being detected as a new encounter thus resulting in an imprecision of the AER. However, in reality, MANET's users rarely move at very high speed (up to 30 m/s ~ 108 km/h) [14], that does not really impact on the application of FIS-MP for controlling HELLO frequency in MANET.

One of the most concerns when changing HELLO frequency is how to accurately discover neighbors appearing, especially when the mobility of the network increases. It is observed that the NTA is very close to that of constant $T_{\text{HELLO}} = 1$ (s) (see Fig. 9), i.e., ~ 90 % and 85 % with $\alpha = 0.7$ and 0.5 respectively.

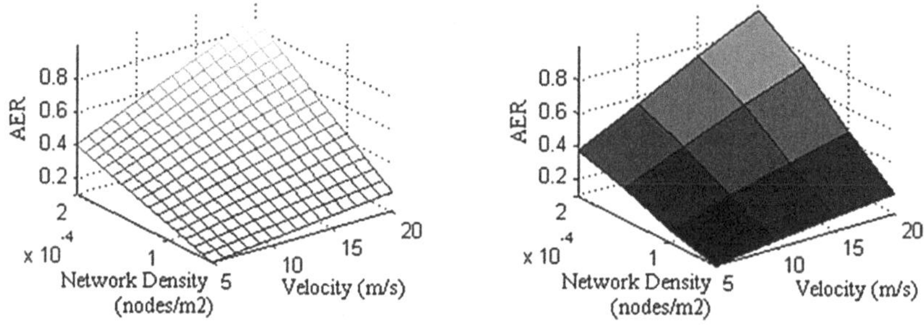

Fig. 7. (a) Theoretical AER analysis based on Eq. (3) (b) Simulation observation.

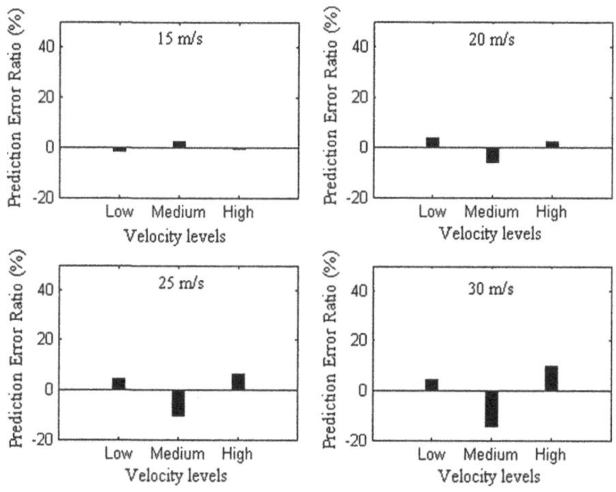

Fig. 8. The prediction errors at different velocities

The main goal of the proposed model is to save energy while still ensuring to accurately discover the number of neighbours. As seen in Fig. 10, energy saves 3 mW/100 s ($\sim$10.8 mW/1 h) and 5 mW/100 s ($\sim$18mW/1 h) with $\alpha = 0.7$ and $\alpha = 0.5$ respectively. These are remarkable improvements for a MANET's node such as a smartphone or a mobile device [15] if it works in a long hours.

From Eq. (14), it can be recognised that the NTA improves when α increases. However, this increment will raise the number of HELLO messages broadcasting across the network, thus occupying a significant bandwidth used for data and wasting a lot of energy (see Fig. 10). In other words, there is a trade-off between the system accuracy and the coefficient α. The study of this trade-off is out of the scope of this paper.

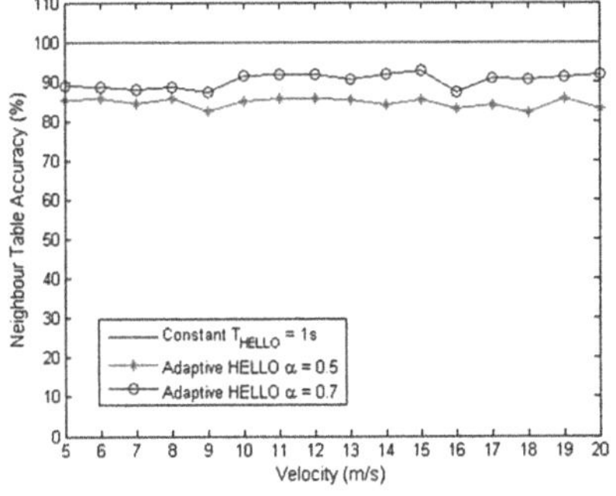

Fig. 9. Neighbour table accuracy

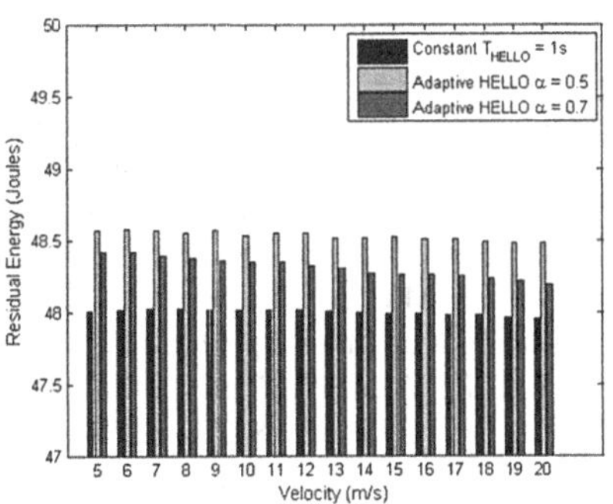

Fig. 10. Residual energy

5 Conclusion

This paper has introduced an HELLO broadcasting scheme for neighbour discovery in MANET. The model uses FIS working as the mobility predictor with two input variables: AER and node degree k to control the HELLO broadcasting frequency. If network mobility is high, the proposed system increases HELLO broadcasting frequency to keep track of quick change of the network topology. In contrast, it decreases HELLO broadcasting frequency to reduce the number of HELLO messages and save energy for network nodes. By using the proposed model, the neighbour table is still updated at a high accuracy ($\sim$90 % and 85 % with $\alpha = 0.7$ and 0.5 respectively) while reducing the redundant HELLO messages. This helps to save 3 mW/100 s ($\sim$10.8 mW/1 h) and 5 mW/100 s ($\sim$18mW/1 h) with $\alpha = 0.7$ and $\alpha = 0.5$ respectively.

References

1. Basagni, S., Conti, M., Giordano, S., Stojmenovic, I.: Mobile Ad hoc Networking. IEEE Press, pp. 1–45 (2004). ISBN 0-471-37313-3
2. Alotaibi, E., Mukherjee, B.: A survey on routing algorithms for wireless Ad-Hoc and mesh networks. Comput. Netw. **56**(2), 940–965 (2012)
3. Boukerche, A., Turgut, B., Aydin, N., Ahmad, M.Z., Bölöni, L., Turgut, D.: Routing protocols in ad hoc networks: A survey. Comput. Netw. **55**(13), 3032–3080 (2011)
4. Sarkar, S.K., Basavaraju, T.G., Puttamadappa, C.: Routing protocols for Ad hoc Wireless Networks. In: Ad Hoc Mobile Wireless Networks: Principles, Protocols, and Applications, pp. 59–112. Taylor & Francis Group, LLC (2008). ISBN 978-1-4200-6221-2
5. Perkins, C., Belding-Royer, E., Das, S.: Ad-hoc On-demand Distance Vector Routing (AODV), July 2003
6. Johnson, D., Hu, Y., Maltz, D.: The Dynamic Source Routing Protocol (DSR) for Mobile Ad Hoc Networks for IPv4, February 2007
7. Clausen, T., Jacquet, P.: Optimized Link State Routing Protocol (OLSR), October 2003
8. Ingelrest, F., Mitton, N., Simplot-Ryl, D.: A turnover based adaptive HELLO protocol for mobile Ad Hoc and sensor networks. In: International Symposium on Modeling, Analysis, and Simulation of Computer and Telecommunication Systems (MASCOTS 2007), Istanbul, Turkey (2007)

9. Ernst, R., Martini, P.: Adaptive HELLO for the neighbour discovery protocol. In: The 37th Annual IEEE Conference on Local Computer Networks (2012)
10. Han, S.Y., Lee, D.: An adaptive hello messaging scheme for neighbor discovery in on-demand MANET routing protocols. IEEE Commun. Lett. **17**(5), 1040–1043 (2013)
11. Son, T.T., Le- Minh, H., Sexton, G., Aslam, N.: A novel encounter-based metric for mobile ad-hoc networks routing. Ad Hoc Netw. **14**, 2–14 (2014)
12. Newman, M.E.J., Strogatz, S.H., Watts, D.J.: Random graphs with arbitrary degree distribution and their applications. Phys. Rev. E **64**(2), 1–17 (2001)
13. Klir, G.J., Yuan, B.: Fuzzy Sets and Fuzzy Logic - Theory and Applications. Prentice Hall PTR (1995). ISBN 0-13-101171-6
14. Perkins, C.E., Royer, E.M., Das, S.R., Marina, M.K.: Performance comparison of two on-demand routing protocols for ad hoc networks. IEEE Pers. Commun. **8**(1), 16–28 (2001)
15. Perrucci, G.P., Fitzek, F.H.P, Widmer, J.: Survey on energy consumption entities on the smartphone platform. In: The IEEE 73rd Vehicular Technology Conference (VTC Spring) (2011)

Resource Allocation for Virtual Service Based on Heterogeneous Shared Hosting Platforms

Nguyen Minh Nhut Pham[1(✉)], Thu Huong Nguyen[1], and Van Son Le[2]

[1] Department of E-commerce, Vietnam Korea Friendship Information Technology College, Danang, Vietnam
{nhutpnm, huongnt}@viethanit.edu.vn
[2] Department of Information, Danang University of Education, Danang University, Danang, Vietnam
levansupham2004@yahoo.com

Abstract. Nowadays, one of the issues in cloud computing is resource optimizing for virtual services to enhance IaaS service performance and meet the requirements of resource exploitation effectively. In this paper, we seek the approach of multi-dimensional resource allocation based on heterogeneous shared hosting platforms for virtual services. We construct the problem as an optimized formulation that uses a linear programming to minimize the number of physical machines. The solution for this formulation is applying the Greedy algorithms to solve and evaluate via emulation-based program.

Keywords: Resource allocation · Greedy algorithm · Shared hosting platforms · Cloud computing · Linear programming

1 Introduction

Cloud computing is built on the achievement of various fields, such as: service-oriented architecture, grid computing, virtualization. In particular, virtualization technology allows partitioning [3] or aggregation [14] the resource of Y ($Y \geq 1$) physical machines into S ($S \geq 1$) virtual machines. A system which consists of multiple physical machines connecting together for sharing resources is called a shared hosting platforms [1, 2, 4–7]. One of the challenges this system faces is to minimize resources of the platform for virtual services while still ensuring QoS requirements.

The resource management in the shared hosting platforms has been investigated in many other studies. For instance, Urgaonkar *et al.* [1] proposed a profiling technique to place applications onto a shared cluster so that maximize resource provider revenue while we focus to minimize the physical resources. Aron *et al.* [2] formulated the resource allocation problem as a constrained optimization problem in which physical machines are considered as a monolithic resource while we consider a multiple resource dimensions. In the documents [4, 6], the authors construct the resource allocation problem for virtual service as the optimal problem. But what was limited to problem in the homogeneous shared hosting platforms, i.e., resource configuration of the physical machines is the same while we consider the hetergeneous shared hosting

N.T. Nguyen et al. (Eds.): ACIIDS 2016, Part II, LNAI 9622, pp. 51–60, 2016.
DOI: 10.1007/978-3-662-49390-8_5

platforms, i.e., the physical machines have the different resource configurations. Mark Stillwell *et al.* [7] referred to resource allocation for virtual services on heterogeneous platforms, but they focused on calculating the optimal performance of system. Nguyen *et al.* [5] also presented the resource model from heterogeneous platforms, but they focused on solving the deadlock handling in resource allocation. One difference with our work is that we focus to minimize the number of physical resources.

The problem of resource allocation is considered in both cases: static and dynamic. In this paper, we only focus on solving the static case, i.e., assuming that workload does not change. The key contributions of the paper are as follows:

(a) Building the system and resource model for virtual service from the heterogeneous shared hosting platforms, ensuring QoS requirements;
(b) Modeling a resource allocation as a linear programming problem and the complexity analysis of problem;
(c) Developing the Greedy algorithms to solve multi-dimensional resource allocation with the goal of minimum number of physical machines.

The rest of the paper is organized as below: Sect. 2 presents a mathematical model of the problem as a linear programming problem and analyzes the complexity of the problem. Section 3 proposes several algorithms to solve the problem. Section 4 follows with experimental results and comparisons in various scenarios. Finally, Sect. 5 concludes the paper and opens some future work.

2 Resource Allocation for Virtual Service

2.1 System and Resource Model

We consider a heterogeneous shared platforms in which a cluster of physical machines having the different resource configuration and being interconnected by a high-speed network devices is deployed for sharing resources to virtual service. Each virtual service consists of a single virtual machine. These virtual machines run on physical machines under the control of a hypersivor [3] and consume resources at different portions. When the system receives a request to provide the resources for virtual service, a resource provider is responsible for making decisions whether to reject or admit a request, and provide resources to virtual service from physical machines.

For each type of resource under consideration a physical machine may have one or more single element resource (i.e., one or more single real CPU, one or more single real memory,…) and aggregate resources. Thus, the resources of a physical machine are represented by an ordered pair of multi-dimensional resource vectors. The elementary resource vector gives the capacity of a single element in each dimension while the aggregate resource vector gives the sum of resource capacity counting all elements.

Similarly, the virtual machines have also the virtual single elements and the virtual aggregate elements. Therefore, the resource needs of virtual service are represented by an ordered pair of multi-dimensional resource needs vectors, including elementary resource needs vector and aggregate resource needs vector. In fact, each virtual service has two kinds of resource needs: rigid needs and fluid needs [4]. A rigid needs represents a specific fraction of required resource. The service cannot benefit from a larger fraction and cannot operate with a smaller fraction than. A fluid needs specifies the additional fraction of a resource that the service could use. The service cannot benefit from a larger fraction, but can operate with a smaller fraction than at the cost of reduced performance.

Thus, the rigid needs of resource type k of service i are given by an ordered vector pair $\left(r_{ik}^{e}, r_{ik}^{a}\right)$, that represents the resource needs to run the service at the acceptable service level. If this resource needs cannot be met, then resource allocation fails. The fluid needs of resource type k of service i are given by a second ordered vector pair (f_{ik}^{e}, f_{ik}^{a}), that represents the additional resources required to run the service at the maximum level of performance. We assume that, if the elementary resource needs of virtual service exceed the elementary resource capacity of physical machine in each dimension, then resource allocation fails.

However, the constraint of QoS requirements should be considered. The QoS attribute describes the necessary characteristics of services to meet the needs of users, such as [15]: time, cost. This constraint is defined by the multiplication of the fluid needs of resource and the QoS requirements of service. In fact, when the CPU resource needs reduced, this leads to the reduction of other resources consumption. Thus, the utilizations of all resources corresponding to fluid needs are linearly correlated. For simplicity, the QoS of all fluid needs can represent the same value. Therefore, resource needs of resource type k of virtual service i on physical machine j with QoS requirements Q_{ij} are given by an ordered vector pair $(r_{ik}^{e} + Q_{ij}f_{ik}^{e}, r_{ik}^{a} + Q_{ij}f_{ik}^{a})$.

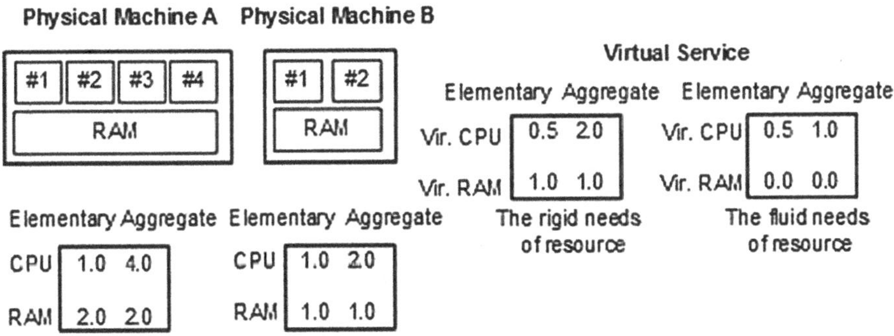

Fig. 1. Example of problem instance with resource of two physical machines and resource needs of one virtual service.

Figure 1 illustrates a simple example about the resources and resource needs model with two physical machines and one virtual service. In particular, the physical machine A comprises 4 CPU cores (i.e., 4 CPU single resource elements) and 1 RAM (i.e., RAM 1 single resource element). Thus, the CPU resource vector pair is (1.0, 4.0) and RAM resource vector pair is (2.0, 2.0). Physical machine B comprises 2 cores and 1 RAM. Thus, the CPU resource vector pair is (1.0, 2.0) and RAM resource vector pair is (1.0, 1.0). For virtual service, the CPU resource needs vector pair of virtual service is (0.5 + QoS × 0.5, 2.0 + QoS × 1.0), and the RAM resource needs vector pair is

(1.0 + QoS × 0.0, 1.0 + QoS × 0.0). If QoS is assumed $Q_{ij} = 1$ then the physical machine A fully satisfied in resource allocation to virtual machine. Otherwise, QoS is assumed $Q_{ij} = 0.5$, the physical machine B is not only fully satisfied in resource allocation but also the resource cost is reduced.

2.2 Objective and Constraints

Assume that each virtual service consists of a single virtual machine which has a fixed resource needs. A static multi-dimention resource allocation problem for virtual service from the heterogeneous shared hosting platforms (HMDRA) is formulated by a Mixed integer linear programming problem (MILP) as follows:

Let S be virtual services, indexed by i $= 1, \ldots, S; S > 0$, Y be physical machines having the different resource configuration, indexed by $j = 1, \ldots, Y; Y > 0$. Each physical machine provides D types of resource, indexed by $k = 1, \ldots, D$. Use r^e_{ik}, r^a_{ik} to denote the elementary rigid needs and aggregate rigid needs of service i for resource type k. f^e_{ik}, f^a_{ik} denote the elementary fluid needs and aggregate fluid needs of service i for resource type k. c^e_{jk},c^a_{jk} denote the elementary resource and aggregate resource of physical machine j for resource type k, and Q_{ij} denote the QoS of service i on physical machine j. A binary variable x_{ij} that is equal to 1 if service i is allocated the resource by physical machine j and 0 otherwise. Finally, a binary variable y_j indicates whether the physical machine j is in use or not for resource allocation to S virtual services. Objective is to minimize the number of used physical machines. Given these notations, a HMDRA problem represented by a MILP problem are as follows:

$$x_{ij} \in \{0, 1\}, Q_{ij} \in [0, 1], \quad \forall i, j \tag{1}$$

$$\sum_j x_{ij} = 1, \quad \forall i \tag{2}$$

$$y_j \geq x_{ij}, \quad \forall i, j \tag{3}$$

$$(r^e_{ik} + Q_{ij} f^e_{ik}) x_{ij} \leq c^e_{jk}, \quad \forall i, j, k \tag{4}$$

$$\sum_i (r^a_{ik} + Q_{ij} f^a_{ik}) x_{ij} \leq c^a_{jk}, \quad \forall j, k \tag{5}$$

and, objective is to

$$min \sum_j y_j \tag{6}$$

Constraint (1) defines the domain of the variables. Constraint (2) determines the state that resources for service i are provided by exactly one physical machine j. Constraint (3) specifies the state which a physical machine j is used, if it provides resource to at least one virtual machine. Constraint (4) states that the elementary resource of the physical machine j does not overcome, and constraint (5) states that the aggregate resource of the physical machine j does not overcome. Finally, constraint (6) is the optimization objective to minimize the number of used physical machines.

2.3 Complexity Analysis

To determine the complexity of HMDRA problem, let's consider HMDRA-Dec, the decision problem associated with HMDRA can be stated as: Is it possible to assign S services, each of which has a r_{ij} resource needs to Y physical machines?

Theorem 1. The decision problem HMDRA-Dec is NP-C.

Proof.

(a) It can be clearly seen that the problem HMDRA-Dec is NP. Because the solution, if it exists, can be verified in polynomial time.

(b) Considering the vector packing problem presented in [8, 9] as NP-C: give a set A consisting of elements of the d-dimensional vector represented by a d-tuple:$(a_1^i, a_2^i, \ldots, a_d^i)$ and set B consisting of elements of the d-dimensional vector represented by d-tuple: (1,1,…,1). Put the elements of the set A into the set B such that $\sum_{i \in B} a_j^i \leq 1, \forall j = 1, \ldots, d$.
The vector packing problem is reduced to the HMDRA problem as follows: let the number of physical machines be B (i.e., $Y = B$), the virtual services be A (i.e., $S = A$), the number of resource types be j (i.e., $k = j$), the resource needs of service i of the resource k be a_j^i (i.e., $r_{ij} = a_j^i$). For each service i, we considered aggregate resource needs and the rigid needs (for the fluid needs, the proof will be similar). Similarly, for the resource of physical machine, we considered the aggregate resource and its value is 1. Clearly, the vector packing problem provides a solution to the HMDRA-Dec problem. In contrast, a solution of the HMDRA-Dec problem will provide a solution to the vector packing problem. From (a) and (b), Theorem 1 has been proved.

3 Solutions for Solving the HMDRA Problem

3.1 Exact Solution

Solving the MILP formulation of HMDRA problem in Sect. 2.2 provides an exact solution. We use publicly solver lp-solver [13] on an Intel Core Duo 1.86 GHz and 2 GB RAM to compute exact solution for small problem instances (few physical machines, few virtual services, few type of resources) in under an hour.

3.2 Solution Based on Greedy Algorithms

The HMDRA is the problem of combinatorial optimization. So far, the methods usually used to solve combinatorial optimization problems, including: heuristic search to find a good enough solution [12]; local search to find local optimal solution [10]; approximate search through natural simulation algorithms [11]. We therefore seek heuristic algorithms that work well in practice. In general greedy algorithms operate by locally optimal decisions and run quickly, but may provide poor results for some problem instances. One way to deal with this issue is to run several greedy algorithms and choose the best solutions from those computed. The pseudo-code of the Greedy algorithm is as Algorithms 1.

Algorithms 1. Greedy Algorithm

Input:

- The number of virtual services S, the types of resources D, the elementary rigid needs r_{ik}^{e}, the aggregate rigid needs r_{ik}^{a}, the elementary fluid needs f_{ik}^{e}, the aggregate fluid needs f_{ik}^{a}, and QoS of service.
- The number of physical machines Y, the elementary resource C_{jk}^{e}, the aggregate resource C_{jk}^{a}.

Output: the number of used physical machines *NumUsedPM*

int i, j, k;
int *ArrayUsedPM[index]* ;
bool *Success* ← **False** ;
$j \leftarrow 1$;
while ($i \leq S$ **and** $j \leq Y$) **do**
  for $i := 1$ **to** S **do**
    for $k := 1$ **to** D **do**
      $Sum_{ik}^{e} \leftarrow r_{ik}^{e} + QoS \times f_{ik}^{e}$;
      $Sum_{ik}^{a} \leftarrow r_{ik}^{a} + QoS \times f_{ik}^{a}$;
      $Load_{ik} \leftarrow 0.0$; $Capacities_{jk} \leftarrow C_{jk}^{a}$;
      if ($C_{jk}^{e} \geq Sum_{ik}^{e}$ **and** $Capacities_{jk} \geq Sum_{ik}^{a}$) **then**
        if ($Capacities_{jk} - Load_{ik} \geq 0$) **then**
          $Load_{ik} \leftarrow Load_{ik} + Sum_{ik}^{a}$;
          $Capacities_{jk} \leftarrow Capacities_{jk} - Sum_{ik}^{a}$;
          $C_{jk}^{a} \leftarrow Capacities_{jk}$;
          Success ← **True** ;
        else
          Success ← **False** ;
        end if
      end if
    end for $k := 1$ **to** D
      if (*Success*)
        ArrayUsedPM[index] ← j;
        $index \leftarrow index + 1$;
        remove i;
      end if
  end for $i := 1$ **to** S
  $j \leftarrow j + 1$;
end while
NumUsedPM ← number of different value elements in *ArrayUsedPM[index]* ;

First, the index of physical machines, virtual services and resource types are declared (line 1). Use an one-dimensional array *ArrayUsedPM[index]* is to store the used physical machine j (line 2). Next, open a physical machine j (line 4) and iterations all resource type k of each service i to be provided resources from the physical machine j (line 5 to 27). Particularly, the elementary resource needs and the aggregate resource needs of resource type k of service i are estimated (line 8 and 9). Initialize the value of total resource needs k should be provided resources from the physical machine j as 0 and the total resources k of the physical machine j as the aggregate resource k of the physical machine j (line 10). Next, if elementary resource needs and aggregate resource needs of physical machine j satisfy the all resource needs k of virtual service i then recalculated the each resource k of the physical machine j after it has provided resources for service i (line 11 to 21). If resource allocaton is success, then value j is stored in the array *ArrayUsedPM[index]* (line 23), increase array index (line 24), and remove service i from S virtual services (line 25). If the current physical machine j does not meet the resources then open a new physical machine (i.e. increasing j) (line 28). Finally, the number of used physical machines are the number of elements which have the different value in the array *ArrayUsedPM[index]* (line 30).

Before applying this algorithm, the S virtual services are sorted in a descending or ascending order based on the resource needs of the following criterias [8, 9]:

(a) *Lexicographical*: given $k \geq 1, \theta^k = \{(r_1, r_2, \ldots r_k), \forall i, 0 \leq r_i \leq 1\}$ and a, $b \in \theta^k$. $a \leq b$ iff $a = b$ or the first nonzero component of $b - a$ is positive

(b) *Maximum Component*: given $k \geq 1, \theta^k = \{(r_1, r_2, \ldots r_k), \forall i, 0 \leq r_i \leq 1\}$ and $a, b \in \theta^k$. $a \leq b$ iff the maximum component in b is not less than the maximum component of a.

(c) *Maximum Sum*: given $k \geq 1, \theta^k = \{(r_1, r_2, \ldots r_k), \forall i, 0 \leq r_i \leq 1\}$ and $a, b \in \theta^k$. $a \leq b$ iff the sum of components of b is not less than the sum of components of a.

The combinatorial algorithms are constructed from Greedy algorithms based on the three criterias, therefore we have 6 algorithms including: *GreedyDecMax*, *GreedyDecLex*, *GreedyDecSum*, *GreedyAscMax*, *GreedyAscLex*, *GreedyAscSum*. If S are the number of virtual services, Y are the number of physical machines, and D resource dimensions are considered as a constant, those algorithms have the computational complexity of O(S.log$S$ + $S.Y$).

4 Numerical Results and Evaluations

4.1 Simulation Setup

To evaluate the proposed algorithms, we used a collection of randomly generated synthetic problem instances as follows: consider S services, Y physical machines and D resource dimensions. For each service, the number of rigid needs are $D/2$ and the number of fluid needs are $D/2$. All elementary resource needs are sampled from a normal probability distribution with mean μ_s and standard deviation σ_s The aggregate resource needs are defined by an elementary resource needs in each resource type k multiplied by a random integer between 1 and 3.

For the resource of physical machines, each physical machine resource meet to the resource of services, the elementary resource is defined by an elementary resource of rigid needs plus by elementary resource of fluid needs in each resource type k. In particular, each elementary resource is sampled from a normal probability distribution with mean μ_p and standard deviation σ_p. The aggregate resource of physical machine is defined by a elementary resource multiplied by a random integer between 3 and 5.

Assuming that the value of the parameters are as follows: QoS = 0.5, the number of services $S = \{32; 64; 128; 256; 512\}$, the number of physical machines $Y = \{32; 64; 128; 256; 512\}$, resource dimensions $D = 6$, $\mu_s = 0.5$, $\sigma_s = \{0.25; 0.5; 0.75\}$, $\mu_p = 0.5$, $\sigma_p = \{0.25; 0.5; 0.75\}$. This setup corresponds to $1 \times 5 \times 1 \times 1 \times 3 \times 1 \times 3 = 45$ scenarios. For each scenario, 100 random samples are generated resulting in a total of 4500 input individual instances used for evaluation.

Two metrics were employed for evaluation: the number of used physical machines and the execution time of the algorithm in seconds. The value of the two metrics are averaged from 900 (i.e., $3 \times 3 \times 100$) experimental instances that corresponding to the 5 values of service $S = \{32; 64; 128; 256; 512\}$. The algorithms were coded in C^{++} language and ran on an Intel Core Duo 1.86 GHz and 2 GB RAM.

4.2 Simulation Results and Evaluations

The execution time of algorithms are presented in Table 1 and Fig. 2(a). The number of used physical machines are presented in Table 2 and Fig. 2(b).

Table 1. The comparison of execution time (s)

Algorithms	Number of services				
	S = 32	S = 64	S = 128	S = 256	S = 512
GreedyDecMax	0.0001471	0.0002294	0.0004000	0.000972	0.002384
GreedyDecLex	0.0001570	0.0001991	0.0003813	0.000864	0.001594
GreedyDecSum	0.0001623	0.0002296	0.0003956	0.000954	0.001794
GreedyAscMax	0.0001828	0.0002112	0.0004045	0.000934	0.001658
GreedyAscLex	0.0001448	0.0002069	0.0003729	0.000863	0.001542
GreedyAscSum	0.0001542	0.0002243	0.0003884	0.000882	0.001742

From results are showed in Tables and Figures. It is clearly seen that with the same set of data, the execution time and the number of used physical machines are linear with the number of virtual services. The execution time is relatively small, thus, ones can be applied in practice. In particular, a *GreedyAscMax* algorithm gives the best number of used physical machines and a *GreedyAscLex* algorithm gives the best execution time.

Table 2. The comparison of used number of physical machines

Algorithms	Number of services				
	S = 32	S = 64	S = 128	S = 256	S = 512
GreedyDecMax	4.150	5.790	8.760	13.11	17.81
GreedyDecLex	4.500	6.090	9.480	12.62	17.83
GreedyDecSum	4.670	6.900	11.45	17.62	27.38
GreedyAscMax	4.130	5.370	7.670	10.45	14.75
GreedyAscLex	4.310	5.730	8.250	10.81	14.73
GreedyAscSum	4.530	6.140	10.17	14.85	22.38

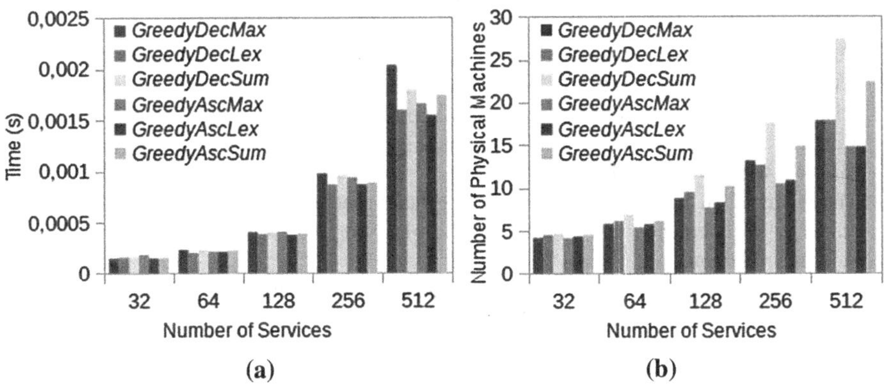

Fig. 2. (a) Execution time (s) (b) The number of used physical machines

5 Conclusion and Future Research

This paper has discussed the static multi-dimention resource allocation based on a hetergeneous shared hosting platforms for virtual services with the optimal constraints and QoS requirements; each service is considered as a single virtual machine. We have developed the formula for resource allocation problem based on linear programming problem with the objective to minimize the number of physical machines; set complexity of problem; and proposed the Greedy algorithms for installation and evaluation with the computational complexity of $O(S.\log S + S.Y)$. For the future work, the proposed model will be extended for dynamic resource allocation.

References

1. Urgaonkar, B., Shenoy, P., Roscoe, T.: Resource overbooking and application profiling in shared hosting platforms. SIGOPS Operating Syst. Rev. **36**(SI), 239–254 (2002). ACM

2. Aron, M., Druschel, P., Zwaenepoel, W.: Cluster reserves: a mechanism for resource management in cluster-based network servers. In: Proceedings of the 2000 ACM SIGMETRICS International Conference on Measurement and Modeling of Computer Systems, SIGMETRICS 2000, vol. 28, no. 1, pp. 90–101. ACM (2000)
3. Uhlig, R., et al.: Intel virtualization technology. IEEE Comput. **38**(5), 48–56 (2005)
4. Stillwell, M., Schanzenbach, D., Vivien, F., Casanova, H.: Resource allocation algorithms for virtualized service hosting platforms. J. Parallel Distrib. Comput. **70**(9), 962–974 (2010). Elsever
5. Nguyen, H.H.C., Dang, H.V., Pham, N.M.N., Le, V.S., Nguyen, T.T.: Deadlock detection for resource allocation in heterogeneous distributed platforms. In: Meesad, P., Boonkrong, S., Unger, H. (eds.) Recent Advances in Information and Communication Technology 2015. AISC, vol. 361, pp. 285–296. Springer, Heidelberg (2015)
6. Pham, N.M.N., Le, V.S.: A genetic algorithm for resource provisioning of virtual service based on homogeneous shared hosting platforms. Int. J. Eng. Res. Technol. (IJERT) **3**(12), 644–649 (2014)
7. Stillwell, M., Vivien, F., Casanova, H.: Virtual machine resource allocation for service hosting on heterogeneous distributed platforms. In: 2012 IEEE 26th International Parallel & Distributed Processing Symposium (IPDPS), pp. 786–797. IEEE (2012)
8. Kou, L.T., Markowsky, G.: Multidimensional bin packing algorithms. IBM J. Res. Dev. **21** (5), 443–448 (1977). IBM Corp
9. Maruyama, K., Chang, S.K., Tang, D.T.: A general packing algorithm for multidimensional resource requirements. Int. J. Comput. Inf. Sci. **6**(2), 131–149 (1977). Kluwer Academic-Plenum
10. Aarts, E., Lenstra, J.K.: Local Search in Combinatorial Optimization. Wiley, Chichester (1997)
11. Graham, R.L., Lawler, E.L., Lenstra, J.K., Rinnooy Kan, A.H.G.: Optimization and approximation in deterministic sequencing and scheduling: a survey. Ann. Discrete Math. **5**, 287–326 (1979)
12. Colornil, A., Dorigo, M., Maffioli, F., Maniezzo, V., Righini, G., Trubian, M.: Heuristics from nature for hard combinatorial optimization problems. Int. Trans. Oper. Res. **3**(1), 1–21 (1996)
13. http://lpsolve.sourceforge.net/5.5/
14. http://www.scalemp.com/technology/versatile-smp-vsmp-architecture/
15. Buyya, R., Yeo, C.S.: Service level agreement based allocation of cluster resources: handling penalty to enhance utility. In: 7th IEEE International Conference on cluster Computing, pp. 1–10. IEEE International, Burlington (2005)

Design and Implementation of Data Synchronization and Offline Capabilities in Native Mobile Apps

Kamoliddin Mavlonov(✉), Tsutomu Inamoto, Yoshinobu Higami, and Shin-Ya Kobayashi

Graduate School of Science and Engineering, Ehime University, 3 Bunkyou-cho, Matsuyama, Ehime 790-8577, Japan
kamol@koblab.cs.ehime-u.ac.jp, {inamoto,higami,kob}@cs.ehime-u.ac.jp

Abstract. This paper describes a solution for data synchronization, mobile offline capabilities, and network bandwidth optimization by utilizing a native smart device app as a distributed storage system. The solution aggregates the best practices in business and academic research to achieve a reduction in redundant data transfer and an ability to work offline in smart devices.

Keywords: Data synchronization · Offline capabilities · Smart devices · Mobile storage

1 Introduction

Currently, we see an explosion of applications (apps) that run on smart devices due to increase in internet usage and penetration of the devices. 7.2 billion People are using the internet from across both developed and developing countries [16]. The number of mobile broadband subscribers was sitting at 1.9 billion by December 2013 [15]. In 2014, the global mobile data traffic grew by 69 percent and reached 2.5 exabytes per month [15]. This increase has a direct relationship with how much electrical energy is used to run the infrastructure that supports the internet. The implication is that, since we cannot afford to reduce internet usage, more effort is needed its efficient usage through network bandwidth optimization. In this paper, we present a data synchronization approach, for apps running on smart devices, as a way of achieving efficient bandwidth usage by users.

We address the apps that behave as thin clients because they store a very minimum amount of remote data locally. Moreover, they largely lack the ability to work offline. This means that those apps regularly and redundantly download data from remote server leading to more packet loss in mobile-access networks [1]. In our study, we aimed at removing this redundancy between Web service and apps to reduce wastage in network bandwidth. In our proposed approach, we treat apps as parts of distributed storage systems. Any localized data that was

N.T. Nguyen et al. (Eds.): ACIIDS 2016, Part II, LNAI 9622, pp. 61–71, 2016.
DOI: 10.1007/978-3-662-49390-8_6

transferred from Web service to an app is not going to be retransferred unless it is outdated or corrupted. We achieve this using a modified version of Vector Clock algorithm [12]. Our focus is only on native apps due to better user experience and better utilization of devices platform [8].

This paper is structured as follows. Section 2 describes our design and implementation of Vector Clock algorithm [12]. Section 3 describes our implementation and outcome, with evaluation of mobile storage, transport protocol, and data interchange format. Section 4 describes our evaluation results. Section 5 describes related work, and we present our conclusion in Sect. 6.

2 Design

2.1 Overview

This section describes our design approach for data synchronization and offline capabilities of smart devices. It is applicable for all apps that require data synchronization by using Web services. Moreover, the design helps in optimizing the network bandwidth by avoiding redundant data transfer between a Web service and smart devices. The main components involved in data synchronization and offline provisioning are Server Synchronizer, Server Database, App Synchronizer and App DB. Section 2.2 gives a brief explanation of each and individual components and their interactions between them.

2.2 Data Synchronization

Figure 1 shows the sequence of interactions between a Web service and a client. Figure 2 shows the detailed overview of the interactions that happen between the server and the client. It also provides data format representation between the server and client interactions. The server and the client consist of a Synchronizer and Database (DB). The Server Synchronizer handles the response for all

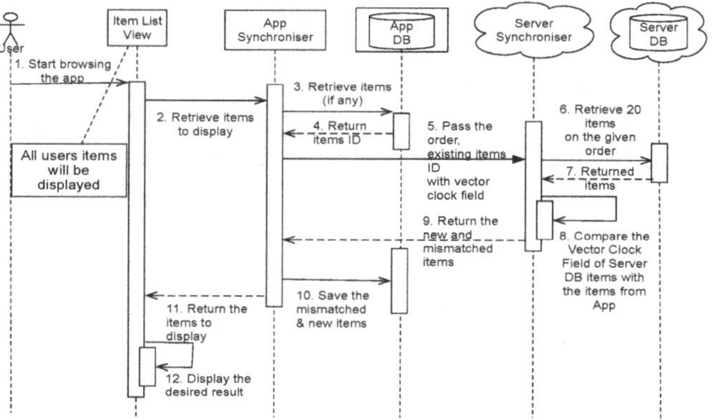

Fig. 1. Sequence diagram for data synchronization

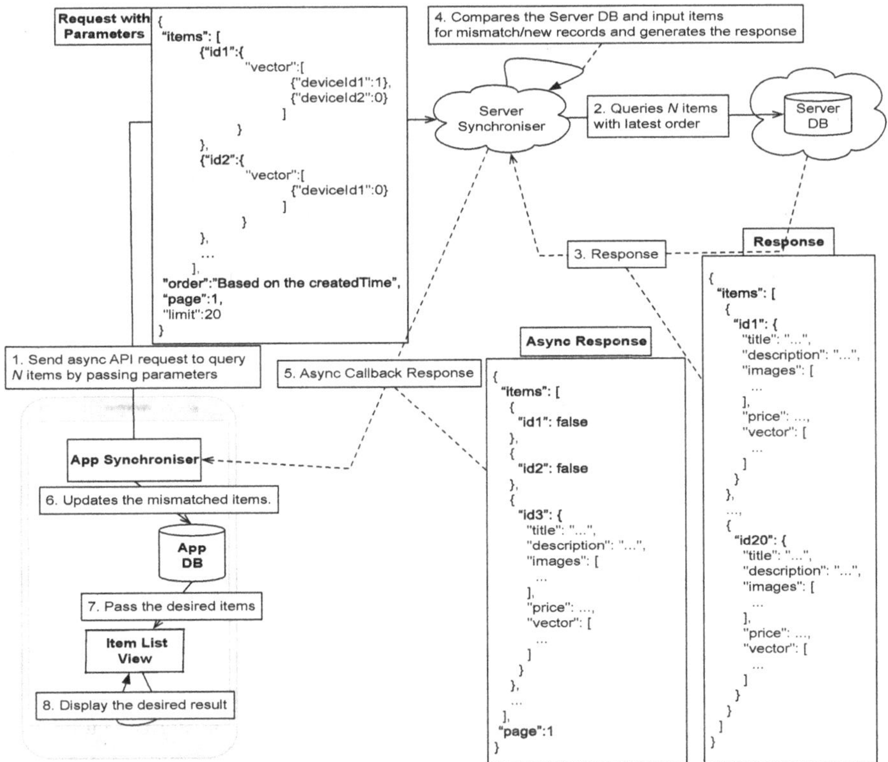

Fig. 2. Detailed overview of the server and client

request from clients. The App Synchronizer requests necessary information to the server and handles the response appropriately. It also acts as the controller for rendering the items to the view. When a user starts to browse Item list view in the app, the App Synchronizer sends an asynchronous request for retrieving N items by passing the items ID, order, page and vector clock field. The vector clock field is used for synchronization purposes and for finding the mismatched items. Section 2.3 describes the vector clock implementation in detail. Order and page parameters are used to retrieve the items in a particular order and corresponding page respectively. In our example, the sort order is based on the created time. The sample request format is shown in Fig. 2. After receiving the request from the App Synchronizer, the Server Synchronizer retrieves the N items from the Server DB for the corresponding page. The sample response from the Server DB to the Server Synchronizer is shown in Fig. 2. The Server Synchronizer compares the items between the Server DB and the client request, and then it sends the mismatched items to the App Synchronizer. The sample response format from Server Synchronizer to App Synchronizer is shown in Fig. 2. If an item is unchanged, then it is represented by a boolean value (*false*). After receiving the response, the App Synchronizer updates the App DB with mismatched items.

Then it passes the resultant items to the Item list view. The Item list view renders the screen by displaying the retrieved items. When the user further scrolls the item list for next page, the above sequence will be followed until all the items are loaded.

We consider that the maximum number of items stored in the App DB is M. If the user continues to browse more than M items, then those are directly fetched from server rather retrieving from the App DB. However, when a user goes offline, the user will be able to see the M items that are being stored in the App DB. Here M can be configurable based on the device available memory.

2.3 Vector Clock Implementation

The vector clock field associated with every item is responsible for the synchronization logic. For our scenario, we use the modified version of Vector Clock algorithm for implementing the data synchronization. The following are the other alternative ways that we could have considered for our synchronization logic.

Using Timestamp. Timestamps does not suit our case because each client device have different times.

Lamport Timestamp. The algorithm of Lamport timestamp [11] is a quite close to vector clock algorithm. This algorithm maintains the logical clock for a total ordering of all the events happenings. It works well when we need to order all the events based on its occurrence. In our app, each and every item is being associated with a particular user. Let us consider, *user1* has logged in three devices. It is enough for the *user1* to keep track of the updates related to his/her associated items. The *user1* does not need to keep track of all the updates of other users items instead it is enough for the *user1* to view the latest items from the other users. But the Lamport timestamps works in such a way that if an event has happened in a process, then it will be notified to all the other processes to keep track of the update happened. In our example, since the user is only interested on his/her item updates, Lamport timestamp is a bit of overhead for our scenario. Thus, we did not choose this approach.

Vector Clock Algorithm. Vector clock algorithm help us to keep track of the updates for which the user wanted, and it helps in data synchronization of the user. Vector clock can be visualized as an array of Lamport timestamp for each user, so the user is aware of a total ordering of all the events for which they are associated. It works in such a way that when a device creates/updates the item, or when a device receives the updated item, the vector clock sequence will be incremented. But for our case we have modified a bit by incrementing the vector fields only when there is an create/update. Let us consider a scenario where a user has been logged into two devices simultaneously and working on both devices at the different times. In this scenario, data synchronization between both the devices play an important role for the user to keep the information up to date. Each item has a vector clock field, which contains the device ID and corresponding counter to each device ID. For example, if an item vector clock fields is {d1:1, d2:2}. Here d1, d2 represent the first and the second device ID respectively. The sequence number that has been mapped to each device ID represents the number of times the create/update has been done by the corresponding device for a particular item. In this case, the d1 has created/modified an item once, and d2 has created/modified an item twice. If a user has another device d3 and if it has not created/modified an item then the d3's sequence number will be zero. If a sequence number for a particular device is zero, it won't be represented in the vector field.

Vector Clock Representation. Let us dive some more into the data synchronization logic using vector clocks. Consider V_i denotes i^{th} version, V_j denotes j^{th} version, D_n denotes n^{th} device ID sequence number. For example, $V_i D_n$ denotes i^{th} version n^{th} device ID sequence number.

Synchronization Logic Representation without Conflict. The following representation provides the synchronization without conflict

$$\forall\ \mathbf{n} \in \mathbf{N}\ \textbf{if}\ \mathbf{V}_i D_n \geq \mathbf{V}_j D_n\ \textbf{then}\ \mathbf{V}_j\ \textbf{will be replaced by}\ \mathbf{V}_i.$$

The above representation states that *for all n in N if i^{th} version is greater than or equal to j^{th} version then j^{th} version (V_j) can be replaced by i^{th} version (V_i).*

Here N represents the natural number. i.e. device IDs of the user for a particular item.

Synchronization Logic Formal Representation with Conflict. The following representation provides the synchronization with conflict

$$\exists\ \mathbf{n} \in \mathbf{N} \mid \textbf{if}\ \mathbf{V}_i D_n > \mathbf{V}_j D_n \wedge \exists\ \mathbf{m} \in \mathbf{N} \mid \textbf{if}\ \mathbf{V}_j D_m > \mathbf{V}_i D_m\ \textbf{then}\ \mathbf{V}_i\ \textbf{conflicts with}\ \mathbf{V}_j$$

The above representation states that *there exists n in N, such that if i^{th} version is greater than j^{th} version and there exists m in N, such that if j^{th} version is greater than i^{th} version, then i^{th} version (V_i) conflicts with j^{th} version (V_j).*

Here also N represents the natural number. i.e. device ID of the user for a particular item.

Synchronization Example. Figure 3 shows the data synchronization with and without conflicts. We determine the conflicts based on the logic representations as per the Sect. 2.3. In Fig. 3, the user has logged-in two devices. In the first device, the user creates an item. The vector clock field for the item has become

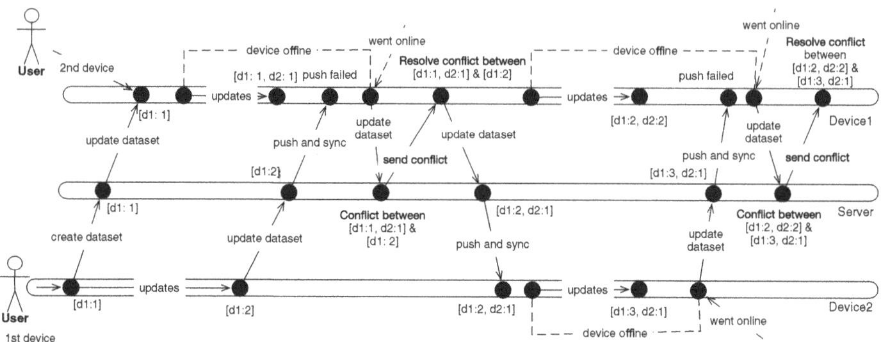

Fig. 3. Data synchronization between two devices

{d1:1}. Here d1 denotes the device ID of the first device. Then the server pushes the item to the second device. The second device vector clock field is empty. So as per the synchronization logic representation 1, the {d1:1, d2:0} is greater than or equal to the {d1:0, d2:0}. So the first device version will be copied in the second device. Later the second device updates its item in offline mode, so the second device vector field has become {d1:1, d2:1}. On the other hand the first device also updates the item, and then its vector clock field has become {d1:2} and pushes its changes to the server. Since the device 2 is offline, the server cannot synchronize the data for device 2. Once the device 2 goes online, it tries to update its data to the server. The server compares the vector field of incoming update {d:1, d2:1} and its own vector field {d1:2}. According to the logic representation 2 the server recognizes the conflict because for n = 1 the vector field {d1:2, d2:0} is greater than {d1:1, d2:1} and for n = 2 the vector field {d1:1, d2:1} is greater than {d1:2, d2:0}. To resolve the conflict, the server pushes the conflicted item to the second device. The user then modifies the conflicted items and updates the desired version to the server. In this case user chooses the first device version, so vector fields are populated in such a way that, the first device vector field {d1:2} is chosen and then the second device vector field is copied and appended to the first device vector. Now the vector field has become {d1:2, d2:1} and updates the server. The same update is also been pushed to the first device. Further, both first device and second device updates the data in offline, their vector fields has become {d1:3, d2:1} and {d1:2, d2:2} respectively. Initially, the first device goes online and updates its item to the server. The server tries to push the update to the second device since the second device is offline, the server push failed. Once the second device also goes online, it tries to update its item to the server. The server compares its vector fields with the incoming second device vector fields. The server recognizes the conflict because, as per the logic representation 2, for n = 1, the server vector field {d1:3, d2:1} is greater than second device vector field {d1:2, d2:2} and for n = 2 the second device vector field {d1:2, d2:2} is greater than server vector field {d1:3, d2:1}. To resolve the conflict, the server again pushes the conflicted item to the first device. The user modifies the item and updates the server with the desired version.

3 Implementation and Outcome

In the implementation, we mainly focus on mobile storage, transport protocol, and data interchange format. Those are our main building blocks of our design.

3.1 Mobile Storage

Most of the widely used mobile datastores are ORM databases. An ORM is Object-relational mapping, an abstraction that maps a relational database to an object-oriented API. ORM relies on a third-party database engine, which is primarily SQLite[1]. Moreover, a lot of mobile database engines are built on top of SQLite source code.

However, we find Realm[2] as the right mobile storage for our scenario. Realm is a database with built-in object relations and an object-oriented API. It has its database engine that fully supports the ACID properties. It does not rely on ORM by providing data as objects. All queries are done by code. By removing ORM, it eliminates the issues related to maintenance and performance. Our performance benchmark of Realm (vs. other ORMs and SQLite) is described in Sect. 3.1.

Benchmark. For the benchmarking purpose we use Nexus 6 device (Android 5.1.1) with the latest available version of each mobile datastores as of 28 Sep 2015. This is a benchmarking suite in a series of common and simple operations:

- Batch Writes: write a bunch of objects in a single transaction
- Simple Writes: write a bunch of objects in separate transactions
- Simple Queries: perform a simple query with filters
- Full Scan: perform the simple query with conditions that ensure a full table scan and get the size of the result
- Count: count the number of entries in a table
- Sum: sum the values of an int field

We run the benchmark with 1000 objects in 50 iterations. The minimum (best) time is taken from each datastore as shown in Table 1. The reason is why we choose the minimum time is because of system overhead. The overhead comes from other activities going on when the measurement is done, despite enabled airplane mode.

[1] http://www.sqlite.org/.

[2] https://realm.io.

Table 1. Benchmarking suite results (milliseconds) to compare Mobile datastores in a series of common and simple operations.

Datastore	Batch Write	Simple Write	Simple Query	Full Scan	Count	Sum
Realm	38.2	0.274	0.22	0.166	0.0229	0.0286
SQLite	36.5	0.662	0.715	0.605	0.311	0.583
OrmLite	93.8	0.828	0.811	0.679	0.158	0.593
Sugar ORM	333	1850	1.16	0.768	0.239	205

3.2 Transport Protocol

As a transport protocol, we choose an HTTP/2 [5]. HTTP/2 is recently approved by The Internet Engineering Task Force (IETF) [6] and is not widely tested in production. But in our scenario, we are using an HTTP client instead of a mobile web browser (Webapp), so it is sufficient for us to utilize the base features of HTTP/2 [2–4] more effectively.

HTTP Header. One of the reasons for choosing HTTP/2 over HTTP/1.1 is because, HTTP/1.1 lacks the header compression. HTTP/1.1 carries header as a plain text for any request/response between apps and Web Services. HTTP/2 addresses this issue, with header compression and binary encoding. By choosing HTTP/2 in our scenario, we have reduced our HTTP header size up to 45 %.

SSL Connection. We understand that there is an increase in the cost of bandwidth and battery utilization due to HTTPS implementation between end-points and the apps [10]. This essential cost of HTTPS implementation can also be reduced slightly by using HTTP/2. Each new SSL connection requires an intensive key pair exchange between device and SSL Termination. However in HTTP/2, only one initial connection is needed instead of multiple connections as in HTTP/1.1.

Single, Persistent Connection. All app's connections are initiated by user's actions whenever there is a need for the user. To overcome many connection initiations from the app, the Server Synchronizer provides the response data on the same connection that has been previously initiated by the user for the first time. Server Synchronizer attempts to guess the app's subsequent request and provides the response in advance.

Server Push. HTTP/2 has Server Push feature, by which it can push data to the HTTP client's cache. In our scenario, during data synchronization, if there is a conflict then Server Push is used to transfer data to the client for resolving the conflict.

3.3 Data Interchange Format

JSON[3] is suited for Webapps [9], however for data interchange format between Web Services and apps, we choose MessagePack[4]. Using MessagePack we can achieve a similar data structure as JSON; however, it is smaller in size. We divide our response data from Web Server into three average sizes as shown in Table 2. In average, we can achieve data reduction of 6–9 % in large, 4–9 % in medium and 29–33 % in small messages.

Table 2. MessagePack and JSON comparison

Message size	JSON (utf-8)	MessagePack (utf-8)	Diff	JSON (ascii)	MessagePack (ascii)	Diff.
Large	4,495 bytes	4,207 bytes	6 %	3,215 bytes	2,927 bytes	9 %
Medium	2,077 bytes	2,003 bytes	4 %	587 bytes	527 bytes	9 %
Small	27 bytes	18 bytes	33 %	42 bytes	30 bytes	29 %

4 Evaluation

For our evaluation, we build two native apps: CurrentApp and NextGenApp. CurrentApp is based on HTTP/1.1 protocol with RESTful API, and local storage is disabled. NextGenApp is based on HTTP/2 protocol with RESTful API, and local storage is enabled with Realm. CurrentApp's and NextGenApp's data interchange formats are JSON and MessagePack respectively. CurrentApp acts as a thin (Webapp) client, and NextGenApp acts as a thick client where we implement our data synchronization design. We run the same amount of requests on both those apps and monitor the Web server HTTP Response (response). Each response is divided by a range of items, e.g. (0; 1], (1; 5], (5; 20] are Small, Medium and Large respectively. Even on worst-case scenario where response

Table 3. Bandwidth utilization in CurrentApp and NextGenApp for uncached Large, Medium and Small responses.

Response	Bandwidth utilization in				
	CurrentApp	NextGenApp			
		Worst-case scenario	Diff.	Best-case scenario	Diff.
Large	13,924 bytes	12,526 bytes	11 %	271 bytes	98 %
Medium	4,389 bytes	3,818 bytes	14 %	74 bytes	98 %
Small	1,154 bytes	802 bytes	31 %	22 bytes	97 %

[3] http://www.json.org/.
[4] http://msgpack.org/.

data is not cached on local datastores, the NextGenApp's bandwidth utilization is lesser than CurrentApp's bandwidth utilization as shown in Table 3. The difference is 11 %, 14 % and 31 % for Large, Medium and Small responses respectively. More we use the NextGenApp, it falls under Medium and Small responses category, thereby saving up to 98 % bandwidth (in best-case scenario) between the Web server and the client.

5 Related Work

Majchrzak [9] has very similar concept. However, Majchrzak's work includes the compilation of requirements for Web apps to work offline. In our case, we focus only on native apps and implementation does not rely on HTML5 or HTML5 storage capabilities.

Lomotey [13] mainly focus on middleware (very similar to our Server Synchronizer). However, Lomotey's mobile device acts as a thin client, and entire implementation of a mobile side of the app is built in HTML5, which similar to Majchrzak [9]. In our case, we also focus on the mobile side, including its native storage, transport protocol and data interchange format as described in Sect. 3.

Xue [14] has shown the similar architecture as ours for data synchronization between mobile devices and server. By assigning a unique identifier (generated by the server) for each mobile device and by using the timestamp, the data synchronization between mobile devices and a server has been accomplished. In our case, we used the Vector clock algorithm and mobile's device id for accomplishing the data synchronization. Section 2.3 describes our implementation in detail.

Couchbase[5] provides Couchbase Mobile, the embedded datastore with built-in data synchronization. It consists of three components. Couchbase Lite - NoSQL [7] datastore on mobile. Couchbase Sync Gateway - a server synchronizer between a mobile device and Couchbase Server and Couchbase Server - NoSQL database. It uses JSON as data interchange format. However, we use MessagePack as described in Sect. 3.3.

6 Conclusion

We presented our novel solution for data synchronization between Web services and mobile devices. Our solution also provides the offline capabilities to enhance the user experience of the apps and also reducing bandwidth usage by removing redundant data requests and response. We were able to present the status quo of mobile storages by benchmarking them. The Vector Clock algorithm in our design solution plays a significant role for data synchronization between the server and the client. Usage of new HTTP protocol (HTTP/2) and binary data interchange format (MessagePack) enhance the performance and reduction in the bandwidth usage respectively.

[5] http://www.couchbase.com/.

References

1. Heikkinen, M.V.J., Berger, A.W.: Comparison of user traffic characteristics on mobile-access versus fixed-access networks. In: Taft, N., Ricciato, F. (eds.) PAM 2012. LNCS, vol. 7192, pp. 32–41. Springer, Heidelberg (2012)
2. Stenberg, D.: HTTP2 explained. SIGCOMM Comput. Commun. Rev. **44**, 120–128 (2014). ACM, New York
3. Grigorik, I.: Making the web faster with HTTP 2.0. Commun. ACM **56**, 42–49 (2013). ACM, New York
4. de Saxce, H., Oprescu, I., Yiping, C.: Is HTTP/2 really faster than HTTP/1.1? In: Computer Communications Workshops (INFOCOM WKSHPS), Hong Kong, pp. 293–299. IEEE (2015)
5. Hypertext Transfer Protocol Version 2 (HTTP/2). https://httpwg.github.io/specs/rfc7540.html
6. The Internet Engineering Task Force. http://www.ietf.org/blog/2015/02/http2-approved/
7. Klein, J., Gorton, I., Ernst, N., Donohoe, P., Pham, K., Matser, C.: Performance evaluation of NoSQL databases: a case study. In: Proceedings of the 1st Workshop on Performance Analysis of Big Data Systems (PABS 2015), pp. 5–10. ACM, New York (2015)
8. Charland, A., Leroux, B.: Mobile application development: web vs. native. Commun. ACM **54**, 49–53 (2011). ACM, New York
9. Majchrzak, T.A., Hillmann, T.: Offline-provisioning and synchronization of content for mobile webapps. In: Proceedings of the 11th International Conference on Web Information Systems and Technologies, pp. 601–612 (2015)
10. Naylor, D., Finamore, A., Leontiadis, I., Grunenberger, Y., Mellia, M., Munafò, M., Papagiannaki, K., Steenkiste, P.: The cost of the "S" in HTTPS. In: Proceedings of the 10th ACM International on Conference on Emerging Networking Experiments and Technologies, pp. 133–140. ACM, New York (2014)
11. Lamport, L.: Time, clocks, and the ordering of events in a distributed system. Commun. ACM **21**, 558–565 (1978)
12. Mattern, F.: Virtual time and global states of distributed systems. In: Cosnard, M., et al. (eds.) Proceedings of the Workshop on Parallel and Distributed Algorithms, pp. 215–226. Elsevier Science Publishers B.V., North-Holland (1989)
13. Lomotey, R., Chai, Y., Ashik, A., Deters, R.: Distributed mobile application for crop farmers. In: Proceedings of the Fifth International Conference on Management of Emergent Digital EcoSystems (MEDES 2013), pp. 135–139. ACM, New York (2013)
14. Xue, Y.: The research on data synchronization of distributed real-time mobile network. In: 2008 International Conference on Computer Science and Software Engineering, Wuhan, vol. 3, pp. 1104–1107. IEEE (2008)
15. ICT Facts and Figures 2005, 2010, 2014, Telecommunication Development Bureau, International Telecommunication Union (ITU). Accessed 24 May 2015
16. ITU Global internet report 2014

Multiple Model Approach to Machine Learning

A Machine Learning Based Technique for Detecting Digital Image Resampling

Hieu Cuong Nguyen(✉)

Faculty of Information Technology, University of Transport and Communications, Hanoi, Vietnam
cuonggt@gmail.com

Abstract. Digital images can easily be tampered because of the popularity and power editing software. In order to create a persuasive forged image, the image is usually exposed to several geometric transformations, such as rescaling and rotating. Since the manipulations require a resampling step, uncovering traces of resampling became an important approach for detecting image forgeries. In this paper, we propose a new technique to reveal image resampling artifacts. The technique employs specific features of the linear dependencies of neighboring image samples for discriminating resampled images from original images. A machine learning method is utilized for classification. Experimental results in a large dataset show that the proposed technique is good in detecting resampled images, even when the manipulated images were slightly transformed.

Keywords: Resampling · Image forensics · SVM · Classification

1 Introduction

Nowadays, digital images are widely used and mostly replace traditional photographs. A significant advantage of digital images is that they can easily be manipulated. Although it is useful to enhance image quality, it can be utilized for illegal purposes. Since image processing tools are popular, even a novice can create forged images without leaving visible evidence. Therefore, the reliability of digital images becomes doubtful and image authentication issued as a crucial problem.

Digital watermarking is a well-known method for image protection and authentication in decades (see, for example [1, 2] for overview). Since watermarking techniques work only for the images where a watermark was embedded, it cannot be applied to authenticate arbitrary images. Recently, image forensics has been regarded as a new approach for revealing image authenticity. Image forensics is a passive method, which can be used to detect image forgeries without using a watermark or any type of fingerprint. It is based on the assumption that although forgeries may leave no obvious evidence, they may alter intrinsic statistics of an image. There are many types of image tampering, which can be exposed by different forensic techniques [3].

To create a convincing forged image, the image is often underwent geometric transformations, which require a step of resampling. Therefore, uncovering resampling artifacts, which have become an effective way for detecting image tampering. They are mostly based on the variances of the second derivatives of image pixels [4–6] or the

N.T. Nguyen et al. (Eds.): ACIIDS 2016, Part II, LNAI 9622, pp. 75–84, 2016.
DOI: 10.1007/978-3-662-49390-8_7

analysis of linear dependencies among neighboring image samples [7–9]. An evaluation for performance and robustness of some widely used resampling detection techniques can be found in [10].

The techniques based on identifying the variances of the second derivatives in images are rather simple and easy to implement but they often suffer from high false positives [6]. Besides, some of them work only for discovering resized images and they are not capable of exposing resampling artifacts for rotated images [4, 5]. Thus, we focus mostly on the techniques [7–9]. The technique of Popescu and Farid [7] is powerful (with high detection capacity and low false positive rate), but it is rather complicated. To overcome the downside of [7], Kirchner proposed a faster resampling detector [8]. In general, the mentioned techniques are good for detecting image resampling. However, a common drawback of the techniques is that they are mostly not possible to detect resampling when the images were slightly manipulated (by very small scaling factors or very small rotation angles). Some studies [8, 9] showed that the probabilities of the correlations of pixels and their neighboring are periodic. Therefore, in this work, based on a reasonable set of predefined weighted values, we first compute the probabilities of image samples, which correlate to their neighbors. Next, the probabilities of all samples of the evaluated image form a probability map. The Radon transform is then applied to the map in order to improve the detection capacity, especially in the case of image rotation.

We extract maximal peaks of the Fourier spectrum of the Radon transformed signals and design specific features, which are subsequently fed to a support vector machine (SVM) based classifier in order to distinguish resampled images from original images. The performance of the proposed technique will be evaluated with a large image dataset. We test the proposed technique and the other related tools [7, 9] under the same condition in order to achieve a fair comparison. Experimental results show that the proposed technique is good for resampling detection and when images were slightly manipulated, it outperforms the other analyzed techniques. Moreover, the robustness against deliberated attacks of the proposed technique is also evaluated.

The rest of this paper is organized as follows: In the next section, we briefly review essential background and related resampling detection techniques and the proposed technique is described in Sect. 3. Experimental results are shown in Sect. 4. Finally, our conclusions are drawn in Sect. 5.

2 Background

2.1 Image Interpolation

Interpolation is the central operation of image resampling for measuring the value of a sample at intermediate position to the original samples. This step is the key to smooth the signal and create a visually appealing image [11].

A p/q resampling of a 1-D discrete sequence x[k] involves the following steps [12]:

- Up-sampling: create a new signal $x_u[k]$ by inserting p-1 zeros after every x[k].
- Interpolating: convolve the sample $x_u[k]$ with h[k], where h[k] is an interpolation filter (e.g. bi-linear and bi-cubic): $x_i[k] = x_u[k] * h[k]$.
- Down-sampling: pick every *qth* sample: $y[k] = x_i[qk]$, k= 0, 1, 2…

Resampling 2-D (e.g. digital images) is straightforward where the aforementioned operations are applied in both spatial directions.

2.2 Resampling Detection

The main part of various resampling detection techniques is to estimate the probability of every image sample, which is correlated to its neighbors. The technique [7] employs a linear predictor to approximate each sample y_i as the weight sum of its neighboring samples:

$$y_i = \sum \alpha_k y_{i+k} + r_i.$$

The residue is modeled as a zero-mean Gaussian random variable in order to compute the correlation probability of the sample. The probability values of all samples of an image make the probability map (p-map). The authors of [7] empirically found that the p-map of a resampled image is periodic and the periodicity can be recognized in the frequency domain. Since this characteristic appeared only in the case of resampled images, it can be used to differentiate resampled images and original images. However, in practice, the neighboring size and the weighted values (α) are not known, so the p-map cannot be directly computed.

In order to compute the p-map, the technique [7] employs the expectation maximization (EM) algorithm and weighted least square (WLS) estimation. Thus, the technique [7] is highly computationally expensive. Kirchner [8] showed that the p-map of a resampled image is always periodic, no matter which set of α would be used. The authors in [9] found many times that using one predefined set of weights for detecting an image, significant peaks can easily be recognized in the transformed p-map, but using another set of α, peaks are not evident (although the periodicity exists in theory).

3 Proposed Technique

A drawback of the mentioned techniques is that they are mostly not possible to detect resampling when the images were slightly manipulated. To overcome the disadvantage, we propose a new technique, which consists of three main steps: (1) computing the probability map of the analyzed image with a predefined set of α (thus, we called it the pseudo probability map or pseudo p-map), (2) applying the Radon transform to the pseudo p-map, and (3) extracting specific features from the transformed map, and employing the features for image classifying.

3.1 Probability Map Computation

For each sample y_i, by using a predefined set of weighted values, the residue r_i is computed as:

$$r_i = \left| y_i - \sum \alpha_k y_{i+k} \right|.$$

The probability can be modeled as a zero-mean Gaussian distribution and the correlation probability of a sample is estimated based on the residue:

$$p_i = \frac{1}{\sigma\sqrt{2\pi}} e^{-r^2/2\sigma^2},$$

where σ is the variance of the Gaussian distribution.

Based on Bayes' rule, the value w_i can be estimated as follows:

$$w_i = \frac{p_i}{p_i + p_0},$$

where p_0 is the reciprocal of the range of the image y. In practice, we set the value of p_0 is equal to 1/max, where max is denoted for the size of the range of possible values for y_i (max = 255 for gray-scale images). The values w_i of all image samples form the pseudo p-map of the image.

3.2 Radon Transform

The Radon transform computes projections of an image along various directions. The transformed version g(r,θ) of an image f(x,y) can be computed as below [13]:

$$g(r, \theta) = \int_{-\infty}^{\infty} \int_{-\infty}^{\infty} f(x, y)\delta(r - x\cos\ \theta - y\sin\theta)\, dxdy,$$

where δ is the sifting property of the impulse function (Dirac function).

Since the Radon transform has robust properties against rotation, scaling and translation (RST) operations as well as noise addition [14], we believe that using the transform can improve the robustness of the detection technique. The Radon transform in this case should be in discrete form [15] because digital images are discrete. In this work, we use the `radon` function of Matlab.

In this technique, we apply the Radon transform with a set of predefined angles to the pseudo p-map of the analyzed image. The transformed result is a set of projected vectors, which are arranged in a matrix. Since our goal is to determine whether the image is being investigated undergone affine transformation, we focus only on the strongest periodic patterns present in the Fourier transform of the auto-covariance of the projected vectors. The patterns are plotted in the Fourier transform spectrum in order to uncover peaks. Due to the periodicity of the pseudo p-maps, significant peaks will be shown in the case of resampled images.

3.3 SVM Based Classification

When using the aforementioned procedures, we found in many times that there are differences for the resampled images and the original images in the Fourier spectrum. For example, the Fourier spectrum of the Radon transform of an original image (Fig. 1 left) is shown in Fig. 1 (right). When the image is resized with the factor of 1.1, the spectrum is changed as shown in Fig. 2 (left). Similarly, if the image is rotated by the angle of 10 degree, the Fourier spectrum is presented as Fig. 2 (right).

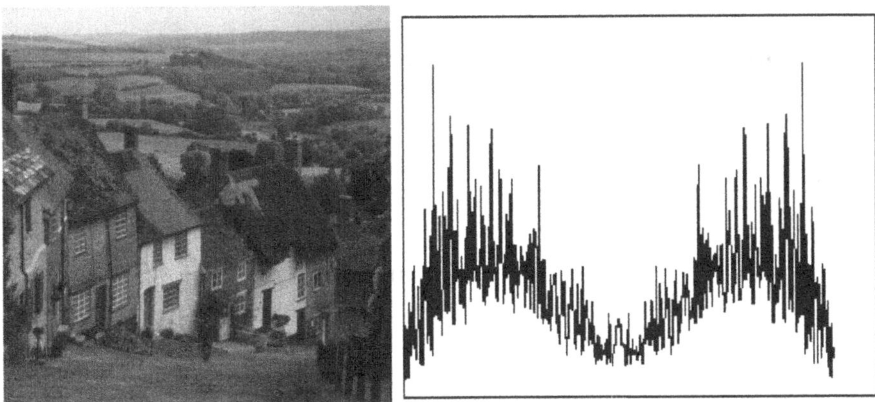

Fig. 1. An original image to be tested (left) and the Fourier spectrum of the Radon transform for the pseudo p-map of the original image (right)

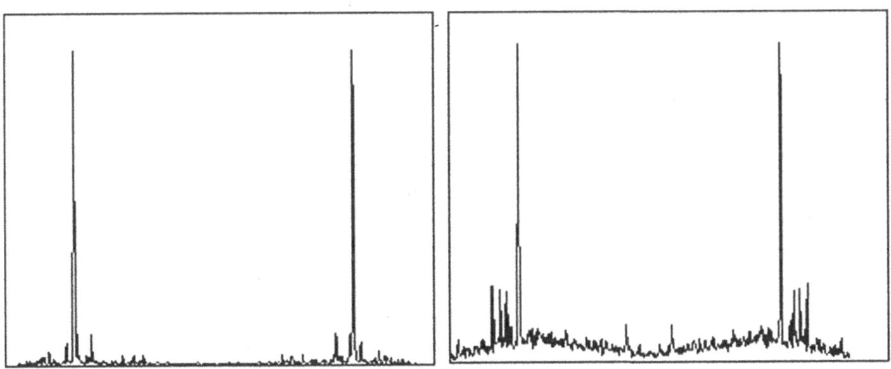

Fig. 2. The Fourier spectrum of the Radon transform for the pseudo p-map of the resized image with the scaling factor is 1.1 (left) and the Fourier spectrum of the Radon transform for the pseudo p-map of the rotated image with the rotation angle is 10° (right)

This can easily be recognized the difference between the spectrums of the original image and its resampled versions, even by human eyes (when observing the significant peaks). However, it seems not enough reliable to give a general conclusion. Some techniques apply a threshold based peak detector searching for local maximum [6, 9]. However, it is difficult to manually choose an appropriate threshold so that the resampling detectors achieve high detection rate and low false positive rate. In this paper, we employ a machine learning based method (SVM) for image classification. Firstly, we analyze images and form their pseudo p-maps (Sect. 3.1). Next, we apply the Radon transform to the pseudo p-maps and extract the values of maximal peaks from the Fourier spectrum (Sect. 3.2). Lastly, in this section, the extracted features are fed to SVM-based classifiers to discriminate resampled images from original images.

4 Experimental Results

Initially, we set values for the parameters for computing the probability map: the variance $\sigma = 0.075$, $p_0 = 1/255$ (since in our experiments, we apply the detection technique to gray-scale images). Following the suggestion in [8, 9], we use a following set of α:

$$\alpha = [-0.25,\ 0.5,\ -0.25,\ 0.5,\ 0,\ 0.5,\ -0.25,\ 0.5,\ -0.25].$$

For the step described in Sect. 3.2, we applied the Radon transform with the angles from 0° to 179° with an incremental step of 1° to the map, which is obtained from the previous procedure.

After that, we prepared two groups of images and extract their specific features, as shown in Sect. 3 for training. Naturally, the more maximal peaks are used, the better quality of classification. However, using more peaks will require more time for forgery detection. After many trials, we found that using 20 peaks is reasonable.

For the first group, we randomly chose 160 original images from the UCID image database [16]. We applied different geometric transformations to the original images in order to create resampled images. As a result, we obtained eight datasets for up-sampling (by scaling factors of 1.05, 1.1, 1.2, and 1.3), down-sampling (by scaling factors of 0.9 and 0.7) and rotating (by rotation angles of 5° and 10°). We selected randomly 20 images from each resampled dataset, therefore, obtained the second group of 160 resampled images. In this case, two groups to train a SVM classifier were used. After training, the classifier can be used to test whether an image has been geometrically transformed.

Next, we randomly selected 200 original images from [16] to test. Several geometric transformations were applied to the original dataset in order to create different resampled image datasets (each dataset consists of 200 resampled images). We rescaled the original images by the scaling factors of 1.01, 1.03, 1.05, 1.1, 1.2, 1.25, 1.3 (up-sampling) and 0.95, 0.9, 0.7, and 0.5 (down-sampling). Next, we rotated the original images by the rotation angles of 1°, 3°, 5°, 10°, 20° and 30°. Consequently, we have total 3400 resampled images to be evaluated.

In order to analyze digital images, we extracted their specific features following the algorithm in Sect. 3. The features were fed to the trained SVM-based classifier in order to discriminate resampled images from original images. In this work, we utilized the machine learning based functions of Matlab (`svmtrain` and `svmclassify`) with the Radial Basic Function (RBF) kernel and other default parameters. The experimental results are shown in Figs. 3 and 4.

We found that, the technique work well to detect up-sampled images by the scaling factor of 1.2 or larger with the detection rates are larger than 80 %. When applying the technique to down-sampled images, we realized that the detection rates were significantly decreased to 50 % or lower. The technique works well in detecting rotated images: it obtains the results about 50 % even the rotation angles are small. With the images, which have been rotated by an angle of 10° or larger, the detection results are about 70 % to 80 %.

In order to compare the proposed technique with other resampling detectors, we selected two other techniques [7, 9] and prepared a common condition for testing. Due to inheritance some important steps from [9], the proposed technique had also low computational complexity and worked fast. To evaluate [7, 9], we use the same dataset as used in testing our proposed technique: a set of 200 original images, and different versions of up-sampled images, down-sampled images, and rotated images (over 3000 resampled images). We adjusted some intrinsic parameters of [7, 9] so that they both had similar false positives to the proposed technique with the rates are lower than 5 %.

The experimental results of three techniques for detecting up-sampled and rotated images are shown in Figs. 3, 4 respectively. We found that, the technique of Popescu and Farid (PF) [7] is the best in detecting up-sampled images with high detection rates, except in the case of very small scaling factor. Although our proposed technique is less powerful than [7] when analyzing up-sampled images, its detection capacity is slightly stronger than the technique of Nguyen and Katzenbeisser (NK) [9]. For exposing resampling artifacts in rotated images, the proposed technique is not as good as [7], but it is usually better than [9]. Our technique works quite well even when the rotation angle is small. Besides, while all three techniques do not work well to detect down-sampled images, with the detection rates are about 50 % or lower, the proposed technique is also better than the others in most cases.

The robustness of an image forensic technique was determined by applying different countermeasures, such as Gaussian noise addition and filtering to resampled images. When an image is down-sampled by a factor of two, no sample in the down-sampled image can be written as a linear combination of its neighbors and traces of resampling will not be noticed [7], the operation can be an effective attack. In order to design a more powerful attack, we use a combination approach. Firstly, the image is up-sampled by a factor of two, then down-sampled by a factor of two. At the end, median filter is applied to the image. We evaluated a dataset of 200 up-sampled images by a factor of 20 %, when no attack was employed, the detection rate of [7] is about 100 % and the detection rate of our technique is about 80 %. However, when we applied the combination attack to the dataset, we found that the detection rate of the approach of [7] reduced impressively to 1 % while the detection rate of the proposed technique remained about 50 %. This implies that our technique is robust against the powerful attack.

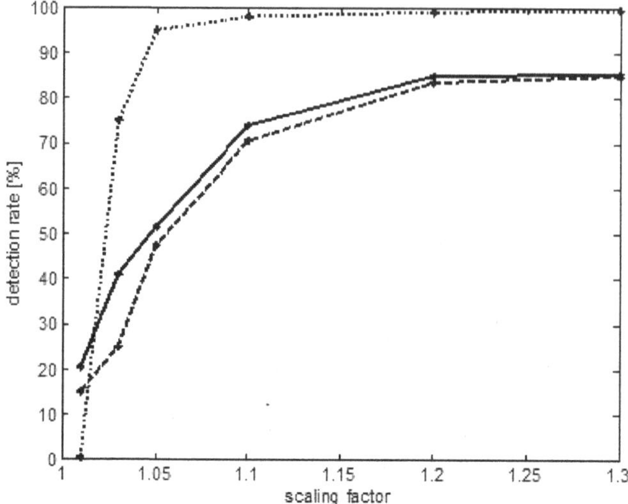

Fig. 3. Detection rates for up-sampled images: dot line for PF's [7], dashed line for NK's technique [9], and solid line for the proposed technique

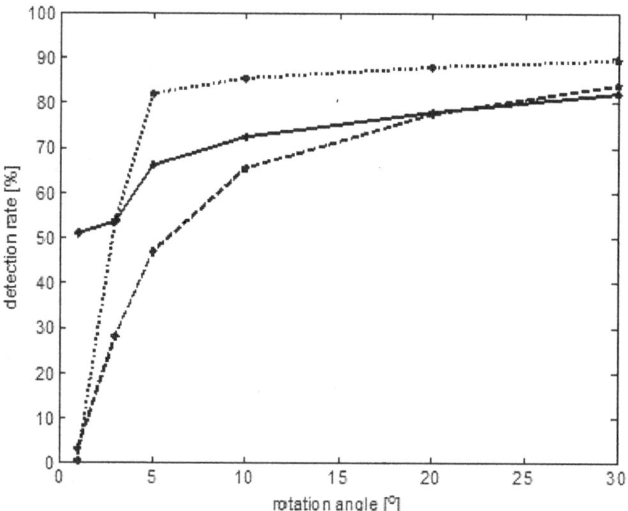

Fig. 4. Detection rates for down-sampled images: dot line for PF's technique [7], dashed line for NK's technique [9], and solid line for the proposed technique

5 Conclusion

In this paper we proposed a new technique for detection traces of resampling in digital images. The technique employed some ideas from [9] for making the probability maps of analyzed images. The maps were then processed by using the Radon transform and the Fourier transform in order to plot distinguishable spectrum. The maximal peaks of the spectrum were used for classifying images by applying SVM. For classifying we designed a binary SVM classifier which was trained by many original images and resampled images in different types of scaling and rotating. In order to get a reliable conclusion, the technique was tested with a large dataset of over 3000 resampled images. The experimental results showed that, the proposed technique work well for detecting up-sampled images by a scaling factor of 1.1 or larger and rotated images by a rotation angle of 5° or larger. To obtain a fair comparison with related techniques, the proposed technique and the techniques [7, 9] were evaluated under the same condition. We found that, in general, the detection capacity of the proposed technique is not as good as [7], but better than [9]. Moreover, the proposed technique outperforms the other techniques in detecting down-sampled images, up-sampled images (with small scaling factors) and rotated images (with small rotation angles). The proposed resampling detection technique is also robust against deliberated attacks.

References

1. Katzenbeisser, S., Petitcolas, F.A.P.: Information Hiding Techniques for Steganography and Digital Watermarking. Artech House, Norwood (2000)
2. Cox, I., Miller, M., Bloom, J., Fridrich, J.: Digital Watermarking and Steganography. Morgan Kaufmann, San Francisco (2009)
3. Piva, A.: An overview on image forensics. ISRN Sig. Process. **2013**, 1–22 (2013). Article ID 496701
4. Prasad, S., Ramakrishnan, K.R.R.: On resampling detection and its application to detect image tampering. In: ICME 2006, pp. 1325–1328 (2006)
5. Gallagher, A.C.: Detection of linear and cubic interpolation in JPEG compressed images. In: The 2nd Canadian Conference on Computer and Robot Vision, pp. 65–72 (2005)
6. Mahdian, B., Saic, S.: Blind authentication using periodic properties of interpolation. IEEE Trans. Inf. Forensics Secur. **3**, 529–538 (2008)
7. Popescu, A.C., Farid, H.: Exposing digital forgeries by detecting traces of re-sampling. IEEE Trans. Signal Process. **53**, 758–767 (2005)
8. Kirchner, M.: Fast and reliable resampling detection by spectral analysis of fixed linear predictor residue. Proceedings of the 10th ACM Workshop on Multimedia and Security, MM&Sec (2008)
9. Nguyen, H.C., Katzenbeisser, S.: Robust resampling detection in digital images. In: Decker, B., Chadwick, D.W. (eds.) CMS 2012. LNCS, vol. 7394, pp. 3–15. Springer, Heidelberg (2012)
10. Nguyen, H.C., Katzenbeisser, S.: Performance and robustness analysis for some re-sampling detection techniques in digital images. In: Shi, Y.Q., Kim, H.-J., Perez-Gonzalez, F. (eds.) IWDW 2011. LNCS, vol. 7128, pp. 387–397. Springer, Heidelberg (2012)
11. Wolberg, G.: Digital Image Warping. IEEE Computer Society Press, Los Alamitos (1994)
12. Oppenheim, A., Schafer, R.: Discrete-Time Signal Processing. Prentice Hall, Englewood Cliffs (1989)
13. Gonzalez, R., Woods, R., Eddins, S.: Digital image processing using Matlab. Gatesmark Publishing, Knoxville (2009)
14. Hoilund, C.: The Radon Transform. Aalborg University (2007)

15. Beylkin, G.: Discrete radon transform. IEEE Trans. Acoust. Speech Signal Process. **35**, 162–172 (1987)
16. Schaefer, G., Stich, M.: UCID: an uncompressed color image database. In: Proceedings of the SPIE, Storage and Retrieval Methods and Applications for Multimedia, San Jose, USA, 2004, pp. 472–480 (2004)

Empirical Study of Social Collaborative Filtering Algorithm

Firas Ben Kharrat[1(✉)], Aymen Elkhlifi[2], and Rim Faiz[1]

[1] LARODEC, University of Carthage-IHEC, 2016 Carthage Présidence, Carthago, Tunisia
{Firas.BenKharrat,Rim.Faiz}@ihec.rnu.tn
[2] LaLic, University Paris-Sorbonne, 75006 Paris, France
Aymen.Elkhlifi@paris.sorbonne.fr

Abstract. In this paper, we propose a new recommender algorithm based on user's social profile and new measurements. It's used by our recommender system that uses external knowledge to solve the cold start problem. Most of Collaborative filtering algorithms are based on user's rating profile, we propose to introduce external resource to create several communities to predict recommendation. These systems are achieving widespread success in E-tourism nowadays. We evaluate our algorithm on tourism dataset and we have shown good results. We compared our algorithm to SVD, Slope One and Weight Slope One. We have obtained an improvement of 8 % in precision and recall as well an improvement of 18 % in RMSE and nDCG.

Keywords: Recommendation system · Tourism recommendation · Collaborative filtering · Machine learning · Social information extraction

1 Introduction

Collaborative filtering *(CF)* recommender algorithms generate personalized recommendation for users based on a set of previously rating items. CF algorithm has been widely applied to various areas: movies, social network, media… The growing popularity of e-commerce brings an increasing interest in recommender systems. When users browse a Website, the system calculated recommendations expose them in the aim to scan their activities around products/services. In this framework, we have introduced a new recommendation system based on new similarity measurement and Facebook user profiles. We have improved the Slope One algorithm by external knowledge and through some experiments results. The approach has been implemented in Java and tested on e-tourism website.

There are two major tasks in our work; the first one is gathering and parsing of data. The second one is using machine learning to come up with relevant suggestions.

The rest of the document is organized as follows: In Sect. (2) we introduce the related works on recommendation systems. In Sect. (3), we report the encountered problems. Then, we present in Sect. (4) our approach using combination of actions to improve recommendation algorithm. The experimentation is described in Sect. (5);

N.T. Nguyen et al. (Eds.): ACIIDS 2016, Part II, LNAI 9622, pp. 85–95, 2016.
DOI: 10.1007/978-3-662-49390-8_8

hence, we evaluate our algorithm. Finally, we conclude with a few notes on future works, in Sect. (6).

2 Related Works

According to [2], recommendation technique has strengths and weaknesses. We categorize the existing method as follows:

Content-based: it recommends items which are similar to previous users [2–4]. Items (products, services or people) are defined by their associated features. User preferences, stored in user profile, considering those associated features, appear in items already rated. According to Schafer et al. [5] a content-based technique, the authors present some works based on it, which are:

- NewsWeeder [6] is a newsgroup filtering system. It recommends unread news for users based on ratings in articles which have been read. The implementations were based on decision trees, neural networks and vector-based representations.
- Pazzani et al. [7] propose a system able to recommend Website based on a user's topic interesting. Their system is based on Bayesian classifier.
- Zhang et al. [8] propose to distinguish amongst relevant documents containing new information. Their system is based on Bayesian system.

Collaborative filtering: CF called people-to-people correlation.

- Ringo [9] recommends music albums and artists based on similarities between user's tastes.
- Tapestry [10] filters electronic documents on a topic that was written by a specific user.
- HOAKS [11, 12] recommends Webpages from news messages. If users are interested, the system find person's contact who posted the message and recommend the webpage.
- Jester [13] is an online joke recommending system. It uses a CF algorithm called Eigentaste and nearest neighbor algorithm for the online phase. Then, recursive rectangular clustering methods for the offline phase.
- GroupLens [14, 15] proposes a system, based on the information filtering, that rates Usenet articles.

Knowledge-based: recommends items based on inferences from user's preferences. The user profile consists of functional knowledge structured and interpreted according to the inference machine:

- Google [16] recommends the most popular links that contain the query provided by users. The system is based on probabilistic models.

Slope One [1] is a CF algorithm, it should deliver reasonably accurate predictions, fast on-line query processing and dynamic updates of the generated predictions when new ratings occur. The algorithm is updatable, and it is efficient at query time. Finally, the Slope One algorithm doesn't need many ratings to generate predictions and it is reasonably accurate.

Similar to Item-based Collaborative Filtering, Slope One uses an item similarity measure. This metric, just like the prediction computation, are different from the variation of Item-based Collaborative Filtering. The idea is to store the differential rating values of item pairs.

We used the following notation in describing schemes:

- u_i is the rating of a user given to item i.
- S(u) is the subset of items consisting of all items rated in u.
- Card(S) is the number of elements in S.
- The average of ratings in an evaluation u is denoted $\bar{u}$.
- The set Si(X) is the set of all evaluations u ∈ X as they contain item i (i ∈ S(u)).
- Given two evaluations u, v, we define the scalar product $(u, v) as \sum_{i \in S(u) \bigcap S(v)} u_i v_i$.

Predictions, P (u), represent a vector where each component is the prediction corresponding to one item: predictions depend implicitly on the training set X.

The similarity in Slope One is defined as the average deviation of items i, j given by:

$$dev(i,j) = \sum\nolimits_{u \in S_{i,j}(X)} \frac{u_i - u_j}{card(S_{i,j}(X))} \tag{1}$$

When new ratings are entered, $\mathrm{dev}(\mathrm{i},\mathrm{j})$ is symmetric and computation time is halved.

$\mathrm{dev}(\mathrm{i},\mathrm{j}) + \mathrm{u_j}$ used as prediction for u_i. Therefore, a usable predictor is the average of all predictions:

$$P^{S1schema}(a)_i = \frac{1}{card(R_j)} \sum\nolimits_{j \in R_j} (dev(i,j) + u_j) \tag{2}$$

With $R_i = \{j | j \in S(u), i \neq j, card(S_{i,j}(X)) > 0\}$.

The authors [1] present an approximation based on a dense dataset, where almost all pairs of items i and j were rated, $\mathrm{R_i} = \mathrm{S(u)}/(i)$. Also in a dense dataset $\sum_{\mathrm{j} \in \mathrm{R_i}} \frac{\mathrm{u_j}}{\mathrm{card(R_j)}}$ is approximately $\bar{\mathrm{u}}$. The approximated prediction is:

$$P^{S1}(u)_j = \bar{u} + \frac{1}{card(R_j)} \sum\nolimits_{i \in R_j} dev_{i,j} \tag{3}$$

3 Problem

Most of online travel booking use different recommender systems, i.e. software agents that elicit the users' preferences to purpose adequate product. Some systems trying to solve the cold start problem, which occurs when no booking or rating history are available. External data from social network, like Facebook, seem promising to overcome this problem. Is this information better or complementary to the information from hotel's description?

4 Our Approach

4.1 General Architecture

Figure 1 shows a general architecture of our recommendation system. It is based on Facebook profile and item's clustering. The proposed system consists of three parts; First, the web service API which interacts with the user's action in the web as prescribed REST description. The action database layer collects and stores all actions tracked. Everyone is stored in the database with additional information like date, time, user and item.

Second, the social database layer extracts data from Facebook for each user separately. We use these data to create a user profile, in Sect. 4.2.

Third, the recommendation layer, where new input is sent to our algorithm. Next, we grouped items in several clusters, described in Sect. 4.3. Then, we calculate the distance between users, we used Weight database, which contains several information about each user. (i.e.: common groups, common page fun, common interests, common activity, common share…), described in Sect. 4.4. Finally, we applied our algorithm, described in Sect. 4.5.

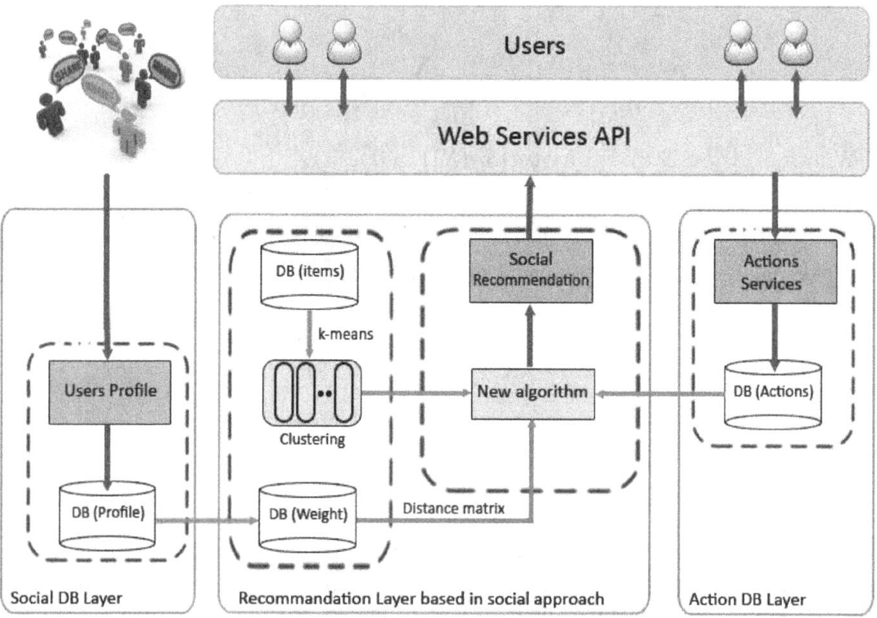

Fig. 1. Architecture of collaborative recommender systems based on Facebook profile and item's clustering.

4.2 Gathering User's Profile

The main confronted problem was the issue of gathering data from social network. We created a script in java to extract all available information based on 'Facebook Graph API', the package is installed in online web site[1] to collect profiles. The dataset used in training and test was a collection of 1 k Facebook profiles, as well as by volunteer students. We make anonymous all users. For privacy issues, sparsity and dynamicity of data were be taken into account when dealing with social network information. In our dataset we have several information about users; such as gender, groups, fun pages, likes mentions, civil status, interests, activities and friends. Finally, all users' profiles were stored in database.

4.3 k-Means Clustering

We used k-means algorithm to partition items into clusters based on mean. In our work, we used a tourism dataset, composed of hotels. For each hotel, we have extracted several users' rate attribute to location, rooms, service, cleanliness and amenities. We proceeded to write a script in java that extracted keywords for each hotel's description after syntactic pretreatment. The attributes of hotels are multidimensional. With this we could use K-means on hotels, using keywords as features, to group hotels together in clusters. Specifically, "k" is the number of clusters we're going to find in our data. In order to use k-means we need to know how many clusters we're looking for at the outset, so we have find the optimize k = 10.

4.4 User's Distance Matrix

The distance matrix containing the distances, taken pairwise, of users. This matrix will have a size of N × N where N is the number of users. We created an algorithm that computes and returns the distance matrix by using the specified formula of distance measure between users. To determine the matrix of distance, we calculated the average of similarity between user's profiles. The objective was to exploit the social links between Facebook users.

The Distance in Rate. We calculated the similarity between users in their rates. We predicted the distance in rate with the following equation:

$$Dist_{Rate}(u,v) = 1 - \sum\nolimits_{i \in S_{u,v}(X)} \frac{|u_i - v_i|}{card(S_{u,v}(X))}.k \tag{4}$$

With k = {0,1} and $Dist_{Rate}(u,v) \epsilon [0..1]$.

[1] www.recommendation-system.com.

The Distance in Gender. We calculated the similarity between users in their gender. We predicted the distance in gender with the following equation:

$$Dist_{Gender}(u, v) = 1 - |gender(u) - gender(v)| \tag{5}$$

With gender(u) return 1 if u is a man else return 0.

The Distance in Civil Status. We calculated the similarity between users in their civil status. We predicted the distance in civil status with the following equation:

$$Dist_{civil}(u, v) = \begin{cases} 1, & civil(u) = civil(v) \\ 0, & else \end{cases} \tag{6}$$

With civil (u) return the civil status of user u.

The Distance in Groups. We calculated the similarity between users in their groups and we predicted the distance in groups with the following equation:

$$Dist_{groups}(u, v) = \frac{1 + Gs_{u,v}}{log(1 + Gd_{u,v})} \tag{7}$$

With: $Gs_{u,v}$ is the number of similar groups between users u and v, $Gd_{u,v}$ is the number of different groups between users u and v.

The Distance in Fun Pages. We calculated the similarity between users in their fun pages and predicted the distance in fun pages with the following equation:

$$Dist_{pages}(u, v) = \frac{1 + Ps_{u,v}}{log(1 + Pd_{u,v})} \tag{8}$$

With: $Ps_{u,v}$ is the number of similar groups between users u and v, $Pd_{u,v}$ is the number of different fun pages between users u and v.

The Distance in Interests. We calculated the similarity between users in their interests. We predicted the distance in interests with the following equation:

$$Dist_{interests}(u, v) = \frac{1 + Is_{u,v}}{log(1 + Id_{u,v})} \tag{9}$$

With: $Is_{u,v}$ is the number of similar interests between users u and v, $Id_{u,v}$ is the number of different interests between users u and v.

The Distance in Likes Mentions. We calculated the similarity between users in their "like" mentions and predicted the distance in "like" mentions with the following equation:

$$Dist_{likes}(u, v) = \frac{1 + Ls_{u,v}}{log(1 + Ld_{u,v})} \tag{10}$$

With: $Ls_{u,v}$ is the number of similar groups between users u and v, $Ld_{u,v}$ is the number of different likes mentions between users u and v.

The Distance in Activity. We calculated the similarity between users in their activity and predicted the distance in activity with the following equation:

$$Dist_{activity}(u,v) = \frac{1 + As_{u,v}}{log(1 + Ad_{u,v})} \tag{11}$$

With: $As_{u,v}$ is the number of similar activity between users u and v, $Ad_{u,v}$ is the number of different activity between users u and v.

The Distance in Friends. We calculated the similarity between users in their friends and predicted the distance in friends with the following equation:

$$Dist_{friends}(u,v) = \frac{1 + Fs_{u,v}}{log(1 + Fd_{u,v})} \tag{12}$$

With: $Fs_{u,v}$ is the number of similar friends between users u and v, $Fd_{u,v}$ is the number of different friends between users u and v.

User's Distance Matrix. We created a matrix that values are the distances for all users and predicted the distance with the following equation:

$$Dist(u,v) = \frac{1 + \sum Dist_i}{\log(1 + \prod Dist_i)} \tag{13}$$

With: i in $\{rate, gender, civil, groups, pages, interests, likes, activity, friends\}$. User's distance matrix are updated when new data is collected.

4.5 Our Algorithm

The new algorithm is defined by an improved Slope One algorithm. We added values of distance's matrix in the deviation.

$$dev(i,j) = \sum\nolimits_{u \in S_{i,j}(X)} \left(\frac{u_i \times u_d - u_j \times u_d}{card(S_{i,j}(X))}\right) \tag{14}$$

With: u_d the distance of current user and predict user in the training set X. Finally, we define our algorithm prediction as the following weighted:

$$P(u)_j = \frac{\sum_{i \in S(u)-\{j\}} (dev_{j,i} + u_i + w_{items}).c_{j,i}.(1 + dis_i)}{\sum_{i \in S(u)-\{j\}} c_{j,i}.(1 + dis_i/c_{j,i})} \tag{15}$$

With: $dis_i = \sum_{u \in S_{j,i}(X)} u_d$; $c_{i,j} = \text{card}(S_{j,i}(X))$.

And $w_{items} = \begin{cases} 0.5, & \textit{if } item_i \textit{ and } item_j \in Cluster_k \\ 0, & else \end{cases}$.

When the user has no rate. The system recommend common best items between 10 users with the highest values of distance: (vi). $item_{vi}$ return the 10 best rated items.

$$P(u) = \bigcap_{i=1}^{10} item_{vi} \tag{16}$$

5 Experimentation and Results

To evaluate our approach, we chose to experiment our system in the tourism domain. We created one benchmark data composed by description of many hotels and we developed an extract program based on 'jaunt API[2]'. It's an open source API to extract data (hotel's description, rank, comments, localization…) from tourism website[3]. TripAdvisor is a famous online website that collects comments, reviews and rank from customers who book hotels all over the world. We have succeeded to extract 1Mof dataset from hotel's description. In addition, we have created an online website to collect users profile from Facebook. In reality, the users of TripAdvisor and Facebook were not the same. The objective of this approach is to recommend the best list of hotels that a user can book. Before buying, users used to view the items many times to check comments and ranks of other users but prediction based on rank is not enough. Hotels booking depend on the user's profile and hotel's cluster. Only users who have rated at least 20 hotels are included in our dataset. Predictions based on new information about users and hotels are important. Finally, we have added several weights into our recommendation approach. The effectiveness of a given CF Systems based on evaluative Slope One algorithm can be measured precisely. To evaluate our approach, we have employed four metrics [18]. RMSE, nDCG, precision and recall measurements [19]. RMSE and nDCG evaluate the prediction of each item recommend. We have found that these two metrics perform quite similarly in our evaluation. To generate training and test sets for different number of ratings from 1 to 19, we first generated a set of test and training splits so that test users have 19 training items using the crossfold strategy. Then, to generate any other number of ratings, we randomly selected a sample of the 19 item training set. This allows us to generate training sets with a given number of ratings for test users, such as each profile size is evaluated with the same test set.

$$RMSE = \sqrt{\frac{1}{n}\sum_{i=1}^{n} (y_i - \hat{y}_i)^2} \tag{17}$$

[2] www.jaunt-api.com.

[3] www.tripadvisor.com.

$$DCG_k = \sum\nolimits_{i=1}^{k} \frac{2^{rel_i} - 1}{log_2(i+1)} \tag{18}$$

With k is the maximum number of entities that can be recommended. $IDCG_k$ is the maximum possible (ideal) DCG for a given set of queries.

$$nDCG_k = \frac{DCG_k}{IDCG_k} \tag{19}$$

We analyzed prediction error of our algorithm called "Slope One Social" compared with several algorithms using the LensKit toolkit [20]. We started to partition the dataset in a 5-fold cross-validation configuration. Users were partitioned into 5 sets; for each user in each partition, we randomly selected 20 % of their ratings as the test, with the remaining ratings from users in the other partitions forming the training set. We then ran five recommender algorithms on the data, and captured the predictions each algorithm made for each test rating. We used the following algorithms, choosing parameters based on prior results in the research literature and experience tuning LensKit for our dataset. Therefore we will compare two Slope One algorithms with our improved Slope One. However, it is interesting to compare with techniques based on matrix factorization as SVD. The high level results of our evaluation are summarized in Fig. 2. Overall we see that our algorithm seems to be the best performing algorithm for cold start use. It was the only algorithm that consistently outperformed the Slope One algorithms on prediction and recommendation items. It is worth noting that, the average improvement of our algorithm over the Slope One recommender represents 8 %of precision and recall; an improvement of 18 % in RMSE and nDCG.

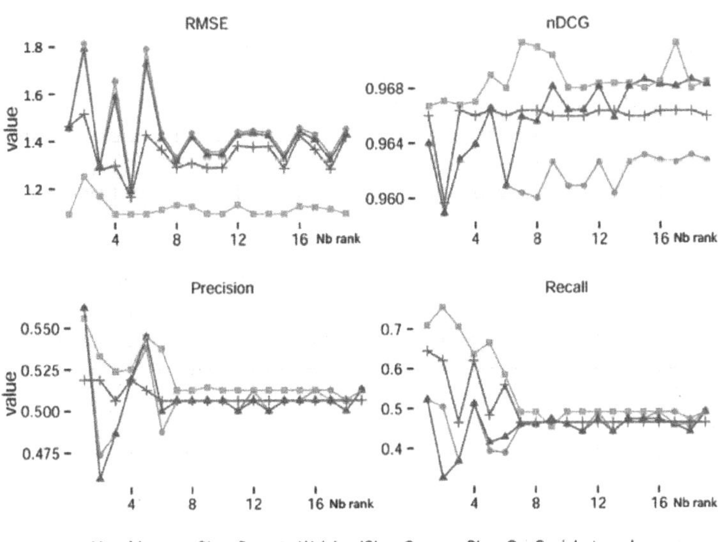

Fig. 2. Results of RMSE, nDCG, Precision and Recall metrics

6 Conclusion and Future Work

In this paper, we have proposed a new approach for the recommendation systems based on users' profiles. The proposed approach is spread over three stages to recommend best items, in a first stage, by the preprocessing that consists in collecting users' profiles from social network. In a second step, we calculated distance matrix and item's cluster. Finally, the recommendation system is predicted with an algorithm based on Slope One. We validated our approach on a touristic benchmark, created from the extraction of data from touristic website. The results obtained are promising compared to those given by the Slope One algorithm. This approach comes within our website of recommendation system; it is being applied in a number of interesting applications like: recommendation restaurants, attractions and combination of different products to give a full package. In our future works, we propose to elaborate user profile using a domain ontology. We aim to extend our algorithm to search several connected communities and we will try also to extract information from other resources such as discussion forums, blogs and search history to enrich knowledge about users.

References

1. Lemire, D., Maclachlan, A.: Slope one prediction for online rating-based collaborative filtering. In: SIAM Data Mining (SDM 2005), Newport Beach, California, 21–23 April 2005 (2005)
2. Burke, R.: Hybrid recommender systems: survey and experiments. User Model. User-Adap. Inter. **12**(4), 331–370 (2002)
3. Adomavicius, G., Tuzhilin, A.: Toward the next generation of recommender systems. IEEE Trans. Knowl. Data Eng. (2005)
4. Perugini, S., Gonçalves, M., Fox, E.A.: Recommender systems research: a connection-centric survey. J. Intell. Inf. Syst. **23**(2), 107–143 (2004)
5. Schafer, J.B., Konstan, J., Riedl, J.: Recommender systems in e-commerce. In: Proceedings of the 1st ACM Conference on Electronic Commerce, EC 1999, p 158. ACM, NY (1999)
6. Lang, K.: Newsweeder: learning to filter netnews. In: Proceedings of the 12th International Conference on Machine Learning, p 331–339. Inc., San Mateo (1995)
7. Pazzani, M., Billsus, D.: Learning and revising user profiles: the identification of interesting web sites. Mach. Learn. **27**(3), 313–331 (1997)
8. Zhang, Y., Callan, J., Minka, T.: Novelty and redundancy detection in adaptive filtering. In: Proceedings of the 25th Annual International ACM SIGIR Conference on Research and Development in Information Retrieval, SIGIR 2002, pp. 81–88. ACM, NY (2002)
9. Shardanand, U., Maes, P.: Social information filtering: algorithms for automating"word of mouth". In: Proceedings of the SIGCHI Conference on Human Factors in Computing Systems, CHI 1995, pp. 210–217. ACM Press/Addison-Wesley Publish, New York, NY (1995)
10. Goldberg, D., Nichols, D., Oki, B., Terry, D.: Using collaborative filtering to weave an information tapestry. Commun. ACM **35**(12), 61–70 (1992)
11. Hill, W., Terveen, L.: Using frequency-of-mention in public conversations for social filtering. In: Proceedings of the 1996 ACM Conference on Computer Supported Cooperative Work, CSCW 1996, pp. 106–112. ACM, New York (1996)

12. Terveen, L., Hill, W., Amento, B., McDonald, D., Creter, J.: Phoaks: a system for sharing recommendations. Commun. ACM **40**(3), 59–62 (1997)
13. Goldberg, K., Roeder, T., Gupta, D., Perkins, C.: Eigentaste: A constant time collaborative filtering algorithm. Inf. Retr. **4**(2), 133–151 (2001)
14. Konstan, J., Miller, B., Maltz, D., Herlocker, J., Gordon, L., Riedl, J.: Grouplens: applying collaborative filtering to usenet news. Commun. ACM **40**(3), 77–87 (1997)
15. Resnick, P., Iacovou, N., Suchak, M., Bergstrom, P., Riedl, J.: Grouplens: an open architecture for collaborative filtering of netnews. In: Proceedings of the 1994, CSCW 1994. ACM (1994)
16. Brin, S., Page, L.: The anatomy of a large-scale hypertextual web search engine. Comput. Netw. ISDN Syst. **30**(1–7), 107–117 (1998)
17. Lindell, J., Haponen, A.: Predicting Movie and TV Preferences from Facebook Profiles: Project in Stanford University (2012)
18. Herlocker J., Konstan J., Borchers A., Riedl A.: An algorithmic framework for performing collaborative filtering. In: Proceedings of Research in Information Retrieval (1999)
19. Fellbaum, C., Grabowski, J., Landes, S.: Performance and confidence in a semantic annotation task, chapter 9, pp. 216–237. The MIT Press, Cambridge (1998)
20. Ekstrand, M.D., Ludwig, M., Konstan, J.A., Riedl, J.T.: Rethinking the recommender research ecosystem: reproducibility, openness, and lenskit. ACM Conference on RS (2011)

Cooperation Prediction in GitHub Developers Network with Restricted Boltzmann Machine

Roman Bartusiak[1(✉)], Tomasz Kajdanowicz[1(✉)], Adam Wierzbicki[1,2], Leszek Bukowski[2], Oskar Jarczyk[2], and Kamil Pawlak[2]

[1] Department of Computational Intelligence, Wrocaw University of Technology, Wyb.Wyspianskiego 27, 50-370 Wroclaw, Poland
{roman.bartusiak,tomasz.kajdanowicz}@pwr.edu.pl

[2] Department of Informatics, Polish-Japanese Academy of Information Technology, Koszykowa 86, 02-008 Warsaw, Poland

Abstract. In order to solve link prediction problem with higher accuracy than achieved by classical supervised approaches we provide a proposal of the method based on information extracted from network using pre-processing done by Restricted Boltzmann Machine (RBM) and statistical inference models. Input space is fed to RBM in order to provide new sparse coded feature space that is used in order to estimate parameters of classical inference models. By accomplishing link prediction with proposed RBM pre-processing noticeable increase of all accuracy related measures was observed in comparison to state-of-the-art approaches.

Keywords: Social networks · GitHub · Link prediction

1 Introduction

Link prediction problem is a quite new and challenging research in network analysis area. Especially, while it tackles strictly social networks with easy application in problem solving. This article considers problem of developers cooperation in GitHub programming repository environment understood as a network. Open source communities, such as these gathered around GitHub platform have been recently rapidly growing. GitHub as code sharing and publishing service with its social networking services for programmers and companies provides a wide variety of advantages exceeding code version control. It gathers a strong attention of code commiters as it allows world-wide collaboration that keeps costs down, improves quality, delivers agility and in general mitigates code development risk. More and more companies are recruiting their programmers based on development activities that can be observed in public software repositories like GitHub. Therefore programmers are willing to be present on GitHub and collaborate with others. Observing and understanding activity of GitHub users is interesting among others for software engineering as well as project management research domains.

N.T. Nguyen et al. (Eds.): ACIIDS 2016, Part II, LNAI 9622, pp. 96–107, 2016.
DOI: 10.1007/978-3-662-49390-8_9

In order to provide a deeper insight on GitHub like collaboration phenomena we provide a proposal for developers collaboration prediction based on information extracted from the platform using network processing and statistical inference models. In complex systems (like GitHub) users interact with each others in different ways. For instance developers can be in relations with others by means of code commits to the same repository. They can also discuss above mentioned commits and related issues staying in the *follow* relation. Such data creates a network of software development community. The proposed collaboration prediction method is using such network data to build a vectorized representation of relations between users. In other words generated vectors describe existing and unexisting edges of the network. Each vector is constructed using centrality measures (e.g. page rank, betweenness, etc.) and clustering coefficients (local and global) of vertices that are on the edge. Additionally each vector is extended with combined measures calculated from degree of particular vertices as well as their neighbours. The proposed methods used statistical inference mechanism based on Restricted Boltzman Machine that is fed with gathered information about edges and utilized for collaboration prediction. Due to the fact that considered GitHub data set is temporal and divided into multiple quarter length time snapshot it is evaluated on future states of developers collaboration. All gathered results confirms that the proposed approach to collaboration prediction is accurate and provides better understanding of users interaction in GitHub service.

Link prediction methods have a lot of real world applications. First of them is in recommendation systems, where newly derived links are used as recommendations (for example: movies). Another application is in reconstruction of networks. We can utilize link prediction methods to reconstruct connections that have been somehow lost (for example: data corruption or lost). Another real world application is knowledge discovery. We can present knowledge as a graph, and then extract new knowledge utilizing existing information [Dunlavy et al. 2011].

2 Related Work

Link prediction problem can be solved using both well known machine learning approaches: supervised and unsupervised. Both approaches have already shown promising results and have their advantages [Davis et al. 2011; Peng et al. 2015; Nickel et al. 2015]. Both of those approaches usually require some vector representation that will be used by inference methods. Majority of recently proposed network vectorization methods were accomplished from the perspective of each single vertex but also for pair of nodes in order to characterize connection. In general we can divide such measures according to the way of computation into multiple groups:

- Neighbourhood based measures. We can try to characterize vertices using their closest neighborhood. This way it will be created a local feature, but obviously it may be extended to further neighbourhood. The most well established and frequently used measures are: *degree* [Newman 2008], *common neighbours* [Newman 2001], *Jacquard coefficient* [Jaccard 1901] (see Eq. 1) or *Adamic/Adar* [Adamic and Adar 2003] (see Eq. 2).

$$J(x, y) = \frac{|N(x) \cap N(y)|}{|N(x) \cup N(y)|} \tag{1}$$

$$A(x, y) = \sum_{z \in N(x) \cap N(y)} \frac{1}{log(D(z))} \tag{2}$$

where $N(x)$ denotes a set of neighbouring nodes of particular vertex x and $D(x)$ means its degree.

- Topology based measures. Another way of describing properties in context of whole network is to take into account its topology. Measures calculated in that way tend to be more global. We can mention here often utilized measures like: *betweenness entrality* [Freeman 1977], *closeness* [Sabidussi 1966], *Katz* [Katz 1953] or *distance*. Additionally to these commonly applied measures there were proposed some other measures that were composed of basic ones. As we can find different type of networks like bipartite, there are also some type specific measures or their modifications to fit into use case.
- Random walk based measures. Due to the fact that networks usually have complex structure, calculating some features can be highly complex or hardly possible. To overcome that issue we can try to approximate some measures using random walk methods. It is worth to mention measures like: *pageRank* [Brin and Page 2012], *hitting time* [Fouss et al. 2007], *PropFlow* [Lichtenwalter et al. 2010].
- Other measures. In addition to above mentioned measures we must remember that there are other ways to extract profile of a node. There can be present some extra attributes that are characterizing nodes (for example: age, gender, nationality), nodes can be accompanied with additional unstructured data (for instance textual data that describes a node can be treated with computational linguistic methods). In this area we can also find application of some social theory like *week ties* [Liu et al. 2013], social balance and microscopic mechanism or *graphlets* [Pržulj et al. 2004].

All measurements presented above can be successfully used in order to build vectorized representation of the network and applied to supervised or unsupervised approach in link prediction. There have been recently published several papers that compared experimentally the usefulness of variety of learning models, e.g. [Peng et al. 2015]. Also experiments on different combinations of features were conducted, thus it can be observed that different sets of attributes are appropriate in distinct problems.

2.1 Problems in Link Prediction

Link prediction in networks is a complex task and depending on the type of the network is related to various problems. Prediction in bipartite networks requires specialized measures that would reflect correctly nodes characteristics, for instance due to the fact that clusters in such networks are not descriptive as well as classical triplets are not formed. Moreover dealing with link prediction is related with issues common with modern machine learning. For sure it is the size of data set that is utilized for models parameters estimation. When our network has enormous size (huge number of nodes, edges or both) some measures will be impossible to calculate. In order to process the network efficiently it might be considered to apply some sampling techniques. However, sampling in the networks itself is not easy task. Simply ignoring parts of network would modify distributions of data and random sampling might not create a robust approximation of input data. Another problem is related to limited availability of real world data sets. Synthetically generated ones are commonly usually used, but newly developed methods can not be evaluated extensively this way. Also we must remember that real world networks are sparse and sparsity problem must be addressed. In cases of time-stamped data time dimension must be also taken into account. It is related to the problem of disappearing edges or connections that exist only during some specific time [Dunlavy et al. 2011].

3 Link Prediction Problem Formulation

Let's consider a social network $G_{t_i}(V, E)$, where V is a set of vertices (nodes), E is a set of edges between vertices, t_i denotes specific unit time of a network. Link prediction problem can be formulated in two folds. Firstly, it is a prediction of links that will be created in future time $t_{i+c} > t_i,\ c > 0$ (see Fig. 1). Future time unit can be understood as following t_i or any other moment that take place after t_i. Secondly, link prediction is to predict invisible (hidden or missing) edges of a given network $G(V, E)$ (see Fig. 2).

(a) $G_{t_1}(V, E_1)$ (b) $G_{t_2}(V, E_1)$

Fig. 1. One network G in different unit of times t_1 and t_2

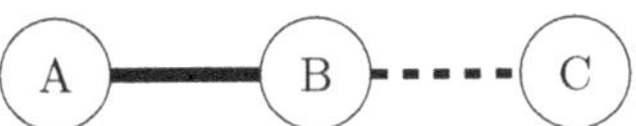

Fig. 2. Network $G(V, E)$ with one edge invisible (B, C)

In both approaches we can clearly see that it can be generalized to a binary classification problem. Edges of network can be classified as *existing* or *nonexistent*. Using that approach we can generalize link prediction problem into a classification task over data set K. Given a data set that is a combination of each pair of vertices $v_1, v_2 \in V$, $v_1 \neq v_2$ what can be written as Cartesian product of vertices set: $K = \{(v_1, v_2) | v_1, v_2 \in V, v_1 \neq v_2\}$ Each data instance from data set K can be classified as an existing edge or nonexistent. Then each vertex pair from set K can be characterized using the already mentioned features. We can calculate indices based on network topology like *common neighbours*, *betweenness*, *closeness*, based on random walking over network like *Katz*, *PageRank* and others. Given the edge description vector expressed with measures it can be used as data for classifier, or we can pre-process it before using more complex models like Restricted Boltzman Machine, or deep neural networks.

4 Link Prediction Using Restricted Boltzman Machine Pre-processing

We are proposing data pre-processing using Restricted Boltzman Machine (RBM) ([Smolensky 1986; Hinton 2002]), to extract better features that are based on common used measures. RBM can be represented as bipartite graph where one set of model nodes creates visible units, and second set hidden units (Fig. 3), which can be denoted as $E(v, h)$, where v represents visible units and h hidden ones.

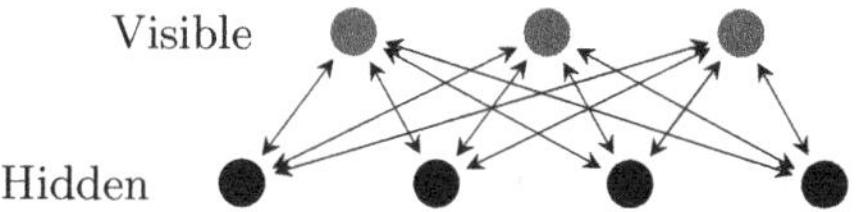

Fig. 3. RBM with 3 visible units and 4 hidden

RBM is an energy based model. It means that there is a scalar *energy* (Eq. 3) associated with each configuration of model.

$$F(v) = -b'v - \sum_i \log \sum_{h_i} e^{h_i(c_i + W_i v)} \tag{3}$$

Where b, c are offsets of visible and hidden units and W are weights of connections.

We can use binary units ($v_i, h_i \in \{0, 1\}$, Eq. 4), which will simplify the whole energy model (Eq. 5)

$$\begin{aligned} P(h_i = 1|v) &= sigm(c_i + W_i v) \\ P(v_j = 1|h) &= sigm(b_j + W_j' h) \end{aligned} \tag{4}$$

$$F(v) = -b'v - \sum_i \log(1 + e^{(c_i + W_i v)}) \tag{5}$$

Utilizing presented equations and assumptions, whole model can be trained using Contrastive Divergence [Carreira-Perpinan and Hinton 2005].

To describe edges of a network we have to compute appropriate features. We have selected features that have different characteristics. RBM is utilized as a graph data pre-processor. Each instance of vector that represents an edge is transformed using pre-trained RBM. Received values of hidden units are used in further data processing. Created pre-processor can be stacked with different machine learning models, what will provide more complex view on effectiveness of that approach (Fig. 4).

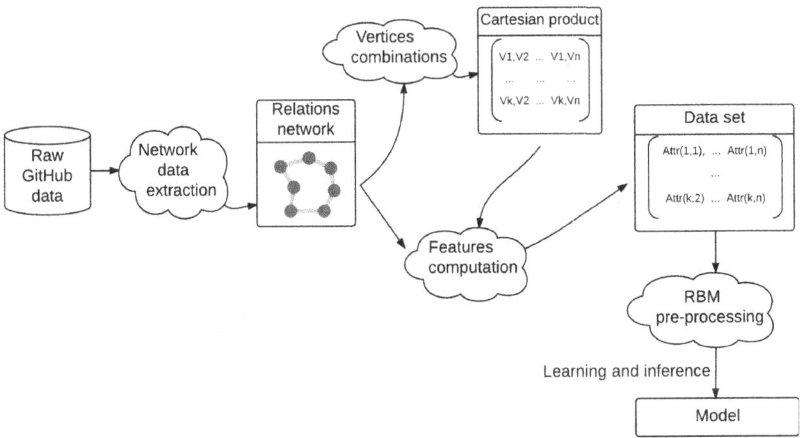

Fig. 4. Data preparation and feature extraction process

4.1 Experiments

The proposed method solves the link prediction in a network created by interaction between users of GitHub. The network considered by us is build by connecting two users with a relation if their are in any one of two interactions. First interaction exsist if they are writing comments in one thread (under same issue, commit and others). Second interaction occurs if two users are active in same repository. As we can see on Fig. 5, *discussion* network has high growth factor. Both, vertices and edges quantity change over time, have *sigmoid* shape. We can assume that it is because of high popularity peek of GitHub in 2012, which was stabilized in 2013.

We have conducted our experiments using data from year 2011. Four quarters were divided into two data sets (first and second quarter; third and fourth quarter) from which training and testing sets were build. The first step of data preparation was to create Cartesian product (as shown above) of all vertices from earlier quarter (first quarter for training set; third quarter for testing set). For each pair of vertices from prepared product, we have computed mentioned earlier measures. Final step of data preparation was identifying if edge described by

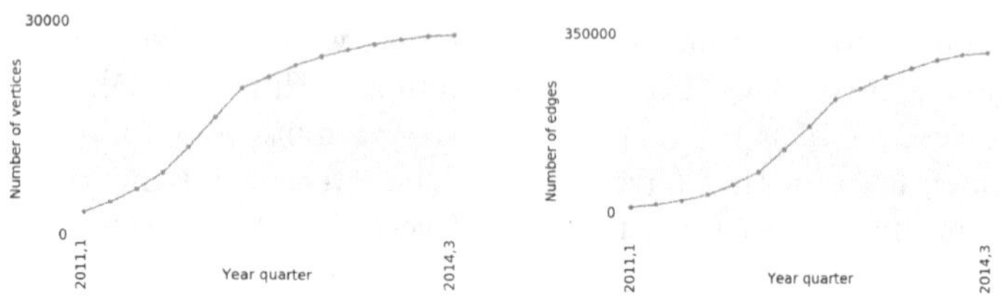

(a) Vertices count change over time (b) Edges count change over time

Fig. 5. *Discussion* network growth over time

computed vector exists in next following quarter (that way second and fourth quarters were only used to identify relation, without measures computation). Data set consists of vectors calculated using first two quarters (first and second) was utilized as training data with 766942 instances (3512 positive ones). Effects of generation based on second two quarters (third and fourth) build testing set with 2692361 instances (8852 positive ones).

In our experiments we have observed well known measures Recall, Precision and F1-Score (Eq. 6). Due to the fact that data sets is extremely unbalanced, less than 10 % relations between users exists comparing to whole possible space, we have observed measures for *edge exists* class with is more important in our research.

$$Recall = \frac{TP}{TP + FN} \quad Precision = \frac{TP}{TP + FP}$$
$$F1_{Score} = \frac{2 * Precision * Recall}{Precision + Recall} \tag{6}$$

where TP - true positive, FN - false negative, FP - false positive

We have conducted experiments over presented data set using two main approaches. The first one consists of classic classification using standard machine learning models. The second approach extends the classic one by pre-processing data using RBM. We have also checked how the number of hidden units in Restricted Boltzman Machine influence quality of classification. Presented results are scores achieved for test set.

4.2 Methods Parameters

Random Forest method was build using ten decision trees. In Random Forest and Decisions Tree based approach Gini impurity was used as a criterion for split. Perceptron had 22 input units, 2 output,and was trained using stochastic gradient descent method. Used by us Naive Bayes is based on Gaussian approach.

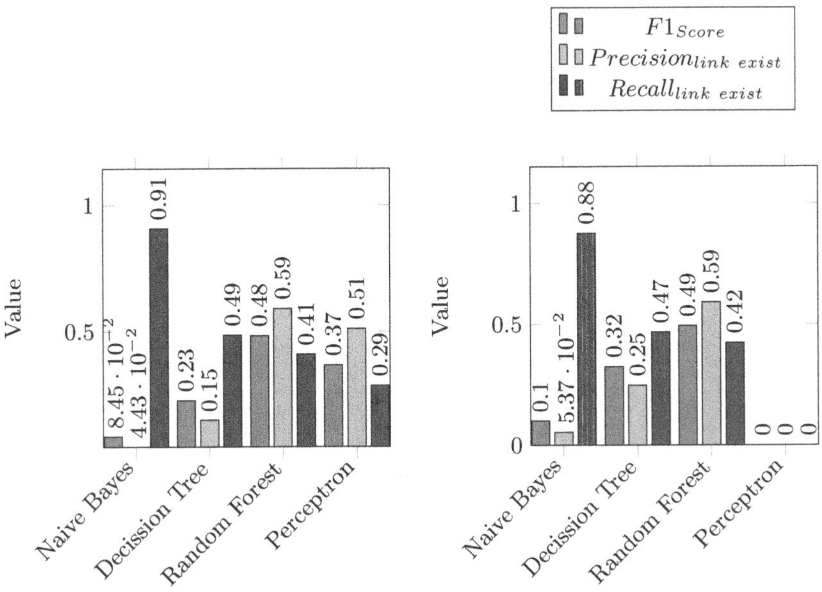

(a) Different learning models on *discussion* network

(b) Best RBM results on *discussion* network

Fig. 6. (a) Different learning models on *discussion* network. (b) Best RBM results on *discussion* network

4.3 Results and Discussion

Classic Approach vs. RBM Pre-processing. First set of experiments shows the difference between link prediction that utilizes advanced RBM as a pre-processor and standard supervised machine learning. On the first plot (Fig. 6) we can see results of classic machine learning based on prepared data set. Clearly the best results are achieved by Random Forest method. The next result was achieved by perceptron. Decission Tree is another method following the F1-Score order. The worst results were obtained by Naive Bayes classifier.

On next bar chart (Fig. 6) we can see what results were achieved by the same classifiers but with RBM used for pre-procesing. Each classifier achieved better classification performance in terms of F1-Score with RBM pre-processing than without it, except perceptron. For Random Forest and Naive Bayes best results were achieved when using with 26 hidden units. Decision Trees achieved highest F1-Score when stacked with RBM with 24 hidden units.

Accuaracy vs. Hidden Units Count. In order to correctly evaluate proposed approach, we have conducted experiments evaluating performance of the method depending on the number of hidden units of RBM. For each model mentioned in earlier experiments, we have evaluated their performance for different number

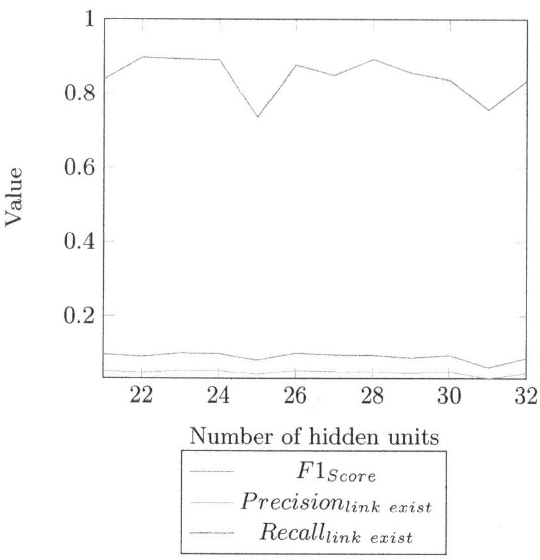

Fig. 7. Measurements for *discussion* layer depending on number of units in hidden layer in RBM in conduction with *Naive Bayes*

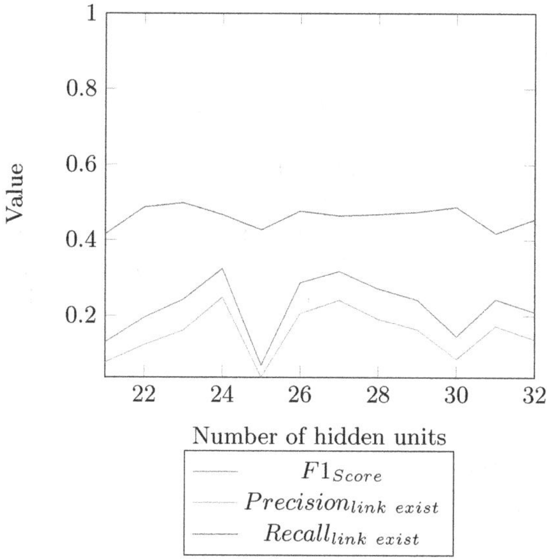

Fig. 8. Measurements for *discussion* layer depending on number of units in hidden layer in RBM in conduction with *Decission Tree*

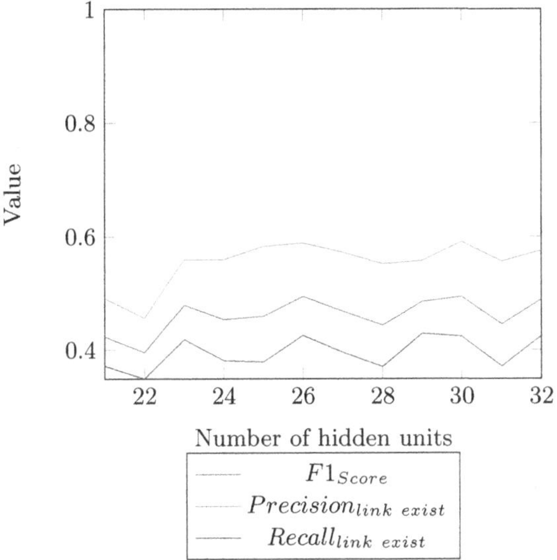

Fig. 9. Measurements for *discussion* layer depending on number of units in hidden layer in RBM in conduction with *Random Forest*

of hidden units of RBM. The experiment was performed with range from 21 hidden units to 32. As in experiments comparing standard supervised learning to learning with RBM pre-processing, we have also observed F1-Score and *edge exists* Precision and Recall.

The first plot (Fig. 7) presents influence of hidden units number in RBM on classification quality of Naive Bayes classifier. We have used classifier with same configuration as in earlier experiments, with means *Gausian Naive Bayes.* Clearly, the number of hidden units do not cause any significant changes in classifier performance.

Second plot (Fig. 8) presents impact of hidden units on performance of *Decission Tree* classifier. Response of classifier performance on hidden unit count is now more noticeable. Unfortunately we can not see any clear trends or dependency. Variance of classification quality is much higher than for *Naive Bayes* based classifier.

In Fig. 9 the last figure, we can see dependency between size of hidden layer of RBM and classification quality of Random Forest based classifier. Despite variance, positive correlation can be observed between variables.

5 Conclusions and Future Work

In this paper we have shown that usage of Restricted Boltzmann Machine combined with classic machine learning methods will increase quality of link prediction task treated as classification problem. We have also shown that Random Forest classifier provides the best results when used for link prediction task.

In further work, evaluation of RBM usage over different data sets should be evaluated. Some data sets can have high number of edges and/or vertices, computation of some metrics can be impossible. Thus, it will be considered computation of network data with high volume and variety characteristic must be solved.

Acknowledgements. The work was partially supported by The National Science Centre, decision no. DEC-2013/09/B/ST6/02317 and the European Commission under the 7th Framework Programme, Coordination and Support Action, Grant Agreement Number 316097, Engine project.

References

Adamic, L.A., Adar, E.: Predicting missing links via local information. Soc. Netw. **25**(3), 211–230 (2003)

Brin, S., Page, L.: Reprint of: the anatomy of a large-scale hypertextual web search engine. Comput. Netw. **56**(18), 3825–3833 (2012)

Carreira-Perpinan, M.A., Hinton, G.E.: On contrastive divergence learning. In: Proceedings of the Tenth International Workshop on Artificial Intelligence and Statistics, pp. 33–40. Citeseer (2005)

Davis, D., Lichtenwalter, R.N., Chawla, N.V.: Multi-relational link prediction in heterogeneous information networks. In: 2011 International Conference on Advances in Social Networks Analysis and Mining, pp. 281–288 (2011)

Dunlavy, D.M., Kolda, T.G., Acar, E.: Temporal link prediction using matrix and tensor factorizations. ACM Trans. Knowl. Discov. Data (TKDD) **5**(2), 10 (2011)

Fouss, F., Pirotte, A., Renders, J.-M., Saerens, M.: Random-walk computation of similarities between nodes of a graph with application to collaborative recommendation. IEEE Trans. Knowl. Data Eng. **19**(3), 355–369 (2007)

Freeman, L.C.: A set of measures of centrality based on betweenness. Sociometry **40**(1), 35–41 (1977)

Hinton, G.E.: Training products of experts by minimizing contrastive divergence. Neural Comput. **14**(8), 1771–1800 (2002)

Jaccard, P.: Etude comparative de la distribution florale dans une portion des Alpes et du Jura. Impr, Corbaz (1901)

Katz, L.: A new status index derived from sociometric analysis. Psychometrika **18**(1), 39–43 (1953)

Lichtenwalter, R.N., Lussier, J.T., Chawla, N.V.: New perspectives and methods in link prediction. In: Proceedings of the 16th ACM SIGKDD International Conference on Knowledge Discovery and Data Mining, pp. 243–252. ACM (2010)

Liu, H., Hu, Z., Haddadi, H., Tian, H.: Hidden link prediction based on node centrality and weak ties. EPL (Europhys. Lett.) **101**(1), 18004 (2013)

Newman, M.E.: The mathematics of networks. New Palgrave Encycl. Econ. **2**(2008), 1–12 (2008)

Newman, M.E.J.: Clustering and preferential attachment in growing networks, 1–13 (2001). http://bit.ly/1OBhDPL

Nickel, M., Murphy, K., Tresp, V., Gabrilovich, E.: A review of relational machine learning for knowledge graphs from multi-relational link prediction to automated knowledge graph construction, pp. 1–18 (2015). http://bit.ly/1OBhyLS

Peng, W., Baowen, X.U., Yurong, W.U., Xiaoyu, Z.: Link prediction in social networks: the state-of-the-art. 8 December 2014, vol. 58, pp. 1–38, January 2015. arXiv:1411.5118v2 [cs.SI]

Pržulj, N., Corneil, D.G., Jurisica, I.: Modeling interactome: scale-free or geometric? Bioinformatics **20**(18), 3508–3515 (2004)

Sabidussi, G.: The centrality index of a graph. Psychometrika **31**(4), 581–603 (1966)

Smolensky, P.: Information Processing in Dynamical Systems: Foundations of Harmony Theory (1986)

Fast and Accurate - Improving Lexicon-Based Sentiment Classification with an Ensemble Methods

Łukasz Augustyniak(✉), Piotr Szymański(✉), Tomasz Kajdanowicz, and Przemysław Kazienko

Department of Computational Intelligence, Wroclaw University of Technology, Wroclaw, Poland
{lukasz.augustyniak,piotr.szymanski,tomasz.kajdanowicz, przemyslaw.kazienko}@pwr.edu.pl

Abstract. A lexicon-based ensemble approach to sentiment analysis that outperforms lexicon-based method is presented in this article. This method consists of two steps. First we employ our own method (called frequentiment) for automatic generation of sentiment lexicons and some of publicly available lexicons. Secondly, an ensemble classification is used to improve the overall accuracy of predictions. Our approach outperforms publicly available sentiment lexicons and automatically generated domain lexicons. We conduct comprehensive analysis based on 10 Amazon review data sets that consist of 4,200,000 reviews.

Keywords: Sentiment analysis · Ensemble classification · Lexicons-based sentiment analysis

1 Introduction

Nowadays a lot of business takes place in the Internet and everybody want to know how their brand is recognizable. Sentiment analysis is techniques, which help in detecting emotions and opinions on social media data. This may help in finding how your brand is seen in the Internet. A social media monitoring shows increased growth, hence an analysis of web written texts (i.e., opinions, reviews) that is fast and accurate is needed. In the past it was possible to read and annotate manually such an opinions. Having enough money, one would hire a group of human annotators, and employ them to read the texts and use their intelligence and knowledge to complete the task. A growth of the Internet usage implies much more opinions and its not possible to conduct it by hands. There appeared need for automatic processing and annotation textual data. For this reason we are witnessing growing popularity of Sentiment Analysis. This kind of analysis in part of Digital Universe of Data. IDC[1] projects that the Digital Universe will reach 40 zettabytes (ZB) by 2020, an amount that exceeds previous

[1] http://www.idc.com.

N.T. Nguyen et al. (Eds.): ACIIDS 2016, Part II, LNAI 9622, pp. 108–116, 2016.
DOI: 10.1007/978-3-662-49390-8_10

forecasts by 5 ZBs, resulting in a 50-fold growth from the beginning of 2010. In terms of sheer volume, 40 ZB of data is equivalent to for example two statistics. First one, there are 700,500,000,000,000,000 grains of sand on all the beaches on earth (or seven quintillion five quadrillion). That means 40 ZB is equal to 57 times the amount of all the grains of sand on all the beaches on earth. The other one, in 2020, 40 ZB will be 5,247 GB per person worldwide. Hence, even small part of such analysis that is for natural language data analysis is outstandingly huge. We need fast, accurate and low memory/processor computation for sentiment analysis alike processing. The approach that provides memory, processor and easy way to count sentiment are lexicons [6,14], but they accuracy is not satisfying. We proposed ensemble-based extension for lexicons and achieved better accuracy while the time and memory complexity of approach stay at level for lexicons.

Sentiment analysis is used in many areas e.g. predicting election outcomes [15], supplying organizations with information on their brands [6], summarizing product in reviews [3], building better recommendation system [5] and even predicting the stock market [2].

In this paper, we compared several different lexicons and fusion classifiers and found out, without much surprise, that combination of various lexicons performs better than any individual sentiment lexicon. The ensemble best benefits from models variability and complementary, thus having diverse set of techniques is desirable. We used several different learners such as Decision Tree, Extra Tree Classifier and AdaBoost. Usage of ensembles in such an approach doesn't appear in literature, hence we wanted to verify our method on big review data (further description of the data in Sect. 4.1).

2 Related Work

In this section we provided some examples related to lexicon-based and ensemble-based approaches to opinion mining tasks.

2.1 Lexicon-Based Approach

The lexicon-based methods assume that sentiment orientation is related to presence of certain words/phrases in a document. The sentiment lexicon is a set of ngrams (one or more consecutive words) with sentiment orientation assigned to these ngrams. Overall sentiment of the document is annotated using these features from the lexicon that are (or are not) present in the document. Sentiment lexicons are used in many sentiment classification tasks. Sentiment words are always divided into at least two classes according to their orientation: positive and negative attitudes. For instance, "good" or "great" are a positive words, and "bad" or "catastrophic" are a negative words. Sentiment words and their weights form the sentiment lexicon [8,11].

2.2 Ensemble Classification Approach

The ensemble techniques are used widely in the literature for classification methods. The ensemble uses a variety of models whose predictions are taken as input to a new model that learns how to combine the predictions into an overall prediction. Whitehead [16] describes ensemble learning as a technique increasing machine learning accuracy with a trade-off of increasing computation time so they are best suited in those domains where computational complexity is relatively unimportant compared to the best possible accuracy.

Lin and Kolcz [9] used Logistic Regression classifiers learned from byte 4-grams (hashed) as features. The 4-grams refers to four characters (and not to four words). They didn't do any processing tasks, not even word tokenization. The ensembles were formed by different models, obtained from different training sets, but with the same learning algorithm that was mentioned above Logistic Regression. Their results show that the ensembles lead to more accurate classifiers. The next approach with ensembles presented Rodriguez et al. [13]. He used classifier ensembles for expression, not char ngrams. In this situation the sentiment orientation label (positive, negative, or neutral) is applied to a phrase or word within the tweet. What is important in such method the sentiment label does not necessarily match the sentiment of the entire tweet.

The class imbalance and the feature space sparsity are big issues in text classification problems. Hassan et al. [7] addressed these problems. They proposed to enrich the corpus using multiple additional datasets related to the task of sentiment classification. The authors used a combination of standard approach with unigrams and bigrams of simple words, part-of-speech, and other semantic features. None of the previous works used AdaBoost [4]. Moreover, lexicon's predictions as a feature space for fusion classifier have not been addressed widely in the literature.

3 Ensembles of Lexicons

Our method consists of two steps:

1. Lexicons-based sentiment classification. We used publicly available lexicons and automatically generated lexicons based on method presented in [1].
2. Ensemble classification (fusion classifier) step.

The main part of our proposed methods is the lexicon ensemble approach. It consists of two stages - building the relevant input space for ensemble classification and learning a fusion classifier based on mentioned input space. This part of our method uses a variety of models (lexicons in this experiment) whose predictions are taken as input to a new model that learns how to combine the predictions into an overall prediction. We built a sentiment polarity matrix $S(\mathcal{L}, \mathcal{D})$ using predictions from sentiment lexicons. Sentiment orientation was obtained for every document $d \in \mathcal{D} = \{d_1, \ldots, d_n\}$ and every lexicon $l \in \mathcal{L} = \{l_1, \ldots, l_n\}$. We denoted the sentiment polarity of a document d using lexicon l as $s_l(d)$ regardless. The sentiment polarity matrix is defined as follows:

$$\mathcal{S}(\mathcal{L},\mathcal{D}) = \begin{pmatrix} s_{l_1}(d_1) & s_{l_1}(d_2) & \cdots & s_{l_1}(d_n) \\ s_{l_2}(d_1) & s_{l_2}(d_2) & \cdots & s_{l_2}(d_n) \\ \vdots & \vdots & \ddots & \vdots \\ s_{l_n}(d_1) & s_{l_n}(d_2) & \cdots & s_{l_m}(d_n) \end{pmatrix} \tag{1}$$

Afterwards, we used such feature space as input for the fusion classifier. We tried couple of a classifiers such as Decision Tree, Extra Tree Classifier, and AdaBoost. The experimental scenario is presented in Fig. 1.

4 Experimental Scenario

In this section the experimental scenario: dataset, text preprocessing with cross-validation division.

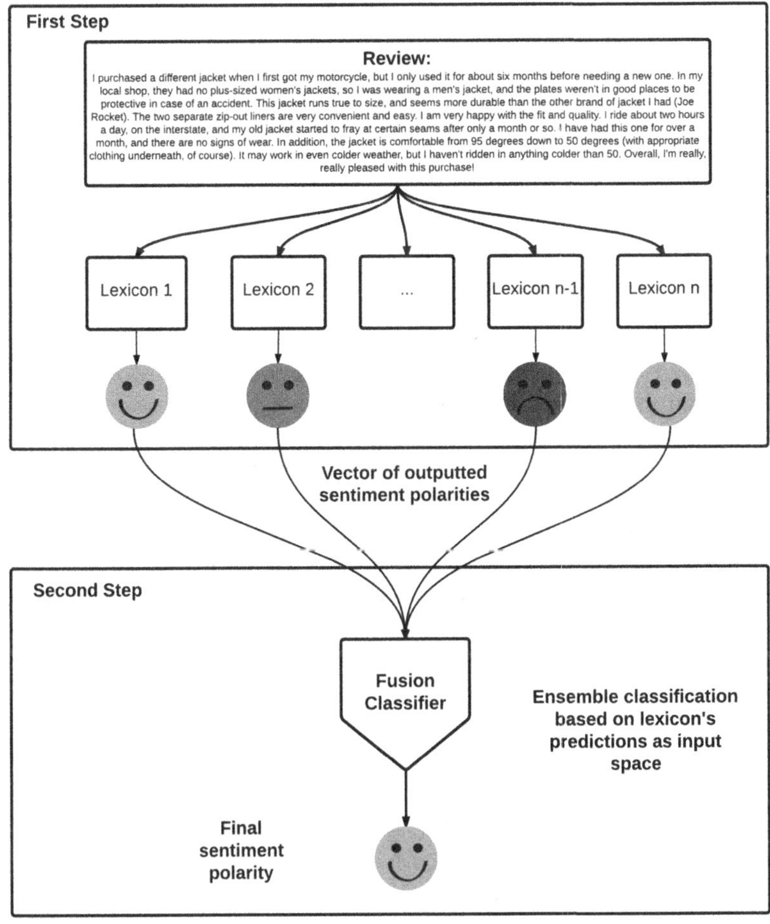

Fig. 1. The concept of the proposed ensemble classification.

4.1 Dataset

We used an Amazon Reviews Dataset published by SNAP [10]. Domains presented in Table 1 were chosen for experimental scenario.

Table 1. Dataset's domains used in experiment

Domain	Number of reviews
Automotive product	188,728
Book	12,886,488
Clothing	581,933
Electronics product	1,241,778
Health product	428,781
Movie TV	7,850,072
Music	6,396,350
Sports Outdoor product	510,991
Toy Game	435,996
Video Game	463,669

4.2 Text Pre-processing

Each of the review consists of Amazon's user opinion (text) and his star score (1–5 scale), where 1 is the worst score and 5 is the best. The review data set was cleaned up from its raw form. All the HTML tags and entities were removed or converted to textual representations using the HTML parser in python library BeautifulSoup4[2]. Next the unicode review texts were decoded to ASCII using the unidecode[3] python library. In addition, all punctuation and numbers were removed. Each of the data sets was divided into a training and test set in 10 cross validations. For tractability, especially with supervised learners, the training data set consisted of 12,000 randomly drawn reviews. Reviews were selected evenly per sentiment, the training set included 2,000 reviews with 1, 2, 4 and 5 stars each and 4,000 labeled with 3 stars. We have thus obtained a balanced set of 4,000 positive, negative and neutral reviews each. The test data set consisted of 30,000 evenly distributed across sentiment labels (distributed analogously to training set). In order to check the accuracy of the proposed methods, the ground truth sentiment was extracted from ratings expressed with stars. Ratings were mapped to text classes "positive", "neutral" and "negative", using 1 and 2 stars, 3 stars, 4 and 5 stars respectively.

[2] https://pypi.python.org/pypi/beautifulsoup4.

[3] https://pypi.python.org/pypi/Unidecode.

4.3 Lexicons

In this paper, we used human generated lexicons, automatic generated lexicon's based on method provided in [1]. In addition, we used Bing Liu's Opinion Lexicon [8], AFINN Lexicons [12] and list of positive/negative words from www.enchantedlearning.com. These lexicons (Table 2) was needed for first step of our method, further description in Sect. 3.

The polarity of the document is calculated based on detecting occurrences of sentiment words from the lexicon in each of the document (some of the lexicons contain weights for sentiment words). We didn't perform any negation handling.

Table 2. Examples of sentiment lexicons

Lexicon	Positive words	Negative words
Simplest (SM)	good	bad
Simple List (SL)	good, awesome, great, fantastic, wonderful	bad, terrible, worst, sucks, awful, dumb
Simple List Plus (SL+)	good, awesome, great, fantastic, wonderful, best, love, excellent	bad, terrible, worst, sucks, awful, dumb, waist, boring, worse
Past and Future (PF)	will, has, must, is	was, would, had, were
Past and Future Plus (PF+)	will, has, must, is, good, awesome, great, fantastic, wonderful, best, love, excellent	was, would, had, were, bad, terrible, worst, sucks, awful, dumb, waist, boring, worse
Bing Liu	2006 words	4783 words
AFINN-96	516 words	965 words
AFINN-111	878 words	1599 words
enchantedlearning	266 words	225 words

5 Results

Table 4 reports the results of each individual model (lexicons and lexicon-based ensemble). The performance was evaluated by F-measure per each method and domain.

5.1 F-Measure Explanation

As the introduction for analyzing the results, the F-measure is described. Firstly, the confusion matrix and two additional measures called precision and recall must be presented.

Let's now define the most basic terms for confusion matrix, which are whole numbers - not any rates (Table 3):

Table 3. Confusion matrix for classification problem of two classes

	Class A	Class B
Classifier says A	true positive (TP)	false positive (FP)
Classifier says B	false negative (FN)	true negative (TN)

- **true positives (TP)** - these are cases when classifier predicted class A, and the classified object do have class A.
- **true negatives (TN)** - classifier predicted class B and the object was classified correctly.
- **false positives (FP)** - classifier predicted class A, but it was mistake, the object should be classified as class B. This is also known as a "Type I error".
- **false negatives (FN)** - classifier predicted class B, but again it was mistake, because it should be class A. It is known as a "Type II error".

$$recall = \frac{TP}{TP + FN} \tag{2}$$

Recall measures the completeness, or sensitivity, of a classifier. A higher recall means less false negatives, while lower recall means more false negatives. Improving recall can often decrease precision because it gets increasingly harder to be precise as the sample space increases.

$$precision = \frac{TP}{TP + FP} \tag{3}$$

Precision measures the exactness of a classifier. A higher precision means less false positives, while a lower precision means more false positives. This is often at odds with recall, as an easy way to improve precision is to decrease recall.

$$F - measure = 2 * \frac{precision * recall}{precision + recall} \tag{4}$$

F-measure is widely used for text classification evaluation, because it uses precision and recall to produce a additional single metric known as F-measure. This metric is the weighted harmonic mean of precision and recall.

5.2 Results Evaluation

We observe immediately that the lexicon-based ensemble with AdaBoost classifier (AB in Table 4) outperformed all lexicons, even this automatically generated for each domain (unigram, bigram and trigram) and other fusion learners. The AdaBoost achieved the highest accuracy for the Toys & Games and the lowest for the Music. The Music domain is an outstanding example, because the overall accuracy is the lowest one across all domains. Some of the tested classifiers presented better than an AdaBoost in specific domain, but the AdaBoost was the most consistent method. The Random Forest classifier performed really well in the Cloths & Accessories (56.9 %) - it was the highest F-measure across all results. However, it's performance in other domains was worse than unigram lexicons. Interestingly, the Cloths & Accessories presents higher accuracy for other fusion classifiers - Decision Tree - 55.4 %, Extra Tree Classifier - 55.5 %.

Table 4. Results for all methods - F-measure

	Method	Auto	Books	C&A	Elect	Health	M&TV	Mus	SP	T&G	VG
Lexicons	SM	0.244	0.227	0.249	0.244	0.245	0.230	0.228	0.245	0.230	0.237
	SL	0.335	0.333	0.341	0.349	0.342	0.382	0.357	0.343	0.344	0.367
	PF	0.352	0.365	0.361	0.348	0.362	0.368	0.340	0.362	0.379	0.357
	SL+	0.351	0.364	0.385	0.364	0.366	0.398	0.362	0.366	0.376	0.395
	AF-111	0.368	0.346	0.364	0.376	0.358	0.370	0.350	0.368	0.359	0.368
	PF+	0.366	0.375	0.411	0.360	0.376	0.389	0.335	0.381	0.387	0.370
	trigr.	0.348	0.395	0.361	0.386	0.380	0.39	0.353	0.366	0.392	0.388
	AF-96	0.390	0.364	0.391	0.401	0.387	0.385	0.371	0.398	0.395	0.388
	EN	0.419	0.389	0.400	0.406	0.411	0.394	0.391	0.410	0.411	0.394
	BL	0.411	0.387	0.421	0.414	0.410	0.406	0.407	0.429	0.439	0.404
	bigr.	0.440	0.461	0.496	0.498	0.503	0.457	0.370	0.472	0.500	0.495
	unigr.	0.500	0.505	0.530	0.508	0.505	0.512	0.435	0.514	0.511	0.499
Ensemble	DT	0.493	0.472	0.554	0.475	0.474	0.476	0.464	0.488	0.504	0.471
	ET	0.495	0.473	0.555	0.474	0.476	0.478	0.465	0.489	0.505	0.474
	RF	0.503	0.482	0.569	0.483	0.485	0.488	0.475	0.499	0.512	0.484
	AB	**0.522**	**0.524**	**0.538**	**0.537**	**0.529**	**0.534**	**0.510**	**0.527**	**0.552**	**0.529**

Domains: Automotive, Books, Clothing & Accessories, Electronics, Health, Movies & TV, Music, Sports & Outdoors, Toys & Games, Video Games. **Methods**: lexicons as described in Table 2 with extension of unigrams, bigrams and trigrams. **Fusion Classifiers**: **DT** - Decision Tree, **AB** - AdaBoost, **ET** - Extra Tree Classifier, and **RF** - Random Forest.

6 Conclusions and Future Work

We have proposed a very simple yet powerful ensemble system for sentiment analysis. We combine lexicon predictions to build more complex and more accuracy sentiment predictor. Each such lexicon contributes to the success of the overall system, outperforming single lexicon approach on Amazon Reviews dataset. We conclude that AdaBoost learner performed the best among all fusion classifiers.

However, we feel that further investigations related to ensembles and feature space for fusion classifier are necessary. Extension of new lexicons could also influence of the performance of the overall method.

Acknowledgment. This work is partially funded by the European Commission under the 7th Framework Programme, Coordination and Support Action, Grant Agreement Number 316097, European research centre of Network intelliGence for INnovation Enhancement (ENGINE).

References

1. Augustyniak, L., Kajdanowicz, T., Szymanski, P., Tuliglowicz, W., Kazienko, P., Alhajj, R., Szymanski, B.K.: Simpler is better? lexicon-based ensemble sentiment classification beats supervised methods. In: 2014 IEEE/ACM International Conference on Advances in Social Networks Analysis and Mining, ASONAM 2014, Beijing, China, 17–20 August 2014, pp. 924–929 (2014)
2. Bollen, J., Mao, H., Zeng, X.: Twitter mood predicts the stock market. J. Comput. Sci. **2**(1), 1–8 (2011)
3. Brody, S., Elhadad, N.: An unsupervised aspect-sentiment model for online reviews. In: Human Language Technologies: The 2010 Annual Conference of the North American Chapter of the Association for Computational Linguistics, HLT 2010, Stroudsburg, PA, USA, pp. 804–812. Association for Computational Linguistics (2010)
4. Freund, Y., Schapire, R.E.: Experiments with a new boosting algorithm (1996)
5. Galitsky, B., McKenna, E.W.: Sentiment extraction from consumer reviews for providing product recommendations, November 12 2009. US Patent App. 12/119,465
6. Ghiassi, M., Skinner, J., Zimbra, D.: Twitter brand sentiment analysis: a hybrid system using n-gram analysis and dynamic artificial neural network. Expert Syst. Appl. **40**(16), 6266–6282 (2013)
7. Hassan, A., Abbasi, A., Zeng, D.: Twitter sentiment analysis: a bootstrap ensemble framework. In: 2013 International Conference on Social Computing (SocialCom), pp. 357–364, September 2013
8. Hu, M., Liu, B.: Mining and summarizing customer reviews. In: Proceedings of the Tenth ACM SIGKDD International Conference on Knowledge Discovery and Data Mining, KDD 2004, pp. 168–177. ACM, New York (2004)
9. Lin, J., Kolcz, A.: Large-scale machine learning at twitter. In: Proceedings of the 2012 ACM SIGMOD International Conference on Management of Data, SIGMOD 2012, pp. 793–804. ACM, New York (2012)
10. McAuley, J., Leskovec, J.: Hidden factors, hidden topics: understanding rating dimensions with review text. In: The 7th ACM Conference on Recommender Systems, pp. 165–172. ACM (2013)
11. Mohammad, S., Dunne, C., Dorr, B.: Generating high-coverage semantic orientation lexicons from overtly marked words, a thesaurus. In: Proceedings of the 2009 Conference on Empirical Methods in Natural Language Processing, EMNLP 2009, Stroudsburg, PA, USA, vol. 2, pp. 599–608. Association for Computational Linguistics (2009)
12. Nielsen, F.Å.: Afinn, March 2011
13. Rodriguez-Penagos, C., Atserias Batalla, J., Codina-Filbà, J., García-Narbona, D., Grivolla, J., Lambert, P., Saurí, R.: FBM: combining lexicon-based ML and heuristics for social media polarities. In: Second Joint Conference on Lexical and Computational Semantics (*SEM). Proceedings of the Seventh International Workshop on Semantic Evaluation (SemEval 2013), vol. 2, pp. 483–489, Association for Computational Linguistics, Atlanta (2013). http://www.aclweb.org/anthology/S13-2080
14. Taboada, M., Brooke, J., Tofiloski, M., Voll, K., Stede, M.: Lexicon-based methods for sentiment analysis. Comput. Linguist. **37**(2), 267–307 (2011)
15. Tumasjan, A., Sprenger, T.O., Sandner, P.G., Welpe, I.M.: Predicting elections with twitter: what 140 characters reveal about political sentiment. In: ICWSM, vol. 10, pp. 178–185 (2010)
16. Whitehead, M., Yaeger, L.: Sentiment mining using ensemble classification models. In: SCSS (1), pp. 509–514. Springer (2008)

Adaptive Ant Clustering Algorithm with Pheromone

Urszula Boryczka(✉) and Jan Kozak

Institute of Computer Science, University of Silesia, Sosnowiec, Poland
urszula.boryczka@us.edu.pl

Abstract. In the midst of data mining tasks, clustering algorithms received special attention, especially when these techniques are bio-inspired and while they use special methods which improve a learning process during clusterization. Most promising among them are ant-based approaches. The process of clustering with colony of virtual ants is emerging and can be an alternative, when the data is complicated. Clustering, based on ant's behavior, was first introduced by Deneubourg et al. in 1991 and this classical proposition still requires investigation to improve stability, scalability and convergence of speed. This investigations will show that we can create a mature tool for clustering. The aim of this research was to examine the execution of a new Ant Clustering Algorithm with a modified scheme of ants' perception and an incorporation of pheromone matrices. To assess the performance of our proposition, certain amount of widely known benchmark data sets were used. Empirical study of our approach shows that the *adACA* performs well when the pheromone matrices influence the behavior of clustering ants and leads to better results.

Keywords: Data mining · Cluster analysis · Ant clustering algorithm

1 Introduction

Data mining denotes withdrawing or extracting lore from significant amount of data. It requires the usage of data examination techniques in order to reveal unfamiliar, significant patterns and affinities in large data sets. This embraces great amount of methods such as clustering, classification, regression, pattern recognition, analyzing changes and detecting anomalies. In recent years, many algorithms established on Swarm Intelligence have been crated to solve data mining tasks, such as classification and decision tree construction [2,3]. Most promising among them are ant-based clustering algorithms. In past twenty years, algorithms that use the ant clustering methodology, has developed promising ideas to base the research on. Two significant sorts of ant-based approaches for clustering are defined: first relies on Ant Colony Optimization (ACO) algorithms, which depend on the foraging behavior of ants ("foraging model"). The second method uses piling ants' corpses and grouping their larvae. In some research articles it has been called "piling model".

N.T. Nguyen et al. (Eds.): ACIIDS 2016, Part II, LNAI 9622, pp. 117–126, 2016.
DOI: 10.1007/978-3-662-49390-8_11

First group of methods, discussed below, represents a hybridization of *ACO* algorithm with some optimization techniques applied in clustering problem. Tsai et al. [6] suggested a new clustering approach: Ant Colony Optimization with Different Favor algorithm (*ACODF*). The *ACODF* has adapted Simulated Annealing process to avoid stacking in local optimal solutions. Chu et al. [8] extended ACO methods and discussed an original data clustering process using Constrained Ant Colony Optimization (*CACO*). Runkler et al. [19] introduced an Ant Clustering method that was similar to the ACO algorithm. This approach shows how objective function can be optimized by ants. Another group of ant-based clustering algorithms represents the combination of issues with classical data mining approaches. Fei Wang et al. [22] created an approach of document clustering originated from *ACO* and Fuzzy C-means clustering (*FCM*). This way the algorithm runs resemble the Ant Colony Algorithm in order to find local extremum. When promising groups are created, Fuzzy C-means clustering is adhibited. C. Fernandes et al. [10] has proposed another approach based on a modification of ACO and Kohonen Self-Organizing Maps. Results show that usage of KohonenAnts (*Kants*) model is justified for clustering and classification tasks. Revised ant clustering algorithm in which k-means were applied, was carried out by Qin Chen and Jinping Mo [5]. It optimizes rules for clustering by ants.

There are interesting methods of clustering that have proposed a new insight of pheromone's additional procedures. Warangkhana Ngenkaew et al. [17] proposed a new way of using different pheromones: trailing and foraging pheromones in ant-based clustering algorithm. Ashid Ghosh et al. [12] presented a new approach for clustering data sets founded on possession of aggreagation pheromone discovered in ants, called *APC*. The motion of given ant is dependent on the amount of pheromone deposited at distinct points of the search space.

Variety of alterations have been proposed to the classical scheme of clustering with ants. We want to focus to the "piling model" of clustering and incorporate pheromone maps, which would improve the quality of the clustering, convergence, stability etc. Our motivation is the scientific work concerning the Adaptive Ant Clustering Algorithm proposed by Vizine [21]. Our paper is divided into 6 parts. Sections 2 and 3 give an expanded illustration of the biological inspirations and original algorithms. Section 4 depicts the modifications of the algorithm. Next unit describes the way experiments have been carried out to set the pheromone maps of *adACA*. The last section contains the summary as well as the future goals of *adACA*.

2 Biological Inspirations and Algorithms

A colony of ants in nature is a compound system in which simple interactions constitute global behavior. This behavior occurs as intelligence when perceiving it from an upper level or scale. Interactions between this insects may be indirect, via local modifications of the environment in which they evolve. Pheromone, as a chemical substance is used by many social insects to communicate with each other. This form of stimulation of the behavior has been called stigmergy. The natural system often exhibits such sophisticated manners of organization at the global scale. The cemetery organization and brood sorting are examples of this kind of phenomena and imitation of those patterns shows some new facilities of ant's application. There are two interesting behaviors in real ants: (a) cemetery organization or corpse clustering and (b) brood sorting. These two phenomena are related and therefore, they should be dealt with together. This two mentioned above occurrences have been intensively analyzed by scientists and observed in different species of ants. The first one was investigated by Chretien [7], who analyzed it in *Lasius Niger*. The confirmation of those observations made by Chretien have been obtained by Deneubourg, who conducted experiments on the ant *Phaidole Pallidula*. Another group of ants - *Laptothorax Unifasciatus* gather larvae in accordance to their size. Franks and Sendova-Franks gave a detailed account of this sorting behavior [9]. They have thoroughly examined the distribution of brood within the brood cluster [11]. Deneubourg et al. [9] defined two mathematical models for discussed phenomena of corpses clustering and larvae sorting in colonies of ants. Those models are alike. Virtual ants-agents move at random on the square grid. Objects were scattered within the grid and could be raised, displaced and finally released by virtual agent-ants. These operations are dependent on the density of objects that were spread in the ants' local vicinity, isolated objects, or those that are surrounded by dissimilar ones are more probable to be moved to a neighborhood of similar. Decision to pick up and drop objects are adapted by probabilities p_p and p_d, presented below:

$$p_p = \left(\frac{k_1}{k_1 + f}\right)^2$$

where f is the detected fraction of objects in the nearness of the ant and k_1 is a threshold value and

$$p_d = \left(\frac{f}{k_2 + f}\right)^2$$

where k_2 is another threshold coefficient.

The neighborhood function $f(i)$ formulation was originally motivated by implementation where simple robots were used. In this approach identical objects are treated as similar while dissimilar ones were not. Gutowitz [13] pursued the dynamics of clustering by using the spatial entropy. This formula is useful for comparing various clustering strategies, and can explain how fast they lead to clusters at different scales. An alternative model of Deneubourg's (BM) was presented by Oprisan et al. [18]. Previously met objects (discounted by a time factor) had impact on this approach. Another type of function, incorporated by Bonabeau, represented the weighted formula [1]. In this approach, the influence of various calculations concerning short-term activation and long-term inhibition were examined and finally, the author found that it is possible to generate reaction-diffusion-like patterns.

The most important modifications for BM were incorporated in an approach of Lumer and Fieta. In this Standard Ant Clustering Algorithm authors introduced four novelties: (a) a dissimilarity-based assessment of local density, (b) unability of communication between virtual ants, (c) ability to only ascertain the similarity of objects in their proximate vicinity, and (d) the notion of short-term memory within each virtual ant. In this way, the ant will most likely proceed in the direction of previously left similar object. Lumer anf Fieta [16] define picking up and dropping probabilities as follows:

$$p_p(o_i) = \left(\frac{k_1}{k_1 + f(o_i)}\right)^2$$

$$p_d(o_i) = \begin{cases} 2f(o_i) \text{ when } f(o_i) < k_2 \\ 1, \quad \text{when } f(o_i) \geq k_2 \end{cases} \tag{1}$$

where k_1, k_2 are two constants which role is originated from the BM.

The algorithm proposed by Lumer and Faieta (hereafter LF) projects the space of attributes onto some lower dimensional space, typically of $z = 2$ dimensions. The assumption for the rule is as follows: an ant is located at site r, where it finds an object o_i at time t. The "local density" $f(o_i)$ for object o_i is given by

$$f(o_i) = \begin{cases} \frac{1}{s^2} \sum_{o_j \in Neigh(s \times s)(r)} [1 - \frac{d(o_i,o_j)}{\alpha}], & \text{when } f > 0 \\ 0, & \text{otherwise} \end{cases}$$

where: $f(o_i)$ is a measure of the average similarity of object o_i to other objects o_j located in the environs of o_i, and α is a coefficient of the dissimilarity scale, establishes if two objects are in proximity.

3 Standard Ant Clustering Algorithm — *ACA*

Two algorithms proposed by Deneubourg (BM) and Lumer and Fieta (LF) have a great influence on the Ant-based clustering algorithms. The main improvement concerns the spatial separation between clusters in the grid. The most interesting novelties were incorporated by Handl, Knowles and Meyer [15]. This modification achieves more satisfying results. A crucial coefficients: parameters of the neighborhood function improve the quality of clustering results. In the discussion concerning the perception radius σ, it is worth mentioning, that when applying more expanded neighborhoods improvement of the quality of the clustering and the distribution on the grid have been observed. It is computationally more demanding and moreover strange creation of clusters during the preliminary stage can be observed. A perception radius, that had been increasing little by little, accelerated the disintegration of initially small clusters. To overcome this problem, "short-term memory" proposed by [16] has been introduced. According to Handl et al. [14] parameter α establishes the fraction of the grid patterns treated as similar. The smaller values of α allows to avoid the creation of disintegrated clusters in the grid. When choosing the large value of this parameter, the fusion of clusters can be observed. Based on these analyses, we proposed an automatic adjustment of α (similarly to Handl et al. [14]) and moreover we examined a new scheme for changing. To establish the probability of picking and dropping objects the equations as follows were proposed:

$$p^*_{pick}(i) = \begin{cases} 1, & \text{if } f^*(i) > 1 \\ \frac{1}{f^*(i)^2}, & \text{otherwise} \end{cases}$$

$$p^*_{drop}(i) = \begin{cases} 1, & \text{if } f^*(i) \geq 1 \\ \frac{1}{f^*(i)^4}, & \text{otherwise,} \end{cases}$$

where $f^*(i)$ constitutes an alteration of Lumer and Faieta's vacinity definition:

- $f^*(i) = \begin{cases} \frac{1}{\sigma^2} \sum_j [1 - \frac{d(i,j)}{\alpha}], & \text{if } f^* > 0 \\ & \text{and } (1 - \frac{d(i,j)}{\alpha}) > 0 \\ 0, & \text{otherwise} \end{cases}$
- $\frac{1}{\sigma^2}$ — a neighborhood scaling parameter,
- α — a parameter scaling the dissimilarities within the neighbourhood function $f^*(i)$,
- $d(i,j)$ — a dissimilarity function.

A self-adaptation of agents activity may be presented as a special type of amended perception represented by parameter α. A population of differential ants is used – with distinct coefficient α. After N_{active} individual moves an adaptation of α is executed. Agent-ants also keep track the unsuccessful dropping operations N_{fail}. It is established by $r_{fail} = \frac{N_{fail}}{N_{active}}$, where N_{active} is equal to 100. The ant's coefficient α in this case is altered by the rule:

$$\alpha = \begin{cases} \alpha + 0.01, & \text{if } r_{fail} > 0.99 \\ \alpha - 0.01, & \text{if } r_{fail} \leq 0.99. \end{cases}$$

High-level specification of the presented algorithm is shown as Algorithm 1, where *randomly* means the special probability function's calculations.

4 Modifications of Classical *ACA*

The main disadvantage in applying ant-clustering algorithms is that agent-ants generate much larger number of clusters than is natural. What is more, the algorithm is very often unstable in a particular distribution of objects. Virtual ants permanently create and destroy clusters through the repeated procedure of clustering. The aim of this stage of the construction of an appropriate algorithm is to recognize the real groups presence in the analyzed data set. The main process of iteratively performed clustering merge the most similar clusters and assess each new distribution of objects. As a result of this operation the finest template for the given data is obtained. For each repetition, the distribution of objects is estimated and confronted against the best template acquired so far. The best distribution of objects is represented by the special kind of swarm

Algorithm 1. Algorithm ACA

```
procedure ACA
    Initialize
    Randomly_spread_objects_on_the_grid
    while termination_condition_not_met do
        Each_ant_randomly_picks_up_one_object
        Each_ant_randomly_places_on_the_grid
        for each_ant (i = 1, ..., n) do
            while ant[i] carries_object do
                ant[i] ← move_randomly_on_the_grid
                if ant[i] decide_to_drop_object then
                    ant[i] ← drop_object
                    Lay_pheromone_on_the_grid
                end if
            end while
        end for
    end while
end procedure
```

intelligence – pheromone updating maps connected with the mentioned before grid file.

To achieve this goal, three changes were incorporated in the classical ant-clustering algorithm ACA: firstly a cooling scheme for the variable that regulates the probability of virtual ants lifting objects from the grid (similar to the cooling procedure in Simulated Annealing algorithm), then a progressive vision field that permits agent-ants to analyze different areas ($k_p \leftarrow 0.98 \cdot k_p$ and $k_{pmin} = 0.001$). Finally the use of pheromone reinforcement mechanism as a method to learn when to drop objects at more or less promissing regions of the grid was proposed. Such alterations promote an adaptive scenario and were firstly presented in [21]. The most complicated modification seems to be the second one. Inappropriate behavior of agent-ants (because of fixed perceptual area) caused that the groups are created from heterogeneous objects without appropriate, separate distances between them. In order to overpass this problem, a progressive (forward-moving) scheme was submitted in accordance to the first scientific work [20]. The same article introduced the interested and useful pheromone function. Based on this analysis described by Sherafat and Vizine [20,21], we incorporate the method of reinforcement learning via pheromone as a two equations of picking and droping the objects:

$$P_{pick}(i) = (1 - Phe(\phi_{min}, \phi_{max}, P, \phi(i))) \cdot \left(\frac{k_p}{k_p + f(i)}\right)^2$$

$$P_{drop}(i) = (1 + Phe(\phi_{min}, \phi_{max}, P, \phi(i))) \cdot \left(\frac{k_p}{k_p + f(i)}\right)^2,$$

where ϕ_{max} expresses the factual greatest value of pheromone and ϕ_{min} the smallest one respectively. The function $Phe(\cdot)$ is calculated in accordance to the proposition firstly presented in [20].

5 Experiments

Great number of tests have been conducted to analyze the execution of the proposed algorithm. This part will investigate an experimental study (see Table 2, Fig. 1) performed for the following adjustments. We performed 30 experiments for each data set. The parameter values chosen for the experiment are given in Table 1. The experiments were carried out on an Intel Core i5 2.27 GHz Computer with 2.9 GB RAM.

In conducted experiments different parameter values have been applied (see Table 2), so that they are better adjusted to the given data set's structure. Thus, for more complex data, more iterations and greater grids as well as perception's area of agent-ants have been examined.

Results for Iris and Wine data sets confirm that, predominantly, the number of clusters is the same as in reality – see Fig. 2. Only for Glass data set less clusters are obtained than expected (in average less than 4, instead 7). Agent–ants do not create correct clusters with very small number of objects. The small value of standard deviations characterizes the rest of the obtained results. This happened every time clusters are being formulated, they consist of almost the same objects. Unfortunately in case of Glass data sets we observed the differencies in objects belonging to the special clusters. Benefit results from the above approach is the representation of ants' activity in form of pheromone maps (see Fig. 1). They represent example of distribution of pheromone trails at the end of algorithm performance. Each acclivity of pheromone represents the cumulation of objects on the grid. In each of the examined data sets the same results have been obtained. This confirms the usability and influence of pheromone trails to the achieved results.

Moreover in Table 3 the indicators of the results obtained for analyzed data sets, during the single algorithm lauch are presented. This analysis confirms that the created clusters are very similar in case of many assessed coefficients. The

Table 1. The parameter values chosen for the experiment

	Data set		
	Glass	Iris	Wine
Iteration No	428 000	300 000	356 000
Grid Size	46	38	42
Ants Radius	65	54	59
No of Ants	10	10	10
Mem. of Ants Size	10	10	10
Fin. radius	5	5	5
Min. No of obj. per cl.	4	4	4
Influence of Pher.	0.4	0.4	0.4
Min. Percept. Size	3–6	3–6	3–6

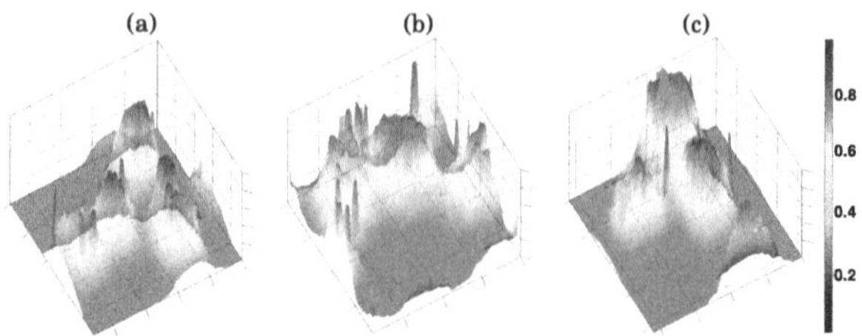

Fig. 1. An example of Pheromone maps for data sets: (a) Glass, (b) Iris, (c) Wine

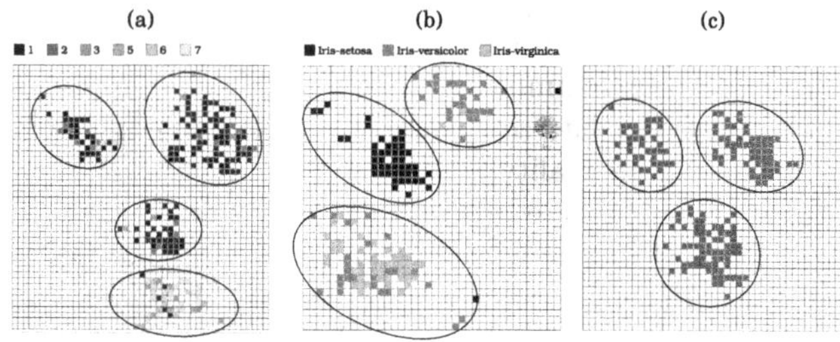

Fig. 2. An example of partitioning objects for data sets: (a) Glass, (b) Iris, (c) Wine

exception constitutes the Glass data set, where the number of clusters is significantly larger than in other data sets. Particularly, the diameter of the created clusters is similar, concerning the given data set. This observation presented above is very interesting because it can be consider as an advantage in case of unsustainable data sets (considering the size of clusters). Simultaneously, we conclude that our approach is not fit to the analysis of data sets with great number of clusters as well as their sizes.

Table 2. Comparative study – standard deviations in parentheses

	ACA alg. – Data set			EM alg. [4] – Data set		
	Glass	Iris	Wine	Glass	Iris	Wine
No of clusters	3.833 (0.59)	3.000 (0.53)	2.900 (0.48)	4.0 (0.0)	4.0 (0.0)	10.0 (0.0)
Diameter between	4.482 (0.13)	3.175 (0.17)	5.577 (0.12)	11.2 (0.6)	3.6 (0.3)	12.3 (0.5)
Inside Diameter	2.819 (0.14)	1.372 (0.24)	3.743 (0.17)	–	–	–
Dunn index	0.022 (0.01)	0.053 (0.05)	0.149 (0.02)	–	–	–
Hubert Gamma	0.362 (0.04)	0.652 (0.12)	0.584 (0.02)	–	–	–
Entropy	1.226 (0.16)	1.020 (0.16)	1.009 (0.17)	–	–	–

Table 3. An example of clusters obtained in our experiments

	Glass			Iris			Wine		
Group	1	2	3	1	2	3	1	2	3
Diameter	39	121	54	50	31	69	71	46	61
Mean Diameter	2.749	2.574	4.829	1.189	1.201	1.381	3.488	3.731	3.799
Med. Distance	2.617	2.136	4.477	0.996	1.150	1.294	3.243	3.548	3.684
Diameter	6.476	11.642	15.767	3.953	3.428	3.721	8.959	8.537	7.927
Separability	0.275	0.275	0.855	0.242	0.312	0.242	1.758	1.337	1.337

The modified ant-based clustering approach developes the classical ant clustering algorithm such as LF method and the ACA, which can operate on different category data sets more swiftly, precisely and efficiently, and maintain the good scalability simultaneously. The actual clustering effectiveness of $adACA$ is expected to be enriched. There exists repeated effervescence in the conveyance process, so that it needs additional improvement that can assure the algorithm single convergence in each conveying process. This process is uniquely observed in big data sets, with unstable number of objects belonging to clusters.

6 Conclusions

The proposed adaptive version of ACA algorithm with pheromone maps was examined to variety of well-known data sets. The new approach has demonstrated a very good robustness and adjustability to the analyzed, sustainable data sets as well as stability of the presented results. Despite the promising results presented here, there are a few directions for future investigations. First of all, the method of governing the parameter values should be incorporated to the presented algorithm. The comparative study with other bio-inspired (*Ant Tree*) clustering algorithms should be performed for better understanding the "piling model" for clustering. Finally, Memetic approach with local search is a good idea to stabilize the results obtained by agent-ants.

References

1. Bonabeau, E.: From classical models of morphogenesis to agent-based models of patern formation. Artif. Life **3**, 191–209 (1997)
2. Boryczka, U., Kozak, J.: Ant colony decision trees – a new method for constructing decision trees based on ant colony optimization. In: Pan, J.-S., Chen, S.-M., Nguyen, N.T. (eds.) ICCCI 2010, Part I. LNCS, vol. 6421, pp. 373–382. Springer, Heidelberg (2010)
3. Boryczka, U., Kozak, J.: Enhancing the effectiveness of ant colony decision tree algorithms by co-learning. Appl. Soft Comput. **30**, 166–178 (2015)
4. Bouckaert, R.R., Frank, E., Hall, M., Kirkby, R., Reutemann, P., Seewald, A., Scuse, D.: Weka manual for version 3-7-10. Nieznane czasopismo (2013)

5. Chen, Q., Mo, J.: Optimizing the ant clustering model based on k-means algorithm. In: Proceeding of the WRI World Congerss on Computer Science and Information Engineering, vol. 3, pp. 699–702 (2009)
6. Tsai, C.W., Tsai, C.-F., Wu, H.-C.: Acof: a novel data clustering approach for data mining in large data bases. J. Syst. Softw. **73**(1), 133–145 (2004)
7. Chretien, L.: Organisation Spatiale du Materiel Provenant de L'excavation du nid chez Messor Barbarus et des Cadavres d'ouvrieres chez Lasius niger (Hymenopterae: Formicidae). Ph.D. thesis, Universite Libre dr Bruxelles (1996)
8. Chu, S.-C., Roddick, J.F., Su, C.-J.: Constrained ant colony optimization for data clustering. J. Artif. Intell., 534–543 (2004)
9. Deneubourg, J.-L., Goss, S., Franks, N., Sendova-Franks, A., Detrain, C., Chretien, L.: The dynamics of collective sorting: robot-like ant and ant-like robot. In: Meyer, J.A., Wilson, S.W. (eds.) First Conference on Simulation of Adaptive Behavior. From Animals to Animats, pp. 356–365. MIT Press, Cambridge (1991)
10. Fernandes, C., Merelo, J.J., Mora, A.M., Ramos, V., Laredo, J.L.J.: Kohonants: a self-organizing algorithm for clustering and pattern classification. Artif. Life XI, 428–435 (2008)
11. Franks, N.R., Sendova-Franks, A.B.: Brood sorting by ants: distributing the workload over the work surface. Behav. Ecol. Sociobiol. **30**, 109–123 (1992)
12. Ghosh, A., Halder, A., Kothari, M., Ghosh, S.: Aggregation pheromone density based data clustering. Inf. Sci. **178**(13), 2816–2831 (2008)
13. Gutowitz, H.: Complexity — seeking ants. Unpublished report (1993)
14. Handl, J., Knowles, J., Dorigo, M.: Ant-based clustering: a comparative study of its relative performance with respect to k-means, average link and id-som. Technical report 24, IRIDIA, Universite Libre de Bruxelles, Belgium (2003)
15. Handl, J., Meyer, B.: Improved ant-based clustering and sorting in a document retrieval interface. In: Guervós, J.J.M., Adamidis, P.A., Beyer, H.-G., Fernández-Villacañas, J.-L., Schwefel, H.-P. (eds.) PPSN VII. LNCS, vol. 2439, pp. 913–923. Springer, Heidelberg (2002)
16. Lumer, E., Faieta, B.: Diversity and adaptation in populations of clustering ants. In: Third International Conference on Simulation of Adaptive Behavior: From Animals to Animats 3, pp. 489–508. MIT Press, Cambridge (1994)
17. Ngenkaew, W., Ono, S., Nakayamo, S.: Multiple pheromone deposition in ant-based clustering as an ant foraging concept. In: Proceedings of the IEEE Symposium on Computational Intelligence in Bioinformatics and Computational Biology, CIBCB 2004, pp. 268–275. IEEE (2004)
18. Oprisan, S.A., Holban, V., Moldoveanu, B.: Functional self-organisation performing wide-sense stochastic processes. Phys. Lett. A **216**, 303–306 (1996)
19. Runkler, T.: Ant colony optimization of clustering model. Int. J. Intell. Syst. **20**(12), 1233–1261 (2005)
20. Sherafat, V., Nunes de Castro, L., Hruschka, E.R.: TermitAnt: an ant clustering algorithm improved by ideas from termite colonies. In: Pal, N.R., Kasabov, N., Mudi, R.K., Pal, S., Parui, S.K. (eds.) ICONIP 2004. LNCS, vol. 3316, pp. 1088–1093. Springer, Heidelberg (2004)
21. Vizine, A., de Castro, L., Hruschka, E.: Towards improving clustering ants: Adaptive clustering algorithm. Inf. J. **29**(2), 143–154 (2005)
22. Wang, F., Zhang, D., Bao, N.: Fuzzy document clustering based on ant colony algorithm. In: Yu, W., He, H., Zhang, N. (eds.) ISNN 2009, Part II. LNCS, vol. 5552, pp. 709–716. Springer, Heidelberg (2009)

Link Prediction in a Semi-bipartite Network for Recommendation

Aastha Nigam[1] and Nitesh V. Chawla[1,2(✉)]

[1] University of Notre Dame, Notre Dame, IN 46556, USA
{anigam,nchawla}@nd.edu
[2] Wroclaw University of Technology, Wroclaw, Poland

Abstract. There is an increasing trend amongst users to consume information from websites and social media. With the huge influx of content it becomes challenging for the consumers to navigate to topics or articles that interest them. Particularly in health care, the content consumed by a user is controlled by various factors such as demographics and lifestyle. In this paper, we use a semi-bipartite network model to capture the interactions between users and health topics that interest them. We use a supervised link prediction approach to recommend topics to users based on their past reading behavior and contextual data associated to a user such as demographics.

1 Introduction

Internet has become the biggest and the most preferred medium for consuming information. As a user we can discover information about a breadth of topics. But, with large amounts of information present across different mediums it becomes extremely laborious for users to find all the content pertinent to them. There are various systems that aim at predicting the rating or preference a user would give to an item. These predictions are based on the user's interaction with the item or similar items.

Reading habits of a user are driven by various factors such as interests, lifestyle, city they are residing in and their age. Particularly, the health content a user consumes is an outcome of many variables. Health content can be defined as information relating to one's health. It could be an article describing the symptoms of a disease or a recipe for healthier eating. Moreover, the articles the user reads may or may not be indicative of an illness they are suffering from or a problem pertaining to them. Given the diversity in the content and other above listed challenges, it becomes difficult to model the user's interest.

Capturing relevance of a particular article for a user becomes extremely granular, therefore we wanted to understand interests of users at a broader level using topics. Every article is associated with a topic or a theme which can be used to cluster them together and be used to understand user behavior. In this paper, we model the interests of users based on the articles they have previously shown interest in and then leveraging other user attributes particularly the city they reside in to predict other topics that would be relevant to them. We propose the

N.T. Nguyen et al. (Eds.): ACIIDS 2016, Part II, LNAI 9622, pp. 127–135, 2016.
DOI: 10.1007/978-3-662-49390-8_12

use of a semi-bipartite network to model this phenomena and identifying missing links in the networks to make recommendations at a user level.

We firstly describe research related to this work in Sect. 2 and then discuss the data used for our model development and validation in Sect. 3. We then present our methodology in Sect. 4 where we discuss the network structure and describe the approach used. Next, we present results obtained by applying our model on the data in Sect. 5. Lastly, we conclude with a discussion in Sect. 6.

2 Related Work

There has been a lot of work on link prediction in general. Liben et al. [1] presented a survey on various methods for link prediction in homogeneous networks where all the nodes are of the same type. They experimented with various measures such as Graph Distance, Common Neighbors, Jaccard's Coefficient, Adamic/Adar Score, Preferential Attachment, Katz, Hitting Time, Page Rank and Sim Rank for predicting new edges in a social network. Backstrom et al. [2] proposed an approach based on supervised random walks that combined information from both nodes and edges. However, many real world systems form complex networks with varied node and interactions types therefore, there has been work on link prediction in heterogeneous network. Davis et al. [3] proposed a supervised approach for link prediction in heterogeneous information networks where they used a modified Adamic Adar measure and compared its performance across various data-sets. They examined the measure on a disease-gene bipartite network. In this paper, we model the health care data using a semi-bipartite network which to the best of our knowledge has not been done before. We then try to understand topical preferences of users using link prediction.

3 Data Description

The data was provided by a digital media company, Everyday Health (EDH)[1] producing content related to health and wellness. EDH owns multiple companies addressing a varied set of topics such as pregnancy, diseases and healthy eating.

Being a web-based platform for health related information they are available across the globe. EDH allows users to sign up to their websites and select the topics they will be most interested in reading about. EDH also sends out weekly newsletters over the email to the users who have signed up and captures the user's reading behavior in terms of when they received the email, when they opened the newsletter, when they read the articles and which health topic category did the each article belong to. Along with health topic information, EDH also collects demographic details for each user such as city they reside in, gender and age group.

In this study, we utilize the data for Saint Joseph County, Indiana from June 2012 to June 2014. Saint Joseph County data comprises of 8 cities: Mishawaka, New Carlisle, South Bend, Osceola, Granger, Walkerton, Lakeville, and North Liberty.

[1] http://www.everydayhealth.com/.

4 Methodology

In this section, we present our method to construct the network and then illustrate the algorithm to perform link prediction.

4.1 Network Model

We propose a semi-bipartite network [4] to model our data for topic recommendation. A semi-bipartite graph can be defined as $G = (V_1, V_2, E_1, E_2)$ where V_1 and V_2 are the two set of nodes, E_1 denotes the edges between V_1 and V_2 whereas E_2 depict the edges (interactions) amongst the nodes in V_1. In our network, the two set of nodes are the user (V_1) and topic (V_2) nodes. As described earlier, the two ways to understand user's interests are firstly when he signs up at the website and selects the topics and secondly when he reads a particular article in the newsletter. Using these two sources, we get an aggregate of which topics the user is interested in. An edge (E_1) between the user and topic node signifies the user's interest. The other set of edges (E_2) in the network are amongst the users. Two users are connected if they come from the same city. This results in cliques of users registering from the same city. This might add noise but we want to leverage this demographic information using topological attributes.

The network was constructed following the above explained methodology. The network was undirected and unweighted. As a result, the network consisted of 4240 nodes and 2,780,874 edges as shown in Fig. 3. As can be seen from Fig. 3, users coming from the same city form cliques and all user cliques are connected to the topics. In the nodes set, 4076 were user nodes and 164 were topic nodes. Similarly, in the edge set, 33,269 edges were between users and topics and rest were between users. Figure 1 illustrates how the users are distributed across the 8 cities in Saint Joseph County. We also study how each city as a whole consumes the EDH content. Figure 2 captures the activity of each city.

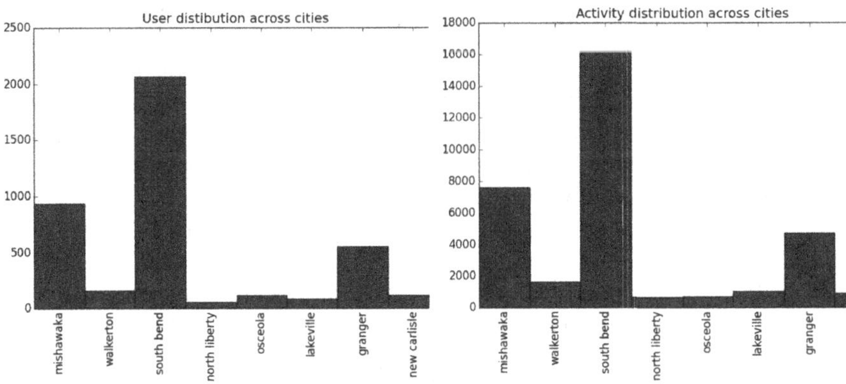

Fig. 1. Captures the user distribution across 8 cities in St. Joseph County.

Fig. 2. Captures how active each city is in terms of consuming information from EDH.

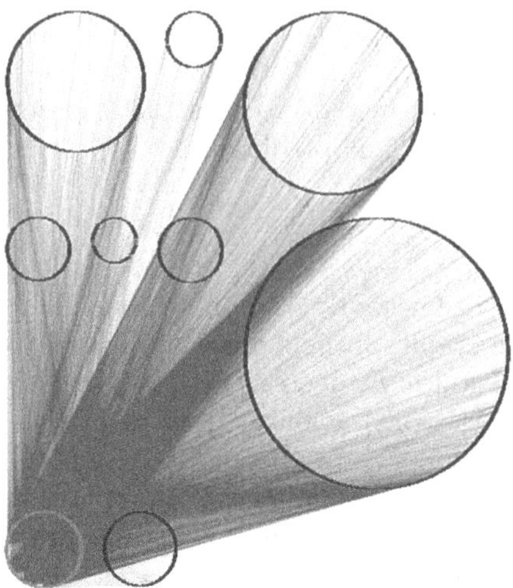

Fig. 3. Representation of the network where the blue nodes signify the user nodes and the red nodes signify the topic nodes. Only the interactions between the user and topics nodes are illustrated here. Since we have 8 cities we have 8 cliques of color blue all connected to the topic nodes (Color figure online).

Activity is calculated based on number of links clicked by the users in their respective cities. It can be seen from Figs. 1 and 2 that the overall activity of users in a city is correlated with the number of users in the city.

Table 1 lists the top 20 topics being consumed in each of the 8 cities. We see that weight management, diet and nutrition and exercise and fitness seem to be popular amongst most cities but articles related to depression seem to be consumed more in North Liberty. Similarly, diabetes seems to be a bigger concern in South Bend, Granger and Walkerton compared to other cities.

4.2 Topic Recommendation

Once the network was constructed, we deal with the problem of topic recommendation using link prediction. The network captures heterogeneous information and we only want to predict links between the user and topic nodes. We calculate various features for links (or nodes) of interest. The topological attributes that we consider can be broadly categorized into two categories of neighborhood methods and path methods. Neighborhood methods are: Common Neighbors, Jaccard's Coefficient, Adamic Adar and Preferential Attachment whereas path methods are: PageRank. For a node u in our network, we denote the set of direct neighbors as $\Gamma(u)$. Using this as the notation, we define the features as follows:

1. **Common Neighbors**: It captures the similarity between the two nodes by identifying the common nodes in their neighborhood [1]. Since, the network is semi-bipartite, the common neighbors can only be the user nodes at path length of 2. It can be calculated using Eq. 1.

Table 1. Top 20 topics being selected by the users across cities

Mishawaka	New carlisle	South bend	Osceola	Granger	Walkerton	Lakeville	North liberty
Weight management	Weight management	Weight management	Weight management	Weight management	Weight management	Weight management	Weight management
Diet and Nutrition	Exercise and Fitness	Diet and Nutrition	Exercise and Fitness	Diet and Nutrition	Diet and Nutrition	Exercise and Fitness	Depression
Exercise and Fitness	Diet and Nutrition	Exercise and Fitness	Diet and Nutrition	Exercise and Fitness	Exercise and Fitness	Diet and Nutrition	Diet and Nutrition
High blood pressure	High blood pressure	Diabetes	Depression	High blood pressure	Depression	High blood pressure	High blood pressure
Depression	Depression	High blood pressure	High blood pressure	Diabetes	Diabetes	Allergies	Exercise and Fitness
Diabetes	Allergies	Depression	Diabetes	Depression	High blood pressure	Diabetes	Diabetes
Heart disease	Diabetes	Heart disease	Allergies	Heart disease	Heart disease	High cholesterol	Arthritis
Allergies	Heart disease	Allergies	Heart disease	Allergies	Arthritis	Depression	Heart disease
High cholesterol	Arthritis	Arthritis	Sleep disorders	Beauty	Pain	Arthritis	High cholesterol
Arthritis	High cholesterol	High cholesterol	High cholesterol	High cholesterol	High cholesterol	Heart disease	Allergies
Anxiety	Menopause	Beauty	Arthritis	Menopause	Anxiety	Migraines	Anxiety
Sleep disorders	Anxiety	Sleep disorders	Anxiety	Arthritis	Allergies	Menopause	Beauty
Beauty	Sleep disorders	Anxiety	Menopause	Anxiety	Sleep disorders	Headache	Menopause
Pain	Beauty	Menopause	Pain	Sleep disorders	Beauty	Beauty	Pain
Menopause	Pain	Diabetes type 2	Beauty	Cancer	Menopause	Sleep disorders	Sleep disorders
Migraines	Cancer	Pain	Migraines	Digestive health	Digestive health	Pain	Digestive health
Diabetes type 2	Migraines	Sexual health	Diabetes type 2	Migraines	Migraines	Sexual health	Sexual health
Cancer	Skin conditions	Migraines	Headache	ADD/ADHD	Osteoarthritis	Anxiety	Osteoperosis
Smoking	Sexual health	Headache	Cancer	Sexual health	Diabetes type 2	Cancer	Cancer
Sexual health	Smoking	Digestive health	Sexual health	Diabetes type 2	Skin conditions	Digestive health	Headache

$$|\Gamma(u) \cap \Gamma(v)| \tag{1}$$

2. **Jaccard's Coefficient**: Number of common neighbors divided by the total combined number of neighbors of both nodes [1]. Instead of considering the raw number, it looks at the ratio of common nodes. It is given by Eq. 2.

$$\frac{|\Gamma(u) \cap \Gamma(v)|}{|\Gamma(u) \cup \Gamma(v)|} \tag{2}$$

3. **Adamic Adar**: It weights the impact of neighbor nodes inversely with respect to their total number of connections [5]. It is based on the assumption that rare relationships are more specific and have more impact on similarity. It can be calculated using Eq. 3.

$$\sum_{z \in \Gamma(u) \cap \Gamma(v)} \frac{1}{log \Gamma(z)} \tag{3}$$

4. **Preferential Attachment**: It emphasizes on the number of neighbors a node has [6]. The higher the degree of a node, the more probable that a new node attaches to it. As shown in Eq. 4, it multiples the number of common neighbors.
$$|\Gamma(u) \times \Gamma(v)| \tag{4}$$
5. **PageRank**: The significance of a node in a network is based on the significance of other nodes that are linked to it. We take the product of the page rank scores for node u and v [7].

We then applied a supervised machine learning approach [8,9]. Each of the above listed attributes contribute to a feature vector. We take the presence of a link as class 1 and the absence of a link as class 0. Since we have a static network, to evaluate the performance of our model we divide the given data into train data and test data. Since there are fewer instances of links compared to absence of links we perform stratified sampling to divide our data set. This ensures that both train and test data sets follow the same class distribution as the original data set. As a result, we had 334,233 train instances which constituted of 317,598 samples from class 0 and 16,635 samples from class 1. Similarly, the test data had 334,231 instances of which 317,597 were class 0 and 16,634 were class 1. This was then evaluated using different machine learning algorithms.

5 Results

We firstly study the degree distributions over the user and topics nodes. The degree distribution of the user nodes was calculated using only the edges between the user and topic nodes. The user-user edges were not considered as they would add noise for the degree distribution. As can be seen from Figs. 4 and 5, they both follow a power law distribution indicating that there are more nodes with fewer links versus fewer nodes with more links. It essentially means that most of the users have presented a very sparse set topic choices whereas there are only a smaller number of users which have indicated their topic interests comprehensively. Similarly, from Fig. 5 we can say that there are lesser topics read by all users but most topics have fewer readers.

In our experiments, we applied three machine learning algorithms: Support Vector Machines (SVM) [10], decision trees [11] and logistic regression [12]. Area under the curve [13] and confusion matrix has been used as the evaluation metric. ROC curves for all three algorithms can be seen in Fig. 6. The area under the curve for SVM, decision trees and logistic regression is 0.5737, 0.5466 and 0.5401 respectively. We can see that SVM performs slightly better than the rest. In Fig. 7, we see we have high number of false positives and false negatives. Due to the inherent imbalance in the dataset, we see that the model is able to predict the absence of the links but due to few positive samples the model suffers.

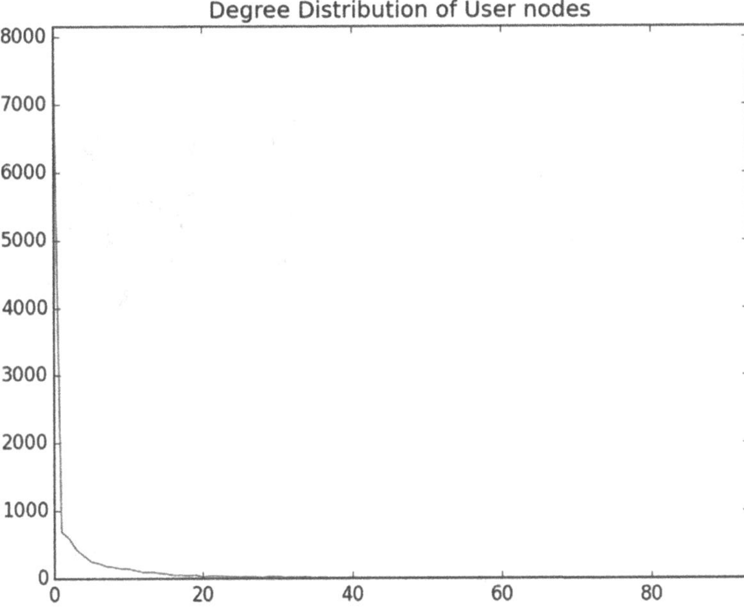

Fig. 4. Captures the degree distribution of user nodes with respect to topic nodes only.

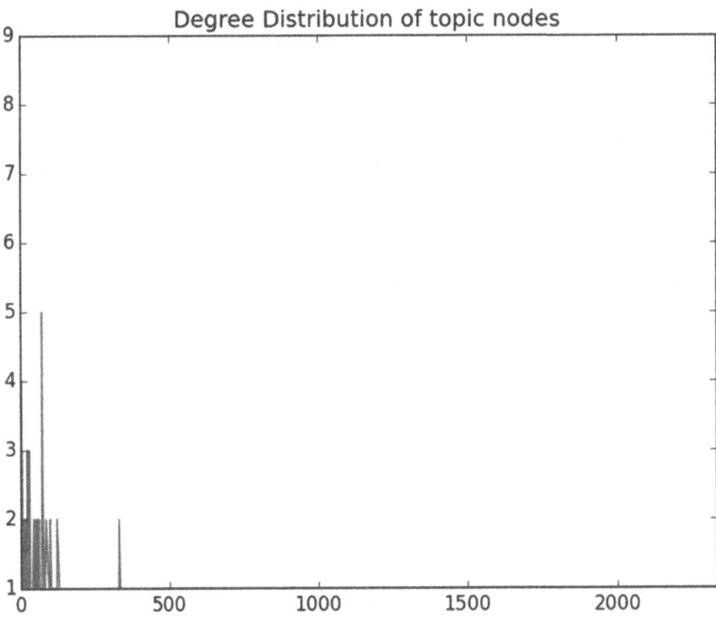

Fig. 5. Captures the degree distribution of topic nodes.

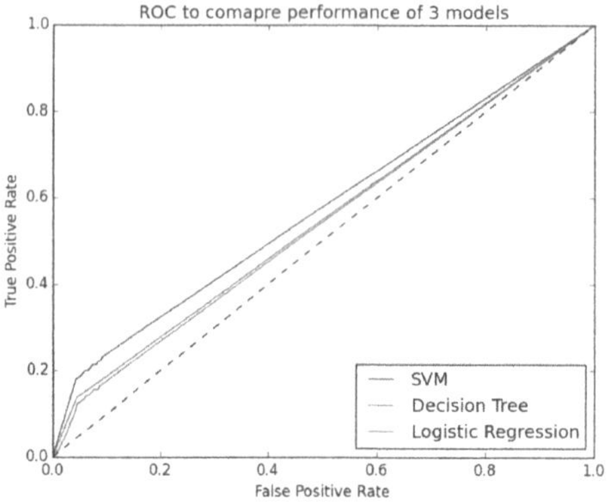

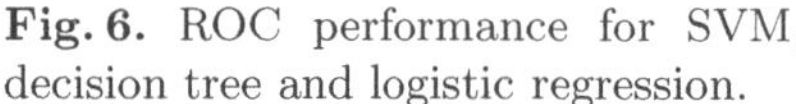

Fig. 6. ROC performance for SVM, decision tree and logistic regression.

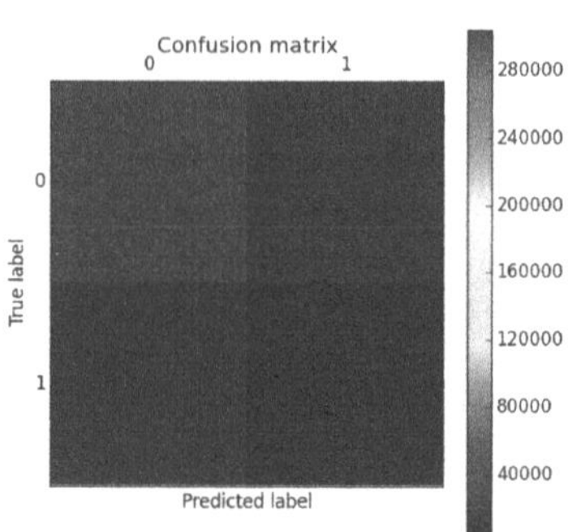

Fig. 7. Confusion matrix for SVM

6 Conclusion

In this work, we have proposed the construction of a semi-bipartite network from the user data. We approach the topic recommendation problem using link prediction in a supervised machine learning framework. We would like to include other features to incorporate the heterogeneity of the data. We evaluated various machine learning algorithms and found that SVMs was most effective on our dataset. It is challenging for a user to navigate through the entire website to find content relevant to him, therefore it becomes important for us to present the user with information he might be most interested in reading about. But many-a-times, data about user and his interests is very sparse. We have tried to capture the user's interest through demographic and reading habits.

As extension to this work we would like to incorporate more features and study the effect of each feature and analyze their effectiveness for predicting a new link in a semo-bipartite network.

Acknowledgements. We would like to thank Everyday Health for providing us their user data.

This research was supported in part by National Science Foundation (NSF) Grant OCI-1029584 and IIS-1447795.

References

1. Liben-Nowell, D., Kleinberg, J.: The link-prediction problem for social networks. J. Am. Soc. Inf. Sci. Technol. **58**(7), 1019–1031 (2007)
2. Backstrom, L., Leskovec, J., Supervised random walks: predicting and recommending links in social networks. In: Proceedings of the Fourth ACM International Conference on Web Search and Data Mining, WSDM 2011, pp. 635–644. ACM, New York, NY, USA (2011)

3. Davis, D., Lichtenwalter, R., Chawla, N.V.: Multi-relational link prediction in heterogeneous information networks. In: Proceedings of the International Conference on Advances in Social Networks Analysis and Mining, ASONAM 2011, pp. 281–288. IEEE Computer Society, Washington, DC, USA (2011)
4. Xu, K., Williams, R., Hong, S.-H., Liu, Q., Zhang, J.: Semi-bipartite graph visualization for gene ontology networks. In: Eppstein, D., Gansner, E.R. (eds.) GD 2009. LNCS, vol. 5849, pp. 244–255. Springer, Heidelberg (2010)
5. Adamic, L.A., Adar, E.: Friends and neighbors on the web. Soc. Netw. **25**, 211–230 (2001)
6. Barabasi, A.-L., Albert, R.: Emergence of scaling in random networks. Science **286**(5439), 509–512 (1999)
7. Page, L., Brin, S., Motwani, R., Winograd, T.: The pagerank citation ranking: bringing order to the web (1999)
8. Benchettara, N., Kanawati, R., Rouveirol, C.: Supervised machine learning applied to link prediction in bipartite social networks. In: Memon, N., Alhajj, R. (eds.) ASONAM, pp. 326–330. IEEE Computer Society (2010)
9. Kunegis, J., De Luca, E.W., Albayrak, S.: The link prediction problem in bipartite networks. In: Hüllermeier, E., Kruse, R., Hoffmann, F. (eds.) IPMU 2010. LNCS, vol. 6178, pp. 380–389. Springer, Heidelberg (2010)
10. Cortes, C., Vapnik, V.: Support-vector networks. Mach. Learn. **20**(3), 273–297 (1995)
11. Quinlan, J.R.: Induction of decision trees. Mach. Learn. **1**(1), 81–106 (1986)
12. Hosmer, D.W., Lemeshow, S.: Applied Logistic Regression (Wiley Series in Probability and Statistics), 2nd edn. Wiley-Interscience Publication, New York (2000)
13. Bradley, A.P.: The use of the area under the roc curve in the evaluation of machine learning algorithms. Pattern Recogn. **30**(7), 1145–1159 (1997)

Hybrid One-Class Ensemble for High-Dimensional Data Classification

Bartosz Krawczyk(✉)

Department of Systems and Computer Networks, Wrocław University of Technology, Wybrzeże Wyspiańskiego 27, 50-370 Wrocław, Poland
bartosz.krawczyk@pwr.edu.pl

Abstract. The advance of high-throughput techniques, such as gene microarrays and protein chips have a major impact on contemporary biology and medicine. Due to the high-dimensionality and complexity of the data, it is impossible to analyze it manually. Therefore machine learning techniques play an important role in dealing with such data. In this paper we propose to use a one-class approach to classifying microarrays. Unlike canonical classifiers, these models rely only on objects coming from single class distributions. They distinguish observations coming from the given class from any other possible states of the object, that were unseen during the classification step. While having less information to dichotomize between classes, one-class models can easily learn the specific properties of a given dataset and are robust to difficulties embedded in the nature of the data. We show, that using one-class ensembles can give as good results as canonical multi-class classifiers, while allowing to deal with imbalanced distribution and unexpected noise in the data. To cope with high dimensionality of the feature space, we propose a novel hybrid one-class ensemble utilizing combination of weighted Bagging and Random Subspaces. Experimental investigations, carried on public datasets, prove the usefulness of the proposed approach.

Keywords: Machine learning · One-class classification · Classifier ensemble · High-dimensional data · Bioinformatics

1 Introduction

Contemporary high-throughput technologies produce massive volumes of biomedical data. Transcriptional research and profiling, with the usage of microarray technologies are powerful tools to gain a deep insight into the pathogenesis of complex diseases that plague modern society, such as cancer. Recent works on cancer profiling showed without a doubt, that gene expression patters can be used for high-quality cancer subtype recognition [25] - leukemias [20], melanoma [18], breast cancer [7] or prostate cancer [14] to name a few.

Identifying cancer properties, based on their distinct expression profiles may provide necessary information for a breakthrough, that is required for patient-tailored therapy. Currently there are no distinct rules on how individuals respond

N.T. Nguyen et al. (Eds.): ACIIDS 2016, Part II, LNAI 9622, pp. 136–144, 2016.
DOI: 10.1007/978-3-662-49390-8_13

to chemotherapy and existing chemotherapies have in most cases severe side-effects with varying medical efficiency.

Due to massive amounts of data generated by microarray experiments and their high complexity and dimensionality, one requires a decision support system to extract the meaningful information from them. Machine learning is widely used for this task [12], with two distinct areas - unsupervised [27] and supervised learning [17]. In this paper we will focus on the latter one.

Supervised machine learning [5] is a promising approach for analyzing microarray results in context of predicting patients outcome. Support Vector Machines are among the most popular classifiers used for this task [1]. Multiple Classifier Systems [29,30], or classifier ensembles [26], have gained an significant attention of the bioinformatics community in recent years. Random Forest [15] and Rotation Forest [13] ensembles have displayed an excellent classification accuracy for small-sample, high dimensionality microarray datasets, outperforming single-model approaches.

Another important issue is the problem of curse of dimensionality. Microarray data suffer from a relatively small number of objects, in comparison to the feature space dimensionality, often reaching several thousands. This causes difficulties for machine learning algorithms, reducing their performance and increasing their computational complexity. Among this data flood a major number of parameters possess small discriminative power and is irrelevant to the classification process, which makes feature selection a crucial step in microarray analysis [9].

Although there are many applications of machine learning-based decision support systems in bioinformatics, there are still many unresolved problems, such as:

- How to integrate heterogeneous data sources to achieve better insight into the mechanism behind complex diseases?
- How to organize, store, analyze and visualize high-dimensionality data obtained from the biomedical data flood?
- How to deal with the problem of high-dimensionality, small sample size, which strongly affects the classification performance and may lead to overfitting, poor generalization and unstable predictors?
- How to cope with difficulties embedded in the nature of microarray data, such as noise or class imbalance, as canonical machine learning classifiers cannot cope with them easily?

In this paper the last two issues are addressed.

We propose to analyze microarray data with the usage of one-class classifiers [28], instead of commonly applied binary ones. To cope with the high dimensionality and complexity of the problem we apply a novel hybrid ensemble approach combining advantages of weighted Bagging and Random Subspaces. It randomly assigns weights to training objects, varying their importance for each base learners. This introduces diversity and assures that each classifier has a different area of competence. To deal with numerous features we propose to use a Random Subspaces method retaining only a small fraction of the original feature space. This reduces the computational complexity of each model while further improving the diversity among ensemble members. Experiments, based on a set of publicly available microarray datasets, show that the proposed approach maintains a good classification accuracy, while displaying an improved robustness to atypical data distribution and prevalent noise.

2 One-Class Classification

The aim of one-class classification (OCC) is to recognize one specific class from the more broad set of classes (e.g., selecting horses from all animals). The given class is known as target class ω_t, while the remaining are denoted as outliers ω_O. During the learning only examples target class (known also as positive examples) are being presented to learner, while it is assumed that during the exploitation phase new, unseen objects from other classes may appear.

OCC problems are common in the real world where positive examples are widely available but negative ones are hard, expensive or even impossible to gather [4]. Let us consider an engine. It is a quite easy and cheap to collect data about its normal work. Collecting observations about failures it is expensive and sometimes impossible, because in this case we would have to spoil the engine.

Such approach is very useful as well for many practical cases especially when the target class is "stable" and outlier one is "unstable". To explain this motivation let us consider a computer security problem as spam filtering or intrusion detection (IDS/IPS) [16].

Among several types of classifiers dedicated to OCC, the most popular is one concentrating on estimation of a closed boundary for given data, assuming that such a boundary will describe sufficiently the target class [23]. The main aim of those methods is to find the optimal size of the volume enclosing given training points. Too small size could lead to overfitting the model, while too big size might lead to extensive acceptance of outliers into the target class. Those methods rely strongly on the distance between objects. Boundary methods require smaller number of objects to properly estimate the decision criterion, which makes them a perfect tool for applications suffering from a small sample size,such as microarrays classification. The well-known boundary methods are one-class support vector machine (OCSVM) [19] and support vector data description (SVDD) [21]. In this work we will use the former one.

3 Proposed One-Class Hybrid Ensemble

None of standard ensembles used in OCC can directly benefit from the usage of Weighted One-Class Support Vector Machines (WOCSVM) [3]. WOCSVMs are much more efficient and robust to internal noise than standard one-class methods, but require dedicated ensemble forming algorithms to being combined in an efficient way. This lead us to the possibility of using a Wagging ensemble approach [10].

Wagging [2] is a variant of Bagging method. It is also known as Weighted Bagging. Here each base classifier is trained on the entire training set, but each objects is randomly assigned a weight.

Bagging can be considered as Wagging with weights drawn from the Poisson distribution, as each instance is represented in the bag a discrete number of times. On the other hand, Wagging often uses an exponential distribution to drawn weights. This is because the exponential distribution is the real-value counterpart of the Poisson distribution.

These properties are very interesting for the OCC task. Canonical Bagging creates random sub-samples of data. However, as one-class methods rely on distance between object using sub-spaces consisting of distant training samples is unadvised. This may lead to training classifiers with excessive size of the classification volume (boundary) and high rate of outliers acceptance Wagging alleviates this drawback of Bagging, as it does not discard the spatial relations between objects.

Wagging offers an interesting way to modify the level of influence of each sample on the classifier's training process by differentiating weights. However, Wagging cannot be directly applied in most of the OCC methods, as they consider each object from the target class to be equally important during the training step. We propose to combine Wagging with WOCSVM classifier and input the drawn weights directly into the WOCSVM training phase.

As we deal with high dimensional data we need a way to reduce the complexity of input space for each learner. That is why we propose to augment Wagging with Random Subspaces [8] method in order to create simplified sub-tasks for each classifier. As each base learner will be trained on a small subset of features it will work in simpler, lower dimensional space. At the same time an ensemble of such learners will cover the entire original feature space.

The pseudo-code for proposed hybrid ensemble for high-dimensional one-class classification (OC-Hyb) is presented in Algorithm 1.

Algorithm 1. One-class hybrid ensemble for high-dimensional data.

Require: WOCSVM training procedure,
number of classifiers L ,
training set $\mathcal{TS}$,
exponential distribution λ,
feature subspace size S

$l \leftarrow 1$
repeat
  $\mathcal{TS}_l \leftarrow \mathcal{TS}$ with random weights drawn from λ
  $\mathcal{FS}_l \leftarrow$ draw subset of S features from the original space
  Train l-th WOCSVM on $\mathcal{TS}_i$ according to weights assigned to each object and using $\mathcal{FS}_l$ feature subset
  $l \leftarrow l + 1$
until $l > L$

The main advantages of the proposed approach are a significant reduction of the training complexity, as weights are given directly by Wagging, instead of calculating them individually, greatly reduced dimensionality of feature space for each classifier and increase of the diversity of ensemble members.

To efficiently combine one-class classifiers we require the knowledge about the values of support functions of each individual classifier from the pool. As WOCSVM work on the basis of distance between the new sample and its decision boundary (known also as reconstruction error) we need a heuristic mapping from distance to support function value:

$$F_{\omega_T}(x) = exp(-dst(x, \omega_T)/s), \tag{1}$$

where $dst(x, \omega_T)$ stands for a distance (usually Euclidean metric is used) between the considered object x and decision boundary for ω_T (this depends on the nature of used classifier, e.g., support vectors or nearest neighbor may be used) and s is the scale parameter that should be fitted to the target class distribution. This scale factor is related to how spread out your data points are. When the distance between objects tend to get very high (e.g., in high dimensional spaces) small value of s is used to control the stability of the mapping. Therefore, in most cases $s = \frac{1}{d}$. This mapping has the advantage that the outputted support value is always bounded between 0 and 1.

In order to combine base classifiers in our OC-Hyb we propose to use product combination of the estimated supports [22], which is expressed by:

$$F_{\omega_T}(x) = \frac{\prod_{l=1}^{L} F_{\omega_T}^{(l)}(x)}{\prod_{l=1}^{L} F_{\omega_T}^{(l)}(x) + \prod_{l=1}^{L} \theta^{(l)}}, \tag{2}$$

where $F_{\omega_T}^{(l)}(x)$ is the support of l-th classifier for object x belonging to the target class and $\theta^{(l)}$ is the classification threshold of l-th classifier for accepting an object as belonging to target class.

For decomposing a binary problem (such as considered here microarray data) we train a separate OC-Hyb on each class. Then the class label for a new object is established by the maximum rule over the outputted decision supports [11].

4 Experimental Investigations

In this section we evaluate the proposed one-class ensemble on the basis of datasets available at[1], whose details are given in Table 1. Four different datasets were used and additional, fifth one, was generated. It was based on the Breast Cancer dataset. To test the performance of classifiers in difficult scenarios we have affected 25 % of objects with Gaussian noise, thus creating in-class outliers in the data.

As base classifier we have used an WOCSVM with RBF kernel [4]. We have trained 100 base classifiers. Size S of feature subspace was set to 5 % of the original feature space.

[1] http://datam.i2r.a-star.edu.sg/datasets/krbd/.

Table 1. Statistics of the datasets used in the experiments.

Dataset	Samples (class 1/class 2)	Features
Breast cancer	78 (34/44)	24481
Breast cancer - noise	78 (34/44)	24481
Central nervous system	60 (21/39)	7129
Colon tumor	62 (22/40)	6500
Lung cancer	181 (31/150)	12533

To put the obtained results into context we have tested the performance of multi-class classifiers used for this task - single SVM (trained with RBF kernel and SMO procedure), Random Forest (consisting of 100 decision trees) and Rotation Forest (consisting of 100 decision trees). Additionally we show the performance of a single OCSVM and one-class Random Forest ensemble (OC-RandF) [6].

Results are based on leave-one-out cross-validation (LOOCV).

All experiments were carried out in the R environment [24], with classification algorithms taken from the dedicated packages, thus ensuring that the results achieved the best possible efficiency and that the performance was not decreased by a bad implementation.

Results with respect to sensitivity and specificity, are given in Table 2.

Table 2. Recognition sensitivity [%] and specificity [%] for examined methods. *RandF* stands for Random Forest, *RotF* for Rotation Forest, $OC - RandF$ for an one-class Random Forest and $OC - Hyb$ for the proposed hybrid one-class ensemble.

Dataset	SVM		RandF		RotF		OCSVM		OC-RandF		OC-Hyb	
	Sens [%]	Spec [%]	Sens [%]	Spec [%]	Sens [%]	Spec [%]	Sens [%]	Spec [%]	Sens [%]	Spec [%]	Sens [%]	Spec [%]
Breast cancer	90.23	91.46	92.32	93.65	92.32	93.65	87.85	90.07	89.15	91.45	93.04	92.89
Breast cancer - noise	74.46	83.59	77.36	84.90	80.05	85.72	75.20	82.98	85.15	87.40	89.01	90.65
Central nervous system	85.60	94.36	88.20	95.90	88.20	95.90	82.95	90.11	85.67	93.15	87.57	93.76
Colon tumor	78.90	91.25	81.35	94.03	82.70	93.90	80.15	92.36	83.85	94.10	86.17	94.45
Lung cancer	61.72	93.05	65.89	95.11	67.00	94.85	69.22	92.08	72.98	94.12	76.03	95.46

From the results one may clearly see, that in case of standard microarray datasets the proposed approach returns both specificity and sensitivity similar to those of the state-of-the-art multi-class models. However in case of noisy (dataset no. 2) and imbalanced (datasets no. 4 and no. 5) our proposed approach is able to outperform significantly the standard classifiers. This happens due to the nature of OCC models - as they are able to learn the distinct properties of the target class, they are able to cope with in-class difficulties.

Ensemble approaches are superior to single-model one-class classifiers. Using decomposition with OCSVMs did not lead to satisfactory results. This can be explained by increasing complexity of data description algorithms in high dimensions. Therefore, without using any approach for reducing the number of features used by each one-class classifier it is impossible to efficiently handle microarray data.

Both OC-RandF and OC-Hyb offer possibility of reducing the size of the feature space. OC-RandF achieves this by using a drawn subset of features in each node of inducted tree. The proposed OC-Hyb trains each base classifier on a small random feature subspace, thus offering a significant dimensionality reduction.

When comparing these two one-class ensembles we may see that our proposal achieves much better performance, especially on noisy and imbalanced data. This is due to the operating modes of these committees. OC-RandF transforms an one-class problem into a binary one by adding artificial counterexamples. Then canonical binary trees are being trained. This does not output a data description, but a standard dichotomization boundary. Therefore, OC-RandF loses the desirable properties of OCC methods and becomes more similar to binary classifiers. The proposed OC-Hyb satisfies the conditions of OCC by working only with objects from the target class.

5 Conclusions

In this paper a novel approach for microarray analysis, based on an ensemble of one-class weighted support vector machines, was presented. To deal with the problem of high dimensionality, which may cause difficulties for one-class models, a novel one-class hybrid ensemble method based on combination of weighted Bagging and Random Subspaces was introduced. Creating base classifiers with the usage of weighted objects and reduced feature space allowed for forming an efficient multiple classifier system that was able to cope with high-dimensionality nature of data and return similar performance as state-of-the-art multi-class methods. The strong points of the proposed method were revealed when dealing with noisy and imbalanced data. In such a case the proposed one-class hybrid ensemble displayed superior quality over its competitors.

The proposed approach may be an attractive tool for bioinformatics decision support systems, in which we deal with uncertain, noisy data or data coming from uneven distributions.

Acknowledgements. This work was partially supported by The Polish National Science Centre under the grant PRELUDIUM number DEC-2013/09/N/ST6/03504.

References

1. Bariamis, D., Maroulis, D., Iakovidis, D.K.: Unsupervised SVM-based gridding for DNA microarray images. Comput. Med. Imaging Graph. **34**(6), 418–425 (2010)

2. Bauer, E., Kohavi, R.: An empirical comparison of voting classification algorithms: bagging, boosting, and variants. Mach. Learn. **36**(1–2), 105–139 (1999)
3. Bicego, M., Figueiredo, M.A.T.: Soft clustering using weighted one-class support vector machines. Pattern Recogn. **42**(1), 27–32 (2009)
4. Cyganek, B.: One-class support vector ensembles for image segmentation and classification. J. Math. Imaging Vis. **42**(2–3), 103–117 (2012)
5. Czarnecki, W.M., Tabor, J.: Two ellipsoid support vector machines. Expert Syst. Appl. **41**(18), 8211–8224 (2014)
6. Desir, C., Bernard, S., Petitjean, C., Heutte, L.: One class random forests. Pattern Recogn. **46**(12), 3490–3506 (2013)
7. Finak, G., Bertos, N., Pepin, F., Sadekova, S., Souleimanova, M., Zhao, H., Chen, H., Omeroglu, G., Meterissian, S., Omeroglu, A., Hallett, M., Park, M.: Stromal gene expression predicts clinical outcome in breast cancer. Nat. Med. **14**(5), 518–527 (2008)
8. Ho, T.K.: The random subspace method for constructing decision forests. IEEE Trans. Pattern Anal. Mach. Intell. **20**, 832–844 (1998)
9. Inza, I., Larraaga, P., Blanco, R., Cerrolaza, A.J.: Filter versus wrapper gene selection approaches in dna microarray domains. Artif. Intell. Med. **31**(2), 91–103 (2004)
10. Krawczyk, B.: Forming ensembles of soft one-class classifiers with weighted bagging. New Gener. Comput. **33**(4), 449–466 (2015)
11. Krawczyk, B., Woźniak, M., Herrera, F.: On the usefulness of one-class classifier ensembles for decomposition of multi-class problems. Pattern Recogn. **48**(12), 3969–3982 (2015)
12. Larranaga, P., Calvo, B., Santana, R., Bielza, C., Galdiano, J., Inza, I., Lozano, J.A., Armananzas, R.: Machine learning in bioinformatics. Briefings Bioinform. **7**(1), 86–112 (2006)
13. Liu, K., Huang, D.: Cancer classification using rotation forest. Comput. Biol. Med. **38**(5), 601–610 (2008)
14. Lynch, C.C., Hikosaka, A., Acuff, H.B., Martin, M.D., Kawai, N., Singh, R.K., Vargo-Gogola, T.C., Begtrup, J.L., Peterson, T.E., Fingleton, B., Shirai, T., Matrisian, L.M., Futakuchi, M.: MMP-7 promotes prostate cancer-induced osteolysis via the solubilization of RANKL. Cancer Cell **7**(5), 485–496 (2005)
15. Moorthy, K., Mohamad, M.S.: Random forest for gene selection and microarray data classification. In: Lukose, D., Ahmad, A.R., Suliman, A. (eds.) KTW 2011. CCIS, vol. 295, pp. 174–183. Springer, Heidelberg (2012)
16. Noto, K., Brodley, C., Slonim, D.: FRaC: a feature-modeling approach for semi-supervised and unsupervised anomaly detection. Data Min. Knowl. Discov. **25**(1), 109–133 (2012)
17. Ringner, M., Peterson, C., Khan, J.: Analyzing array data using supervised methods. Pharmacogenomics **3**(3), 403–415 (2002). Cited By (since 1996): 43
18. Schatton, T., Murphy, G.F., Frank, N.Y., Yamaura, K., Waaga-Gasser, A.M., Gasser, M., Zhan, Q., Jordan, S., Duncan, L.M., Weishaupt, C., Fuhlbrigge, R.C., Kupper, T.S., Sayegh, M.H., Frank, M.H.: Identification of cells initiating human melanomas. Nature **451**(7176), 345–349 (2008)
19. Schölkopf, B., Smola, A.J.: Learning with Kernels: Support Vector Machines, Regularization, Optimization, and Beyond. Adaptive Computation and Machine Learning. MIT Press, Cambridge (2002)
20. Silveira, V.S., Scrideli, C.A., Moreno, D.A., Yunes, J.A., Queiroz, R.G.P., Toledo, S.C., Lee, M.L.M., Petrilli, A.S., Brandalise, S.R., Tone, L.G.: Gene expression pattern contributing to prognostic factors in childhood acute lymphoblastic leukemia. Leukemia Lymphoma **54**(2), 310–314 (2013)

21. Tax, D.M.J., Duin, R.P.W.: Support vector data description. Mach. Learn. **54**(1), 45–66 (2004)
22. Tax, D.M.J., Duin, R.P.W.: Combining one-class classifiers. In: Kittler, J., Roli, F. (eds.) MCS 2001. LNCS, vol. 2096, p. 299. Springer, Heidelberg (2001)
23. Tax, D.M.J., Juszczak, P., Pekalska, E., Duin, R.P.W.: Outlier detection using ball descriptions with adjustable metric. In: Yeung, D.-Y., Kwok, J.T., Fred, A., Roli, F., de Ridder, D. (eds.) SSPR 2006 and SPR 2006. LNCS, vol. 4109, pp. 587–595. Springer, Heidelberg (2006)
24. R Development Core Team: R: a language and environment for statistical computing. R Foundation for Statistical Computing, Vienna, Austria (2008)
25. Tinker, A.V., Boussioutas, A., Bowtell, D.D.L.: The challenges of gene expression microarrays for the study of human cancer. Cancer Cell **9**(5), 333–339 (2006)
26. Trawiński, B.: Evolutionary fuzzy system ensemble approach to model real estate market based on data stream exploration. J. UCS **19**(4), 539–562 (2013)
27. Wang, Y., Yu, Z., Anh, V.: Fuzzy C-means method with empirical mode decomposition for clustering microarray data. Int. J. Data Min. Bioinf. **7**(2), 103–117 (2013)
28. Wilk, T., Woźniak, M.: Soft computing methods applied to combination of one-class classifiers. Neurocomputing **75**, 185–193 (2012)
29. Woźniak, M., Grana, M., Corchado, E.: A survey of multiple classifier systems as hybrid systems. Inf. Fusion **16**(1), 3–17 (2014)
30. Woźniak, M., Zmyślony, M.: Chosen problems of designing effective multiple classifier systems. In: International Conference on Computer Information Systems and Industrial Management Applications, CISIM, Krakow, Poland, 8–10 October 2010, pp. 42–47 (2010)

Advanced Data Mining Techniques and Applications

A Lossless Representation for Association Rules Satisfying Multiple Evaluation Criteria

Marzena Kryszkiewicz(✉)

Institute of Computer Science, Warsaw University of Technology, Nowowiejska 15/19, 00-665 Warsaw, Poland
mkr@ii.pw.edu.pl

Abstract. A lot of data mining literature is devoted to association rules and their evaluation. As the number of discovered rules is often huge, their direct usage by a human being may be infeasible. In the case of classical strong association rules, which are defined as rules supported by sufficiently large fraction of data and having sufficient confidence, a number of their concise representations have been proposed. However, as indicated in the literature, support and confidence measures of association rules seem not to cover many aspects that could be of interest to a user. In consequence, many other measures have been proposed to evaluate association rules. In this paper, we identify an important and wide class of rule *ACBC*-evaluation measures and offer a lossless representation of association rules satisfying constraints for any set of evaluation measures from this class. A number of properties of the representation is derived as well.

1 Introduction

A lot of data mining literature is devoted to association rules and their evaluation. As the number of discovered rules is often huge, their direct usage by a human being may be infeasible. In the case of classical strong association rules, which are defined as rules supported by sufficiently large fraction of data and having sufficient confidence, a number of their concise representations have been proposed. Most of them consist of association rules built from so called generators and/or closed itemsets [2, 4, 6–8, 17, 18]. Though these representations have proved their value, as indicated in the literature, support and confidence measures of association rules seem not to cover many aspects that could have been of interest to a user such as unexpectedness of the knowledge presented in the form of an association rule. In consequence, many other measures have been proposed to evaluate association rules (see e.g. [3, 5, 9–16]). Nevertheless, the representations of association rules that satisfy solely support and confidence criteria are not adjusted to other evaluation measures.

In this paper, we define a new notion of an *ACBC* (Antecedent-Consequent-Base-Constants)–evaluation measure, which covers many rule evaluation measures, including, for example, the novelty, lift, certainty factor, dependence factor, cosine, Jaccard, accuracy, F-score, but also classical support and confidence. Next, we introduce and examine a new notion of a rule template which we define as a pair of rules: one built from generators and another one built from closed itemsets. Based on the

N.T. Nguyen et al. (Eds.): ACIIDS 2016, Part II, LNAI 9622, pp. 147–158, 2016.
DOI: 10.1007/978-3-662-49390-8_14

introduced notions, we offer a lossless representation of association rules satisfying constraints for any set of *ACBC*-evaluation measures. We prove that for any set M of *ACBC*-evaluation measures, the proposed representation is lossless, correctly distinguish between association rules satisfying multiple evaluation criteria based on M and association rules not satisfying them, and enables correct determination of values of measures M for rules satisfying given criteria. We also discuss possible ways of compact storing of our representation.

Our paper has the following layout. In Sect. 2, we briefly recall basic notions of itemsets and (association) rules, support, confidence as well as recall the notions and properties of two particular types of itemsets: generators and closed itemsets. Our new contribution is presented in Sects. 3–4. In Sect. 3, we introduce a definition of a class of *ACBC*-evaluation measures and provide examples of such measures. In Sect. 4, we introduce and examine notions of rule templates and, based on them, a lossless representation of association rules satisfying criteria expressed in terms of *ACBC*-evaluation measures. There, we also propose a few ways for compact storing of rule template representations. Section 5 summarizes our work.

2 Basic Notions and Properties

In this section, we provide definitions and properties related to *rules* and their specific type called *association rules* [1]. Let I be a set of *items* (products, features,…). Any set $X \subseteq \mathrm{I}$ is called an *itemset*. A *transaction dataset* is denoted by D and is defined as a set of itemsets. Each itemset T in D is called a *transaction*. By T_X we will denote the set of transactions in D that contain X.

Let X and Y be any itemsets. An expression $X \rightarrow Y$ is called a *rule*. If additionally:

$$X \cap Y = \emptyset$$

then $X \rightarrow Y$ is called an *association rule*.

Itemsets X and Y, which occur in a rule $X \rightarrow Y$, are called its *antecedent* and *consequent*, respectively. The itemset $Z = X \cup Y$ is called the *base* of $X \rightarrow Y$.

Itemsets and rules are typically characterized by *support* (or *relative support*) and *confidence* as follows. *Support* of an itemset X is denoted by $sup(X)$ and is defined as the number of transactions in D that contain X; that is:

$$sup(X) = |\mathrm{T}_X|.$$

Alternatively, instead of a support, a notion of a relative support is used: *Relative support* of an itemset X is defined as $sup(X)/|\mathrm{D}|$. The relative support of itemset X can be understood as the probability of the occurrence of X in a transaction. The relative support of X will be denoted by $P(X)$.

Clearly, if $X \subseteq Y$, then $\mathrm{T}_X \supseteq \mathrm{T}_Y$, $sup(X) \geq sup(Y)$ and $P(X) \geq P(Y)$.

An itemset is called *frequent* if its (relative) support is not less than a user specified minimum (relative) support threshold.

Support of a rule $X \rightarrow Y$ is denoted by $sup(X \rightarrow Y)$ and is defined as the support of $X \cup Y$; that is,

$$sup(X \rightarrow Y) = sup(X \cup Y).$$

The *confidence* of a rule $X \rightarrow Y$ is denoted by $conf(X \rightarrow Y)$ and is defined as the conditional probability that Y occurs in a transaction provided X occurs in the transaction; that is:

$$conf(X \rightarrow Y) = sup(X \rightarrow Y)/sup(X) = P(XY)/P(X).$$

A rule is regarded as *strong* if its (relative) support and confidence are greater than or equal to user defined minimum (relative) support and minimum confidence thresholds, respectively.

A number of concise lossless representations of strong association rules have been proposed (see e.g. [7, 8] for an overview). Such representations consist typically of rules built from particular itemsets called *generators* and/or *closed itemsets*. For example, in the case of representative association rules [6–8] and informative basis [2], antecedents of rules are generators, whereas their consequents are set theoretical differences between closed itemsets and rules' antecedents. The number of rules in those representations is typically by 1–3 orders of magnitude less than the number of all (strong) association rules. Beneath we recall definitions and/or properties of itemsets, closed itemsets and generators to which we will refer later in the paper.

Property 1. Let $X \subseteq U \subseteq Z$ and $T_X = T_Z$. Then $T_X = T_U = T_Z$ and $P(X) = P(U) = P(Z)$.

Proof. Since $X \subseteq U \subseteq Z$, then $T_X \supseteq T_U \supseteq T_Z$. However, taking into account the fact that $T_X = T_Z$, one may conclude that $T_X = T_U = T_Z$ and, in consequence, that $P(X) = P(U) = P(Z)$. □

Closure operator for itemset X is denoted by $\gamma(X)$ and is defined as the intersection of all transactions in T_X if T_X is not empty, and as I otherwise. Hence, $\gamma(X)$ is the greatest superset (w.r.t. subset inclusion) of X such that $T_X = T_{\gamma(X)}$. As follows from the definition of $\gamma(X)$, an itemset has exactly one closure. An itemset X is called *closed* if $\gamma(X) = X$; that is, if $\forall Y \supset X$, $T_Y \subset T_X$.

An itemset Y is called a *generator of* X if Y is a minimal subset (w.r.t. subset inclusion) of X such that $T_Y = T_X$. Clearly, each itemset has at least one generator. An itemset X is called a *key generator* if X has itself as its (only) generator; that is, if $\forall Y \subset X$, $T_Y \supset T_X$.

The number of (frequent) closed itemsets as well as the number of all (frequent) key generators is typically by a few orders of magnitude less than the number of all (frequent) itemsets.

Example 1. Let D be a transaction dataset from Table 1 and the minimum support threshold equals 1. Then there are 132 frequent itemsets for this dataset. In Table 2, we list the set of all frequent closed itemsets and all frequent key generators. Please note that their number is much less than the number of frequent itemsets; namely, there are only 7 frequent closed itemsets and 11 frequent key generators. Clearly, all frequent itemsets have their closures and generators only among frequent closed itemsets and frequent key generators, respectively.

Let us consider a frequent closed itemset $\{bcdehi\}$ and its generators $\{h\}$ and $\{i\}$ (see Table 2). Clearly, these three itemsets occur in the same set of transactions (namely, transaction #3 and transaction #5) and so, have the same support (2) and relative support (2/5). By Property 1, each itemset U such that $\{h\} \subseteq U \subseteq \{bcdehi\}$ (or such that $\{i\} \subseteq U \subseteq \{bcdehi\}$) will also occur in transactions #3 and #5 and will have relative support equal to 2/5. Note that there are $2^{|\{bcdehi\}/\{h\}|} = 2^5 = 32$ itemsets U such that $\{h\} \subseteq U \subseteq \{bcdehi\}$, which are losslessly represented by a pair of the itemsets: (key generator $\{h\}$, closed itemset $\{bcdehi\}$). □

Table 1. Example transaction dataset D

Transaction Id	Transaction
#1	$\{abcde\}$
#2	$\{abcdef\}$
#3	$\{abcdehi\}$
#4	$\{abe\}$
#5	$\{bcdehi\}$

Table 2. Frequent closed itemsets and frequent key generators from dataset D in Table 1

Frequent closed itemsets	Frequent key generators	Relative support
$\{be\}$	$\emptyset$	5/5
$\{abe\}$	$\{a\}$	4/5
$\{bcde\}$	$\{c\}$, $\{d\}$	4/5
$\{abcde\}$	$\{ac\}$, $\{ad\}$	3/5
$\{bcdehi\}$	$\{h\}$, $\{i\}$	2/5
$\{bef\}$	$\{f\}$	1/5
$\{abcdehi\}$	$\{ah\}$, $\{ai\}$	1/5

3 Rules Satisfying Multiple *ACBC*-Evaluation Criteria

We find that a large number of evaluation measures of rules, say $X \rightarrow Y$, considered in the literature can be defined in terms of at most the following four components:

- the probability $P(X)$ of rule's antecedent X,
- the probability $P(Y)$ of rule's consequent Y,
- the probability $P(Z)$ of its base $Z = X \cup Y$,
- constants.

Each rule evaluation measure that can be defined in this way, we will call an *ACBC-evaluation measure*.

In Table 3, we list example *ACBC*-evaluation measures. In fact, all or almost all evaluation measures listed in [3, 5, 9–16] are *ACBC*-evaluation measures. Clearly, *ACBC*-evaluation measures include also the support and confidence.

Note that marginal probabilities $P(\bar{X})$, $P(\bar{Y})$ and joint probabilites $P(\bar{X}Y)$, $P(X\bar{Y})$ and $P(\bar{X}\bar{Y})$ are derivable from $P(X)$, $P(Y)$ and $P(XY)$; namely:

- $P(\bar{X}) = 1 - P(X)$, $P(\bar{Y}) = 1 - P(Y)$,

Table 3. Example *ACBC*-evaluation measures of (association) rules

measure	definition
sup(X→Y)	$P(XY) \times \vert D \vert$
conf(X→Y)	$P(XY) / P(X)$
novelty(X→Y)	$P(XY) - P(X) \times P(Y)$
lift(X→Y)	$P(XY) / (P(X) \times P(Y))$
certaintyFactor(X→Y)	$\begin{cases} \frac{conf(X \to Y) - P(Y)}{1 - P(Y)} & \text{if } conf(X \to Y) > P(Y) \\ 0 & \text{if } conf(X \to Y) = P(Y) \\ -\frac{P(Y) - conf(X \to Y)}{P(Y) - 0} & \text{if } conf(X \to Y) < P(Y) \end{cases} = \begin{cases} \frac{P(XY) - P(X) \times P(Y)}{P(X) - P(X) \times P(Y)} & \text{if } P(XY) > P(X) \times P(Y) \\ 0 & \text{if } P(XY) = P(X) \times P(Y) \\ -\frac{P(X) \times P(Y) - P(XY)}{P(X) \times P(Y) - 0} & \text{if } P(XY) < P(X) \times P(Y) \end{cases}$
dependenceFactor(X→Y)	$\begin{cases} \frac{conf(X \to Y) - P(Y)}{(\min\{P(X), P(Y)\}/P(X)) - P(Y)} & \text{if } conf(X \to Y) > P(Y) \\ 0 & \text{if } conf(X \to Y) = P(Y) \\ -\frac{P(Y) - conf(X \to Y)}{P(Y) - (\max\{0, P(X) + P(Y) - 1\}/P(X))} & \text{if } conf(X \to Y) < P(Y) \end{cases}$ $= \begin{cases} \frac{P(XY) - P(X) \times P(Y)}{\min\{P(X), P(Y)\} - P(X) \times P(Y)} & \text{if } P(XY) > P(X) \times P(Y) \\ 0 & \text{if } P(XY) = P(X) \times P(Y) \\ -\frac{P(X) \times P(Y) - P(XY)}{P(X) \times P(Y) - \max\{0, P(X) + P(Y) - 1\}} & \text{if } P(XY) < P(X) \times P(Y) \end{cases}$
cosine(X→Y)	$P(XY) / \sqrt{P(X) \times P(Y)}$
Jaccard(X→Y)	$P(XY) / (P(X) + P(Y) - P(XY))$
accuracy(X→Y)	$P(XY) + P(\bar{X}\bar{Y}) = 1 + 2P(XY) - P(X) - P(Y)$
F-score(X→Y)	$\frac{(P(XY)/P(X)) \times (P(XY)/P(Y))}{((P(XY)/P(X)) + (P(XY)/P(Y)))/2} = \frac{2P(XY)}{P(X) + P(Y)}$

- $P(\bar{X}Y) = P(Y) - P(XY)$, $P(X\bar{Y}) = P(X) - P(XY)$,
- $P(\bar{X}\bar{Y}) = P(\bar{X}) - P(\bar{X}Y) = 1 - P(X) - P(Y) + P(XY)$.

Hence, any measure of a rule $X \to Y$ that is specified in terms of some marginal probabilities $P(X)$, $P(Y)$, $P(\bar{X})$, $P(\bar{Y})$, joint probabilities $P(XY)$, $P(\bar{X}Y)$, $P(X\bar{Y})$, $P(\bar{X}\bar{Y})$ and eventually constants is, in fact, an *ACBC*-evaluation measure as it can be expressed only in terms of $P(X)$, $P(Y)$, $P(XY)$ and constants.

Proposition 1. Let $X \to Y$ and $Z \to V$ be rules such that $P(X) = P(Z)$, $P(Y) = P(V)$, $P(XY) = P(ZV)$ and μ be an *ACBC*-evaluation measure. Then:

$$\mu(X \to Y) = \mu(Z \to V).$$

In the remainder of the paper, we assume that a set $\mathrm{M} = \{\mu_1, \ldots, \mu_n\}$ of *ACBC*-evaluation measures is used to evaluate (association) rules w.r.t. their corresponding threshold values $E = \{\varepsilon_1, \ldots, \varepsilon_n\}$. A rule $X \to Y$ will be called an (M, *E*)-*rule* if:

$$\forall \mu_i \in \mathrm{M}(\mu_i(X \to Y) \geq \varepsilon_i).$$

4 Using Rule Templates to Represent Association Rules that Satisfy Multiple *ACBC*-Evaluation Criteria

In this section, we will first introduce a new notion of a rule template and examine its properties and relationships with rules, and in particular, with association rules. Then, based on rule templates, we will offer, examine and illustrate a lossless representation of association rules satisfying multiple *ACBC*-evaluation criteria.

4.1 Rule Templates, Rules and Association Rules

Any pair of two rules ($X \rightarrow Y$, $Z \rightarrow V$) such that X and Y are key generators, $X \cap Y = \emptyset$, Z and V are closed itemsets, $Z = \gamma(X)$ and $V = \gamma(Y)$ will be called a *rule template*.

Let ($X \rightarrow Y$, $Z \rightarrow V$) be a rule template. Then, $X \rightarrow Y$ is called its *lower rule*, while $Z \rightarrow V$ is called its *upper rule*. Please note that by definition of a rule template, a lower rule $X \rightarrow Y$ is an association one ($X \cap Y = \emptyset$), while upper rule $Z \rightarrow V$ does not need to be association one (Z and V may share some items).

Proposition 2. Let ($X \rightarrow Y$, $Z \rightarrow V$) be a rule template. Then:

(a) $P(X) = P(Z)$.
(b) $P(Y) = P(V)$.
(c) $P(XY) = P(ZV)$.

Proof. Since ($X \rightarrow Y$, $Z \rightarrow V$) is a rule template, then Z is the closure of X and V is the closure of Y. Hence, $T_Z = T_X$ (*) and $T_V = T_Y$ (**).

Ad a) By (*), $|T_Z| = |T_X|$, so $P(Z) = P(X)$.
Ad b) By (**), $|T_V| = |T_Y|$, so $P(V) = P(Y)$.
Ad c) $T_{Z \cup V} = T_Z \cap T_V$ = /* by (*) and (**) */ = $T_X \cap T_Y = T_{X \cup Y}$. Hence, $|T_{Z \cup V}| = |T_{X \cup Y}|$, so $P(ZV) = P(XY)$. □

Proposition 3. Let ($X \rightarrow Y$, $Z \rightarrow V$) be a rule template and μ be an *ACBC*-evaluation measure. Then:

$$\mu(X \rightarrow Y) = \mu(Z \rightarrow V).$$

Proof. By Propositions 1–2. □

Thus, a value of any *ACBC*–evaluation measure μ is the same for both lower rule and upper rule of a rule template. This property allows us to define an *evaluation measure of rule template* as beneath:

Let μ be an *ACBC*-evaluation measure of a rule. Then, the *ACBC-evaluation measure* μ *of a rule template* ($X \rightarrow Y$, $Z \rightarrow V$) is denoted by $\mu(X \rightarrow Y, Z \rightarrow V)$ and is defined as follows:

$$\mu(X \to Y, Z \to V) = \mu(X \to Y)$$
$$\text{(or equivalently as : } \mu(X \to Y, Z \to V) = \mu(Z \to V)).$$

Proposition 4. Let $(X \to Y, Z \to V)$ be a rule template and $U \to W$ be a rule such that $X \subseteq U \subseteq Z$ and $Y \subseteq W \subseteq V$. Then:

(a) $P(X) = P(U) = P(Z)$.
(b) $P(Y) = P(W) = P(V)$.
(c) $P(XY) = P(UW) = P(ZV)$.

Proof. By Proposition 2 and Property 1. □

Theorem 1. Let μ be an *ACBC*-evaluation measure, $(X \to Y, Z \to V)$ be a rule template and $U \to W$ be a rule such that $X \subseteq U \subseteq Z$ and $Y \subseteq W \subseteq V$. Then:

$$\mu(U \to W) = \mu(X \to Y) = \mu(Z \to V) = \mu(X \to Y, Z \to V).$$

Proof. By Proposition 4 and Proposition 1. □

In the remainder of the paper, we will say that *a rule template* $(X \to Y, Z \to V)$ *covers a rule* $U \to W$ (or equivalently, *rule* $U \to W$ *is covered by rule template* $(X \to Y, Z \to V)$) if $X \subseteq U \subseteq Z$ and $Y \subseteq W \subseteq V$.

Theorem 2. Each association rule is covered by at least one rule template.

Proof. Let $U \to W$ be an association rule. Then $U \cap W = \emptyset$. Let X be a generator of U and Z be the closure of U. Then, $X \subseteq U \subseteq Z$ and $Z = \gamma(X)$ (*). Now, let Y be a generator of W and V be the closure of W. Then, $Y \subseteq W \subseteq V$ and $V = \gamma(Y)$ (**). Since sets U and W are mutually exclusive, then their subsets X and Y are also mutually exclusive (***). Thus, by (*), (**) and (***), $U \to W$ is covered by rule template $(X \to Y, Z \to V)$. □

Example 2. Let $(\{\boldsymbol{a}\} \to \{\boldsymbol{e}\}, \{\boldsymbol{abc}\} \to \{\boldsymbol{cde}\})$ be a rule template. Each association rule, say r_1, covered by this rule template will have $\boldsymbol{a}$ in its antecedent and $\boldsymbol{e}$ in its consequent; item b may, but does not have to, occur in r_1's antecedent; item d may, but does not have to, occur in r_1's consequent; item c, which is common to $\{abc\}$ and $\{cde\}$, may occur at any side of rule r_1 or may not occur in r_1 at all. As a result, $(\{\boldsymbol{a}\} \to \{\boldsymbol{e}\}, \{\boldsymbol{abc}\} \to \{\boldsymbol{cde}\})$ covers $2^{|\{b\}|} \times 2^{|\{d\}|} \times 3^{|\{c\}|} = 2^1 \times 2^1 \times 3^1 = 12$ association rules: $\{a\} \to \{e\}$, $\{ab\} \to \{e\}$, $\{a\} \to \{de\}$, $\{ab\} \to \{de\}$, $\{ac\} \to \{e\}$, $\{abc\} \to \{e\}$, $\{ac\} \to \{de\}$, $\{abc\} \to \{de\}$, $\{a\} \to \{ce\}$, $\{ab\} \to \{ce\}$, $\{a\} \to \{cde\}$, $\{ab\} \to \{cde\}$.

Now, let us consider a rule template $(\{\boldsymbol{a}\} \to \{\boldsymbol{ij}\}, \{\boldsymbol{abcdef}\} \to \{\boldsymbol{efghij}\})$. Then each association rule, say r_2, covered by this rule template will have $\boldsymbol{a}$ in r_2's antecedent, $\boldsymbol{i}$ and $\boldsymbol{j}$ in r_2's consequent; items b, c, and d may, but do not have to, occur in r_2's antecedent; items g and h may, but do not have to, occur in r_2's consequent; items e and f, which are shared by $\{abcdef\}$ and $\{efghij\}$, may occur at any side of r_2 or may not occur in r_2 at all. As a result, $(\{a\} \rightarrow \{ij\}, \{abcdef\} \rightarrow \{efghij\})$ covers $2^{|\{bcd\}|} \times 2^{|\{gh\}|} \times 3^{|\{ef\}|} = 2^3 \times 2^2 \times 3^2 = 288$ association rules. □

Beneath, we formalize the calculation of the number of association rules covered by a single rule template.

Theorem 3. Let $(X \rightarrow Y, Z \rightarrow V)$ be a rule template. Then, $(X \rightarrow Y, Z \rightarrow V)$ covers $2^m \times 2^n \times 3^k$ distinct association rules, where:

- m is the number of items in the antecedent of the upper rule of $(X \rightarrow Y, \mathbf{Z} \rightarrow V)$ that occur neither in the antecedent of the lower rule nor in the consequent of the upper rule; that is, $m = |\mathbf{Z} \setminus (X \cup V)|$;
- n is the number of items in the consequent of the upper rule of $(X \rightarrow Y, Z \rightarrow \mathbf{V})$ that occur neither in the consequent of the lower rule nor in the antecedent of the upper rule; that is, $n = |\mathbf{V} \setminus (Y \cup Z)|$.
- k is the number of items common to both parts of the upper rule of $(X \rightarrow Y, \mathbf{Z} \rightarrow \mathbf{V})$; that is, $k = |\mathbf{Z} \cap \mathbf{V}|$.

Proof. Let $U \rightarrow W$ be an association rule that is covered by $(X \rightarrow Y, Z \rightarrow V)$. Then, the antecedent of $U \rightarrow W$ must contain X and cannot contain Y, while the consequent of $U \rightarrow W$ must contain Y and cannot contain X. Now, each item from outside $X \cup Y$ that is common to Z and V (that is, belongs to $Z \cap V$) may occur either in the antecedent of $U \rightarrow W$ or in its consequent or may not occur in $U \rightarrow W$ at all, which gives 3 possibilities per each such item. Each remaining item in Z that is from outside $X \cup Y \cup V$ (that is, belongs to $Z \setminus (X \cup V)$) may, but does not have to, occur in the antecedent of $U \rightarrow W$, which gives 2 possibilities per each such item. Analogously, each remaining item in V that is from outside $X \cup Y \cup Z$ (that is, belongs to $V \setminus (Y \cup Z)$) may, but does not have to occur in the consequent of $U \rightarrow W$, which gives 2 possibilities per each such item. □

4.2 (M, E)–Rule Templates as a Lossless Representation of Association Rules

Let M be a set of *ACBC*-evaluation measures. Then, a rule template $(X \rightarrow Y, Z \rightarrow V)$ is called an (M, E)*–rule template* if:

$$\forall \mu_i \in \mathrm{M}(\mu_i((X \rightarrow Y, Z \rightarrow V)) \geq \varepsilon_i).$$

The set of all (M, E)–rule templates will be denoted by $RT_{(\mathrm{M},E)}$.

Theorem 4. Let M be a set of *ACBC*-evaluation measures and $U \rightarrow W$ be an association rule. If there is a rule template in $RT_{(\mathrm{M},E)}$ covering $U \rightarrow W$, then $U \rightarrow W$ is an (M, E)–association rule and $\forall \mu_i \in \mathrm{M}\ (\mu_i(U \rightarrow W) = \mu_i(t)$, where t is any rule template in $RT_{(\mathrm{M},E)}$ that covers $U \rightarrow W$. Otherwise, $U \rightarrow W$ is not an (M, E)–association rule.

Proof. Let $U \rightarrow W$ be an association rule. By Theorem 2, it is covered by at least one rule template, say t, and, by Theorem 1, $\forall \mu_i \in \mathrm{M}\ (\mu_i(U \rightarrow W) = \mu_i(t)$. If the rule template t belongs to $RT_{(\mathrm{M},E)}$, then $\forall \mu_i \in \mathrm{M}\ \mu_i(t) \geq \varepsilon_i$, so, $\forall \mu_i \in \mathrm{M}\ (\mu_i(U \rightarrow W) \geq \varepsilon_i$ and thus,

$U \rightarrow W$ is an (M, E)–association rule. Otherwise, the rule template t does not have sufficiently high value for at least one measure in M, say measure μ_i. Then, $\mu_i(U \rightarrow W) = \mu_i(t) < \varepsilon_i$, and so, $U \rightarrow W$ is not an (M, E)–association rule. □

In the beneath example, we illustrate the conciseness of the (M, E)–association rules' representation consisting of (M, E)–rule templates.

Example 3. Let us consider the transaction dataset D from Table 1. Let the set of evaluation measure M = {*sup*, *novelty*, *lift*, *certaintyFactor*} and the set of their corresponding threshold values E = {1, 0.1, 2.0, 0.3}. Table 4 shows all (M, E)–rule templates found for these constraints. The found six (M, E)–rule templates cover 486 (M, E)–association rules. □

Table 4. (M, E)–rule templates for dataset D from Table 1, where M = {*sup*, *novelty*, *lift*, *certaintyFactor*} and E = {1, 0.1, 2.0, 0.3}

(M, E)–rule template	*sup*	*novelty*	*lift*	*certaintyFactor*
$(\{h\} \rightarrow \{i\}, \{bcdehi\} \rightarrow \{bcdehi\})$	2	0.24	2.5	1
$(\{h\} \rightarrow \{ai\}, \{bcdehi\} \rightarrow \{abcdehi\})$	1	0.12	2.5	0.375
$(\{i\} \rightarrow \{h\}, \{bcdehi\} \rightarrow \{bcdehi\})$	2	0.24	2.5	1
$(\{i\} \rightarrow \{ah\}, \{bcdehi\} \rightarrow \{abcdehi\})$	1	0.12	2.5	0.375
$(\{ah\} \rightarrow \{i\}, \{abcdehi\} \rightarrow \{bcdehi\})$	1	0.12	2.5	1
$(\{ai\} \rightarrow \{h\}, \{abcdehi\} \rightarrow \{bcdehi\})$	1	0.12	2.5	1

4.3 Storage Issues

Our proposed representation of (M, E)–association rules can be stored in several compact ways. For example, a rule template $(X \rightarrow Y, Z \rightarrow V)$ may be stored as a triple: $(X \rightarrow Y, M \rightarrow N, K)$, where $K = Z \cap V$, $M = Z \setminus (X \cup K))$ and $N = V \setminus (Y \cup K)$. Clearly, upper rule $Z \rightarrow V$ is retrievable from $(X \rightarrow Y, M \rightarrow N, K)$ as $Z = X \cup M \cup K$, while $V = Y \cup N \cup K$. In addition, instead of storing values of potentially many *ACBC*-evaluation measures M, one may store the probabilities of antecedents, consequents and bases of lower rules of rule templates, based on which all *ACBC*-evaluation measures may be calculated. Table 5 shows the result of applying this approach to storing rule templates from Table 4.

Table 5. Reducing storage of (M, E)–rule templates by split of their upper rules

$(X \rightarrow Y, M \rightarrow N, K)$	$P(XY)$	$P(X)$	$P(Y)$
$(\{h\} \rightarrow \{i\}, \{i\} \rightarrow \{h\}, \{bcde\})$	2/5	2/5	2/5
$(\{h\} \rightarrow \{ai\}, \{i\} \rightarrow \{h\}, \{bcde\})$	1/5	2/5	1/5
$(\{i\} \rightarrow \{h\}, \{h\} \rightarrow \{i\}, \{bcde\})$	2/5	2/5	2/5
$(\{i\} \rightarrow \{ah\}, \{h\} \rightarrow \{i\}, \{bcde\})$	1/5	2/5	1/5
$(\{ah\} \rightarrow \{i\}, \{i\} \rightarrow \{h\}, \{bcde\})$	1/5	1/5	2/5
$(\{ai\} \rightarrow \{h\}, \{h\} \rightarrow \{i\}, \{bcde\})$	1/5	1/5	2/5

Another alternative (or additional) approach to storing (M, E)–rule templates in a compact way is to keep only upper rules (without duplication) and generators of closed itemsets that occurred in their antecedents and/or consequents. Let us consider for instance the following rule templates:

$(\{a\} \rightarrow \{b\}, \{abcd\} \rightarrow \{abcd\})$, $(\{a\} \rightarrow \{cd\}, \{abcd\} \rightarrow \{abcd\})$, $(\{b\} \rightarrow \{a\}, \{abcd\} \rightarrow \{abcd\})$, $(\{b\} \rightarrow \{cd\}, \{abcd\} \rightarrow \{abcd\})$, $(\{cd\} \rightarrow \{a\}, \{abcd\} \rightarrow \{abcd\})$, $(\{cd\} \rightarrow \{b\}, \{abcd\} \rightarrow \{abcd\})$, $(\{e\} \rightarrow \{a\}, \{efg\} \rightarrow \{abcd\})$, $(\{e\} \rightarrow \{b\}, \{efg\} \rightarrow \{abcd\})$, $(\{e\} \rightarrow \{cd\}, \{efg\} \rightarrow \{abcd\})$.

These nine rule templates have only two different upper rules; namely, $(\{abcd\} \rightarrow \{abcd\})$ and $(\{efg\} \rightarrow \{abcd\})$. Table 6 presents generators of closed itemsets $\{abcd\}$ and $\{efg\}$, which occur in antecedents and/or consequents of the upper rules. The information about the upper rules and generators of their antecedents and consequents is sufficient to retrieve the lower rules if needed – the lower rules are derivable from the upper rules by replacing the antecedents and consequents of the upper rules with their generators. This solution seems to be particularly useful when closed itemsets have many generators.

Table 6. Associating closed itemsets from upper rules with their generators from lower rules

Closed itemset	Generators
$\{abcd\}$	$\{a\}$, $\{b\}$, $\{cd\}$
$\{efg\}$	$\{e\}$

Finally, we would like to note that our $RT(\mathrm{M}_{,E})$ representation of (M, E)–association rules can be reduced by half in the case when M consists of only $ACBC$–evaluation measures μ that are symmetric (that is, such that $\mu(X \rightarrow Y) = \mu(Y \rightarrow X)$). In such cases, if $(X \rightarrow Y, Z \rightarrow V)$ is an (M, E)–rule template, then $(Y \rightarrow X, V \rightarrow Z)$ is an (M, E)–rule template. Then it is sufficient to store either $(X \rightarrow Y, Z \rightarrow V)$ or $(Y \rightarrow X, V \rightarrow Z)$ in $RT(\mathrm{M}_{,E})$, rather than both rule templates. Please note that in Table 3 all measures except for confidence and certainty factor are symmetric.

5 Summary

In this paper, we identified an important and wide generic class of $ACBC$-rule evaluation measures, which can be formulated at most in terms of the probability of the antecedent of a rule, the probability of its consequent, the probability of its base and constants. Then we proposed the lossless representation of association rules satisfying multiple evaluation criteria expressible in terms of $ACBC$-evaluation measures. Our proposed representation is based on the new notion of rule templates each of which consists of two rules: a lower rule built from disjoint key generators and an upper rule built from closed itemsets – their closures. We have proved that a rule that is covered by a given rule template has the same values of $ACBC$-evaluation measures as the covering rule template. Hence, evaluation of an association rule, say r, by means of our representation requires looking for just one covering rule template, even if there are many rule templates covering rule r. This property is not guaranteed for representations of strong association rules offered so far in the literature [2, 4, 6–8, 17, 18] – in general, for each of these representations all its rules have to be examined in order to determine exact values of support and confidence of an evaluated strong association rule. We have also shown how many association rules are covered by a single rule template. The theoretical result we obtained is promising. The examples considered in the paper showed that the number of rule templates may be by 2–3 orders of magnitude less numerous than the number of rules satisfying given *ACBC*-criteria. Finally, we suggested a few ways of storing our representation in a compact way. We anticipate that they may provide an additional insight into knowledge hidden in data and rules. In particular, groups of rule templates sharing some common characteristics (say, *K* or upper rules) may enable identification of groups of transactions and their properties that could be of interest to a user.

References

1. Agrawal, R., Imielinski, T., Swami, A.N.: Mining association rules between sets of items in large databases. In: ACM SIGMOD International Conference on Management of Data, pp. 207–216 (1993)
2. Bastide, Y., Pasquier, N., Taouil, R., Stumme, G., Lakhal, L.: Mining minimal non-redundant association rules using frequent closed itemsets. In: Palamidessi, C., Moniz Pereira, L., Lloyd, J.W., Dahl, V., Furbach, U., Kerber, M., Lau, K.-K., Sagiv, Y., Stuckey, P.J. (eds.) CL 2000. LNCS (LNAI), vol. 1861, pp. 972–986. Springer, Heidelberg (2000)
3. Brin, S., Motwani, R., Ullman, J.D., Tsur, S.: Dynamic itemset counting and implication rules for market basket data. In: ACM SIGMOD 1997 International Conference on Management of Data, pp. 255–264 (1997)
4. Hamrouni, T., Yahia, S.B., Nguifo, E.M.: Succinct minimal generators: theoretical foundations and applications. Int. J. Found. Comput. Sci. **19**(2), 271–296 (2008)
5. Hilderman, R.J., Hamilton, H.J.: Evaluation of interestingness measures for ranking discovered knowledge. In: Cheung, D., Williams, G.J., Li, Q. (eds.) PAKDD 2001. LNCS (LNAI), vol. 2035, pp. 247–259. Springer, Heidelberg (2001)
6. Kryszkiewicz, M.: Closed set based discovery of representative association rules. In: Hoffmann, F., Adams, N., Fisher, D., Guimaraes, G., Hand, D.J. (eds.) IDA 2001. LNCS, vol. 2189, pp. 350–359. Springer, Heidelberg (2001)
7. Kryszkiewicz M.: Concise Representations of Frequent Patterns and Association Rules, Prace Naukowe Politechniki Warszawskiej. Elektronika, no. 142 (2002)
8. Kryszkiewicz, M.: Concise Representations of Association Rules. In: Hand, D.J., Adams, N. M., Bolton, R.J. (eds.) Pattern Detection and Discovery. LNCS (LNAI), vol. 2447, pp. 92–109. Springer, Heidelberg (2002)
9. Kryszkiewicz, M.: Dependence factor for association rules. In: Nguyen, N.T., Trawiński, B., Kosala, R. (eds.) ACIIDS 2015. LNCS, vol. 9012, pp. 135–145. Springer, Heidelberg (2015)
10. Kryszkiewicz, M.: Dependence factor as a rule evaluation measure. In: Matwin, S., Mielniczuk, J. (eds.) Challenges in Computational Statistics and Data Mining, Studies in Computational Intelligence, vol. 605, pp. 205–223. Springer, Switzerland (2016)
11. Lavrač, N., Flach, P.A., Zupan, B.: Rule evaluation measures: a unifying view. In: Džeroski, S., Flach, P.A. (eds.) ILP 1999. LNCS (LNAI), vol. 1634, pp. 174–185. Springer, Heidelberg (1999)

12. Lenca, P., Meyer, P., Vaillant, B., Lallich, S.: On selecting interestingness measures for association rules: user oriented description and multiple criteria decision aid. Eur. J. Oper. Res. **184**, 610–626 (2008). Elsevier
13. Piatetsky-Shapiro, G.: Gregory piatetsky-shapiro: discovery, analysis, and presentation of strong rules. In: Knowledge Discovery in Databases, pp. 229–248. AAAI/MIT Press (1991)
14. Sheikh, L.M., Tanveer, B., Hamdani, S.M.A.: Interesting measures for mining association rules. In: Proceedings of INMIC 2004, IEEE 2004 (2004)
15. Shortliffe, E., Buchanan, B.: A model of inexact reasoning in medicine. Math. Biosci. **23**, 351–379 (1975)
16. Suzuki, E.: Interestingness measures - limits, desiderata, and recent results. In: Lenca, P., Lallich, S. (Eds.) QIMIE/PAKDD 2009 (2009)
17. Stumme, G., Taouil, R., Bastide, Y., Pasquier, N., Lakhal, L.: Intelligent structuring and reducing of association rules with formal concept analysis. In: Baader, F., Brewka, G., Eiter, T. (eds.) KI 2001. LNCS (LNAI), vol. 2174, pp. 335–350. Springer, Heidelberg (2001)
18. Zaki, M.J.: Generating non-redundant association rules. In: 6th ACM SIGKDD (2000)

Learning Algorithms Aimed at Collinear Patterns

Leon Bobrowski[1,2] and Paweł Zabielski[1(✉)]

[1] Faculty of Computer Science, Bialystok University of Technology, Bialystok, Poland
{l.bobrowski,p.zabielski}@pb.edu.pl
[2] Institute of Biocybernetics and Biomedical Engineering, PAS, Warsaw, Poland

Abstract. Collinear (flat) pattern appears in a given set of multidimensional feature vectors when many of these vectors are located on (or near) some plane in the feature space. Flat pattern discovered in a given data set can give indications for creating a model of interaction between selected features. Patterns located on planes can be discovered even in large and multidimensional data sets through minimization of the convex and piecewise linear (*CPL*) criterion functions. Discovering flat patterns can be based on the search for degenerated vertices in the parameter space. The possibility of using learning algorithms for this purpose is examined in this paper.

Keywords: Data mining · Flat patterns · *CPL* criterion functions · Learning algorithms

1 Introduction

Data sets composed of feature vectors of the same dimensionality can be represented as clouds of points in the feature space. Data mining tools are used for discovering useful patterns in such data sets [1, 2]. The word *pattern* means a specific type of regularity such as clusters, prognostic trends, or models of interactions which have been discovered in a given dataset. The overall goal of the data mining process is to obtain useful knowledge on the basis of extracted patterns.

A large number of feature vectors located on and around a plane in a given feature space form the so called collinear (*flat*) pattern. Flat patterns allow to define the central planes and can be used in the *K-planes* algorithm [3, 4]. Discovered collinear pattern can be used also among others for creating model of interaction between a number of features (genes) contained in the selected subset.

Flat patterns can be selected from data sets through minimization of the convex and piecewise linear (*CPL*) criterion functions []. The basis exchange algorithms can be used for this purpose. The possibility of using a special type of learning algorithms for the purpose of the flat patterns discovering is examined in the presented paper.

N.T. Nguyen et al. (Eds.): ACIIDS 2016, Part II, LNAI 9622, pp. 159–168, 2016.
DOI: 10.1007/978-3-662-49390-8_15

2 Hyperplanes and Vertices in the Parameter Space

Let us consider the data set C which contains m feature vectors $\mathbf{x}_j = [x_{j,1},\ldots,x_{j,n}]^T$ belonging to a given n-dimensional feature space $F[n]$ ($\mathbf{x}_j \in F[n]$):

$$C = \{\mathbf{x}_j : j = 1, \ldots, m\} \tag{1}$$

Components $x_{j,i}$ of the feature vector $\mathbf{x}_j$ can be treated as the numerical results of n standardized examinations of the j-th object O_j, where we assume that $x_{j,i} \in R$ or $x_{j,i} \in \{0,1\}$.

Each of m feature vector $\mathbf{x}_j$ from the set C (1) defines the below (*dual*) hyperplane h_j in the parameter space R^n [6]:

$$(\forall \mathbf{x}_j \in C) h_j = \left\{\mathbf{w} : \mathbf{x}_j^T \mathbf{w} = 1\right\} \tag{2}$$

where $\mathbf{w} = [w_1,\ldots,w_n]^T$ is the parameter (*weight*) vector ($\mathbf{w} \in R^n$).

Each of n unit vectors $\mathbf{e}_i = \mathbf{e}_i[n] = [0,\ldots,1,\ldots,0]^T$ defines the below hyperplane h_i^0 in the parameter space R^n:

$$(\forall i \in \{1, \ldots, n\}) \quad h_i^0 = \left\{\mathbf{w} : \mathbf{e}_i^T \mathbf{w} = 0\right\} = \{\mathbf{w} : \ w_i = 0\} \tag{3}$$

Let us consider the set S_k of n linearly independent feature vectors $\mathbf{x}_j$ ($j \in J_k$) and unit vectors $\mathbf{e}_i$ ($i \in I_k$).

$$S_k = \{\mathbf{x}_j : j \in J_k\} \cup \{\mathbf{e}_i : i \in I_k\} \tag{4}$$

where J_k is the k-th subset of the r_k ($1 \leq r_k \leq n$) indices j, and I_k is the k-th subset of the $n - r_k$ indices i.

The k-th *vertex* $\mathbf{w}_k$ in the parameter space R^n is the intersection point of r_k hyperplanes h_j (2) defined by the vectors $\mathbf{x}_j$ ($j \in J_k$) and $n - r_k$ hyperplanes h_i^0 (3) defined by the vectors $\mathbf{e}_i$ ($i \in I_k$) constituting the set S_k (4). The intersection point $\mathbf{w}_k$ is defined by the below linear equations:

$$(\forall j \in J_k) \quad \mathbf{w}_k^T \mathbf{x}_j = 1 \tag{5}$$

And

$$(\forall i \in I_k) \quad \mathbf{w}_k^T \mathbf{e}_i = 0 \tag{6}$$

The Eqs. (5) and (6) can be represented in the following matrix form:

$$\mathbf{B}_k \mathbf{w}_k = \mathbf{1}' = [1, \ldots, 1, 0, \ldots, 0]^T \tag{7}$$

where $\mathbf{B}_k$ is the nonsingular n x n matrix (the k-th *basis*) linked to the vertex $\mathbf{w}_k$ defined by the vectors $\mathbf{x}_j$ and $\mathbf{e}_i$ from the set S_k (4).

$$\mathbf{B}_k = \left[\mathbf{x}_{j(1)}, \ldots, \mathbf{x}_{j(rk)}, \mathbf{e}_{i(rk+1)}, \ldots, \mathbf{e}_{i(n)}\right]^T \tag{8}$$

and

$$\mathbf{w}_k = \mathbf{B}_k^{-1}\mathbf{1}' \tag{9}$$

Definition 1: The *rank* r_k ($1 \le r_k \le n$) of the vertex $\mathbf{w}_k$ (9) is defined as the number of the (linearly independent) feature vectors $\mathbf{x}_j$ ($j \in J_k$ (5)) contained in the set S_k (4).

It can be noted that the *rank* r_k of the vertex $\mathbf{w}_k = [w_{k,1},\ldots,w_{k,n}]^T$ (9) is also equal to the number of its nonzero components $w_{k,i}$ ($w_{k,i} \neq 0$). The vertex $\mathbf{w}_k$ is *degenerated* if it can be determined (9) by more than one basis $\mathbf{B}_k$

Definition 2: The *degree of degeneration* d_k of the vertex $\mathbf{w}_k$ (9) of the rank r_k is defined as the number $d_k = m_k - r_k$, where m_k is the number of such feature vectors $\mathbf{x}_j$ ($\mathbf{x}_j \in C$) from the set C (1), which define hyperplanes h_j (2) passing through this vertex ($\mathbf{w}_k^T\mathbf{x}_j = 1$).

3 Vertexical Planes in the Feature Space

The hyperplane $H(\mathbf{w}, \theta)$ in the feature space $F[n]$ is defined as follows:

$$H(\mathbf{w}, \theta) = \{\mathbf{x} : \mathbf{w}^T\mathbf{x} = \theta\} \tag{10}$$

where $\mathbf{w}$ is the *weight vector* ($\mathbf{w} \in R^n$) and θ is the *threshold* ($\theta \in R^1$).

The vertex $\mathbf{w}_k$ (9) of the rank r_k ($r_k > 1$) (*Def.* 1) allows to define the (r_k - 1) - dimensional *vertexical plane* $P_k(\mathbf{x}_{j(1)},\ldots,\mathbf{x}_{j(rk)})$ in the feature space $F[n]$ as the linear combination of r_k feature vectors $\mathbf{x}_{j(i)}$ ($j(i) \in J_k$) (5) belonging to the set S_k (4) [5]:

$$P_k\left(\mathbf{x}_{j(1)}, \ldots, \mathbf{x}_{j(rk)}\right) = \{\mathbf{x} \in F[n] : \mathbf{x} = \alpha_1\mathbf{x}_{j(1)} + \ldots + \alpha_k\mathbf{x}_{j(rk)}\} \tag{11}$$

where $j(i) \in J_k$ (5) and the parameters α_i ($\alpha_i \in R^1$) satisfy the following equation:

$$\alpha_1 + \ldots + \alpha_{rk} = 1 \tag{12}$$

Remark 1: If the vertex $\mathbf{w}_k$ (9) has the rank $r_k = n$, then the vertexical plane (11), (12) has the dimension equal to (n - 1) similarly as the hyperplane $H(\mathbf{w},\theta)$ (10).

Remark 2: None of the *vertexical planes* $P_k(\mathbf{x}_{j(1)},\ldots, \mathbf{x}_{j(rk)})$ (11), (12) passes through the point zero $\mathbf{0}[n]$ (the *origin*) [5].

Theorem *1:* The feature vector $\mathbf{x}_j$ defines the hyperplane h_j (2) which passes through the vertex $\mathbf{w}_k$ (9) of the rank r_k if and only if the vector $\mathbf{x}_j$ is situated on the (r_k - 1)-dimensional *vertexical plane* $P_k(\mathbf{x}_{j(1)},\ldots,\mathbf{x}_{j(rk)})$ (11), where $j(i) \in J_k$ (5).

The proof of this Theorem can be found in the paper [5].

4 Convex and Piecewise Linear (*CPL*) Penalty and Criterion Functions

Let us consider convex and piecewise linear (*CPL*) penalty functions $\varphi_j(\mathbf{w})$ defined in the below manner on the feature vectors $\mathbf{x}_j$ from the data subset $C_k \subset C$ (1) [5]:

$$(\forall \mathbf{x}_j \in C_k) \quad \varphi(\mathbf{x}_j;\mathbf{w}) = \begin{cases} \delta_j^- - \mathbf{w}^T\mathbf{x}_j & \text{if} \quad \mathbf{w}^T\mathbf{x}_j \le \delta_j^- \\ 0 & \text{if} \quad \delta_j^- < \mathbf{w}^T\mathbf{x}_j < \delta_j^+ \\ \mathbf{w}^T\mathbf{x}_j - \delta_j^+ & \text{if} \quad \mathbf{w}^T\mathbf{x}_j \ge \delta_j^+ \end{cases} \tag{13}$$

where δ_j^- ($\delta_j^- \in R^1$) is the *bottom margin* and δ_j^+ ($\delta_j^+ \in R^1$) is the is the *top margin* ($\delta_j^- \le \delta_j^+$) (Fig. 1).

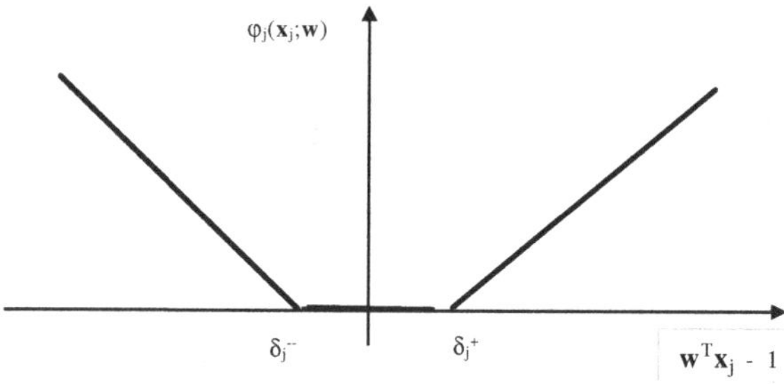

Fig. 1. The penalty functions $\varphi(\mathbf{x}_j;\mathbf{w})$ (13) with the bottom margin δ_j^- and the top margin δ_j^+.

For the purpose of the flat patterns discovering we pay particular attention to the penalty functions $\varphi_j(\mathbf{w})$ (13) with the constant values of the margins δ_j^- and δ_j^+:

$$(\forall j \in \{1,\ldots,m\}) \quad \delta_j^- = 1 - \varepsilon \quad and \quad \delta_j^+ = 1 + \varepsilon \tag{14}$$

where ε is a small, positive parameter ($\varepsilon \ge 0$).

We can remark that the *perceptron penalty functions* $\varphi_j^+(\mathbf{w})$ or $\varphi_j^-(\mathbf{w})$ can be treated as a special cases of the penalty functions $\varphi_j(\mathbf{w})$ (13) with the particular choice of the margins ($\delta_j^- = -\infty$ and $\delta_j^+ = 1$ for the positive penalty function $\varphi_j^+(\mathbf{w})$) and ($\delta_j^- = -1$ and $\delta_j^+ = \infty$ for the negative penalty function $\varphi_j^-(\mathbf{w})$) [5].

The criterion function Φ_k (1) ($\mathbf{w}$) is defined as the weighted sum of the penalty functions $\varphi_j(\mathbf{w})$ (13) defined on m_k feature vectors $\mathbf{x}_j$ defining the subset $C_k \subset C$ (1):

$$\Phi_k(\mathbf{w}) = \Sigma_j \alpha_j \varphi(\mathbf{x}_j;\mathbf{w}) \tag{15}$$

where the positive parameters α_j ($\alpha_j > 0$) can be treated as the *prices* of particular feature vectors $\mathbf{x}_j$. The standard choice of the parameters α_j values is given below:

$$(\forall j \in \{1, \ldots, m_k\}) \quad \alpha_j = 1/m_k \tag{16}$$

The criterion function $\Phi(\mathbf{w})$ (15) is convex and piecewise linear (*CPL*) as the sums of the *CPL* functions $\alpha_j\ \varphi(\mathbf{x}_j;\mathbf{w})$ (13). It can be proved that the minimal value of the function $\Phi(\mathbf{w})$ can be found in one of the vertices $\mathbf{w}^*_{k'}$ (9) [6]:

$$(\exists \mathbf{w}^*_{k'}) \quad (\forall \mathbf{w}) \quad \Phi_k(\mathbf{w}) \geq \Phi_k(\mathbf{w}^*_{k'}) = \Phi^*_k \geq 0 \tag{17}$$

The basis exchange algorithms which are similar to the linear programming allow to find efficiently the minimal value Φ^*_k of the criterion functions $\Phi_k(\mathbf{w})$ (15) and the optimal vertex $\mathbf{w}^*_{k'}$ (9) even in the case of large, multidimensional data sets C (1) [7].

Theorem 2: The minimal value Φ^*_k (17) of the criterion function $\Phi_k(\mathbf{w})$ (15) with $\delta^-_j = \delta^+_j = 1$ (13) is equal to zero $\Phi^*_k = 0$, if and only if all the feature vectors $\mathbf{x}_j$ from the subset $C_k \subset C$ (1) are situated on a certain hyperplane $H(\mathbf{w},\theta)$ (10) with $\theta \neq 0$.

The proof of this Theorem can be found in the paper [5]. The above Theorem can be formulated in a slightly more general form for the case, when the feature vectors $\mathbf{x}_j$ from the subset $C_k \subset C$ (1) are situated on and around certain hyperplane $H(\mathbf{w},\theta)$ (10) with $\theta \neq 0$.

5 Learning Algorithms

The minimal value Φ^* (17) of the criterion function $\Phi(\mathbf{w})$ (15) and the optimal vertex $\mathbf{w}^*_k$ (9) can be found also by using adequate learning algorithms. The learning algorithms aimed at minimization of the criterion function $\Phi(\mathbf{w})$ (15) can be obtained by using the Robbins–Monro procedure in the framework of the stochastic approximation [8]. For this purpose it is convenient to represent the criterion function $\Phi(\mathbf{w})$ (15) as the regression function in the below manner:

$$\Phi(\mathbf{w}) = \sum_i \mathrm{p}(\mathbf{x}_j)\varphi_j(\mathbf{x}_j;\mathbf{w}) \tag{18}$$

where $\mathrm{p}(\mathbf{x}_j)$ is probability distribution defined on elements $\mathbf{x}_j$ of the data set C:

$$(\forall j \in \{1, \ldots, m\}) \quad \mathrm{p}(\mathbf{x}_j) = 1/m \tag{19}$$

The uniform probability distribution $\mathrm{p}(\mathbf{x}_j)$ is used in generation of the below learning sequence $\{\mathbf{x}(k)\}$:

$$\mathbf{x}(1), \mathbf{x}(2), \ldots, \mathbf{x}(k), \ldots. \tag{20}$$

The elements $\mathbf{x}(k)$ of the above sequence are generated in each learning step k independently in accordance with the probability distribution $p(\mathbf{x}_j)$ (20) defined on elements $\mathbf{x}_j$ of the data set C (1).

The Robbins – Monro procedure aimed at minimization of the regression function $\Phi(\mathbf{w})$ (19) can be given in the below manner:

$$\mathbf{w}(k+1) = \mathbf{w}(k) - \alpha_k \nabla_w \varphi(\mathbf{x}(k); \mathbf{w}(k)) \tag{21}$$

where $\mathbf{w}(k)$ is the value of the weight vector during k-th learning step, and $\{\alpha_k\}$ is a sequence of decreasing, positive parameters α_k, e.g. $\alpha_k = 1/k$ which fulfill the below conditions:

$$\Sigma_k \alpha_k = \infty \textit{ and } \Sigma_k \alpha_k^2 < \infty \tag{22}$$

The Robbins – Monro procedure (21) applied to the regression function $\Phi(w)$ (18) with the margins $\delta^- = 1 - \varepsilon$ and $\delta^+ = 1 + \varepsilon$ (14) allows to propose the below learning algorithm:

$$\begin{aligned}
&(\forall k = 1, 2, \ldots)\\
&\textit{if } \mathbf{w}(k)^T\mathbf{x}(k) < 1 - \varepsilon, \textit{ then } \mathbf{w}(k+1) = \mathbf{w}(k) + \alpha_k \mathbf{x}(k), \textit{ and}\\
&\textit{if } \mathbf{w}(k)^T\mathbf{x}(k) > 1 + \varepsilon, \textit{ then } \mathbf{w}(k+1) = \mathbf{w}(k) - \alpha_k \mathbf{x}(k), \textit{ and}\\
&\textit{if } 1 - \varepsilon \leq w(k)^T x(k) \leq 1 + \varepsilon, \textit{ then } w(k+1) = w(k)
\end{aligned} \tag{23}$$

The learning algorithm (23) can be treated as a modification of the famous *error-correction algorithm* used in the *Perceptron* [1].

Theorem 3: If the learning sequence $\{\mathbf{x}(k)\}$ (20) is generated in accordance with the probability distribution $p(x_j)$ (19), then the learning algorithm (23) with the condition (22) generates such sequence $\{\mathbf{w}(k)\}$ of the weight vectors $\mathbf{w}(k)$ which converges to the optimal vertex $\mathbf{w}_{k'}^*$ (17) which constitutes the minimum of the criterion function $\Phi(\mathbf{w})$ (15) defined on the data set C (1) with the margins δ_j^- and δ_j^+ (14).

$$\mathbf{w}(k) \underset{k \to \infty}{\rightarrow} \mathbf{w}_{k'}^* \tag{24}$$

where the stationary point $\mathbf{w}_k^*$ satisfies the condition (17).

The proof of this theorem can be given using basic theorems of the stochastic approximation method [6].

6 Experimental Results

The learning algorithm (23) was implemented in C++ and run on an Intel Core i7, 1.73 GHz CPU (turbo boost to 2.93 GHz).

Two synthetic data sets D_1 and D_2, each with $m = 100$ feature vectors $\mathbf{x}_j$ has been created for the purpose of the computational experiments. The set D_1 contained 100 two-dimensional feature vectors $\mathbf{x}_j$ ($\mathbf{x}_j \in R^2$). The set D_2 contained 100 three-dimensional feature vectors $\mathbf{x}_j$ ($\mathbf{x}_j \in R^3$). The data sets D_1 and D_2 were ***collinear***. It means in this case, that elements $\mathbf{x}_j$ of each set D_k ($k = 1, 2$) has been located on the vertexical line $L_k(\mathbf{x}_{j(1)},\mathbf{x}_{j(2)})$ (11) defined by two basic feature vectors $\mathbf{x}_{j(1)}$ and $\mathbf{x}_{j(2)}$ contained in the basis $\mathbf{B}_k$ (8):

$$L_k\left(\mathbf{x}_{j(1)}, \mathbf{x}_{j(2)}\right) = \{\mathbf{x} : \mathbf{x} = \alpha_1 \mathbf{x}_{j(1)} + \alpha_2 \mathbf{x}_{j(2)}\} \tag{25}$$

where the parameters α_i ($\alpha_i \in R^1$) satisfy the following Eq. (12):

$$\alpha_1 + \alpha_2 = 1 \tag{26}$$

In the case of the two-dimensional feature space R^2 ($\mathbf{x}_j \in R^2$) the basis $\mathbf{B}_k$ (8) has been constituted by two basic feature vectors $\mathbf{x}_{j(1)}$ and $\mathbf{x}_{j(2)}$. In the case of the three-dimensional feature space R^3 ($\mathbf{x}_j \in R^3$) the basis $\mathbf{B}_k$ (8) has been constituted by two basic feature vectors $\mathbf{x}_{j(1)}$ and $\mathbf{x}_{j(2)}$ and by the basic unit vector $\mathbf{e}_{i(3)}$.

The collinear datasets D_1 and D_2 were generated randomly by using two pre-selected basic feature vectors $\mathbf{x}_{j(1)}$ and $\mathbf{x}_{j(2)}$ (25) and the Eq. (26):

$$(\forall \mathbf{x}_j \in D_k) \quad \mathbf{x}_j = \alpha_{j,1} \mathbf{x}_{j(1)} + \alpha_{j,2} \mathbf{x}_{j(2)} \tag{27}$$

were $\alpha_{j,1} + \alpha_{j,2} = 1$ (26).

The parameters $\alpha_{j,1}$ were generated randomly in accordance with the normal distribution $N(0, 1)$ with the mean value equal to 0 and the variance equal to 1 ($\alpha_{j,1} \sim N(0, 1)$). The parameters $\alpha_{j,2}$ were computed an the basis of the Eq. (26) as: ($\forall j = 1, 2,\ldots, 100$) $\alpha_{j,2} = 1 - \alpha_{j,1}$.

The basic feature vectors $\mathbf{x}_{j(1)}$ and $\mathbf{x}_{j(2)}$ (25) were pre-selected as:

$$\begin{aligned} &D_1 : \mathbf{x}_{j(1)} = [1, 0]^{\mathrm{T}} \quad \textit{and} \quad \mathbf{x}_{j(2)} = [0, 1]^{\mathrm{T}} \\ &D_2 : \mathbf{x}_{j(1)} = [1, 1, 0]^{\mathrm{T}} \colon \textit{and} \quad \mathbf{x}_{j(2)} = [0, 1, 1]^{\mathrm{T}} \end{aligned} \tag{28}$$

The computational experiments were carried out both on the collinear data sets D_k ($k = 1, 2$) as well as on the sets D_k with added *outliers*. The term *outliers* means here such additional feature vectors $\mathbf{x}_j$ which were not located on the vertexical line $L_k(\mathbf{x}_{j(1)}, \mathbf{x}_{j(2)})$ (25). The *outlier* feature vectors $\mathbf{x}_j$ were generated in accordance with the normal distribution $N_2(\mathbf{0}, \boldsymbol{I})$ with the unit covariance matrix $\boldsymbol{I}$ in the case of the two-dimensional feature space R^2 or with the normal distribution $N_3(\mathbf{0}, \boldsymbol{I})$ in the case of the three-dimensional feature space R^3.

Several simulations of the learning process were carried out with different values of the parameter ε in the algorithm (23). The results of the learning process based on the algorithm (23) with the parameter $\varepsilon = 0.00001$ (due to the fact of) limitations of the

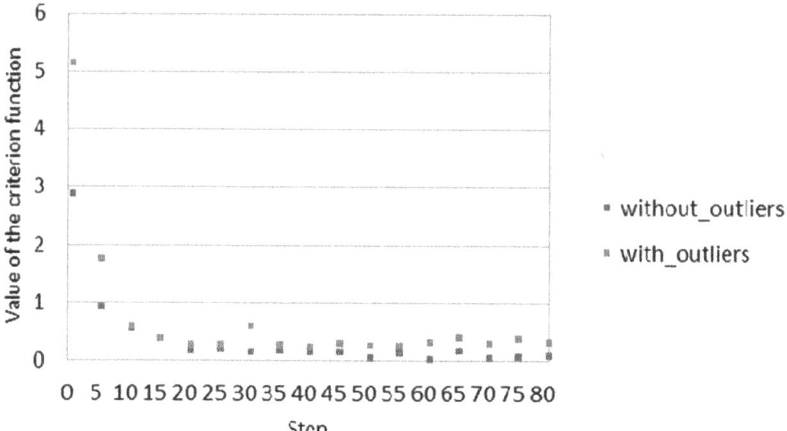

Fig. 2. The value of the criterion function $\Phi(\mathbf{w})$ (15) during successive steps k with the two - dimensional learning set D_1 (28). Two series: data without outliers and with added 40 % of outliers.

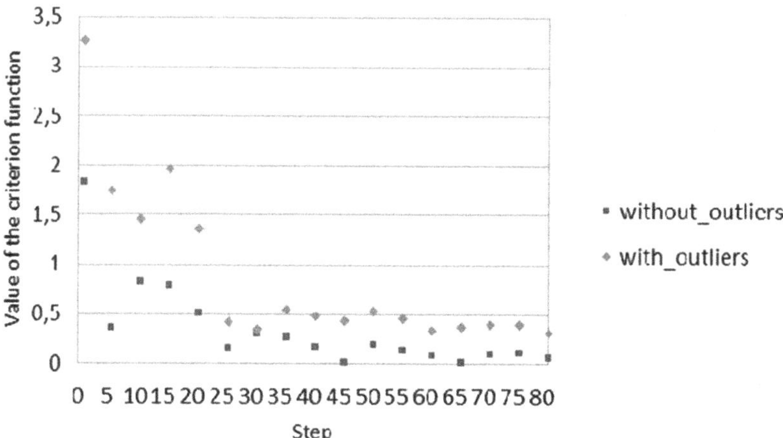

Fig. 3. The value of the criterion function $\Phi(\mathbf{w})$ (15) during successive steps k with the three - dimensional learning set D_2 (28). Two series: data without outliers and with added 40 % of outliers.

precision of the machine in case of real numbers are presented on Figs. 2 and 3. The learning algorithm (23) was stopped if the condition: $1 - \varepsilon \leq \mathbf{w}(k)^T\mathbf{x}(k) \leq 1 + \varepsilon$ was reached.

Figure 2 shows results of the learning process with the algorithms (23) as the values of the criterion function $\Phi(\mathbf{w})$ (15) defined on the two-dimensional learning data sets D_1 firstly without outliers and then with added 40 % of outliers.

As shown by those figures the algorithm converges after about 40 steps. However, in case of data with outliers the method takes longer than in case of data without outliers. Algorithm ends after 283 steps in case of data without outliers and after 375 in the case of data sets with outliers. The execution time for data without outliers (two-dimensional set 0.000016 s, three-dimensional set 0.000021 s) is about two times shorter than in case of data with outliers (two-dimensional set 0.000031 s, three-dimensional set 0.000047 s). This time is calculated as an average from 10 runs.

The Fig. 3 shows the results of the learning algorithm (23) by the values of the criterion function $\Phi(\mathbf{w})$ (15) defined on the three-dimensional learning data sets D_2 firstly without outliers and then with added 40 % of outliers. The results of the experiment in the three-dimensional feature space are similar to the results obtained in the two-dimensional feature space.

The values of the criterion function $\Phi(\mathbf{w})$ (15) defined on the learning data set D_k without outliers converges to zero during the learning process with the algorithm (23) both in the case of the two- as well the three-dimensional feature space.

7 Concluding Remarks

Flat (collinear) patterns made of a large number of feature vectors $\mathbf{x}_j$ allow for a basis for modeling linear interactions between selected features x_i. The data set C (1), can be decomposed on the basis of flat patterns into family of subsets C_m ($C_m \subset C$) each of which may be modeled by another linear model of interactions and represent another category.

Collinear patterns can be discovered in large data sets through minimization of the convex and piecewise linear (*CPL*) criterion functions [5]. Extraction flat patterns can be also carried out during learning process by using the proposed here learning algorithm (23).

Discovering collinear patterns can be linked to the search of the degenerated vertices (9) in an oriented graph [4]. This relationship would create a bridge between data mining methods and graph methods. Such relationship could be useful, for example, in modeling social networks.

Discovering collinear patterns on the basis of the CPL functions can be compared also with the methods based on the Hough transformation used in computer vision for detection lines and curves in pictures [7, 8].

Acknowledgments. This work was supported by the project S/WI/2/2015 from the Białystok University of Technology, Poland.

References

1. Duda, O.R., Hart, P.E., Stork, D.G.: Pattern Classification. J. Wiley, New York (2001)
2. Hand, D., Smyth, P., Mannila, H.: Principles of Data Mining. MIT Press, Cambridge (2001)
3. Bradley, P.S., Mangasarian, O.L.: K-plane clustering. J. Glob. Optim. **16**, 23–32 (2000)
4. Bobrowski, L.: K-lines clustering with convex and piecewise linear (CPL) functions. MATHMOD, Vienna (2012)

5. Bobrowski, L.: Discovering main vertexical planes in a multivariate data space by using CPL functions. In: Perner, P. (ed.) ICDM 2014. LNCS, vol. 8557, pp. 200–213. Springer, Heidelberg (2014)
6. Kushner, H.J., Yin, G.G.: Stochastic Approximation Algorithms and Applications. Springer Verlag, Berlin (1997)
7. Duda, O.R., Hart, P.E.: Use of the hough transformation to detect lines and curves in pictures. Commun. Assoc. Comput. Mach. **15**(1), 11–15 (1972)
8. Ballard, D.H.: Generalizing the hough transform to detect arbitrary shapes. Pattern Recogn. **13** (2), 111–122 (1981)

Fast Human Activity Recognition Based on a Massively Parallel Implementation of Random Forest

Jan Janoušek[1,2(✉)], Petr Gajdoš[1,2], Pavel Dohnálek[1,2], and Michal Radecký[1]

[1] Department of Computer Science, FEECS, VŠB - Technical Univesity of Ostrava, 17. listopadu 15, 708 33 Ostrava, Czech Republic
[2] IT4innovations, Centre of Excellence, VŠB - Technical Univesity of Ostrava, 17. listopadu 15, 708 33 Ostrava, Czech Republic
{jan.janousek,petr.gajdos}@vsb.cz

Abstract. This article elaborates on the task of Human Activity Recognition being solved with the Random Forest algorithm. A performance measure is provided in terms of both recognition accuracy and computation speed. In addition, the Random Forest algorithm was implemented using CUDA, a technology providing options for massively parallel computations on low-cost hardware. The results suggest that Random Forest is a suitable and highly reliable technique for recognising human activities and that Graphics Processing Units can significantly improve the computation times of this otherwise rather time-consuming algorithm.

Keywords: Random forests · Classification · GPU · CUDA · Parallelisation

1 Introduction

Human Activity Recognition (HAR) is a modern and extensively studied research area with new datasets emerging and being made available rather frequently. This allows researchers to discover ways of improving the reliability and/or speed of HAR-solving techniques, be it by means of physical sensors placed on the subject's body or analysing data from contactless devices such as cameras, physical quantity sensors, spatial scanners and others. These two means of sensing human activities are actually two distinct disciplines in the area: sensor-based and vision-based HAR. Sensor-based HAR generally refers to detecting activities from wearable sensor data, while vision-based systems rely most often on cameras, less often on various sensing apparatuses.

Many HAR datasets exist with several of them being freely available in the UCI Machine Learning Database. Their focus and methods vary from sensor arrays placed around the experimental surroundings and the subjects themselves (the OPPORTUNITY Activity Recognition dataset, [22]), through a small number of wearable sensors placed on the body (the PAMAP2 Physical Activity Monitoring dataset, [21]), through not using a specialised sensor at all, to

N.T. Nguyen et al. (Eds.): ACIIDS 2016, Part II, LNAI 9622, pp. 169–178, 2016.
DOI: 10.1007/978-3-662-49390-8_16

utilising devices that people use and wear on their own, such as smartphones (Human Activity Recognition Using Smartphones, [1]). While methods for capturing activity data are diverse, methods for processing this data are abundant. Generally speaking, recognising activities is a classification task, therefore classic algorithms such as k-Nearest Neighbors [14], Support Vector Machine [1], Nearest Centroid Classifier [8] and others are all well suited for application in this research branch. More complex algorithms can also be used. To name a few, temporal templates [4], multidimensional indexing [3] or R-transform [25] were all successfully applied. While some papers suggest that sensor-based systems can be used with nearly perfect reliability, a great challenge of robustness of the methods still remains. Several papers address this issue presenting various levels of success [23].

In this paper, we focus on one of the new HAR datasets available in the UCI Machine Learning Repository and present a method that improves both recognition accuracy and computation speeds.

2 Data Collection

The number of wearable devices on the market has rapidly increased in the last few years. These devices usually contain a large number of various sensors, which allows collecting information about their environment and the individuals who use them. HAR is one of the possible ways of utilisation of this data. For this paper's experiment purposes the HAR (PUC-Rio) dataset freely available on the UCI Machine Learning Database [13] was used. With it's publication year of 2013, this dataset is one of the most recent and contains an 8 hour record of 4 subjects performing activities captured as 165642 vectors. Each vector contains 18 attributes (user, gender, age, height, weight, BMI and x, y, z values from accelerometers placed around the waist, left thigh, right arm and right ankle). These values were divided into 5 classes corresponding to *sitting*, *sitting down*, *standing*, *standing up* and *walking*.

3 The Ensemble Approach

In the paper we utilise a *group model* (also known as an *ensemble*) algorithm. The group model category of algorithms clusters algorithms together to form a larger algorithmic structure that often provides better results at the cost of higher computational times. Classifiers used in the group models are known as weak learners [2]. Generally, a weak learner is characterised by low bias and high variance. We can illustrate the effect of the group models with a simple case of interpolation. Figure 1 shows a dashed and a solid line that illustratively express the interpolation for two different training sets using a non-ensemble technique. Both splines interleave all points, but this interpolation does not reflect the variability of the group where each observation may not be accurately measured. In this type of models, over-fitting often occurs. The variance of different types of datasets is high. In case of low variance with high bias, the model usually

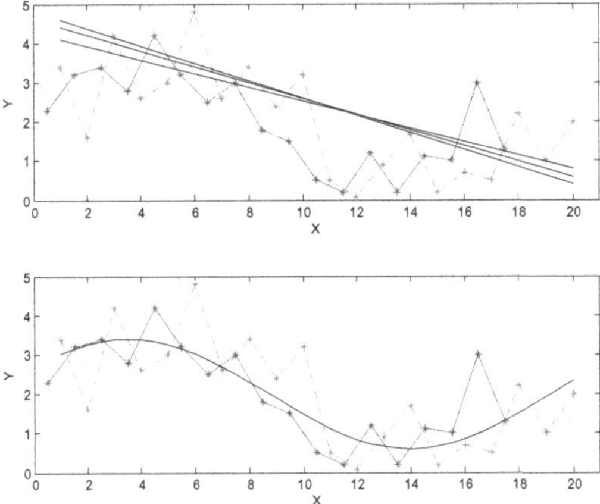

Fig. 1. Model with low bias and high variance (fitted by red and blue splines) and model with high bias and low variance (fitted by black lines - the top half) and Averaged weak models (the bottom half) (Color figure online)

performs poorly in terms of accuracy and is considered to be an under-fit. Such a model is illustrated by the three black solid lines in the upper figure of Fig. 1. It is necessary to find a model that has both bias and variance as low as possible. Both of these errors can be reduced by using a combination of different models, leading researchers to the ensemble approach. Let us consider splitting the training set into random parts and iterating the interpolation procedure N times, shuffling the training data in each iteration. Then, N weak models will be created. If the splines are averaged, a much more suitable solution with low bias and variance such as that illustrated in the lower figure of Fig. 1 can be obtained. A good explanation and justification of the claim that ensemble techniques can be more successful can be found, for example, in the much cited [10].

3.1 Random Forests

The Random Forest (RF) method belongs to the group model category of algorithms [5,7,9]. Generally there are many ways of determining the final answer from individual votes. For instance, regression often uses averaging of the results while classification simply picks the answer that was most frequent. Thanks to its underlying structure, the Classification and Regression Tree (CART) [12], the RF can work in both the regression and classification modes. Since the classification mode is used in this paper, the most frequent answer approach, also known as *majority voting*, was selected.

With CARTs being purely deterministic it is obvious that training each CART in the forest with the same dataset would result in the same answer with the same error. Therefore, the training dataset needs to be randomised so that each CART handles a different set of information. This process is called bagging [16].

As is the case with the CART, the RF algorithm was also designed by Breiman with [6] being the original paper where the entire algorithm was introduced and described. Each RF consists of N CARTs. Any such predictive forest can be expressed as $h(X_1, \Theta_1), \ldots, h(X_N, \Theta_N)$, where h is a function and Θ_N are the predictors. After splitting the input dataset into the testing and training part, bootstrap selections from the training set take place. A bootstrap selection is a repeated random selection of the same number of values of a given observation. The selections are performed independently of each other, leading to some of them being present in training sets for more than one individual CART while others may not be included in any training set at all. Selections not included in the training sets are used to estimate the tree error (the out-of-bag estimation). The classification result of the entire forest is given by voting that can be expressed as:

$$\hat{C}_{rf} = voting\{\hat{C}(x_i)\}_1^N \tag{1}$$

where $\hat{C}_{rf}$ is the classification answer of the entire forest, x_i is the i-th observation being classified and N is the number of trees in the forest.

In the beginning of the algorithm, a specific number of randomly selected predictors is chosen and a tree is built. This procedure is repeated for each bootstrap file. At the end, the final classification is evaluated by a vote. In order for the algorithm to execute, it is necessary to determine both the number of trees from which the forest will consist and the number of variables. To the best of our knowledge, the best values cannot be precomputed and need to be discovered experimentally. We can also explain how a combination of existing models reduces the total classification or regression error. There are three basic types of errors incurred during the classification task.

1. Noise - quantifies the output deviation from the optimal model.
2. Bias - defined as the average error of the model with regards to the optimal model.
3. Variance - error that describes the error rate of the model during training on various learning sets.

The total model error is given as the sum of the above mentioned errors. As describing the solution to minimise this sum would be lengthy and was done in the past, the readers are encouraged to further study the matter in [6,17,19] and others.

4 GPU and CUDA

Modern graphics hardware plays an important role in the area of high performance parallel computing. Graphics cards have been primarily used to accelerate gaming and 3D graphics applications, but today they are also used to accelerate computations in relatively distant research areas such as remote sensing, environmental monitoring, business forecasting, medical applications, physical simulations, etc.

The architecture of Graphics Processing Units (GPUs) is suitable for vector and matrix algebraic operations. That leads to the wide usage of GPUs in the area of information retrieval, data mining, image processing, data compression, etc. [15].

CUDA (Compute Unified Device Architecture) is a general purpose parallel computing architecture. GPUs utilised in our experiments are based on the Kepler architecture [18], which still belongs to the most common GPU architecture since the original G80. Currently, the new architecture called Maxwell has been introduced by NVIDIA.

The main advantage of CUDA technology lies in the power of different architecture of graphics processing units. There exists a number of tasks that were solved on GPU rather than CPU such as Fourier transform and convolution, matrix multiplication, neural network, data mining algorithms, etc. We refer to [15] for more information.

4.1 Parallel Random Forest

In our CUDA implementation of the RF, the algorithm was divided into six kernels. The first kernel takes care of initialising the data structures on the GPU and is executed only once at the beginning of the tree build process. The following five kernels are triggered iteratively with each iteration corresponding to a single node creation.

The first of these five kernels, *generateFunctionCandidates*, is responsible for the selection of function candidates for the current dataset. The number of features that will be selected is one of the configuration parameters for this subroutine. The threshold for each feature is chosen as the feature value of a randomly selected vector that belongs to the currently processed dataset.

The second kernel, *evaluateFunctionCandidates*, gets function candidates from the first kernel, computes the information gain of a particular division and selects the best one.

For each input vector, the *updateMarkers* kernel creates a vector of binary marks based on the best discovered function candidate from *evaluateFunctionCandidates*. If a vector feature value is lower than the threshold, the corresponding mark is set to 0, otherwise it is set to 1.

The next kernel named *reorderData* reorders the input vectors according to the markers from the previous step. This is a performance-favouring step as it allows to determine the most efficient order of swap operations in the global memory.

The last kernel is *growTrees*. This kernel takes care of creating a new tree node. The nodes of a tree are generated in the depth first order. Since every kernel needs to be configured with a thread block and block grid sizes, the configurations were set according to the best practices in CUDA programming. Thread blocks were set to the size divisible by warp size (32) and the number of blocks in the grid corresponded to the number of trees in the forest. Since building each tree is an operation fully independent on any other tree, this implementation has a significant potential for execution on virtually any number of GPUs.

5 Experimental Setup and Results

In our experiments, the following settings were experimentally determined as the most successful: RF contained 128 trees, each of them trained on 43 % of the vectors. The limit for tree growth termination was set as the minimal error mitigation of 0.0005. All experiments on CUDA were carried out on an NVIDIA GeForce GTX TITAN with 6 GB of global memory. Experiments on CPU were performed on an Intel Core i5-2500K processor clocked at 3.3 GHz with 8 GB RAM and 64-bit Windows 7.

To verify the quality of classification, 10-fold cross validation was used. Results were compared with the original results for this dataset published in [24]. Results obtained with the proposed algorithm are presented in Table 1. To make the comparison easier, Table 2 presents the results obtained with C4.5 [20] using AdaBoost [11] proposed and presented in the original paper. For both algorithms, the better values are highlighted in bold. The tables show that the number of incorrect classifications using our implementation of the RF is lower by 351 misses when compared with C4.5 with AdaBoost.

Table 1. Random Forest confusion matrix

	Sitting	Sitting down	Standing	Standing up	Walking
Sitting	**50619**	**2**	0	**14**	**0**
Sitting down	82	**12148**	46	**18**	121
Standing	0	**0**	47270	20	84
Standing up	61	**42**	**5**	11681	**39**
Walking	1	28	40	**16**	**43305**

Table 2. C4.5 with AdaBoost confusion matrix

	Sitting	Sitting down	Standing	Standing up	Walking
Sitting	50601	9	0	20	1
Sitting down	**10**	11484	**29**	297	**7**
Standing	0	4	**47342**	**11**	**13**
Standing up	**14**	351	24	**11940**	85
Walking	**0**	**8**	**27**	60	43295

Tables 3 and 4 present statistical comparisons for both algorithms. Again, better values are highlighted. From the tables it can bee seen that the RF has better sensitivity in almost all classes. Specificity and accuracy for each single class seems better when using C4.5 since more values are highlighted. However, when averaged, RF is shown to be superior across all metrics, showing that while there are cases where RF performs worse, these cases are fully compensated by performing much better in the remaining tasks.

Table 3. Statistical results for Random Forests

	TP	FP	Precision	Sensitivity	Specificity	Accuracy
Sitting	50619	144	0.9972	**0.9997**	0.9987	0.9990
Sitting down	12148	72	0.9941	**0.9785**	**0.9995**	**0.9980**
Standing	47270	91	0.9981	0.9978	0.9992	0.9988
Standing up	11681	68	0.9942	**0.9876**	**0.9996**	**0.9987**
Walking	43305	244	0.9944	**0.9980**	0.9980	0.9980
Average	**33004.6**	**123.8**	**0.9956**	**0.9923**	**0.9990**	**0.9985**

Table 4. Statistical results for C4.5 with AdaBoost

	TP	FP	Precision	Sensitivity	Specificity	Accuracy
Sitting	50601	24	0.9995	0.9994	**0.9998**	**0.9997**
Sitting down	11484	372	0.9686	0.9710	0.9976	0.9957
Standing	47342	80	0.9983	**0.9994**	**0.9993**	**0.9993**
Standing up	11940	388	0.9685	0.9618	0.9975	0.9948
Walking	43295	106	0.9976	0.9978	**0.9991**	**0.9988**
Average	32932.4	194	0.9865	0.9859	0.9987	0.9977

Not surprisingly, the quality of classification depends on the number of trees in the forest. The graph of the number of misclassifications based on the number of trees is shown in Fig. 2.

Figure 3 shows a comparison of time required for building the RF classifier based on the number of trees for the serial (C) and parallel (CUDA) implementation. The graph shows that the serial version of the algorithm is faster for about 20 trees up. It is a common occurrence in parallel programming that serial versions of an algorithm are faster for smaller problem sizes due to the extra one-time overhead requirements in parallel processing. In CUDA, a prominent

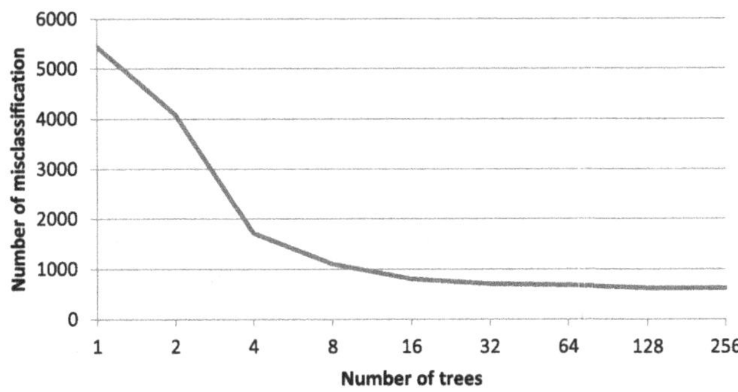

Fig. 2. Number of misclassification based on the number of trees

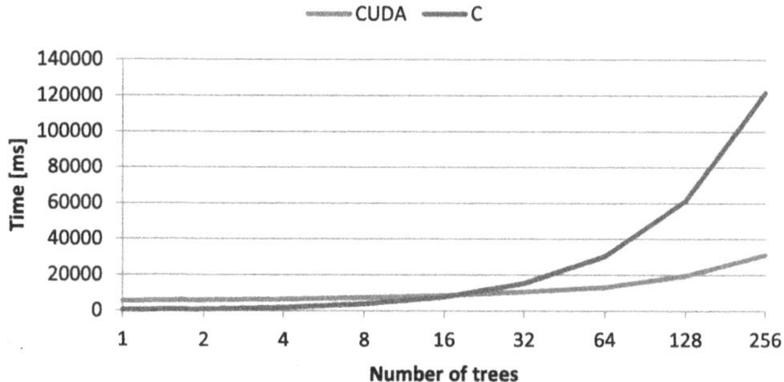

Fig. 3. Comparsion of CUDA and C version

overhead is data transfers between host memory (RAM) and GPU memory. In order to fully utilise the GPU computation power, enough parallel computations need to be enqueued for processing. Figure 3 confirms this widely-known phenomenon.

6 Conclusion and Future Work

In this paper, a reliable human activity recognition with high classification success rate was presented. The method was an improvement over the original proposition made by the authors of the HAR dataset the method was tested on. A massively parallel implementation showing significant speed-up of the computation was also presented. While many HAR approaches are sufficiently fast to provide a real-time activity recognition for a single subject, our implementation shows potential for executing on multiple hardware units simultaneously, thus allowing either greater speeds for a single subject or grouping computations for several subjects into one application. This could be useful in applications involving collaboration between subjects in team efforts such as team games and sports or supervising team work (especially in manual labour such as construction). With sufficiently accurate sensors, a digital assistant for precision-heavy tasks can be built. Future research focuses include utilising the scalability potential of the implementation, improving its robustness in different sensor setups and further optimising the algorithm for more accuracy and speed, both in terms of its theoretical background and practical implementation for future state-of-the-art hardware.

Acknowledgement. This work was supported by the IT4Innovations Centre of Excellence project (CZ.1.05/1.1.00/02.0070), funded by the European Regional Development Fund and the national budget of the Czech Republic via the Research and Development for Innovations Operational Programme and by Project SP2015/105 "DPDM - Database of Performance and Dependability Models" of the Student Grand System, VŠB Technical University of Ostrava and by Project SP2015/146 "Parallel processing of Big data 2" of the Student Grand System, VŠB Technical University of Ostrava.

References

1. Anguita, D., Ghio, A., Oneto, L., Parra, X., Reyes-Ortiz, J.L.: Human activity recognition on smartphones using a multiclass hardware-friendly support vector machine. In: Bravo, J., Hervás, R., Rodríguez, M. (eds.) IWAAL 2012. LNCS, vol. 7657, pp. 216–223. Springer, Heidelberg (2012)
2. Bańczyk, K., Kempa, O., Lasota, T., Trawiński, B.: Empirical comparison of bagging ensembles created using weak learners for a regression problem. In: Nguyen, N.T., Kim, C.-G., Janiak, A. (eds.) ACIIDS 2011, Part II. LNCS, vol. 6592, pp. 312–322. Springer, Heidelberg (2011)
3. Ben-Arie, J., Wang, Z., Pandit, P., Rajaram, S.: Human activity recognition using multidimensional indexing. IEEE Trans. Pattern Anal. Mach. Intell. **24**(8), 1091–1104 (2002)
4. Bobick, A.F., Davis, J.W.: The recognition of human movement using temporal templates. IEEE Trans. Pattern Anal. Mach. Intell. **23**(3), 257–267 (2001)
5. Breiman, L.: Random forests. Mach. Learn. **45**(1), 5–32 (2001). http://dx.doi.org/10.1023/A%3A1010933404324
6. Breiman, L.: Random forests. Mach. Learn. **45**(1), 5–32 (2001)
7. Cuzzocrea, A., Francis, S.L., Gaber, M.M.: An information-theoretic approach for setting the optimal number of decision trees in random forests. In: SMC, pp. 1013–1019 (2013)
8. Dabney, A.R.: Classification of microarrays to nearest centroids. Bioinformatics **21**(22), 4148–4154 (2005)
9. Denil, M., Matheson, D., de Freitas, N.: Narrowing the gap: Random forests in theory and in practice. CoRR abs/1310.1415 (2013)
10. Dietterich, T.G.: Ensemble methods in machine learning. In: Kittler, J., Roli, F. (eds.) MCS 2000. LNCS, vol. 1857, pp. 1–15. Springer, Heidelberg (2000)
11. Freund, Y.: An adaptive version of the boost by majority algorithm. Mach. Learn. **43**(3), 293–318 (2001). http://dx.doi.org/10.1023/A%3A1010852229904
12. Gehrke, J.: Classification and regression trees. In: Encyclopedia of Data Warehousing and Mining, pp. 192–195 (2009)
13. Groupware@LES, R.G: Wearable computing: Classification of body postures and movements data set (2013). http://groupware.les.inf.puc-rio.br/har
14. Hastie, T., Tibshirani, R.: Discriminant adaptive nearest neighbor classification. IEEE Trans. Pattern Anal. Mach. Intell. **18**(6), 607–616 (1996)
15. Kirk, D., Hwu, W.: Programming Massively Parallel Processors: A Hands-on Approach. Applications of GPU Computing Series, Elsevier Science (2010). http://books.google.cz/books?id=qW1mncii_6EC
16. Kotsiantis, S.B.: Bagging and boosting variants for handling classifications problems: a survey. Knowl. Eng. Rev. **29**(1), 78–100 (2014)
17. Loh, W.Y.: Classification and regression trees. Wiley Interdisc. Rev. Data Min. Knowl. Discov. **1**(1), 14–23 (2011)
18. nVIDIA: Nvidia kepler gk110 architecture whitepaper (2014). http://www.nvidia.com/content/PDF/kepler/NVIDIA-Kepler-GK110-Architecture-Whitepaper.pdf
19. Breiman, L., Friedman, J., Stone, C.J., Olshen, R.A.: Classification and Regression Trees. Wadsworth International Group, California (1984)
20. Quinlan, J.R.: C4.5: Programs for Machine Learning. Morgan Kaufmann Publishers Inc., San Francisco (1993)
21. Reiss, A., Stricker, D.: Creating and benchmarking a new dataset for physical activity monitoring. In: The 5th Workshop on Affect and Behaviour Related Assistance (ABRA) (2012)

22. Roggen, D., Calatroni, A., Rossi, M., Holleczek, T., Forster, K., Troster, G., Lukowicz, P., Bannach, D., Pirkl, G., Ferscha, A., Doppler, J., Holzmann, C., Kurz, M., Holl, G., Chavarriaga, R., Sagha, H., Bayati, H., Creatura, M., del R Millan, J.: Collecting complex activity datasets in highly rich networked sensor environments. In: 2010 Seventh International Conference on Networked Sensing Systems (INSS), pp. 233–240, June 2010
23. Sagha, H., Digumarti, S.T., del R Millan, J., Chavarriaga, R., Calatroni, A., Roggen, D., Troster, G.: Benchmarking classification techniques using the opportunity human activity dataset. In: 2011 IEEE International Conference on Systems, Man, and Cybernetics (SMC), pp. 36–40. IEEE (2011)
24. Ugulino, W., Cardador, D., Vega, K., Velloso, E., Milidiú, R., Fuks, H.: Wearable computing: accelerometers' data classification of body postures and movements. In: Barros, L.N., Finger, M., Pozo, A.T., Gimenénez-Lugo, G.A., Castilho, M. (eds.) SBIA 2012. LNCS, vol. 7589, pp. 52–61. Springer, Heidelberg (2012)
25. Wang, Y., Huang, K., Tan, T.: Human activity recognition based on r transform. In: IEEE Conference on Computer Vision and Pattern Recognition, 2007. CVPR 2007, pp. 1–8. IEEE (2007)

An Item-Based Music Recommender System Using Music Content Similarity

Ja-Hwung Su[1(✉)] and Ting-Wei Chiu[2]

[1] Department of Information Management, Cheng Shiu University, Kaohsiung, Taiwan
bb0820@ms22.hinet.net

[2] Department of Information Management, Kainan University, Taoyuan, Taiwan

Abstract. Nowadays, music data grows rapidly because of the advanced multimedia technology. People are always spending much time to listen to music. This incurs a hot research issue for how to discover the users' interested music preferences from a large amount of music data. To deal with this issue, the music recommender system has been a solution that can infer the users' musical interests by a set of learning methods. However, recent music recommender systems encounter problems of new item and data sparsity. To alleviate these problems, in this paper, we propose a new recommender system that fuses user ratings and music low-level features to enhance the recommendation quality. The experimental results show that our proposed recommender system outperforms other well-known music recommender systems.

Keywords: Music recommender system · Item-based collaborative filtering · Music content · New item · Data sparsity

1 Introduction

Recent advanced multimedia technology makes the music data available at an explosive rate. For an online customer, it is not easy to obtain the interested music from an enormous music data set. Hence how to provide the user with the preferred music effectively has been becoming a hot topic over the past few decades. To cope with this issue, the music recommender system is employed as a solution, and a lot of past studies have been made on this topic. Traditionally, the user interest for a music piece (also called item in this paper) is shown in a numeric rating ranged from 1 to 5, where the larger the score, the higher the interest. Then, the rating matrix is conducted by the user-to-item ratings.

Table 1 is an example of a rating matrix which is composed of 6 items and 6 users. In this table, a number of ratings (also called unknown ratings) are zero, which represents the items never rated by the users. Based on the rating table, the recommendation procedure can typically be divided into two stages, namely rating prediction and item selection.

N.T. Nguyen et al. (Eds.): ACIIDS 2016, Part II, LNAI 9622, pp. 179–190, 2016.
DOI: 10.1007/978-3-662-49390-8_17

Table 1. Example of a rating matrix.

	$item_1$	$item_2$	$item_3$	$item_4$	$item_5$	$item_6$
User1	0	0	0	0	0	5
User2	0	2	0	0	3	0
User3	0	0	0	2	0	0
User4	2	0	0	0	0	0
User5	3	0	0	1	0	4
User6	0	2	0	0	3	0

I. Rating prediction: The aim of this stage is to predict the active user's unknown ratings for items.
II. Item selection: According to the predicted ratings, the un-rated items are sorted and the top q un-rated items are thereby recommended to the active user.

Actually, there have been a number of past studies were conducted on the first stage because the second stage is straightforward and cheaper. Nevertheless, the recommender systems based on only the rating matrix cannot earn the users' satisfaction because of the following problems.

I. New item: This problem indicates the items never rated by any users. For example, the third item $item_3$ is a new item in Table 1. Hence, it is difficult to predict ratings of new items because no related information in the matrix can be used.
II. Data sparsity: This problem indicates that, the rating density of the rating matrix is very low. For example, there are 10 ratings in Table 1 and the rating density is $10/36 = 28\ \%$. Because the information is insufficient, the unknown ratings are hard to infer effectively.

To aim at the problems above, in this paper, a new music recommender system is proposed, which integrates information of music content and ratings to improve quality of the rating prediction. In the proposed recommender system, the method of item-based collaborative filtering is performed to predict the user ratings by using the music content similarity instead of the music rating similarity. Behind the proposed method, our intent is to predict the users' preferences on music even facing very sparse data. The empirical evaluations reveal that the proposed approach can effectively predict the unknown ratings and outperform other well-known recommender systems on real data. The rest of this paper is organized as follows. In Sect. 2, a brief review of previous work is shown. In Sect. 3, we demonstrate our proposed method for predicting the unknown ratings in great detail. Experimental evaluations of the proposed methods are illustrated in Sect. 4. Finally, conclusions and future work are stated in Sect. 5.

2 Related Work

In principal, the music recommender system refers a set of predictive algorithms that infer the users' preferences by learning from the users' implicit or explicit behaviors. Collaborative Filtering called CF is the well-known recommender system which predicts the user preferences based on the user ratings. In fact, a considerable number of CF-based recommender systems have been conducted on music recommendation over the past few years. Although these predecessors have been shown to be effective, there are some problems unsettled. In the followings, the review of past CFs is briefly described by categories.

I. Memory-based CF. This is a traditional recommendation paradigm that infers the unknown ratings using the user/item similarity on ratings. It can further be divided into two well-known categories, namely user-based CF and item-based CF. In terms of user-based recommender systems [5, 8], the item ratings are predicted by the most-relevant users on similar ratings. In contrast, item-based recommender systems [4, 9] predict the item ratings by the most-relevant items on similar ratings. In order to attack the individual lacks of above methods, Wang *et al.* [10] proposed an algorithm to unify the user-based and item-based collaborative filtering. Another method similar to user-based CF using significances of the users and items is proposed by Bobadilla *et al.* [1]. In fact, the significances are still calculated by ratings. As mentioned in Sect. 1, this type of recommender systems considering only ratings suffers from problems of data sparsity and new item.

II. Model-based CF. Also on the basis of the ratings, the main goal of model-based CF is to model the behaviors by machine learning techniques. Through learning the behaviors from the users' rating logs, the user's preferences hidden in the rating behaviors are therefore implied. SVM (Support Vector Machine), Decision Tree and Bayesian are the most popular solutions to recognize patterns for classification, which were adopted as the rating classifiers by [2, 7, 11], respectively. Lin *et al.* [6] found the overlaps of several users' tastes to match the active user's taste by utilizing the discovered user associations and article associations. The item-to-item regression model derived from item-to-item correlations and item-to-item cosine similarities has been proposed by Chuan et al. [3]. However, the effectiveness of model-based CF is limited in the rating space that incurs problems of data sparsity and new item.

3 Proposed Method

3.1 Basic Idea

As mentioned in Sect. 2, most existing music recommender systems predict the unknown ratings based on the rating similarities. This type of recommender systems suffers from the problem that, the user and item similarities based on the ratings are not robust enough to imply the users' preferences by the similar users and items, respectively. Let us take Table 1 as an example to illustrate this problem. In Table 1, assume that, user1 is the active user and itm_2 is the target item to predict. For traditional item-based recommender systems, the most similar item to $item_2$ is $item_5$ because only $item_5$ and $item_2$ have the ratings by the same users {user2, user6}. Similarly, the most similar user to user1 is user5 for the user-based recommender systems because only user1 and user5 have the ratings for the same item {itm_6}. From this example, it is obvious that, the rating of the target item $item_2$ cannot be predicted whether using itm_5's or user5's ratings. Indeed, it is caused by problem of data sparsity, that is, the known ratings are too few to predict the known ratings.

Another problem called new item in this example is that, the rating of $item_3$ cannot be predicted because no related rating can be employed to detect the most similar items to $item_3$. Unfortunately, such above problems occur frequently in real applications. To avoid these problems, in this paper, we propose a new recommender system to predict the unknown ratings by the music content. The major intention behind our proposed method is to calculate the item similarities based on the music low-level features. Without ratings, the most similar items can be detected to facilitate the predictions of the unknown ratings. In the succeeding sections, we will give the details of how to achieve the item-based music recommender system using music low-level features instead of ratings.

3.2 Overview of the Proposed Method

Although the existing music recommender systems using ratings can predict the users' preferences effectively, the problems mentioned in above are not easy to solve. Hence, the goal of our proposed method is to achieve higher quality of music recommendations by fusing information of ratings and music low-level features. To reach this goal, as shown in Fig. 1, the proposed recommender system can be decomposed into two phases, namely offline preprocessing and online prediction.

I. Offline preprocessing phase. The primary goal of this phase is to accelerate the online prediction. To achieve this goal, the item similarity matrix has to be conducted in this phase. In this phase, features of all music items have to be extracted first. Then, through the item similarity calculations, the item-to-item similarity matrix is generated.
II. Online prediction phase. Based on the item-to-item similarity matrix, first, the most similar items to the target item can be determined without calculating the item similarities in this phase. Then, the unknown ratings of the target item can be computed by using item-based CF.

3.3 Audio Features

Actually, feature extraction plays a basic role for the whole recommender system. Therefore, before showing how to perform the two main phases, the feature extraction is described in this subsection. As shown in Fig. 2, each music piece is divided into a set of frames and each frame is described by a number of low-level features {a_1, a_2, …., a_n}. In this paper, Mel-scale Frequency Cepstral Coefficients (MFCCs) are served as the major audio features.

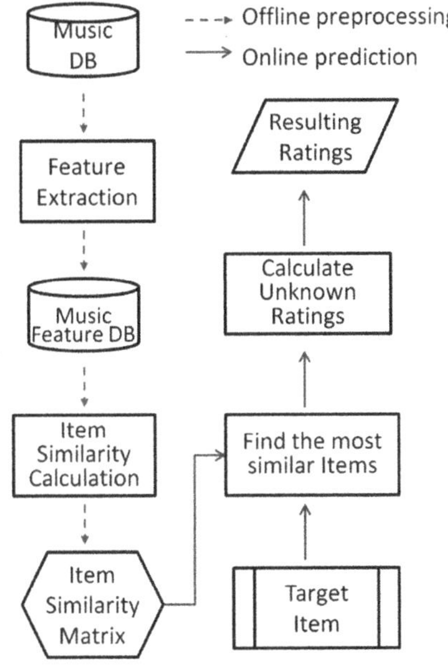

Fig. 1. The framework of the proposed method.

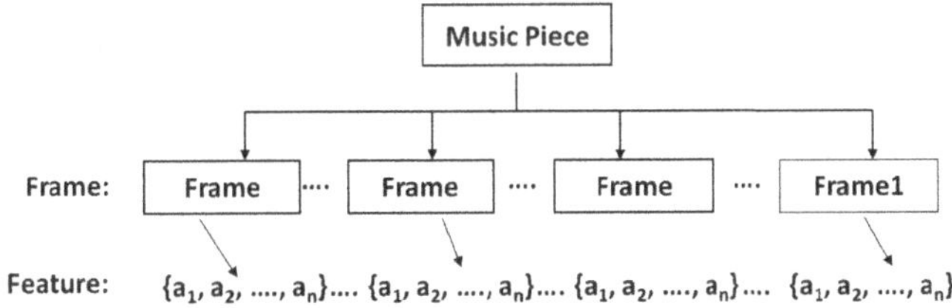

Fig. 2. Data structure of audio features.

3.4 Offline Preprocessing

Basically, the offline preprocessing can be regarded as the support of the online prediction. That is, the online prediction will be inefficient if lacking offline preprocessing. In this phase, low-level features of all music pieces in the database have to be extracted first. Then, the similarities of every two items are calculated according to Definition 1 and the item-to-item similarity matrix is thereupon made. Table 2 is an example of the resulting item-to-item similarity matrix. From this table, we can obtain the similarities of every two items and this information will increase the efficiency of online prediction significantly.

Table 2. Example of the item-to-item similarity matrix.

	itm_1	itm_2	itm_3	itm_4	itm_5	itm_6
itm_1	1	0.85	0.9	0.75	0.65	0.8
itm_2	–	1	0.8	0.9	0.7	0.85
itm_3	–	–	1	0.9	0.85	0.8
itm_4	–	–	–	1	0.7	0.8
itm_5	–	–	–	–	1	0.9
itm_6	–	–	–	–	–	1

Definition 1. Assume a feature vector contains n dimensions for a music piece where the dimension denotes the mean of frame dimension features. Given two music pieces x and y with two vectors $\{a_{x,1}\, a_{x,2}, \ldots, a_{x,n}\}$ and $\{a_{y,1}\, a_{y,2}, \ldots, a_{y,n}\}$, respectively, the similarity between x and y is defined as:

$$sim_{x,y} = \frac{\sum_{0<i\leq n} a_{x,i} * a_{y,i}}{\sqrt{\sum_{0<i\leq n} \left(a_{x,i}\right)^2 * \sum_{0<i\leq n} \left(a_{y,i}\right)^2}}, \quad (1)$$

where $a_{x,i}$ and $a_{y,i}$ are the ith dimensions of music pieces x and y, respectively.

3.5 Online Prediction

In general, this phase is triggered by an active user's visit. Once an active user visits the recommender system, all of her/his unknown ratings have to be predicted one by one. For each unknown rating, the most similar items to the referred target item have to be determined first. Next, the rating of the referred target item is derived by Definition 2.

Definition 2. Based on Definition 1, assume there is a set IM of unique music items in the database and there is a user-to-item rating matrix generated in the offline preprocessing phase, which contains a set of users and items. Assume the most relevant item set to the target item itm_i for an active user u_z is $U_z = \cup itm_p$, where $itm_p \in IM$. Thereby the rating v of un-rated itm_i for u_z is defined as:

$$v_i^z = \frac{\sum_{itm_p \in U_z} sim_{i,p} * v_p^z}{\sum_{itm_p \in U_z} sim_{i,p}}. \quad (2)$$

where v_p^z indicates the rating of itm_p for user u_z.

Let us take a simple example to explain predictions of the unknown ratings. Following above examples Tables 1 and 2, assume the active user is u_1 and itm_2 is the target item. That is, the target unknown rating to predict is v_2^1. First, according to Table 2, the most k similar items to itm_2 are itm_4 and itm_6, where $k = 2$. Then, the rating v_2^1 of itm_2 for u_1 is:

$$v_2^1 = \frac{0.9 * 0 + 0.85 * 5}{0.9 + 0.85} = 2.4.$$

Similarly, the other unknown ratings for u_1 can be calculated by Eq. 2 and finally the top q items can be recommended by sorting the predicted ratings.

4 Empirical Study

In the above section, we have presented our proposed music recommender system in great detail. In this section, the proposed method will be evaluated by a set of experiments. Overall, the whole experiments were conducted on two aspects: (1) comparisons with other well-known recommender systems in terms of RMSE (Root Mean Square Error) and (2) comparisons with other well-known recommender systems in terms of numbers of un-predicted items. Through the experimental evaluations, the technical contributions for solving problems mentioned above can be justified clearly.

Table 3. Details of experimental settings.

Description	Value
#Users	20
#Items	400
#Known rating	1099
Music genre	20
#music pieces of each genre	20
#music pieces	400

4.1 Experimental Settings

The music data was gathered from the collections of music websites and CDs. To make the evaluations more solid, the experimental data contains two types of music, namely instrument music data and voice music data. Each of instrument and voice music data consists of 10 genres and a genre is composed of 20 music pieces, i.e., total of 400 music pieces. In the experiments, we constructed a voting website and invited 20 users to give the ratings for the items randomly selected from the music data. For each user, she/he rated at most 60 items. That is, 3 items of each genre are randomly selected as the ones to rate. After collecting the ratings, there are 1099 ratings in the user-to-item rating matrix initially. On the whole, the density of the rating matrix is 13.7 %, which is very spare. From 1099 ratings, 10 ratings of each user were selected as the testing (unknown) ratings. The details of experimental settings and music data are shown in Tables 3 and 4. In the experiments, two CF methods including item-based and user-based CFs using only ratings were adopted as the compared methods. Additionally, to clarify the effectiveness of different similarity functions, in this paper, three similarity functions were employed to compute the item and user similarities. The details of compared methods are depicted in Table 5.

Table 4. Details of music data.

Music Type	Music Genre
Instrument	Piano
	Saxophone
	Erhu Fiddle
	Flute
	Harmonica
	Guitar
	Symphony
	Techno
	Zither/Guzheng
	Violin
Voice	Rock&Metal
	Classical Chorus
	Rap
	Disco Dance
	Vocal
	Country
	Pop
	Latin
	Folk
	Jazz&Blues

Table 5. Details of compared methods.

Approach	Similarity function	Similarity Source	Terminology
Item-based CF	Adjusted cosine similarity	Rating	IAC
Item-based CF	Cosine similarity	Rating	IC
Item-based CF	Correlation similarity	Rating	ICR
User-based CF	Cosine similarity	Rating	UC
Item-based CF	Cosine similarity	Content	CC (proposed)

To analyze the effectiveness of our proposed music recommender system, the popular criterion, namely Root Mean Square Error (RMSE), was employed to measure the related experimental performances. It is defined as:

$$RMSE = \sqrt{\frac{\sum_{0<i\leq|test|}(v_i - \hat{v}_i)^2}{|test|}} \tag{3}$$

where v_i stands for the ground truth of the ith test rating, $\hat{v}_i$ stands for the predicted value of the ith test rating and *test* stands for the testing data set. Generally, RMSE shows the error variance. That is, the lower the RMSE, the lower the error, the higher the precision, the better the recommendation.

4.2 Comparisons with Other Well-Known Methods in Terms of RMSE

From the rating density, we can know that the experimental data is very sparse. Accordingly, the goal of the experiments is to reveal the effectiveness of our proposed method on such sparse data. The first experimental result we want to show is the RMSE comparisons between our proposed method and other existing methods. Figure 3 depicts the experimental comparisons on RMSE. From Fig. 3, we can obtain some discovery. First, the impact of k is so unclear that the best k is recommended as 5 for this data. Second, the worst one is UC and IC is better than UC. It says that, by ratings, the item-based CF is more robust than user-based CF. Third, IAC is worse than IC and ICR. Moreover, results of IC and ICR are pretty close. These results deliver that, Adjusted Cosine Similarity is not robust. The effectiveness of Cosine Similarity and Correlation Similarity are almost the same. Fourth, our proposed method CC performs better than all compared methods. That is, the proposed method can really achieve higher quality of music recommendation.

4.3 Comparisons with Other Well-Known Methods in Terms of Numbers of Un-Predicted Items

The goal of the final experimental result is to show if the proposed method can solve the problems of new item and data sparsity. In the experiments, there are 200 ratings

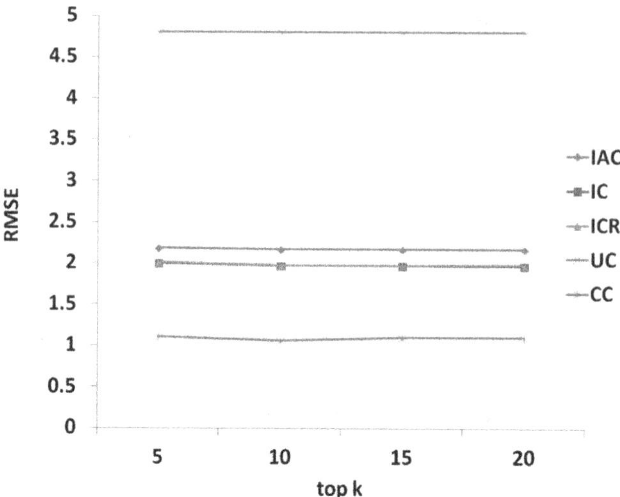

Fig. 3. RMSE comparisons of compared methods.

adopted as the testing (unknown) ratings. Table 6 is the experimental result showing the numbers of un-predicted ratings of all methods. There are some important points in Table 6. First, UC is the worst. Second, IAC is worse than IC and ICR. Third, results of IC and ICR are the same. Fourth, our proposed method is the best, which can predict all unknown ratings successfully. The result can be regarded as an echo of Fig. 3. On the whole, through music content similarity, our proposed method can really avoid problems mentioned in above.

Table 6. Comparisons of numbers of un-predicted ratings of compared methods.

Approach	#un-predicted ratings
IAC	28
IC	21
ICR	21
UC	183
CC	0

5 Conclusion and Future Work

So far, there is very few music recommender systems succeeding in predicting the user preferences due to problems of new item and data sparsity. To deal with these problems, in this paper, we propose a new item-based method that integrates information of music contents and ratings. On one hand, the un-robust item similarity using ratings is replaced with the content similarity. Therefore, the rating predictions can be enhanced. On the other hand, even though the new items with no ratings, the referred ratings can still be predicted. The experimental results reveal that, our proposed recommender system indeed provides lower errors of preference predictions than compared methods. It tells the truth that, the problems mentioned in this paper can be alleviated successfully.

In the future, there remain some issues to investigate further. First, we will enlarge the music data and users to realize the effectiveness of the proposed method under large data. Second, more information will be hybridized with music low-level features to attack the lack of music content. Third, the main idea will be applied to other kinds of multimedia data.

Acknowledgements. This research was supported by Ministry of Science and Technology, Taiwan, R.O.C. under grant no. MOST 104-2632-S-424-001 and MOST 104-2221-E-230-019.

References

1. Bobadilla, J., Hernando, A., Ortega, F., Gutiérrez, A.: Collaborative filtering based on significances. Inf. Sci. **185**(1), 1–17 (2012)
2. Breese, J.S., Heckerman, D., Kadie, C.: Empirical analysis of predictive algorithms for collaborative filtering. In: Proceedings of the International Conference on Uncertainty in Artificial Intelligence, pp. 43–52 (1998)
3. Chuan, Y.U., Jieping, X.U., Xiaoyong, D.U.: Recommendation algorithm combining the user-based classified regression and the item-based filtering. In: Proceedings of the International Conference on Electronic Commerce, Fredericton, pp. 574–578, New Brunswick, Canada (2006)
4. Deshpande, M., Karypis, G.: Item-based top-*N* recommendation algorithms. ACM Transactions on Information Systems. **22**(1), 143–177 (2004)
5. Herlocker, J.L., Konstan, J.A., Borchers, A., Riedl, J.: An algorithmic framework for performing collaborative filtering. In: Proceedings of the International ACM SIGIR Conference on Research and Development in Information Retrieval, pp. 230–237, Berkeley, USA (1999)
6. Lin, W., Alvarez, S.A., Ruiz, C.: Collaborative recommendation via adaptive association rule mining. In: Proceedings of the ACM SIGKDD International Conference on Knowledge Discovery and Data Mining (2002)
7. Nikovski, D., Kulev, V.: Induction of Compact Decision Trees for Personalized Recommendation. In: Proceedings of the International Conference on the ACM Symposium on Applied Computing, pp. 575–581 (2006)
8. Resnick, P., Iacovou, N., Suchak, M., Bergstrom, P., Riedl, J.: GroupLens: an open architecture for collaborative filtering of netnews. In: Proceedings of the International Conference on ACM Computer Supported Cooperative Work, pp. 175–186 (1994)
9. Sarwar, B.M., Karypis, G., Konstan, J., Riedl, J.: Item-based collaborative filtering recommendation algorithms. In: Proceedings of the International Conference on World Wide Web, pp. 285–295, Hong Kong (2001)

10. Wang, J., de Vries, A.P., Reinders, M.J.T.: Unifying user-based and item-based collaborative filtering approaches by similarity fusion. In: Proceedings of the International Conference on Special Interest Group on Information Retrieval, pp. 501–508 (2006)
11. Xu, J.A., Araki, K.: A SVM-based personal recommendation system for TV programs. In: Proceedings of the International Conference on Multi-Media Modeling Conference, pp. 401–404 (2006)

Efficient Mining of Fuzzy Frequent Itemsets with Type-2 Membership Functions

Jerry Chun-Wei Lin[1(✉)], Xianbiao Lv[1], Philippe Fournier-Viger[2], Tsu-Yang Wu[1], and Tzung-Pei Hong[3,4]

[1] School of Computer Science and Technology, Harbin Institute of Technology Shenzhen Graduate School, Harbin, China
jerrylin@ieee.org, {lxbiao209,wutsuyang}@gmail.com

[2] School of Natural Sciences and Humanities, Harbin Institute of Technology Shenzhen Graduate School, Harbin, China
phifv@hitsz.edu.cn

[3] Department of Computer Science and Information Engineering, National University of Kaohsiung, Taiwan, China
tphong@nuk.edu.tw

[4] Department of Computer Science and Engineering, National Sun Yat-sen University, Taiwan, China

Abstract. In the past, the Apriori-based algorithm with fuzzy type-2 membership functions was designed for discovering fuzzy association rules, which is very time-consuming to generate-and-test candidates in a level-wise way. In this paper, we present a list-based fuzzy mining algorithm to mine the fuzzy frequent itemsets with fuzzy type-2 membership functions. A fuzzy-list structure and an efficient pruning strategy are respectively designed to speed up the mining process of fuzzy frequent itemsets. Several experiments are carried to verify the efficiency and effectiveness of the designed algorithm compared to the state-of-the-art Apriori-based algorithm in terms of runtime and number of traversal nodes (candidates).

Keywords: Type 2 fuzzy-sets · Fuzzy frequent itemsets · List structure · Data mining · Fuzzy association rules

1 Introduction

Association-rule mining (ARM) plays an important issue in data mining since it can reveal the relationships of items in the binary databases [1,4]. Apriori [3] is the first algorithm to address this issue and has two-phases to derive the required association rules based on minimum support and minimum confidence thresholds. To speed up the performance of Apriori algorithm, Han et al. presented the FP-growth algorithm with a FP-tree structure for mining the frequent itemsets without candidate generation [6]. Most previous works [1,3,4,6] whether ARM or frequent itemset mining (FIM) aims at handling the binary databases. In real-life applications, several items may purchase together, thus an item may attach

N.T. Nguyen et al. (Eds.): ACIIDS 2016, Part II, LNAI 9622, pp. 191–200, 2016.
DOI: 10.1007/978-3-662-49390-8_18

with its purchase quantity in the transactional database. In recent decades, the fuzzy-set theory [16] has been frequently used in intelligent system because its simplicity and similarity to human reasoning. Hong et al. designed an Apriori-like algorithm to level-wisely mine the fuzzy assassination rules [7]. Lin et al. developed three pattern-growth approaches namely FFPT [10], CFFPT [11], and UBFFPT [12], to mine the fuzzy frequent itemsets.

The above approaches focus on mining the fuzzy frequent itemsets with the membership functions of type-1 fuzzy sets. Mendel and John designed the type-2 fuzzy sets to consider the measurement of uncertainty [13]. Several algorithms have been extensively presented to adopt the fuzzy type-2 theory in different applications [8, 14]. Chen et al. first developed an algorithm for mining fuzzy association rules with fuzzy type-2 membership functions in a level-wise way [5]. In this paper, we first design an efficient list structure to discover the fuzzy frequent itemsets with type-2 membership functions. An efficient pruning is also developed to early prune the unpromising candidates for discovering the actual fuzzy frequent itemsets. Extensive experiments were conducted on both real-life and synthetic datasets to evaluate the performance of the proposed approach compared to the state-of-the-art fuzzy association-rule mining with type-2 membership functions [5] in terms of execution time and number of traversal nodes.

2 Related Work

Mining association rules is the fundamental task in data mining [1,3,4]. In recent decades, fuzzy-set theory [16] was adopted to handle the quantitative database for mining the required information. Kuok et al. introduced a fuzzy mining algorithm to handle the quantitative attributes [9]. Hong et al. also proposed a fuzzy mining algorithm to mine fuzzy rules from quantitative databases based on the generate-and-test mechanism [7]. Lin et al. presented the fuzzy frequent pattern (FFP)-tree [10], the compressed fuzzy frequent pattern (CFFP)-tree [11], and the upper bound fuzzy frequent pattern tree (UBFFPT) [12] algorithms to respectively mine the fuzzy frequent itemsets based on their designed specific tree structures.

The above studies focus on the type-1 fuzzy sets without considering the uncertain factor. The membership functions used in the type-1 fuzzy sets are totally crisp, which is insufficient to handle the uncertain models. In this situation, Mendel and John developed the type-2 fuzzy sets for dealing the uncertainty problems [13]. Chen et al. first presented the state-of-the-art approach to mine the fuzzy association rules with the type-2 membership functions [5]. This approach is, however, very time-consuming since the amount of candidates are required to be generated and examined at each k-level ($k \geq 2$).

3 Preliminaries and Problem Statement

3.1 Preliminaries

Let $I = \{i_1, i_2, \ldots, i_m\}$ be a finite set of m distinct items. A quantitative database is a set of transactions $D = \{T_1, T_2, \ldots, T_n\}$, where each transaction

$T_q \in D (1 \leq q \leq m)$ is a subset of I and has a unique identifier q, called its *TID*. Besides, each item i_j in a transaction T_q has a purchase quantity denoted as $q(i_j, T_q)$. A set of k distinct items $X = \{i_1, i_2, \ldots, i_k\}$ such that $X \subseteq I$ is said to be a k-itemset, where k is the length of the itemset. An itemset X is said to be contained in a transaction T_q if $X \subseteq T_q$. A minimum support threshold is set as δ according to users' preference. An illustrated example is shown in Table 1 as the running example in this paper.

Table 1. A quantitative database.

TID	Transaction (item, quantity)
1	A:4, C:1, E:5
2	C:1, E:3
3	C:2, D:2, E:1
4	A:1, C:3, F:2
5	A:5, B:3, C:3, E:4, F:5
6	B:1, C:4, F:4
7	A:4, C:2
8	C:4, D:3, F:4
9	A:4, C:2, D:1, F:3
10	C:5, D:5

A type-2 membership functions ($= \mu$) used in this paper are illustrated in Fig. 1.

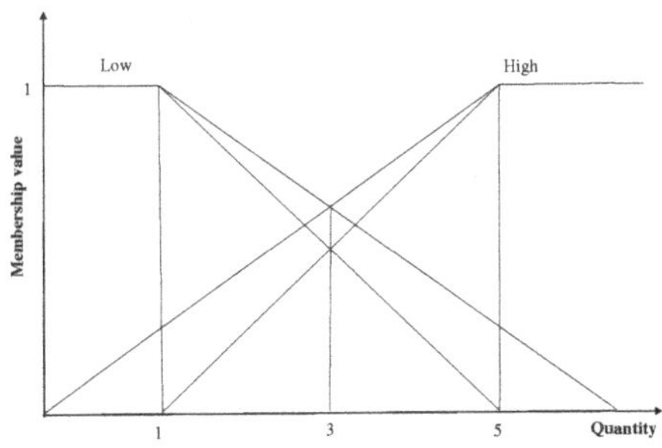

Fig. 1. Type-2 membership functions of linguistic 2-terms.

Definition 1. The linguistic variable i ($i \subset I$) is an attribute of a quantitative database whose value is the set of fuzzy terms represented in natural language as $(R_{i1}, R_{i2}, \ldots, R_{ih})$ and can be defined in the membership functions.

Definition 2. The quantitative value of i denoted as v_{iq}, is the quantitative of the item (linguistic variable) i in transaction T_q.

Definition 3. The fuzzy set, denoted as f_{iq}, is the set of fuzzy terms with their membership degrees (fuzzy values) transformed from the quantitative value v_{iq} of the linguistic variable i by the membership functions μ as:

$$f_{iq} = \mu_i(v_{iq})(= \frac{(fv_{iq1}^{lower}, fv_{iq1}^{upper})}{R_{i1}} + \frac{(fv_{iq2}^{lower}, fv_{iq2}^{upper})}{R_{i2}} + \cdots + \frac{(fv_{iqh}^{lower}, fv_{iqh}^{upper})}{R_{ih}}), \tag{1}$$

where h is the number of fuzzy terms of i transformed by μ, R_{il} is the l-th fuzzy terms of i, fv_{iql}^{lower} is the lower membership degree (fuzzy value) of v_{iq} of i in the l-th fuzzy terms R_{il}, fv_{iql}^{upper} is the upper membership degree (fuzzy value) of v_{iq} of i in the l-th fuzzy terms R_{il}, $fv_{iql}^{lower} \leq fv_{iql}^{upper}$, and $fv_{iql}^{lower}, fv_{iql}^{upper} \subseteq [0, 1]$.

In the past, Chen et al. first designed an Apriori-based algorithm to level-wise mine the fuzzy association rules (FARs) with fuzzy type-2 membership functions [5]. Based on this methodology, the problem statement is stated below.

3.2 Problem Statement

The problem of fuzzy frequent itemset mining (FFIM) is to discover the complete set of fuzzy frequent itemsets from the quantitative databases as:

$$FFIs \leftarrow \{sup(X) \geq \delta \times |D|\}, \tag{2}$$

in which X is a fuzzy linguistic term (can also be considered as an itemset), $sup(X)$ is the support count of X, δ is the minimum support threshold, and $|D|$ is the size of the quantitative database.

4 Proposed Fuzzy-List Structure and Mining Algorithm

4.1 Fuzzy-List Construction

To construct the fuzzy-list structure, the items in the quantitative database are first transformed into several linguistic terms based on the pre-defined fuzzy type-2 membership functions. After that, the type-2 fuzzy values of a linguistic term is then reduced as a type-1 fuzzy value by centroid type-reduction method with the given split number of point [5]. The linguistic term is thus kept and concerned as the fuzzy frequent 1-itemsets (1-FFIs) if its fuzzy (cardinality) value is no less than the pre-defined minimum support count. The database is then revisited to remove the non-fuzzy-frequent 1-itemsets for constructing the fuzzy-list structure. Each remaining 1-FFI is required to build its related fuzzy-list structure for keeping the necessary information.

Definition 4. Let X be the set of the linguistic terms, and let T be a transaction with the condition as $X \subseteq T$. The set of all fuzzy terms after X in T is denoted as T/X.

Definition 5. Let i be the fuzzy linguistic term of X with the condition as $i \subseteq T/X$. The fuzzy value of i in T can be represented as $fv(i, T)$.

Definition 6. The maximum remaining fuzzy value of X in T, denoted as $mrfv(X, T)$, is the maximum fuzzy membership value of all terms in T/X as: $mrfv(X, T) = max\{fv(i, T/X)\}$. Each element in the fuzzy list of term X contains three ordered fields as: *tid*, fv, and $mrfv$.

- *tid* indicates the term X is contained in transaction T.
- *fv* indicates the fuzzy membership value of X in T.
- *mrfv* indicates the maximum remaining fuzzy membership value after X in T.

The constructed fuzzy-list structure from the given example is thus shown in Fig. 2.

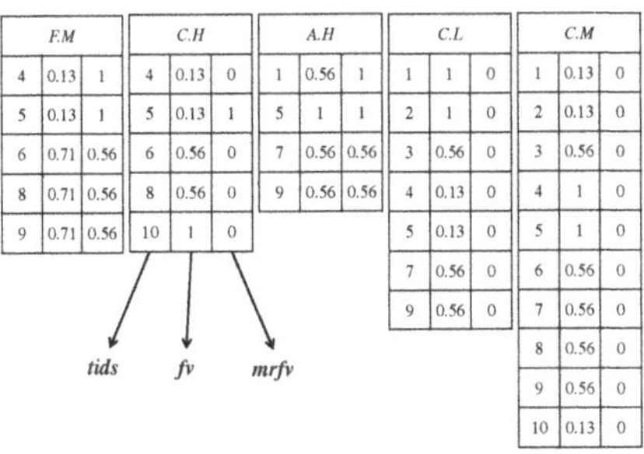

Fig. 2. The initial constructed fuzzy-list structure.

F.M			C.H			A.H			C.L			C.M		
4	0.13	1	4	0.13	0	1	0.56	1	1	1	0	1	0.13	0
5	0.13	1	5	0.13	1	5	1	1	2	1	0	2	0.13	0
6	0.71	0.56	6	0.56	0	7	0.56	0.56	3	0.56	0	3	0.56	0
8	0.71	0.56	8	0.56	0	9	0.56	0.56	4	0.13	0	4	1	0
9	0.71	0.56	10	1	0				5	0.13	0	5	1	0
									7	0.56	0	6	0.56	0
									9	0.56	0	7	0.56	0
												8	0.56	0
												9	0.56	0
												10	0.13	0

For example, in Fig. 2, the fuzzy term $\{F.M\}$ exists in transactions T_4, T_5, T_6, T_8, and T_9; its fuzzy membership value in T_4 is $fv(\{F.M\}, T_4)(= 0.13)$, and its maximum remaining fuzzy membership value is $mrfv\{(0.13,\ 1,\ 1)\}(= 1)$.

For the combination process of k-itemsets ($k \geq 2$), the tids of (k-1)-itemsets are required to check for discovering the same transaction IDs by intersection operation. Besides, the same linguistic variable of the fuzzy terms cannot be used for generating the fuzzy-list structure. Details of the fuzzy-list construction algorithm is shown in Algorithm 1.

In Algorithm 1, suppose two terms of P_x and P_y are used to generate the fuzzy-list structure of P_{xy}. First, the fuzzy terms are then checked to determine whether their linguistic variable is the same or not for generating the valid P_{xy} fuzzy-list structures as $P_{xy}.FL$. If the two terms of P_x and P_y have the same *transaction IDs*, the intersection operation is processed to find every fv in T. Besides, the maximum operation is also adopted to find the *maximum remaining fv* ($mrfv$) of the P_{xy} in T. Note that if the sum of fv is no larger than the pre-defined minimum support count, it is not considered as the fuzzy frequent itemset and will be directly removed. This process is recursively performed until no candidates can be combined.

Algorithm 1. fuzzy-list construction

Input: $P_x.FL$, the fuzzy list of P_x; $P_y.FL$, the fuzzy list of P_y.
Output: $P_{xy}.FL$, the fuzzy list of $P_x y$.

set $P_{xy}.FL \leftarrow null$;
if *x, y are not from the same i* **then**
  for *each element* $E \in P_x.FL$ **do**
    if $\exists E_y \in P_y.FL \wedge E_x.id == E_y.id$ **then**
      $E_{xy} \leftarrow (E_x.tid, min(E_x.fv, E_y.fv), E_y.mrfv)$;
      $P_{xy}.FL \leftarrow P_{xy}.FL \bigcup E_{xy}$;
  return $P_{xy}.FL$;

4.2 Pruning Strategy

The search space of the designed algorithm is based on the enumeration tree. The remaining 1-FFIs are then sorted in ascending order by their support counts (the sum of the *fv* values of a linguistic term). An example of the enumeration tree is shown in Fig. 3.

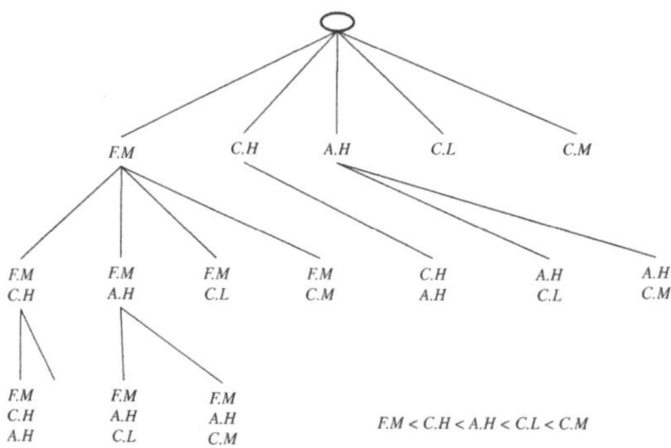

Fig. 3. The search space of the designed algorithm.

Since the search space to discover the promising fuzzy frequent itemsets is very huge in the enumeration, a pruning strategy is designed to early reduce the unpromising candidates, thus speeding up the computations.

Theorem 1. *For a fuzzy frequent itemset X in the enumeration tree, if the sum of the X.mrfv is less than the minimum support count, any supersets (extensions) of X cannot be considered as the promising fuzzy frequent itemsets and can be early pruned.*

Based on this pruning strategy, it can greatly reduce the search space for mining the fuzzy frequent itemsets. The designed algorithm for mining fuzzy frequent itemsets is described in Algorithm 2.

Algorithm 2. The designed mining algorithm

Input: *FLs*, the fuzzy list of 1-itemsets; δ, the minimum support threshold.
Output: *FFIs*, the set of the discovered fuzzy frequent itemsets.

for *each list* $X \in FLs$ **do**
  if $\sum_{X \subset T_q \wedge T_q \in D} fv(X) \geq \delta \times |D|$ **then**
    $FFIs \leftarrow FFIs \bigcup X$;
    if $\sum_{X \subset T_q \wedge T_q \in D} mrfv(X) \geq \delta \times |D|$ **then**
      $exFLS \leftarrow null$;
      for *each fuzzy list* Y *after* X *in* FLs **do**
        $exFLs = exFLs + Constrcut(X, Y)$;
      $mining(exFLs, \delta)$;

return *FFIs*;

5 Experimental Results

In this section, the proposed algorithm is evaluated compared to the state-of-the-art Apriori-based algorithm [5] in several datasets. Three real-life chess [15], mushroom [15], foodmart [15] datasets and one synthetic T10I4D100K dataset [2] were used in the experiments. The quantitative value of each item in the datasets is first transformed by the pre-defined fuzzy type-2 membership functions shown in Fig. 1. The characteristics of these datasets are shown in Table 2.

Table 2. Characteristics of used datasets.

Dataset	#**|D|**	#**|I|**	AvgLen	MaxLen	Type
Chess	3196	75	37	37	condense
Mushroom	8124	119	23	23	condense
Foodmart	21556	1559	4	11	sparse
T10I4D100K	100000	942	10.1	29	sparse

5.1 Runtime

In this section, the runtime of the compared algorithms for 2-terms membership functions w.r.t. different minimum support thresholds is shown in Fig. 4. From the results in Fig. 4, it can be observed that the designed algorithm has better results in term of runtime compared to the state-of-the-art Apriori-based algorithm [5] for mining fuzzy frequent itemsets based on the fuzzy type-2 membership functions in all datasets. When the minimum support threshold is increased, the runtime of the compared algorithms is decreased. This is reasonable since when the minimum support threshold is set higher, the number of the discovered rules or information is thus reduced. For example in Fig. 4(c), when the minimum support threshold are respectively set as 13 % and 16 %, the number of the discovered fuzzy frequent itemsets are respectively 2,171 and 917; the runtime of the designed algorithm takes 76.0 and 17.2 s but the state-of-the-art Apriori algorithm requires 1,094 and 188 s based on its designed level-wise approach. Thanks to the designed list structure, the necessary information can be kept without candidate generation in a level-wise way.

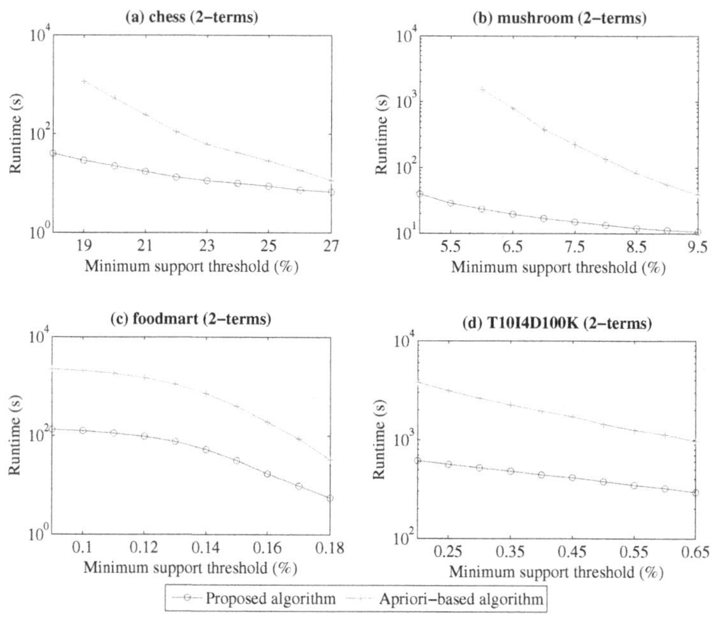

Fig. 4. Runtime w.r.t. variants of minimum support thresholds.

5.2 Number of Traversal Nodes

The number of generated candidates of the Apriori-based algorithm [5] and the number of traversal nodes in the enumeration tree of the designed algorithm is then compared to show the efficiency of the designed pruning strategy. The results for fuzzy 2-terms membership functions are shown in Fig. 5. From Fig. 5, we can observe that the traversal nodes of the designed algorithm in the enumeration tree is less than that of the Apriori-based algorithm. For example in Fig. 5(c), when the minimum support threshold is set as 0.1 %, the traversal nodes of the designed algorithm is 4,290,569 and the Apriori-based algorithm requires to determine 4,316,185 candidates in the level-wise way. For more sparse datasets, the proposed algorithm still performs well in the foodmart and T10I4D100K datasets and has nearly the same number of determined candidates shown in Fig. 5(c) and (d) especially when the minimum support threshold is set lower. The reason is that when the minimum support threshold is set lower or in a very sparse dataset, more information is necessary to be kept in the list structure for speeding up the computations based on the designed algorithm.

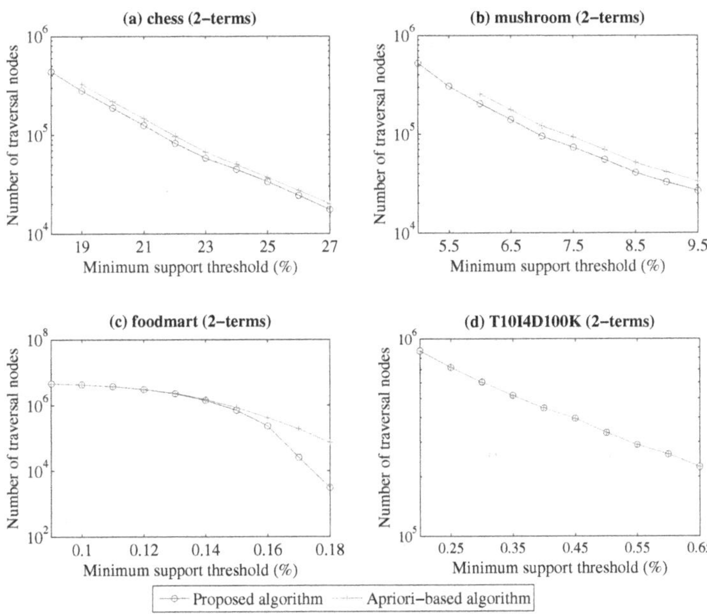

Fig. 5. Number of the traversal nodes (candidates) w.r.t. variants of minimum support thresholds.

6 Conclusion

In this paper, we present an efficient algorithm to mine the fuzzy type-2 frequent itemsets based the designed fuzzy-list structure. A pruning strategy is also developed to reduce the size of search space for early pruning the unpromising candidates of the fuzzy frequent itemsets. From the experimental results, it can be observed that the designed algorithm has better performance in terms of the runtime and the number of traversal nodes compared to the state-of-the-art Apriori-based algorithm.

Acknowledgements. This research was partially supported by the National Natural Science Foundation of China (NSFC) under grant No.61503092, by the Tencent Project under grant CCF-TencentRAGR20140114, and by the Shenzhen Strategic Emerging Industries Program under grant ZDSY20120613125016389.

References

1. Agrawal, R., Imieliski, T., Swami, A.: Mining association rules between sets of items in large databases. ACM SIGMOD Record, pp. 207–216 (1993)
2. Agrawal, R., Srikant, R.: Quest synthetic data generator (1994). http://www.Almaden.ibm.com/cs/quest/syndata.html
3. Agrawal, R., Srikant, R.: Fast algorithms for mining association rules in large databases. In: The International Conference on Very Large Databases, pp. 487–499 (1994)
4. Chen, M.S., Han, J., Yu, P.S.: Data mining: an overview from a database perspective. IEEE Trans. Knowl. Data Eng. **8**(6), 866–883 (1996)
5. Chen, C.-H., Hong, T.-P., Li, Y.: Fuzzy association rule mining with type-2 membership functions. In: Nguyen, N.T., Trawiński, B., Kosala, R. (eds.) ACIIDS 2015. LNCS, vol. 9012, pp. 128–134. Springer, Heidelberg (2015)
6. Han, J., Pei, J., Yin, Y., Mao, R.: Mining frequent patterns without candidate generation: a frequent-pattern tree approach. Data Min. Knowl. Discov. **8**, 53–87 (2004)
7. Hong, T.P., Kuo, C.S., Chi, S.C.: Mining association rules from quantitative data. Intell. Data Anal. **3**, 363–376 (1999)
8. Karnik, N.N., Mendel, J.M.: Introduction to type-2 fuzzy logic systems. In: International Conference on Fuzzy Systems, pp. 915–920 (1998)
9. Kuok, C.M., Fu, A., Wong, M.H.: Mining fuzzy association rules in databases. ACM SIGMOD Record **27**, 41–46 (1998)
10. Lin, C.W., Hong, T.P., Lu, W.H.: Linguistic data mining with fuzzy FP-trees. Expert Syst. Appl. **37**, 4560–4567 (2010)
11. Lin, J.C.W., Hong, T.P., Lin, T.C.: A CMFFP-tree algorithm to mine complete multiple fuzzy frequent itemsets. Appl. Soft Comput. **28**, 431–439 (2015)
12. Lin, C.W., Hong, T.P.: Mining fuzzy frequent itemsets based on UBFFP trees. J. Intell. Fuzzy Syst. **27**, 535–548 (2014)
13. Mendel, J.M., John, R.I.B.: Type-2 fuzzy sets made simple. IEEE Trans. Fuzzy Syst. **10**, 117–127 (2002)
14. Melin, P., Castillo, O.: A review on the applications of type-2 fuzzy logic in classification and pattern recognition. Expert Syst. Appl. **40**, 5413–5423 (2013)
15. SPMF: an open-source data mining library. http://www.philippe-fournier-viger.com/spmf/index.php
16. Zadeh, L.A.: Fuzzy sets. Inf. Control **8**, 338–353 (1965)

Improving the Performance of Collaborative Filtering with Category-Specific Neighborhood

Karnam Dileep Kumar[1(✉)], Polepalli Krishna Reddy[1], Pailla Balakrishna Reddy[1], and Longbing Cao[2]

[1] International Institute of Information Technology Hyderabad, Hyderabad, Telangana, India
karnamdileep1993@gmail.com, pkreddy@iiit.ac.in, balakrishnar@gmail.com
[2] Advanced Analytics Institute, University of Technology, Sydney, Australia
Longbing.Cao@uts.edu.au

Abstract. Recommender system (RS) helps customers to select appropriate products from millions of products and has become a key component in e-commerce systems. Collaborative filtering (CF) based approaches are widely employed to build RSs. In CF, recommendation to the target user is computed after forming the corresponding neighbourhood of users. Neighborhood of a target user is extracted based on the similarity between the product rating vector of the target user and the product rating vectors of individual users. In CF, the methodology employed for neighborhood formation influences the performance. In this paper, we have made an effort to improve the performance of CF by proposing a different approach to compute recommendations by considering two kinds of neighborhood. One is the neighborhood by considering the product ratings of the user as a single vector and the other is based on the neighborhood of the corresponding virtual users. For the target user, the virtual users are formed by dividing the ratings based on the category of products. We have proposed a combined approach to compute better recommendations by considering both kinds of neighborhoods. The experiments results on real world MovieLens dataset show that the proposed approach improves the performance over CF.

Keywords: Data mining · Recommender systems · Collaborative filtering

1 Introduction

Recommender system (RS) technology has emerged to help customers to select appropriate products from millions of products in e-commerce environment [12]. RS provides personalized information to customers and reduces the information overload. In the e-commerce environment, RS receives information from the customer about which products he/she is interested in, and recommends products that are likely to fit his/her needs. In the modern internet era, almost

N.T. Nguyen et al. (Eds.): ACIIDS 2016, Part II, LNAI 9622, pp. 201–210, 2016.
DOI: 10.1007/978-3-662-49390-8_19

all businesses are providing their services online and employing RS to recommend products and services. Improving the efficiency of RSs is an active area of research [1,18].

In the literature, collaborative filtering (CF) approach is widely employed in building RS [14]. There are two types of CF approaches: model-based and memory-based. Model-based CF [5,8] uses user-item ratings to build a predictive model. Memory-based CF [12,13] relies on user's history to predict items of interest to the user. In this paper, we have made an effort to improve the performance of memory-based CF.

In CF, recommendations of the target user are predicted by forming the corresponding neighborhood which is computed based on the similarity between the product rating vector of the target user and the product rating vectors of individual users. Neighborhood computation is the key step in CF which influences the performance. In CF, it can be observed that the neighborhood of the target user is computed by considering all the ratings of a user as a single unit. If the user rates certain number of products, equal weightage is given for each product rating value of the vector for computing the neighbourhood. However, it can be observed that the products could be grouped into categories and user may rate different number of products in each category. For example, consider that a user rates more movies in *comedy* category and very few movies in each of the other categories. The performance could be improved, if we give more weightage to the movies rated in *comedy* category over other categories while computing neighborhood. In the literature, the category-based collaborative filtering (CCF) approach [15] has been proposed by employing the notion of category-specific neighbourhood. In CCF, virtual users are formed by dividing the ratings of the user based on the category of product. The recommendations for the target user are computed based on the neighborhood of the corresponding virtual users.

It can be observed that the recommendations to the target user are computed using different kinds of neighbourhood in CF and CCF. Normally, user likes products with varying degree of aspects. So, user may like some products recommended by CF and some more products recommended by CCF. In this paper, rather than recommending products by using either CF or CCF approach, we propose that it is possible to obtain the better performance by combining the recommendations of both approaches. The experimental results on real world MovieLens dataset show that the proposed approach improves the performance over CF.

The rest of the paper is organized as follows. In Sect. 2, we explain the related work. In Sect. 3, we explain CF and CCF. In Sect. 4, we explain the proposed approach. We present experimental results in Sect. 5. In the last section we provide summary and conclusions.

2 Related Work

In [3], CF is introduced to provide suggestions to the members of closely related community. A survey [1] contains various CF approaches along with their limitations and possible extensions. Several types of CF based social recommender systems are discussed in [18].

In user-based CF [12], recommendations for the target user are computed based on the products rated by neighborhood users. In [2,13], item-based CF algorithms are discussed. In this model, similarity between items is taken as a measure to form neighborhood for an item. Recommendations for a user are computed using neighborhoods of products rated by the user. An approach, [4] proposing adjustments to CF with application of weight on similarity coefficient on target user rating using previous ratings of the user.

In [6], Koren proposes improvements to two models of CF. Extensions to latent vector model is proposed to integrate feedback and the neighborhood model is changed to optimize a global cost function. These improvements enables the integration of both models to maximize the performance. There are efforts [7, 10] to cope with the changes in user interests and trust over time. In these models, users and products pattern of ratings are tracked through a time period to provide recommendations. A group recommender system model is proposed in [11], in which, contextual information from users and items is used along with constraining similarity to factors like demographic information, interactions, etc. An evolutionary approach to automate the choice of various techniques available in CF depending on the variation in the data is proposed in [16]. A recursive prediction algorithm is proposed to overcome the data sparseness problem in [19] which uses second level neighbors i.e., neighbors to nearest neighbors to influence recommendations of items that are not rated by nearest neighbors.

There are research efforts to improve the performance of RS by combing the strengths of multiple methods. A unifying model of user-based and item-based CF has been proposed in [17]. In this model, unknown rating of a product by a user is predicted using the product ratings of the user and ratings of users similar to the target user ratings on items which are similar to the target item. An approach using a hybrid model of global and local similarity measures is proposed in [9] by employing transitive property to improve the user's neighborhood. An approach to exploit categories of products is proposed in [15]. In this approach, the target user is divided into as many virtual users as the categories of products he/she rated, recommendations for each virtual user are calculated by employing CF and the results from all the virtual users are combined to give recommendations to target user.

In this paper we have proposed an improved approach by combining CF and the approach proposed by exploiting category of products. The proposed approach is different from other hybrid approaches as it exploits the positive aspects of CF and CF with category-specific neighbourhood.

3 Overview of CF and Category-Based CF

In this section we briefly explain the CF [12] and category-based CF [15].

3.1 Collaborative Filtering

User based CF works on the idea that user likes the products that are liked by the users who are similar to him/her. Similarity measures like cosine similarity, pearson correlation, etc., can be employed to find similarity between a pair users given their product rating vectors. Neighborhood for the given user is formed by computing similarity value with other users. Recommendations are computed from the products rated by the users in neighborhood. The CF approach contains three steps: i.e., data representation, neighborhood formation and generation of recommendations.

i. **Representation:** For m users and n products, user-product rating data is represented by a two dimensional matrix R_{CF} of size $m \times n$. Rating of user i for a product j is represented as $r_{i,j}$ (value in ith row and jth column of matrix R_{CF}).
ii. **Neighbourhood Formation:** For each user i, we calculate similarity with each user j and compute similarity matrix S_{CF}. For a user i, we calculate the neighbourhood by selecting top-K similar users.
iii. **Recommendations:** For a user i, we compute the support of each item from the neighborhood. The top-N frequent items that are not yet rated by the target user are given as recommendations. This is called *most frequent item method* of recommendation generation.

3.2 Category-Based CF

The basic idea of category-based CF (CCF) is as follows. Normally, a typical user rates more number of products in certain categories and rates few products among other categories. In this model, user is divided into multiple virtual users depending on the category of products. For each virtual user, neighborhood is formed by computing the similarity between the target virtual user product ratings and the product ratings of the corresponding virtual users. The recommendation to the target user are computed as follows. First, recommendations for each virtual user are computed based on the corresponding neighborhood. The final recommendations are computed by combining the recommendations made for the corresponding virtual users. The steps of CCF are as follows.

1. **Representation:** Suppose there are m users, n products, and c product categories. For a category l ($0 < \mathrm{l} <= \mathrm{c}$), virtual user of a user i is created if the user rated any products belonging to the category l. Maximum c virtual users are created for a user. Let the total number of virtual users of all users be v. The virtual user-product rating data is represented by a two dimensional matrix R_{CCF} of size $v \times n$.
2. **Neighborhood Formation:** For each virtual user $v_{i,l}$, we calculate similarity with every other virtual user and produce similarity matrix S_{CCF}. For virtual user, we calculate the neighbourhood of by selecting top K similar virtual users.

3. **Recommendations:** Recommendations for each virtual user is formed using *most frequent item recommendation* method in virtual neighborhood. For a real user i, recommendations from the corresponding virtual users are merged to form the final recommendations. As proposed in [15], we can use random approach or ranking approach. In addition, we can also employ the support driven merging approach where the final recommendations are given based on the support (frequency count) of recommendations in virtual user neighborhood.

4 Proposed Approach

User likes products due to different aspects or criteria. A typical user may like a few products by considering one aspect and few other products by considering another aspect. If all recommendations are made by computing based one aspect, less quality recommendation could be included.

It can be observed that the recommendations made using CF captures the one aspect how target user likes the products. The CF approach tries to recommend the products to the target user if they are liked by other users who rate all products in a similar manner as that of the target user. It can also be observed that the recommendations made using CCF captures another aspect on how the target user likes the products. The CCF approach tries to recommend the target user if they are liked by other users who rate certain category of products in a similar manner as that of the target user.

Each of the CF and CCF approaches has the potential to recommend certain number of quality recommendations. As the recommendations from both CF and CCF are formed by different kinds of neighbourhood, even though some recommendations are common, there is a possibility that each approach extracts distinct recommendations. So, there is an opportunity to improve the performance if we combine the strengths of both CF and CCF.

In the proposed approach, we first compute the recommendations to the target user by following CF as well as CCF. Next, we compute quality recommendations by taking recommendation sets of CF and CCF as input.

4.1 Combined Approach

The proposed approach consists of three steps.

i. **Data Representation:** For m users, n products, and c product categories, we represent the data in a matrix R of size m $\times$ n where a product j is represented in a column as $< l, j >$ where l represent category of the product j, i.e., $r_{i,<l,j>}$ denotes the product rating of user i for product j, which belongs to category l.
ii. **Neighbourhood Formation:** In case of CF, we consider each row of R as a vector and compute similarity rating matrix S_{CF}. By using S_{CF}, the neighbourhood of target user is computed by identifying the required number of similar users.

In case of CCF, by treating user ratings of the products in that category x as vectors, we form user similarity matrix S_x. By using S_x, the neighbourhood of size K for a target virtual user is computed by identifying the top similar virtual users.

iii. **Generation of Recommendations:** The steps to compute final recommendations are as follows.

For a target user u_i, we compute the recommendations using CF and corresponding support of each recommendation j as $\mathbf{SCF}_{i,j}$. For u_i, we also compute recommendations using CCF and support of each recommendation as the frequency count of a recommendation in the neighborhood of the corresponding virtual user ($\mathbf{SCCF}_{i,j}$).

We propose two approaches to combine the recommendations of CF(u_i) and CCF(u_i): Maximum Combined Approach (Maximum-CA) and Weighted Combined Approach (Weighted-CA). Top-N recommendations from each of CF and CCF are given as input. Let N represent the required number of recommendations.

- **Maximum-CA:** We consider recommendations is of high quality based on the support value of recommendations computed through CF and CCF. If an item belongs to both sets, we assign the maximum of the two supports. Top-N recommendations are provided in the decreasing order of support value.
- **Weighted-CA:** We compute the recommendations based on the combined support value in CF(u_i) and CCF(u_i). We introduce a weight b ($0 \leq b \leq 1$) and compute the final support for the recommendation j as $\boldsymbol{b}$***($\mathbf{SCCF}_{i,j}$) + (1-$\boldsymbol{b}$)*($\mathbf{SCF}_{i,j}$)**. Top-N recommendations are provided in decreasing order of the combined support.

5 Experimental Evaluation

In this section we explain about the data set, evaluation metrics and present the results. We conducted experiments on real world MovieLens dataset[1]. The dataset consists of 100,000 ratings by 943 users on 1682 movies. A total of 18 genres serve as categories. Five fold cross validation is done where the dataset is divided into five folds and four folds are used as training set and results of experiments are tested against the unknown fifth fold (test data).

For a user in training set we obtain final *top-N* (where ***N*** is the number of recommendations) recommendations. The ***test set*** represents user ratings in test data. The ***hit Set*** (HS) represents the intersection of items between *top-N* set i.e., $Hitset = Test \bigcap top - N$.

We employ precision, recall and F1 score as evaluation metrics. Precision is defined as ratio of *hit set* size to the *top-N* set size and equal to $\frac{|HS|}{|N|}$. Recall is defined as ratio of hit set size to the test set size and equal to $\frac{|HS|}{|testset|}$. F1 Score is determined as the combination of precision and recall and equal to

[1] http://grouplens.org/datasets/movielens/.

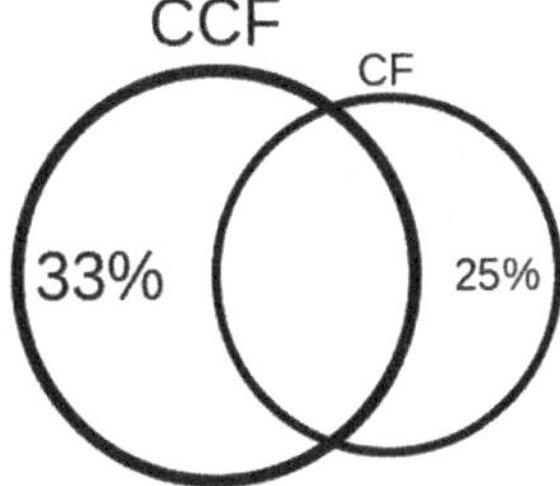

Fig. 1. Depiction of distinct and common hits

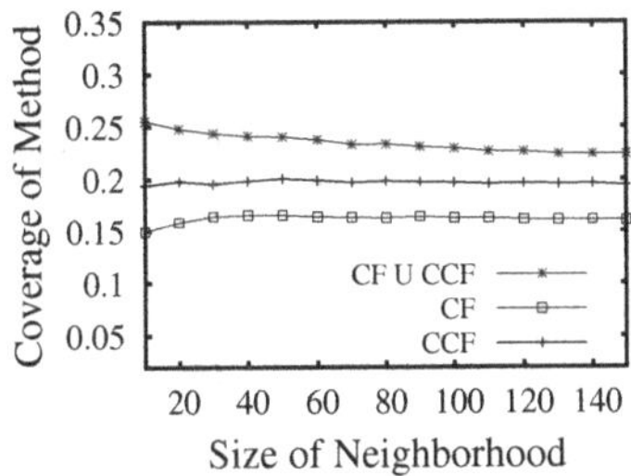

Fig. 2. Coverage

$\frac{2*recall*precision}{precision+recall}$. For a given approach and the test set, we use the notion of ***Coverage*** which is equal to the percentage of relevant movies recommended/number of movies in the test set.

The number of recommendations is fixed at $N = 10$. For CF and CCF, Center-based method is used for the extraction of neighborhood. For CF, *most frequent item recommendation* is used to generate recommendations. In the recommendations step of implementation of CCF, we used support of items in virtual neighborhood to provide recommendations. Recommendations of all virtual users are taken in a set and sorted in decreasing order of support and top-N recommendations are taken as the final set of recommendations rather than taking recommendations at random as in [15].

Analysis of Hit Sets of CF and CCF: Figure 1 depicts the percentage of distinct hits and common hits of both approaches. We have conducted experiments to compute the recommendations and *hit set* for every user by fixing the size of neighborhood at $K = 70$ users in CF. For each virtual user of CCF also, we fix $K = 70$ virtual users. Cosine based similarity is used to compute similarity. In this experiment, it can be observed that 33 % of hits by CCF are distinct and 25 % hits of CF are distinct.

Figure 2 shows the coverage of CF, CCF and combination of CF and CCF which is the union of recommendations formed through CF and CCF (CF ∪ CCF). From the graph it can be observed that CCF coverage is greater to that of CF indicating that CCF could perform better than CF. If CCF is improving CF by adding the new recommendations, coverage of CCF ∪ CF should be the same as that of CCF. However, CF ∪ CCF coverage is greater than either of CF or CCF. This shows that the recommendations formed through CCF are distinct from that of CF due to the differences in the methodology of computing the neighborhood. These results also show that there is a scope to improve performance by combining CF and CCF.

Performance of CF, CCF and Proposed Approach: We have conducted the experiments by varying the size of neighborhood K. Figure 3(a) shows the precision performance of CF, CCF and Maximum-CA approaches. It can be observed that the performance of Maximum-CA approach is improved significantly over both CF and CCF. This is due to the fact that the proposed

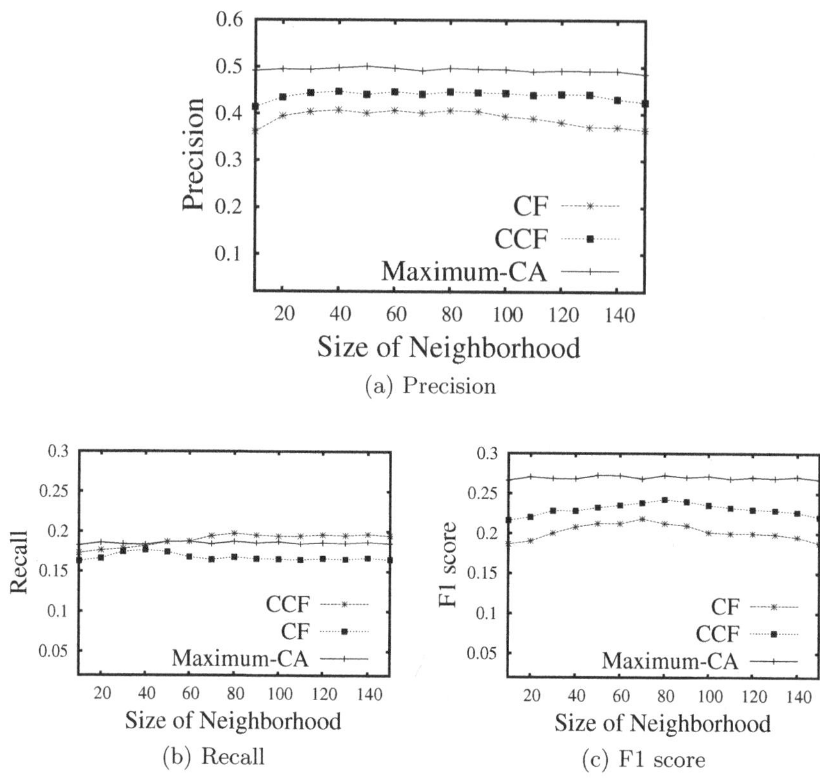

(a) Precision

(b) Recall

(c) F1 score

Fig. 3. Performance

approach combines the strengths of both approaches and increases the size of the hit set. Figure 3(b) and (c) shows the recall and f1 score performance respectively. It can be observed that, CCF performs slightly better than the proposed approach in terms of recall which may be due to the disproportionate selection of CF recommendations over CCF recommendations. Weighted-CA uses proportionality factor to solve this problem. Overall it can be observed that the maximum-CA improved f1 score performance significantly.

Weighted-CA Approach: We conducted experiments by fixing $K = 70$ and varying the weight b. Figure 4(a) shows precision performance of CF, CCF, Maximum-CA and Weighted-CA approaches. It can be noted that Maximum-CA performs better than both CF or CCF. It can be observed that Weighted-CA performs slightly better than Maximum-CA at $b = 0.5$. This is because Weighted-CA allows some new recommendations that are not top-N in either of CF or CCF but have considerable support in both, such that sum of the support allows them in the set of final N recommendations. The performance of recall in Fig. 4(b) and f1-score in Fig. 4(c) follow the similar trend. For the movielens dataset, we obtained the maximum performance at $b \simeq 0.5$, but this may vary with datasets which will be investigated as a part of future work.

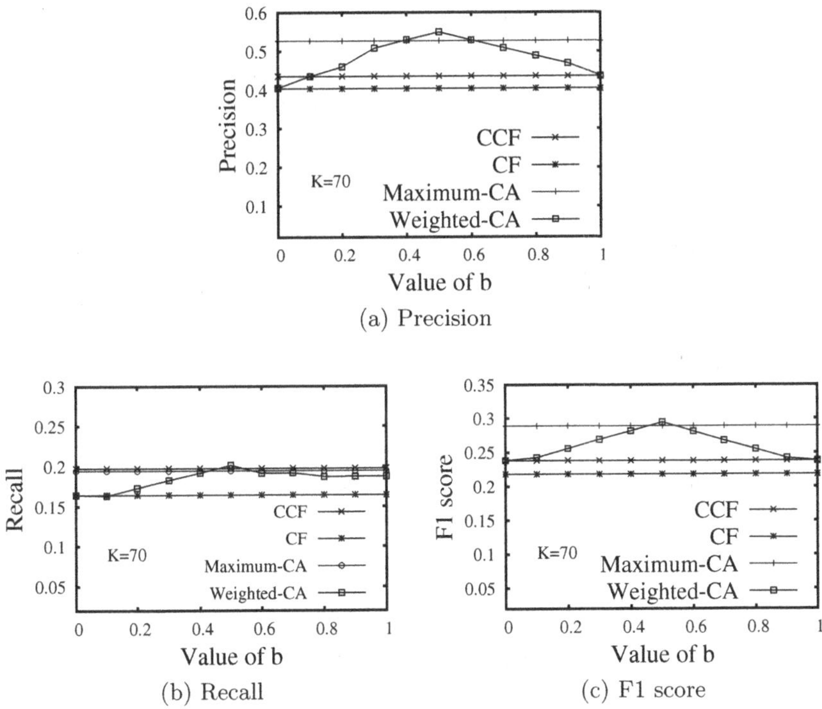

(a) Precision

(b) Recall

(c) F1 score

Fig. 4. Performance with variation of *b*

6 Conclusion

The collaborative filtering is one of the widely used methods in building recommender systems. By extending the notion of category-specific neighbourhood, we proposed an approach to improve the performance of CF. By combining the strengths of both CF and CF with category-specific neighbourhood, the proposed approach improves the recommendation performance. Through experiments results on Movielens data, it has been shown that it is possible to improve the performance of CF using the proposed approach.

As a part of future work, we are planning to conduct extensive experiments by considering different kinds of data sets. We explore how notion of category-specific neighbourhood works by considering concept hierarchy of the products. We also plan to explore how the notion of category-specific neighbourhood works to item-based and model based variations of collaborative filtering approaches.

References

1. Adomavicius, G., Tuzhilin, A.: Toward the next generation of recommender systems: a survey of the state-of-the-art and possible extensions. TKDE **17**(6), 734–749 (2005)

2. Deshpande, M., Karypis, G.: Item-based top-n recommendation algorithms. TOIS **22**(1), 143–177 (2004)
3. Goldberg, D., Nichols, D., Oki, B.M., Terry, D.: Using collaborative filtering to weave an information tapestry. ACM Commun. **35**(12), 61–70 (1992)
4. Herlocker, J.L.: Understanding and Improving Automated Collaborative Filtering Systems. Ph.D. thesis (2000)
5. Huang, Q.H., Ouyang, W.M.: Fuzzy collaborative filtering with multiple agents. J. Shanghai Univ. **11**, 290–295 (2007)
6. Koren, Y.: Factorization meets the neighborhood: a multifaceted collaborative filtering model. In: SIGKDD, pp. 426–434. ACM (2008)
7. Koren, Y.: Collaborative filtering with temporal dynamics. ACM Commun. **53**(4), 89–97 (2010)
8. Li, F., Xu, G., Cao, L.: Coupled item-based matrix factorization. In: Benatallah, B., Bestavros, A., Manolopoulos, Y., Vakali, A., Zhang, Y. (eds.) WISE 2014, Part I. LNCS, vol. 8786, pp. 1–14. Springer, Heidelberg (2014)
9. Lopes, A.R., Prudencio, R.B., Bezerra, B.L.: A collaborative filtering framework based on local and global similarities with similarity tie-breaking criteria. In: International Joint Conference on Neural Networks (IJCNN), pp. 2887–2893. IEEE (2014)
10. Min, S.H., Han, I.: Detection of the customer time-variant pattern for improving recommender systems. Expert Syst. Appl. **28**(2), 189–199 (2005)
11. Santos Jr., E.B., Goularte, R., Manzato, M.G., et al.: Personalized collaborative filtering: a neighborhood model based on contextual constraints. In: 29th Symposium on Applied Computing. ACM (2014)
12. Sarwar, B., Karypis, G., Konstan, J., Riedl, J.: Analysis of recommendation algorithms for e-commerce. In: 2nd Conference on Electronic Commerce, pp. 158–167. ACM (2000)
13. Sarwar, B., Karypis, G., Konstan, J., Riedl, J.: Item-based collaborative filtering recommendation algorithms. In: The 10th International Conference on World Wide Web, pp. 285–295. ACM (2001)
14. Schafer, J.B., Konstan, J., Riedl, J.: Recommender systems in e-commerce. In: Conference on Electronic Commerce, pp. 158–166. ACM (1999)
15. Sharma, M., Reddy, P.K., Kiran, R.U., Ragunathan, T.: Improving the performance of recommender system by exploiting the categories of products. In: Kikuchi, S., Madaan, A., Sachdeva, S., Bhalla, S. (eds.) DNIS 2011. LNCS, vol. 7108, pp. 137–146. Springer, Heidelberg (2011)
16. da Silva, E.Q., Camilo Junior, C.G., Pascoal, L.M.L., Rosa, T.C.: An evolutionary approach for combining results of recommender systems techniques based on collaborative filtering. In: Congress on Evolutionary Computation (CEC), pp. 959–966. IEEE (2014)
17. Wang, J., De Vries, A.P., Reinders, M.J.: Unifying user-based and item-based collaborative filtering approaches by similarity fusion. In: SIGIR, pp. 501–508. ACM (2006)
18. Yang, X., Guo, Y., Liu, Y., Steck, H.: A survey of collaborative filtering based social recommender systems. Comput. Commun. **41**, 1–10 (2014)
19. Zhang, J., Pu, P.: A recursive prediction algorithm for collaborative filtering recommender systems. In: Conference on Recommender Systems, pp. 57–64. ACM (2007)

Mining Drift of Fuzzy Membership Functions

Tzung-Pei Hong[1,2(✉)], Min-Thai Wu[1], Yan-Kang Li[1], and Chun-Hao Chen[3]

[1] Department of Computer Science and Information Engineering, National University of Kaohsiung, Kaohsiung City, Taiwan
tphong@nuk.edu.tw, wmt@wmt35.idv.tw, m1025506@mail.nuk.edu.tw

[2] Department of Computer Science and Engineering, National Sun Yat-sen University, Kaohsiung City, Taiwan

[3] Department of Computer Science and Information Engineering, Tamkang University, New Taipei, Taiwan
chchen6814@gmail.com

Abstract. In this paper, the fuzzy c-means (FCM) clustering approach is adopted to find concept drift of fuzzy membership functions. The proposed algorithm is divided into two stages. In the first stage, the FCM approach is used to find appropriate fuzzy membership functions at different periods or at different places. Then in the second stage, the proposed algorithm compares the results in the first stage to find different types of drift of fuzzy membership functions. Experiments are also made to show the performance of the proposed approach.

Keywords: Concept drift · Data mining · Fuzzy C-means · Membership functions

1 Introduction

Fuzzy data mining uses fuzzy membership functions to derive fuzzy frequent itemsets and association rules from quantitative data [20]. As the result, fuzzy membership functions play a crucial role that affects the quality of the results. In the past, the membership functions were usually static or manually defined. Some meta-heuristic approaches have thus been recently proposed to automatically obtain appropriate membership functions for mining. For example, Hong et al. proposed a genetic fuzzy mining algorithm to train the membership functions and attain association rules from quantitative data [1]. Yang et al. used a self-organizing feature map to generate fuzzy membership functions with unsupervised learning [2].

In this paper, we discuss the concept-drift problem of fuzzy membership functions. Membership functions may be modified due to the change of environments. The drift of membership functions can provide some useful information to decision makers. We thus design an approach with three criteria to judge the happening of the drift. First, the fuzzy c-means clustering approach is adopted to get the membership functions of each item in two databases at different time or different places. The results from the two different databases are then compared to derive possible concept-drift patterns.

N.T. Nguyen et al. (Eds.): ACIIDS 2016, Part II, LNAI 9622, pp. 211–218, 2016.
DOI: 10.1007/978-3-662-49390-8_20

2 Related Works

Concept drift becomes increasingly important in these years since it can provide decision makers varying information [4–7]. Tsymbal defined concept drift as finding patterns which changed over time in unexpected ways [3]. Mukkavilli et al. proposed a method to detect network attacks based on the pattern drift [6]. Hayat et al. then designed a model based on it to filter junk mails [5]. Lee et al. also proposed an approach with decision trees to mine concept-drift rules [8]. The concept-drift patterns were used to data classification and data streams in [7, 9–14]. Song et al. also defined three types of concept-drift patterns in association-rule mining [15].

In this paper, we try to find the concept drift of membership functions of items for fuzzy association rules. For data mining with quantitative transactions, Srikant and Agrawal proposed a method to handle quantitative transactions by partitioning possible values of each attribute [19]. Then some approaches [17, 18, 20] based on the fuzzy set theory [16] were then proposed to solve it. Since membership functions will greatly affect the results of fuzzy mining, Hong et al. thus proposed a genetic fuzzy approach to automatically derive them [1].

FCM is a popular method for clustering data [21]. It is based on the fuzzy set theory and allows one piece of data to belong to more than one cluster. FCM is based on the minimization of the following objective function:

$$J(U, c_1, c_2, \ldots, c_c) = \sum_{i=1}^{c} \sum_{j=1}^{n} (u_{ij})^m \left\| c_i - x_j \right\|^2,$$

where m is a parameter which is greater than 1, u_{ij} is the fuzzy value of a data point x_j in cluster i, c_i is the center of cluster i. Euclidean distance is commonly used. FCM uses an iterative procedure to update centers and membership functions until they converge.

3 Drift of Fuzzy Membership Functions

As mentioned above, the proposed algorithm is divided into two stages. In the first stage, the FCM approach is used to find appropriate fuzzy membership functions; in the second stage, the proposed algorithm compares the results to find drift. The two stages are stated below, respectively.

3.1 Generating Fuzzy Membership Functions

In this paper, we propose a simple method to generate a set of membership functions for each item by FCM. Each membership function is designed as an isosceles triangle and represented by a pair (c, w) as used in [1]. The peak of the triangle is located at c and the distance between c and its endpoints is w. If n membership functions are generated for each item, then the proposed algorithm will obtain n cluster centers by using FCM. Each center will be the peak c of a certain membership function. The span w is calculated as the distance between the current center and the one of the previous membership function. The w value of the first membership function is the distance between the peak and 0.

Membership functions play a critical role in converting the purchasing behavior of commodity items into semantic representation which is similar to human understanding. Figure 1 shows a simple example for apples purchased. It consists of three membership functions representing low, medium and high for different purchased amounts. If a customer buys five apples, the fuzzy value for "low" amount is 0.4, for "medium" amount is 0.6, and for "high" amount is 0.

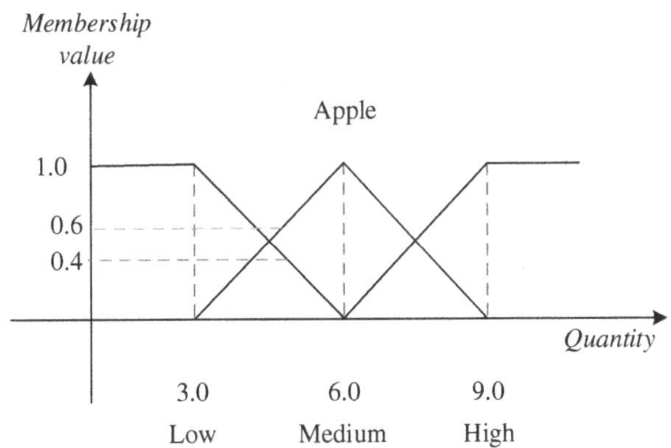

Fig. 1. A simple example of membership functions for the amounts of apples purchased.

3.2 Deriving Drift of Fuzzy Membership Functions

When membership functions of an item change significantly at different times or places, it usually means the customer behavior for the item has changed as well. The drift of fuzzy membership functions can thus provide some useful information for decision makers. In this section, we propose three different kinds of drifts of fuzzy membership functions. They are explained below.

(a) **The change of the representative value for a linguistic term.** Assume Fig. 1 shows the membership functions of the purchased amount of apples per transaction over the last year and Fig. 2 shows those for this year. When comparing the two figures, it can be easily seen that for the linguistic term of "low", the representative value, which is the center of the membership function for "low", reduces from three to two. On the contrary, for the linguistic term of "high", the representative value increases from nine to ten. It represents the concepts of the "low" and the "high" linguistic terms have already changed. In the example, the representative value of the "medium" linguistic term retains the same as the original one.
The change degree of the representative value (cr) of a linguistic term at different time or places is then measured as follows:

$$cr = \frac{\left| c_{nm}^{t} - c_{nm}^{t+k} \right|}{w_{nm}^{t}},$$

where c_{nm}^{t} and c_{nm}^{t+k} are the centers of the m-th linguistic term (membership function) of the n-th item for two different databases at time t and time $t + k$, respectively, and w_{nm}^{t} are the span of the m-th linguistic term of the n-th item for the database at time t. When the value of cr is larger than a predefined threshold, a drift occurs.

(b) **The change of the span for a linguistic term.** The range of a membership function is also important to a linguistic term. For the above example, although the representative value of the "medium" linguistic term has not been changed in Figs. 2 and 3, the range of the membership function for this year is larger than that in the previous year. The influence of the "medium" linguistic term in this year is then bigger than before. The change degree of the span (cs) of a linguistic term at different time or places is then measured as follows:

$$cs = \frac{\left| w_{nm}^{t} - w_{nm}^{t+k} \right|}{\left| c_{nM}^{t} - c_{n1}^{t} \right| / (M - 1)},$$

where M is the number of linguistic terms for an item, w_{nm}^{t} and w_{nm}^{t+k} are the spans of the m-th linguistic term (membership function) of the n-th item for two different databases at time t and time $t + k$, respectively, c_{nM}^{t} and c_{n1}^{t} are the centers of the last and the first linguistic terms of the n-th item for the database at time t. When the value of cs is larger than a predefined threshold, a drift occurs.

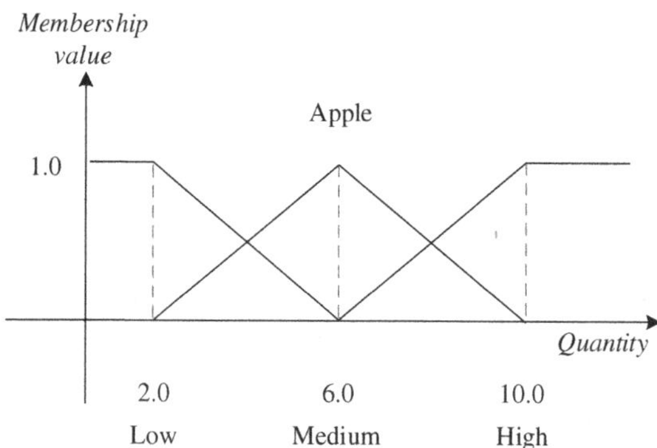

Fig. 2. Membership functions for the amounts of apples purchased at different time.

(c) **The change of the fuzzy support for a linguistic term.** The fuzzy support of a linguistic term is the summation of the membership values of the individual purchased amounts for an item in all the transactions from a database. A change in fuzzy support represents the group size significantly evolves for this linguistic term.

Algorithm: Finding Drift of Membership Functions (FDMF)

Input:

D^t, D^{t+k}: two databases at t and $t+k$;
N: the number of items;
M: the number of linguistic terms;
α: the threshold for the change of the representative value;
β: the threshold for the change of the span,
γ: the threshold for the change of the fuzzy support value.

Output: The three kinds of drift of membership functions.

Step 1: Generate fuzzy membership functions for each item from D^t and D^{t+k}, respectively, by the following sub-steps.

Step 1.1: Set $i = 1$, where i is used to keep the identity number of the current item processed.

Step 1.2: Tune the center points of M clusters by the fuzzy c-means algorithm for the i-th item and set the final centers as the centers of the M membership functions.

Step 1.3: Calculate the spans by the distance from the centers.

Step 1.4: Set $i = i + 1$. If $i \leq M$, go to **Step 1.2**; else, go to **Step 2**.

Step 2: Find the drift of fuzzy membership functions between D^t and D^{t+k} by the following sub-steps.

Step 2.1: Set the initial drift set S as ϕ.

Step 2.2: Set $i = 1$, where i is used to keep the identity number of the current item.

Step 2.3: Calculate the change degree (cr) of the representative value for the i-th linguistic term. If cr of a linguistic term is larger than α, then mark the drift as the first type and put it in S.

Step 2.4: Calculate the change degree (cs) of the span for the i-th linguistic term. If cs of a linguistic term is larger than β, then mark the drift as the second type and put it in S.

Step 2.5: Calculate the change degree (cf) of the fuzzy support for the i-th linguistic term. If cfs of a linguistic term is larger than γ, then mark the drift as the third type and put it in S.

Step 2.6: Set $i = i + 1$. If $i \leq M$, go to **Step 2.3**; else, go to **Step 3**.

Step 3: Output S.

Fig. 3. The algorithm of finding drift of membership functions.

The change degree of the fuzzy support (cf) of a linguistic term at different time or places is then measured as follows:

$$cf = \frac{|sup^t_{nm} - sup^{t+k}_{nm}|}{sup^t_{nm}},$$

where sup^t_{nm} and sup^{t+k}_{nm} are the fuzzy support values of the m-th linguistic term (membership function) of the n-th item for two different databases at time t and time $t + k$, respectively. When the value of cf is larger than a predefined threshold,

a drift occurs. This type of drift can easily represent change of customer behavior. For example, the following pattern may be formed: more people buy expensive mobile phones this year than last year.
The proposed algorithm will compare the membership functions of each item at different databases with predefined thresholds to find the drift. If any of the three kinds of change degree is larger than their individual thresholds, the related concept-drift patterns are immediately generated. The detailed algorithm is described in Fig. 3.

4 Experimental Results

In this section, a simulation dataset containing 1,559 items and 21,556 transactions was used in the experiments. In the data set, the number of purchased items in transactions was first randomly generated and the purchased items and their quantities in each transaction were then generated. Each transaction was also assigned a date in one year. The cluster size was set at 3, the parameter m in FCM was set at 2, the threshold α was set at 2, β was set at 1, and γ was set 1. The following three scenarios were used, with different pairs of databases:

1. We divided the first half of the whole year in the database as the first database and the second half as the second one.
2. We selected the transactions in a random month as a database and compared it with the original database (the whole year).
3. We selected the transactions in two random months as two databases.

Table 1 shows the number of drift patterns derived by the proposed approach in different scenarios. From the experimental results, it can be observed that the number of patterns in the first scenario is larger than that of the second scenario, which is then larger than that of the third scenario. It was because a short-term database usually contained more special rules, so when we compared membership functions from two short-term databases, more drift patterns could be found. On the contrast, since long-term databases tended to be stable, less drift patterns occurred. Thus, analyzing a long-term database could get more reliable drift patterns.

Table 1. The numbers of the three types of drift in different scenarios.

	Representative value	Span	Fuzzy support
First half with second half of a year	9	39	234
A random month with a whole year	40	112	274
Two random months	112	236	441

5 Conclusion

In the paper, a concept-drift approach for fuzzy membership functions is proposed for mining concept-drift patterns in different situations. Three types of patterns are defined and experiments are made to show the drift behavior under three scenarios.

In the future, we will design more effective ways to decrease computing time and combine the proposed concept-drift patterns with fuzzy association rules.

References

1. Hong, T.P., Chen, C.H., Wu, Y.L., Lee, Y.C.: A GA-based fuzzy mining approach to achieve a trade-off between number of rules and suitability of membership functions. Soft. Comput. **10**, 1091–1101 (2006)
2. Yang, C.C., Bose, N.: Generating fuzzy membership function with self-organizing feature map. Pattern Recogn. Lett. **27**, 356–365 (2006)
3. Tsymbal, A.: The problem of concept drift: definitions and related work. Technical Report (2004)
4. Widyantoro, D.H., Yen, J.: Relevant data expansion for learning concept drift from sparsely labeled data. IEEE Trans. Knowl. Data Eng. **17**, 401–412 (2005)
5. Hayat, M.Z., Basiri, J., Seyedhossein, L., Shakery, A.: Content-based concept drift detection for email spam filtering. In: International Symposium on Telecommunications, pp. 531–536 (2010)
6. Mukkavilli, S.K., Shetty, S.: Mining concept drifting network traffic in cloud computing environments. In: IEEE/ACM International Symposium on Cluster, Cloud and Grid Computing, pp. 721–722 (2012)
7. Sun, J., Li, H., Adeli, H.: Concept drift-oriented adaptive and dynamic support vector machine ensemble with time window in corporate financial risk prediction. IEEE Trans. Syst. Man Cybern. Syst. **43**, 801–813 (2013)
8. Lee, C.I., Tsai, C.J., Wu, J.H., Yang, W.P.: A decision tree-based approach to mining the rules of concept drift. In: Fuzzy Systems and Knowledge Discovery, pp. 639–643 (2007)
9. Thuraisingham, B.: Data mining for security applications: Mining concept-drifting data streams to detect peer to peer botnet traffic. In: Intelligence and Security Informatics (2008)
10. Hayat, M.Z., Hashemi, M.R.: A dct based approach for detecting novelty and concept drift in data streams. In: Soft Computing and Pattern Recognition, pp. 373–378 (2010)
11. Shetty, S., Mukkavilli, S.K., Keel, L.: An integrated machine learning and control theoretic model for mining concept-drifting data streams. In: Technologies for Homeland Security, pp. 75–80 (2011)
12. Patil, P.D., Kulkarni, P.: Adaptive supervised learning model for training set selection under concept drift data streams. In: Cloud & Ubiquitous Computing & Emerging Technologies, pp. 36–41 (2013)
13. Dongre, P.B., Malik, L.G.: A review on real time data stream classification and adapting to various concept drift scenarios. In: IEEE International Advance Computing Conference, pp. 533–537 (2014)
14. Padmalatha, E., Reddy, C., Rani, B.P.: Classification of concept drift data streams. In: Information Science and Applications, pp. 1–5 (2014)
15. Song, H.S., Kim, S.H.: Mining the change of customer behavior in an internet shopping mall. Expert Syst. Appl. **21**, 157–168 (2001)

16. Zadeh, L.A.: Fuzzy sets. Inf. Control **8**, 338–353 (1965)
17. Hong, T.-P., Lin, K.-Y., Wang, S.-L.: Fuzzy data mining for interesting generalized association rules. Fuzzy Sets Syst. **138**, 255–269 (2003)
18. Hüllermeier, E.: Fuzzy sets in machine learning and data mining. Appl. Soft Comput. **11**, 1493–1505 (2011)
19. Srikant, R., Agrawal, R.: Mining quantitative association rules in large relational tables. ACM Special Interest Group on Management Of Data, pp. 1–12 (1996)
20. Hong, T.P., Kuo, C.S., Chi, S.C.: Mining association rules from quantitative data. Intell. Data Anal. **3**, 363–376 (1999)
21. Bezdek, J.C.: Pattern recognition with fuzzy objective function algorithms. Springer Science & Business Media, New York (2013)

Mining Discriminative High Utility Patterns

Jerry Chun-Wei Lin[1(✉)], Wensheng Gan[1], Philippe Fournier-Viger[2], and Tzung-Pei Hong[3,4]

[1] School of Computer Science and Technology, Harbin Institute of Technology Shenzhen Graduate School, Shenzhen, People's Republic of China
jerrylin@ieee.org, wsgan001@gmail.com

[2] School of Natural Sciences and Humanities, Harbin Institute of Technology Shenzhen Graduate School, Shenzhen, People's Republic of China
phifv@hitsz.edu.cn

[3] Department of Computer Science and Information Engineering, National University of Kaohsiung, Kaohsiung, Taiwan, ROC

[4] Department of Computer Science and Engineering, National Sun Yat-sen University, Kaohsiung, Taiwan, ROC
tphong@nuk.edu.tw

Abstract. Recently, many approaches for high utility pattern mining (HUPM) have been proposed, but most of them aim at mining high-utility patterns (HUPs) instead of frequent ones. The major drawback is that any combination of a low-utility item with a very high utility pattern is regarded as a HUP, even if this combination is infrequent and contains items that rarely co-occur. Thus, the HUIPM algorithm was proposed to derive high utility interesting patterns (HUIPs) with strong frequency affinity. However, it recursively constructs a series of conditional trees to produce candidates, and then derive the HUIPs. It is time-consuming and may lead to a combinatorial explosion. In this paper, a Fast algorithm for mining Discriminative High Utility Patterns with strong frequency affinity (FDHUP) is proposed by considering both the utility and frequency affinity constraints. Two compact structures named EI-table and FU-table, and two pruning strategies are designed to reduce the search space, and efficiently and effectively discover DHUPs. Experimental results show that the proposed FDHUP algorithm considerably outperforms the state-of-the-art HUIPM algorithm in all datasets.

Keywords: Utility mining · Frequency affinity · Discriminative high utility patterns · Pruning strategies

1 Introduction

Frequent Itemset Mining (FIM) [2,3,8] has become an important data mining task having a wide range of real-world applications. However, an important limitation of FIM is that it considers the frequencies of items/itemsets in a transactional database but not other implicit factors such as their weight, interest, risk or profit. To address this issue, the problem of high utility pattern mining

N.T. Nguyen et al. (Eds.): ACIIDS 2016, Part II, LNAI 9622, pp. 219–229, 2016.
DOI: 10.1007/978-3-662-49390-8_21

(HUPM) [5,7,12,13,17] was proposed to discover profitable patterns. It consists of discovering high utility patterns (HUPs) in quantitative transactional databases by considering both the purchase quantities (internal utility) and unit profits (external utility) of items. The problem of HUPM in quantitative databases was introduced by Chan et al. [7]. Yao et al. then proposed a strict unified framework for mining HUPs [17]. In HUPM, the utility represents how "useful" or "profitable" a pattern is to users. The goal of HUPM is to find the complete set of HUPs having utilities no less than a user-specified minimum utility threshold. HUPM is an emerging topic in data mining, which has many practical applications, many algorithms have been proposed [5,7,9–13,15–17].

However, traditional HUPM algorithms for discovering HUPs without considering the frequency and affinity of items in patterns. Hence, HUPM algorithms may discover many highly profitable patterns that are infrequent and contain items that rarely co-occur. For example, the item "diamond" has a very high utility, any pattern containing "diamond" would be regarded as a HUP in traditional HUPM. But most of these patterns are meaningless, redundant or non-discriminative (have happened by chance). Hence, it is a critical issue for real-world applications to design efficient algorithms to discover discriminative HUPs based on the utility and affinity of items. In the past, an algorithm named HUIPM was proposed to discover high utility interesting patterns (HUIPs) with strong frequency affinity [6]. To the best of our knowledge, it is the first work that considers the frequency affinity of items in HUPM. The HUIPM algorithm recursively constructs a series of conditional trees to produce candidates, and then filters candidates to obtain the HUIPs. However, this procedure is very time-consuming and may lead to a combinatorial explosion.

In this paper, an efficient enumeration-tree-based algorithm named Fast algorithm for mining Discriminative High Utility Patterns (FDHUP) is proposed to efficiently identify DHUPs containing items having strong frequency affinity. The key contributions of this paper are summarized as follows:

- An efficient FDHUP algorithm is proposed to ensure that discovered patterns satisfy not only the user-specified minimum utility threshold, but also have a high frequency affinity. Hence, many redundant and non-interesting patterns can be filtered to reveal more meaningful and useful discriminative HUPs.
- Two compact data structures, called element-information table (EI-table) and frequency-utility table (FU-table), are designed to capture knowledge on frequency affinity from a database by performing two database scans. The FDHUP algorithm employs them to efficiently discover the DHUPs from a proposed frequency-utility tree (FU-tree) without candidate generation.
- Two efficient pruning strategies are further developed. They rely on a new upper-bound that is tighter than the KWU used in the HUIPM algorithm.
- An extensive experiments carried on both real-life and synthetic datasets show that the proposed algorithm outperforms the state-of-the-art HUIPM algorithm in terms of runtime and memory consumption in all datasets.

2 Preliminaries and Problem Statement

Let $I = \{i_1, i_2, \ldots, i_m\}$ be a finite set of m distinct items appearing in a transactional database $D = \{T_1, T_2, \ldots, T_n\}$, where each transaction $T_q \in D$ is a subset of I, and has a unique identifier called *TID*. An unit profit $pr(i_j)$ is assigned to each item $i_j \in I$, which represents its profit (e.g. interest, risk). Unit profits for all items are stored in a profit table $ptable = \{pr(i_1), pr(i_2), \ldots, pr(i_m)\}$. A pattern $X \subseteq I$ with k distinct items $\{i_1, i_2, \ldots, i_k\}$ is called a k-pattern. As a running example, Table 1 shows a transactional quantitative database containing 10 transactions and 5 items, and the profit table is set as $ptable = \{pr(a){:}6, pr(b){:}2, pr(c){:}15, pr(d){:}4, pr(e){:}9\}$.

Table 1. An example database

TID	Transaction	*tu*
T_1	$(a,2);(c,4);(e,1)$	81
T_2	$(d,3);(e,3)$	39
T_3	$(a,5);(b,2);(c,1);(d,2)$	57
T_4	$(c,3);(d,2)$	53
T_5	$(b,2);(c,1);(e,2)$	37
T_6	$(b,4);(d,3)$	20
T_7	$(a,1);(b,3);(c,2);(d,5);(e,1)$	71
T_8	$(b,5);(c,3);(e,4)$	91
T_9	$(c,2);(d,1)$	34
T_{10}	$(a,3);(b,1);(d,2);(e,4)$	64

Table 2. Derived DHUPs

Pattern	*u*	*af*	*au*
(c)	240	16	240
(e)	135	15	135
(ac)	153	4	84
(bc)	129	7	119
(be)	121	8	88
(cd)	160	6	114
(ce)	222	6	144
(bce)	173	5	130

Definition 1. The utility of an item i_j in T_q is $u(i_j, T_q) = q(i_j) \times pr(i_j)$.

Definition 2. The utility of a pattern X in a transaction T_q is denoted as $u(X, T_q)$, and defined as: $u(X, T_q) = \sum_{i_j \in X \wedge X \subseteq T_q} u(i_j, T_q)$.

Definition 3. The utility of a pattern X in a database D is denoted as $u(X)$ and is defined as: $u(X) = \sum_{X \subseteq T_q \wedge T_q \in D} u(X, T_q)$.

Definition 4. The utility of a transaction T_q is denoted as $tu(T_q)$, and defined as: $tu(T_q) = \sum_{i_j \in X \wedge X \subseteq T_q} u(i_j, T_q)$. where m is the number of items in T_q.

Definition 5. The total utility in a database D is denoted as TU, and defined as the sum of all transaction utilities: $TU = \sum_{T_q \in D} tu(T_q)$.

For example, $tu(T_1) = u(a, T_1) + u(c, T_1) + u(e, T_1) = 12 + 60 + 9 = 81$. And the transaction utilities of T_1 to T_{10} are respectively shown in Table 1.

Definition 6. A pattern X in a database D is said to be a high utility pattern (HUP) if it satisfies $HUP \leftarrow \{X | u(X) \geq \delta \times TU\}$, where δ is the user-specified minimum utility threshold.

A new measure called frequency affinity [6] was proposed to mine high utility interesting patterns (HUIPs). The concept of affinitive frequency and affinitive utility are defined below, and are more precise than that in HUIPM [6].

Definition 7. The affinitive frequency of a pattern X in a transaction T_q is denoted as $af(X, T_q)$ and defined as the minimum frequency value for items of X in T_q, that is: $af(X, T_q) = min\{q(i_1, T_q), q(i_2, T_q), ..., q(i_j, T_q)\}, i_j \in X$.

For example in Table 1, $af(ab, T_3) = min\{q(a, T_3), q(b, T_3)\} = min\{5, 2\} = 2$, and $af(ab, T_7) = min\{q(a, T_7), q(b, T_7)\} = min\{1, 3\} = 1$.

Definition 8. The affinitive utility of a pattern X in T_q is denoted as $au(X, T_q)$, and defined as: $au(X, T_q) = eu(X) \times af(X, T_q)$, where $eu(X) = \sum_{i_j \in X} pr(i_j)$.

For example, $au(ab, T_3) = eu(ab, T_3) \times af(ab, T_3) = (6 + 2) \times 2 = 16$.

Definition 9. The affinitive utility of a pattern X in a database D is denoted as $au(X)$, and defined as: $au(X) = \sum_{X \subseteq T_q \wedge T_q \in D} au(X, T_q)$.

Definition 10. A pattern X in a database D is said to be a discriminative high utility patterns with strong frequency affinity ($DHUP$) if it satisfies $DHUP \leftarrow \{X | au(X) \geq \delta \times TU\}$, where δ is the user-specified minimum utility threshold.

The problem of discriminative high utility pattern mining with strong frequency affinity (DHUPM) is to find the complete set of DHUPs. When δ is set to 15 %, the derived DHUPs in the running example is shown in Table 2.

3 Proposed FDHUP Algorithm for Mining DHUPs

To facilitate the mining process and avoid performing multiple database scans, two compact data structures, called element-information table (EI-table) and frequency-utility table (FU-table), are designed in the proposed algorithm.

Definition 11. Assume that patterns and transactions are sorted according to a total order $\prec$ on items (e.g. the lexicographic order). The affinitive utility of an item i_j under a pattern X in T_q is denoted as $au(i_j|_X, T_q)$ and defined as the external utility of i_j multiplied by the minimum frequency between X and i_j:

$$au(i_j|_X, T_q) = min\{af(X, T_q), q(i_j, T_q)\} \times pr(i_j), i_j \in X \subseteq T_q \wedge X \prec i_j.$$

For example, since $af(ab, T_3) = 2$, the affinitive utility of (ab) in T_3 is calculated as: $au(c|_{ab}, T_3) = min\{af(ab, T_3), q(c, T_3)\} \times pr(c) = min\{2, 1\} \times 15 = 15$; and $au(d|_{ab}, T_3) = min\{af(ab, T_3), q(d, T_3)\} \times pr(d) = min\{2, 2\} \times 4 = 8$.

Definition 12. The remaining affinitive utility of a pattern X in a transaction T_q is denoted as $rau(X, T_q)$ and defined as the sum of the affinitive utilities values of each item appearing after X in T_q according to the total order $\prec$, that is: $rau(X, T_q) = \sum_{i \in X \subseteq T_q \wedge X \prec i_j \wedge T_q \in D} au(i_j|_x, T_q)$.

For example in Table 1, $rau(ab, T_3) = au(c|_{ab}, T_3) + au(d|_{ab}, T_3) = 15 + 8 = 23$; and $rau(ab, T_{10}) = au(d|_{ab}, T_{10}) + au(e|_{ab}, T_{10}) = 4 + 9 = 13$.

Definition 13 (EI-Table). The element-information table (EI-table) of a pattern X contains a set of elements corresponding to the transactions where the pattern X appears. An element consists of three fields: the TID of a transaction T_q containing X ($tid, X \subseteq T_q \wedge T_q \in D$), the affinitive frequency of X in T_q ($af(X, T_q)$), and the remaining affinitive utility of X in T_q ($rau(X, T_q)$).

Note that the total order $\prec$ used in the FDHUP algorithm adopts TWU-ascending order of 1-patterns. Since the TWU values of 1-items are $\{a : 273, b : 340, c : 424, d : 338, e : 383\}$, the TWU-ascending order is $\{a \prec d \prec b \prec e \prec c\}$. To discover DHUPs, the FDHUP algorithm performs a single database scan to create the EI-tables of all 1-patterns, as shown in Fig. 1.

(a)		
tid	*af*	*rau*
1	2	39
3	5	27
7	1	30
10	3	37

(d)		
tid	*af*	*rau*
2	3	27
3	2	19
4	2	30
6	3	6
7	5	45
9	1	15
10	2	20

(b)		
tid	*af*	*rau*
3	2	15
5	2	33
6	4	0
7	3	39
8	5	81
10	1	9

(e)		
tid	*af*	*rau*
1	1	15
2	3	0
5	2	15
7	1	15
8	4	45
10	4	0

(c)		
tid	*af*	*rau*
1	4	0
3	1	0
4	3	0
5	1	0
7	2	0
8	3	0
9	2	0

Fig. 1. The built EI-tables of 1-items.

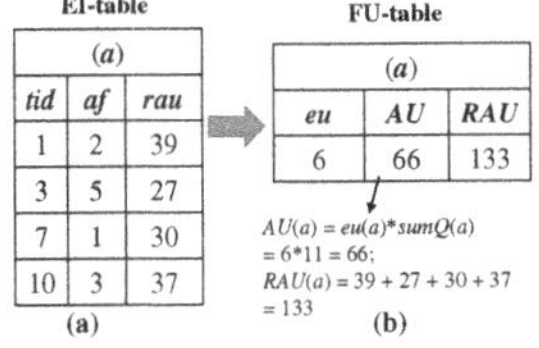

Fig. 2. The constructed FU-table of item (a).

The sum of the affinitive quantities of a pattern X in a database D can be efficiently calculated by adding the affinitive frequencies of all elements in the EI-table of X. In fact, five important types of information can be obtained as: (1) the name of X; (2) the TID of transactions where X appears; (3) the sum of the affinitive frequencies (purchase quantities) of X in D; (4) the actual affinitive utility of X in D; and (5) the total remaining affinitive utility of X in D. Details are described in the following definitions and the concept of FU-Table (Fig. 2).

Definition 14. The summation of affinitive quantities of a pattern X in D is denoted as $sumQ(X)$ and defined as: $sumQ(X) = \sum_{X \subseteq T_q \wedge T_q \in D} af(X, T_q)$.

Definition 15. The summation of the external utilities of a pattern X in a database D is denoted as $eu(X)$ and is defined as: $eu(X) = \sum_{i_j \in X} pr(i_j)$.

Definition 16. The summation of the utilities of a pattern X in the entire database according to its EI-table is denoted as $AU(X)$, and can be determined as: $AU(X) = \sum_{X \subseteq T_q \wedge T_q \in D} au(X, T_q) = sumQ(X) \times eu(X)$.

Definition 17. The summation of the remaining affinitive utilities of a pattern X in the entire database is denoted as $RAU(X)$, and can be determined as: $RAU(X) = \sum_{X \subseteq T_q \wedge T_q \in D} rau(X, T_q)$.

Definition 18 (FU-Table). The frequency-utility table (FU-table) of a pattern X contains four parts: the name of X; the external utility of X ($\boldsymbol{eu}$); the summation of the utilities of X in D ($\boldsymbol{AU}$); and the summation of the remaining affinitive utilities of X in D ($\boldsymbol{RAU}$).

The FU-table of a pattern X is built after the construction of its EI-table, and it stores the necessary information, as shown in Table 2. Details of the construction procedure of EI-table is shown in Algorithm 1. Note that the search space of the proposed FDHUP algorithm can be represented as a set-enumeration tree [14], denoted as the frequency-utility tree (FU-tree), details of FU-tree and the proof of Lemma 3 are skipped due to the limited space of the paper.

Input: X, X_a, X_b $(a \neq b)$.
Output: X_{ab} which having EI-table and FU-table as $X_{ab}.EIT$ and $X_{ab}.FUT$.

set $X_{ab}.EIT \leftarrow \emptyset$;
for *each element* $E_a \in X_a.EIT$ **do**
    if $\exists E_a \in X_b.EIT \wedge E_a.tid == E_b.tid$ **then**
        if $X.EIT \neq \emptyset$ **then**
            search for element $E \in X.EIT$, such that $E.tid = E_a.tid$;
            $E_{ab} \leftarrow < E_a.tid, min\{E_a.af, E_b.af\}, E_b.rau >$;
        else
            $E_{ab} \leftarrow < E_a.tid, min\{E_a.af, E_b.af\}, E_b.rau >$;
        $X_{ab}.EIT \leftarrow X_{ab}.EIT \cup E_{ab}$;
return X_{ab};

Algorithm 1. Construction(X, X_a, X_b)

Lemma 1. *In the FU-tree, the affinitive frequency of a node is always no less than the affinitive frequency of any child nodes of this node.*

Proof. Let the node corresponding to X in the FU-tree be denoted as X^{k-1}, and any children of X^{k-1} be denoted as X^k. For all transaction $X^k \subseteq T_q$, we have: $af(X^{k-1}, T_q) = min\{f(i_1, T_q), ..., f(i_j, T_q)\}, j = k-1 \wedge i_j \in X^{k-1}$; $af(X^k, T_q) = min\{f(i_1, T_q), ..., f(i_j, T_q), f(i_{j+1}, T_q)\}, j+1 = k \wedge i_j \in X^k$.

Hence, in T_q, $af(X^{k-1}, T_q) \geq af(X^k, T_q)$; in D, $sumQ(X^{k-1}) = \sum_{X^{k-1} \subseteq T_q \wedge T_q \in D} af(X^{k-1}, T_q) \geq \sum_{X^k \subseteq T_q \wedge T_q \subseteq D} af(X^k, T_q) = sumQ(X^k)$.

Thus, $sumQ(X^{k-1}) \geq sumQ(X^k)$, and Lemma 1 holds.

Lemma 2. *Based on the designed DHUPM framework, the transaction-weighted downward closure property (TWDC) of the TWU model [13] can still be used to filter unpromising patterns, i.e., if a pattern X is not a high transaction-weighted utilization pattern (HTWUP), then X and all its supersets are not DHUP.*

Proof. Let X be a pattern. Based on above definitions, we have that:

(1) $au(X, T_q) = eu(X) \times af(X, T_q) = \sum_{i_j \in X} pr(i_j) \times af(X, T_q) = \sum_{i_j \in X}(pr(i_j) \times af(X, T_q))$, and $tu(T_q) = \sum_{j=1}^{m} u(i_j, T_q) = \sum_{j=1}^{m}(pr(i_j) \times q(i_j, T_q)$;

(2) Since $af(X, T_q) = min\{f(i_1, T_q), ..., f(i_j, T_q)\}$, $i_j \in X$. Thus: $au(X) = \sum_{X \subseteq T_q \wedge T_q \in D} au(X, T_q) \leq \sum_{X \subseteq T_q \wedge T_q \in D} tu(T_q) = TWU(X)$. Hence, $au(X) \leq TWU(X)$, this lemma holds.

Lemma 3. *Let X be a node (a pattern) in the FU-tree. The sum of $AU(X)$ and $RAU(X)$ is always no less than the affinitive utility of X and its children.*

Pruning Strategy 1. After scanning the database, the TWU value of each 1-pattern can be used to filter unpromising patterns. If the TWU value of a pattern X is less than $(\delta \times TU)$, X and all its supersets can be discarded.

Pruning Strategy 2. When traversing the search space of the FU-tree using a depth-first search, if the sum of $AU(X)$ and $RAU(X)$ for a node corresponding to a pattern X is less than the user-specified minimum utility count according to the nodes FU-table, then no children of that node is a DHUP.

Details of the proposed FDHUP algorithm (Algorithm 2) and the **DHUP-Mining** mining procedure (Algorithm 3) are described below.

Input: D, a transaction quantitative database; *ptable*, the user-specified profit table; δ, the user-specified minimum utility threshold.
Output: The complete set of DHUPs.
scan D to calculate the TWU value of each item $i \in I$ and the TU of D;
find $I^* \leftarrow \{i \in I | TWU(i) \geq \delta \times TU\}$;
sort I^* in the designed total order $\prec$ (w.r.t. TWU-ascending order);
scan D once to build the EI-table and FU-table for each 1-item $i \in I^*$;
call DHUP-Mining(ϕ, I^*, δ);
return $DHUPs$.

Algorithm 2. FDHUP algorithm

4 Experiments

We performed extensive experiments in terms of runtime, memory consumption and the number of patterns to evaluate the proposed algorithm with the state-of-the-art HUIPM algorithm [6]. Experiments are conducted on four datasets including both real-world datasets (kosarak, retail and mushroom) [1] and synthetic dataset (T10I4D100K) generated using the IBM Quest Synthetic Data Generator [4]. The purchased quantity and unit profit of each item in a database was randomly generated using the same way as the previous studies [9,10,16]. Characteristics of the three real-life datasets can be referred to [1], and the T10I4D100K has 100,000 transactions with 870 distinct items, and the average length of transactions in it is 10.1.

Input: X, a pattern; $extenEITOfX$, a set of extension EI-tables of X; δ.
Input: The set of DHUPs
for *each pattern* $X_a \in extendOfX$ **do**
    obtain $AU(X_a)$ and $RAU(X_a)$ from the built $X_a.FUT$;
    if $AU(X_a) \geq \delta \times TU$ **then**
        $DHUPs \leftarrow DHUPs \cup X_a$;
    if $AU(X_a) + RAU(X_a) \geq \delta \times TU$ **then**
        $extenEITOfX_a \leftarrow \emptyset$;
        for *each pattern* X_b *after* X_a *in* $extenEITOfX$ **do**
            $X_{ab} \leftarrow X_a \cup X_b$;
            call Construct(X, X_a, X_b) to build $X_{ab}.EIT$ and $X_{ab}.FUT$ for X_{ab};
            $extenEITOfX_a \leftarrow extenEITOfX_a \cup X_{ab}$;
    call DHUP-Mining$(X_a, extenEITOfX_a, \delta)$;
return $DHUPs$.

Algorithm 3. DHUP-Mining$(X, extenEITOfX, \delta)$

4.1 Runtime

The runtime of the two algorithms are compared under various minimum utility thresholds. Results on four datasets are shown in Fig. 3.

From Fig. 3, it can be seen that the proposed FDHUP algorithm has better performance than the HUIPM algorithm on all datasets, and almost faster than the HUIPM up to one or two order of magnitude. This result is reasonable since the HUIPM algorithms has to construct a large number of condition trees for retaining HKWUIPs before deriving DHUPs, that takes up much mining time. This limits the application of this kind of algorithms to real situations.

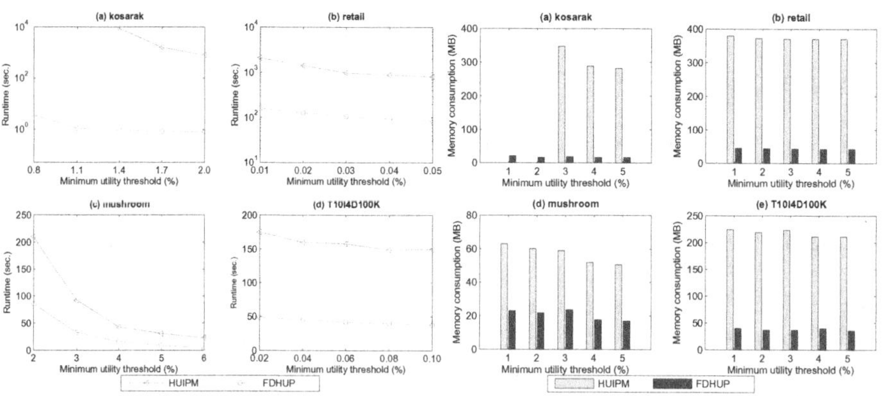

Fig. 3. Runtime under various minimum utility thresholds.

Fig. 4. Memory consumption under various minimum utility thresholds.

While the FDHUP algorithm use two pruning strategies to effectively prune the unpromising patterns early, this makes the search space much smaller and thus reduce much execution time. Thus, the FDHUP algorithm is more effective and outperforms than the state-of-the-art algorithm.

4.2 Memory Consumption

The memory consumption of the two algorithms are also compared under various minimum utility thresholds, as shown in Fig. 4. From Fig. 4, it can be noticed that the proposed FDHUP algorithm with different pruning strategies always consumes less memory than the compared HUIPM algorithm on all datasets. As mentioned previously, the performance advantage of FDHUP algorithm is that it easily maintains the necessary information about the processing database by constructing a series of EI-tables and FU-tables. Without multiple datasets scanning and candidates generation-and-test, it can significantly save memory usage. Thus, the proposed FDHUP algorithm with two compact data structures and different pruning strategies always consumed less memory than the state-of-the-art HUIPM algorithm.

4.3 Pattern Analysis

The number patterns including HKWUIPs, HUPs and DHUPs which are respectively derived by the HUIPM algorithm, traditional HUPM algorithm, and the proposed FDHUP algorithm, are further evaluated to analysis the relationship between the DHUPM and traditional HUPM. Let ***reduce ratio1*** $= \frac{(HKWUIPs - DHUPs)}{HKWUIPs} \times 100\,\%$, and ***reduce ratio2*** $= \frac{(HUPs - DHUPs)}{HUPs} \times 100\,\%$. Selected each "middle" threshold in Fig. 3, results of patterns are shown in Table 3. It can be seen that the number of DHUPs is always smaller than that of HUPs in all datasets, and the number of HKWUIPs is quite large than that of DHUPs. It indicates that numerous of redundant patterns (HUPs) are generated in the traditional HUPM algorithm, rare of them has the highly frequency affinity among the items inside the combinational patterns. By considering frequency affinity factor, fewer but more useful information could be discovered by the FDHUP algorithm for the precision marketing. The proposed algorithm for mining discriminate high utility patterns is reasonable and acceptable in real-world applications.

Table 3. The number of different type patterns.

Dataset	#HKWUIPs	#HUPs	#DHUPs	*reduce ratio1*	*reduce ratio2*
kosarak	28903	29	**13**	99.955 %	55.172 %
retail	55779	5237	**23**	99.959 %	99.561 %
mushroom	555673	379732	**6211**	98.882 %	98.364 %
T10I4D100K	31103	13722	**4434**	85.744 %	67.687 %

5 Conclusion

In this paper, an efficient algorithm called fast algorithm for mining discriminative high utility patterns (FDHUP) has been proposed, it not only consider the utility measure, but also measure the frequency affinity among the items inside the patterns. Two compact data structures EI-table and FU-table are designed to capture knowledge, and two efficient pruning strategies are further developed. Experimental results also show that the performance of the proposed algorithm outperforms than HUIPM algorithm in terms of runtime, memory consumption. Moreover, the derived DHUPs can provide fewer but more precise information instead of mass redundant information.

References

1. Frequent itemset mining dataset repository. http://fimi.ua.ac.be/data/
2. Agrawal, R., Imielinski, T., Swami, A.: Database mining: a performance perspective. IEEE Trans. Knowl. Data Eng. **5**(6), 914–925 (1993)
3. Agrawal, R., Srikant, R.: Fast algorithms for mining association rules in large databases. In: The Intentional Conference on Very Large Data Bases, pp. 487–499 (1994)
4. Agrawal, R., Srikant, R.: Quest synthetic data generator. http://www.Almaden.ibm.com/cs/quest/syndata.html
5. Ahmed, C.F., Tanbeer, S.K., Jeong, B.S., Le, H.J.: Efficient tree structures for high utility pattern mining in incremental databases. IEEE Trans. Knowl. Data Eng. **21**(12), 1708–1721 (2009)
6. Ahmed, C.F., Tanbeer, S.K., Jeong, B.S., Choi, Y.K.: A framework for mining interesting high utility patterns with a strong frequency affinity. Inf. Sci. **181**(21), 4878–4894 (2011)
7. Chan, R., Yang, Q., Shen, D.: Minging high utility itemsets. In: IEEE International Conference on Data Mining, pp. 19–26 (2003)
8. Chen, M.S., Han, J., Yu, P.S.: Data mining: an overview from a database perspective. IEEE Trans. Knowl. Data Eng. **8**(6), 866–883 (1996)
9. Fournier-Viger, P., Wu, C.-W., Zida, S., Tseng, V.S.: FHM: faster high-utility itemset mining using estimated utility co-occurrence pruning. In: Andreasen, T., Christiansen, H., Cubero, J.-C., Raś, Z.W. (eds.) ISMIS 2014. LNCS, vol. 8502, pp. 83–92. Springer, Heidelberg (2014)
10. Lin, J.C.-W., Gan, W., Hong, T.-P., Pan, J.-S.: Incrementally updating high-utility itemsets with transaction insertion. In: Luo, X., Yu, J.X., Li, Z. (eds.) ADMA 2014. LNCS, vol. 8933, pp. 44–56. Springer, Heidelberg (2014)
11. Lin, J.C.W., Gan, W., Fournier-Viger, P., Hong, T.P.: Mining high-utility itemsets with multiple minimum utility thresholds. In: The International C* Conference on Computer Science and Software Engineering, pp. 9–17 (2015)
12. Liu, M., Qu, J.: Mining high utility itemsets without candidate generation. In: ACM International Conference on Information and Knowledge Management, pp. 55–64 (2012)
13. Liu, Y., Liao, W., Choudhary, A.K.: A two-phase algorithm for fast discovery of high utility itemsets. In: Ho, T.-B., Cheung, D., Liu, H. (eds.) PAKDD 2005. LNCS (LNAI), vol. 3518, pp. 689–695. Springer, Heidelberg (2005)

14. Rymon, R.: Search through systematic set enumeration. Technical reports, 297 (1992)
15. Tseng, V.S., Wu, C.W., Shie, B.E., Yu, P.S.: UP-Growth: an efficient algorithm for high utility itemset mining. In: ACM SIGKDD International Conference on Knowledge Discovery and Data Mining, pp. 253–262 (2010)
16. Tseng, V.S., Wu, C.W., Shie, B.E., Yu, P.S.: Efficient algorithms for mining high utility itemsets from transactional databases. IEEE Trans. Knowl. Data Eng. **25**(8), 1772–1786 (2013)
17. Yao, H., Hamilton, H.J., Butz, C.J.: A foundational approach to mining itemset utilities from databases. In: SIAM International Conference on Data Mining, pp. 211–225 (2004)

An Experimental Study on Cholera Modeling in Hanoi

Ngoc-Anh Thi Le[1], Thi-Oanh Ngo[2], Huyen-Trang Thi Lai[2], Hoang-Quynh Le[2], Hai-Chau Nguyen[2], and Quang-Thuy Ha[2(✉)]

[1] Hanoi Medical University, No.1, Ton That Tung Street, Hanoi, Vietnam
lengocanh@hmu.edu.vn

[2] VNU-University of Engineering and Technology (UET), 144, Xuan Thuy, Cau Giay, Hanoi, Vietnam
{oanhnt_57, tranglth_57, QuynhLH, ChauNH, thuyhq}@vnu.edu.vn

Abstract. Cholera modeling for prediction Cholera state is the research topic is specially considered by researchers. An experimental frame for Cholera modeling based on using data mining techniques is showed in this paper. In the Data Preparation Phase, two versions of data presentation and a solution for feature selection are considered. Moreover, both cases of two-value Cholera state and three-value Cholera state also investigated. Experimental results show that the global presentation is better than the local presentation, F1 measures are between 0.79 and 0.86, and the target variable has correlation with climate condition variables in some cases.

Keywords: Cholera modeling · Feature selection · Global data preparation · Local data preparation · Prediction cholera state

1 Introduction

Cholera has been showed as one of most serious disease with the seven Cholera pandemics occurred in the world. Cholera modeling for prediction Cholera state is the research topic is specially considered by researching community in the world. Many works on topic of Cholera modeling are done. There are three categories of methods on Cholera modeling. The first categories consists methods based on using mathematical epidemiology, such as [1, 6, 9, 12, 14]. The second categories consists methods based on using data mining techniques, such as [2, 8, 13, 16]. Some other methods based on multi-agent systems, such as [4].

Cholera is one of the most sensitive disease with climate change. Rita R. Colwel [3] remarked that cholera offers an excellent example of how information concerning environmental factors permits better understanding of disease-not only virulence, but equally important, transmission and epidemiology and is regarded as a model of the impact of climate change to the epidemic. There are many research focus for modeling for climate-related infectious diseases, such as [10, 12–14, 16].

In this work, a Frame of Modeling Cholera in Hanoi has been proposed. In this model, we describe clearly three phases in the modeling process.

N.T. Nguyen et al. (Eds.): ACIIDS 2016, Part II, LNAI 9622, pp. 230–240, 2016.
DOI: 10.1007/978-3-662-49390-8_22

This study is the first study on Cholera modeling for prediction Cholera state in Hanoi based on using data mining techniques, and it makes the following contributions:

- A feature selection solution is applied to improve effect of results model,
- Two versions of data presentations are determined. Experiments show the global data representation is better than the local data representation,
- The mutual relations between the target variable (Cholera value at time-point t) and climate condition variables (climate values at time-points t-1 and t-2) is investigated.

The rest of this article is organized as following. In the next section, a Frame of Cholera modeling in Hanoi is showed. This Frame concludes three components of Data Preparation, Modeling, and Using Outcomes. Experiments and remarks are described in the third section. In fourth section, related works are introduced to show the importance of the Feature Selection in Cholera modeling in Hanoi. Conclusions are shown in the last section.

2 Our Approach

2.1 The Problem

The paper focuses to propose a frame of cholera modeling in Hanoi. In this paper, the problem is described as follows.

Let QH be set of districts in Hanoi. There are twenty ninth districts in Hanoi and let QH_i ($i = 1, 2, \ldots, 29$) be the ith district. For each district QH_i, let $LCQH_i$ be the set of districts, which are neighbors of QH_i.

Let $DT_{i,t}$ be the Cholera state in QH_i at time-point t, let $DTLC_{i,t}$ be the set of Cholera states in districts belonged to $LCQH_i$ at time-point t, let DT_t be the set (vector) of Cholera states in all districts in Hanoi at time-point t.

Let $KH_{i,t}$ be climate feature values in QH_i at the time-point t, let $KHLC_{i,t}$ be the set of climate feature values in districts belonged to $LCQH_i$ at time-point t. There are five weather stations (Ba Vi, Son Tay, Lang, Hoai Duc, Ha Dong) and three hydrological stations (Ha Noi, Son Tay, Thuong Cat). The parameters in $KH_{i,t}$ are determined based on data from the nearest weather station and the nearest hydrological station. Let KH_t be set (vector) of climate feature values from five weather stations and three hydrological stations.

In this study, we consider two kinds of predictive models, related with two cases of data presentations. In the local predictive model for a district, a data-point is presented as a vector ($KH_{i,t-2}$, $KH_{i,t-1}$, $KHLC_{i,t-1}$, $KHLC_{i,t-2}$, $DT_{i,t-1}$, $DT_{i,t-2}$, $DTLC_{i,t-1}$, $DTLC_{i,t-2}$, $DT_{i,t}$), where $DT_{i,t}$ is the target variable and the other variables are condition variables. In the global predictive model, a data-point is presented as a vector (KH_{t-2}, KH_{t-1}, DT_{t-1}, DT_{t-2}, DT_t) where DT_t is the target vector and the other parameters are condition variables.

2.2 A Frame of Cholera Modeling in Hanoi

Figure 1 describes our frame of Cholera modeling in Hanoi. The frame includes three parts of Data Preparation, Modeling, and Using Outcomes.

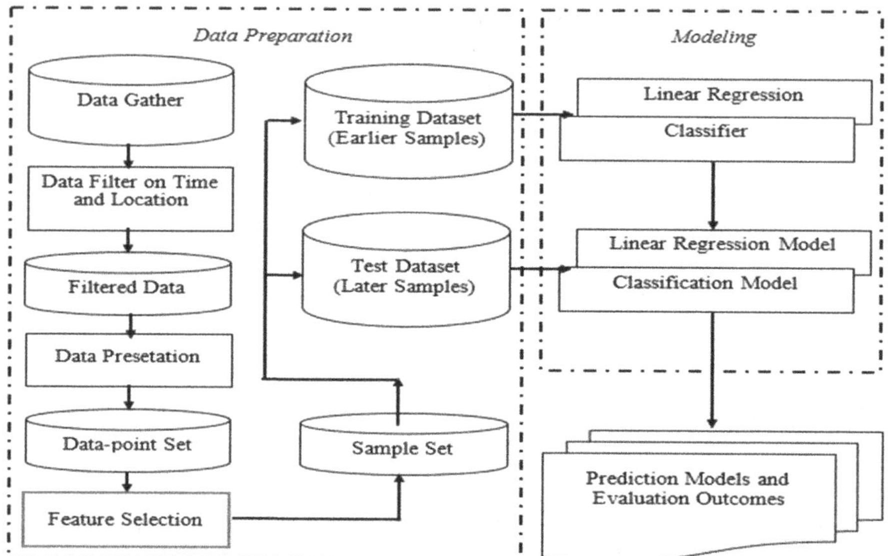

Fig. 1. A frame of Cholera modeling in Hanoi

2.3 Data Preparation Phase

Because the gathered dataset is uncompleted that it should be filtered based on removing some parts caused incomplete.

There are two versions of data presentation, the local version and the global version. In the local version, the data points are presented locally for each district. The target variable is the cholera value in the district at the time-point t. Condition variables are cholera values and climate values in the district and in its surrounding districts at time-points t-1 and t-2. Climate values are chosen from the nearest station. The local version is similarity with data presentation in [16].

In the global version, the data points are presented global. The target variables is the cholera values in twenty ninth districts at the time-point t. Condition variables are cholera values in all of the districts at time-points t-1 and t-2, and climate values in stations at time-points t-1 and t-2.

Example 1. Climate features include the average value, the maximum value, and the minimum value of air temperature, rainfall, air humidity, sunny hours, wind speed, and Red river height. Red river height values are measured at three stations of Ha Noi, Son Tay, and Thuong Cat. The other climate feature are measured at five stations Ba Vi, Son Tay, Hoai Duc, Lang, and Ha Dong. Cholera features are valued either {Cholera state 0: "no cholera", Cholera state 1: "cholera"} or {Cholera state 0: "no cholera", Cholera state 1: "cholera", Cholera state 2: "heavy cholera" (number of Cholera cases is greater than 20)}.

We would like to apply a solution for feature selection in this study. In the feature selection step, some weak-correlative features should be deleted. After deleting the weak-correlated features, we have the Sample Set. After that, the sample set are divided into the Training Dataset included earlier samples and the Test Dataset included later samples.

2.4 Modeling Phase

In the modeling phase, data mining algorithms Linear Regression, Naïve Bayes, SVM, and Random Tree are applied. The earlier sample set is used for modeling and the later sample set is used for evaluation the result model.

As above description, two cases of the local version and the global version of data presentations are examined for comparison. To evaluate the effect of applying solutions of feature selection, two cases of input of origin data presentation and chosen data presentation are investigated. To determine relationships between climate factors and Cholera, the cases of condition variables with only climate factors are examined. Both of cases of Cholera features valued in {0, 1} and {0, 1, 2} are investigated.

2.5 Evaluation Phase

The effect of output models are evaluated in some measures of Mean absolute error (MAE), Root mean squared error (RMSE), Precision, Recall, F-Measure. The Correlation coefficient (CC) measure indicates the correlation between target variable (the cholera value at the time-point t) and condition variables (cholera values and climate values at the time-points t-1 and t-2). The evaluation is implemented on the test dataset.

Assume that the test dataset consists n data-points. Let $a_1, a_2, \ldots, a_n$ be the actual of target variable in the test dataset and $p_1, p_2, \ldots, p_n$ be the predicted values by the model, corresponding. The computing formulas for Mean absolute error (MAE), Root mean squared error (RMSE), and Correlation coefficient (CC) as followings:

$$MAE = \frac{\sum_{i=1}^{n} |p_i - a_i|}{n}, \qquad RMSE = \sqrt{\frac{\sum_{i=1}^{n} (p_i - a_i)^2}{n}},$$

$$CC = \frac{S_{PA}}{\sqrt{S_P S_A}}, \text{ where } S_{PA} = \frac{(p_i - \overline{p})(a_i - \overline{a})}{n-1}, \quad S_P = \frac{\sum_{i=1}^{n} (p_i - \overline{p})^2}{n-1},$$

$$S_A = \frac{\sum_{i=1}^{n} (a_i - \overline{a})^2}{n-1}, \quad \overline{p} = \frac{\sum_{i=1}^{n} p_i}{n}, \text{ and } \overline{a} = \frac{\sum_{i=1}^{n} a_i}{n}$$

Let confusion matrix be described as follows:

	Predicted class		
Actual class		Yes	No
	Yes	TP	FN
	No	FP	TN

where, TP (TN) is the number of data-points in the test sample set actually (do not) belongs to the class as same as the prediction by the model, FP (FN) is number of confusion cases of actual belonging (actual not-belonging) in the class. The computing formulas for Recall, Precision, F-Measure as follows:

$$Recall = \frac{TP}{TP + FN}, \quad Precision = \frac{TP}{TP + FP}, \text{ and}$$

$$F - Measure = \frac{2 \times recall \times precision}{recall + precision}.$$

3 Experiments and Results

3.1 The Data

The gathered dataset is described as follows:

- The Cholera cases in years of 2004, 2007-2010. After transformation into the number of Cholera cases in months, this data has been transformed into either {0: "no cholera", 1: "cholera"} or {0: "no cholera", 1: "cholera", 2: "heavy cholera"},
- The climate data consists the average value, the maximum value, and the minimum value of air temperature, rainfall, air humidity, sunny hours, wind speed in years 2001–2012,
- The Red river height data at the Ha Noi station in years 1960-2012, at Son Tay and Thuong Cat stations in years 1960-2013.

Because the Cholera data recorded completely in years 2007-2010 then the dataset in the period of 2007-2010 has been kept for the study. Two cases of data presentations has been made.

3.2 Experiments

For feature selection, the related function of STATISTICA [17] has been used then weak-correlative has been deleted. In the modeling phase, this study also used data mining algorithms Linear Regression, Naïve Bayes, SVM (C-SVC), and Random Tree from WEKA [18].

3.2.1 The Case of Local Data Presentation

In this case, prediction models has been made for all of twenty ninth districts in Hanoi. The experiment results for two example districts are described in the Table 1. F1 measure values seem low in range between 0.6 and 0.758. The Correlation coefficient values seem separated. In some case, the absolute value is very small, which indicates there are no correlation between the target variable and condition variables. In some other case, the absolute value is so high, which indicates there are correlation between the target variable and condition variables.

Table 1. The case of local data presentation in districts Dong Da and Hoang Mai

District	Measures	Linear regression	Naive Bayes	LibSVM	Random tree
Dong Da	CC	**–0.0713**			
	MAE	22.8332	0.2504	0.2222	0.3333
	RMSE	26.5469	0.4741	0.4714	0.5774
	Precision		0.5830	0.4440	0.7220
	Recall		0.6670	0.6670	0.5000
	F		**0.6110**	0.5330	0.5280
Hoang Mai	CC	**0.5317**			
	MAE	12.7367	0.2227	0.2222	0.2222
	RMSE	13.8483	0.4530	0.4714	0.4714
	Precision		0.4440	0.4440	0.5830
	Recall		0.6670	0.6670	0.6670
	F		0.5330	0.5330	**0.6110**

3.2.2 The Case of Global Data Presentation

In this case, two kinds of experiments had been done. In the first kind, all condition variables are used for modeling, and in the second kind, only climate condition variables used for modeling. In each kind, the size of time-points in the past also is investigated. In following tables, "t-12" indicates the size by 2 (condition variables

Table 2. All condition variables are used for modeling: Linear Regression algorithm

Measure \ Size of time-point	t-12	t-1
Correlation coefficient	**0.4857**	**0.5197**
Mean absolute error	2.8307	2.6200
Root mean squared error	5.1425	5.0157

cover at the time-points t-1 and t-2), and "t-1" indicates the size by 1 (condition variables cover at the time-point t-1).

The results in experiments for cases of all condition variables are used for modeling

Table 3. All condition variables are used for modeling: Naïve Bayes, SVM, and Random Tree algorithms

No. of classes	Measures	Naive bayes		LibSVM		Random tree	
		t-12	t-1	t-12	t-1	t-12	t-1
Two classes {0,1}	MAE	0.1406	**0.0958**	**0.0958**	**0.0958**	0.1145	0.1860
	RMSE	0.2913	0.3095	0.3095	0.3095	0.3082	0.3935
	Precision	0.7190	0.7330	0.7330	0.7330	0.7330	0.7190
	Recall	0.7010	0.8560	0.8560	0.8560	0.8560	0.7010
	F	0.7100	**0.7900**	**0.7900**	**0.7900**	**0.7900**	0.7100
Three classes {0,1,2}	MAE	0.1558	**0.1437**	**0.1437**	**0.1437**	0.1718	0.2354
	RMSE	0.3791	0.3790	0.3790	0.3790	0.3775	0.4223
	Precision	0.7330	0.7330	0.7330	0.7330	0.7330	0.8890
	Recall	0.8560	0.8560	0.8560	0.8560	0.8560	0.7760
	F	0.7900	0.7900	0.7900	0.7900	0.7900	**0.8060**

are showed in the Table 2 (Linear Regression algorithm) and the Table 3 (Naïve Bayes, SVM (C-SVC), and Random Tree algorithms). In the both cases, the Correlation coefficient values are around 0.5, the F-measure values are around 0.8, and the MAE values are from 0.1 to 0.2. The Random Tree algorithm is the best in three-class experiments.

The results in experiments for cases of all condition variables are used for modeling

Table 4. Only climate condition variables are used for modeling: Linear Regression algorithm

Size of time-point / Measure	t-12	t-1
Correlation coefficient	**-0.0179**	**0.4699**
Mean absolute error	8.2019	12.3981
Root mean squared error	10.5614	14.4287

are showed in the Table 4 (Linear Regression algorithm) and the Table 5 (Naïve Bayes, SVM (C-SVC), and Random Tree algorithms). The Correlation coefficient value in the t-12 case is –0.0179 (no correlation) and the Correlation coefficient value in the t-1 case is 0.4699 (medium correlation). Most of the F-measure values are near 0.8. The Random Tree algorithm is the worst in three-class experiments.

3.2.3 Discussions

There are some findings from the results showed in Tables 2, 3, 4, and 5 as follows:

Table 5. Only climate condition variables are used for modeling: Naïve Bayes, SVM, and Random Tree algorithms

No. of classes	Measures	Naive bayes		LibSVM		Random tree	
		t-12	t-1	t-12	t-1	t-12	t-1
Two classes {0,1}	MAE	**0.0958**	**0.0975**	**0.0958**	**0.0958**	0.1315	0.1900
	RMSE	0.3095	0.3093	0.3095	0.3095	0.3261	0.3973
	Precision	0.7330	0.7330	0.7330	0.7330	0.7330	0.7190
	Recall	0.8560	0.8560	0.8560	0.8560	0.8560	0.7010
	F	**0.7900**	**0.7900**	**0.7900**	**0.7900**	**0.7900**	0.7100
Three classes {0,1,2}	MAE	0.1437	0.1437	0.1437	0.1437	0.3363	0.5660
	RMSE	0.3790	0.3790	0.3790	0.3790	0.5322	0.7178
	Precision	0.7330	0.7330	0.7330	0.7330	0.7200	0.6330
	Recall	0.8560	0.8560	0.8560	0.8560	0.7010	0.3790
	F	**0.7900**	**0.7900**	**0.7900**	**0.7900**	0.7100	0.4700

- The global data representation is better than the local data representation,
- The effect of models in the global data representation is similarity with the effect of model in [Reiner12],
- There exist correlation between the target variable and climate condition variables.

4 Related Work

Robert C. Reiner et al. [13] proposed a probabilistic spatial model for prediction Cholera state in Darka (Bangladesh). The region consists 21 sub-regions (thana), whereby the prediction model should determine the level of export cholera each region by month. In the study, there are three values of Cholera state, the state 0 corresponds to no Cholera, the state 1 to low cholera, and state 2 to high cholera. Based on assuming that the cholera state of each thana will change stochastically from one month to the next according to transition probabilities that depend on the current cholera state of the thana, the cholera states of its neighbors, the season, and the value of the climate covariates, the authors establish the Climate-Independent Model, Climate-Dependent Model, then make Alternative Parameterization for these models. Then the model is fitted by maximizing the likelihood. The model captures the inter-annual and intra-annual dynamics of cholera in each thana. To assess result model's prediction performance, the authors took a 12-cross-validation approach, corresponding with 12 months in a year and the model's 11-mo predictions' sum of squared error (SSE) is 0.79.

Based on using linear regression analysis, Yujuan Yue et al. [16] focus on the relationships between positive rate of *V.cholerae* and climate factors were obtained through analyzing the data from 2008 to 2009 in the Pearl River estuary, Guangzhou, China. The study found that the positive rate of *V.cholerae* was correlated closely with monthly climate factors of water temperature and air temperature, respectively in 2009. Positive correlations between *V.cholerae* and water temperature is highest (r = 0.746, p = 0.003) and negative correlation between *V.cholerae* and air pressure is highest (r = –0.765, p = 0.002). Moreover, there existed seasonal characteristic for *V.cholerae*.

Analyzing 70 *V.cholerae* isolates which were collected from patients and the physical environment in northern Vietnam during three cholera outbreaks between November 2007 and February 2008 [15], Binh Minh Nguyen et al. [11] detected that Vibrio cholera causing by *V.cholerae O1 Ogawa serotype*. The finding indicates special vaccines prevent three cholera outbreaks in Vietnam and in other south and Southeast Asia. Dang Duc Anh et al. [5] studied the using of oral Cholera vaccines (OCV) in an outbreak of three waves in Hanoi, Vietnam from 28 October 2007 to 15 July 2008. They found that the killed OCV provided protection against this new strain suggesting that there may be a role for reactive use of the killed OCV in future cholera outbreaks.

Michael Emch et al. [7] announced studying results on the impact of local environmental factors to cholera in Bangladesh and Vietnam. Their findings provide further evidence of the formation of cholera and cholera levels are related to the local environment. Particularly in Vietnam, the increase in Sea surface temperature has the greatest influence in Hue (in factors of River height 2-month lag, Sea surface temperature, Sea surface height, Sea surface height 2-month lag, Rainfall 2-month lag, Temperature, Ocean chlorophyll concentrations, Overall Wald χ^2), while increasing the height of Dinh River and Cai River has an important role in Nha Trang (Monthly rainfall, Cai River Height, Dinh River Height, Dinh River Height 2-month lag, Dinh River discharge, Sea surface temperature, Sea surface, Ocean chlorophyll concentrations, Temperature, Overall Wald χ^2).

Our study is first study on Cholera modeling for prediction Cholera state in Hanoi based on using data mining techniques. The results show that the global presentation is better than the local presentation. In most cases, the outcome model based on using Random Tree algorithm is showed as the best model reached the F1 measure be between 0.79 and 0.91. This study also investigates the mutual relations between the target variable (Cholera value at time-point t) and climate condition variables (climate values at time-points t-1 and t-2) also had been showed and in the some cases, the correlation coefficient reached at level about 0.50.

5 Conclusions

In this paper, we have showed an experimental study on Cholera modeling for prediction Cholera state in Hanoi. This is the first study on Cholera modeling for prediction Cholera state in Hanoi based on using data mining techniques. Two versions of data presentations of local data presentation and global data presentation had been investigated. Some methods of Linear Regression, Naïve Bayes, SVM, and Random Tree had been applied. The results show that the global presentation is better than the local presentation. In most cases, the outcome model is showed the F1 measure be between 0.79 and 0.86. The mutual relations between the target variable (Cholera value at time-point t) and climate condition variables (climate values at time-points t-1 and t-2) also had been showed and in the some cases, the correlation coefficient reached at level about 0.50.

Because the Red river height values had been recorded monthly then the time-step in this study is a month. Some investigations with time-step of a week should be done. In some time periods of Cholera Outbreak in Hanoi, such as three Cholera outbreak waves in the time period from October 2007 to December 2007, some investigations with time-step of a day for each Cholera outbreak wave should be considered. Moreover, ensemble methods such as Average or Advisor Perceptron should be investigated.

Acknowledgments. This work was supported in part by Hanoi Grant 01C-08/-8-2014-2, VNU Grant QG-15-22.

References

1. Andrews, J.R., Basu, S.: Transmission dynamics and control of cholera in Haiti: an epidemic model. Lancet **377**(9773), 1248–1255 (2011)
2. Chunara, R., Andrews, J.R., Brownstein, J.S.: Social and news media enable estimation of epidemiological patterns early in the 2010 Haitian cholera outbreak. Am. J. Trop. Med. Hyg. **86**(1), 39–45 (2012)
3. Colwell, R.R.: Global climate and infectious disease: the cholera paradigm. Science **274** (5295), 2025–2031 (1996)
4. Crooks, A.T., Hailegiorgis, A.B.: An agent-based modeling approach applied to the spread of cholera. Environ. Modell. Softw. **62**(2014), 164–177 (2014)
5. Anh, D.D., Lopez, A.L., Thiem, V.D., Grahek, S.L., Duong, T.N., Park, J.K., Kwon, H.J., Favorov, M., Hien, N.T., Clemens, J.D.: Use of oral cholera vaccines in an outbreak in Vietnam: a case control study. PLoS Neglected Trop. Dis. **5**(1), e1006 (2011)
6. Eisenberga, M.C., Kujbida, G., Tuite, A.R., Fisman, D.N., Tien, J.H.: Examining rainfall and cholera dynamics in Haiti using statistical and dynamic modeling approaches. Epidemics **5**(2013), 197–207 (2013)
7. Emch, M., Feldacker, C., Yunus, M., Streatfield, P.K., DinhThiem, V., Canh do, G., Ali, M.: Local environmental predictors of cholera in Bangladesh and Vietnam. Am. J. Trop. Med. Hyg. **78**(5), 823–832 (2008)
8. Ginsberg, J., Mohebbi, M.H., Patel, R.S., Brammer, L., Smolinski, M.S., Brilliant, L.: Detecting influenza epidemics using search engine query data. Nature **457**, 1012–1014 (2009)
9. Koepke, A.A., Longini Jr., I.M., Elizabeth Halloran, M., Wakeeld, J., Minin, V.N.: Predictive modeling of cholera outbreaks in Bangladesh, 3 February 2014. arXiv: http://arxiv.org/abs/1312.4414 [stat.AP]
10. Lipp, E., Huq, A., Colwell, R.: Effects of global climate on infectious disease: the cholera model. Clin. Microbiol. Rev. **15**(4), 757–770 (2002)
11. Nguyen, B.M., Lee, J.H., Cuong, N.T., Choi, S.Y., Hien, N.T., Anh, D.D., Lee, H.R., Ansaruzzaman, M., Endtz, H.P., Chun, J., Lopez, A.L., Czerkinsky, C., Clemens, J.D., Kim, D.W.: Cholera outbreaks caused by an altered Vibrio cholerae O1 El Tor biotype strain producing classical cholera toxin B in Vietnam in 2007 to 2008. J. Clin. Microbiol. **47**(5), 1568–1571 (2009)

12. Posny, D., Wang, J.: Modelling cholera in periodic environments. J. Biol. Dyn. **8**(1), 1–19 (2014). doi:10.1080/17513758.2014.896482
13. Reiner, R.C., King, A.A., Emch, M., Yunus, M., Faruque, A.S.G., Pascual, M.: Highly localized sensitivity to climate forcing drives endemic cholera in a megacity. Proc. Nat. Acad. Sci. USA **109**, 2033–2036 (2012)
14. Wang, J., Liao, S.: A generalized cholera model and epidemic – endemic analysis. J. Biol. Dyn. **6**(2), 568–589 (2012)
15. World Health Organization: severe acute watery diarrhoea with cases positive for Vibrio cholerae, Viet Nam. Wkly. Epidemiol. Rec. **83**(18), 157–168 (2008)
16. Yue, Y., Gong, J., Wang, D., Kan, B., Li, B., Ke, C.: Influence of climate factors on Vibrio cholerae dynamics in the Pearl River estuary, South China. World J. Microbiol. Biotechnol. (2014). doi:10.1007/s11274-014-1604-5
17. http://www.statsoft.com/Products/STATISTICA/Product-Index
18. http://sourceforge.net/projects/weka/

On Velocity-Preserving Trajectory Simplification

Josh Jia-Ching Ying[1(✉)] and Ja-Hwung Su[2]

[1] Department of Computer Science, National Chiao Tung University, Hsinchu, Taiwan
jashying@gmail.com

[2] Department of Information Management, Cheng Shiu University, Kaohsiung, Taiwan
bb0820@ms22.hinet.net

Abstract. Trajectory data plays crucial role in many real-world applications with moving objects. The size of trajectory dataset is always very huge because of high sampling rate. Therefore, it is desired to simplify each trajectory before it is stored and processed. As the result, many trajectory simplification notions have been proposed. However, existing studies on trajectory simplification more or less rely on geometric-preserving manner (e.g., minimizing position-based or direction-based errors). These manners directly avoid effectiveness of velocity in many real-world applications. Actually, the velocity of a moving object is very important in many real-world applications, such as map-matching, mobility prediction, moving pattern mining, etc. In this paper, we propose a novel trajectory simplification, velocity-preserving trajectory simplification (VPTS), which minimize both geometric error and velocity error. We present an efficient algorithm for optimal velocity-preserving trajectory simplification. Through a series of experimental evaluation with real trajectory data, we examine the benefit of our proposed velocity-preserving trajectory simplification.

Keywords: Velocity-based error · Trajectory simplification · Moving object · Data mining

1 Introduction

With the widely development of market of GPS-embedded devices (e.g., smart phones and navigation system), trajectory data is becoming easily collected. Thus, users' movement analysis has been addressed comprehensively in the past decades in the literature of trajectory analysis [13, 15]. Since the trajectory data is usually very large, simplifying trajectory data plays crucial role in many real-world application. For example, consider a city with 1 million smart phone users. Suppose that we set the sampling rate as once per minute to track users' movement. The size of the collected trajectories for just one day is more than 30 GB.

Actually, most existing query processing and data mining algorithms on trajectory data are very time-consuming and thus cannot be used with whole collected trajectory data that is too huge. A possible way to solve the problem of huge size is sampling less frequently to reduce the size of the trajectory data. However, in real-world application, moving objects have great variance in their velocities. For example, a smartphone user

N. T. Nguyen et al. (Eds.): ACIIDS 2016, Part II, LNAI 9622, pp. 241–250, 2016.
DOI: 10.1007/978-3-662-49390-8_23

might stay in a certain location for a while then go another place by a bus. Obviously, we require less frequent observations when the user stays than when he moves. Therefore, standard practice is to oversample initially, and then to simplify by eliminating some unnecessary GPS point from a collected GPS trajectory.

To do so, several algorithms have been proposed for simplifying GPS trajectory data [6, 8, 11, 12, 14]. However, these algorithms assume that the simplified trajectories should preserve the position or direction information, which is similar to the position or direction information captured in the original trajectories. We can call them geometric-preserving trajectory simplification algorithms. However, these geometric-preserving trajectory simplification algorithms can not preserve users' moving behavior. Take Fig. 1 as an example. The Trajectory_1 is original trajectory. Suppose the sample rate is fixed. In other words, the time interval between each pair of contiguous GPS point in Trajectory_1 is fixed. Trajectory_2 is metric-preserving simplified trajectory, and Trajectory_3 is velocity-preserving simplified trajectory. We can see that the Trajectory_3 is different from Trajectory_2. Trajectory_3 preserve GPS point p_3, p_4, and p_8 such that we can aware that the velocity form p_3 to p_4 is very different from that form p_4 to p_7.

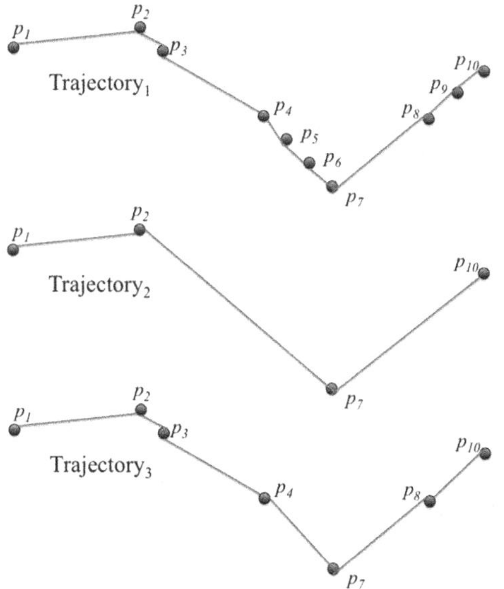

Fig. 1. An example of geometric-preserving trajectory simplification and velocity-preserving trajectory simplification

In this paper, we propose a new trajectory simplification mechanism called Velocity-Preserving Trajectory Simplification (VPTS) such that the velocity of each segment of trajectory could be significantly preserved. Meanwhile, we propose a velocity-based measurement V_t, which is defined to measure the error of a simplified trajectory in terms of the velocity information. Let *Tr* be a trajectory and *Tr'* be a simplification of *Tr*. The simplification error of *Tr'* under V_t, denoted by *err*(*Tr'*), is equal to the maximum difference of velocity between *Tr* and t *Tr'*. Then, the problem of VPTS is to simplify a given GPS trajectory such that its size is minimized under a acceptable simplification error (i.e., *err*(*Tr'*)) by a given error tolerance *max_err*, here the error tolerance *max_err* is user-specified and belongs to [0, π).

In summary, in this study we have a number of contributions as follows:

- We propose a novel notion of velocity-preserving trajectory simplification (VPTS), which can completely preserve smartphone users' moving and staying behaviors for real-world applications on trajectory data.
- We show that VPTS not only preserves velocity information, but also preserves geometric information, such that the simplified trajectory can support a wide range of applications.
- We adopt a common dynamic programming (DP) technique for VPTS. To solve inefficient problem of dynamic programming, we propose an efficient algorithm called approximated VPTS for VPTS. SP is an approximate algorithm for solving VPTS.
- We use a real-world trajectory dataset obtained from WnSN, SJTU [16] to evaluate the performance of our proposal. The results show that our proposed approach not only are faster than state-of-the-art trajectory simplification, DPTS [4, 5], over 100 times but also outperforms other trajectory simplification techniques [6] in terms of accuracy.

For the rest of this paper, we first briefly review the related work, and then describe our proposed velocity-preserving trajectory simplification. Next, we present dynamic programming for VPTS and an efficient algorithm for seeking approximation of VPTS. Finally, we evaluate the performance of different components of our map-matching system and compare our approach with the existing map-matching method.

2 Related Work

Many trajectory simplification techniques have been proposed. We categorize them by the main idea employed in the algorithm as follows. They are Split [6, 11], Merge [12, 14], Greedy [8, 11] and Dead-Reckoning [10]. Split is an approach which finds a position in a given trajectory, according to the heuristic value of the position, to split the whole trajectory into two sub-trajectories and continues the process iteratively on each of the split sub-trajectories which can- not be approximated by a line segment connecting its start position and its end position. Merge is an approach which finds two adjacent segments in a given trajectory, according to the heuristic value computed from these two adjacent segments, discards the position p bridging these two segments, and create a segment connecting the non-bridging end position of one segment and the non-bridging end position of the other segment. It continues the process iteratively until discarding any position p violates the error tolerance. Greedy is an approach which finds a sequence of the greatest number of consecutive segments to be discarded and create a segment connecting the two end positions of this sequence iteratively until discarding any sequence of 2 consecutive segments violates the tolerance constraint. Dead-Reckoning is an online algorithm which reads each position sequentially and determines whether this position is discarded or not according to a heuristic criterion.

Other related studies include [1] which studies the error bounds of several queries on the simplified trajectories with bounded simplification errors mainly measured by the position information, [3] which studies the trajectory simplification problem with the consideration of the shape and also the semantic meanings of the trajectory, [2] which introduces a multi-resolution polygonal curve approximation (also called line simplification) algorithm for trajectory simplification, and [7] which studies the trajectory simplification problem where the trajectories are constrained to a road network. Long *et al.* [4, 5] focus on direction information for trajectory simplification which can preserve the direction information in the simplified trajectory. None of these studies pay attention to the velocity information for trajectory simplification.

3 Velocity-Preserving Trajectory Simplification

To our best knowledge, existing trajectory simplification methods always consider the geometric limitation of GPS points for measuring the probability of eliminating a GPS point from a trajectory. However, people always do not move fast when they are staying somewhere such that the velocities of trajectory could not achieve the high speed. Besides, when a user stays in a certain place, the direction or position of a trajectory might violently be changed because of GPS error. Accordingly, the traditional Geometric-Preserving trajectory simplification methods probably preserve many redundant GPS points and eliminate necessary GPS points. In other word, the traditional trajectory simplification methods cannot work well based only on the geometric information such as direction and position. Therefore, we propose a novel velocity error measurement instead of the geometric error measurement for trajectory simplification.

Before introducing the velocity error measurement, we first describe the formal definitions for illustrating the velocity error measurement metric:

Definition 1 (GPS Trajectory). A GPS trajectory (Tr) is a sequence of GPS points, each of which contains a latitude ($p_i.Lat$), longitude ($p_i.Lngt$) and timestamp ($p_i.T$). Thus, $Tr = p_1 \rightarrow p_2 \rightarrow \ldots \rightarrow p_n$, where $p_i.T < p_{i+1}.T$.

Definition 2 (Trajectory Traffic Velocity). Given two GPS points, p_i and p_{i+k}, of a GPS trajectory (Tr), the Trajectory Traffic Velocity, denoted as $TTV(p_i, p_{i+k})$, is defined as the average velocity of the moving object which moves along with the Tr from p_i to p_{i+k}. The $TTV(p_i, p_{i+k} \mid Tr)$ is formally defined as follows:

$$TTV(p_i, p_{i+k}|Tr) = \frac{\sum_{n=1}^{k} |p_{i+n} - p_{i+n-1}|}{p_{i+k}.T - p_i.T}, \tag{1}$$

where $|p_{i+n} - p_{i+n-1}|$ indicates the geometric distance between p_{i+n} and p_{i+n-1}.

Definition 3 (Average Traffic Velocity). Given two GPS points, p_i and p_{i+k}, of a GPS trajectory (*Tr*), the Average Traffic Velocity, denoted as $ATV(p_i, p_{i+k})$, is defined as the average velocity of the moving object which moves along with straight line from p_i to p_{i+k}. The $ATV(p_i, p_{i+k})$ is formally defined as follows:

$$ATV(p_i, p_{i+k}) = \frac{|p_{i+k} - p_i|}{p_{i+k}.T - p_i.T}, \quad (2)$$

where $|p_{i+k} - p_i|$ indicates the geometric distance between p_{i+k} and p_i.

Definition 4 (Velocity-Based Error Measurement). Given a GPS trajectory (*Tr*) and its simplified GPS trajectory (*Tr'*), the velocity-based error measurement, denoted as $V_t(Tr' \mid Tr)$, is defined as the average relative difference between Trajectory Traffic Velocity and Average Traffic Velocity. The $V_t(Tr' \mid Tr)$ is formally defined as follows:

$$V_t(Tr'|Tr) = \frac{1}{|Tr'|} \sum_{p \in Tr'} \frac{|ATV(p, N(p, Tr')) - TTV(p, N(p, Tr')|Tr)|}{TTV(p, N(p, Tr')|Tr)}, \quad (3)$$

where $N(p,Tr')$ indicates next GPS point of p in *Tr'*

Definition 5 (Velocity-Preserving Simplified Trajectory). Given a GPS trajectory (*Tr*) and a tolerance *e*, velocity-preserving simplified trajectory (*Tr'*) is defined as the $V_t(Tr' \mid Tr) > e$ and including least GPS points.

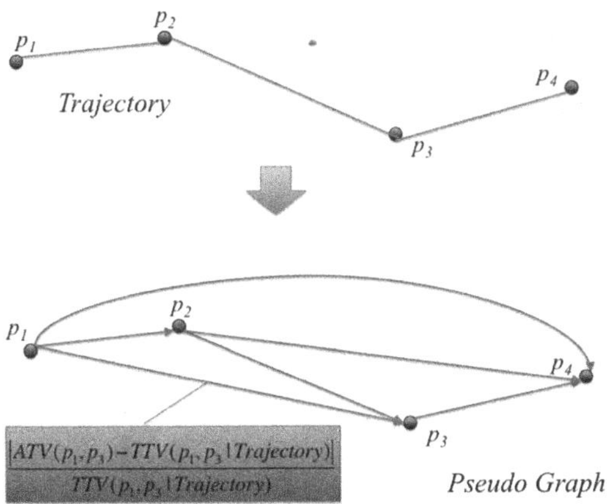

Fig. 2. An example of pseudo graph

According to above-mentioned definition, the velocity-preserving trajectory simplification could be formulated as a shortest path search problem. We can build a pseudo graph, as shown in Fig. 3, in which a node represents a GPS point, and each edge is weighted by velocity-based error. Therefore, we can utilize the Dynamic Programming to find shortest path with the velocity-based error constraint from the start and destination of the trajectory as the simplified trajectory (Fig. 2).

```
Input: tolerance e, A trajectory Tr = p1 → p2 → ... → pn
Output: A simplified trajectory Tr'
ps ← p1
err ← 0
Tr' ← ∅
for i from 2 to n
    err ← |ATV(ps, pi)- TTV(ps, pi|Tr)| / TTV(ps, pi|Tr)
    if err > e then
        err ← 0
        Tr' ← Tr' ∪ ps
        ps ← pi
    end
end
return root
```

Fig. 3. Approximation Algorithm of VPTS

4 Approximation of VPTS

Since the Dynamic Programming requires the velocity-based error between each pair of GPS points, the time complexity of the algorithm should be O(n2). That is, if a trajectory consists of large number of GPS point, the execution time should be very long. Accordingly, we prefer to find an approximated solution and bound the algorithm by O(n). To do so, we propose an algorithm shown in Fig. 4 to find Approximation of VPTS.

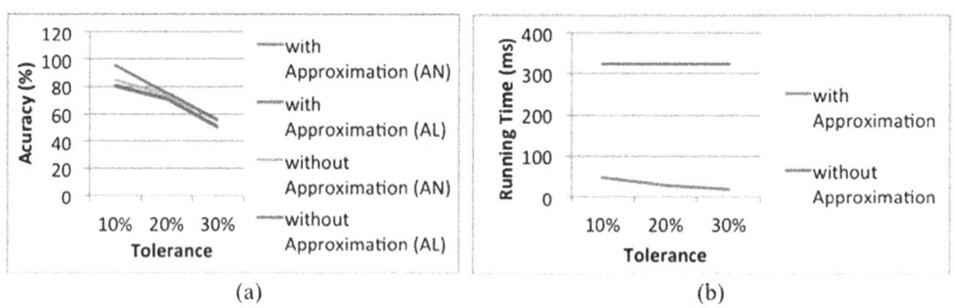

Fig. 4. Impact of Road Network Decomposition under Various Maximum Tolerance

5 Experiments

In this section, we conduct a series of experiments to evaluate the performance for the proposed VPSF using a real trajectory set obtained from WnSN, SJTU [16]. All the experiments are implemented in Java JDK 1.6 on an Intel Xeon CPU W3520 2.67 GHz machine with 24 GB of memory running Microsoft Windows win7. We first describe the data preparation on the real trajectory set obtained from WnSN, SJTU [16], and the evaluation methodology. Then, we show our experimental results for following discussions.

5.1 Evaluation Dataset and Methodology

We use the trajectory data from GPS-equipped taxis of Shanghai city, which were collected by Wireless and Sensor networks Lab (WnSN), Shanghai Jiao Tong University [16]. However, the GPS points of trajectories are not labeled whether they are redundant, i.e., there is no ground truth for trajectory simplification. As the result, we use a vehicle movement simulation model [9] to insert the redundant GPS points of trajectory.

The follows are the main measurements for the experimental evaluations. The *Accuracy by Number* (*AN*) and *Accuracy by Length* (*AL*) are defined as Eqs. (4) and (5).

$$AN = \frac{\#\,\text{correctly eliminated}\ GPS\ \text{points}}{\#\,\text{redundant}\ GPS\ \text{points}} \tag{4}$$

$$AL = \frac{\textit{total length of}\ \text{correctly eliminated GPS}\ \textit{segments}}{\textit{total length of}\ \text{redundant GPS points}} \tag{5}$$

5.2 Experimental Results

We divide the experiment into two parts: (1) impact of Approximated Algorithm and (2) comparison of existing Trajectory Simplification. We examine the impact of Approximated Algorithm in terms of running time, Accuracy by Number and Accuracy by Length. For the comparison of existing Trajectory Simplification, we compare our method with several state-of-the-art map-matching methods, DPTS [5] and Douglas-Peucker algorithm [6], in terms of running time, Accuracy by Number and Accuracy by Length.

5.2.1 Impact of Road Network Decomposition

Figure 4(a) shows the Accuracy by Number and Accuracy by Length of our trajectory simplification method with/without approximated algorithm under various maximum tolerance. We can observe that all results of our trajectory simplification without approximated algorithm are slightly accurate than that with approximated algorithm. The reason is that the method without approximated algorithm consider whole trajectory for trajectory simplification task. Therefore, it would be more accurate. However, it just achieves about 5 % to 10 % accuracy higher than the method with approximated algorithm. But, if we consider the running time shown in Fig. 4(b), we can find that such tiny accuracy gain would cost about 6 times as much running time. Accordingly, our approximated algorithm is still an acceptable and necessary component in trajectory simplification method.

Besides, we also can see that the tolerance setting would affect our method. The reason is that our velocity of moving object is always changed such that the optimum solution might be effect by the tolerance setting significantly. It does not easily distinguish the redundant GPS points form the candidate set.

5.2.2 Comparison of Existing Map-Matching Method

Figure 5 shows the Accuracy by Number and Accuracy by Length of our VPTS method and several state-of-the-art trajectory simplification methods, DPTS [5] and *Douglas-Peucker* algorithm [6]. We can observe that all results of our VPTS method are slightly more accurate than DPTS [5] and *Douglas-Peucker* algorithm [6]. The reason is that the all these state-of-the-art trajectory simplification methods partially consider spatial-temporal information to help improve accuracy of trajectory simplification task. Our VPTS method fully and deeply addresses spatial-temporal information, i.e., velocity, for improving accuracy of trajectory simplification task. Therefore, our method would be more accurate than the state-the-art trajectory simplification method.

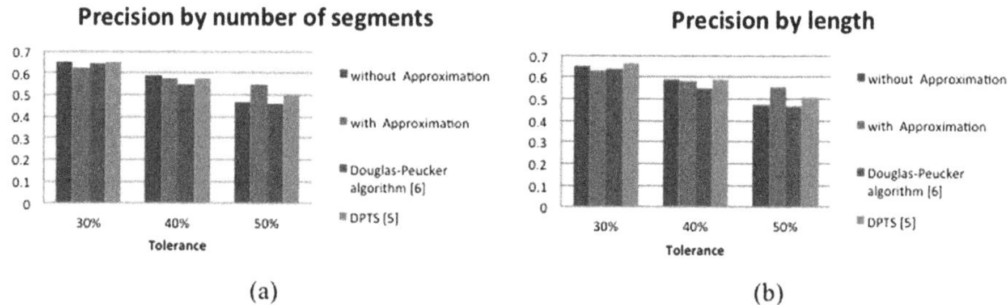

Fig. 5. Comparison with Existing Trajectory Simplification Methods in Terms of Accuracy

However, as shown in Fig. 6, we can find that our method is eleven times slower than DPTS [5] and *Douglas-Peucker* algorithm [6]. The reason is that our method involves the velocity error which cost a lot of computation time.

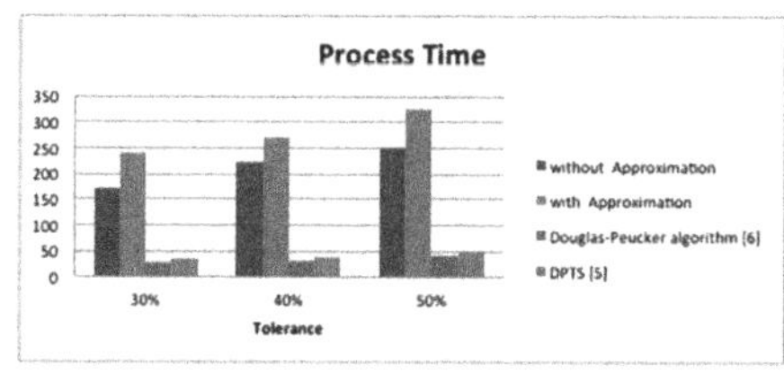

Fig. 6. Comparison with Existing Trajectory Simplification Methods in Terms of Execution Time

6 Conclusions and Future Work

In this paper, we propose a new breed of trajectory simplification called Velocity-Preserving trajectory simplification (*VPTS*) to simplify the GPS trajectory data. The trajectory simplification method fully and deeply addresses velocity information preserving for improving accuracy of trajectory simplification task. Based on preserved velocity information, we propose an approximation for speeding up the trajectory simplification task. The experiment results demonstrate that our *VPTS* significantly outperforms state-of-the-art trajectory simplification methods, DPTS [5] and *Douglas-Peucker* algorithm [6], in terms of accuracy and running time. In our future work, we plan to deal with the problem that trajectory simplification algorithm is sensitive to the GPS sampling rate.

Acknowledgement. This research was partially supported by Ministry of Science and Technology, Taiwan, R.O.C. under grant no. MOST 104-2632-S-424-001 and MOST 104-2221-E-230-019.

References

1. Cao, H., Wolfson, O., Trajcevski, G.: Spatio-temporal data reduction with deterministic error bounds. VLDB J. **15**(3), 211–228 (2006)
2. Chen, M., Xu, M., Fränti, P.: A fast o(n) multiresolution polygonal approximation algorithm for GPS trajectory simplification. IEEE Trans. Image Process. **21**(5), 2770–2785 (2012)
3. Chen, Y., Jiang, K., Zheng, Y., Li, C., Yu, N.: Trajectory simplification method for location-based social networking services. In: Proceedings of the 2009 International Workshop on Location Based Social Networks, pp. 33–40. ACM (2009)
4. Long, C., Wong, R.C.-W., Jagadish, H.V.: Trajectory simplification: On minimizing the direction-based error. In: The 41st International Conference on Very Large Data Bases (VLDB 2015), Kohala Coast, Hawaii, USA, 31 August-4 September 2015 (2015)
5. Long, C., Wong, R.C.-W., Jagadish, H.V.: Direction-preserving trajectory simplification. In: The 39th International Conference on Very Large Data Bases (VLDB 2013), Riva del Garda, Trento, Italy, 26-30 August 2013 (2013)
6. Douglas, D., Peucker, T.: Algorithms for the reduction of the number of points required to represent a digitized line or its caricature. Can. Cartographer **11**(2), 112–122 (1973)
7. Kellaris, G., Pelekis, N., Theodoridis, Y.: Map-matched trajectory compression. J. Syst. Softw. **86**, 1566–1579 (2013)
8. Kolesnikov, A.: Efficient online algorithms for the polygonal approximation of trajectory data. In: MDM 2011, pp. 49–57 (2011)
9. Lan, K.-C., Chou, C.-M.: Realistic mobility models for vehicular ad hoc network (VANET) simulations. In: IEEE ICST 2008, October 2008
10. Lange, R., Farrell, T., Durr, F., Rothermel, K.: Remote real-time trajectory simplification. In: PerComm 2009, pp. 1–10 (2009)
11. Meratnia, N., de By, R.: Spatiotemporal compression techniques for moving point objects. In: Bertino, E., Christodoulakis, S., Plexousakis, D., Christophides, V., Koubarakis, M., Böhm, K. (eds.) EDBT 2004. LNCS, vol. 2992, pp. 765–782. Springer, Heidelberg (2004)
12. Muckell, J., Hwang, J.H., Patil, V., Lawson, C.T., Ping, F., Ravi, S.: Squish: an online approach for gps trajectory compression. In: COM.Geo 2011, pp. 13:1–13:8 (2011)

13. Patel, D., Sheng, C., Hsu, W., Lee, M.L.: Incorporating duration information for trajectory classification. In: ICDE 2012, pp. 1132–1143 (2012)
14. Potamias, M., Patroumpas, K., Sellis, T.: Sampling trajectory streams with spatiotemporal criteria. In: SSDBM 2006, pp. 275–284 (2006)
15. Singh, M., Zhu, Q., Jagadish, H.: Swst: A disk based index for sliding window spatio-temporal data. In: ICDE 2012, pp. 342–353 (2012)
16. SUVnet-Trace data: http://wirelesslab.sjtu.edu.cn

Computational Intelligence in Data Mining for Complex Problems

Sink Toward Source Algorithm Finding Maximal Flows on Extended Mixed Networks

Viet Tran Ngoc[1(✉)], Hung Hoang Bao[1], Chien Tran Quoc[2], Thanh Le Manh[3], and Van Hoang Thi Khanh[1]

[1] Lecturer of Computer Science Department, Vietnam Korea Friendship IT College Hoaquy, Danang, Vietnam
{viettn,hunghb,vanhtk}@viethanit.edu.vn

[2] Lecturer of Computer Science Department, The University of Danang, 459 Ton Duc Thang, Danang, Vietnam
tqchien@dce.udn.vn

[3] Lecturer of Computer Science Department, Hue University, 77 Nguyen Hue, Hue, Vietnam
lmthanh1953@yahoo.com

Abstract. Graph is a powerful mathematical tool applied in many fields as transportation, communication, informatics, economy, … In ordinary graph the weights of edges and vertexes are considered independently where the length of a path is the sum of weights of the edges and the vertexes on this path. However, in many practical problems, weights at a vertex are not the same for all paths passing this vertex, but depend on coming and leaving edges. The paper develops a model of extended network that can be applied to modelling many practical problems more exactly and effectively. The main contribution of this paper is sink toward source algorithm finding maximal flows on extended mixed networks.

Keywords: Graph · Extended mixed networks · Network · Flow · Maximal flow · Algorithm

1 Introduction

Graph is a powerful mathematical tool applied in many fields as transportation, communication, informatics, economy, … In ordinary graph the weights of edges and vertexes are considered independently where the length of a path is simply the sum of weights of the edges and the vertexes on this path. However, in many practical problems, weights at a vertex are not the same for all paths passing this vertex, but depend on coming and leaving edges. Therefore, a more general type of weighted graphs, called extended weighted graph, is defined in this work. The paper develops a model of extended mixed network that can be applied to modelling many practical problems more exactly and effectively. Therefore, necessary to build a model of the extended network so that the stylization of practical problems can be applied more accurately and effectively. Based on the results of the study of the problem regarding finding the maximum flow [1, 2] and extended graphs [3, 6], the main contribution of this paper is the sink toward source algorithm finding maximal flows on extended mixed networks and improving computing performance.

N.T. Nguyen et al. (Eds.): ACIIDS 2016, Part II, LNAI 9622, pp. 253–260, 2016.
DOI: 10.1007/978-3-662-49390-8_24

2 Extended Mixed Network

A network is a mixed graph of the traffic $G = (V, E)$, circles V and roads E. Roads can be classified as either direction or non-direction. There are many sorts of means of transportation on the network. The non-direction shows two-way roads while the direction shows one-way roads. Given a group of the functions on the network as follows:

+ The function of the route circulation possibility c_E: $E \rightarrow R^*$, $c_E(e)$ the route circulation possibility $e \in E$.
+ The function of the circle circulation possibility $c_V : V \rightarrow R^*$, $c_V(u)$ the circle circulation possibility $u \in V$.
+ $G = (V, E, c_E, c_V)$: extended mixed network.

3 Flow of the Extended Mixed Network

Given an extended mixed network $G = (V, E, c_E, c_V)$, a source point a and a sink point z.
Set: $\{f(x,y) \mid (x,y) \in E\}$, is called the flow of network G if the requirements are met:

(i) $0 \le f(x,y) \le c_E(x,y) \; \forall (x,y) \in E$
(ii) Any value of point r is referring to neither a sourse point nor a sink point

$$\sum_{(v,r)\in E} f(v,r) = \sum_{(r,v)\in E} f(r,v)$$

(iii) Any value of point r is referring to neither a sourse point nor a sink point

$$\sum_{(v,r)\in E} f(v,r) \le c_V(r)$$

Expression: $v(F) = \sum_{(a,v)\in E} f(a,v)$, is called the value of flow F.

- ***The maximum problem:*** Given an extended mixed network $G = (V, E, c_E, c_V)$, a source point a and a sink point z. The task required by the problem is finding the flow which has a maximum value. The flow value is limited by the total amount of the circulation possibility on the roads starting from source points. As a result of this, there could be a confirmation on the following theorem.

Theorem 1. Given an extended mixed network $G = (V, E, c_E, c_V)$, a source point a and a sink point z, then exist is the maximal flow [1].

4 Sink Toward Source Algorithm

+ ***Input:*** Given an extended mixed network $G = (V, E, c_E, c_V)$, a source point a and a sink point z [2, 6]. The points in graph G are arranged in a certain order.
+ ***Output:*** Maximal flow $F = \{f(x,y) \mid (x,y) \in E\}$.

(1) ***Start:***

The departure flow: $f(x,y) := 0,\ \forall (x,y) \in E$.

Points from the sink points will gradually be labelled L_1 for the first time including 5 components.

Form backward label:

$L_1(v) = [\downarrow, prev_1(v), c_1(v), d_1(v), bit_1(v)]$ and can be label $(\downarrow)$ for the second time $L_2(v) = [\downarrow, prev_2(v), c_2(v), d_2(v), bit_2(v)]$.

Put labeling $(\downarrow)$ for sink point:

$$z[\downarrow, \phi, \infty, \infty, 1]$$

The set T comprises the points which have already been labelled $(\downarrow)$ but are not used to label $(\downarrow)$, T' is the point set labelled $(\downarrow)$ based on the points of the set T. Begin $T: = \{z\}$, $T': = \phi$

(2) ***Backward label generate***

(2.1) Choose backward label point:

- Case $T \neq \phi$: Choose the point $v \in T$ of a minimum value. Remove the v from the set T, $T: = T\backslash\{v\}$. Assuming that the backward label of v is $[\downarrow, prev_i(v), c_i(t), d_i(t), bit_i(t)]$, $i = 1$ or 2. B is the set of the points which are not backward label time and adjacent to the backward label point v. Step (2.2).
- Case $T = \phi$ and $T' \neq \phi$: Assign $T: = T'$, $T': = \phi$. Return to step (2.1).
- Case $T = \phi$ and $T' = \phi$: The flow F is the maximum. End.

(2.2) Backward label the points which are not backward label and are adjacent to the backward label points v

- Case $B = \phi$: Return to step (2.1).
- Case $B \neq \phi$: Choose $t \in B$ of a minimum value. Remove the t from the set B, $B: = B\backslash\{t\}$. Assign backward labeled point t:

If $(t,v) \in E, f(t,v) < c_E(t,v),\ bit_i(v) = 1$ put backward label point t: $prev_j(t) := v$;
$c_j(t): = \min\{c_i(v), c_E(t,v) - f(t,v)\}$, if $d_i(v) = 0$,
$c_j(t): = \min\{c_i(v), c_E(t,v) - f(t,v), d_i(v)\}$, if $d_i(v) > 0$;
$d_j(t) := c_V(t) - \sum_{(i,t)\in E} f(i,t)$;
$bit_j(t): = 1$, if $d_j(t) > 0$,
$bit_j(t): = 0$, if $d_j(t) = 0$.
If $(v,t) \in E, f(v,t) > 0$ put backward label point t: $prev_j(t) := v$;
$c_j(t): = \min\{c_i(v), f(v,t)\}$,
$d_j(t) := c_V(t) - \sum_{(i,t)\in E} f(i,t)$; $bit_j(t): = 1$.
If t is not backward label, then return to step (2.2).

If t is backward label and $t = a$, then making adjustments in increase of the flow. Step (3).
If t is backward label and $t \neq a$, then add t to T', $T' := T' \cup \{t\}$, and return to step (2.2).

(3) ***Making adjustments in increase of the flow***
Suppose t is backward label [$\downarrow$, $prev_i(t)$, $c_i(t)$, $d_i(t)$, $bit_i(t)$]:

(3.1) Adjustment made from t back to z according to forward label

(3.1.1) Start
$x := a$, $y := prev_1(a)$, $\delta := c_1(a)$.

(3.1.2) Making adjustments

(i) Case (x, y) the road section whose direction runs from x to y: put $f(x,y) := f(x,y) + \delta$.

(ii) Case (y, x) the road section whose direction runs from y to x: put $f(y,x) := f(y,x) - \delta$.

(iii) Case (x, y) non-direction roads:

If $f(x,y) \geq 0$ and $f(y,x) = 0$ then put $f(x,y) := f(x,y) + \delta$.

If $f(y,x) > 0$ then put $f(y,x) := f(y,x) - \delta$.

(3.1.3) Moving

(i) Case $x = z$, then step (3.2).

(ii) Case $x \neq z$, put $x := y$ and $y := k$, k is the second component of the backward labeled point x. Then return to step (3.1.2).

(3.2) Remove all the labels of the network points, except for the sink point z. Return to step (2).

Theorem 2. If the value of the route circulation possibility and the circle circulation possibility are integers, then after a limited number of steps, the processing of the maximum network problem will end.

Proof. According to Theorem 1, after each time of making adjustment of the flow, the flow will be increased with certain units (due to c_E is a whole number, c_V is a whole number, and δ is, therefore, a positive whole number). On the other hand, the value of the flow is limited above by the total amount of the circulation possibility at roads leaving the source points. So, after a limited number of steps, the processing of the maximum network problem will end.

Theorem 3. Given an F = $\{f(x,y) \mid (x,y) \in E\}$ is the flow on extended mixed network *G*, a source point *a* and a sink point *z*:

$$\sum_{(a,x)\in E} f(a,x) = \sum_{(x,z)\in E} f(x,z)$$

Proof. The points of the set *V*. If *x,y* is not previous, assign $f(x,y) = 0$

$$\sum_{y\in V}\sum_{x\in V} f(x,y) = \sum_{y\in V}\sum_{x\in V} f(y,x)$$

$$\Leftrightarrow \sum_{y\in V}\left(\sum_{x\in V} f(x,y) - \sum_{x\in V} f(y,x)\right) = 0$$

$$\Leftrightarrow \sum_{y\in V\backslash\{a,z\}}\left(\sum_{x\in V} f(x,y) - \sum_{x\in V} f(y,x)\right) + \left(\sum_{x\in V} f(x,z) - \sum_{x\in V} f(z,x)\right) + \left(\sum_{x\in V} f(x,a) - \sum_{x\in V} f(a,x)\right) = 0$$

$$-\sum_{(a,x)\in E} f(a,x) + \sum_{(x,z)\in E} f(x,z) = 0 \Leftrightarrow \sum_{(a,x)\in E} f(a,x) = \sum_{(x,z)\in E} f(x,z)$$

- ***The complexity of the algorithm:*** It is assumed that the road circulation possibility and the point circulation possibility are whole integer. After each round step, to find the roads to increase the amount of circulation on the flow, we have to approve to pass $|E|$ roads in maximum, and in order to adjust the flow we have to approve to pass $2.|V|$ roads, in maximum. As a result, the complexity of each time of increasing the flow is $O(|E| + 2.|V|)$. Mark v^* is the value of the maximum flow. The number of times to increase the flow in maximum is v^*. So the complexity of the algorithm is $O(v^*(|E| + 2.|V|))$.

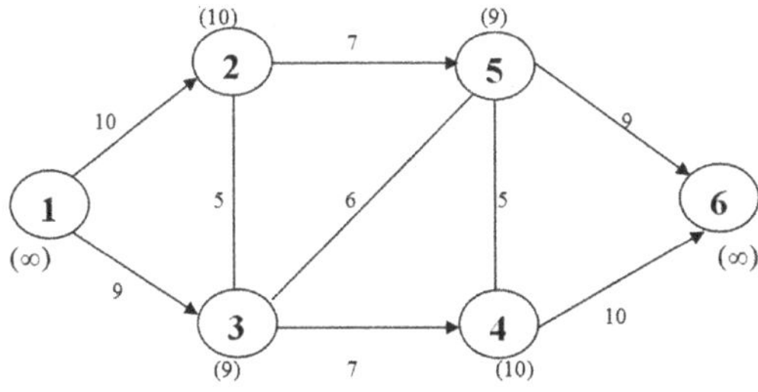

Fig. 1. Extended mixed network

5 Result of the Experiment

Given an extended mixed network graph Fig. 1. The network has six circles, six direction roads and three non-direction roads. The road circulation possibility c_E and the circle circulation possibility c_V. The source point is 1, the sink point is 6.

+ Result of the first backward label (Fig. 2):
  Sink point is **6**: backward label [↓, ϕ, ∞, ∞, 1]
  Point **5**: backward label [↓, 6, 9, 9, 1]
  Point **4**: backward label [↓, 6, 10, 10, 1]
  Point **3**: backward label [↓, 4, 7, 9, 1]
  Point **2**: backward label [↓, 5, 7, 10, 1]
  Point **1**: backward label [↓, 3, 7, ∞, 1]

Result of the flow increasing adjustment in Fig. 3 and the value of the increase $v(F) = 7$

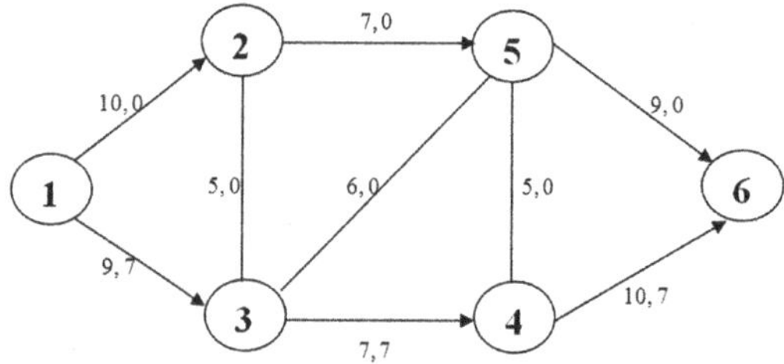

Fig. 3. The value of the increase $v(F) = 7$

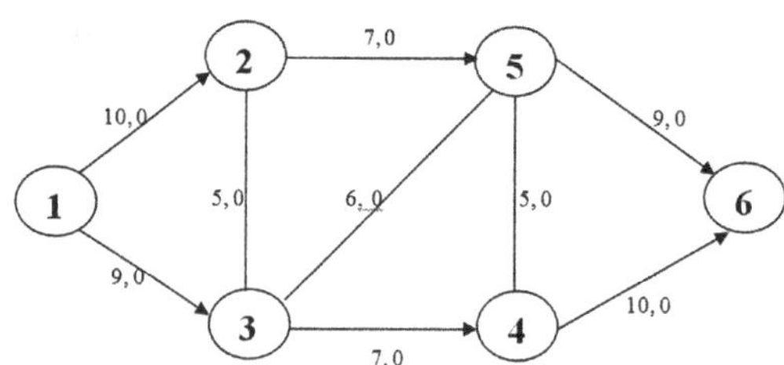

Fig. 2. The departure flow 0

+ Result of the second backward label:
  Sink point is **6**: backward label [↓, ϕ, ∞, ∞, 1]
  Point **5**: backward label [↓, 6, 9, 9, 1]
  Point **4**: backward label [↓, 5, 5, 3, 1]
  Point **3**: backward label [↓, 5, 6, 2, 1]
  Point **2**: backward label [↓, 5, 7, 10, 1]

Point **1**: backward label [↓, 2, 7, ∞, 1]
Result of the flow increasing adjustment in Fig. 4 and the value of the increase $v(F) = 14$

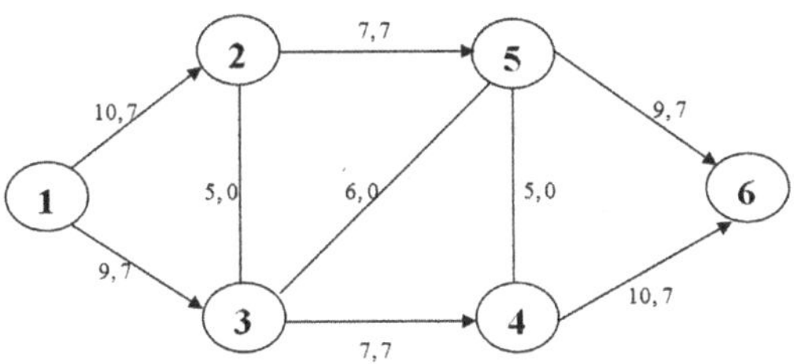

Fig. 4. The value of the increase $v(F) = 14$

+ Result of third backward label:
  Sink point is **6**: backward label [↓, ϕ, ∞, ∞, 1]
  Point **5**: backward label [↓, 6, 2, 2, 1]
  Point **4**: backward label [↓, 6, 3, 3, 1]
  Point **3**: backward label [↓, 5, 2, 2, 1]
  Point **2**: backward label [↓, 3, 2, 3, 1]
  Point **1**: backward label [↓, 2, 2, ∞, 1]
  Result of the flow increasing adjustment in Fig. 5 and the value of the increase $v(F) = 16$.

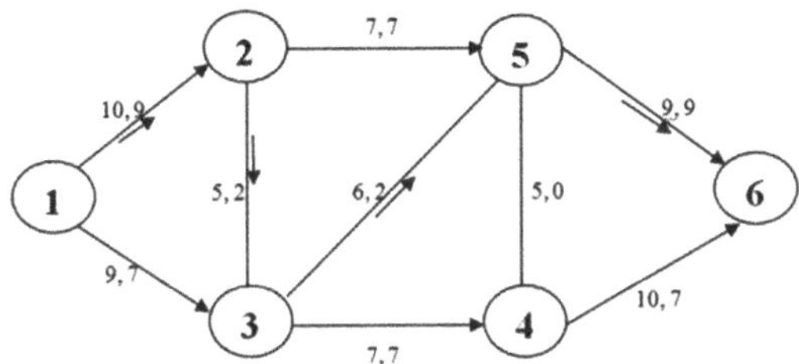

Fig. 5. The value of the increase $v(F) = 16$

This is the maximum flow, because in the following backward label is not labelled - Source point is **1**.

6 Conclusion

The article regarding building a model of an extended mixed network so that the stylization of practical problems can be applied more accurately and effectively. Next, sink toward source algorithm finding maximal flows on extended mixed networks is being built. Finally, a concrete example is presented to illustrate sink toward source algorithm.

References

1. Chien, T.Q.: The problem to find the maximum flow on the network. The research project on the Ministry level, B2005-16-34, Vietnam (2005)
2. Chien, T.Q.: Weighted source-sink alternative algorithm. Univ. Danang-Vietnam J. Sci. Technol. **3**(26), 99–105 (2008)
3. Chien, T.Q., Tue, N.M., Viet, T.N.: Shortest path algorithm on extended graphs. In: Proceedings of the 6th National Conference on Fundamental and Applied Information Technology Research (FAIR), Vietnam, pp. 522–527 (2013)
4. Taylor, M.A.P. (ed.): Transportation and traffic theory in the 21st Century. Pergamon Press Amsterdam, The Netherlands (2002)
5. Johnson, E.L., Committee Chair, Nemhauser, G.L.: Shortest paths and multicommodity network flow (2003)
6. Ngoc, V.T., Quoc, C.T., Van, T.N.: Improving computing performance for algorithm finding maximal flows on extended mixed networks. J. Inf. Comput. Sci. 075–080, England, UK (2015)
7. Trevisan, L.: Generalizations of the maximum flow problem. Stanford University-CS261: Optimization (2011)
8. Madry, A.: Faster approximation schemes for fractional multicommodity flow problems via dynamic graph algorithms. Massachusetts Institute of Technology (2009)

An Efficient Algorithm for a New Constrained LCS Problem

Daxin Zhu[1], Yingjie Wu[2], and Xiaodong Wang[3](✉)

[1] Quanzhou Normal University, Quanzhou 362000, China
[2] Fuzhou University, Fuzhou 350002, China
[3] Fujian University of Technology, Fuzhou 350108, China
wangxd135@139.com

Abstract. The solution to a generalized longest common subsequence problem is addressed in this paper. In this problem, two given sequences of their length s and t must be included as the subsequences of the two main sequences A and B, and the result subsequences must be the longest. If the lengths of the two main sequences A and B are n and m, and the lengths of their two constrained sequences are s and t respectively, the time complexity of the dynamic programming algorithm presented in this paper to solve the new generalized LCS problem must be $O(nmst)$.

Keywords: Generalized LCS problem · Similarity · Constrained sequences · Time complexity, Dynamic programming algorithm

1 Introduction

LCS problem (the longest common subsequence) can be declared as follows. For the given two sequences A and B, we want to find a common subsequence of the main sequences A and B of the longest length.

This problem is a well-known problem for measuring the similarity of two sequences. It seems to have a lot of applications in various apparently unrelated fields, such as computer science, mathematics, molecular biology, file comparison, pattern matching, speech recognition, gas chromatography [3,4,8,9].

For some biological applications the new variants of the LCS problems are considered. In such problems, some constraints can be applied to the longest common subsequence problem, and the different kinds of variant of the longest common subsequence problem are commonly called the constrained LCS (CLCS) problem [2,4,6,7,10,11]. Recently, a new kind of generalized forms of the constrained LCS problem, the generalized constrained LCS (GC-LCS) problem was proposed by Chen and Chao [1]. For the two input sequences A and B with their lengths m and n, respectively, and a constrained string U of length r, the generalized constrained LCS problem can be formulated as a set of four problems to find the longest common subsequence of A and B including/excluding U as a subsequence or a substring, respectively.

N.T. Nguyen et al. (Eds.): ACIIDS 2016, Part II, LNAI 9622, pp. 261–267, 2016.
DOI: 10.1007/978-3-662-49390-8_25

The purpose of this paper is to consider another more general constrained LCS problem called SEQ-IC-SEQ-IC-LCS, in which two constrained sequences with their lengths s and t must be included as the subsequences of the two main sequences A and B, and the length of the resulting subsequences must be maximal. A first efficient dynamic programming algorithm will be presented to solve this problem.

We will organize our paper as follows.

Our main result on the original presented dynamic programming algorithm to solve the SEQ-IC-SEQ-IC-LCS problem will be described in the following 3 sections.

In Sect. 2, we will discuss the preliminary knowledge to present our algorithm for the newborn SEQ-IC-SEQ-IC-LCS problem. In Sect. 3, the new dynamic programming algorithm to solve the SEQ-IC-SEQ-IC-LCS problem is presented, and its give time complexity $O(nmst)$ is given, where n and m are their lengths of the two main input sequences A and B, and the lengths of the two constrained sequences are s and t, respectively. In Sect. 4, we give some concluding comments.

2 Some Characterizations to the New SEQ-IC-SEQ-IC-LCS Problem

A sequence can be characterized as a string of characters over some alphabet $\sum$. A subsequence is obtained by deleting some characters(not necessarily contiguous),zero or more, from the original sequence A. A substring is a subsequence of A and its characters are successive within A.

For a given sequence $A = a_1a_2 \cdots a_n$ with length n, its ith character can be denoted as $a_i \in \sum$ for any $i = 1, \cdots, n$. $A[i : j] = a_ia_{i+1} \cdots a_j$ is used to denote the substring of A from position i to j. The substring $A[i : j] = a_ia_{i+1} \cdots a_j$ is designated as a proper substring of A, if $i \neq 1$ or $j \neq n$. If $i = 1$ or $j = n$, then the substring $A[i : j] = a_ia_{i+1} \cdots a_j$ is referred to as a prefix of A or a suffix of A, respectively.

An appearance of sequence $A = a_1a_2 \cdots a_n$ in sequence $B = b_1b_2 \cdots b_m$, for any given sequences A and B, beginning from the position j is a strictly increasing indexes sequence $i_1, i_2, \cdots, i_n$, such that $i_1 = j$, and $A = b_{i_1}, b_{i_2}, \cdots, b_{i_n}$. A compact appearance of A in B beginning from the position j is its appearance with the smallest last index i_n.

For the two input sequences $A = a_1a_2 \cdots a_n$ and $B = b_1b_2 \cdots b_m$ with their lengths m and n, respectively, and the two constrained sequences $U = u_1u_2 \cdots u_s$ and $V = v_1v_2 \cdots v_t$ with their lengths s and t, the SEQ-IC-SEQ-IC-LCS problem is defined as finding a constrained longest common subsequence of A and B including U and V as its subsequences.

Definition 1. *Let $L(i, j, k, r)$ be the set of all longest common subsequences of $A[1 : i]$ and $B[1 : j]$ such that for each $c \in L(i, j, k, r)$, c includes $U[1 : k]$ and $V[1 : r]$ as subsequences, where $0 \leq r \leq t$, and $1 \leq i \leq n, 1 \leq j \leq m, 0 \leq k \leq s$. The length of a longest common subsequence in $L(i, j, k, r)$ is denoted as $g(i, j, k, r)$.*

To compute the longest common subsequences in $L(i,j,k,r)$, for any $1 \leq i \leq n, 1 \leq j \leq m, 0 \leq k \leq s$, and $0 \leq r \leq t$, we will use the following theorem to characterize the structure of the optimal solution based on the properties of the optimal solutions to their subproblems.

Theorem 1. *If $C[1:l] = c_1, c_2, \cdots, c_l \in L(i,j,k,r)$, then we can build the following conditions to the optimal solutions:*

1. *If $k > 0, r > 0, a_i = b_j = u_k = v_r$, then $c_l = a_i$ and $C[1:l-1] \in L(i-1, j-1, k-1, r-1)$.*
2. *If $k > 0, a_i = b_j = u_k$ and $(r = 0 \vee a_i \neq v_r)$, then $c_l = a_i$ and $C[1:l-1] \in L(i-1, j-1, k-1, r)$.*
3. *If $r > 0, a_i = b_j = v_r$ and $(k = 0 \vee a_i \neq u_k)$, then $c_l = a_i$ and $C[1:l-1] \in L(i-1, j-1, k, r-1)$.*
4. *If $a_i = b_j$ and $(k = 0 \vee a_i \neq u_k)$ and $(r = 0 \vee a_i \neq v_r)$, then $c_l = a_i$ and $C[1:l-1] \in L(i-1, j-1, k, r)$.*
5. *If $a_i \neq b_j$, then $c_l \neq a_i$ implies $C[1:l] \in L(i-1, j, k, r)$.*
6. *If $a_i \neq b_j$, then $c_l \neq b_j$ implies $C[1:l] \in L(i, j-1, k, r)$.*

Proof.

1. In this first case, we can add a_i to $C[1:l-1]$ such that $a_i = c_l$, and thus $C[1:l-1]$ must be an LCS of $A[1:i-1]$ and $B[1:j-1]$ including $U[1:k-1]$ and $V[1:r-1]$ as subsequences, i.e. $C[1:l-1] \in L(i-1, j-1, k-1, r-1)$.
2. In this second case, we have no constraints on V, and thus a_i cab be added to $C[1:l-1]$ and we can conclude that $C[1:l-1]$ is a common subsequence of $A[1:i-1]$ and $B[1:j-1]$ including $U[1:k-1]$ as a subsequence and including $V[1:r]$ as a subsequence. It can be seen further that $C[1:l-1]$ is an LCS of $A[1:i-1]$ and $B[1:j-1]$ including $U[1:k-1]$ as a subsequence and including $V[1:r]$ as a subsequence. If there is a common subsequence a of $A[1:i-1]$ and $B[1:j-1]$, assumed by contradiction, including $U[1:k-1]$ as a subsequence and including $V[1:r]$ as a subsequence, and its length must be greater than $l-1$. Then we can concatenate a and a_i to form a common subsequence of $A[1:i]$ and $B[1:j]$ including $U[1:k]$ and $V[1:r]$ as subsequences, and its length will be greater than l. This will be a contradiction.
3. In this third case, we have no constraints on U, and thus a_i cab be added to $C[1:l-1]$ and we can conclude that $C[1:l-1]$ is a common subsequence of $A[1:i-1]$ and $B[1:j-1]$ including $U[1:k]$ as a subsequence and including $V[1:r-1]$ as a subsequence. It can be seen further that $C[1:l-1]$ is an LCS of $A[1:i-1]$ and $B[1:j-1]$ including $U[1:k]$ as a subsequence and including $V[1:r-1]$ as a subsequence. If there is a common subsequence a of $A[1:i-1]$ and $B[1:j-1]$, assumed by contradiction, including $U[1:k]$ as a subsequence and including $V[1:r-1]$ as a subsequence, and its length must be greater than $l-1$. Then we can concatenate a and a_i to form a common subsequence of $A[1:i]$ and $B[1:j]$ including $U[1:k]$ and $V[1:r]$ as subsequences, and its length will be greater than l. This will be a contradiction.

4. In this fourth case, we have no constraints on U and V, and thus a_i cab be added to $C[1:l-1]$ and we can conclude that $C[1:l-1]$ is a common subsequence of $A[1:i-1]$ and $B[1:j-1]$ including $U[1:k]$ and $V[1:r]$ as subsequences. It can be seen further that $C[1:l-1]$ is an LCS of $A[1:i-1]$ and $B[1:j-1]$ including $U[1:k]$ and $V[1:r]$ as subsequences. If there is a common subsequence a of $A[1:i-1]$ and $B[1:j-1]$, assumed by contradiction, including $U[1:k]$ and $V[1:r]$ as subsequences, and its length must be greater than $l-1$. Then we can concatenate a and a_i to form a common subsequence of $A[1:i]$ and $B[1:j]$ including $U[1:k]$ and $V[1:r]$ as subsequences, and its length will be greater than l. This will be a contradiction.
5. Since $a_i \neq b_j$ and $c_l \neq a_i$, we can conclude that $C[1:l]$ is a common subsequence of $A[1:i-1]$ and $B[1:j]$ including $U[1:k]$ and $V[1:r]$ as subsequences. It is obvious that $C[1:l]$ is also an LCS of $A[1:i-1]$ and $B[1:j]$ including $U[1:k]$ and $V[1:r]$ as subsequences.
6. Since $a_i \neq b_j$ and $c_l \neq b_j$, we can conclude that $C[1:l]$ is a common subsequence of $A[1:i]$ and $B[1:j-1]$ including $U[1:k]$ and $V[1:r]$ as subsequences. It is obvious that $C[1:l]$ is also an LCS of $A[1:i]$ and $B[1:j-1]$ including $U[1:k]$ and $V[1:r]$ as subsequences.

The proof is now completed. □

3 An Efficient Dynamic Programming Solution

Suppose the length of a longest common subsequence in $L(i,j,k,r)$ be denoted as $g(i,j,k,r)$. By Theorem 1 we know that the SEQ-IC-SEQ-IC-LCS problem will have the optimal substructure properties. Therefore, following recursive formula can be built to compute $g(i,j,k,r)$. For any $1 \leq i \leq n, 1 \leq j \leq m, 0 \leq k \leq s$, and $0 \leq r \leq t$, following recursive formula (1) can be used to compute the values of $g(i,j,k,r)$.

$$g(i,j,k,r) = \begin{cases} \max\{g(i-1,j,k,r), g(i,j-1,k,r)\} & \text{if } a_i \neq b_j \\ 1 + g(i-1,j-1,k-1,r-1) & \\ \quad \text{if } k>0, r>0, a_i = b_j = u_k = v_r & \\ 1 + g(i-1,j-1,k-1,r) & \\ \quad \text{if } k>0, a_i = b_j = u_k \wedge (r=0 \vee a_i \neq v_r) & \\ 1 + g(i-1,j-1,k,r-1) & \\ \quad \text{if } r>0, a_i = b_j = v_r \wedge (k=0 \vee a_i \neq u_k) & \\ 1 + g(i-1,j-1,k,r) & \\ \quad \text{if } a_i = b_j \wedge (k=0 \vee a_i \neq u_k) \wedge (r=0 \vee a_i \neq v_r) & \end{cases} \tag{1}$$

The boundary conditions to use this recursive formula read as follows.

$g(i,0,0,0) = g(0,j,0,0) = 0$ and $g(i,0,k,r) = g(0,j,k,r) = -\infty$ for any $0 \leq i \leq n, 0 \leq j \leq m, 0 \leq k \leq s$, and $0 \leq r \leq t$.

Our new algorithm to compute $g(i,j,k,r)$ is a standard dynamic programming algorithm based on above formula. The dynamic programming algorithm to compute $g(i,j,k,r)$ can be drawn up as the following Algorithm 1.

Algorithm 1. SEQ-IC-SEQ-IC-LCS

Input: Strings $A = a_1 \cdots a_n$, $B = b_1 \cdots b_m$ with their lengths n and m, respectively, and the two constrained sequences $U = u_1 u_2 \cdots u_s$ and $V = v_1 v_2 \cdots v_t$ with their lengths s and t

Output: $g(i, j, k, r)$, the length of a longest common subsequence of $A[1:i]$ and $B[1:j]$ including $U[1:k]$ and $V[1:r]$ as subsequences, for all $1 \le i \le n, 1 \le j \le m, 0 \le k \le s$, and $0 \le r \le t$.

```
for all i, j, k, r , 0 ≤ i ≤ n, 0 ≤ j ≤ m, 0 ≤ k ≤ s and 0 ≤ r ≤ t do
   g(i, 0, k, r), g(0, j, k, r) ← −∞, g(i, 0, 0, 0), g(0, j, 0, 0) ← 0 {boundary condition}
end for
for all i, j, k, r , 1 ≤ i ≤ n, 1 ≤ j ≤ m, 0 ≤ k ≤ s and 0 ≤ r ≤ t do
   if a_i ≠ b_j then
      g(i, j, k, r) ← max{g(i − 1, j, k, r), g(i, j − 1, k, r)}
   else if k > 0 and a_i = u_k then
      if r > 0 and a_i = v_r then
         g(i, j, k, r) ← 1 + g(i − 1, j − 1, k − 1, r − 1)
      else
         g(i, j, k, r) ← 1 + g(i − 1, j − 1, k − 1, r)
      end if
   else if r > 0 and a_i = v_r then
      g(i, j, k, r) ← 1 + g(i − 1, j − 1, k, r − 1)
   else
      g(i, j, k, r) ← 1 + g(i − 1, j − 1, k, r)
   end if
end for
```

It can be seen obviously that time and space complexities of the algorithm are both $O(nmst)$. The corresponding LCS of $A[1:i]$ and $B[1:j]$ including $U[1:k]$ and $V[1:r]$ as subsequences, for each value of $g(i, j, k, r)$ computed by algorithm $Suffix$, can be constructed by a backtracking algorithm pass through the paths from the computation (i, j, k, r) to $(0, 0, 0, 0)$. The following algorithm $back(i, j, k, r)$ can be used as the backtracking algorithm to construct the longest common subsequence, not only its length. It is obvious that the time required by the algorithm $back(i, j, k, r)$ must be $O(n + m)$.

4 Conclusions

We have proposed an efficient dynamic programming algorithm to solve the further generalized constrained LCS problem SEQ-IC-SEQ-IC-LCS. The worst case time complexity of the new dynamic programming algorithm is $O(nmst)$, where s, t, n, m are their lengths of the four given input sequences respectively. The space complexity of this problem can also be reduced to $O(\min(n, m)st)$, by using the technique based on Hirschbergs Algorithm [5].

There are many other generalized constrained LCS problems (GC-LCS) with similar structures of the problem dealt within this paper. It is not clear whether our technique developed in this paper can also be applied to these analogous

Algorithm 2. *back*(i, j, k, r)

Input: Integers i, j, k, r

Output: The LCS of $A[1:i]$ and $B[1:j]$ including $U[1:k]$ as a subsequence and $V[1:r]$ as a suffix

```
if i < 1 or j < 1 then
    return
end if
if a_i ≠ b_j then
    if g(i − 1, j, k, r) > g(i, j − 1, k, r) then
        back(i − 1, j, k, r)
    else
        back(i, j − 1, k, r)
    end if
else if k > 0 and a_i = u_k then
    if r > 0 and a_i = v_r then
        back(i − 1, j − 1, k − 1, r − 1)
        print a_i
    else
        back(i − 1, j − 1, k − 1, r)
        print a_i
    end if
else if r > 0 and a_i = v_r then
    back(i − 1, j − 1, k, r − 1)
    print a_i
else
    back(i − 1, j − 1, k, r)
    print a_i
end if
```

problems to achieve the corresponding efficient algorithms. We will explore these problems further.

Acknowledgments. The work on this paper was supported by Data-Intensive Computing Key Laboratory of Fujian Province, and the Quanzhou Foundation of Science and Technology under Grant No.2013Z38.

References

1. Chen, Y.C., Chao, K.M.: On the generalized constrained longest common subsequence problems. J. Comb. Optim. **21**(3), 383–392 (2011)
2. Baeza-Yates, R.A.: Searching subsequences. Theoret. Comput. Sci. **78**, 363–376 (1991)
3. Deorowicz, S.: Quadratic-time algorithm for a string constrained LCS problem. Inf. Process. Lett. **112**(11), 423–426 (2012)
4. Deorowicz, S.: Bit-parallel algorithm for the constrained longest common subsequence problem. Fundamenta Informaticae **99**(4), 409–433 (2010)

5. Hirschberg, D.S.: Algorithms for the longest common subsequence problem. J. ACM **24**, 664–675 (1977)
6. Gotthilf, Z., Hermelin, D., Lewenstein, M.: Constrained LCS: hardness and approximation. In: Ferragina, P., Landau, G.M. (eds.) CPM 2008. LNCS, vol. 5029, pp. 255–262. Springer, Heidelberg (2008)
7. Gotthilf, Z., Hermelin, D., Landau, G.M., Lewenstein, M.: Restricted LCS. In: Chavez, E., Lonardi, S. (eds.) SPIRE 2010. LNCS, vol. 6393, pp. 250–257. Springer, Heidelberg (2010)
8. Masek, W.J., Paterson, M.S.: A faster algorithm computing string edit distances. J. Comput. Syst. Sci. **20**(1), 18–31 (1980)
9. Peng, Y.H., Yang, C.B., Huang, K.S., Tseng, K.T.: An algorithm and applications to sequence alignment with weighted constraints. Int. J. Found. Comput. Sci. **21**(1), 51–59 (2010)
10. Tang, C.Y., Lu, C.L.: Constrained multiple sequence alignment tool development and its application to RNase family alignment. J. Bioinform. Comput. Biol. **1**, 267–287 (2003)
11. Tsai, Y.T.: The constrained common subsequence problem. Inf. Process. Lett. **88**, 173–176 (2003)

A New Betweenness Centrality Algorithm with Local Search for Community Detection in Complex Network

Youcef Belkhiri(✉), Nadjet Kamel, and Habiba Drias

University of Science and Technology Houari Boumediene, Bab Ezzouar, Algeria
belkhiri.youcef@gmail.com, nkamel@univ-setif.dz, hdrias@usthb.dz

Abstract. Community structure identification in complex networks has been an important research topic in recent years. In this paper, a new between-ness centrality algorithm with local search called BCALS in short, is proposed as an effective optimization technique to solve the community detection problem with the advantage that the number of communities is automatically determined in the process. BCALS selects at first, leaders according to their measure of between-ness centrality, then it selects randomly a node and calculates its local function for all communities and assigns it to the community that optimizes its local function. Experiments show that BCALS gets effective results compared to other detection community algorithms found in the literature.

Keywords: Network · Community detection · Between-ness centrality · Modularity Q

1 Introduction

In the real world, complex networks allow to describing complex systems stemming from different fields, such as the web (set of web pages handling the same theme), biological networks and social networks (identifying groups of friends), etc. The structure of relations between their entities represents common characteristics, thus a common study regardless of their background. The community structure is a property in the complex network that is often defined as a set of nodes strongly connected together but weakly connected to the rest of network nodes [1].

There are many works in the literature on the study of community structure in networks. The issue is closely related to graph partitioning [2]. Several algorithms have been proposed in order to find communities. We distinguish two main categories: heuristic based on intuitive hypothesis [3–5], and optimization methods based on objective function such as Modularity Q [6], external optimization, simulated annealing, and genetic algorithms (GA) [7]. The approach we suggest aims at improving Modularity Q.

This paper is organized as follow: in Sect. 2 we define the community detection problem in the complex networks as an optimization problem. Then, we discuss some related works in Sect. 3. In Sect. 4 we describe our proposed algorithm. Section 5 presents the obtained results using real life network as data sets.

N.T. Nguyen et al. (Eds.): ACIIDS 2016, Part II, LNAI 9622, pp. 268–276, 2016.
DOI: 10.1007/978-3-662-49390-8_26

2 Community Detection Problem

The aim of community discovery problem is to detect and interpret community structures in complex networks. The community detection is generally based on the idea of optimizing modularity function Q, which was a formal point of view. The purpose of community detection in a graph $G = (V, E)$, where V is the set of vertices and E is the set of edges connecting pairs of vertices, is to find a partition $P = \{C1, ..., Ck\}$ of the set V of vertices such as $Q(P)$ is maximal. Note that the information about the number and size of this partition is unknown a priori.

Let us Consider the following hypotheses:

- G = (V, E) an unweighted and undirected network,
- A_{ij} is the adjacency matrix of the graph.

then :

- $K_i = \sum_i A_{ij}$ is the degree of a vertex i,
- $m = \frac{1}{2} \sum_{ij} A_{ij}$ is a number of total edges,
- the functions (u, v) is equal to 1 if $u = v$, and 0 otherwise. c_i denotes the community to which vertex i has been assigned.

The modularity function were introduced by Newman and Girvan, it is defined as [8]:

$$Q = \frac{1}{2m} \sum_{ij} (A_{ij} - \frac{k_i k_j}{2m}) s(c_i, c_j) \tag{1}$$

If we rewrite (1) into (2), function Q can be expressed as the sum of function f of all vertices. As we can see, from the angle of each vertex, here function f denotes the actual number of edges of a vertex within community, minus the expected value of the same quantity if edges fall at random without regard for the community structure. Thus, it can be also regarded as a quality metric for communities representing the same meaning as function Q, while in terms of the local view of each vertex.

$$Q = \frac{1}{2m} \sum_i f(i), \qquad f(i) = \sum_{j \in c_i} (A_{ij} - \frac{k_i k_j}{2m}) \tag{2}$$

Following the analyzes of [9],$\forall i \in V$ the overall function is monotonically increasing with the local function f for each vertex. From this Eq. (2). We use Eq. 2 in our approach and we perform local search for each vertex $f(i)$ aiming at finding a node to assign to a given community. Quality functions typically lead to relevant actual partitions on graphs. The major inconvenient of these criteria is that, their optimization on the set of vertices is a NP-hard problem [10].

Thus, the optimum is hard to calculate in a reasonable time, since the size of the graph to be treated exceeds a few hundred nodes. It is then necessary to resort to the methods of optimization.

3 Related Works

In the literature we find a wide range of different community detection algorithms that follow different strategies. The most common family in the community detection that is based on maximizing modularity [6] among them the clustering technique as algorithm based on K-means [11] and hierarchical clustering (CDHC Algorithm) [12]. We are interested in Distance Centrality based on Detection Community (DCDC) [13] proposed by Wu et al. Longju in 2013 FSKD conference and Top Leader approach [14] proposed by Reihaneh Rabbany Khorasgani et al. in 2010 KDD conference. Both methods are based on the K-means algorithm.

The DCDC consists in selecting center nodes based on the distance between nodes to find the initial communities, then in calculating the similarity between the center nodes and each other nodes in order to assign each node to the most similar community. This approach requires a processing order, starting with the nodes that are directly linked with the central nodes. This processing may not provide a good solution because the order of affected node has a very important role in the quality result and influences the partitioning graph.

The Top Leader algorithm selects at first K representative nodes called "leaders" in the network. Then the association of a node to its leader takes into account the number of nodes that share, and assign it to the one that has more nodes in the intersection. The drawback with this algorithm relies in the fact it requires defining the number and size of the community a priori, which is not always obvious to obtain in advance.

The most positive point of our approach is that it considers the number of communities and their size as indefinite a priori. It calculates between-ness centrality for each node in order to define leader nodes. The remaining nodes are chosen randomly so as, to be affected to the appropriate community considering the local function $f(i)$ that allows the optimization of the overall function Q of the network.

4 Our Approach

4.1 Selection of Leader Nodes

For the selection of leader nodes, we opted for calculating between-ness centrality measure for each node. It represents a measure of quality of nodes centrality that quantifies the number of times a node acts as a bridge along the shortest path between two other nodes, introduced by Linton Freeman [15]. The between-ness of a vertex v in a graph $G = (V, E)$ with V vertices is computed as follows:

- For each pair of vertices (s, t), compute the shortest paths between them.
- For each pair of vertices (s,t), determine the fraction of shortest paths that pass through the vertex v.
- Sum this fraction over all pairs of vertices (s,t).

Concretely, Formula 3 allows the computation of this measure, It is defined as [13]:

$$C_B(v) = \sum_{s \neq v \neq t \in v} \frac{\sigma_s t(v)}{\sigma_s t} \tag{3}$$

Where $\sigma_s t$ is the total number of shortest paths from node s to node t and $\sigma_s t(v)$ is the number of those paths that pass through V.

Once we calculate the between-ness centrality for all nodes, we have to consider a certain number of those nodes as leader nodes. For this purpose, we select the node X that has the maximum value of between-ness centralities as the first leader (Center Nodes). Then, we select the additional leaders, which measure more than 50 percent of the between-ness centrality of node X. Those nodes form the initial communities of the network.

4.2 Associating the Remaining Nodes to Communities

The remaining nodes are the nodes that have their measure of between-ness centrality less than 50 percent of the maximal measure. These nodes have to be assigned to the initial communities, but we are faced to the processing order problem, because we have to choose a certain sort of nodes to be affected to their communities. This association falls into three main strategies:

- Increasing sort of remaining nodes by their betweenness centrality, it affects directly the value of the local function $f(i)$ because of the number of edges of the sorted node and its links with the initial communities, which means that the association could not be pertinent.
- Decreasing sort of remaining nodes by their between-ness centrality, the same problem occurs with it.
- Random sort of remaining nodes.

Starting from the fact that our approach is based on the random sort of remaining nodes and the modularity function (2). In this state we have n number of nodes and m number of communities. For each n number of nodes we must calculate the local function $f(i)$ related to each m number of communities and also to calculate the same function for it. After that, the strategy takes all values of local function for each communities and compare all of them to take the maximum value and to decide which community receives the appropriate node.

Now if its own local function $f(i)$ is maximal compared with all other values f(i) of communities, the node is not related to any community and becomes a leader node and we create a new community. We must repeat this process until there is no more nodes to be affected.

When the number of nodes in the community is less than 3, we should repeat the same process with $f(i)$ and change the node to the other appropriate community.

Finally when all communities are created, we take them one by one all communities. For each community we repeat the same strategy in order to discover if the community is going to be divided into other communities. If the modularity function of the new partitioning is higher than the previous one we maintain this partitioning; otherwise we pass over this partitioning.

4.3 BCALS Algorithm

Based on the above discussion, we describe in this section the Betweenness Centrality Algorithm with Local Search Affectation called BCALS. Its framework is illustrated in Algorithm 1.

```
Procedure BCALS
    Data: Network G = (V, E)
    Result: C //network community Structure
    1 - Calculate betweenness centrality for each node;
    2 - Consider max the maximum measure of all betweenness centralities;
    3 - Select top leader nodes where their measure is more than the half of
    max and initialize communities;
    4 - while there is node to be affected do
        - Pick randomly a node v from the remaining nodes;
        - Calculate local function f1 of modularity of v for each community
        and select the community c that has the highest value;
        - Calculate its local function f2 of modularity of v alone;
        if f1 > f2 then
            assign v to c;
        else
            create a new community led by v;
        end
    end
    if there is a community having less than 3 nodes then
        -consider those nodes unaffected and go to step (4);
    else
        -For each resulted community c_i, calculate modularity Q_1(c_i), then
        repeat the process from step (1);
        -Calculate the new Q_2(c_i) ;
        if if Q_2(c_i) > Q_1(c_i) then
            maintain the new resulted communities C;
        end
    end
```

Algorithm 1. BCALS Algorithm

BCALS relies on Betweenness Centrality notion of nodes and deals with large data set. The basic idea is as follow: At the beginning, it calculates the measure of between-ness centrality for each vertex, and defines the initial communities. Firstly, it selects the node with the maximal value as the first leader. After that, it considers other nodes as leaders according to their measure, which must be more than 50 percent than the first leader value. Afterwards, it draws randomly a vertex V and calculates its local function $f(i)$ for all communities. It then assigns it to the community that gives $f(i)$ maximal. On the contrary, if the possibility to form its own community gives $f(i)$ maximal than the communities already created, it creates a new community to be led by this vertex v. This step stops when there is no vertex to be affected in the network.

In order to further improve the performance for the BCALS, the nodes that are affected in a community, which has less than three nodes, must be assigned to other communities. After that, we consider each community as a new network and we execute the three following operations: between-ness centrality, selecting leader nodes and associating the remaining nodes to populate communities. Next, we apply this new configuration of this community if its global new function is better than the old one. Finally, we update the value of the global function and the partitioning scheme of original network.

This process will be iteratively executed K iterations Q. Then the best solution found across all runs according to the objective function is returned.

5 Experiments

We have established a series of experiments on binary networks (i.e., with undirected and unweighted links) by using a real world network according to modularity function defined before. It is compared with the following five representatives as well as efficient community detection algorithms: Fast Modularity, CFinder, Scan, Top Leader and Distance Centrality Detection Community. We used the programming environment Java 7.

5.1 The Tested Data Sets

We apply BCALS on three real life networks, the Zachary Karate Club, the Bottlenose Dolphin and American College football networks which are well studied in the literature.

Karate Club Benchmark: The Karate Club data set was first analyzed in [16], contains the community structure of a karate club. The network consists of 34 nodes. Due to a conflict between the club president and the karate instructor, the network is divided in two groups almost of the same size. The network consists of 34 vertices and 78 edges. The numerical results of BCALS and other algorithms represent the Q value and are shown in Table 1.

In fact, over 10 runs, as shown in Table 1, BCALS obtained a modularity value Q equal to 0.471, which divides the network into four communities as shown in Fig. 1. At first, the leaders are nodes 1 and 34, and by choosing randomly other nodes, new leaders appeared, they are nodes 7 and 26. Compared to Distance Centrality, Top leader and other algorithms, BCALS gets the best results on the Karate Club data set.

Table 1. The BCALS results on Karate Club network and other algorithms

Algorithm	Q-value
BCALS	0.4197
DC	0.371
Top leader (2)	0.371
Top leader (3)	0.374
Top leader (4)	0.361
Fast modularity	0.380
CFinder	0.182
Scan	0.312

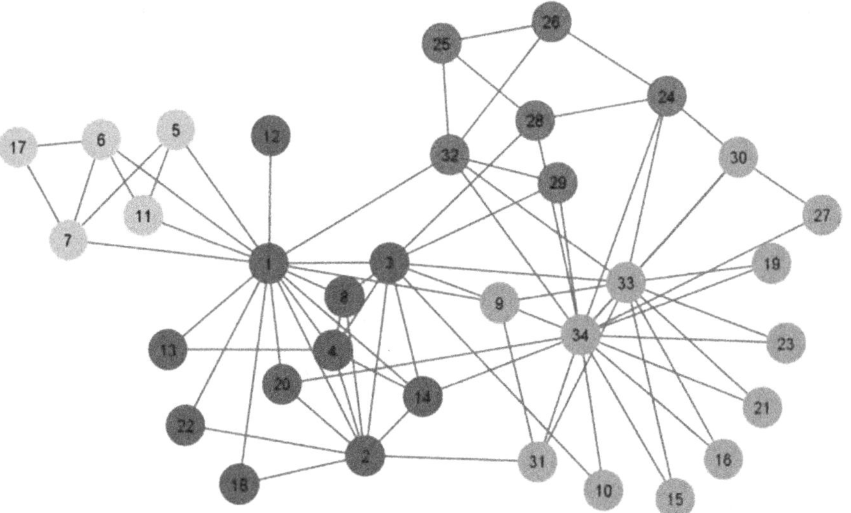

Fig. 1. The results of BCALS on the Zachary Karate Club

The Dolphin Network: The Bottlenose Dolphin network was compiled by Lusseau [17]. It is based on observations over a period of seven years of the behavior of 62 bottlenose dolphins living in Doubtful Sound, New Zealand. The network splits naturally into two large groups. There are 62 nodes and 159 edges in this network. The network has been divided into 6 communities in BCALS. On the contrary the results of DC Algorithm on doplhin network found 5 communities. Consequently the Q value of BCALS is better than Distance Centrality as shown in Table 2.

The Football Network: American College football network [1] represents football games between American colleges during a regular season in Fall 2000. Nodes in the graph represent teams, and edges represent regular-season games between

Table 2. The BCALS results on Dolphin network and other algorithms

Algorithm	Q-value
BCALS	0.5207
DC	0.4967

Table 3. The BCALS results on Football network and other algorithms

Algorithm	Q-value
BCALS	0.5925
Top leader (7)	0.394
Top leader (11)	0.513
Top leader (12)	0.511
Fast modularity	0.567
CFinder	0.532
Scan	0.501

the two teams they connect. What makes this network interesting is that it incorporates a known community structure. The teams are divided into conferences containing around 812 teams each. Games are more frequent between members of the same conference than between members of different conferences. The network is divided into 12 conferences.

As we can see in Table 3, our solution with less prior information offers the best Q value compared to all top leaders with different k number of communities, fast modularity, cfinder and scan as well. The number of obtained communities is 12.

6 Conclusion

In this paper, a new algorithm BCALS for community detection has been proposed. Its originality lies in the fact it uses the measure of between-ness centrality to pick leader nodes in network and selects randomly the remaining nodes to affect them based on the local function of modularity, so that each node could optimize its local function. Moreover, the algorithm considers each community as a network and executes the same process. In other words, it tries to divide community into other communities and optimizes the global function Q. The results demonstrate that the performance of the proposed algorithm are highly effective and efficient for discovering communities.

As a future work, we intend to continue working on the between-ness centrality measure as it yielded interesting results, but using data mining techniques with unsupervised learning instead of optimization techniques.

References

1. Newman, M.E.J., Girvan, M.: Community structure in social and biological networks. Proc. Natl. Acad. Sci **99**, 7821–7826 (2002)
2. Johnson, D.S., Garey, M.R.: Computers and intractability: a guide to the theory of np-completeness. Wiley Computer Publishing, (Freeman, San Francisco) (1979)
3. Farkas, I., Palla, G., Derenyi, I., Vicsek, T.: Uncovering the overlapping community structures of complex networks in nature and society. Nature **435**, 814–818 (2005)
4. Cheung, W.K., Yang, B., Liu, J.: Community mining from signedsocial networks. IEEE Trans. Knowl. Data Eng. **17**, 1333–1348 (2007)
5. Albert, R., Raghavan, U.N., Kumara, S.: Near linear-time algorithm to detect community structures in large-scale networks. Phys. Rev. E **76**, 036–106 (2007)
6. Newman, M.E.J., Girvan, M.: Finding and evaluating community structure in networks. Phys. Rev. E **69**, 026–113 (2004)
7. Fortunato, S.: Community detection in graphs. Phys. Rep. **486**, 75–174 (2010)
8. Clauset, A., Newman, M.E.J., Moore, C.: Finding community structure in very large networks. Phys. Rev. E **70**, 066–111 (2004)
9. Liu, D., Jin, D., He, D., Baquero, C.: Genetic algorithm with local search for community mining in complex networks. In: Proceedings of the 22th International Conference on Tools with Artificial Intelligence (ICTAI 2010), pp. 105–112 (2010)
10. Gaertler, M., Gorke, R., Hoefer, M., Nikoloski, Z., Brandes, U., Delling, D., Wagner, D.: On modularity clustering (2008)
11. Wang, Y.: An improved complex network community detection algorithm based on k-means. Adv. Intell. Soft Comput. **160**, 243–248 (2012)
12. Chen, H., Zhang, B., Yin, C., Zhu, S., David, B.: A method for community detection of complex networks based on hierarchical clustering. Int. J. Distrib. Sens. Netw. **2015**, 9 (2015)
13. Wang, Z., Wang, L., Hu, Y., Wu, L., Bai, T., Ji, J.: A new community detection algorithm based on distance centrality. In: 10th International Conference on Fuzzy Systems and Knowledge Discovery (FSKD) 2013, pp. 898–902 (2013)
14. Khorasgani, O.R.R., Chen, J., Zaane, R.: Top leaders community detection approach in information networks. In: Proceedings of the 2010 International Conference on Knowledge Discovery, Data Mining (KDD 2010), Washington DC, USA (2010)
15. Freeman, L.: A set of measures of centrality based upon betweenness. Sociometry **40**, 35–41 (1977)
16. Zachary, W.W.: An information flow model for conflict and fission in small groups. J. Anthropol. Res. **33**, 452–473 (1977)
17. Lusseau, D.: The emergent properties of dolphin social network. In: Reaka-Kulda, M.L., Wilson, D.E., Wilson, E.O. (eds.) Proceedings of the Royal Society of London. Series B: Biological Sciences, vol. 270, pp. 186–188 (2003)

Meta-Apriori: A New Algorithm for Frequent Pattern Detection

Neyla Cherifa Benhamouda(✉), Habiba Drias, and Célia Hirèche

Computer Science Department, LRIA, USTHB, Algiers, Algeria
benhamoudaneyla@gmail.com, {hdrias, chireche}@usthb.dz

Abstract. Frequent patterns mining is one of the most important data mining techniques, used for extracting information and knowledge from ordinary data. In this paper, we are interested by improving Apriori, which is one of the most used algorithm for extracting frequent patterns. First, we propose some enhancements for the Apriori algorithm. Then we develop Meta-Apriori a new recursive algorithm based on Apriori. As we know, the major drawback of Apriori is its temporal complexity, which makes it difficult to practice. The aim of our algorithm is to reduce substantially the runtime so as, to increase its efficiency while preserving its effectiveness. The main idea is to use the "divide and conquer" technique, which consists in partitioning the whole database into small ones and then applying Meta-Apriori if the database is huge or Apriori if it is of reasonable size. By merging the achieved results, we obtain the outcomes for the whole database.

Keywords: Frequent patterns mining · Apriori · Transactions · Fusion algorithm · And efficiency

1 Introduction

Data analysis is one of the oldest and most important science that exists because of the importance of data and knowledge. Several techniques have been developed in order to analyze, decrypt and extract information previously unknown. These techniques are well established in the domain of data mining (DM). Nevertheless, many efforts remain to undertake in order to reduce the disadvantages of certain methods.

One important data mining task is dicovering frequent patterns from a database. Frequent patterns are items that appear frequently in a dataset and are used for instance to generate association rules and unsupervised learning. One of the most used frequent patterns detecting algorithms is Apriori [1]. Proposed by R. Agrawal and R. Srikant, Apriori is basic, very used and especially revisited and improved in order to yield a reduced time.

In our work, we are interested in improving Apriori by decreasing its execution time. For this purpose, we propose an algorithm called Meta-Apriori based on a split and merge operations. Concretely, it consists of three steps, which are:

- Splitting the database into smaller ones in order to reduce its size.
- Apply Apriori on each sub-database if the latter is of a small size or Meta-Apriori.
- Merge the results of the second step.

N.T. Nguyen et al. (Eds.): ACIIDS 2016, Part II, LNAI 9622, pp. 277–285, 2016.
DOI: 10.1007/978-3-662-49390-8_27

The remainder of this document is organized as follows. The next section presents some works performed around the Apriori algorithm. Then the Apriori algorithm is described in Sect. 3. In Sect. 4, an improved version of Apriori as well as the new algorithm Meta-Apriori algorithm are proposed. In Sect. 5, the experimental evaluation is presented, Conclusions are finally summarized and some perspectives are suggested.

2 Related Works

Apriori has stimulated the interest of researchers for many years. Several improvements of the algorithm were proposed in the literature. Some of these efforts are reported in the following:

Akshita Bhandari, Ashutosh Gupta and Debasis Das considered in [2] the item with the minimum support, to minimize the database scans and hence to reduce the runtime. They also used the FP-Tree algorithm in order to reduce the memory space.

V. B. Nikam and B. B. Meshram proposed in [3] a GPU parallel algorithm and provided interesting experimentations. To obtain this results they match between high performance parallel and/or distributed computing.

In [4] Feng Wang and Yong-Hua Li used a matrix indicating to limit the database scaning by using “AND operation” to create the largest frequent itemset and others The database scans and the number of candidates of frequent itemsets are reduced in this way.

YE Yanbin and Chiang Chia-Chu proposed an improvement by reading transactions with a parallel computer [5].

Mahesh Balaji and G Subrahmanya VRK Rao adapted Apriori for a Cloud Environment, using the approach of mining the frequent itemset to minimize response time [6].

All these studies focused on reducing the number of database scans using some algorithmic tricks and most of the time parallelism. Some of them invoke FP-Tree in order to reduce the memory space. In this paper, we propose an original algorithm based on a ‘divide and conquer’ technique using split and fusion operations.

3 The Apriori Algorithm

Before describing Apriori Algorithm, we have to define some preliminaries such as the minimum Support (MinSup), which is the minimal frequency of an itemset.

The Apriori algorithm [7] consists in determining from a set of transactions, items that are repeated the most. The set of these items is called *Frequent ItemSet* and its frequency is greater than or equal to the given minimum support. The algorithm is iterative and consists in finding, in a first time, the frequent itemset with a single item and on each iteration, an itemset with (k + 1) items by joining k-ItemSet Frequent with itself. A validation of reliable itemsets is performed afterwards.

In summary, Apriori algorithm consists of two parts:

- A join step, where a k-Itemset self-join is done, to release a (k + 1) Itemset candidate.
- A validation step that checks if the Itemset-candidates are frequent (>=MinSup). This step is called prune step.

The Apriori algorithm is outlined in Algorithm 1.

```
Algorithm
Ci : ith itemsets candidates
i,j : indices
F: frequency
|><| : join
Input :
TB: transaction base
MinSUp: minimum support
Output:
FPB: frequent patterns base
Procedure Read(TB): C1, C2
   1.#For (i :0 →TB length)
      1.1 Extract 1-Itemset + Extract all 2-Itemset
   2.#Return C1, C2
Procedure Validation(Ci): Li
   1.#For (j=0 → Ci length)
      1.1Fj= occurrence of itemset in TB: course entire base
      1.2If Fj >= MinSup
         1.2.1 Li [j] ← Ci [j]
   2.#Return Li
Method:
   1.#Read (TB)
   2.#Validation(C1),Validation(C2)
   3.#i = 2
   4.#While(Li<>∅ )
      4.1Ci+1=Li |><| Li
      4.2Validation(Ci+1)
      4.3I=i+1
   5.#Return E={L1, L2, L3,…}
```

3.1 Complexity of the Algorithm

Let us consider the following variables:

- k be the number of itemsets (maximal length of an itemset)
- m the number of transactions
- n the number of items (variables)
- l the maximal length of a transaction (the length of the longest transaction).

In the generation step, the number of itemsets is:
for k = 1, n Itemsets.
For k = 2, n(n-1) (n-2)

$$\begin{cases} n : \textit{number of } k-1 \textit{frequent itemset} \\ n(n-1) : \textit{self join} \\ n-2 : \textit{to check if the other subitemset are generated} \end{cases}$$
$$= n^3 - 3n^2 + 2n = n^3$$

…

for k = l, $\prod_{i=1}^{L-1} n^3$.

For k = 1… l, $n + \sum_{k=2}^{L} \prod_{i=1}^{k-1} n^3 = n^{\max(n,m)}$

In the pruning step:

For each itemset we scan the transaction base of m

$$n^{\max(n,m)} * m$$

The complexity is then O $(n^{\max(n,m)} * m)$.

Two issues arise from this algorithm:

- The space problem provoked by the storage of the transaction base.
- The runtime problem because of the important number of scans.

To fix these limitations, we proposed an improvement of the Apriori algorithm that we integrated next in the Meta-Apriori algorithm.

4 A Simple Improved Apriori Algorithm

In this section, we introduce an improvement of Apriori algorithm to reduce the scope of the drawbacks seen previously.

First, we adopt a representation of the database that reduces the number of scans and hence reduce the temporal complexity. Given a table of transactions, where a transaction is a set of items, we convert the table into a vertical structure. The entry of this structure is an item and the contents is the set of transactions where the item appears. Figure 1 illustrates this representation of the data.

We also adopt this representation for the table of itemsets. We associate for each item all the frequent itemsets containing it instead of working with a table of itemsets.

We also undertake another operation to reduce the time through the validation step. Instead of scanning all the transactions to determine whether an itemset is frequent or not, we add a restriction in the validation step as follows: For an itemset candidate C,

- Use the item that appears the least in the database. An itemset is frequent when its frequency is greater than or equal to the minimum support. It is then sufficient to check on a minimal number of transactions.
- Eliminate items that appear in less than MinSup transactions.
- Eliminate all transactions with a number of items less than k, k being the number of items in the current itemset. If a transaction contains less than 5 items for example, it is a waste of time to validate a 6- itemset on this transaction.
- Stop the scan when the minimum support condition is satisfied. If the minimum support is 100, why continue to 150 when we can validate it at 100.

5 Meta Apriori

All the improvements presented previously allow a time saving. Nevertheless, we considered them as minim; we therefore design an original and more powerful algorithm that we called Meta-Apriori. Of course, all the improvements previously mentioned, will be integrated in the new algorithm, which includes three steps.

- Partitioning step: we divide the transaction base into two clusters, using two methods
  1. Random portioning: which is an intuitive clustering that put a transaction in a cluster, the next in the other, and so on. Figure 2 depicts the process of distributing transactions over clusters.
  2. Frequency clustering that is based on the dispersion of the items in the transaction base. The objective is to have the same frequency of items in both clusters to have the best reduced representation of the database on each cluster. Figure 3 illustrates this technique.
- Apriori Step where we apply the improved Apriori to determine the frequent patterns of both of the two clusters in parallel. If the size of the sub-bases is still huge, we execute Meta-Apriori.
- A fusion step where we join the itemsets of the two clusters.

The framework of Meta-Apriori is shown in Fig. 4 and the algorithm in Algorithm 2.

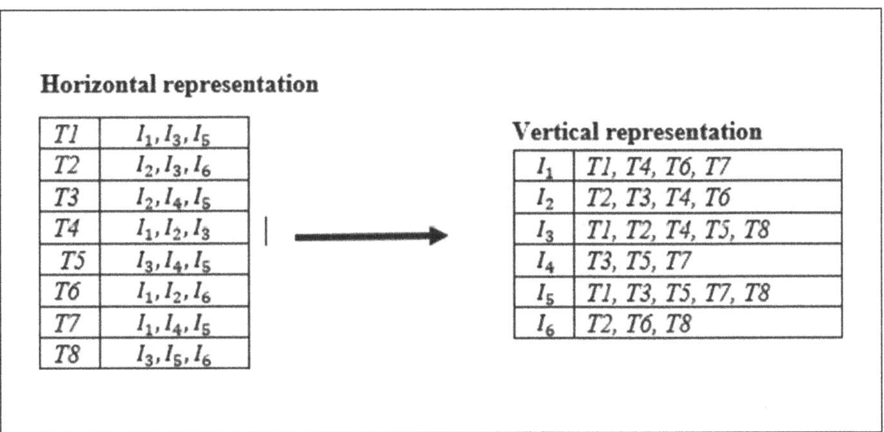

Fig. 1. From horizontal to vertical representation.

5.1 Algorithm

This algorithm is almost the same compared to the Apriori algorithm (1) the only difference is in the validation procedure.

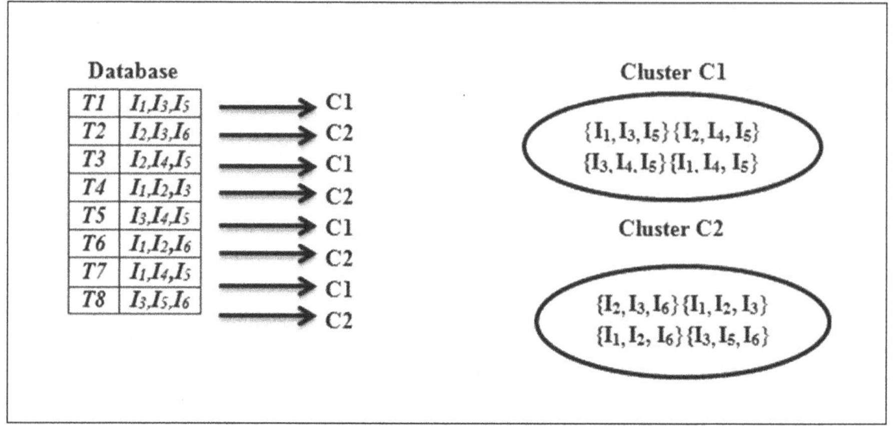

Fig. 2. Random clustering

```
Procedure Validation(Ci) : Li
   1. For (j=0 → Ci length)
         1.1 Fj= 0; k=0;
         1.2 While(k<TB length  and  Fj < MinSup ) : Stop
         when the condition is satisfied
            1.2.1 Fj ← frequency ; k=k+1
         1.3 If Fj >= MinSup
            1.3.1 Li [j] ← Ci [j]
   2. Return Li
```

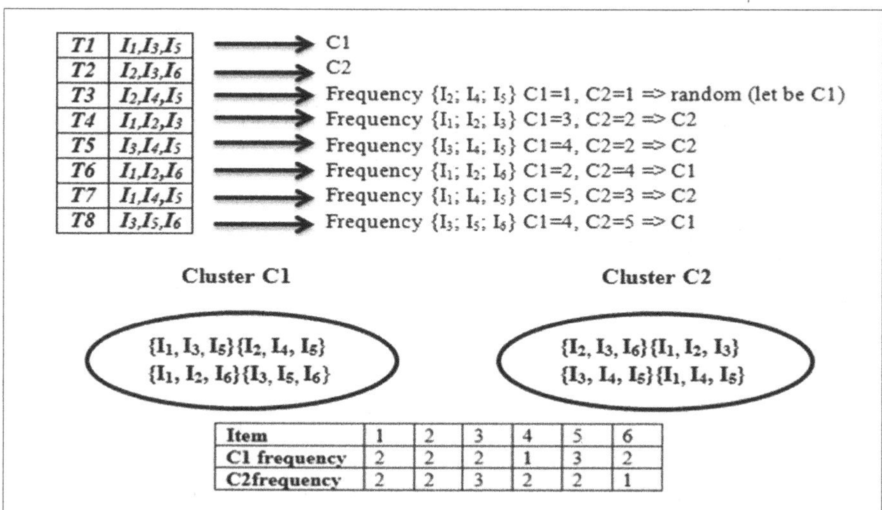

Item	1	2	3	4	5	6
C1 frequency	2	2	2	1	3	2
C2frequency	2	2	3	2	2	1

Fig. 3. Frequency clustering

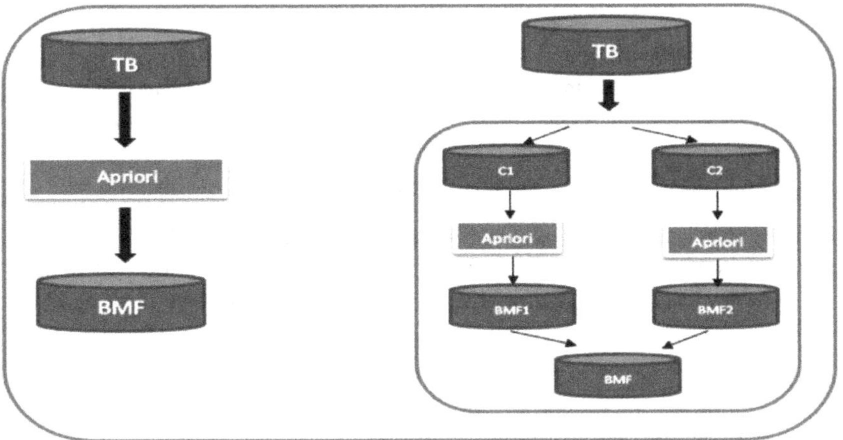

Fig. 4. Apriori vs Meta Apriori

6 Experimental Results

The experiments were conducted on the following benchmarks:

- Ibm2 (Bounded Model Checking) [8]
- Ibm7 (Bounded Model Checking) [8]
- T10I4D100 K [9]

Table 1. Benchmark's description

Benchmark name	Number of items(variables)	Number of transactions
IBM2	2810	11266
IBM7	8710	39374
T10I4D100 K(Fimi)	999	100000

Table 1 describes the characteristics of the datasets that are selected according to their size. We tested our algorithms on a small, a medium and large datasets in order to observe the impact of the data size.

The experimentations were performed on an i7 2.40 Ghz 4Go and the implementation on Microsoft visual studio C# 2013.

Table 2. Apriori – Ameliorate Apriori results

Benchmark name	Apriori		Ameliorate apriori	
	Itemset number	Time(s)	Itemset number	Time(s)
Ibm2	3824	1,803305	3824	0,3300005
Ibm7	11510	55,05679	11510	9,7610137
Fimi	27521	4340,222	27521	2008,42724

Table 2 shows the results obtained by the improved Apriori. We notice that the algorithm is as effective as Apriori since it generates the same number of Itemset as Apriori. Further, it is more efficient than Apriori since it runs faster than the latter.

Table 3. Meta Apriori results

Benchmark name	Meta apriori random		Meta apriori frequentiel	
	Itemset number	Time(s)	Itemset number	Time(s)
Ibm2	3824	0,2053877	3824	0,202613
Ibm7	11510	5,8213687	11510	8,076405
Fimi	27521	1980,064836	27521	1995,84

Table 3 shows the results obtained by the execution of Meta-Apriori. The computed number of itemsets is the same. However, the runtime is shorter than that of the other algorithms. We can conclude that based on the tested benchmarks, Meta-Apriori as effective as the other algorithms but the most efficient (Fig. 5).

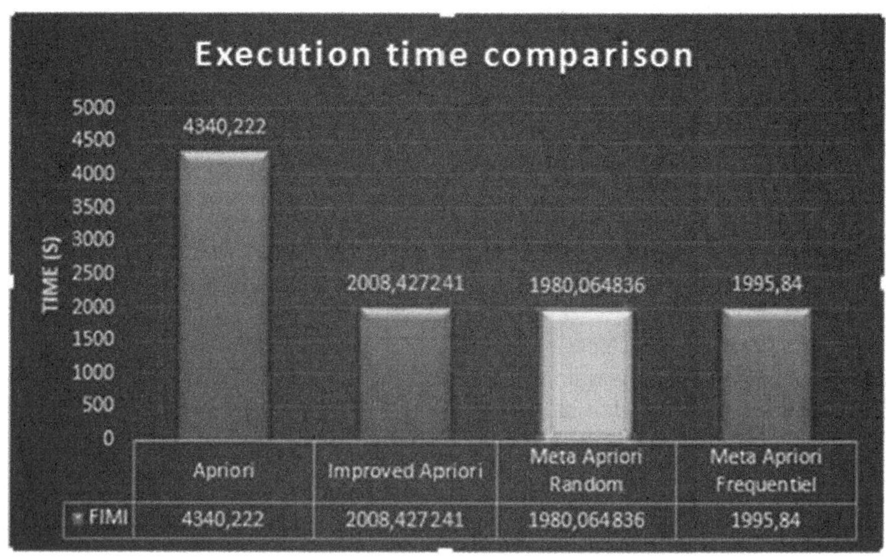

Fig. 5. Time saving representation: FIMI.

This histogram attests the efficiency of the improved Apriori and the Meta Apriori algorithm. It also reveal that Meta Apriori is the best one in time saving.

7 Conclusion

In this paper, we first proposed an improved version of Apriori algorithm by adopting a judicious data structure and ameliorating the prune step to reduce the number of scans. These enhancements yielded better outcomes than those of apriori algorithm. Then we designed a new algorithm called Meta-Apriori, which uses the divide and conquer technique. More precisely, it consists in splitting data into groups and applying meta-Apriori or Apriori on the latters. Two manners of splitting were used, a random one, and a frequency-based one that represents effectively the database. This algorithm provides the best results in terms of efficiency.

On the other hand, what we notice about our approach is that it is parallelizable. Therefore, as the first perspective, we intend to study the best way to parallelize the proposed algorithms in order to increase their efficiency. Also we can think about splitting the database into several small clusters of reasonable size and then apply Apriori on them just in one iteration. The fusion process in this case has to be investigated.

References

1. Agrawal, R., Srikant, R.: Fast algorithms for mining association rules. In: Proceedings of 1994 International Conference on Very Large Data Bases (VLDB 1994), Santiago, Chile, pp. 487–499, September 1994
2. Bhandari, A., Gupta, A., Das, D.: Improvised apriori algorithm using frequent pattern tree for real time applications in data mining. In: Procedia Computer Science Vol. 46, International Conference on Information and Communication Technologies (ICICT 2014), pp. 644–651 (2015)
3. Nikam, V.B., Meshram, B.B.: Scalable frequent itemset mining using heterogeneous: para-priori algorithm. Int. J. Distrib. Parallel Syst. (IJDPS) **5**(5), 1–11 (2014)
4. Feng, W., Yong-hua, L.: An improved apriori algorithm based on the matrix. In: International Seminar on Future BioMedical Information Engineering, FBIE, pp. 152–155 (2008)
5. Yanbin, Y., Chia-Chu, C.: A parallel apriori algorithm for frequent itemsets mining. In: Fourth International Conference on Software Engineering Research, Management and Applications (SERA 2006), pp. 87–94 (2006)
6. Mahesh, B., Rao, VRK, Subrahmanya, G: An adaptive implementation case study of apriori algorithm for a retail scenario in a cloud environment. In: 2013 13th IEEE/ACM International Symposium on Cluster, Cloud, and Grid Computing, CCGRID, pp. 625–629 (2013)
7. Han, J., Kamber, M., Pei, J.: Data mining: Concepts and techniques, 3rd edn. Morgan KaufmannPublisher, San Francisco (2012)
8. Biere, A., Cimatti, A., Clarke, E., Zhu, Y.: Symbolic model checking without BDDs. In: Cleaveland, W. (ed.) TACAS 1999. LNCS, vol. 1579, pp. 193–207. Springer, Heidelberg (1999)
9. Frequent Itemset Mining Dataset Repository. http://fimi.ua.ac.be/data/

Solving a Malleable Jobs Scheduling Problem to Minimize Total Weighted Completion Times by Mixed Integer Linear Programming Models

Nhan-Quy Nguyen[1(✉)], Farouk Yalaoui[1], Lionel Amodeo[1], Hicham Chehade[2], and Pascal Toggenburger[3]

[1] ICD, LOSI, Universite de Technologie de Troyes, UMR 6281, CNRS, 12 Rue Marie Curie, BP 2060, 10010 Troyes, France
nhan_quy.nguyen@utt.fr
[2] Opta-LP, 2 Rue Gustave Eiffel, 10430 Rosieres-pres-Troyes, France
[3] Park'nPlug, 13 Rue Marie Curie, 10430 Rosieres-pres-Troyes, France

Abstract. In this paper we report two mixed integer linear programming models to resolve the malleable jobs scheduling problem with single resource. Jobs' release dates and deadlines are taken into account. The total amount of available resource of the system is variable at different times. Numerical experimentation is conducted to evaluate the performance variability between two introduced models. The objective of this optimization problem is to minimize the total weighted completion time.

Keywords: Mixed integer linear programming · Malleable jobs · Scheduling · Total weighted completion times minimization

1 Introduction

We study in this paper a scheduling problem with jobs' variable processing times. Our problem can be considered as a relaxed version of both of the problems in the literature namely *discrete-continuous scheduling problem (DCSP)* and *malleable parallel jobs scheduling problem (MJSP)*. In the DSCP, n jobs have to be scheduled in m machines with an additional continuous and divisible resource. At each time, one machine can process at most one job and the resource amount allotted to this job decides its processing rate by a non-decreasing function, i.e. processing rate/speed function, which leads the job faster or slower to the desire final state. Formal definitions, examples and resolution methods of this problem can be found in [4,5,10]. In the MJSP, the additional resources are not taken in account. The only discrete resource for jobs to proceed is the total available identical machines. At a given time, a job can be processed on several machines. The more machines used to execute a job at a time, the faster this job reaches its completion state [3,6,7,9]. According to our knowledge, in most of the prior works the objective function is to minimize the scheduling length when the models do not take into account release dates and deadlines of tasks.

N.T. Nguyen et al. (Eds.): ACIIDS 2016, Part II, LNAI 9622, pp. 286–295, 2016.
DOI: 10.1007/978-3-662-49390-8_28

There are few papers which considered the time windows constraints and the total (weighted) completion times minimization. In [14], Rozycki and Weglarz introduce a new general methodology with classical problem decomposition onto tasks sequencing and resource allocating while adding optimal properties to deal with tasks release dates and deadlines. For the total (weighted) completion time minimizing problem, many dominance properties can be found in [15,18]. In our recent work [13], we attempted to solve the total completion time minimization problem using a MILP and strip-packing inspired heuristics. To approach the resolution of the considered problem, we describe our optimization problem by Mixed-Integer Linear Programming (MILP, or MIP) models. The important numbers of methods of solving and decomposing MIP problems in the literature [8,16,17] have motivated us to chose MIP model to formulate our problem. General principles, examples and case studies can be found in [1,2,11,12]. The MIP models are solved by IBM ILOG CPLEX solver in the numerical tests section.

Our paper is organized as follows: the problem formulation is introduced in Sect. 2. Two MIP models are formulated in Sect. 3. Numerical result of conducted tests are presented in Sect. 4 and some conclusions are drawn in Sect. 5.

2 Problem Description

Given a set of n jobs to be scheduled in H decision intervals, each interval lasts 1 unit of times. Let k be the interval time index, $k \in 0, ..., H$. Jobs are non-preemptable, each job i has a release date r_i when it arrives into the system and a due date d_i when it has to leave. During interval $[k, k+1[$ job i consumes an amount of resource $u_{i,k}$. The bigger amount of resource a job receives during an interval, the faster it can reach to its desired final state. This final state can be named also the *workload* of a job, denoted by $\tilde{x}_i$. Let $x_{i,k}$, namely state-of-job i, be the total resource amount that job i has consumed up to time k. The state-of-job is defined recursively as shown in (1) with the initial conditions $x_{i0} = 0 \quad \forall i$.

$$x_{i,k} = x_{i,k-1} + u_{i,k-1} \qquad \forall i \quad k = 1, ..., H \tag{1}$$

By that definition, the amount of resource attributed to a job at a time $u_{i,k}$ can be considered also as the *processing rate* of a job because $u_{i,k} = x_{i,k+1} - x_{i,k} = \dot{x}_{i,k}$ which is the changing rate of state of job between two consecutive intervals. Let C_i denote the completion time of job i therefore $x_{i,k} \geq \tilde{x}_i \quad \forall k \geq C_i$. The workload of job i can also be expressed in the following equation:

$$\tilde{x}_i \leq \sum_{k=1}^{C_i} u_{i,k} = x_{i,C_i} \tag{2}$$

An inequality is used instead of an equation to define the job's workload because of the existence of the resource consumption upper and lower limits of a job. If a job i is executing at time k, the amount of resource consumed is bounded by u_i^{min} and u_i^{max} otherwise $u_{i,k} = 0$. Therefore

$$u_{i,k} \in [u_i^{min}, u_i^{max}] \cup \{0\} \quad \forall i, k \tag{3}$$

For instance, at the beginning of the interval where a task receives its last amount of resource before it could be completed, i.e. at time $C_i - 1$, the amount of resource left to consume to reach the final state is less than the lower resource consumption limit, job i has to take an amount u_i^{min} which is more than necessary. If $\tilde{x}_i - x_{i,C_i-1} < u_i^{min}$ then $x_{i,C_i} = x_{i,C_i-1} + u_i^{min} > \tilde{x}_i$. Furthermore, jobs are non-preemptable which means that once a job i has started to process and has not completed yet, the resource allotted to it cannot be reduced to less than u_i^{min}. Jobs also have different weights which are denoted by $w_i; i = 1, ..., n$. In addition, the total amount of available resource at time k is denoted by U_k. Let $\mathcal{J} = 1, 2, ..., n$ be the set of jobs and $\mathcal{T} = 0, 1, ..., H$ be the set of time periods which is commonly based on the time discretization of the scheduling horizon (hourly, every 15 min, ...). We aim to minimize the total weighted completion time $\sum_{i \in \mathcal{J}} w_i C_i$.

Thus, in the parallel jobs processing perspective, our considered problem is a relaxed problem of the classical malleable jobs scheduling problem where the number of processors is very large that one can assume this large set of identical machines as a continuous additional resource to execute tasks. In another point of view, one can consider also this problem as a sub-problem of the DCSP when the number of jobs is less than or equal to the number of machines.

3 MIP Models

The formulation of the MIP models is based on the interval times indexed model [13]. Unlike the classical scheduling problem where the processing times of jobs are constant, the difficulty of this problem lies on the variation of the processing time according to the allocation of resource to jobs. Because of that reason, we introduced two MIP models. The main difference between the two models is in the formulation of the completion and the processing state of jobs. The two models are named $MIPw - C$ and $MIPw - \alpha\beta$. In addition to the difference in modeling the completions states and processing states of jobs, those two models share two common variables which are $x_{i,k}$ defined in (1) and $u_{i,k}$ constrained by (3). In order to add more bounds for variables which will be introduced shortly, two more parameters are defined in Eq. (4) for each job, namely earliest completion time e_i and latest starting time l_i.

$$e_i = r_i + \lceil \frac{\tilde{x}_i}{u_i^{max}} \rceil; \quad l_i = d_i - \lceil \frac{\tilde{x}_i}{u_i^{max}} \rceil \quad \forall i \in \mathcal{J} \tag{4}$$

Therefore, the starting time of a job $i \in \mathcal{J}$ is bounded in $[r_i, l_i]$ and the completion time of this job is bounded in $[e_i, d_i]$.

3.1 MIPw-C Model Formulation

For the first model we introduce the following variables:

- $x_{i,k}$ the state of job $i \in \mathcal{J}$ at time $k \in \mathcal{T}$

- $u_{i,k}$ decision variable of the amount of resource job $i \in \mathcal{J}$ consumes at time $k \in \mathcal{T}$
- $c_{i,k}$ completion indication variables, 1 if job $i \in \mathcal{J}$ already completed at time $k \in \mathcal{T}$, 0 otherwise.
- $\mu_{i,k}$ processing state indication variables, 1 if job $i \in \mathcal{J}$ is processing at time $k \in \mathcal{T}$, 0 otherwise.

The resource distribution variable and the processing state indication are fundamentals for the formulation since the state of jobs at a moment k is defined recursively and depends only on the amount of resource already alloted to it up to k. The completion indication variable is used to formulate the completion time of a job. By the definition of $c_{i,k}$, completion time of job i can be formulated as

$$C_i = H + 1 - \sum_{k \in \mathcal{T}} c_{i,k}$$

Therefore

$$\sum_{i \in \mathcal{J}} w_i C_i = \sum_{i \in \mathcal{J}} w_i (H + 1 - \sum_{k \in \mathcal{T}} c_{i,k}) = \sum_{i \in \mathcal{J}} w_i (H+1) - \sum_{i \in \mathcal{J}} \sum_{k \in \mathcal{T}} w_i c_{i,k} \quad (5)$$

The $MIPw - C$ formulation is given as follows:

$$\begin{aligned}
Minimize \quad & \sum_{i \in \mathcal{J}} w_i(H+1) - \sum_{i \in \mathcal{J}} \sum_{k \in \mathcal{T}} w_{i c_{i,k}} && && (6)\\
s.t. \quad & x_{i,k} = x_{i,k-1} + u_{i,k-1} && \forall i \in \mathcal{J}, k \in \mathcal{T} : k \geq 1 && (7)\\
& x_{i,0} = 0; u_{i,0} = 0 && \forall i \in \mathcal{J} && (8)\\
& \mu_{i,k} = 0 && \forall i \in \mathcal{J}, k \in \mathcal{T} \setminus \{r_i, ..., d_i\} && (9)\\
& \mu_{i,k} \in \{0, 1\} && \forall i \in \mathcal{J}, k \in \{r_i, ..., d_i\} && (10)\\
& x_{i,k} \geq \tilde{x}_i && \forall i \in \mathcal{J}, k \in \mathcal{T} : k \geq d_i && (11)\\
& u_{i,k} \leq u_i^{max} \mu_{i,k} && \forall i \in \mathcal{J}, k \in \mathcal{T} && (12)\\
& u_{i,k} \geq u_i^{min} \mu_{i,k} && \forall i \in \mathcal{J}, k \in \mathcal{T} && (13)\\
& \sum_{i \in \mathcal{J}} u_{i,k} \leq U_k && \forall k \in \mathcal{T} && (14)\\
& \tilde{x}_i c_{i,k} \leq x_{i,k} && \forall i \in \mathcal{J}, k \in \mathcal{T} && (15)\\
& c_{i,k} = 0 && \forall i \in \mathcal{J}, k \in \mathcal{T} : k < e_i && (16)\\
& c_{i,k} \in \{0, 1\} && \forall i \in \mathcal{J}, k \in \mathcal{T} : k \geq e_i && (17)\\
& \mu_{i,k} \geq \mu_{i,k-1} - c_{i,k} && \forall i \in \mathcal{J}, k \in \mathcal{T} && (18)\\
& \mu_{i,k} \leq 1 - c_{i,k} && \forall i \in \mathcal{J}, k \in \mathcal{T} && (19)\\
& u_{i,k}, x_{i,k} \geq 0 && \forall i \in \mathcal{J}, k \in \mathcal{T} && (20)
\end{aligned}$$

The objective function (6) and the set of constraints (7), (8) have been already introduced in (1) and (5). Time windows constraint (9) prohibits jobs to be executed before their release dates and after their due date while Eq. (10)

defines the processing indication variables. In this problem, jobs are supposed to have strict deadline. Therefore, after the deadlines, all jobs have to be finished (11). This latter constraint can be relaxed in case of job only having non-strict due date. Constraints (12) and (13) limit the value of $u_{i,k}$ according to Eq. (3). System capacity constraint (14) forbids jobs' over consumption of resources at every period. With the total completion time formulated in (6), the minimization of the total weighted completion time is equivalent to the maximization of the total weighted $c_{i,k}$ so this variable "has to turn" to 1 as much as possible. Constraint (15) allows $c_{i,k}$ turning to 1 only if job i has already reached its final state before or at time k. Constraints (16) and (17) bounds the value of $c_{i,k}$ because the completion indicator cannot turn on before the earliest completion time. The non-preemptions constraint (18) prevents an already started job at time k, $\mu_{i,k-1}$=1 but has not finished yet, $c_{i,k} = 0$ to turn off $\mu_{i,k} \geq 1 - 0 = 1$. Otherwise, when $u_{i,k-1} = 0$ and $c_{i,k} = 0$ (has not started, has not finished), or $u_{i,k-1} = 1$ and $c_{i,k} = 1$ (already started and already finished): $\mu_{i,k} \geq 0 - 0 = 0$ or $\mu_{i,k} \geq 1 - 1 = 0$. Finally, $\mu_{i,k}$ has to turn off once job i has finished as shown in Eq. (19). Evidently, job's resource allocations and states have to be non-negative (21).

3.2 MIPw-$\alpha\beta$ Model Formulation

We create more than one model to investigate the performance variation between different MIP models and to have more choices for latter development of decomposition and solving method. In this second model, beside the commonly used variables $x_{i,k}$ and $u_{i,k}$, we introduce two different variables to model the starting and the finishing state of jobs:

- $\alpha_{i,k}$ job's starting state variables, 1 if job $i \in \mathcal{J}$ starts to process at time $k \in \mathcal{T}$, 0 otherwise.
- $\beta_{i,k}$ job's finishing state variables, 1 if job $i \in \mathcal{J}$ finishes at time $k \in \mathcal{T}$, 0 otherwise.

Remark 1. *For each $i \in \mathcal{J}, k \in \mathcal{T}$, the variable $\mu_{i,k}$ in $MIPw - C$ model can be re-formulated with two variables $\alpha_{i,k}$ and $\beta_{i,k}$ in the following way*

$$\mu_{i,k} = \sum_{\tau=0}^{k} (\alpha_{i,\tau} - \beta_{i,\tau}). \tag{21}$$

Proof. If job i is being executed at time k so $\exists \tau_1 \leq k : \alpha_{i,\tau_1} = 1$ and $\nexists \tau_2 \leq k : \beta_{i,\tau_2} = 1$ since the processing time has to be in between starting time and completion time. Thus $\sum_{\tau=0}^{k} \alpha_{i,\tau} = 1$ and $\sum_{\tau=0}^{k} \beta_{i,\tau} = 0$ so by (21) $\mu_{i,k} = 1 - 0 = 1$.

In the other case, if job i has not started to process at time k yet so $\nexists \tau_1 \leq k : \alpha_{i,\tau_1} = 1$ and $\nexists \tau_2 \leq k : \beta_{i,\tau_2} = 1$ since the job has not been started yet so it cannot be ended neither. Thus, by (21) $\mu_{i,k} = 0 - 0 = 0$.

Otherwise, job i has already completed at time k so $\exists \tau_1 \le k : \alpha_{i,\tau_1} = 1$ and $\exists \tau_2 \le k : \beta_{i,\tau_2} = 1$ since the starting time and the completion have happened already before k. Therefore by (21) $\mu_{i,k} = 1 - 1 = 0$.

Furthermore the completion time of a job is now formulated as shown in Eq. (22)

$$C_i = \sum_{k \in \mathcal{T}} (k\beta_{i,k}) \quad \forall i \in \mathcal{J}. \tag{22}$$

Time-windows Constraints. With starting and finishing state variables, one can formulate the time windows constraints in this way:

$$\alpha_{i,k} = 0 \quad \forall i \in \mathcal{J}, k \in \mathcal{T} \setminus \{r_i, ..., l_i\} \tag{23}$$

$$\alpha_{i,k} \in \{0,1\} \quad \forall i \in \mathcal{J}, k \in \{r_i, ..., l_i\} \tag{24}$$

$$\beta_{i,k} = 0 \quad \forall i \in \mathcal{J}, k \in \mathcal{T} \setminus \{e_i, ..., d_i\} \tag{25}$$

$$\beta_{i,k} \in \{0,1\} \quad \forall i \in \mathcal{J}, k \in \{e_i, ..., d_i\}. \tag{26}$$

Observation 1. *With the formulations of the set of constraints, all the bounds e_i and l_i are used in $MIPw - \alpha\beta$ while the equivalent set of constraints in $MIPw - C$ model can only use one bound e_i.*

Non-preemption Constraints. Since jobs are non-preemptable, only one start and one end are permitted for each job's processing. Furthermore, the finishing time must be after the starting time. Thus, the set of constraints is given by

$$\sum_{k=r_i}^{l_i} \alpha_{i,k} = 1 \quad \forall i \in \mathcal{J} \tag{27}$$

$$\sum_{k=e_i}^{d_i} \beta_{i,k} = 1 \quad \forall i \in \mathcal{J} \tag{28}$$

$$\beta_{i,k} \le \sum_{\tau=r_i}^{k} \alpha_{i,\tau} \quad \forall i \in \mathcal{J}, k \in \{e_i, ..., d_i\}. \tag{29}$$

Resource Consumption Constraints. With the Remark 1, resource consumption constraints can be re-written as follows:

$$u_{i,k} \le u_i^{max} \sum_{\tau=0}^{k} (\alpha_{i,\tau} - \beta_{i,\tau}) \quad \forall i \in \mathcal{J}, k \in \mathcal{T} \tag{30}$$

$$u_{i,k} \ge u_i^{min} \sum_{\tau=0}^{k} (\alpha_{i,\tau} - \beta_{i,\tau}) \quad \forall i \in \mathcal{J}, k \in \mathcal{T}. \tag{31}$$

Task Completions Constraints. The variable $\beta_{i,k}$ can only turn to 1 only if $x_{i,k} \ge \tilde{x}_i$ at a time k.

$$\beta_{i,k}.\tilde{x}_i \le x_{i,k} \quad \forall i \in \mathcal{J}, k \in \mathcal{T} \tag{32}$$

The $MIPw - \alpha\beta$ model formulation

$$\begin{aligned} Minimize \quad & \sum_{i \in \mathcal{J}} \sum_{k \in \mathcal{T}} (w_i.k.\beta_{i,k}) \\ s.t. \quad & (7), (8), (14), (21) \\ & (23 - 32). \end{aligned} \tag{33}$$

4 Computational Experiments

The number of tasks to be tested is $n \in \{10, 20\}$. We aim to resolve the online dynamic scheduling problem because the duration of a decision intervals in actual situation is block of 5 or 10 min in a 12 h scheduling horizon. Therefore the scheduling horizon in this test is $H \in \{72, 144\}$. With each pair of (n, H), 20 random instances will be generated by the test instances random generator described in [13]. The way values are generated can be briefly described as follow: $u_i^{min} := \mathcal{N}(12, 3.6)$, $u_i^{max} := \mathcal{N}(180, 72)$, $r_i := \mathcal{U}(0, \lfloor 0.4 \times H \rfloor)$ with the referencing elements generated as $u_i^{avg} := \mathcal{N}(90, 27)$, $p_i^{avg} := \mathcal{U}(\lfloor 0.4 \times H \rfloor, \lfloor 0.7 \times H \rfloor)$ then $d_i := r_i + \mathcal{U}(\lfloor 1.2 \times p_i^{avg} \rfloor, \lfloor 1.7 \times p_i^{avg} \rfloor)$ and $\tilde{x}_i = u_i^{avg} \times p_i^{avg}$. $\mathcal{N}(\mu, \sigma)$ is the normal distribution random generator function and $\mathcal{U}(a, b)$ is the discrete uniform distribution random generator. Because of the short interval times, the solver has to resolve quickly the problem. For that reason, the time limit for the solver is set to 60 s. All runs were performed on PC with CPU Intel core i5 3.20 GHz with 8 GB RAM. The random generator is executed on Matlab 2009 and the MIP models are solved in IBM ILOG CPLEX 12.6.1.0. With each test instance, the execution time (s), the gap (%) between the lower bound found by LP relaxation and the objective value of incumbent integer solution are logged and shown in Tables 1 and 2. If CPLEX found optimal solutions for an instance, the gap will be zero. *#Op* denotes the numbers of optimal solution found for each instances set.

According to the results of tests having small size $H = 72$ both of the models prove efficiency in solving the generated instances with a rate of finding optimal maintaining nearly 100 %. Furthermore, the $MIPw - \alpha\beta$ model is proved more efficient in term of execution times in compared to the $MIPw - C$ counterpart while the gaps found by CPLEX using both models do not have significant differences.

During the first tests with $n = 10$ and $H = 144$ the two models were always stable in resolving problem with average gaps maintaining around 1 %. In coping with the large instances ($n = 20, H = 144$) the solutions quality deteriorates more quickly with the first model, where the average gap goes up to 14 % and a little more slower in the second model where the average gap is at 6 %. While observing elaborately the result of the $MIPw - \alpha\beta$ we can find 65 % of the solutions founds having $gap \leq 2\%$; 15 % of the solutions found having gap in between 4 % and 5 %. 10 % of the solutions have significant outsider gap with the rate of 45 % and 47 % while $MIPw - C$ can find better solutions in those latter cases. We assume that those two models search differently for the solutions so

Table 1. Percentage of gap and time elapsed (seconds) for CPLEX to execute the $MIPw - C$ and $MIPw - \alpha\beta$ models for test instances having $n = 10$. Better results found are in bold.

#	n=10; H=72				n=10; H=144			
	$MIPw - C$		$MIPw - \alpha\beta$		$MIPw - C$		$MIPw - \alpha\beta$	
	Gap	Time	Gap	Time	Gap	Time	Gap	Time
1	**0 %**	18.2	**0 %**	**7.9**	**0 %**	19.4	**0 %**	**18.4**
2	**0 %**	1.9	**0 %**	**1.0**	7 %	60.0	**4 %**	60.0
3	**0 %**	23.3	**0 %**	**11.1**	**2 %**	60.0	3 %	60.0
4	**0 %**	1.1	**0 %**	**0.8**	**0 %**	20.6	**0 %**	**14.0**
5	**0 %**	**2.7**	**0 %**	5.0	**1 %**	60.0	**1 %**	60.0
6	**0 %**	23.9	**0 %**	**11.5**	**0 %**	35.0	**0 %**	**11.3**
7	**0 %**	3.8	**0 %**	**1.5**	**0 %**	60.0	**0 %**	**28.8**
8	**0 %**	1.9	**0 %**	**1.4**	**0 %**	33.3	**0 %**	**26.5**
9	**0 %**	**2.3**	**0 %**	2.6	**0 %**	11.0	**0 %**	**5.4**
10	**0 %**	0.7	**0 %**	**0.6**	2 %	60.0	**1 %**	60.0
11	**0 %**	1.5	**0 %**	**0.8**	**0 %**	26.9	**0 %**	**15.3**
12	**0 %**	2.2	**0 %**	**0.9**	**0 %**	6.4	**0 %**	**2.8**
13	**0 %**	16.2	**0 %**	**8.2**	**0 %**	45.7	**0 %**	**27.8**
14	**0 %**	2.6	**0 %**	**1.0**	**1 %**	60.0	**1 %**	60.0
15	**0 %**	6.9	**0 %**	**6.0**	**0 %**	56.5	1 %	60.0
16	**0 %**	1.0	**0 %**	**0.6**	2 %	60.0	**1 %**	60.0
17	**0 %**	37.1	**0 %**	**11.3**	**1 %**	60.0	2 %	60.0
18	**0 %**	**3.3**	**0 %**	3.7	**0 %**	22.5	**0 %**	**15.0**
19	**0 %**	1.7	**0 %**	**1.0**	**0 %**	36.1	**0 %**	**18.7**
20	**0 %**	10.5	**0 %**	**6.4**	**6 %**	60.0	10 %	60.0
avg	**0 %**	**8**	**0 %**	4	**1 %**	**43**	**1 %**	**36**
#Op	**20**		**20**		**12**		**11**	

in some cases the gap of solution found by $MIPw - C$ dominates the gap found by $MIPw - \alpha\beta$ and vice versa. Those first results imply that $MIPw - \alpha\beta$ has a better chance finding good solutions in a shorter execution time, not inevitably faster. Further tests should be conducted to understand fully the reasons of those outsider gaps leading to bad integer solutions and the searching mechanics of those two MIP models when they are run by CPLEX.

Empirically, the latter introduced model $MIPw - \alpha\beta$ proved an out-performance to its counterpart $MIPw - C$ in this test. However, during this empirical evaluations, both of the models have shown its good performance while dealing with randomly generated instances where the time could be discretized down to 5 min per interval, during a 12 h-scheduling plan.

Table 2. Percentage of gap and time elapsed (seconds) for CPLEX to execute the $MIPw - C$ and $MIPw - \alpha\beta$ models for test instances having $n = 20$. Better results found are in bold.

#	n=20; H=72				n=20; H=144			
	$MIPw - C$		$MIPw - \alpha\beta$		$MIPw - C$		$MIPw - \alpha\beta$	
	Gap	Time	Gap	Time	Gap	Time	Gap	Time
1	**0 %**	34.8	**0 %**	**43.4**	**3 %**	60.0	4 %	60.0
2	**0 %**	35.0	**0 %**	**28.0**	2 %	60.0	**1 %**	60.0
3	**0 %**	20.4	**0 %**	**21.7**	43 %	60.0	2 %	60.0
4	**0 %**	60.0	**0 %**	**25.6**	3 %	60.0	**2 %**	60.0
5	**1 %**	**60.0**	**0 %**	60.0	**1 %**	60.0	**1 %**	60.0
6	**0 %**	60.0	**0 %**	**60.0**	43 %	60.0	**2 %**	60.0
7	**0 %**	18.0	**0 %**	**14.2**	45 %	60.0	**45 %**	60.0
8	**0 %**	18.3	**0 %**	**8.0**	**1 %**	60.0	**1 %**	60.0
9	**0 %**	**14.9**	**0 %**	10.1	41 %	60.0	**4 %**	60.0
10	**0 %**	7.0	**0 %**	**9.1**	**0 %**	**4.1**	**0 %**	10.0
11	**0 %**	60.0	**0 %**	**60.0**	5 %	60.0	**1 %**	60.0
12	**0 %**	60.0	**0 %**	**60.0**	**9 %**	60.0	47 %	60.0
13	**0 %**	4.3	**0 %**	**4.9**	41 %	60.0	**1 %**	60.0
14	**0 %**	16.3	**0 %**	**10.8**	**2 %**	60.0	**2 %**	60.0
15	**0 %**	60.0	**0 %**	**60.0**	**0 %**	53.9	**0 %**	60.0
16	**0 %**	31.6	**0 %**	**17.4**	**0 %**	60.0	5 %	60.0
17	**0 %**	60.0	**0 %**	**60.0**	**3 %**	60.0	**3 %**	60.0
18	**0 %**	**13.0**	**0 %**	9.6	40 %	60.0	**1 %**	60.0
19	**0 %**	13.6	**0 %**	**7.2**	3 %	60.0	**2 %**	60.0
20	**0 %**	47.2	**0 %**	**38.3**	3 %	60.0	**1 %**	60.1
AVG	**0 %**	**35**	**0 %**	**30**	**14 %**	**57**	**6 %**	**58**
#Op	**19**		**20**		**3**		**2**	

5 Conclusion

In this paper, two MIP models are introduced to solve a specific MJSP with time windows constraints and variant resource availability. The numerical results prove a good performance of the two models while dealing with random generated instances and the out-performance of $MIPw - \alpha\beta$ to the $MIPw - C$. Further researches should be made to find a more efficient algorithm for solving this problem to find more qualified solutions while assuring the short execution time in treating big size instances.

Acknowledgments. This research has been supported by ANRT (Association Nationale de la Recherche et de la Technologie, France).

References

1. Atakan, S., Lulli, G., Sen, S.: An improved mip formulation for the unit commitment problem (2015)
2. Bixby, R.E., Fenelon, M., Gu, Z., Rothberg, E., Wunderling, R.: Mixed integer programming: a progress report. The sharpest cut: The impact of Manfred Padberg and his work, MPS-SIAM Series on Optimization, vol. 4, pp. 309–326 (2004)
3. Blazewicz, J., Kovalyov, M.Y., Machowiak, M., Trystram, D., Weglarz, J.: Preemptable malleable task scheduling problem. IEEE Trans. Comput. **55**(4), 486–490 (2006)
4. Blazewicz, P.D.J., Ecker, P.D.K.H., Pesch, P.D.E., Schmidt, P.D.G., Weglarz, P.D.J.: Scheduling under resource constraints. In: Scheduling Computer and Manufacturing Processes, pp. 317–365. Springer, Heidelberg (2001)
5. Dror, M., Stern, H.I., Lenstra, J.K.: Parallel machine scheduling: processing rates dependent on number of jobs in operation. Manage. Sci. **33**(8), 1001–1009 (1987)
6. Dutot, P.-F., Mounié, G., Trystram, D.: Scheduling parallel tasks: approximation algorithms. Handbook of Scheduling: Algorithms, Models, and Performance Analysis, pp. 1–26 (2004)
7. Fan, L., Zhang, F., Wang, G., Liu, Z.: An effective approximation algorithm for the malleable parallel task scheduling problem. J. Parallel Distrib. Comput. **72**(5), 693–704 (2012)
8. Hooker, J.N.: Planning and scheduling by logic-based benders decomposition. Oper. Res. **55**(3), 588–602 (2007)
9. Jansen, K., Zhang, H.: Scheduling malleable tasks with precedence constraints. J. Comput. Syst. Sci. **78**(1), 245–259 (2012)
10. Jedrzejowicz, P., Skakovski, A.: Population learning with differential evolution for the discrete-continuous scheduling with continuous resource discretisation. In: 2013 IEEE International Conference on Cybernetics (CYBCONF), pp. 92–97 (2013)
11. Lima, R.M., Grossmann, I.E.: Computational advances in solving mixed integer linear programming problems (2011)
12. Lodi, A.: 50 Years of Integer Programming 1958–2008. Mixed integer programming computation, pp. 619–645. Springer, Heidelberg (2010)
13. Nguyen, N.-Q., Yalaoui, F., Amodeo, L., Chehade, H., Toggenburger, P.: Total completion time minimization for machine scheduling problem under time windows constraints with jobs' linear processing rate function. Journal of Scheduling, manuscript submitted for publication (2015)
14. Rozycki, R., Weglarz, J.: Power-aware scheduling of preemptable jobs on identical parallel processors to meet deadlines. Eur. J. Oper. Res. **218**(1), 68–75 (2012)
15. Sadykov, R.: A dominant class of schedules for malleable jobs in the problem to minimize the total weighted completion time. Comput. Oper. Res. **39**(6), 1265–1270 (2012)
16. Sirikum, J., Techanitisawad, A., Kachitvichyanukul, V.: A new efficient ga-benders' decomposition method: for power generation expansion planning with emission controls. IEEE Trans. Power Syst. **22**(3), 1092–1100 (2007)
17. Tahar, D.N., Yalaoui, F., Chu, C., Amodeo, L.: A linear programming approach for identical parallel machine scheduling with job splitting and sequence-dependent setup times. Int. J. Prod. Econ. **99**(1), 63–73 (2006)
18. Yalaoui, F., Chu, C.: New exact method to solve the $P_m/r_j/\sum C_j$ schedule problem. Int. J. Prod. Econ. **100**(1), 168–179 (2006)

Theoretical Analysis of Workload Imbalance Minimization Problem on Identical Parallel Machines

Yassine Ouazene[1](✉), Farouk Yalaoui[1], Alice Yalaoui[1], and Hicham Chehade[2]

[1] Institut Charles Delaunay, Laboratoire d'Optimisation des Systèmes Industriels (UMR-CNRS 6281), Université de Technologie de Troyes, 12 rue Marie Curie, CS 42060, 10004 Troyes, France
yassine.ouazene@utt.fr

[2] Opta-Lp, 2 Rue Gustave Eiffel, 10430 Rosières-prés-Troyes, France
chehade@opta-lp.com

Abstract. This paper considers the problem of assigning N non-preemptive jobs to M identical parallel machines or processors as equally as possible. This problem is known as workload imbalance minimization problem. First, we establish that this problem can be formulated as the difference between the maximum and minimum workloads. In other words, it is defined as the minimization of the difference between the workload of the bottleneck machine and the workload of the fastest machine.

Then, we present comparative analysis between this criterion and other criteria proposed in the literature such as: The average absolute deviation from the mean value of the total workload and Normalized Sum of Square for Workload Deviations ($NSSWD$) criteria proposed in the literature.

Keywords: Parallel machines · Workload balancing · Normalized sum of square for workload deviations · Mean absolute error

1 Introduction

Minimizing workload imbalance among parallel resources is important in many service and production environments. For example, in manufacturing industry, balancing the workload among the machines is important to reduce the idle times and work-in-process. It helps also to remove bottlenecks in manufacturing systems. In this paper, we address the classical problem of assigning a set of jobs to identical parallel machines with the objective to balance their workloads as equally as possible.

In the literature, the workload balancing problem has been associated with different scheduling criteria in different ways, even by considering the workload imbalance as a constraint or as an objective. For example,

N.T. Nguyen et al. (Eds.): ACIIDS 2016, Part II, LNAI 9622, pp. 296–303, 2016.
DOI: 10.1007/978-3-662-49390-8_29

Ouazene et al. [8] studied the identical parallel machine scheduling problem with minimizing simultaneously total tardiness and workload imbalance. The authors developed a mathematical formulation and a genetic algorithm based on the aggregation of the two objective functions. Yildirim et al. [15] addressed a mathematical formulation and an approximate resolution method based on some heuristics and a genetic algorithm for dealing with the context of semi-related machines, sequence dependent setups and load balancing constraints. Keskinturk et al. [6] presented a non linear mathematical model for a parallel machine problem with sequence-dependent setups with the objective of minimizing the total relative imbalance. They proposed an ant colony optimization algorithm and a genetic algorithm for the approached resolution of the problem.

Raghavendra et al. [12] proposed a genetic algorithm based approach with SPT and LPT rules for reducing the imbalance among identical parallel machines. The authors have concluded that their genetic algorithm provides better solutions than the strategies proposed in their first paper [10]. The same genetic based heuristics algorithm was compared with other approximate approaches proposed in the literature. Indeed, Raghavendra and Murthy [11] presented a comparative study among different test examples to illustrate the efficiency of their algorithm. The authors have shown that their algorithm outperforms the different heuristics proposed by Heinrich [4] and the genetic and simulated annealing algorithms proposed by Liu and Wu [7]. Recently, Ouazene et al. [9] proposed a mathematical programming approach for solving optimally these different instances previously used in the literature.

As mentioned by Cossari et al. [2], there is no established measure of performance in the scheduling literature to characterize workload balance. But some different criteria have been identified. The first criterion is the relative percentage of imbalances (RPI) introduced by Rajakumar et al. [13,14]. Ho et al. [5] proved that, in the case of identical parallel machine, the relative percentage of imbalances (RPI) depends solely of the minimization of maximum of workloads and does not assess workload balancing.

In this paper, we are interested on the analysis of a fundamental criterion proposed by [5]. This criterion, based on square errors, is called the Normalized Sum of Square for Workload Deviations ($NSSWD$). The authors proved that $NSSWD$-minimization problem implies the maximum completion time minimization. They proposed an approached resolution which bases on existing algorithm for minimizing maximum completion time. Later, Cossari et al. [2,3] addressed the same criterion and other criteria directly derived from $NSSWD$-minimization problem.

The remainder of this paper is organized as follows. The next section introduces the problem considered is this study with the different notations. In this section, we also establish that the workload imbalance minimization problem can be formulated as the minimization of the difference between the maximum and the minimum of machines completion times. Section 3, presents a comparative study between this criterion other criteria issued from the literature. Finally, Sect. 4 summarizes the contribution of this paper.

2 Problem Discerption

The problem considered in this paper can be formally described as follows: a set of N independent jobs $\{J_1, J_2, ..., J_N\}$ are to be scheduled on M identical parallel machines. The objective is to balance the workloads of the different machines as equally as possible.

Each job J_j has a deterministic processing time p_j. The jobs may be assigned to any one of the machines. A machine can process only one job at once and no preemption is allowed. All jobs are available at time zero.

The different notation adopted in this paper are listed below.

N	Total number of jobs
M	Total number of machines
m	Machine index
p_j	Processing time of job j
S_m	Set of jobs scheduled on the machine m
$C_m = \sum_{j \in S_m} p_j$	Completion time or workload of machine m
C_{max}	Maximum of workloads (completion times)
C_{min}	Minimum of workloads (completion times).

2.1 $(C_{max} - C_{min})$ Criterion to Minimize Workload Imbalance

If we consider the analogy with number partitioning, we can define the workload imbalance minimization problem as the minimization of the difference between the greatest and the least completion times $(C_{max} - C_{min})$. This can be easily established.

Let us consider a set of N jobs to be scheduled on M parallel machines $(M < N)$. Since the machines are identical, we can suppose, without loss of generality, that the machines indexes are rearranged in an increasing way according to the machines completion times $(C_1 \leq C_2 \leq C_3 \leq ... \leq C_M)$. So, $C_1 = C_{min}$ and $C_M = C_{max}$.

Let be $\delta_m = C_{m+1} - C_m, \forall m = 1...M-1$ the workload imbalance between two successive machines and $\eta_{m,k} = C_m - C_k$ the workload imbalance between the machine m and another non successive machine k $(m > k)$. The different notations and definitions are illustrated in Fig. 1.

We notice that $\eta_{m,k}$ can be expressed as a linear combination of the workloads imbalances between successive machines as follows.

$$\begin{aligned}
\eta_{m,k} &= C_m - C_k \\
&= (C_m - C_{m-1}) + (C_{m-1} - C_{m-2}) + ... + (C_{k+2} - C_{k+1}) + (C_{k+1} - C_k) \\
&= \delta_{m-1} + \delta_{m-2} + \delta_{m-3} + ... + \delta_k \\
&= \sum_{i=k}^{m-1} \delta_i
\end{aligned}$$

If we define the workload balancing problem as the minimization of the maximum workload imbalance, we can write:

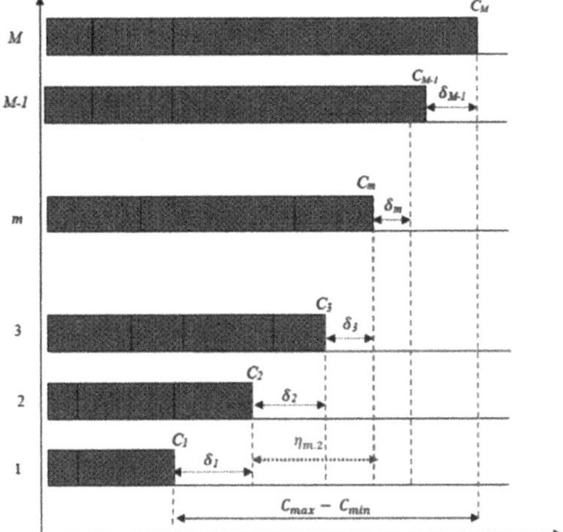

Fig. 1. Workload imbalance illustration using the workloads imbalances between successive machines

$$\begin{aligned}
\min WI &\iff \min\max \eta_{m,k}; m,k \in \{1...M\}, m \neq k \\
&\iff \min\max \textstyle\sum_{i=k}^{m-1} \delta_i; m,k \in \{1...M\}, m \neq k \\
&\iff \min \textstyle\sum_{i=1}^{M-1} \delta_i \\
&\iff \min \delta_{M-1} + \delta_{m-2} + \delta_{m-3} + ... + \delta_1 \\
&\iff \min(C_M - C_{M-1}) + (C_{M-1} - C_{M-2}) + ... + (C_2 - C_1) \\
&\iff \min(C_M - C_1) \\
&\iff \min(C_{max} - C_{min}).
\end{aligned}$$

3 Correlation Between $NSSWD$ and $(C_{max} - C_{min})$ Criteria

The Normalized Sum of Square for Workload Deviations ($NSSWD$) has been introduced by Ho et al. [5]. This criterion is based on the sum of squares principle known in measuring variability in statistics and serves as a precise measurement criterion. $NSSWD$ is defined as follows.

$$NSSWD = \frac{1}{\mu} \times \sqrt{\sum_{m=1}^{M} (C_m - \mu)^2}$$

Where:

$$\mu = \frac{1}{M} \times \sum_{m=1}^{M} C_m = \frac{1}{M} \times \sum_{j=1}^{N} p_j$$

Cossari et al. [3] showed that: $NSSWD = \sqrt{M} \times \frac{\sqrt{\frac{1}{M} \times \sum_{m=1}^{M} (C_m - \mu)^2}}{\mu}$, where $\sqrt{\frac{1}{M} \times \sum_{m=1}^{M} (C_m - \mu)^2}$ is the standard deviation and $\frac{\sqrt{\frac{1}{M} \times \sum_{m=1}^{M} (C_m - \mu)^2}}{\mu}$ is the coefficient of variation.

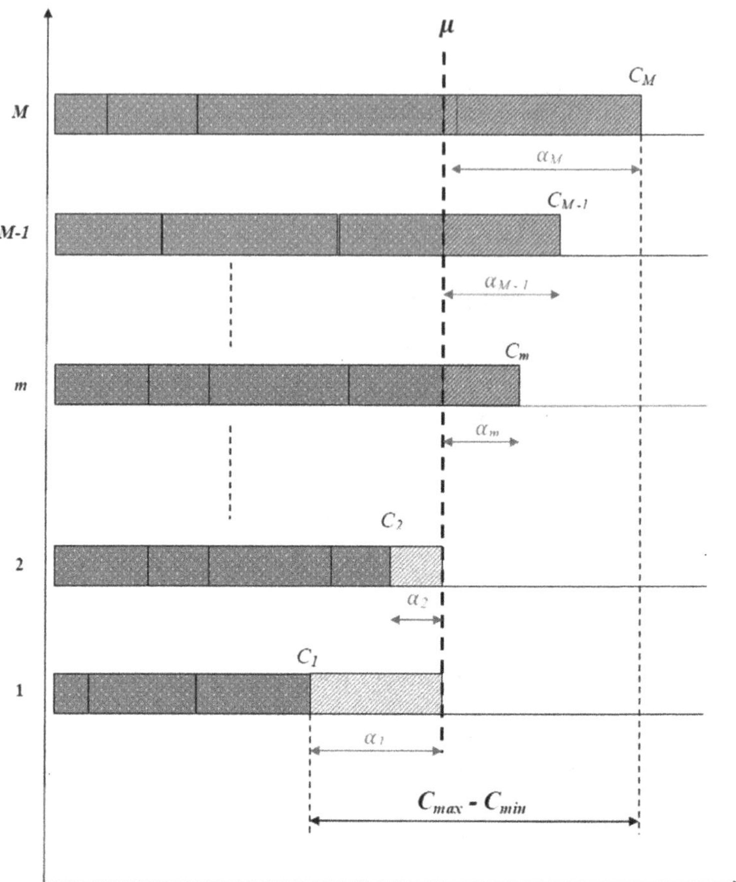

Fig. 2. Workload imbalance illustration using the deviations from the mean value of the total workload

So, the authors concluded that $NSSWD$ criterion is equivalent to the coefficient of variation and both of these two criteria may be used as performance measures.

This section demonstrates some proprieties of the Normalized Sum of Square for Workload Deviations ($NSSWD$) criterion and establishes that minimizing the difference between the maximum and the minimum of workload implies the minimization of the $NSSWD$.

Proposition 1. *The average absolute deviation from the mean value of the total workload imbalance is bounded as follows:*

$$\frac{1}{M} \times (C_{max} - C_{min}) \leq \frac{\sum_{m=1}^{M} |C_m - \mu|}{M} \leq \frac{max\{l, M-l\}}{M} \times (C_{max} - C_{min}).$$

Proof. Without loss of generality, we consider that the machines indexes are rearranged in an increasing way according to the machines completion times ($C_1 \leq C_2 \leq C_3 \leq ... \leq C_M$) as illustrated in Fig. 2. Based on these notations, the average deviation from the mean value of the total workload imbalance minimization problem can be written as follows.

$$\min \sum_{m=1}^{M} |C_m - \mu| \Longleftrightarrow \min \sum_{m=1}^{M} \alpha_m$$

Where,

$$\alpha_m = \begin{cases} \mu - C_m \text{ if } C_m \leq \mu \\ C_m - \mu \text{ if } C_m \geq \mu \end{cases}$$

Let be l the index of the machine corresponding to: $C_l \leq \mu$ and $C_{l+1} > \mu$. The first bound is obtained as follows. we have:

$$\begin{aligned} \sum_{m=1}^{M} \alpha_m &= \alpha_1 + \alpha_2 + ... + \alpha_M \\ &\geq \alpha_1 + \alpha_M \\ &\geq (\mu - C_1) + (C_M - \mu) \\ &\geq C_{max} - C_{min}) \end{aligned}$$

The second bound is based on the following:

$$\forall m = 1...l; \alpha_m \leq \alpha_1$$

Because ($C_1 \leq C_2 \leq ... \leq C_l \Longrightarrow \mu - C_1 \geq \mu - C_2 \geq ... \geq \mu - C_l$)

$$\forall m = l...M; \alpha_m \leq \alpha_M$$

Because ($C_l \leq C_{l+1} \leq ... \leq C_M \Longrightarrow C_l - \mu \leq C_{l+1} - \mu \leq ... \leq C_M - \mu$)
So,

$$\begin{aligned} \sum_{m=1}^{M} \alpha_m &= \sum_{m=1}^{l} \alpha_m + \sum_{m=l+1}^{M} \alpha_m \\ &\leq \sum_{m=1}^{l} \alpha_1 + \sum_{m=l+1}^{M} \alpha_M \\ &\leq l \times \alpha_1 + (M - l) \times \alpha_M \\ &\leq \max\{l, M - l\} \times (\alpha_1 + \alpha_M) \\ &\leq \max\{l, M - l\} \times ([\mu - C_1] + [C_M - \mu]) \\ &\leq \max\{l, M - l\} \times (C_{max} - C_{min}) \end{aligned}$$

Based on the two bounds developed above, we can establish that:

$$(C_{max} - C_{min}) \leq \sum_{m=1}^{M} \alpha_m \leq max\{l, M - l\} \times (C_{max} - C_{min})$$

Since max $\{l, M - l\}$ is a constant, we have demonstrated that minimizing ($C_{max} - C_{min}$) implies the minimization of dispersion around the mean $\sum_{m=1}^{M} |C_m - \mu|$.

Proposition 2. *The ratio between* ($NSSWD$) *and* ($C_{max} - C_{min}$) *criteria is:*

$$\frac{NSSWD}{C_{max} - C_{min}} \leq \frac{M^2}{\sum_{j=1}^{n} p_j}.$$

Proof. As a preliminary step, before dealing with the demonstration, we remind the following result (see Al-Saleh and Yousif [1] for the details):

$$\sqrt{\sum_{m=1}^{M}(C_m-\mu)^2} \leq \sum_{m=1}^{M}|C_m-\mu|.$$

$$\sqrt{\sum_{m=1}^{M}(C_m-\mu)^2} \leq \sum_{m=1}^{M}|C_m-\mu| \Longrightarrow \frac{1}{\mu}\times\sqrt{\sum_{m=1}^{M}(C_m-\mu)^2} \leq \frac{1}{\mu}\times\sum_{m=1}^{M}|C_m-\mu|$$
$$\Longrightarrow NSSWD \leq \frac{1}{\mu}\times\sum_{m=1}^{M}|C_m-\mu|$$

Or based on Proposition 1, it is established that:

$$\sum_{m=1}^{M}|C_m-\mu| \leq \max\{l, M-l\}\times(C_{max}-C_{min}).$$

So we can conclude that:

$$\begin{aligned}
\sqrt{\sum_{m=1}^{M}(C_m-\mu)^2} \leq \sum_{m=1}^{M}|C_m-\mu| &\Longrightarrow \frac{1}{\mu}\times\sqrt{\sum_{m=1}^{M}(C_m-\mu)^2} \leq \frac{1}{\mu}\times\sum_{m=1}^{M}|C_m-\mu| \\
&\Longrightarrow NSSWD \leq \frac{1}{\mu}\times\sum_{m=1}^{M}|C_m-\mu| \\
&\Longrightarrow NSSWD \leq \frac{\max\{l,M-l\}}{\mu}\times(C_{max}-C_{min}) \\
&\Longrightarrow \frac{NSSWD}{(C_{max}-C_{min})} \leq \frac{\max\{l,M-l\}}{\mu} \\
&\Longrightarrow \frac{NSSWD}{(C_{max}-C_{min})} \leq \frac{M}{\mu} \\
&\Longrightarrow \frac{NSSWD}{(C_{max}-C_{min})} \leq \frac{M}{\frac{1}{M}\times\sum_{j=1}^{N}p_j} \\
&\Longrightarrow \frac{NSSWD}{(C_{max}-C_{min})} \leq \frac{M^2}{\sum_{j=1}^{N}p_j}.
\end{aligned}$$

4 Conclusion

In many organizations, distributing workload as equally as possible among a group of persons or machines is a real challenge and some times it is considered as a strong constraint especially when it concerns human resources. This why workload balancing is an important optimization problem.

In this paper, we studied the different criteria proposed in the literature to deal with this problem in the case of identical parallel resources. Based on this study, we have established that the workload balancing problem can be formulated as a problem of minimizing the difference between the workload of the bottleneck machine and the workload of the fastest machine. Some important theoretical proprieties have been also presented. A possible extension of this research is to consider this new formulation of the workload imbalance minimization problem in the definition to develop exact and approached optimization methods.

References

1. Al-Saleh, M.F., Yousif, A.E.: Properties of the standard deviation that are rarely mentioned in classrooms. Austrian J. Stat. **38**(3), 193–202 (2009)

2. Cossari, A., Ho, J.C., Paletta, G., Ruiz-Torres, A.J.: Minimizing workload balancing criteria on identical parallel machines. J. Ind. Prod. Eng. **30**(3), 160–172 (2013)
3. Cossari, A., Ho, J.C., Paletta, G., Ruiz-Torres, A.J.: A new heuristic for workload balancing on identical parallel machines and a statistical perspective on the workload balancing criteria. Comput. Oper. Res. **39**(7), 1382–1393 (2012)
4. Heinrich, K.: A heuristic algorithm for the loading problem in flexible manufacturing systems. Int. J. Flex. Manuf. Syst. **7**, 229–254 (1995)
5. Ho, J.C., Tseng, T.L., Ruiz-Torres, A.J., López, F.J.: Minimizing the normalized sum of square for workload deviations on m parallel processors. Comput. Indus. Eng. **56**(1), 186–192 (2009)
6. Keskinturk, T., Yildirim, M.B., Barut, M.: An ant colony optimization algorithm for load balancing in parallel machines with sequence-dependent setup times. Comput. Oper. Res. **39**(6), 1225–1235 (2012)
7. Liu, M., Wu, C.: A genetic algorithm for minimizing the makespan in the case of scheduling identical parallel machines. Artif. Intell. Eng. **13**(4), 399–403 (1999)
8. Ouazene, Y., Hnaien, F., Yalaoui, F., Amodeo, L.: The joint load balancing and parallel machine scheduling problem. In: Bo, H., Morasch, K., Pickl, S., Siegle, M. (eds.) Operations Research Proceedings 2010. Operations Research Proceedings, pp. 497–502. Springer, Heidelberg (2011)
9. Ouazene, Y., Yalaoui, F., Chehade, H., Yalaoui, A.: Workload balancing in identical parallel machine scheduling using a mathematical programming method. Int. J. Comput. Intell. Syst. **7**(sup1), 58–67 (2014)
10. Raghavendra, B.V., Murthy, A.N.N.: Some solution approaches to reduce the imbalance of workload in parallel machines while planning in flexible manufacturing system. Int. J. Eng. Sci. Technol. **2**(5), 724–730 (2010)
11. Raghavendra, B.V., Murthy, A.N.N.: Workload balancing in identical parallel machine scheduling while planning in flexible manufacturing system using genetic algorithm. ARPN J. Eng. Appl. Sci. **6**(1), 49–55 (2011)
12. Raghavendra, B.V., Murthy, A.N.N.: Some solution approaches to reduce the imbalance of workload in parallel machines while planning in flexible manufacturing system through genetic algorithm. Int. J. Eng. Sci. Technol. **2**(5), 724–730 (2010)
13. Rajakumar, S., Arunachalam, V.P., Selladurai, V.: Workflow balancing strategies in parallel machine scheduling. Int. J. Adv. Manufact. **23**, 366–374 (2004)
14. Rajakumar, S., Arunachalam, V.P., Selladurai, V.: Workflow balancing in parallel machines through genetic algorithm. Int. J. Adv. Manuf. Technol. **33**, 1212–1221 (2007)
15. Yildirim, M.B., Duman, E., Krishna, K., Senniappan, K.: Parallel machine scheduling with load balancing and sequence dependent setups. Int. J. Oper. Res. **4**(1), 42–49 (2007)

Collective Intelligence for Service Innovation, Technology Opportunity, E-Learning and Fuzzy Intelligent Systems

A New Cosine Similarity Matching Model for Interior Design Drawing Case Reasoning

Kuo-Sui Lin(✉) and Ming-Chang Ke

Department of Information Management, Aletheia University, New Taipei City, Taiwan, ROC
{au4234, au1130}@mail.au.edu.tw

Abstract. In our previous study, we proposed a recommender system for interior design drawing retrieval. In that paper, a cosine similarity matching function is used for measuring binary bit string similarity between two design cases. After wider practical applications, we found that the design features on the interior design drawings could be with mixed interval, nominal, ordinal or ratio measurement scales. We further found that a case-based reasoning system is more suitable for interior design drawing retrieval than a recommender system, because the case-based reasoning system can begin with a few number of reference cases and allows the case database to be developed incrementally. Therefore, the objective of this study is to propose a new cosine similarity matching model for interior design drawing case retrieval in a case-based reasoning system, in which mixed measurement scales are considered and applied. Finally, a numerical case study is carried out to demonstrate the effectiveness and capabilities of the proposed cosine similarity matching model for the case-based reasoning system.

Keywords: Case-based reasoning system · Cosine similarity matching · Fuzzy measurement scale · Interior design drawing

1 Introduction

An interior design prototype drawing is better to be created earlier during requirements analysis stage; thereafter, any adjustment, correction, or addition to the design prototype drawing can be implemented, bridging the design requirements gap between the consigner and the designer. In our previous study [1], we proposed a virtue reality based recommender system to retrieve an interior design drawing from a historical drawings database. The binary bit measurement scale is used in that paper and the cosine similarity measure function is used for binary string pattern matching between two sequences of 0's and 1's in the retrieval engine of the recommender system, 0 for absence of a specific design feature and 1for presence of a specific design feature. In further practical similarity measurement, we found that features could be interval, nominal, ordinal and/or ratio scale types. Since cosine similarity measure is the most commonly used similarity measure when dealing with numerical features, this measure fails as the Euclidean distance is not defined for features of ordinal, nominal or mixed scale types. No studies is reported in the literature on the use of cosine similarity

N.T. Nguyen et al. (Eds.): ACIIDS 2016, Part II, LNAI 9622, pp. 307–318, 2016.
DOI: 10.1007/978-3-662-49390-8_30

measure function in developing a retrieval engine for dealing with mixed measurement scales of interval, nominal, ordinal and ratio scale types at the same time. Further, the revising and learning characteristics of a case-based reasoning (CBR) are useful when the new query is hard to articulate their requirement. Thus, a CBR system can begin with a few number of reference cases and allows the case database to be developed incrementally, regarded as an initial version that could be revised, retained and restored to increase the number of reference cases. Therefore, the objective of this study has been two fold. The first goal was to propose a new cosine similarity matching model for case reasoning, in which mixed measurement scales are considered. The second goal was to propose a CBR system and apply the new cosine similarity matching model to the retrieval engine of the CBR system for interior design drawing retrieval problems in the interior design industry.

2 Case-Based Reasoning System

Case-based reasoning is a problem solving methodology that deals with a new problem by first retrieving a past similar case, and then reusing and adapting that case for solving the new problem [2]. For A CBR System, the four main phases of the CBR 4 R cycle are iteratively performed: retrieval, reuse, revision and retention. Firstly, previous cases are stored in the case database. Typically, each case contains a description of the problem, plus a solution and/or the outcome. A new case is the description of a new problem to be solved and is an input to the system. In the retrieval phase, cases that are to be retrieved from the case base will be matched with the new case based on several features. The most similar case to the target problem from the case base is retrieved. In the reuse phase, the retrieved case is suggested and reused to solve the new problem. In the revise phase, if there is a very similar past case whose solution needs little adjustment, revise or adapt the proposed solution for the new problem if necessary. In the retain phase, the revised or adapted case for future problem solving in the next retrieval cycle is retained as part of a new case.

3 The New Fuzzy Similarity Matching Model

The most important part of a CBR system is the retrieval engine. Thus, improving the performance efficiency of the retrieval engine is critical for the CBR system [3]. A commonly used similarity measure for the retrieval engine is cosine similarity [4, 5]. For improving the performance efficiency of the retrieval engine, we proposed a new cosine similarity matching model, which is described as follows.

Step1: Translating Various Measurement Scales into Fuzzy Measurement Scales. From a practical perspective, four different formats of measurement scales are considered for interior design drawing case retrieval in a CBR system. Thus, we need to translate the four traditional measurement scales into fuzzy measurement scales, including nominal, ordinal, interval and ratio measurement scales. The stored old cases and query case which are rated by various numerical or categorical scales will then be translated into their corresponding fuzzy linguistic terms. Then, the translated fuzzy linguistic terms will be transformed into their corresponding membership functions. Delphi technique can be used for membership functions construction. By referring to several types of triangular fuzzy numbers in linguistic terms [6], we parameterized these linguistic terms with Triangular Fuzzy Numbers.

Step 2: Computing Local Similarity Measure. The local similarity is the similarity measure on each feature of the new query vector Q and the stored case vector C_i. Let $sim(f_j^Q, f_j^{Ci})$ denote the local similarities (feature similarities) between the stored case vector $\boldsymbol{C_i}$ and the new query vector $\boldsymbol{Q}$ with regard to feature F_j. f_j^Q *and* f_j^{Ci} is the rating of F_j in the query case Q and the historical case C_i, respectively. Depending on different measurement scales, the local similarity can be calculated as follows.

(1) **Cosine Similarity Measure for Fuzzy Nominal Scale.** Let $sim(f_j^Q, f_j^{Ci})$ denote the feature similarity measure between historical case C_i and query case Q with regard to the feature F_j. If the feature F_j is a nominal variable, the local similarity between two j th feature ratings f_j^Q and f_j^{Ci} is given by:

$$sim\left(f_j^Q, f_j^{Ci}\right) = 1, \text{ if } f_j^Q = f_j^{Ci};\ sim\left(f_j^Q, f_j^{Ci}\right) = 0, \text{ if } f_j^Q \neq f_j^{Ci}$$

It assigns value 1 if the two features match and value 0 otherwise.

(2) **Cosine Similarity Measure for Fuzzy Ordinal, Interval and Ratio Scale.** The ordinal, interval and ratio scale can be translated into corresponding triangular fuzzy number. Thus, a cosine similarity measure for triangular fuzzy numbers is proposed in an analogous manner to the cosine similarity measure (angular coefficient) between fuzzy sets [7, 8]. Let $\boldsymbol{A} = (\mu_A(x_1), \mu_A(x_2), \mu_A(x_3))$ be a triangular fuzzy number in the set of real numbers R, the three parameters in $\boldsymbol{A}$ can be considered as a vector representation with the three elements. Assume that there are two triangular fuzzy numbers $\boldsymbol{A} = (\mu_A(x_1), \mu_A(x_2), \mu_A(x_3))$ and $\boldsymbol{B} = (\mu_B(x_1), \mu_B(x_2), \mu_B(x_3))$ in the set of real numbers R. Based on the extension of the cosine similarity measure for fuzzy sets, a cosine similarity measure between $\boldsymbol{A}$ and $\boldsymbol{B}$ is proposed as follows:

Cosine Similarity$(\boldsymbol{A}, \boldsymbol{B})$ = Dot $(\boldsymbol{A}, \boldsymbol{B})/\|\boldsymbol{A}\|\|\boldsymbol{B}\| = \boldsymbol{A} \bullet \boldsymbol{B}/\|\boldsymbol{A}\|\|\boldsymbol{B}\| = \Sigma_{k=1}^3\ (\mu_A(x_k)$ $*\mu_B(x_k))/(\text{sqrt}(\Sigma_{k=1}^3\ \mu_A(x_k)\ \wedge\ 2) \times \text{sqrt}(\Sigma_{k=1}^3\ \mu_B(x_k)\wedge 2)) = (\mu_A(x_1)\ *\mu_B(x_1) + \mu_A(x_2)$ $*\mu_B(x_2) + \mu_A(x_3)*\mu_B(x_3))/(\text{sqrt}(\mu_A(x_1)\wedge 2 + \mu_A(x_2)\wedge 2 + \mu_A(x_3)\wedge 2) \times \text{sqrt}(\mu_B(x_1)\wedge 2$ $+\mu_B(x_2)\wedge 2 + \mu_B(x_3)\wedge 2))$

Step 3: Computing Global Similarity Measure. To retrieve proper historical case, it is necessary to measure the global similarities. Thus, the local similarities (feature similarities) can be aggregated into the global similarity. The global similarity is the similarity measure between each of historical cases in the case database to the new query case. Let $Sim(Q, C_i)$ denote the global similarity between the historical case C_i and the query case Q, then a global similarity measure can be derived by the weighted summation of the local similarity matching measures: $Sim(Q, C_i) = (\Sigma_{j=1}^{n} w_j \times sim(f_j^Q, f_j^{Ci}))/\Sigma_{j=1}^{1} w_j$, where w_j, are the local weights assumed to be a quantity reflecting importance of the corresponding feature.

4 The Architecture of the Proposed CBR System

After proposed the cosine similarity matching model, we further developed a virtual reality based CBR system for interior design prototype drawing retrieval. As shown in Fig. 1, the proposed CBR system consists of a case base, a virtual reality based display module, a query module and a retrieval engine. The case base stores a group of historical interior design drawings drawn by major interior design software tools. The retrieval engine calculates the similarity of the cases in the case base to the query using the new similarity matching model and retrieves the case which is most suitable to the user' request and match his design preferences. The case retrieval phase is typically the main step of the CBR cycle and the majority of CBR systems can be described as sophisticated retrieval engines. The query module generates the preference query from the user and transfers the user query into a query profile. The display module provides a virtual reality based platform for displaying retrieved design drawings by using FancyDesigner® interior design drawing tool [9].

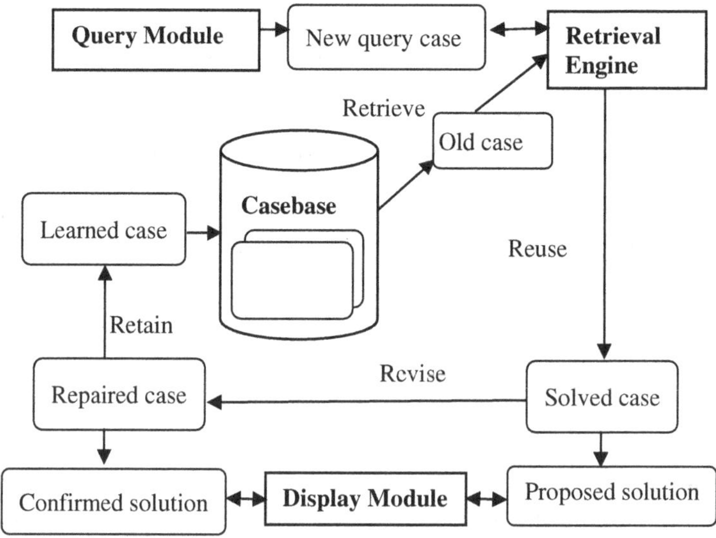

Fig. 1. The Architecture of the proposed CBR system (modified after Aamodt & Plaza, 1994)

5 A Numerical Case Study

To show the effectiveness of the proposed cosine similarity matching model, we apply the proposed model to the proposed CBR system for interior design drawing retrieval. When a user needs to find an interior design drawing that matches the user's design requirements, the user inputs some preference ratings about the design features of the design drawing through a questionnaire query module. The user's query is then taken to retrieve a most suitable case and displayed to the user through a display module. If the user is not satisfied with the retrieved case, the user can modify and adapt the case to form a new case to satisfy the current query and store the new case to the case base. The case study of the CBR system consists of the following steps.

Step 1: Extracting Features of Cases. In the present study, the authors used modified Delphi technique to develop a consensus list of features that characterize the cases. The modified Delphi technique utilized within this study was modified from the classical Delphi technique by including an initial list of features, which will supplied to the design experts instead of beginning the classical Delphi technique with a blank page. Firstly, in the data gathering phase of the modified Delphi technique, the initial and comprehensive list of features related to interior design drawings can be gathered from different information sources that include design handbook, relevant industrial codes and standards, and Web pages. Secondly, in the development phase of the modified Delphi technique, group creativity techniques, such as brainstorming, nominal group techniques, idea mapping, affinity diagram, etc., are used to clarify, comment on the features, brainstorm additional features and compile the list of features. The compiled list of features formed the questionnaire for the ranking phase of the modified Delphi round. Finally, in the consensus phase of the modified Delphi technique, the above mentioned questionnaires were sent by mail or email to a Delphi expert panel consisting of 11 design experts (participants). The expert panel was invited to score each and every research question on a numeric scale. The appropriateness of each features (questions) was rated on a five-point Likert scale that ranged from 1 (not important at all) to 5 (very important), with one equivocal point 3, and two "grey" points 2 and 4. Each feature was considered consensus if the expert's opinion rating was 4 or 5 for more than 70 % (8 of 11 experts) of the experts. Details on the modified Delphi technique described here to gather initial list of features, develop and compile the list, as well as consent and prioritize the list will be reported in another research paper.

Let n be the number of extracted feature set F. $F = \{F_1, F_2,\ldots,F_n\}$ and F_j be the j-th feature ($j = 1,2,\ldots,n$). These features will be used for case based reasoning in the next phase. A set of 10 extracted design features have been developed for classifying the interior design drawings, which are: $F = \{F_1, F_2,\ldots, F_{10}\}$ = {Budget, Working Duration, Design Style, Design Theme, Design Hue, Brightness, Saturation, Color temperature, Daylighting source, Lighting type}.

Step 2: Collecting Historical Cases to the Case Base. Before proceeding, several historical cases of interior design drawings have been pre-loaded to the case database, or case base, of the CBR system. In most situations, cases are represented as feature-value pairs. It is a group of related records of data set. Each record is a group of related design features of drawings. A feature is a distinctive property (or characteristic, attribute) used to distinguish two objects. Feature values are ratings of features that are provided by the users. An individual case C_i is assumed to be comprised of a finite list of feature-value pairs: $C_i = \{(F_1, f_1^{Ci}), (F_2, f_2^{Ci})...(F_j, f_j^{Ci})...(F_n, f_n^{Ci})\}$. Let the case base, CB, be a finite set of individual cases, $CB = \{C_1, C_2,...,C_m\}$ and C_i be the i-th case ($i = 1,2...m$), where m is the number of cases.

Step 3: Establishing Case Profiles and Query Profile. Both feature ratings of the historical cases and the user's questionnaire query are transformed into a set of old case-feature-rating vectors (case profiles) and user's query-feature-rating vector (query profile). A set of old case-feature-rating vectors is represented as a utility matrix.

(1) **Collecting case-feature rating matrix and query-feature rating vector.** The old case profiles can be represented by a case–feature-rating matrix. Each row in the matrix is the vector representation of a design case. As shown in Fig. 2, case-feature-rating entries usually are described as an $m \times n$ ratings matrix $\boldsymbol{R_{mn}}$, where each entry f_{ij} ($1 \leqq i \leqq m$, $1 \leqq j \leqq n$) means the rating assigned to the case i on the feature j; the row represents m cases and the column represents n features. The user's preference query can be described as the following feature-value pairs: $Q = \{(F_1, f_1^Q), (F_2, f_2^Q),..., (F_j, f_j^Q),..., (F_n, f_n^Q)\}$.
As shown in Table 1, example of values of the case–feature-rating matrix and query-feature-rating vector can be nominal, ordinal, interval or ratio scale.

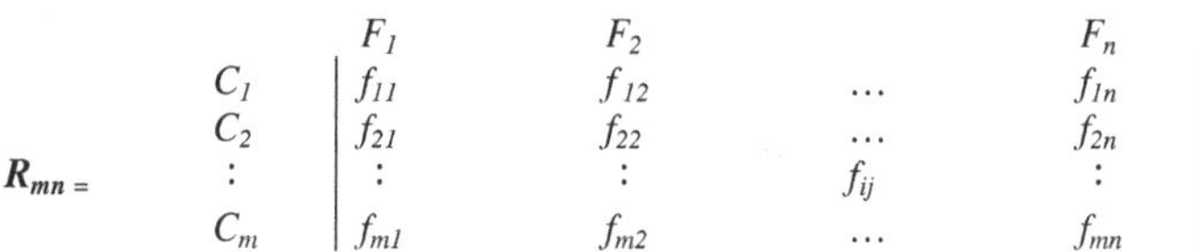

$$\boldsymbol{R_{mn}} = \begin{array}{c|ccccc|} & F_1 & F_2 & & F_n \\ C_1 & f_{11} & f_{12} & \cdots & f_{1n} \\ C_2 & f_{21} & f_{22} & \cdots & f_{2n} \\ \vdots & \vdots & \vdots & f_{ij} & \vdots \\ C_m & f_{m1} & f_{m2} & \cdots & f_{mn} \end{array}$$

Fig. 2. A case–feature rating matrix

(2) **Translating traditional measurement scales into fuzzy measurement scales.** A linguistic term can be defined by a membership function. Based on the available a priori information or the phenomenon intending to describe, a membership function may take many forms. Typical shapes for membership functions include triangular, trapezoid and Gaussian membership functions. The shape of each membership function used in this study is triangular due to its popularity in specifying fuzzy sets [5].

Because we have more knowledge about Budget and Working Duration than other rational features in this study, we defined five linguistic terms. The semantics of the linguistic terms are given using fuzzy numbers. By referring to [6], we further translated them into five triangular fuzzy numbers, ranging between zero and one (Tables 2, 3 and Fig. 3).

For translating feature of Design Style, Design Theme and Color Hue, we use binary bit string of nominal scale for measurement (Tables 4, 5 and 6).

Table 1. Collected content profiles and Query Profile

	F_1 \$ Budget	F_2 Days Working Duration	F_3 Design Style	F_4 Design Theme	F_5 Color Hue	F_6 Brightness	F_7 Saturation	F_8 Color Temperature	F_9 Daylighting Source	F_{10} Lighting Type
	Ratio	Ratio	Nominal	Nominal	Nominal	Ordinal	Ordinal	Ordinal	Ordinal	Nominal
Case 1	10,000	12	100000	00100000	000010	B	M	W	H	010
Case 2	20,000	14	001000	00100000	000010	M	M	M	M	100
Case 3	20,000	14	000100	000100000	000010	D	D	C	L	001
Case 4	60,000	21	000001	000100000	000010	B	I	W	M	001
Case 5	50,000	21	000010	00100000	000010	B	M	W	H	100
Case 6	80,000	42	010000	00100000	000010	D	D	C	L	100
•										
Query	30,000	20	010000	00100000	000010	B	I	M	H	001

For translating feature of Brightness, Saturation, Color Temperature and Daylighting Source, we categorized three linguistic terms of ordinal scale and the range is defined between zero and one (Table 7 and Fig. 4).

For translating feature of Lighting Type, we use binary string of nominal measurement scale (Table 8). Lighting Type is classified as task, accent, or general lighting, depending largely on the distribution of the light produced by the fixture.

Table 2. Budget

Description (days)	Linguistic term	Triangular fuzzy number
Budget ≤ 10	Very Low(V L)	TFN(0, 0, 0.25)
11 ≤ Budget ≤ 30	Low(L)	TFN(0, 0.25, 0.5)
31 ≤ Budget ≤ 50	Medium(M)	TFN(0.25, 0.5, 0.75)
51 ≤ Budget ≤ 70	High(H)	TFN(0.5, 0.75, 1)
71 ≤ Budget	Very High (VH)	TFN(0.75, 1, 1)

Table 3. Working duration

Description (calendar days)	Linguistic term	Triangular fuzzy number
Duration ≤ 10	Very Short(VS)	TFN(0, 0, 0.25)
11 ≤ Duration ≤ 20	Short(S)	TFN(0, 0.25, 0.5)
21 ≤ Duration ≤ 30	Medium(M)	TFN(0.25, 0.5, 0.75)
31 ≤ Duration ≤ 40	Long(L)	TFN(0.5, 0.75, 1)
41 ≤ Duration	Very Long(VL)	TFN(0.75, 1, 1)

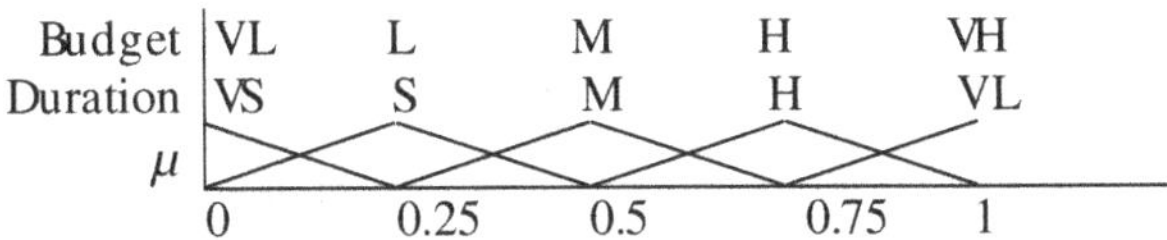

Fig. 3. Membership functions for features with nominal scales

Table 4. Binary string of design style

Linguistic term	Binary string
Royal	100000
Zen	010000
Nautical/Coastal	001000
High Tech	000100
Modern	000010
Country	000001

Table 5. Design theme

Linguistic term	Binary string
Greece	10000000
Middle East	01000000
European	00100000
American	00010000
Taiwanese	00001000
Chinese	00000100
Japanese	00000010
Korean	00000001

Table 6. Color Hue

Linguistic term	Binary string
Red	100000
Yellow	010000
Green	001000
Cyan	000100
Blue	000010
Magenta	000001

Table 7. Brightness, saturation, color temperature and daylighting source

Description	Linguistic term	Linguistic term	Linguistic term	Linguistic term	Triangular fuzzy number
ordinal #1	Dark	Cool	Dull	Low	TFN(0, 0, 0.5)
ordinal #2	Medium	Medium	Medium	Medium	TFN(0, 0.5, 1)
ordinal #3	Bright	Warm	Intense	High	TFN(0.5, 1, 1)

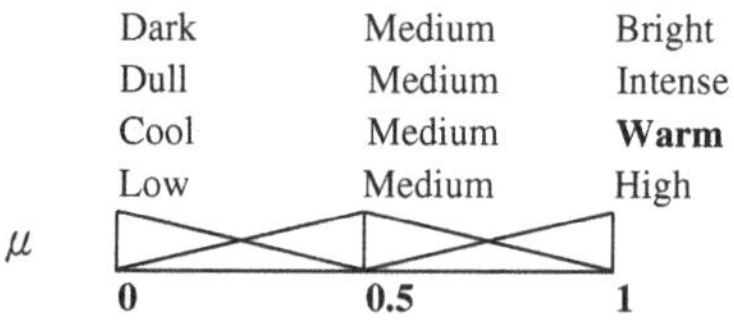

Fig. 4. Membership functions for features with ordinal scale

Table 8. Lighting type

Linguistic term	Binary string
General lighting	100
Accent lighting	010
Task lighting	001

Example of calculation sheet for translating measurement scales into linguistic terms is shown in Table 9.

Step 4: Computing Similarity Measures

(1) **Computing local similarity measures on features between a case and a query.** The local similarity calculates the similarity for each feature j of the case C_i and the query Q. By applying the local similarity measure $sim(f_j^Q, f_j^{Ci})$, the similarity on feature j between old case C_i in the casebase and the new query Q can be calculated. Taking query Q and case C_6 for examples, calculation sheet for translated fuzzy number and calculated local similarity measures and global similarity measure is shown in Table 10.

(2) **Computing global similarity measure between a case and a query.** The global similarity function measures similarity between each of historical cases $C_{i,}$ $i = 1 \ldots m_,$ in the casebase to the new query case Q. Suppose the weightings are equally weighted, each being equal to 1. Then the global similarity between the historical case C_6 and the query case Q is calculated as: Sim (Q, C_6) = $\left(\Sigma_{j=1}^{n} w_j \times sim\left(f_j^Q, f_j^{C1}\right)\right)/\Sigma_{j=1}^{n}$ w_j = 1* (0.84 + 0.84 + 1+0 + 1 +1 + 0.89 + 0.89 + 1+0)/10 = 0.75.

Step 5: Retrieving the Case. In this step, the query case and the historical cases are fed into the retrieval engine to retrieve the most similar case to meet the user's preferences. The transformed query vector and each case vector of the case-feature-rating matrix is compared through the new fuzzy similarity matching model. According to the calculation results of the matching model, the global similarity values in descending order are derived and the case with the highest degree of similarity matching is retrieved.

Table 9. Example of calculation sheet for translating measurement scales into linguistic terms

	F_1 Budget	F_2 Working Duration	F_3 Design Style	F_4 Design Theme	F_5 Color Hue	F_6 Brightness	F_7 Saturation	F_8 Color Temperature	F_9 Daylighting Source	F_{10} Lighting Type
Case 1	VL	S	100000	00100000	000010	3	2	3	3	010
Case 2	L	S	001000	00100000	000010	2	2	2	2	100
Case 3	L	S	000100	000100000	000010	1	1	1	1	001
Case 4	H	M	000001	000100000	000010	3	3	3	2	001
Case 5	M	M	000010	00100000	000010	3	2	3	3	100
Case 6	VH	VL	010000	00100000	000010	1	1	1	1	100
□										
Query	L	S	100000	00100000	000010	3	3	2	3	011

Table 10. Example of calculation sheet for translated fuzzy number and calculated local similarity measures

	F_1 budget	F_2 working duration	F_3 design style	F_4 design theme	F_5 design hue	F_6 brightness	F_7 saturation	F_8 color temperature	F_9 daylighting source	F_{10} lighting type
C_6=	VH	VL	100000	00100000	000010	#3	#2	#3	#3	010
Q =	M	M	100000	00100000	000010	#3	#3	#2	#3	001
C_1=	TFN(0.75, 1, 1)	TFN(0.75, 1, 1)	000110	00100000	000010	TFN (0.5,1,1)	TFN (0,0.5,1)	TFN(0,0,0.5)	TFN(0.5,1,1)	010
Q =	TFN (0,0.25,0.5)	TFN (0,0.25,0.5)	100000	01000000	000010	TFN (0.5,1,1)	TFN (0.5,1,1)	TFN(0,0.5,1)	TFN(0.5,1,1)	001
$sim(f_j^Q, f_j^{C1})$ =	0.84	0.84	1	0	1	1	0.89	0.89	1	0

Step 6: Revising and Retaining Case. The target design drawing which is associated with the retrieved case is then triggered and displayed to the user as a prototype drawing through a FancyDesigner® based virtue reality platform. Thereafter, necessary review/revision to the retrieved drawing can be made and stored into the casebase as a new case.

Step 7: Evaluating Retrieval Performance. In this study, *Precision* is defined as the measure of how many correct cases ($N_{correct}$) the system hits in the total retrieved cases ($N_{retrieved}$): $Precision = N_{correct}/(N_{correct} + N_{false}) = N_{correct}/N_{retrieved}$. *Precision* is a vital measure for evaluating retrieval performance as it is a direct measurement of the quality and user satisfaction of the retrieval process. Compared with previous study, this case study showed that the similarity matching efficiency improved significantly and demonstrated its ability with 73 % *Precision*.

6 Conclusions and Future Works

We have proposed a new fuzzy cosine similarity matching model for case interior design drawing retrieval in a case-based reasoning system, in which mixed measurement scales are considered and applied to the case-based reasoning. A numerical case study of the CBR system was made. This study has demonstrated the effectiveness and capabilities of the proposed similarity matching model. In the future, with increased number of case queries, more cases will be collected and stored in the casebase, so that the availability of choices on cases can be increased. Besides, more design features will be extracted from more data sources. Thus, with increased number of stored cases and extracted features, the next step of this study will be further to enhance more accurate retrieval performance and better user satisfaction.

References

1. Lin, K.-S., Ke, M.-C.: A virtual reality based recommender system for interior design prototype drawing retrieval. In: Barbucha, D., Nguyen, N.T., Batubara, J. (eds.) New Trends in Intelligent Information and Database Systems. SCI, vol. 598, pp. 141–150. Springer, Heidelberg (2015)
2. Aamodt, A., Plaza, E.: Case-based reasoning: foundational issues, methodological variations, and system approaches. AI Commun. **7**(1), 39–59 (1994)
3. Leake, D.B.: CBR in context. The present and future. In: Leake, D.B. (ed.) Case-Based Reasoning: Experiences, Lessons, and Future Directions, pp. 3–30. American Association for Artificial Intelligence, Menlo Park (1996)
4. Larsen, B., Aone, C.: Fast and effective text mining using linear-time document clustering. In: Proceedings of the Fifth ACM SIGKDD International Conference on Knowledge Discovery and Data Mining, San Diego, California, pp. 16–22 (1999)
5. Nahm, U.Y., Bilenko, M., Mooney, R.J.: Two approaches to handling noisy variation in text mining. In: Proceedings of the ICML-2002 Workshop on Text Learning, Sydney, Australia, pp. 18–27 (2002)

6. Chen, T.Y., Ku, T.C.: Importance-assessing method with fuzzy number-valued fuzzy measures and discussions on TFNs and TrFNs. Int. J. Fuzzy Syst. **10**(2), 92–103 (2008)
7. Salton, G., McGill, M.J.: Introduction to Modern Information Retrieval. McGraw-Hill, Auckland (1983)
8. Ye, J.: Multicriteria decision-making method based on a cosine similarity measure between trapezoidal fuzzy numbers. Int. J. Eng. Sci. Technol. **3**(1), 272–278 (2011)
9. FancyDesigner Homepage. http://www.fancydesigner.com.tw/bbs/modules/wordpress/

The Framework of Discovery Early Adopters' Incipient Innovative Ideas

Chao-Fu Hong[1(✉)], Mu-Hua Lin[2], and Woo-Tsong Lin[1,2]

[1] Department of Information Management, Aletheia University, Taipei, Taiwan
au4076@au.edu.tw, lin@mis.nccu.edu.tw
[2] Management Information Systems,
National Chengchi University, Taipei, Taiwan
95356503@nccu.edu.tw

Abstract. Crossing the chasm between early adopters and early majority in the market is not only an important issue for innovation diffusion, but also important information for firms to have the chance to occupy position and get great business success. Additionally, consumers can easily share their consumer-related articles through various IT blogs with Web 2.0, hence there is a big consuming data on the Internet. This research tried to discover incipient innovative ideas from early adopters to help firms to win the business. A new textual association analysis (Term Frequency - Inverse Clusters Frequency, TF-ICF) framework is a methodology to discover the more rare and useful ideas for designing future innovative business service. In the present study, TF-ICF methodology does not only find what instant foods or entertainment are needed for the passengers on travelling vehicles, but also reveal that the moisturizing emulation is another possible need of theirs. The results show that the TF-ICF method is useful to discover early adopters' incipient innovative ideas.

Keywords: Innovation diffusion · Rare association · Social influence · Chance discovery

1 Introduction

From Rogers' Innovation Diffusion Model (IDM), it shows that the earlier adopters' account is 12.5 % of all consumers. Therefore, the information that the earlier adopters uploaded onto the Internet must be rare. In other word, the uploaded information is easily flooded by early majority (34 %) and late majority (34 %). As results, there is the bottleneck, which has kept Rogers' Innovation Diffusion Model from success. Hong (2009) verified that the useful innovations can be accepted by early majority as long as the innovative ideas developed by early adopters can be obtained and passed on to early majority. It means that the small number of early adopters had not only the power of observe and sort out the useful innovative products among a majority of innovative products but also the acuity of lead users to find the features of being useful and easy to use in these innovative products (von Hippel 1986; Morrison et al. 2000). The abilities can help facilitate the process, which has been illustrated using the analogy of a

N.T. Nguyen et al. (Eds.): ACIIDS 2016, Part II, LNAI 9622, pp. 319–327, 2016.
DOI: 10.1007/978-3-662-49390-8_31

bowling alley to cross the chasm between early adopters and early majority simply by increasing the latter's level of acceptability toward innovative products (Moore, 1991).

In order to fasten the process of obtaining the innovative ideas which developed by early adopters, an algorithm framework called the inverse clusters frequency (ICF) was developed to discover the connection between rare nodes, which are words with low-frequency. In this network, a rare node connects to words clusters indirectly through its connected rare node, so that early adopters' primary innovations can be discovered without the intermediate of many clusters. In the beginning of this paper, the framework and the ICF methodology is presented. A literature review will be discussed next. Finally, a case study is used to verify the methodology.

2 Literature Reviews

2.1 Informational Social Influence, Grounded Theory (GT) and Chance Discovery

"Lead user" shares their ideas with others, as a 'tribal identity', to attract people to participate in the consuming tribe (Muniz and O'Guinn, 2001). Only few users are lead users who are able to create new ways of using innovative products that lead to the popularity of such products (Price and Ridgway, 1982). For example, in Morrison's experiment (2000), only about 26 % of the consumers have the "leading edge status" and in-house technical capabilities for them to freely share their innovations with others. The characteristics of these consumers so called lead users are similar to the innovators and early adopters mentioned by Rogers and they occupy only about 20 % of all consumers (Rogers, 2003). When a firm decides to involve in innovation/knowledge creation, how to sift out the lead users and bring them on board could be the critical key to a successful business.

Rare association analysis can be used to identify purchased goods with low support and high confidence. It can also be used to solve the problem of low frequency, but it still fails to link to future trend analysis, which is shown in Fig. 1a. Ohsawa and Nels (1998) proposed the KeyGraph algorithm, which is used to calculate both the frequencies of a node and the co-occurrence of two nodes in a shopping cart, in which strong clusters are expected to emerge, which is shown in Fig. 1b. Next, the key-value derived from all the nodes, which are connected to each other, is calculated. Among the strong clusters, some chance nodes with high key values and low frequency may be found. The researcher may integrate the hints given by the chance nodes and strong clusters to build a chance scenario, which is called chance discovery.

There have been many people attempting to overcome the bottleneck of extracting innovative ideas generated by early adopters, who entail information technologies to extract rare and useful information from a huge amount of information. Recently, Liu et al. (1999) developed a rare association analysis, and Ohsawa and Nels (1998) performed link analysis to analyze the data. They presented reasonable ways to make good use of networks, which are built up by rare nodes and the words clusters connect to them to explore the ranking of importance of innovations. However, these methods all have their own weaknesses. For instance, there are only the connections between the

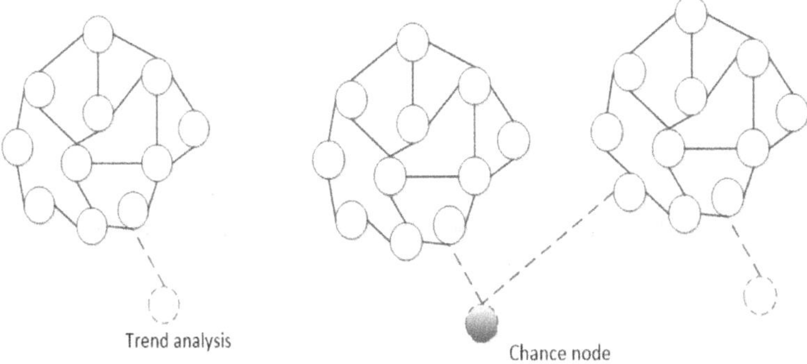

Fig. 1. The conceptual diagraph of (a) trend analysis and (b) chance node

rare nodes and their connected clusters, which are full of nodes and the frequencies of the nodes are all higher than the threshold frequency. This means that all these nodes will soon be known or accepted by majority. Since they cannot be qualifies as rare nodes, they are not real future chances for company to create new businesses.

2.2 Knowledge Creation from the Internet/Big Data

Nowadays, Web 2.0 is the most popular Internet technology and it helps consumers easily share what consuming values they write on the Internet (O' Reilly, 2007). Furthermore, a report about big data in 2011 by the McKinsey Global Institute reported that the was data increase 7000 PB (petabyte) in 2010, so they predicted that the 2020 data would be 44 times of the scale in 2010. Big data includes the data of browsing websites, virtual social communication, weblog articles, weather data and transaction data etc.

In Lin and Hong (2011) innovation model, they used the Google blog search to sift out related consumer's data, and they also used grounded theory (GT) based co-occurrence analysis to extract innovative consuming values from the collected data to influence the majority to accept a new product (Hong, 2009). In the model, a new innovative product was successfully created. They used scenario analysis to explain business processes, how the consumers used the innovative product to bring the good service to improve their lives. As previously discussion, a bottleneck will be eventually met at some point. The new innovations created from such a model will be known by the majority of public soon, because they are created from clusters full of strong nodes, which are nodes with high frequency. So the company will have no time or little chance to enter the innovative business. Although the model in 2011 could not find rare nodes, it evidenced that there are early adopters' ideas in the articles of weblog on the internet.

In this paper, a model has been developed to use weblog's documents to find the rarer nodes, which just newly listed by early adopters and will become new innovative ideas for new products. The company will have enough time to create their new products and business model and publish the innovative idea to people and then earn money.

3 ICF Methodology

The problem of the previous models focused the rare nodes and their connected clusters. As a result, the innovation ideas are based on the nodes in the clusters, but the frequencies of these nodes are higher than the threshold frequency. This means the innovation ideas are that new and will soon be known or accepted by the majority of markets, so that they may not be the real future chance for companies.

The new model in this paper is designed to find more rare nodes, which will lead us to some useful innovative ideas originally from early adopters and discover chances. In Fig. 2, a brief process of the model to find more rare notes is shown. In Fig. 2(a), word clusters are built. In Fig. 2(b), based on the sentences of the nodes in the clusters, the nodes may extend and connect to some other rare nodes. In Fig. 2(c), based on the sentences of these rare nodes, more rare nodes such as node A and node B may be found. In Fig. 2(d), after the network of the notes and the clusters is set up, the ICF value of all the rare nodes can be calculated.

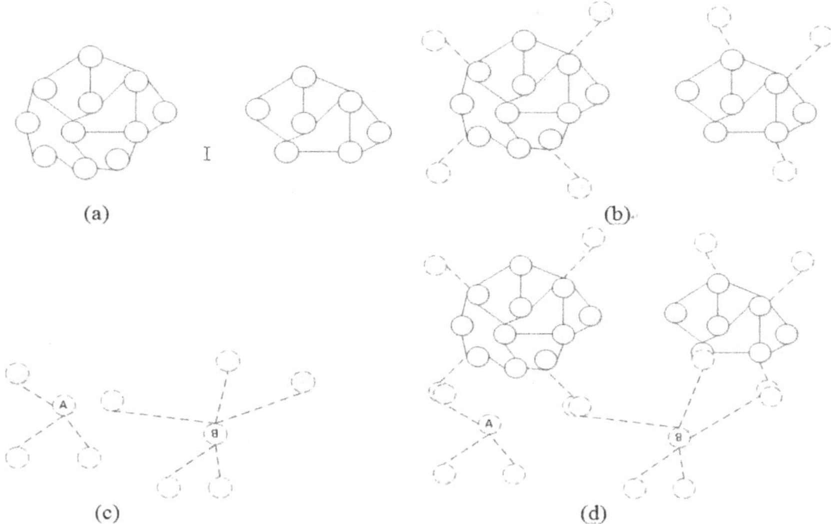

Fig. 2. The conceptual diagrams of ICF

3.1 TF-ICF Algorithm

In Rogers' IDM, it claims that the early adopters create innovative ways of using the products and pass on these novel ways to the early majority to motivate them to accept the products. "Here, The uses of innovative products, which are conceptualized as a consumer's receptivity/attraction to and creativity with using innovative products in new ways" (Price and Ridgway, 1982). According to Rogers' IDM, if the early majority could not obtain or accept the uses of the innovative products of the early adopters, a chasm will exist between them (Moore, 1991). This means whether the early majority uses of innovative products or not is the key factor of social inference to decide that the majority will cross the chasm. In this paper the GT and a text-mining method are employed to develop a TF-ICF Computing System. The steps of the research process of this paper are as followed.

Phase 1: Preparing data and initial clustering analysis. To create an innovation diffusion scenario, the researcher first collects data of the innovative uses of a product to conduct preprocessing on the data and then combines GT with text mining to process the data. The detailed process is listed as follows: H labels indicate the steps to be done by experts, and C labels indicate the steps to be done by the system.

Step 0: Data preprocess.
(0-1) The researcher defines the domain and relevant key words he/she intends to study.
(0-2) The researcher sifts out the data which correspond to keywords from the Internet.
(0-3) Based on his/her domain knowledge, the researcher interprets the texts, and at the same time, segments texts into words, removes useless words, and marks meaningful words as conceptual labels.

Step 2: Word co-occurrence analysis to emerge the clusters.
(1-1) Use Eq. (1) to calculate the association values of all words as below:

$$\begin{aligned}&N \textit{ is all words}\\&i = 1 \textit{ to } N-1\\&\quad j = i+1 \textit{ to } N\\&\qquad assoc(w_i, w_j) = \sum_{s \in allD} \min(|w_i|_s, |w_j|_s)\end{aligned} \tag{1}$$

where s represents the co-occurrence of words in the sentence, and D represents all textual data.
(1-2) Decide the threshold value of word frequency and link value to remove rare words and low linkage between word and word.
(1-3) Search the linking each other words to emerge the clusters.

Phase 2: finding related rare nodes based on the sentences of the nodes in the clusters.

(2-1) Use cluster's words to collect the related sentences as below:

$$\begin{aligned}
& n \text{ is all sentences} \\
& m \text{ is all clusters} \\
& l \text{ is all words in sentence} \\
& i = 1 \text{ to } m \\
& \quad j = 1 \text{ to } n \\
& \qquad k = 1 \text{ to } l \\
& \qquad\quad \text{if } w_i = w_k \text{ than } (w)_{s_j} \text{ put in } (w)_{ci}
\end{aligned} \tag{2}$$

(2-2) Decide extend rare words set in cluster.

Phase 3: finding more related rare nodes cluster's based on the sentences of the rare nodes.

(3-1) Use rare word to collect the related sentences as below:

$$\begin{aligned}
& n \text{ is all sentences} \\
& w_{ri} \text{ is ith rare word} \\
& l \text{ is all words in sentence} \\
& i = 1 \text{ to } n \\
& \quad k = 1 \text{ to } l \\
& \qquad \text{if } w_{ri} = w_k \text{ than } (w)_{s_i} \text{ put in } (w)_{ri}; break;
\end{aligned} \tag{3}$$

(3-2) Repeat 3-1 until all rare words find their word set.

Phase 4: calculating the ICF values of the rare notes.

(3-1) Use $(w)_{ri}$ and $(w)_{ci}$ to identify how many clusters is connected with $(w)_{ri}$ by directly or indirectly linkage as below:

$$\begin{aligned}
& (w)_{ci} \text{ is } ith\ cluster\ words_set \\
& (w)_{rj} \text{ is } jth\ rare_words\ set \\
& i = 1 \text{ to } m \\
& \quad \text{if } (w)_{rj} \cap (w_k)_{ci} \neq \phi \text{ than } Connect_j = connect_j + 1;
\end{aligned} \tag{4}$$

(3-2) *ith* rare word's ICF is identified as below:

$$(w_{ri})_{ICF} = \frac{m}{connect_i}$$

(3-3) Sort the TF-ICF value of all rare words to find incipient innovation of early adopter.

4 A Case Study

Since 2006, more and more Taiwanese wanted to go abroad and enjoy their leisure lives, which means the number of passengers are increasing. Therefore, some questions arose: What services does passages' want? What would be the important business problem for each airline? The research purpose of this paper is to find out more useful service for passengers, and how airlines could supply theses good services for passengers to win the business.

4.1 Data Resources

Six months after notebooks were newly introduced to the market, ranging from October 1, 2008 to October 31, the data posted on blogs relevant to the use of notebooks was collected. Searching keywords, travel + flight + train + kill-time on Google blogs (http://blogsearch.google.com/blogsearch), the researchers obtained 330 blog articles. After these articles were carefully read and some of them with no innovative uses were removed. There were 48 articles with new ideas about using the notebook from early adopters.

4.2 Experimental Results and Discussion

The data is collected from the Internet, after the Eqs. 1, 2, 3 and 4 calculating, the TF-ICF and TF-CF results are shown in Table 1. The features of TF-CF are very close the concept of association analysis, which is based on low support but high confidence. Additionally, the TF-CF does not only include connection between the rare node and clusters, but also include the connection between rare nodes. From Table 1, the words with high TF-CF values also have high TF-ICF values. On the contrary, the words with high TF-ICF values does always have high TF-CF values. Especially, the Top 5 words have big difference.

In Table 1, top ten words of TF-CF present the most important behaviors including eating delicious foods and enjoying movie. On the other hand, top ten words of TF-ICF present the most important behaviors which are chatting with friends, enjoying instant noodle and coffee and keeping skin moisture in the travelling vehicle. Amongst these two groups of behaviors, 'how to keep skin moisture in the flight or train' is the most unique behavior. This means passengers would like to keep their skin looking good when travelling. These results tell us that providing only food, movie and music or upgrading these services is not good enough for passengers, how to help them look good during a long distance travelling is promising topic for people who are looking for new business opportunities.

As previous discussion, the TF-ICF methodology does not only find instant foods or entertainment needed when passenger are travelling by vehicles, but also find the moisturizing emulation being another need of the passengers. This result points out that the TF-ICF methodology could find the more rare and useful words as future innovative and useful ideas.

Table 1. The values of TF-ICF and TF-CF

	TF	TF-ICF		TF	TF-CF
Chatting	1	786	foods	1	0.320611
Coffee	1	786	Entertainment	2	0.274809
Instant noodle	2	655	Car window	3	0.229008
Eye	3	524	cards	4	0.21374
Skin	4	393	snacks	4	0.21374
Moisture emulsion	4	393	space	5	0.19084
Blanket	4	393	Magazine	5	0.19084
Air	4	393	Beautiful scenery	6	0.183206
Lunch	5	327.5	map	7	0.160305
dinner	6	327.5	Inflight Entertainment	7	0.160305
map	6	305.6667	body	8	0.152672
Inflight Entertainment	6	305.6667	newspaper		0.152672
Remote control	7	262	computer		0.152672
Wireless	7	262	Remote control	9	0.137405
PDA	7	262	wireless	9	0.137405
IPOD	7	262	Japan Drama	10	0.122137
battery	7	262	films	11	0.114504
cards	8	229.25	Free seat	11	0.114504
snacks	9	229.25	enjoyment	12	0.091603
films	10	218.3333	feeling	12	0.091603
Tablecloth	11	196.5	APPLE_IPOD_TOUCH	12	0.091603
Eye mask	11	196.5	charger	12	0.091603
enjoyment	12	174.6667	reviews	12	0.091603
feeling	12	174.6667	problem	12	0.091603
APPLE_IPOD_TOUCH	12	174.6667	evaluation	12	0.091603
charger	12	174.6667	lunch	13	0.076336
			dinner	14	0.076336

5 Conclusion

In this paper the TF-ICF framework are used to find out the rare words and rare words' linkages, and then the rare word's linkages also are used to discover more rare and useful words as the future innovative ideas. Therefore, these future innovative ideas will be an innovative chance for crossing the chasm between early adopters and early majority. In the market it means that the future innovative and useful ideas is not only an important issue for innovative service diffusion, but also important information for firms to have the chance to occupy position and get great success in business. Such as in the present study, the most unique behavior on the top 10 lists in Table 1 is how to keep skin moisture in flights or trains. This means the passengers want to keep their appearance looking good when arriving at the tourism place. This experimental result shows that the moisturizing emulation is needed for the passengers, and this service, passengers to maintain their appearance, will be a new business opportunity for firms. The result points out that the TF-ICF methodology could find the more rare and useful words as future innovative and useful ideas.

References

Hong, C.F.: Qualitative chance discovery: Extracting competitive advantages. Inf. Sci. **179**(11), 1570–1583 (2009)

Lin, M.H., Hong, C.F.: Opportunities for crossing the chasm between early adopters and the early majority through new uses of innovative products. Rev. Socionetwork Strat. **5**(2), 27–42 (2011)

Liu, B., Hsu, W., and Ma, Y.: Mining association rules with multiple minimum supports. In: KDD 1999 Proceedings of the Fifth ACM SIGKDD International Conference on Knowledge Discovery and Data Mining. ACM Press, New York, pp.337–341 (1999)

Moore, G.A.: Crossing the Chasm - Marketing and Selling Technology Products to Mainstream Customers. HarperCollins, New York (1991)

Morrison, P.D., Roberts, J.H., von Hippel, E.: Determinants of user innovation and innovation sharing in a local market. Manag. Sci. **46**(12), 1513–1527 (2000)

Muniz, A.M., O'Guinn, T.: Brand community. J. Consum. Res. **27**, 412–432 (2001)

O' Reilly, T.: What is web 2.0: design patterns and business models for the next generation of software. Commun. Strat. **65**, 17–37 (2007)

Ohsawa, Y., Nels, E.B., Yachida, M.: KeyGraph: Automatic indexing by co-occurrence graph based on building construction metaphor. In: ADL 1998: Proceedings of the Advances in Digital Libraries Conference, pp.12–18. IEEE Computer Society, Washington DC (1998)

Price, L.L., Ridgway, N.M.: Use innovativeness, vicarious exploration and purchase exploration: three facets of consumer varied behavior. In: AMA Educator's Conference Proceedings, pp. 56–60 (1982)

Rogers, E.M.: Diffusion of Innovations. Free Press, New York (2003)

von Hippel, E.: Lead users: a source of novel product concepts. Manag. Sci. **32**(7), 791–805 (1986)

Collaborative Learning in Teacher Education Community for Pre-service Teacher Practice Learning

Chia-Ling Hsu(✉), Yi-Fang Chang, Chien-Han Chen, and Huey-Fang Ju

Center for Teacher Education, Tamkang University, 151, Ying-Chuan Road, Tamsui, New Taipei City 25137, Taiwan
{clhsu,yifangchang,chienhanc,fueyfj}@mail.tku.edu.tw

Abstract. Pre-service teachers take courses in the university in order to become good teachers in the near future. However, during the training period, the pre-service teachers learn about the teaching theories and skills from the instructors but lack of the field teaching. Therefore, this study forms a collaborative learning community for pre-service teachers to learn from the secondary school teachers' experiences. Through this collaborative learning community, the pre-service teachers work with their instructors and with the secondary school teachers while they are taking courses in the university. The purpose of this study is to investigate how the collaborative learning community works. In other words, what kind of activities would fit for the collaborative learning community and what the roles would the members paly in the group? Of course, what the benefits for the participants is concerned. That is, how does the professional development for each kind of participants? The results indicated that each of different kind of participants did the different work together and had growth in the collaborative learning community.

Keywords: Collaborative learning · Teacher education · Collaborative learning community · Higher education · Pedagogical teaching

1 Introduction

The collaborative learning community is an innovation in educational system especially in teacher education. It is popular to use computer technology to form collaborative learning community (Hidayanto & Setyady, 2014; Hanewald, 2013) [1, 2]. However, these studies did not focus on how the computer technology using in collaborative learning community. Some studies emphasize on students evaluation in the collective work and focus on in-service teachers' professional development by using collective learning community (Erickson, 1991; Butler & Schnellert, 2012) [3, 4]. Those studies indicate that the collective learning community is essential in educational system.

The questions of this study are to understand how the pre-service teachers get benefits when they participated in collaborative learning community and what the roles the members play in each different groups. In other words, the researchers would like to know as below:

N.T. Nguyen et al. (Eds.): ACIIDS 2016, Part II, LNAI 9622, pp. 328–335, 2016.
DOI: 10.1007/978-3-662-49390-8_32

- How did the pre-service teachers self-report what they had learned?
- What were the pre-service teachers' jobs?
- What were the secondary school teachers' jobs?
- What were the professors' jobs?

The importance of this study is to form a collaborative learning community for difference groups to work together since this is an innovation for froming different groups in practical educational system. The groups are the pre-service teachers, the secondary school teachers and the professors in the university. Usually these people in different groups would not work together. Now this study is going to form these different people in one group and working together. This is a new experience for these people. Therefore, this study would like to check how the collaborative learning community operated. If the collaborative learning community would show its function, the people in the community would be benefits and would achieve the professional development for each group members.

Collaborative learning is not a new theory in education but to form different groups into a collaborative learning group is very difficult especially each member has his/her own schedule and some members did not use the computer technology. Therefore, this study is trail experiment to combine three different types members working together to achieve the goal of collaborative learning.

1.1 Research Purpose and Significance

The purposes of this study are to integrate how the collaborative learning community works and what the benefits the group members can get. In detail, the purposes below:

- Find the appropriate frequency the collaborative learning community meeting.
- Find the content in each community meeting.
- Find the benefits to the pre-service teachers in teacher education program.
- Find the benefits to the secondary school teachers.
- Find the benefits to the professors in the university.

The purposes of this study are to try on a collaborative learning community with different interested members working together and to help each member paly their roles in the group also getting assistance from each other.

The significance of this study is trying to provide an opportunity for pre-service teachers to understand the real school setting through this collaborative learning community. The pre-service teachers have a chance to contact with the secondary school teachers and to watch the pedagogical teaching. This kind of training may enhance the pre-service teachers feel comfortable when they are in internship in the future.

2 Relative Literature

This study is based on the collaborative learning. The reflective model is also important. Although this study hopes that the collaborative learning community can help each member, the main focus on the pre-service teachers. Therefore the practical learning is also discussed. So, the literature review related to the collaborative learning, reflective model, and practical learning is described as belows.

2.1 Collaborative Learning

Dr. Hanewald (2013) found the definition of collaborative learning was that collaborative learning as two or more people learning or trying to learn something together by using each other's knowledge, skills, experience and resources [2]. Hidayanto and Setyady (2014) indicated that doing the task in group requires an ability to manage and adjust time each other, so they can do the discussion which lead them to solve the problem in the assignment [5]. Hence, this study tries to form a collaborative learning community to achieve the goal of the collaborative learning.

2.2 Reflective Model

Thinking about value was to decide what you want and then to figure out how you can get it. So, it was a nature way for alternative-focus thinking. Hsu, Hong, Wang, Chiu, and Chang (2009) used the VFT model in instruction design to improving the teaching quality [6]. No matter what approach the research used would lead to a reflection of instruction [7, 8, 9, 10, 11]. The Kurt Lewin's Model (2013), in Fig. 1, was the recursive model for reflective [12].

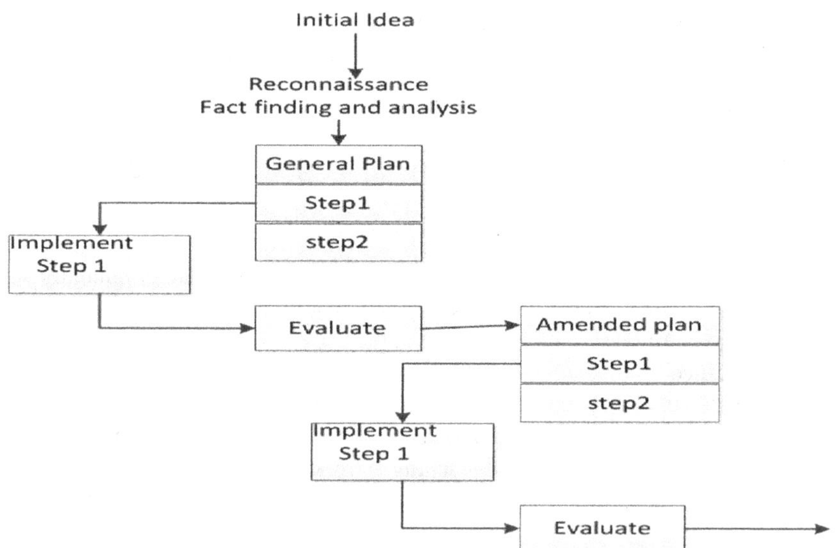

Fig. 1. The Kurt Lewin's Model

2.3 Practical Learning

Many studies indicate that practical learning is powerful for student learning. Moeed's (2014) research showed that students do learn and develop science understanding through engaging in practical experiences [13]. Moran (2014) provides a new model and suggests that while the program can not negate all the difficulties associated with ensuring quality placements, it does provide some solutions that assist in improving the professional experiences of pre-service teachers [14].

From the literature review, this study hops to form a collaborative learning community to help the pre-service teachers learn.

3 Research Method

This study is to organize a collaborative learning community in order to help the pre-service teachers having opportunities to contact with the secondary school teachers. The purposes of this study are to try on a collaborative learning community with different interested members working together and to help each member paly their roles in the group also getting assistance from each other. Therefore, the Kurt Lewin's Model was modified for the operation of the collaborative learning community. The research procedure bellows in Fig. 2.

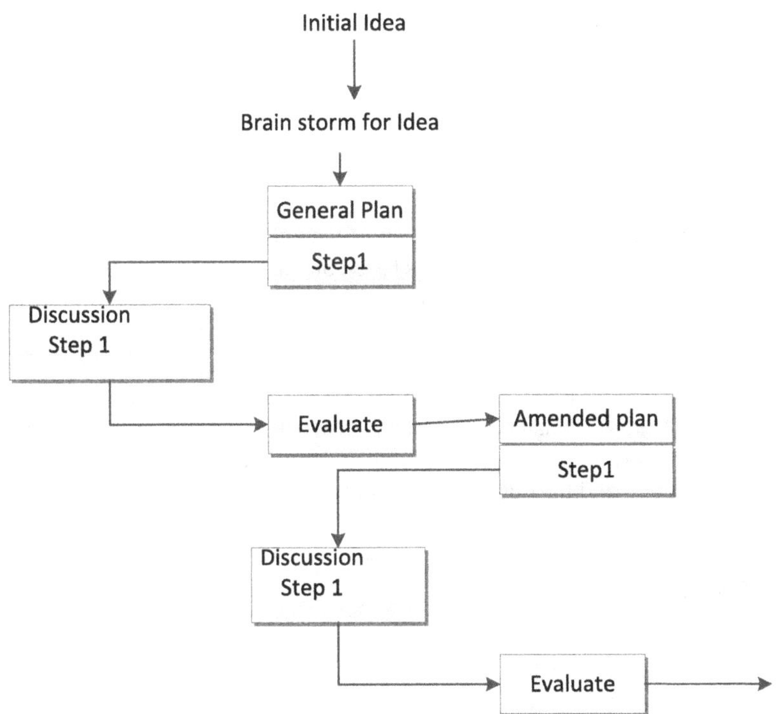

Fig. 2. The research procedure

3.1 Participants

The participants in this study were the members of the collaborative learning community. The participants contained six pre-service teachers, two secondary school teachers and three professors in the university.

The pre-service teachers were all in the second year of the teacher education program. Four of the pre-service teachers were graduate students and two were undergraduate students. They were all in the Department of Chinese Literature.

Two secondary school teachers all had over 20 years teaching experiences. One was a teacher with administration job and one is a class instructor.

Two of the three professors were from teacher education program and one was the subject matter form the Department of Chinese Literature.

3.2 Data Collection

The data collections used in this study were observation, interview, and reflection. The observation data contain the meeting and teaching data. Each member in the collaborative learning community was doing brain storming in each meeting. After the progress of the brain storming, the community came out a teaching plan or an idea for designing working sheet. The observation teaching sheet was collected for each time when the members observed the school teacher's teaching.

The interview was held with the secondary school teacher.

The reflection was collected after the meeting and class observation.

4 Results

The results in this study were the observation of the collaborative learning community meeting. The frequency of the meeting was based on the task. The overall working plan meeting was once a month. The courses preparing meeting was three to four times for one unit. The Fig. 3 indicated the overall working plan meeting. The Fig. 4 showed the courses preparing meeting.

During the class observation, 36 junior high students were involved. The observation sheet contained "class", "material", "date", "teacher", and "observer" items. The observers not only wrote down the data but also described how the teacher's behaver with students reaction and the observation feedbacks. Figure 5 showed the class observation.

The interview with secondary school teacher Su and Wu was showed as bellows.

Interview with teacher Su:

> "*...The collaborative learning community stimulates my teaching a lot. However, I still think that the traditional teaching method also good for me....*"

> "*...The cooperative learning takes too much time in class. The working sheet, I fell, is not help students learn because they are used to copy to each other.....*"

Fig. 3. The overall working plan meeting

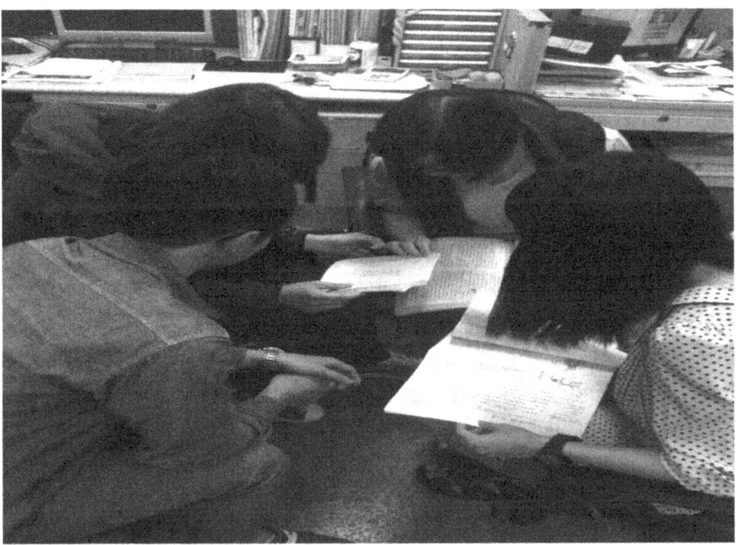

Fig. 4. The course preparing meeting

"*....Since the cooperative learning is the policy in education reform. I will try it.....*"

Interview with teacher Wu:

"*....Because I also have administration work, I have different classes for the pre-service teachers to observe. It will help them to know the different class atmosphere......*".

"*....I am not against the new teaching method; nevertheless, it take too much time to finish one unit. I have much material needed to be taught....*"

Fig. 5. The class observation

> *".....The collaborative learning community helps me a lot because of the pre-service teachers' help......".*

From the interview, the data indicated that the collaborative learning community can help the secondary school teachers developing their professional skill. Since the pre-service teachers involve with the preparing course, the school teachers felt the collaborative learning was a good method to working together.

The reflation of observations indicated that the collaborative learning method make each student involve in class activities. The pre-service teachers learned how to design the work sheet for students. They absorbed the school teachers' teaching style and reflected to think how they will teach in the future.

5 Conclusion and Suggestion

The purposes of this study are to try on a collaborative learning community with different interested members working together and to help each member paly their roles in the group also getting assistance from each other. The results indicated that the pre-service teachers and the secondary school teachers got the benefits from the collaborative learning community. Moreover, the reform of the teaching skills was also put into practice in the secondary school. In addition, the process would help the pre-service teachers to observe the class teaching and to prepare the course material before their internship.

5.1 Suggestion

Some suggestions are provided from this study.

First, for the educational administration, the collaborative learning is accepted by the teachers in this study, but may not be practice in the real setting because of the time consuming. Therefore, the school administration may select some topics for collaborative learning and the rest topics remain the traditional teaching.

Second, for the teacher education administration, the results indicated that the pre-service teachers would learn a lot in collaborative learning community. Therefore, the program of teacher education may consider adding the participating collaborative learning community as requirement for each pre-service teacher.

Third, for the researchers, because of the limitation of this study, the shortage of time and human resource, the connection among the collaborative learning community is not good enough. Hence, to communicate with three parts members is important. If it is possible to use some internet technology may help. In addition, more research is needed.

Acknowledgments. We would like to thank the Zheng De Junior High School and the teacher Su and teacher Wu.

References

1. Hidayanto, A.N., Setyady, S.T.: Impact of collaborative tools utilization on group performance in university students. Turk. Online J. Edu. Technol. - TOJET **13**(2), 88–98 (2014)
2. Hanewald, R.: Learners and collaborative learning in virtual worlds: a review of the literature. Turk. Online J. Distance Edu. **14**(2), 233–247 (2013)
3. Erickson, G.L.: Collaborative inquiry and the professional development of science teachers. J. Edu. Thought/Rev. De La Pensee Educative **25**(3), 228–45 (1991)
4. Butler, D.L., Schnellert, L.: Collaborative inquiry in teacher professional development. Teach. Teach. Edu.: Int. J. Res. Stud. **28**(8), 1206–1220 (2012)
5. Hidayanto, A.N., Setyady, S.T.: Impact of collaborative tools utilization on group performance in university students. Turk. Online J. Edu. Technol. - TOJET **13**(2), 88–98 (2014)
6. Hsu, C.L., Hong, C.F., Wang, A.L., Chiu, T.F., Chang, Y.F.: Value Focused Association Map (VFAM) - an alternative learning outcomes presenting, word conference on educational multimedia, hypermedia & telecommunications, pp. 3270–3275, Hawaii, USA, 22–26 June 2009
7. Hsu, C.L., Chang, Y.F.: Study of the relationship with the media material and the students' learning motivation. J. Edu. Study **116**, 64–76 (2003)
8. Hsu, C.L.: E-CAI case study. Edu. Technol. Media **33**, 28–35 (1997)
9. Hsu, C.L. Kuo, C.H.: Study of e-learning material technology. In: 2000 e-Learning Theory and Practice Conference, pp. 61–65. National Chiao Tung University, Shin-Chu (2000)
10. Hsu, C.C., Wang, L.H., Hong, C.F. Sung, M.Y., Tasi, P.H.: The KeyGraph perspective in ARCS motivation model. In: The 6th IEEE International Conference on Advanced Learning Technologies, pp. 970–974. IEEE Press, New York (2006)
11. Hsu, C.C. Wang, L.H. Hong, C.F.: Understanding students' conceptions and providing scaffold teaching activities. In: International Conference of Teaching and Learning for Excellence, pp. 166–175, Tamsui (2007)
12. McNiff, J.: Action Research Principles and Practice, Routledge, USA
13. Moeed, A.: Successful science learning from practical work. Sch. Sci. Rev. **93**(343), 121–126 (2011)
14. Moran, W.: Enhancing understanding of teaching and the profession through school innovation rounds. Aust. J. Teach. Edu. **39**(3) (2014)

Mobile-Assisted Model of Teaching and Learning English for IT Students

Ivana Simonova(✉) and Petra Poulova

Faculty of Informatics and Management, University of Hradec Králové, Hradec Králové, Czech Republic
{ivana.simonova,petra.poulova}@uhk.cz

Abstract. The paper introduces the model of hybrid flexible model of teaching/learning English for students of Informatics-related study programmes. Based on the theoretical feedback the model includes the traditional face-to-face lessons, work in the online course and autonomous activities available through mobile devices. For the purpose of efficient use of the model, a survey on mobile devices availability and their use was conducted in the group of 312 students of the Faculty of Informatics and Management, University of Hradec Kralove. After the three-month exploitation of the model students' feedback was collected. Finally, several recommendations for designing the mobile-assisted (language) learning are provided.

Keywords: Case study · Cloud computing · Database · Education · E-learning · M-learning

1 Introduction

There is no doubt knowledge of foreign language/s belongs to key competences, as defined in numerous European documents, e.g. EU 1993, 1995, 1997, 2001, 2006, 2009, 2010 etc. [1–7]. Particularly English is not a foreign language but a basic skill, Graddol said [8]. Reflecting the latest trends in technical and technological development, a shift was detected from using 'traditional' (i.e. immobile) electronic devices to mobile ones [9]. Their exploitation was detected in both private and professional spheres, e.g. in e-banking, e-shopping, and in education. Until recently the traditional e-learning, using mostly non-portable, immobile devices, was widely implemented into the education in the Czech Republic reflecting the fact that mobile devices for reasonable prices were not available to a large extent. Within a few years this situation changed substantially and a wide choice of various price levels and types of mobile devices is offered on the market. Thus mobile devices could have become standard didactic means on all levels of education and in nearly all subjects, including foreign languages [9]. Moreover, foreign languages could be even said to be the pioneers in mobile-assisted learning in general – been used for listening to music first, which is the entertainment widely spread among young people, including higher education students, then their exploitation spread to younger and older users and other fields and services, including foreign languages, particularly English.

N. T. Nguyen et al. (Eds.): ACIIDS 2016, Part II, LNAI 9622, pp. 336–345, 2016.
DOI: 10.1007/978-3-662-49390-8_33

The ownership of mobile devices and adequate skills developed within a wide use for private purposes were two main reasons why mobile devices can be used in education. Moreover, among the students of study programmes relating to information technology (IT students) the interest in mobile devices and services was even higher compared to students of other fields [10]. Currently, the mobile-assisted learning can be applied on all levels of Czech education system, gradually moving from small-scale, short-term trials to larger, more sustained and blended deployment.

Reflecting the fact that the use of mobile devices in the whole Czech education system, including in teaching/learning English for IT students, had been rather new, following questions were set to be researched:

1. Are students sufficiently equipped with mobile devices?
2. What purposes do students use the mobile devices?
3. What is the students' feedback after mobile-assisted language learning (MALL)?

The third question was connected to the new model of teaching/learning English for IT students which was applied at the Faculty of Informatics and Management (FIM), University of Hradec Kralove (UHK), Czech Republic. This model was piloted within two semesters in the 2014/15 academic year. Thus *the main objective of this paper is to introduce how the model was designed and present results on how it worked*, i.e. what learners' experience and opinions after each semester were.

2 The Hybrid Flexible Model of Mobile-Assisted Teaching/ Learning English for IT Students

The model of mobile-assisted teaching/learning English for IT students was designed applying the hybrid approach and flexibility. The 'hybrid' feature of the model, also called 'blended' by some authors (e.g. [11], is ensured by the combination of the traditional face-to-face lessons with the ICT-enhanced teaching/learning; the 'flexibility' is enabled by learners' choice between immobile and mobile devices and activities accessible through them to conduct the process of learning. The model includes three approaches: (1) traditional face-to-face approach, (2) e-learning running within online courses in LMS and (3) mobile-assisted learning in which two mobile applications designed by IT students for their peers and many others according to learners choice can be exploited on mobile devices. A brief summary of helpful mobile applications is available e.g. [12].

The process of mobile-assisted teaching/learning English for IT students was held in online courses in the LMS which could be accessed from immobile computers, mobile notebooks, netbooks, tablets and other mobile devices. The direct communication language skills were practiced in face-to-face lessons (held six times per 90 min in one semester). Moreover, the LMS is a high quality environment designed for education purposes, so it provides all tools necessary for efficient simulation of all phases of the process of instruction. In mobile-assisted teaching/learning English for IT students, following tools from the version for mobile devices were exploited:

- the learning content was available to students in the form of study materials containing texts, figures, images, animations;
- tests of new vocabulary and grammar structures and tests on listening/reading comprehension were available in online courses (but not all types of tests worked in the mobile version compared to traditional LMS),
- listening/reading files in the podcasting form were provided,
- discussions were held, both in the written form on mobile devices in discussion forums and orally via Skype, which was linked with the LMS.

Above all, for practicing (revising) *vocabulary and grammar* two applications were designed, working in three steps: (1) after 'click 1' a word, phrase or short sentence appeared on the screen; (2) students wrote the answer (i.e. translated the item) and after 'click 2' the notice appeared saying whether the reply was correct, or not; (3) after 'click 3' the correct answer was displayed on the screen. Both Czech/English and English/Czech versions were prepared. The online courses also contained short texts and animations to explain single grammar items.

Moreover, the *listening/reading comprehension* skills were developed through reading a textbook – collection of professional texts with professional vocabulary translated (using the Insert, Comments tool) and recordings in mp3 format called 'English Reader for IT and Management students'. This is also an output of students' project work for their peers. Both these activities enable students to show what they have learned from the studied field and to reflect and share their professional experience, which strengthens their motivation to further study.

3 Theoretical Background

According to Roschelle or Liang, in this research mobile devices are defined as small items to accompany users anytime and anywhere, autonomous from the electrical supply [13, 14]. They vary significantly in their abilities, sizes and prices but the common ability is their easy mobility/portability and wireless connections. The main types of mobile devices used in the education process are notebooks, netbooks, tablets, Personal Digital Assistants (PDA), cellular (mobile) phones and smartphones.

The hybrid model of flexible mobile-assisted teaching/learning English for IT students includes (instead of others) the mobile-assisted language learning (MALL) approach which was defined and used by Ally [15]. M-learning is understood as "learning across multiple contexts, through social and content interactions, using personal electronic devices" [16]; it is concerned with a society on the move, particularly with the education of "… how the mobility of learners augmented by personal and public technology can contribute to the process of gaining new knowledge, skills and experience" [17].

Additionally to Ally, Kearney designed pedagogical framework to mobile learning from socio-cultural perspective [18] and proved the positive impact on individualization (83 %), collaboration (74 %) and authenticity (73 %) within learning, whereas technology was detected a stronger culture-shaping factor than inherited cultural environment or age and learner's gender was identified to be a predictor of differences in students' attitudes to MALL.

Relating to our research topic, Chen [19] ran a similar survey and identified tablets as an ideal tool for interactive, collaborative and ubiquitous environment for independent informal language learning supported by students' positive attitudes towards their usability, efficiency and satisfaction for the MALL purposes.

The end-users' perception of MALL through cross-cultural analyses in seven countries and regions was investigated by Hsu [20]. He discovered that significant differences still exist but all respondents agreed MALL provided potential for learning English as a foreign language.

Numerous related works have been published e.g. by Viberg, Gronlund [21]. They focused on monitoring attitudes towards the use of mobile devices in the second and foreign language learning in higher education from the point of cross-cultural approach.

On the other side, de la Fuente [22] focused on the field of knowledge development, particularly on the aural input. He indicated significantly higher levels of bottom-up comprehension and top-down overall comprehension with learners in the MALL group in instructor-directed language learning. Compared to this result, Golonka et al. [23] dealt with the effectiveness of technology use in foreign language learning and teaching. He analyzed studies comparing the use of new technologies (mobile and portable devices, network-based social computing) to traditional methods and material (PC); he discovered limited efficacy of mobile technologies but strong impact within computer-assisted pronunciation training, particularly automatic speech recognition.

The only remarkable research on mobile devices in the Czech Republic was published in 2011 by Lorenz [24]. He ran a quantitative research in a group of 274 students of IT in a library services study programme at Masaryk University, Brno, Czech Republic. Following the Corbeil and Valdes-Corbeil design [25] he analyzed the concept of mobile education within the changing university environment focusing on the process of learning and the support which library services can provide, both to teachers and learners. Despite the Lorenz's study was carried in 2010, the data were expected to be changed within the four-year-long period. This was one of the reasons why our study was conducted.

4 Research Sample

The mobile-assisted language learning (MALL) approach was piloted in the sample group of 312 students of FIM UHK who enrolled in part-time bachelor IT study programmes (Applied Informatics; Information Management). The structure of research sample was as follows:

- 60 % of male students;
- 72 % of them 19–24 years old, 13 % of 25–29-year-old ones, 11 % of 30–39-year-old and 4 % of 40+-year-old,
- they attended the 1st, 2nd or 3rd year of the study in 2014/15 academic year.

5 Methods and Tools

To monitor students' mobile equipment before the exploitation of the hybrid flexible model of mobile-assisted teaching/learning English for IT students, Questionnaire 1 was applied; students' feedback after studying in the model courses was reflected by Questionnaire 2.

Questionnaire 1 (Q1) consisted of 12 multiple-choice questions focusing on sources of information which student use for their university study, frequency of private and education-related communication through mobile device, the extent to which the devices are used for MALL and other subjects etc. The collected data were structured under three criteria: what mobile devices students possess, what purpose they exploit them: private or educational, to what extent students used mobile devices for MALL compared to other subjects.

Students' feedback after studying in the model courses was collected by *Questionnaire 2 (Q2)*. Students expressed their opinions in MALL. The questionnaire which was available online in each particular course in LMS after the course was finished. The questionnaire consisted of 22 statements evaluated on the six-point scale from 1 (completely agree) to 6 (completely disagree) point scale. Both questionnaires were considered by the team of five experts and approved for application within this research.

6 Results

All data were processed by the NCSS2007 statistic software and the method of frequency analysis was applied. Results are presented in tables and figures, interpreted and discussed. The findings are structured into three parts according to the research questions.

First, questionnaire Q1 focused on mobile devices from two points of views: (a) what mobile devices student possess, (b) what purposes they exploit them – for learning English and/or other subjects. Results are displayed in Fig. 1.

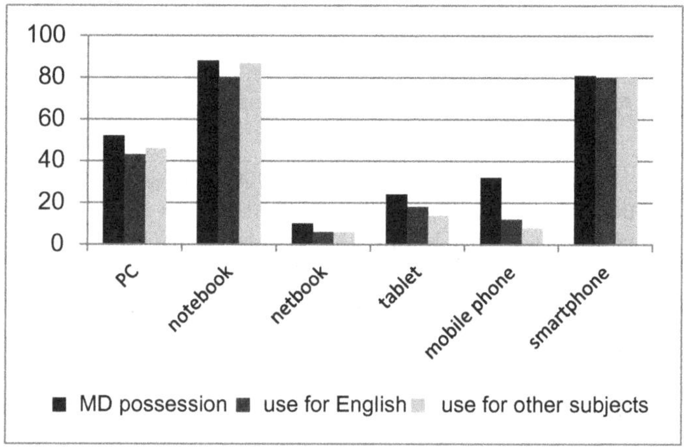

Fig. 1. Q1 – Mobile devices: possession and exploitation in ESP and other subjects (%)

The data show notebooks are currently the most frequently *possessed* mobile devices (88 % of respondents possess them), followed by smartphones (81 %), PCs (52 %), mobile phones (32 %), tablets (24 %) and netbooks (10 %). As expected, students did not declare the possession of one type of mobile devices only, but they simultaneously own PCs, notebooks, netbooks, tablets, smartphones etc.

Respondents declared the use of mobile devices *for different purposes*. For teaching/learning English mostly notebooks (80 %) and smartphones (80 %) are used, followed by PCs (43 %), mobile phones (12 %), tablets (18 %) and netbooks (6 %). What was not expected, that respondents declared higher frequency of small-size mobile devices use (tablets – 18 % and mobile phones – 12 %) for learning English than for other subjects compared to large ones – notebooks (80 %) and PCs (43 %). In other subjects within the university study notebooks were detected (87 %), smartphones (80 %), PCs (46 %), tablets (14 %), mobile phones (8 %) and netbooks (6 %).

Students' feedback after hybrid flexible model of mobile-assisted teaching/learning English for IT students was collected by Questionnaire 2. The selected items out of 22 items describing students' MALL-related feedback are displayed in Table 1.

Table 1. Selected items of students' feedback

	Statement
1	I consider this blended model helpful for the process of ESP learning
2	I consider this blended model significantly helpful for the process of ESP learning
3	I worked with recommended operational system and devices (if you did not, list the operational systems and devices you used)
4	I did not have any technical problems (if you did so, please explain)
5	The Internet access was as displayed in Internet signal maps
6	Would you appreciate using mobile devices in learning English in the future?
7	Would you use mobile devices in learning other subjects in the future? (if yes, provide examples, please)
8	Total evaluation of the model

Evaluation of single items is displayed in Fig. 2. The results show highly positive feedback, which was surprising. The highest score was detected in statement 4 (monitoring technical problems) and statement 6 (appreciation of mobile devices implemented in English instruction). The lower score was with statement 7 (the use of mobile devices in other subjects) and relatively 'worst' score (2.1 out of 6) was with statement 1 (how helpful the model was to students). Totally, the hybrid flexible model of mobile-assisted teaching/learning English for IT student was highly appreciated as most learners considered it highly helpful, or helpful for them – the total evaluation score was 1.8.

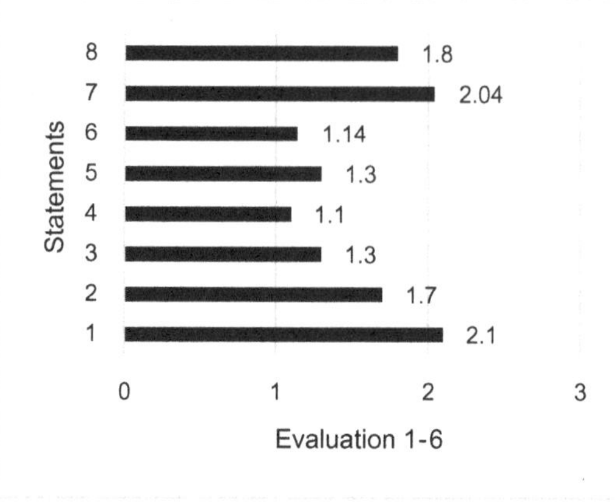

Fig. 2. Q2 – Students' feedback on MALL-related items (1- complete agree; 6 – complete disagree)

7 Discussions and Conclusions

The mobile devices market has grown at amazing rate and is expected to grow on. Over recent years information about designing and developing mobile devices for education purposes has increased substantially [26] and the development of m-learning is a great challenge. The future will bring even more mobile devices and the 'traditional' (i.e. immobile) device is not likely to occur. It is difficult to create a design that will work across all devices, however there are some best practices which can help:

Briefly summarized, the state of the matter has two main features: (1) the long-time research data in the field of mobile devices and their exploitation in (higher) education from the Czech Republic still are not available, (2) results collected in our and Lorenz' researches correlate with the state in the world [18, 20, 21].

In the current globalized world differences in availability of technologies are diminishing and various scientifically-verified methodology (didactics) are available. To list any of them is impossible within this paper; see e.g. [27].

In general, when planning the m-learning design, the FRAME model should be followed. The author, M. Koole [28] asked several questions before planning the model: How can the learner take full advantage of the mobile experience? How can practitioners design materials and activities appropriate for mobile access? How can mobile learning be effectively implemented in both formal and informal learning? The FRAME model [28, p. 27] offers answers to these questions. It describes such an approach to learning where learners are moving from real to virtual environments and situations, they interact there with other people, information or systems, all doing anywhere, anytime. The technology is the mediator of interaction with information. The FRAME model is expressed through a Venn diagram. Three aspects (circles) representing the device (D), learner (L) and social aspects (S) intersect there. The device usability (DL) and social technology (DS) intersections describe the availability (which is called the possession in our research) of mobile technology. Instructional and learning theories emphasizing social constructivism are displayed in the intersection labeled interaction learning (LS). In the center of the Venn diagram all three aspects overlap at the primary intersection (DLS), thus defining an ideal mobile learning situation. The FRAME model also considers the technical characteristics of mobile devices and social and personal aspects of learning, which refers to concepts found in psychological theories e.g. by Vygotsky [29] on mediation and the zone of proximal development. In the FRAME model, the mobile device is an active component of learning and social processes, paying more emphasis on constructivism: the word 'rational' refers to the "belief that reason is the primary source of knowledge and that reality is constructed rather than discovered" [30, p. 15].

Moreover, following didactic recommendations should be applied in designing the meaningful mobile learning [31]:

- Confirm that mobile delivery makes sense.
- Understand the targeted end-users and their contexts.
- Plan for the disconnected mobile user.
- Know the limitations and capabilities of the technologies involved, i.e. think about user's technological limitations and the role of WIFI.
- If developing native apps for iOS, make sure you are familiar with Apple's conventions before you start. They will be only accepted if meeting their standards.
- Warn BYOD (Bring Your Own Device) users of large data downloads, if applied.
- And last but not least – remember that tablets are more appropriate when data require a larger display. This is especially true when users share the data with others. Within this field following aspects should be considered:
  (1) Reflect the device, i.e.
    - keep the specific device and its services in mind when preparing the m-learning model (iPad, Android device, iPhone, etc.) and beware of formats that are not compatible across all devices,
    - reflect user experience in m-learning (whereas the use of mobile devices for private purposes (entertainment) is usually on high level,
    - keep course size small (approx. 10 min per topic) thus reflecting the quality of Internet access, as well as keep the tests short and concise, restrict options to three or four as maximum,
    - avoid scrolling as it is rather difficult on the small screen.
  (2) Use appropriate imagery and font, i.e.
    - place bold images on white backgrounds,
    - use clear and simple images and do not use text in images,
    - provide strong visual cues (bright colors, descriptive icons, etc.),
    - maximize the use of icons and graphics to represent content instead of text.
    - make your default font size at least 14pt.

Mobile technologies' affordances in specific learning activities and learning contents (subjects, topics), difficult financial situation in the field of education in last years, ethical problems relating to the use of mobile devices – these are the hot topics which should be researched and solved in the near future. The didactic design of mobile devices implementation in MALL (as well as in other subjects to some extent) should also consider the problem of unlimited availability of the teacher, which is a strongly appreciated as a positive feature from learners' side, but logically very conflicted from teachers' view. The ethical feeling of non-/contacting the teacher is connected to general level of behaviour and good manners of learners. And, current learners, being allowed to feel free from childish age, sometimes disrupt this gentle border line, expecting teacher's support really on the 7/24/365 principle.

Despite this and as most results show, vast majority of both students and teachers are ready for efficient use of mobile devices, as proved in this model. Didactic recommendations from experienced users are and would be appreciated.

Acknowledgment. The paper is supported by the IGA project Didactic aspects of Blackboard Mobile Learn implementation into instruction at FIM.

References

1. European Commission. Growth, competitiveness, and employment. The challenges and ways forward into the 21st century. Brussels: European Commission (1993)
2. European Commission. White Paper on Education and Training - Teaching and Learning - Towards the Learning Society. Brussel: European Commission (1995)
3. European Commission. Towards a Europe of knowledge (1997). http://europa.eu/legislation_summaries/other/c11040_en.htm
4. European Commission. The eLearning Action Plan: Designing tomorrow's education. Brussel: European Commission (2001)
5. ETUC. The European Union's Lisbon Strategy. ETUC European Trade Union Confederation (2006). http://www.etuc.org/a/652
6. ET2020. Education and Training 2020. Europa (2009). http://europa.eu/legislation_summaries/education_training_youth/general_framework/ef0016_en.htm
7. EU2020. About the Europe 2020 Strategy. EU2020 (2010). http://ec.europa.eu/europe2020/index_en.htm
8. Graddol, D.: English next. http://www.britishcouncil.org/learningenglish-next.pdf
9. Traxler, J.: Current state of mobile learning. In: Ally, M. (ed.) Mobile Learning. Transforming the Delivery of Education and Training, pp. 9–24. AU Press, Athabasca (2009)
10. Simonova, I., Poulova, P.: Learning Style Reflection Within Tertiary e-education. WAMAK, Hradec Kralove (2012)
11. Allen, E., et al.: (2007). https://www.uwb.edu/learningtech/elearning/hybrid-and-online-learning/hybrid-learning/about-hybrid-learning/definition-hybrid-learning
12. Simonova, I.: Mobile-assisted ESP learning in technical education (to be published)
13. Roschelle, J.: Unlocking the learning value of wireless mobile devices. J. Comput. Assist. Learn. **19**, 260–272 (2003)
14. Liang, L., Liu, T., Wang, H., Chang, B., Deng, Y., Yang, J., Chou, C., Ko, H., Yang, S., Chan, T.: A few design perspectives on one-on-one digital classroom environment. J. Comput. Assist. Learn. **21**(3), 181–189 (2005)
15. Ally, M. (ed.): Mobile Learning. Transforming the Delivery of Education and Training. AU Press, Athabasca University (2009)
16. Crompton, H.: A historical overview of mobile learning: toward learner-centered education. In: Berge, Z.L., Muilenburg, L.Y. (eds.) Handbook of Mobile Learning. Routledge, Florence (2013)

17. Sharples, M., Sánchez, I.A., Milrad, M., Vavoula, G.: Mobile learning: Small devices, Big issues. http://www.uio.no/studier/emner/matnat/ifi/INF5790/v12/undervisningsmateriale/articles/KAL_Legacy_Mobile_Learning_(001143v1).pdf
18. Kearney, M., Schuck, S., Burden, K., Aubusson, P.: Viewing mobile learning from a pedagogical perspective. Res. Learn. Technol. **20**, 1–17 (2012)
19. Chen, X.B.: Tablets for informal language learning: student usage and attitudes. Lang. Learn. Technol. **17**(1), 20–37 (2013)
20. Hsu, L.W.: English as a foreign language learners' perception of mobile assisted language learning: a cross-national study. Comput. Assist. Lang. Learn. **26**(3), 197–213 (2013)
21. Viberg, O., Gronlund, A.: Cross-cultural analysis of users' attitudes towards the use of mobile devices in second and foreign language learning in higher education: a case study from Sweden and China. Comput. Educ. **69**, 169–180 (2013)
22. de la Fuente, M.J.: Learners' attention to input during focus on form listening tasks: the role of mobile technology in the second language classroom. Comput. Assist. Lang. Learn. **27**(3), 261–276 (2014)
23. Golonka, E.M., Bowles, A.R., Frank, V.M., Richardson, D.L., Freynik, S.: Technologies for foreign language learning. Comput. Assist. Lang. Learn. **27**(1), 70–105 (2014)
24. Lorenz, M.: Kde nechala škola díru: m-learning aneb Vzdělání pro záškoláky [Where school let a hole yawn: m-learning or education for truants] (2011). http://pro.inflow.cz/kde-nechala-skola-diru-m-learning-aneb-vzdelani-pro-zaskolaky
25. Corbeil, J.R., Valdes-Corbeil, M.E.: Are you ready for mobile learning? Educause Q. **30**(2), 51–58 (2007)
26. Wasiluk, G.: Best Practices for Designing Mobile Learning. https://community.articulate.com/articles/design-mobile-learning-like-a-pro-best-practices-for-mlearning
27. Palalas, A.: Mobile-assisted language learning: designing for your students. In: Thouësny, S., Bradley, L. (eds.) Second Language Teaching and Learning with Technology: Views of Emergent Researchers, pp. 71–94. Research-publishing.net, Dublin (2011)
28. Koole, M.L.: A model for framing mobile learning. In: Ally, M. (ed.) Mobile Learning. Transforming the Delivery of Education and Training, pp. 25–48. AU Press, Athabasca (2009)
29. Vygotsky, L.: Mind in Society: The Development of Higher Psychological Processes. Harvard University Press, Cambridge (1978)
30. Smith, P., Ragan, T.: Instructional Design. Wiley, Toronto (1999)
31. Mobile learning handbook. sites.google.com/a/adlnet.gov/mobile-learning-guide/best-practices#TOC-Planning-for-Design

IPO and Financial News

Jia-Lang Seng(✉), Pi-Hua Yang, and Hsiao-Fang Yang

Department of Accounting, College of Commerce,
National Chengechi University, Taipei 11605, Taiwan, ROC
jia.lang.seng@gmail.com, yangbihua624@gmail.com,
hfyang.wang@gmail.com

Abstract. This paper explores the relation between the sentiment of the media and IPOs underpricing which means whether the media were responsible for the phenomenal rise and fall in the market value of IPOs shares from 2010 to 2014. We focus on the effect of pre-IPOs media on the first day's return and we use sentiment analysis to calculate the media sentiment and use the sentimental words to classify the news. We found that the sentiment of media is significantly related to initial return. This paper also indicates significantly results that initial return is related to market return than sentiment of media in non-electronics industry. Furthermore, as our expected, the number of positive (negative) news has a positive (negative) influence on the initial return in the electronics industry.

Keywords: IPOs underpricing · Media sentiment · Sentiment analysis

1 Introduction

Corporations may raise capital in the primary market by way of an initial public offerings (IPOs). The IPOs[1] is a type of public offering in which shares of stock in a company usually are sold to institutional investors that in turn, sell to the general public, on a securities exchange, for the first time. The main IPOs method are book building, auction method, and public offer. Most companies in Taiwan securities market are using the combination way of book-building and public offer. The book building models beginning with [9, 10] argue that the underwriter's control of both price and allocations may be used to induce investors to reveal their private information. The book building bids and allocation data are not publicly accessible. It is difficult for us to observe how the final shares are allocated. Hence, it is difficult to research on the implications of book building model. Accordingly, we try to use the information flows as the key factor to observe this pricing process. [2] suggest that individual investors are usually uninformed, poorly trained, and inclined to rely on noise in the market. In addition, emerging markets are usually operated in an environment of information opaqueness as pointed out by [25], stock prices in these markets are quite sensitive to political events and rumors. Unlike United States, there is no restriction on news announcements of issuers around IPOs in Taiwan. It also allows IPOs firms to release news that can generate favorable views at their discretion.

[1] https://en.wikipedia.org/wiki/Initial_public_offering.

N.T. Nguyen et al. (Eds.): ACIIDS 2016, Part II, LNAI 9622, pp. 346–353, 2016.
DOI: 10.1007/978-3-662-49390-8_34

Therefore, examining financial news of IPOs firms in Taiwan allows us to investigate whether the investors are influenced by media sentiment. This paper motivated by the recent literature that investigates the role of investor sentiment [4, 27]. We focus on whether the IPOs firms can engage in other qualitative types of reporting to influence investors' perception, and their demand for the new issue shares. Thus, we collect the news from the Internet news database and categorize them. [3] argue that through the sentiment analysis process, information is converted from a textual form to a numeric form. Thus, we use sentiment analysis to convert the qualitative news into quantitative measures. This paper is related to [20], we categorize the news as positive, negative, and neutral. We investigates whether the media sentiment is related with IPOs underpricing. The empirical results show that the media sentiment is association with IPOs underpricing. Furthermore, the number of positive (negative) news has a positive (negative) influence on the initial return in the electronics industry. The remainder of this paper is organized as follows. Section 2 introduces relevant literatures and theories. Section 3 introduces the research method, the data set and all the variables used in the model. In Sect. 4, it presents the empirical results and the relation between the media coverage and underpricing. Finally, Sect. 5 concludes.

2 Literature Review

2.1 IPOs Underpricing, Asymmetric Information, Signaling Theory, Information Production Theory, and Optimistic Investors

[6] uses a model of underpricing where issuers delegate the pricing decision to underwriters. [6] presents that the investment banker is better informed about the capital market than is the issuer, and the issuer cannot observe the distribution effort expended by the banker. Most models of IPOs underpricing are based on asymmetric information. A known study of asymmetric information about the IPOs underpricing is modeled by [28]. [28] presents the privileged investors whose information is superior to the issuer as well as other investors. In the model of Rock is that the issuer himself is also uninformed. This means that the offer price is not informative to any investor in the IPOs, so that uninformed investors face severe adverse selection in the [28]. [28] theorizes "winner's curse", "informed investors" like investment institution have superior information about true value to the stock after the listing; on the contrary, "uninformed investors" have more probability of buying the inferior offering. [8] extend [28] theory and find evidence that risk does affect after-market returns. [29] show that the effect of media on initial returns should be asymmetric and should be magnified by uncertainty.

Past studies [1, 17, 32] formalize the notion that good firms underprice to "leave a good taste in investors' mouths." [32] model implies the high-quality firms underprice at the IPOs in order to obtain a higher price at a seconded offering. [17] two-parameter signaling model indicates the firm value and the degree of underpricing are positively related. [1] confirms that underpricing can signal favorable prospects for the firm, and that it is temporary, industry-specific, and associated with improvements in the profitability of entry. [31] state that in fixed-price offerings, the issuer can maintain investors' propensity to produce information by appropriately adjusting the offering price even if information costs are high.

[13] find in the French IPO market that book-built issues attract more press than auctions, but only after the book building route is selected. [14, 21] suggest that the issuer and regular customers of an investment banker benefit from the presence of sentiment investors, neither model considers that an investment banker might promote an issue in such a way as to induce sentiment investors into the market for an IPOs. [11] imply that initial IPOs returns are positively correlated with pre-issue publicity. [5] find that retail investors are more likely to purchase attention-grabbing stocks. Similarly, [30] finds that either the media report investor sentiment before the sentiment is fully incorporated into market prices or the media directly influence investors' attitudes toward securities. [15] determine that individual investors are more likely to hold stock in highly visible companies. Above discussions suggest that an investment banker's efforts to promote a IPOs through increased media increase retail interest in that stock.

2.2 Sentiment Analysis

To easily define sentiment analysis, this paper quotes from the [26, 33] entitled their paper title. Their paper may explain the sentiment analysis among communities self-identified as focused on natural language processing. A large number of papers mentioning sentiment analysis focus on the specific application of classifying reviews as to their polarity, a fact that appears to have caused some authors to suggest that the phrase refers specifically to this narrowly defined task. However, nowadays many construe the term more broadly to mean the computational treatment of opinion, sentiment, and subjectivity in text. Thus, when broad interpretations are applied, sentiment analysis and opinion mining denote the same field of study. [18] quantifies tone using computer-based analytical tools to measure the frequency of positive and negative words found in earnings press releases and concludes the tone of earnings press releases influences investors' reaction to earnings. [12] find a significant positive association between levels of optimistic tone in earnings press releases and both future return-on-assets and the initial market response. [30] concludes that high values of media pessimism induce downward pressure on market prices; unusually high or low values of pessimism lead to temporarily high market trading volume. This is consistent with sentiment theories under the assumption that media content is linked to the behavior of individual investors, who own a disproportionate fraction of small stocks.

3 Research Method

3.1 Hyphothesis, Data Collection and Description

We propose two hypotheses: (H1) The media sentiment is related to IPO underpricing (initial return); (H2) The number of positive (negative) news positively (negatively) relates to the initial return of IPO. In order to test H1 and H2, we collect the IPOs firms which listed on TWSE or the over-the-counter GTSM for the first time from 2010 to 2014. Excluding IPOs of financial firms, Taiwan depositary receipts. This paper identifies and extract the sample companies from the list of the Taiwan Economic Journal database (TEJ), and the number of final sample IPOs firms listed on TWSE and GTSM are 75 and 166, respectively. This paper removes one firm that lacks news and two firms that company name changed. All the public offering data, financial and return data obtained from the TEJ database. The news resources collected from the KMW[2] database. This paper searches the news by the company abbreviation name, and there are 5,876 news. We follow [20] using the one-month window before the issue day. The number of firms is concentrated from 2011 to 2013, the proportion of firms in these three years is 68.05 %. The breakdown of the 241 IPOs exhibits an unevenly distributed pattern across 24 industries[3] classified by the TSE industrial classification code. As expected, electronics industry related firms dominate the Taiwanese IPO market with 60.17 % of the total sample. We divide the news[4] into three types: (1) positive, (2) neutral, and (3) negative.

3.2 Research Design

Using a dictionary to compile sentiment words is an obvious approach because most dictionaries [24] list synonyms and antonyms for each word. Three graduate students read the financial news and extract sentiment words. The way of building sentimental lexicons is based on [19]. We classified news by four rules. Rule one and rule two: when the number of positive (negative) words is more than the number of negative (positive) words, the news classified into positive (negative). Rule three: when the number of negative words is more than the number of positive words and the number of negative words is more than six, the news classified into negative. Rule four: when the number of positive words equal to the number of negative words and the number of negative words is less than six, the news classified into neutral. In order to test our hypothesis, we use IR, TONE_RATIO, TONE, NUM_POS, NUM_NEG, TLHIT, LN_FIRMSIZE, LN_OFFERSIZE, IPO_PERIOD, SCALED_OFFER, MKTRET as our research variables (shown in Table 1). The dependent variable IR is firm i's initial return which calculated as (CPi – OPi)/OPi. The CPi is firm i's closing price of the issue day, the OPi is the offering price of firm i. The independent variables TONE_RATIO and TONE are calculated which according to [7, 16] studies. The PW (NW) is the number of positive (negative) news of firm i during the period from the one month prior to the issue day. We use different models shown in Table 2 to test our hypothesis between two different periods.

[2] Knowledge Management Winner, http://kmw.chinatimes.com/.

[3] Biotechnology & medical care, optoelectronic, electronic parts & components, semiconductor, communications & Internet, electric machinery, electronics & peripheral equipment, information service, tourism, other electronic, trading and consumers' goods, shipping & transportation, chemical, cultural & creative, textiles, building material & construction, plastics, electrical & cable, iron & steel, automobile, glass & ceramics, foods, electronic products distribution, other.

[4] News that is considered to have a positive (negative) impact on IPO firms. News is considered to have a trivial impact is classified as neutral news.

Table 1. Variables description

Variables	Description
PWij (NWij)	The count of positive (negative) words appears in the news j of the firm i
POS_DIC (NEG_DIC)	The total number of positive (negative) words appears in dictionary
LN_FIRMSIZE	the natural logarithm of the market value of the IPO frim at the offering and is noted in millions dollars in Taiwan
IPO_PERIOD	the natural logarithm of difference in the number of days between the date of listing on emerging stock market and the issue date
SCALED_OFFER	Dummy variable where a SCALED_OFFER is given a value of 1 if revision is more than 1 or a 0 if a revision is less than or equal to 0
MKTRET	Market return for 30 trading days prior to issue day

Table 2. Regression models

Model	Independent variables	Control variables
(1)	TONE_RATIO	TLHIT[a], IPO_PERIOD, LN_FIRMSIZE, SCALED_OFFER, MKTRET[b]
(2)		TLHIT, IPO_PERIOD, LN_OFFERSIZE, SCALED_OFFER, MKTRET
(3)	TONE	TLHIT, IPO_PERIOD, LN_FIRMSIZE, SCALED_OFFER, MKTRET
(4)		TLHIT, IPO_PERIOD, LN_OFFERSIZE, SCALED_OFFER, MKTRET
(5)	NUM_POS, NUM_NEG	IPO_PERIOD, LN_OFFERSIZE, SCALED_OFFER, MKTRET

[a] [23] implies a negative correlation between issue size and price update, and initial return.
[b] [22] finds a positive relationship between pre-IPO market return and IPO first-day return.

$$\mathrm{TONE\_RATIO}_{ij} = (PW_{ij}/POS\_DIC) - (NW_{ij}/NEG\_DIC) \tag{1}$$

$$TONE_{ij} = (PW_{ij} - NW_{ij})/(PW_{ij} + NW_{ij}) \tag{2}$$

4 Research Results and Findings

On average, our sample has initial return of 29.29 % and the first quartile shows 7.93 % representing that at least 75 % of IPO initial returns are larger than 7.93 %. It shows that the underpricing phenomenon is quite prevailing in Taiwan. The 239 samples is divided into two subsamples[5]. We using two time windows[6] and four models to test hypothesis H1. For all sample, the minimum of TONE_RATIO and TONE are −0.011 and −0.023 that consists with the direction of IR (−18.488) in TW2. The empirical results indicate that IR is significantly positive association with the TONE_RATIO and TONE in TW1 and TW2. The SCALED_OFFER, MKTRET, IPO_PERIOD is significantly positive association with IR in model (1) to model (4) which means the investors' demand is higher when IR is going up, the market return can predict future underpricing in the hot market, and the longer TW1, the higher visibility in the media is. In addition, the result shows that the model (1) and is better than model (3) and model (2) is better than model (4) because of the adjusted R2 (0.21 > 0.11, 0.17 > 0.14) which means the TONE_RATIO is fit in our research. For electronics and non-electronics' industries, we only use model (1) and model (2) to test our hypothesis because of the model (1) and model (2) has higher adjusted R2. The IR has a significantly positive association with the TONE_RATIO in all situations. The adjusted R2 of model (1) is higher than model (2) in two industries (in TW1: 0.24 > 0.18, in TW2: 0.16 > 0.14). About hypothesis H2, we use model (5) to test it. The empirical result indicates that IR has significantly positive association with the NUM_POS. The association of underpricing and the sentiment of media is significantly difference between the electronics and non-electronics. The results indicate that the IR has significantly positive (negative) association with the NUM_POS/NUM_NEG in electronics industry which means that the IR react to positive (negative) news. Interestingly, the MKTRET is statistically significant in non-electronics.

5 Conclusion and Discussion

This paper focuses on the relation between the sentiment of the media and IPO underpricing based on the IPOs firm listing on TWSE corporations or GTSM corporations. This paper finds a positive coefficient on the positive sentiment score of media coverage and statistically significant association, indicating that the positive sentiment of media coverage are more likely to influence on IPO underpricing. The interesting results that the association of underpricing and the sentiment of media is significantly difference between the companies in electronics and those in non-electronics industry. The result indicates that initial return has a significantly positive (negative) association with the number of positive (negative) news in electronics industry. On the contrary, the initial return is highly related to market return in non-electronics industry. The media is a market signal for investors, issuers and underwriters. In Taiwan, pre-IPO market that has an organized trading platform – the Emerging Stock Market in Taiwan. The pre-market plays an important of price discovery, and the pre-market price should be useful, thus if we may analyze the media coverage into different categories, such as earnings and financials, strategy and policy, and so on. It may help us to further discuss the content of the media and to reach a detailed and consistent point.

[5] Electronics: 106 and Non-electronics: 133.

[6] TW1: pre-IPO period, TW2: one month prior to the issue day.

Acknowledgement. This research is supported by NSC 102-2627-E-004 -001, MOST 103-2627-E-004 -001, MOST 104-2627-E-004-001.

References

1. Allen, F., Faulhaber, G.R.: Signaling by underpricing in the IPO Market. J. Financ. Econ. **23**(2), 303–323 (1989)
2. Arbel, A., Strebel, P.: Pay attention to neglected firms!*. J. Portfolio Manag. **9**(2), 37–42 (1983)
3. Azar, P.D.: Sentiment Analysis in Financial News, Harvard University (2009)
4. Baker, M., Stein, J.C., Wurgler, J.: When does the market matter? stock prices and the investment of equity-dependent firms. Q. J. Econ. **118**(3), 969–1005 (2002)
5. Barber, B.M., Odean, T.: All that glitters: the effect of attention and news on the buying behavior of individual and institutional investors. Rev. Financ. Stud. **21**(2), 785–818 (2008)
6. Baron, D.P.: A model of the demand for investment banking advising and distribution services for new issues. J. Finan. **37**(4), 955–976 (1982)
7. Bautin, M., Vijayarenu, L., Skiena, S.: International sentiment analysis for news and blogs. In: Proceedings of the International Conference on Weblogs and Social Media (2008)
8. Beatty, R.P., Ritter, J.R.: Investment banking, reputation, and the underpricing of initial public offerings. J. Finan. Econ. **15**(1), 213–232 (1986)
9. Benveniste, L.M., Spindt, P.A.: How investment bankers determine the offer price and allocation of new issues. J. Financ. Econ. **24**(2), 343–361 (1989)
10. Benveniste, L.M., Wilhelm, W.J.: A comparative analysis of IPO proceeds under alternative regulatory environments. J. Financ. Econ. **28**(1), 173–207 (1990)
11. Cook, D.O., Kieschnick, R., Van Ness, R.A.: On the marketing of IPOs. J. Financ. Econ. **82**(1), 35–61 (2006)
12. Davis, A.K., Piger, J.M., Sedor, L.M.: Beyond the Numbers: An Analysis of Optimistic and Pessimistic Language in Earnings Press Releases (2006)
13. Degeorge, F., Derrien, F., Womack, K.L.: Quid Pro Quo in IPOs: Why Book-Building is Dominating Auctions (2004)
14. Derrien, F.: IPO pricing in "Hot" market conditions: Who leaves money on the table? J. Finan. **60**(1), 487–521 (2005)
15. Frieder, L., Subrahmanyam, A.: Brand perceptions and the market for common stock. J. Finan. Quant. Anal. **40**(1), 57–85 (2005)
16. Godbole, N., Srinivasaiah, M., Skiena, S.: Large-scale sentiment analysis for news and blogs. In: International Conference on Weblogs and Social Media (ICWSM 2007), Denver CO, 26-28 March 2007 (2007)
17. Grinblatt, M., Hwang, C.Y.: Signalling and the pricing of new issues. J. Finan. **44**(2), 393–420 (1989)
18. Henry, E.: Are investors influenced by how earnings press releases are written? J. Bus. Commun. **45**(4), 363–407 (2008)
19. Lin, I.H., Chen, K.T.: Creating and Verifying Sentiment Dictionary of Finance and Economics via Financial News (2013)
20. Liu, L.X., Sherman, A.E., Zhang, Y.: Media Coverage and IPO Underpricing (2009)
21. Ljungqvist, A., Nanda, V., Singh, R.: Hot markets, investor sentiment, and IPO pricing. J. Bus. **79**(4), 1667–1702 (2006)
22. Logue, D.E.: On the pricing of unseasoned equity issues: 1965–1969. J. Finan. Quant. Anal. **8**(1), 91–103 (1973)

23. Lowry, M., Schwert, G.W.: Is the IPO pricing process efficient? J. Finan. Econ. **71**(1), 3–26 (2004)
24. Miller, G.A., Beckwith, R., Fellbum, C., Gross, D., Miller, K.J.: Introduction to WordNet: an On-line lexical database. Int. J. Lexicography **3**(4), 235–312 (1990)
25. Morck, R., Yeung, B., Yu, W.: The information content of stock markets: Why do emerging markets have synchronous stock price movements? J. Finan. Econ. **58**(1), 215–260 (2000)
26. Nasukawa, T., Yi, J.: Sentiment analysis: Capturing favorability using natural language processing. In: Proceedings of the 2nd International Conference on Knowledge Capture, pp. 70–77 (2003)
27. Neal, R., Wheatley, S.M.: Do measures of investor sentiment predict returns? J. Finan. Quant. Anal. **33**(4), 523–547 (1998)
28. Rock, K.: Why new issues are underpriced. J. Finan. Econ. **15**(1–2), 187–212 (1986)
29. Sherman, A.E., Titman, S.: Building the IPO order book: underpricing and participation limits with costly information. J. Finan. Econ. **65**(1), 3–29 (2002)
30. Tetlock, P.C.: Giving content to investor sentiment: the role of media in the stock market. J. Finan. **62**(3), 1139–1168 (2007)
31. Trauten, A., Langer, T.: Information production and bidding in IPOs: an experimental analysis of auctions and fixed-price offerings. Zeitschrift für Betriebswirtschaft **82**(4), 361–388 (2012)
32. Welch, I.: Seasoned offerings, imitation costs, and the underpricing of initial public offerings. J. Finan. **44**(2), 421–449 (1989)
33. Yi, J., Nasukawa, T., Bunescu, R., Niblack, W.: Sentiment analyzer: extracting sentiments about a given topic using natural language processing techniques. In: Third IEEE International Conference on Data Mining (ICDM), pp. 427–434 (2003)

An Evaluation of the Conversation Agent System

Ong Sing Goh, Yogan Jaya Kumar(✉), Ngo Hea Choon,
Pui Huang Leong, and Mohammad Safar

Faculty of Information and Communication Technology,
Universiti Teknikal Malaysia Melaka, 76100 Melaka, Malaysia
{goh, yogan, heachoon}@utem.edu.my,
phuang1122@gmail.com, neel2292@gmail.com

Abstract. In this paper, we have highlighted the increasing need for standard metrics to assess and measure the quality of responses produced by conversational agent systems based on different approaches and domains. To demonstrate the approach, we will present the data and results obtained through an evaluation performed on three different conversation systems, namely Eliza (Rogerian Psychotherapist), ALICE (Artificial Linguistic Internet Computer Entity) and AINI (Artificial Intelligent Natural-Language Identity).

Keywords: Conversation Agent (CA) · Artificial Intelligence (AI) · AINI · Evaluation

1 Introduction

The growth of conversation systems is limited by a lack of evaluation, especially of the performance and the quality of responses from the conversation agents (CAs). The Turing Test, for instance, is a proposal to test a machine's capability to demonstrate intelligence in five minutes. In the Loebner Prize contest, judges evaluate which of the CAs entered is the most humanlike in ten minutes. In this paper, a novel methodology based on quantitative and qualitative approaches is proposed. It is apparent that in evaluating a CAs framework, it is required to decide whether a development is a step forward, and, whether the development is worth the effort.

2 Natural Language Query

In AINI, the communication with users takes place through typed text messages and is processed based on natural language query. AINI's engine implements its decision making network based on the information it encounters in the six levels of natural language modules, as have been discussed in references [1, 2, 13]. The input and output of each module is an XML-encoded data structure that keeps track of the current computational state. The knowledge modules can be considered as transformations over this XML data structure. The system accepts queries from the users and it processes the queries based on the information contained in AINI's knowledge bases.

N.T. Nguyen et al. (Eds.): ACIIDS 2016, Part II, LNAI 9622, pp. 354–365, 2016.
DOI: 10.1007/978-3-662-49390-8_35

The system is implemented by open-source architecture based on LAMP (Linux, Apache, MySQL, PHP) solution and knowledge bases stored in a MySQL server. All the domain services are written in the Perl scripting language. Perl has been chosen because it has advantages such as its use of the concepts of objects, modular, arbitrary data structures, classes, methods, and inheritance.

In this paper, we only investigate and evaluate the NL-Query of the agent brain. The agent brain tier handles the process of the queries or business logic. Here, one or more domain service tiers are configured to compute the dialogue logic through the multilevel natural language query algorithm. In this tier, it is based on a goal-driven or top-down natural language query (NL-Query) approach, which is similar to the way that humans process their language. As indicated by literature in the field of Natural Language Processing (NLP), the top-down approach is by far the best approach. Mentalese, or 'language of thought', and conceptual representation support the ideas of a top-down approach [3]. This was also supported by research in generation schemas [4], rhetorical structure theory [5], summarisation [6], plan-based approaches, [7], and SHRDLU [8] the first CA to use NLU, are examples of top-down approaches. Therefore, AINI's agent brain uses a top-down NL-query approach to simulate human conversation. However, in the robotic design, the MIT Cog Robot research fervently supports the bottom-up approach when modelling the human brain [9].

3 Natural Language Understanding and Reasoning (NLUR)

Natural Language Understanding and Reasoning (NLUR) in Level 1 is the most important level of the AINI system. It refers to the process of constructing machine understandable meaning representations from natural language inputs. Preliminary definitions of what "understanding" natural language could imply as introduced by [10], that suggests that "an intelligent person or program should be able to answer [...] questions based on the information in [...] (a) story". According to Hubert Dreyfus [11] in his controversial book "What Computers Cannot Do" (and the revised version "What Computers Still Cannot Do"), the fundamental reason why computers cannot achieve human level intelligence, including the ability to understand human language, is that computers cannot use any formal symbolic system to adequately model the vast background knowledge which humans take for granted in interpreting and reasoning. Since the publication of Dreyfus's book, research of AI in general has gradually shifted from trying to come up with a general problem solver [12], to solving specific problems in narrowly restricted domains. SHRDLU[1] [8] is a classical natural language understanding written in MacLisp at the M.I.T. Artificial Intelligence Laboratory in 1968-70. SHRDLU uses a top-down, left-to-right parser that analyses a pattern, identifies its structure, and recognises its relevant features and grammar. But the system can only answer simple queries about the current state of its toy block world. SHRDLU demonstrated the promising future of NLU research at that time.

[1] http://hci.stanford.edu/~winograd/shrdlu/.

In previous papers [1, 2], NL-query is based on NLUR comprising three parts, namely (a) understanding documents to produce facts which will be integrated into the knowledge base, (b) understanding questions and finally, (c) reasoning using facts and rules to look for answers from the knowledge base. The practice of natural language understanding is widely reflected through the use of understanding modules for both the question and information source. The design of the NLU mechanism took into consideration the various levels of analysis up to the discourse level [20]. Although there are existing concepts or techniques out there for various stages of analysis in NLU, they are mostly studied separately without regard for compatibility of the algorithms which are required to be integrated for full natural language understanding. Hence, for this research, a series of algorithms have been proposed based on actual theories for various stages of analysis that were designed to work seamlessly together. In syntax analysis, an existing external module for sentence parsing called X-MINIPAR is used. X-MINIPAR was a modified version of the off-shelf MINIPAR [14]. X-MINIPAR has been enhanced to allow the parser to load the hash tables once and stay resident (as a background daemon process) so that the parser can parse multiple sentences without having to re-load the hash tables each time. The original MINIPAR has been patented in the United States of America[2] and is a broad-coverage parser for the English language. An evaluation with the SUSANNE corpus parses newspaper text at about 500 words per second, MINIPAR achieves about 88 % precision and 80 % recall with respect to dependency relationships.

AINI[3] conversation robot used typical full-discourse NLUR system consists basically of two subsystems, namely NLU and network-based advanced reasoning system. The NLU subsystem is responsible for reading and understanding two things: questions from users, and sentences of processed news articles from a news repository. The process is carried out in four phases by four natural language processing modules, namely (a) sentence parsing, (b) named-entity recognition, (c) relation inference, and (d) discourse integration.

However, the network-based advanced reasoning subsystem is responsible for discovering the valid answer and generating an unambiguous answer or explanation in response to users' questions [15]. The process is executed in five phases by five modules, namely (a) network-to-path reduction, (b) selective path matching, (c) relaxation of event constraint, (d) explanation on failure, and (e) template-based response generation.

4 An Evaluation of the Parsers

Off-shelf modified version of MINIPAR [14] called X-MINIPAR was used as the parser in the AINI's NL-Query module. A comparison of X-MINIPAR was made with other popular parsers. In this study, X-MINIPAR was compared to the CMU Link

---

[2] http://www.patentstorm.us/patents/7146308-description.html.

[3] AINI's research on "Intelligent Agent Technology in E-commerce", *Intelligent Data Engineering and Automated Learning*, vol. 2690, pp 10-17, 2003 has been cited by US Pattent examiner for the Patent no US8639638, 2014, http://www.google.com/patents/US8639638.

Grammar parser [16] and the Stanford parser [17]. These three parsers represent a cross-section of approaches to producing dependency analyses: X-MINIPAR uses a constituency grammar internally before converting the result to a dependency tree. CMU Link Grammar is based on link grammar, and the Stanford Parser is a lexicalised statistical syntactic parser. With the emergence of broad-coverage parsers, quantitative evaluation of parsers becomes increasingly more important. Firstly, such an evaluation scheme is necessary to quantitatively measure the progress in the field of broad-coverage parsing and to compare and evaluate different parsing techniques. Secondly, in the development of a broad-coverage parser, it is usually very difficult to predict the consequences of a change to the parser or the grammar. An attempt to extend the coverage that is motivated by a few examples may well cause the parser to over generate and/or lose coverage in other areas. This will make the evaluation biased and difficult to compare. Thirdly, efficiency and coverage are often conflicting goals for a parser. A meaningful trade-off can only be arrived at if both of them can be measured precisely. Finally, quantitative evaluation may provide crucial information for determining the suitability of a parser in a particular domain and/or for a particular task.

4.1 Performance of the Parsers

In order to evaluate the comparative performance of the parsers, 1428 uncategorised pandemic bird flu web documents are extracted and wrapped using Google API[4] and BootCAT Toolkit [18]. Regular expressions are used to filter out the HTML information and to extract well-formed sentences. From the 1428 pages, the first sentence of each of the first 150 pages are extracted and parsed by X-MINIPAR, CMU Link Grammar and the Stanford parser. Each sentence has an average of 50 words. A uniform policy facilitates a fair comparison between the parsing techniques. In this experiment, the composition or syntactic structure of these sentences are the main focuses. The performance of all of these three systems varied across different argument types. CMU Link Grammar took an average of 0.7 s to parse a sentence, Stanford Parser 0.5 s and X-MINIPAR 0.2 s[5]. Refer to Table 1. It is expected that X-MINIPAR yields the highest performance because it was the fastest. This result was comparable to the original MINIPAR evaluation with the SUSANNE corpus, which is able to parse newspaper text at about 500 words per second on a Pentium-III(tm) 700 MHz with 500 MB memory [21].

4.2 Accuracy of the Parsers

The accuracy of a parser depends on the formalisms they use to model language and the corresponding outputs they produce. Dependency parsers model language is a set of relationships between words, and they do not make widespread use of concepts like 'phrase' or 'clause'. Dependency parsers are popular in the applied NLP circles. The

[4] http://www.google.com/apis.

[5] In this experiment, Dell Precision PWS380 Server 3 GH with 1 GB of memory was used.

Table 1. Performance Test for X-MINIPAR, CMU Link Grammar and the Stanford Parser

Number of sentence extracted from web documents	Stanford Parser (seconds)	CMU Link Grammar (seconds)	X-MINIPAR (seconds)
150	75	105	30
Average per sentence	0.5	0.7	0.2

grammatical relationships that dependency parsers specify are similar to the semantic relationships encoding logical predicates of which NLP developers use to reduce a sentence. From the parsed output, dependency graphs[6] can be created representing how the words in the sentences governed or depended on one another.

For this accuracy test, the most frequently asked questions (FAQs) obtained from the "who.int" and "pandemicflu.gov" websites are collected and applied to the three parsers. There were 158 FAQ questions used in this evaluation. An example is shown below.

"Bird flu did occur in which countries?"

Results from the CMU Link Grammar parse output shown in Fig. 1 show that the second-last word has been left untagged. This figure also shows the constituent output of the parse, and Fig. 2 depicts the dependency graph showing the part-of-speech of each word, if there is any. It can be seen that the word "*which*" failed to be tagged due to the null-links feature of the parser.

```
[(bird.n)(flu.n)(did.v)(occur.v)(in)([which])(countries.n)]
[[0 1 0 (AN)][1 2 0 (Ss)][2 3 0 (I*d)][3 4 0 (MVp)][4 6 0
(Jp)]]
[0]
```

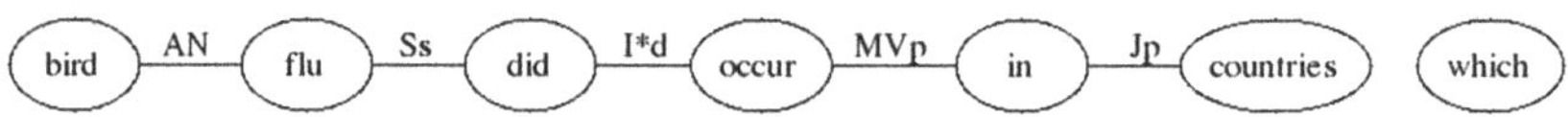

Fig. 1. Parse output and dependency graphs generated by CMU Link Grammar

In the Fig. 3 example, X-MINIPAR and Stanford parser correctly parse every word. The constituents can be easily produced from the grammatical relationships listed in Fig. 3. X-MINIPAR and Stanford parser use different types of grammar (rule-based and principle-based); therefore they produce different types of output. According to Klein [17] and Lin [14], both the Stanford parser and X-MINIPAR have been used successfully in the past and proven in this study.

6 The dependency graphs generated using MINIPAR, CMU Link Grammar and Stanford parser parse visualisation tool which can be obtained at http://cgi.stanford.edu.

```
((
E2 (() U * )
E0 (() fin C E2 )
1 (Bird ~ N 2 nn (gov flu))
2 (flu ~ N 4 s (gov occur))
3 (did do Aux 4 aux (gov occur))
4 (occur ~ V E0 i (gov fin))
E3 (() flu N 4 subj (gov occur)
(antecedent 2))
5 (in ~ Prep E2 p)
6 (which ~ Det 7 det (gov country))
7 (countries country N 5 pcomp-n
(gov in))
)
```

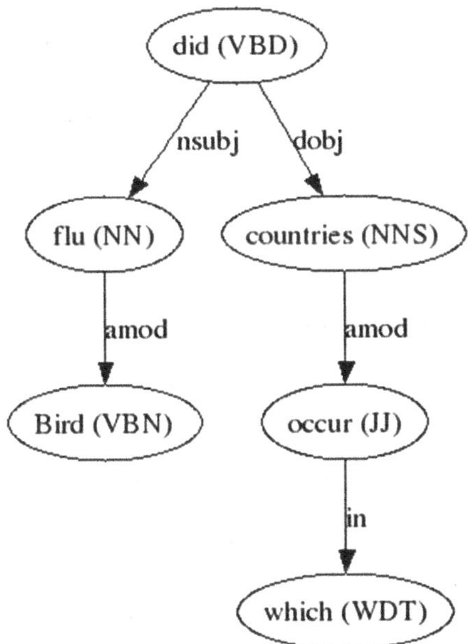

Fig. 2. Parse output and dependency graphs generated by Stanford Parser

However, using two parsers means the system requires an additional component to reconcile them sufficiently to parse the sentences. Building a CA system is not straightforward; using an incremental implementation should remove some of the complexity. The use of multiple parsers is one aspect that makes the system more complex, therefore the first implementation increment of the system may be better served with the use of only one parser.

In this case, X-MINIPAR added richness to the representation making it possible for the CA system to generate a more accurate machine-readable sentence in reply to

```
(ROOT
  (SINV
    (FRAG
      (NP (NN bird) (NN flu)))
    (VP (VBD did)
      (S
        (VP (VB occur)
          (FRAG
            (WHPP (IN in)
              (WHNP (WDT which)))))))
    (NP (NNS countries))))
```

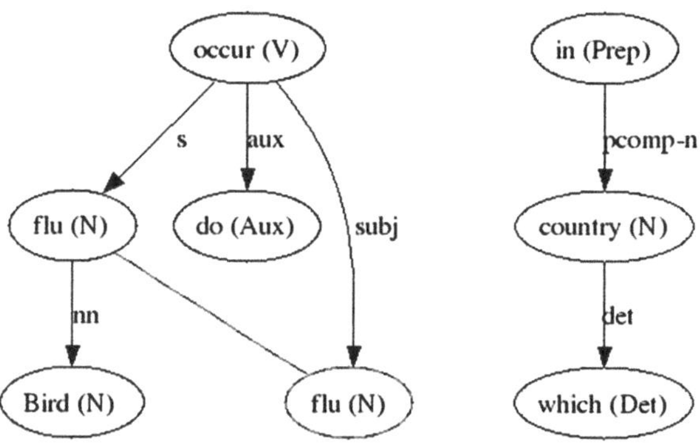

Fig. 3. Parse output and dependency graphs generated by X-MINIPAR

the database language query. X-MINIPAR codes were optimised with 90 lines, compared to CMU Link Grammar with around 300 lines of codes, in order to extract the syntactic categories and tokenisation for visual scrutiny as shown in Figs. 1 and 3. Moreover, the morphological roots of verbs and nouns from X-MINIPAR output could be obtained. This eliminates the need for a morphology analyser. In addition, the X-MINIPAR path is both shorter and simpler for the same predicate-argument relationship, and could be encoded in various ways that take advantage of the additional semantic and lexical information that is provided. Using both a syntactic and semantic grammar parser in series should also improve the richness of the interpretation of the natural language which was implemented in the natural language understanding and reasoning of the AINI conversation system. Most of the performance tests seem attributable to the modified version of X-MINIPAR.

5 An Evaluation of the Performance Conversation Agents

At the moment of this evaluation has been carried out, no other research has reported about the performance evaluation of the CA, except for question-answering systems. However, the results of previous evaluation by researchers of question-answering system such as AnswerBus [19] and AINI [15] are used in this evaluation as a benchmark for ensuring that the results from this evaluation do not deviate too significantly. In this quantitative approach for performance evaluation on the AINI conversation system, a set of 98 stimulus questions extracted from three-time Loebner Prize winner ALICE's transcripts [20] was used. This set of stimulus questions was collected from conversations between four judges and ALICE in the 2001 contest at Science Museum, London. These judges' questions were used to simulate the conversation with AINI, ALICE, ALICE Silver Edition and ELIZA remotely over the World Wide Web.

The maximum, minimum and average response time and also the standard deviation obtained from this evaluation are displayed in Table 2. The response time for each question submitted to these four CAs is recorded, collected and analysed for average and standard deviation. The pattern of the response times for the three systems is depicted in the following graph in Fig. 4.

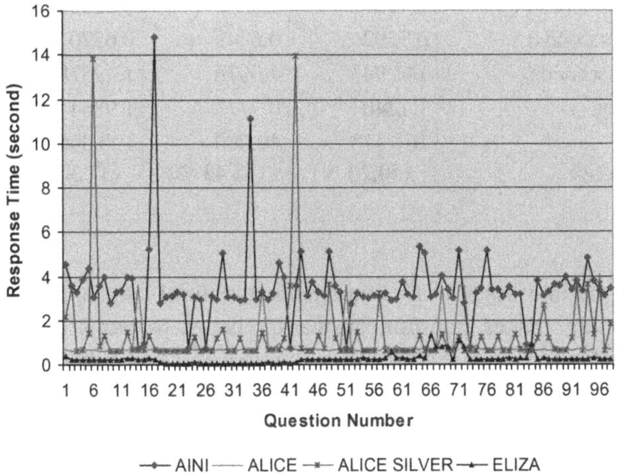

Fig. 4. Response times for AINI, ALICE, ALICE Silver edition and ELIZA

The response time from the evaluation is shown to have similar outliers to the results in previous research [15]. It is expected that the response times for each query will depend on the population of the response categories of the CAs' knowledge base. For instance, the total number of AINI stimulus-response categories was 161,473, whereas the original ELIZA has only 200 stimulus response categories [21]. ALICE standard edition [22], which was ranked the "most human" computer, has about 40,000 response categories from their AAA, and ALICE Silver Edition [23] has about 120,000 response categories, where another 80,000 response categories were taken from MindPixel.

As shown in Table 2, ELIZA's response times will be better than other CAs, followed by ALICE, ALICE Silver Edition and AINI, based on the number of stimulus-response categories in their knowledge bases. However, the response time for

AINI was comparable with ALICE Silver Edition. Although AINI's stimulus-response categories knowledge have 12.9 % more than ALICE Silver Edition, AINI's response time was within the range of 0.7292 to 14.7943 s, compared to ALICE with the range 0.6370 to 13.9578 s. The standard deviations also exhibit some similarity between the two CAs - AINI is 1.7240 s and ALICE Silver Edition is 1.9864 s.

The only major difference is the average time that might be caused by the number of stimulus-response categories for each of the CAs. Although AINI's (1.4527 s) average response time was double that of ALICE Silver Edition (0.72482 s), this was compensated by the fact that the total number of AINI's stimulus-categories is about 13 % more than ALICE Silver Edition.

Table 2. Average response time and standard deviation for AINI, ALICE, ALICE Silver Edition and ELIZA

	AINI	ALICE	ALICE Silver Edition	ELIZA
Average time (seconds)	1.4527	0.3688	0.72482	0.1061
Minimum time (seconds)	0.7292	0.6349	0.6370	0.0699
Maximum time (seconds)	14.7943	4.0976	13.9578	1.2986
Standard deviation	1.7240	0.7773	1.9864	0.21023
Knowledge base (Stimulus-response categories)	161,473 (50.20 %)	40,000 (12.43 %)	120,000 (37.30 %)	200 (0.06 %)

In addition, the introduction of components in natural language understanding and advanced reasoning makes question answering better in terms of response quality without compromising response time. These components can either originate from new ideas or innovative use of existing concepts. Referring to Table 3, the complexity and the demanding nature of the system increases as more components are included, but at the same time, the quality of responses produced also improves. This can be attributed

Table 3. Natural language query components in AINI compare to ELIZA and ALICE conversation systems

Components	Conversation system		
	ELIZA	ALICE/ALICE Silver Edition	AINI
Spelling checker			√
Pattern matching	√	√	√
Case-base reasoning	√	√	√
Index search			√
Natural language understanding and reasoning			√
Dynamic response generation		√	√
Supervised learning			√

to the fact that more and more computation is performed on a decreasing amount of information in an attempt to exploit more aspects of natural language to achieve richer meaning representation.

Here this may provide an explanation to the results. ELIZA is a system of an entirely different class to ALICE and AINI. While AINI appears to be slower, the system has to go through a larger amount of processing as shown in Table 3. In general, the graphs have revealed that the response time of AINI is actually similar to other systems which appear to include less demanding resources and processing. In addition, the results have also shown that the response time of AINI is consistent despite the uncertainty in the type of questions. This is illustrated with the relatively low standard deviation as compared to ALICE Silver Edition. This is important as AINI was designed to handle questions of open-domain and domain-specific nature. This is unlike other existing CA systems that are only restricted to a specific domain such as the virtual therapist in ELIZA, or, unrestricted domain as with ALICE.

6 Conclusions

In this session, we have presented an evaluation of the CA system, AINI, using quantitative and qualitative approaches in laboratory experiments. It is not practical to develop a CA from scratch; therefore, based on the performance and accuracy of the natural language parser, such as CMU Link Grammar and Stanford parser, X-MINIPAR has been selected and integrated into the NLUR component in AINI's framework.

Using the available natural language parser, the AnswerBus question-answering systems were compared with the AINI conversation system in a form of quality evaluation. Initial results have shown that AINI is comparatively better in terms of the quality of responses generated. One of the criteria that have contributed to the higher score of AINI is the capacity to generate useful responses dynamically using two advanced reasoning components, namely explanation on failure and dynamic answer generation to cater for the condition when no answers are available.

In the quantitative approach for performance evaluation on the CA, AINI was compared with other three CAs, namely ELIZA, ALICE and ALICE Silver Edition. Due to its simplistic design, it is expected, ELIZA's response time will be better than other CAs. This is followed by ALICE, ALICE Silver Edition and AINI. This can be attributed to the fact that more computation is performed by AINI. Although AINI's stimulus-response categories knowledge have 12.9 % more than ALICE Silver Edition, AINI's slower response time was due to the introduction of the NLUR components which improve the quality of the responses.

Acknowledgements. This research is supported and funded through Top Down PJP/2014/FTMK(11B)/S01371 grant, awarded by the Centre for Research and Innovation Management (CRIM), Universiti Teknikal Malaysia Melaka (UTeM). The MeetInventor™ was first trademark (2015002886) and copyright (No LY2015000274) under the Intellectual Property Corporation of Malaysia, 2015. AINI Conversation Agent can be accessed from http://meetinventor.utem.edu.my.

References

1. Goh, O.S., Abd Ghani, M.K., Yogan, J.K., Choo, Y.H., Muda, A.K.: Massive Open Online Course (MOOC) with learning objects and intelligent agent technologies. In: IEEE International Conference on IT Convergence and Security, pp. 125–128, Beijing (2014)
2. Goh, O.S., Depickere, A., Fung, C.C.: Domain knowledge query conversation bots in instant messaging. Knowl.-Based Syst. **21**, 681–691 (2008)
3. Fodor, J.A.: Elm and the Expert: An Introduction to Mentalese and Its Semantics. Cambridge University Press, Cambridge (1994)
4. McKeown, K.R.: Text Generation. Cambridge University Press, Cambridge (1995)
5. Moore, J.D., Paris, C.L.: Planning text for advisory dialogues: capturing intentional and rhetorical information. Comput. Linguist. **19**, 651–695 (1992)
6. Radev, D., McKeown, K.: Generating natural language summaries from multiple on-line sources. Comput. Linguist. **24**, 469–500 (1998)
7. Reiter, E., Dale, R.: Building Natural Language Generation Systems. Cambridge University Press, Cambridge (2000)
8. Winograd, T.: Procedures as a representation for data in a computer program for understanding natural language. Cogn. Psychol. **3**(1), 1–191 (1972)
9. Brooks, R.A., Breazeal, C., Marjanovic, M., Scassellati, B., Williamson, M.M.: The cog project: building a humanoid robot. In: Nehaniv, C.L. (ed.) CMAA 1998. LNCS (LNAI), vol. 1562, pp. 52–87. Springer, Heidelberg (1999)
10. McCarthy, J.: An example for natural language understanding and the AI problems it raises. Stanford University, 17 November 2015. http://jmc.standford.edu/articles/mrhug.html
11. Dreyfus, H.L.: What computers still can't do – a critique of artificial reason. MIT Press, New York (1992)
12. Newell, A., Shaw, J.C., Simon, H.A.: Report on a general problem–solving program. In: International Conference on Infomnation Processing, pp. 256–264. NESCO House, Paris (1959)
13. Goh, O.S., Depickere, A., Fung, C.C., Wong, W.: Top-down natural language query approach for embodied conversational agent. In: International Multi Conference of Engineers and Computer Scientists (IMECS 2006), Hong Kong (2006)
14. Lin, D.: Dependency-based evaluation of MINIPAR. In: Abeillé, A. (ed.) 1st International Conference on Language Resources and Evaluation. Springer, The Netherlands (1998)
15. Goh, O.S., Fung, C.C., Wong, W., Depickere, A.: An embodied conversational agent for intelligent web interaction on pandemic crisis communication. In: IEEE/WIC/ACM International Conference on Web Intelligence and Intelligent Agent Technology (WI-IATW 2006), Hong Kong (2006)
16. Sleator, D., Temperley, D.: Parsing English with a link grammar. Carnegie Mellon University Computer Science Technical report CMU-CS-91-196 (1991)
17. Klein, D., Manning, C.: Accurate unlexicalized parsing. In: 41st Annual Meeting of the Association for Computational Linguistics, pp. 423–430 (2003)
18. Baroni, M., Ueyama, M.: BootCaT: bootstrapping corpora and terms from the web. In: Fourth Language Resources and Evaluation Conference (2004)
19. Zheng, Z.: AnswerBus question answering system. In: Conference on Human Language Technology (2002)
20. ALICE: ALICE Loebner prize winner transcripts. Loebner Prize Homepage, 17 November 2014. http://loebner.net/Prizef/2005_Contest/Alice/
21. Weizenbaum, J.: ELIZA—a computer program for the study of natural language communication between man and machine. Commun. ACM **9**, 36–45 (1966)

22. Wallace, R.S.: The anatomy of A.L.I.C.E. Artificial Intelligence Foundation, Inc., 17 November 2014. http://www.alicebot.org/anatomy.html
23. ALICE: ALICE Silver Edition. ALICE Foundation, 20 August 2014. http://www.alicebot.org/join.html

Simulation of Each Type of Defibrillation Impulses by Using LabVIEW

Lukas Peter(✉) and Radek Osmancik

Department of Cybernetics and Biomedical Engineering, VSB – Technical University of Ostrava, 17. listopadu 15, Ostrava, Czech Republic
{lukas.peter,radek.osmancik}@vsb.cz

Abstract. Defibrillation is the only known method of treatment of cardiac arrhythmias, especially during full cardiac arrest. The success of defibrillation is dependent on many factors which influence defibrillation. These factors include the impedance of the patient's chest, elapsed time from cardiac arrest until start of applying the defibrillation's discharges, the type of defibrillation pulse, which is defined by the shape, amplitude, initial phase and also the pulse length and ratio of phases. It is not currently possible to try apply defibrillation pulses with different parameters so that we get feedback about success of the applied defibrillation pulses. This issue of applying defibrillation pulses, a brief description and user's feedback has been processed in the program created in LabVIEW environment which is described in this article.

Keywords: Defibrillation · Impulses · Education · LabVIEW

1 Introduction

The heart works as a pump which pumps the blood into the blood circulation through the contraction of the heart muscle. These contractions related to heart's ventricles and atria. The cardiac cycle can be deflected from the regularity or speed of the optimal heart rhythm by a number of factors that cause inactivity ventricles or atria. In this case we are talking about cardiac arrhythmias. [1]

Fibrillation is one of the most occurred arrhythmia among people. It occurs mainly in the elderly population, although this is not condition. Fibrillation is represented by chaotic vibrations of ventricular or atrial cardiac fibers. The muscle's fibers work on the basis of electric potential of myocardium. So we can chaotic vibrations return to optimal synchronous state just by electric pulse – defibrillation. Defibrillation resets all fibers of the heart muscle at the same time by its current flow. This is the time during which there is a large chance that the heart muscle starts again to work synchronously. If defibrillation shock is not applied on time than the heart muscle vibration stops and this is called isoline (Fig. 1). Figure 1 is representing the complete arrest of the heart muscle which does not produce electrical activity at the end. We cannot apply defibrillation in this state. For reintegration of fibrillation and defibrillation use is necessary to use indirect heart massage. [2, 3]

N.T. Nguyen et al. (Eds.): ACIIDS 2016, Part II, LNAI 9622, pp. 366–373, 2016.
DOI: 10.1007/978-3-662-49390-8_36

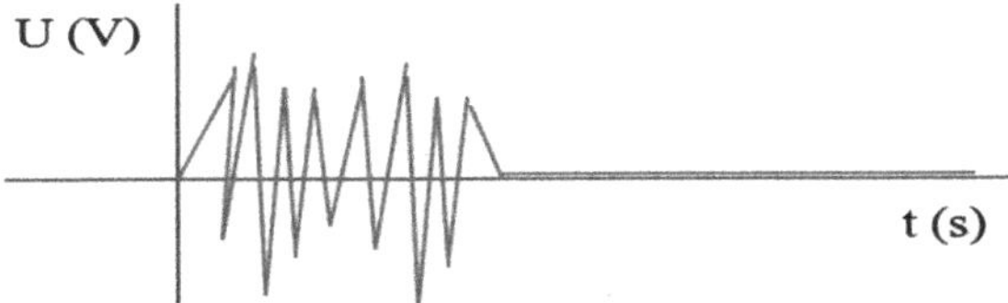

Fig. 1. From fibrillation to complete arrest. The picture shows signal during fibrillation in time.

Untreated cardiac fibrillation is a common cause of death and that is why the correct use and select of a defibrillation pulse is the mainstay of treatment by electrical impulse. Practical setting and its application to various types of arrhythmia was simulated in LabVIEW where the user can select his parameters of defibrillation pulse. Then the user applies the pulse and gets feedback about its success. [4]

2 Problem Definition

The biggest problem regarding defibrillation pulse is their practical training. Practical tool on which we should be able to set the shape of the pulse, its parameters and thereby improve knowledge about dependency of the adjustable parameters on success rate of an electric shock. [5, 6]

This insufficiency is reflecting in health professions education of students who acquire this knowledge only in its practice. So my main goal was creation of software used for teaching issues of defibrillation pulse. The program will have available choice of arrhythmia and choice of different type of shape defibrillation pulse. There are also settings of suitable parameters and applications selected defibrillation pulse on the arrhythmia included in the application. After shock application comes part which informs the user about the accuracy of settings parameters and success of discharge. [7]

3 Description of Application

The overall program should have a structure which offers the choice of educational path through the program for inexperienced users. Users should have a clue about the type of cardiac arrhythmia, defibrillation and choosing the suitable defibrillation parameters after program completion (Fig. 2).

If the user is already experienced he does not have to loose time with educational path and has a choice of direct path to the main program. The main program includes all the settings of mentioned elements such as cardiac arrhythmias, type of defibrillation pulses and mode selection (customization parameters by user or the parameters according to energy). The application also includes a part with graphical display divided into two parts. The first one reflects the heartbeat and the other separate section shows the settings of the defibrillation pulse.

Buttons for program management will ensure easy control. Use of these buttons is particularly in stopping the program and further analysis of the heart rhythm. There is also “Help” section for easier user’s control located at the top right corner. It describes individual sections of the program, their function and usage.

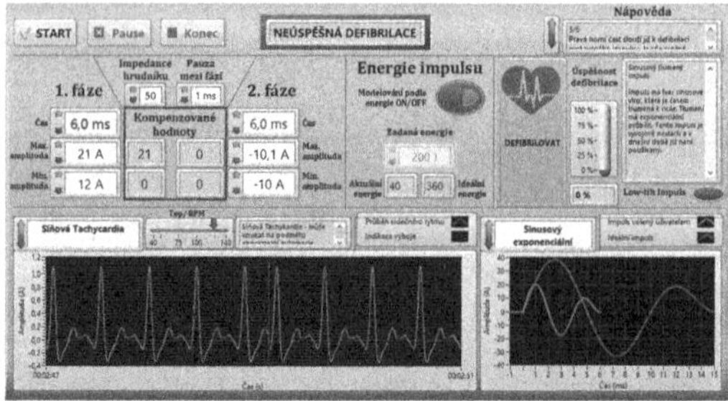

Fig. 2. Intuitive user interface of main application. There is possible to set all of parameters of each type of defibrillation impulses.

The most important is informative part of program consisting of a graphic pointer for success rate, result description, a status indication of low-tilt pulse and it also indicates if defibrillation pulse was or was not successful after application. The text section is connected with parameter settings of defibrillation pulse and that is why it adds, changes or removes information from the text box when the conditions are fulfilled. So that user information is always up to date and correct. For example, the status of low-tilt defibrillation pulse illuminates the indicator and adds notice to text box explaining what low-tilt defibrillation pulse means.

4 Implementation of New Solution

The software was developed in LabVIEW which are used primarily to simulate and test processes in real time. This environment uses the principle of graphical modeling.

The software development had the character of gradual development components with final connection between them. At first, we had to obtain information on which parameters defibrillation pulse is based. Such information may contain the amplitude, duration, threshold voltage values, types of cardiac arrhythmias and the success of various types of defibrillation pulses. This part of the program development is also the understanding of evolution and innovation defibrillation pulse from the history to present with look into the near future. We also used a research about triphasic and quadriphasic pulses. Searching information was done by research of articles in scientific libraries.

Simulation in LabVIEW follows after gathering sufficient amount of information. The first normal heart function was modeled. I have realized it by simulator with predefined default ECG signal in LabVIEW. The simulator has also the option of atrial and ventricular tachycardia and AV block. Fibrillation was set by coordinate points which formed a continuous signal periodically repeated. The simulator is able to set the speed of the heartbeat. The choice of this value is implemented by a sliding bar which also serves as an indicator. Individual simulators have been implemented in case structure serving as a tool to switch heart rhythms.

Modeling of defibrillation pulse was also implemented into case structure. All types of pulses were modeled in the same way using combinations of signal patterns which were set one by one and then ranked by instrument called "insert into array" so that the final signal gives the desired shape of the defibrillation pulse. The time base of the final signal is defined by the 100 points. 100 points is equal to 1 ms. Inputs of signal pattern are connected with numerical controllers for users. Through user sets these controllers as the duration, amplitude and shape of the defibrillation pulse. The summary of amplitude for each of the defibrillation pulse is shown in "Table 1". Pulse energy is automatically calculated from set parameters. If the user does not want set parameters through controllers than he has the option to switch to the automatic rendering of the defibrillation pulse. This option locks all numerical controllers and the user selects only required energy which the pulse must have. The program continuously counts the pulse energy after selection. When there is difference between required energy and current energy the program automatically continuously changes the value of amplitude by small steps until the energies are equal. Only the amplitude is changing continuously because the time base has the ideal value in this case.

Table 1. Parameters of each type of defibrillation impulses.

Pulses	Peak current [A]	Energy [J]	Pulse duration [ms]	Effectiveness [%]
Monophasic	28	360	18	63–93
Biphasic	20	130	10–12	90–95
Modified ZOLL	14	120	10–12	99
Exponential	45	200	18	60–93

Once defining rhythms and cardiac defibrillation pulse was finished the development of algorithm for calculating the success rate of applied defibrillation pulses followed. There are lots of methods to calculate success rate. I chose method of averaging all of the parts which affect defibrillation pulse. See a more detailed description in the Sect. 5.

A large part of the program is applying pulses to cardiac arrhythmia with interaction depending on the value of successful of defibrillation shock. The first part of the application is the detection correct locus of cardiac rhythm for applying the pulse to avoid applying shocks to the T-wave and conversely to ensure discharge as soon as possible after the R-peak. It is implemented by module which detects R-peaks. Subsequently the position is delayed by a few milliseconds representing the device delay and evaluates the status as a logical true value.

Generation of the random number and comparing with value of the success of a defibrillation shock followed. If the comparison comes out positively it will generates again a logical true value that comes along with the logical value of detection to "AND" logical operation. This output affects operation that determines whether the selected cardiac rhythm remains the same or changes to the optimum sinus rhythm. There is also provided a condition for applying a discharge on a healthy normal sinus rhythm. Sinus rhythm changes to cardiac fibrillation with equal percentage chance as successful defibrillation in this case. Application locus of discharge is shown by a red vertical line to give possibility for program to make analysis in future (Fig. 3).

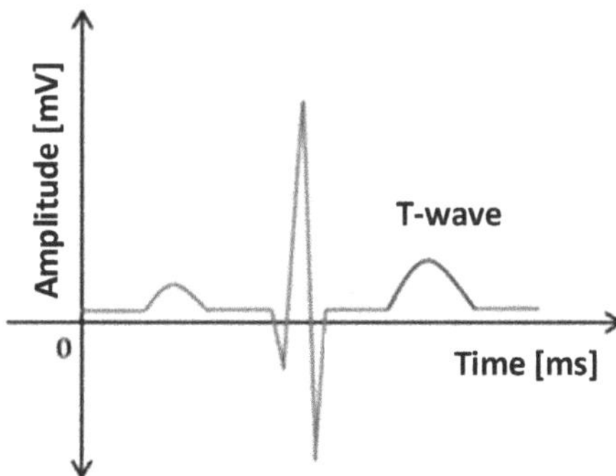

Fig. 3. Typical ECG wave with highlighted T-wave (Color figure online).

Final program adjustment includes a design modification of user interface. Program has to be clear, legible, easy to use and visually affable. In addition, the whole program had be properly tested for the exceptional states which may not be implemented. The conditions are implemented because of these states. For example, program includes the status when the time base of discharge is unrealistically small and amplitude of discharge is unrealistically high. Pulse energy is the same as the correct momentum and success of discharge indicates about 80 % which is nonsense. Finally, all errors had to be fixed with the goal of smooth running program.

5 Development of Application

LabVIEW project is divided into three parts. Project includes a part for simulating cardiac arrhythmias and cardiac rhythm, the second part for defining parameters of defibrillation shape and the third part for applying and evaluating a defibrillation shock. All these sections are independent.

5.1 Simulation of Cardiac Arrhythmias and Cardiac Rhythm

Simulation of the heartbeat was built by using predefined rhythms and cardiac arrhythmias which were placed into a while loop. The choice of first parameters like heart frequency or cardiac arrhythmia is included in this block. The options are for example atrial fibrillation, atrial and ventricular tachycardia or AV block.

5.2 Modeling of the Defibrillation Pulse

Section for the modeling the defibrillation pulse is based on sort of individual impulse consecutively. The user selected them gradually. The user have to select the type of pulse from the menu which includes monophasic, biphasic, biphasic Zoll and sinus exponential pulse. After that, user selects the main parameters of the pulses which are the duration of the phases, the maximum and minimum amplitude, the time interval between phases and potential patient impedance (Fig. 4).

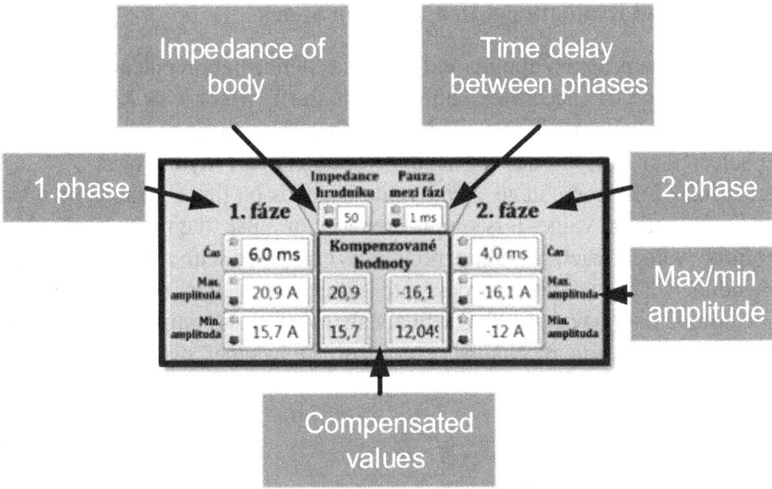

Fig. 4. Numeric controllers for define shape of defibrillation impulse.

Patient impedance has a significant influence on the shape of the impulse. Depending on the deflection from the optimal values impedance app. 50 Ω the pulse shape is compensated so that we have to always keep the same value of the energy which is calculated from (1).

$$E = \int I^2 \cdot R\,dt \tag{1}$$

Compensating constant is used for the calculation with which one the defibrillation pulse is multiplied. Once the reference of impedance is 50 Ω than the compensation constant is equal to 1. Constant is expressed by the formula (2) which is derived from (1) (Fig. 5).

$$comp.constant = \sqrt{\frac{R_{Chest}}{R_{ref}}} \tag{2}$$

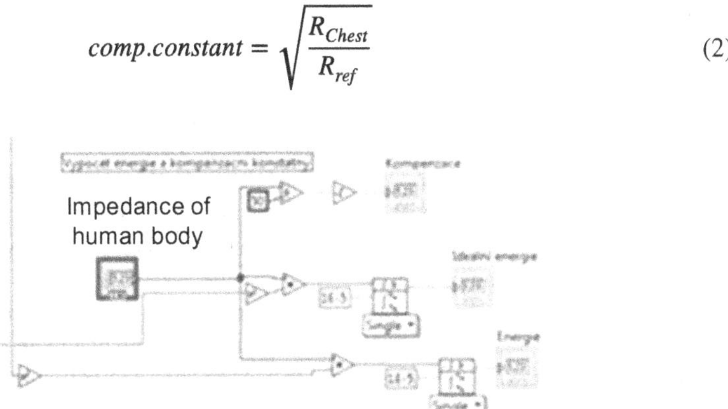

Fig. 5. Construction of calculating of compensating constant in LabVIEW

5.3 Applying and Evaluating a Defibrillation Shock

The next part of the program applies and evaluates defibrillation pulse. Firstly, it calculates the percentage of success of defibrillation shock which was set. After that the percentage value is compared to a random number. The result of comparison decides whether selected an arrhythmia returns to the optimal heart rhythm or not.

Calculation of the percentage success rate depends on three parameters which are compared with optimal value. It is supplied energy of pulse, the time of pulse duration and amplitude values. Each percent parameter is compared with reference value which had been defined. The evaluated success rate is then averaged by number of parameters on which the calculation was made. The final value of success compares with the random number generated during discharge. The result is whether discharge is successful (Fig. 6).

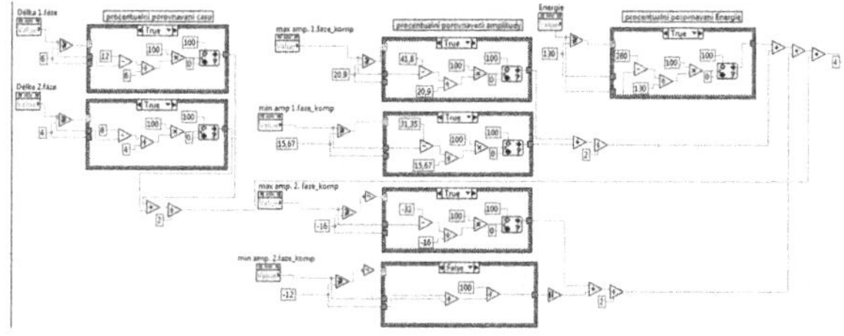

Fig. 6. Calculation of the final percentage value of the success of defibrillation discharge.

6 Conclusion

The software which is aimed at improving the education of students of health profession was implemented. Users can try to setup and apply the defibrillation shock in the program. Training starts already with setting up parameter where each adjustable parameter informs the user about the ideal values. In addition it contains their influence on the shape and function of the pulse. Tutorial part goes on in the main part of the program where the user gets familiar with cardiac rhythms, their differences and the importance of various parameters on discharge. Indications and explanations of low-tilt states defibrillation pulse are also included in the tutorial section.

Users can track changes of parameters of defibrillation pulse or heart rhythm in real time on a graphical waveform. If the user does not want to directly specify the value of the pulse can be switched to the mode in which the shapes of defibrillation pulse are defined only by the value of energy.

Although there are many known types of arrhythmia. The software offers the choice of only three types. Then the software is able to provide an adequate feedback on these arrhythmias. Adding other types of arrhythmia is subject to software's future development.

User interface offers limitless possibilities for students in which they can gain experience how their set parameters affecting the shape and effectiveness of the defibrillation shock.

Acknowledgment. The work and the contributions were supported by the project SP2015/179 'Biomedicínské inženýrské systémy XI', and this paper has been elaborated in the framework of the project "Support research and development in the Moravi-an-Silesian Region 2014 DT 1 - Research Teams" (RRC/07/2014). Financed from the budget of the Moravian-Silesian Region and also from project FRVS2015/204 'Modelling and Analysis of Cardiovascular System'.

References

1. Adgey, A.A.J., Spence, M.S., Walsh, S.J.: Theory and practice of defibrillation:(2) defibrillation for ventricular fibrillation. Heart **91**(1), 118–125 (2005)
2. Bennett, J.R., et al.: Low-tilt monophasic and biphasic waveforms compared with standard biphasic waveforms in the transvenous defibrillation of ventricular fibrillation. Pacing Clin. Electrophysiol. **37**(3), 279–289 (2014)
3. Guan, D., Powell, C., Malkin, R.: Defibrillation impedance: including an inductive element. In: Computers in Cardiology 2000, pp. 549–552. IEEE (2000)
4. Huang, J., et al.: Ascending-ramp biphasic waveform has a lower defibrillation threshold and releases less troponin I than a truncated exponential biphasic waveform. Circulation **126**(11), 1328–1333 (2012)
5. Kodoth, V., et al.: Waveform optimization for internal cardioversion of atrial fibrillation. J. Electrocardiol. **44**(6), 689–693 (2011)
6. Lung, D., Bašić, I.: Digital simulating a wave shape of current impulse appearing on the heart during defibrillation. In: Engineering in Medicine and Biology Society, Proceedings of the Annual International Conference of the IEEE, vol. 13, pp. 753–755. IEEE (1991)
7. Song, H., et al.: Decreasing the defibrillation energy by optimizing the pulse duration of electrical shock. In: The 2nd International Conference on Bioinformatics and Biomedical Engineering, ICBBE 2008, pp. 1606–1608. IEEE (2008)

Analysis of Image, Video and Motion Data in Life Sciences

Building the Facial Expressions Recognition System Based on RGB-D Images in High Performance

Trung Truong[1](✉) and Ngoc Ly[2]

[1] Faculty of Information Technology, University of Science, VNU-HCMC, Ho Chi Minh City, Vietnam
sirquangtrung@gmail.com

[2] Faculty of Information Technology, Computer Vision and Robotics, University of Science, VNU-HCMC, Ho Chi Minh City, Vietnam
lqngoc@fit.hcmus.edu.vn

Abstract. In this paper, we propose a novel idea for automatic facial expression analysis with the aim of resolving the existing challenges in 2D images. The subtle combination of the geometry-based method with the appearance-based features in depth and color images contributes to increasing in distinguishable features among various facial expressions. Particular functions are utilised to calculate the correlation between expressions in order to determine the exact facial expression. Our approach consists of a sequence of steps including estimating the normal vector of facial surface, then extracting the geometric features such as the orientation of normal vector in the point cloud. The useful color information is known as LBP. According to the result of the experiment, we demonstrate that the effective fusion scheme of texture and shape feature on color and depth images. In comparison with the non fusion scheme, our fusion scheme has resulted in the increase of recognition under low and high illuminated light, about 19.84 % and 1.59 %, respectively.

Keywords: Facial expression recognition · Local binary pattern · Normal vector · Covariance matrix · RGBD datasets

1 Introduction

Facial expression analysis plays an important role in daily communication, which contributes to the success in conveying semantic information of the speaker's face when chatting. Therefore, the development of automatic facial expression recognition will be extremely useful. For instance, it will facilitate the human-machine interaction, security and health care, etc. However, there are still many challenging problems in recognizing facial expressions are not yet completely solved such as illumination, pose, occlusion. Therefore, in this paper, we apply new algorithm to solve part of the problems.

Two popular approaches for facial expression in static image can be classiffied as geometric feature-based method and appearance feature-based method.

N.T. Nguyen et al. (Eds.): ACIIDS 2016, Part II, LNAI 9622, pp. 377–387, 2016.
DOI: 10.1007/978-3-662-49390-8_37

Specifically, geometry-based method is the shape feature, which involves calculating the distance, the angle of the particular vector as normal vector on the surface of the face. Recently, geometric feature-based method has been utilized as ASM proposed by Shbib et al. [17]. Besides, the appearance-based method represents the texture feature of another change of facial expression. They are not lighting and direction invariant features, but increase discriminative characteristics in analyzing the expression in the case of good lighting conditions. Appearance-based method includes LBP like S. Guo et al. [6]. Features on dynamic images or video have tended to exact feature which typically rely on motion feature or moving space-time information such as Liu et al. [10].

Limitations of the above methods only evaluate facial expression feature on particular lighting condition, except when various challenges in case of lighting changes or pose will be no longer in the effective area. The data are collected according to traditional methods caused in the loss of the 3D surface information. Therefore it is difficult for many computer vision tasks such as object recognition, detection, identification, tracking, scene reconstruction, and motion analysis. The techniques used for exploiting depth map are geometric feature-based method and appearance feature-based method. Geometry-based method has also been studied by Malawski et al. [11]. Nonetheless, it has faced with the noise and low resolution from dataset induced to fail approach proposed. In terms of a specific application of MS-LNP on 3D models as Hengliang Tang et al. [8], they exploited multiple global histograms of local normal patterns from multiple normal components and multiple binary encoding scales for the classification. However, the 3D model is relatively less noisy, more smooth compared with the point cloud set derived from depth map.

In this work, we introduce a novel algorithm for automatic human facial expression recognition. There are limitations in our approach since it has not been done yet to conduct the experiments in the conditions of occlusion and pose. The main contributions of this paper are summarized as follows:

- We empirically evaluate facial representation based on statistical features such as LBP and normal information derived 3D facial surface. We focus on exploring the characteristic of normal vector to address problems that geometric feature does not affect to the challenges such as color channels. This will help to distinguish the changing expression shape.
- We address many challenges: diverse age, gender, skins, lighting changes even dramatically illumination. Our experimental result proves the truth that in good illumination the texture feature performs the recognition of human facial expression best. However, we strongly show under different or bad lighting conditions that we fuse both features of color and depth image, the features derived from the depth image will contribute to the ineffectively recognized color features to increase the facial expression recognition result.
- Utilizing covariance matrix for feature representation will bring promising results to contribute to increasing the discrimination and emerging the different characteristic in numeric data that is strong feature for analyzing the facial expression.

The remainder of the paper is organized as follows. Section 2 reviews related works. Our approach will be presented in details in Sect. 3. Section 4 will be devoted to experimental results that show the high performance of our proposed method. In Sect. 5, we draw conclusions of our work and indicate our future works.

2 Related Works

Facial expression recognition has been an active research topic during the last few decades. However, there are still scientific challenges and potential applications. In pursuit of using color images, several feature extraction methods used in the face image are given such as Gabor filter Y. Pang et al. [14], LBP [6]. Among these features, [14] proposed Gabor filter for feature extraction in face recognition. It produces a large number of features from extracting multi-scale and multi-orientation coefficients. However, the computational complexity is expensive due to face images with multi-banks of Gabor filter. Some researchers have improved LBP features in order to reduce the dimension computational time and to achieve higher accuracy than Gabor feature such as [6].

However, robust recognition with conventional 2D cameras is still not possible in real conditions, in the presence of unexpected illumination, occlusion and pose variations. To deal with these problems, some researchers proposed a histogram of mesh gradient (HoG) and histogram of shape index (HoS) [9], the Scale-invariant feature transform (SIFT) [1], Local normal pattern (LNP) [8] following the 3D face models with the support of expensive specialized sensor. These challenges are completely solved because the approach to 3D model is not affected by these problems in comparison with those of the 2D face image. In addition, [9] suggested an idea to calculate the weighted statistical distributions of surface differential quantities, including HoG and HoS based on curvature estimation method. However, the fusion schemes are quite simple and a set of landmarks is selected by manual landmarks. Therefore, it is quite expensive and manually selected landmarks in the database that leads to higher accuracy of the experiment. [8] proposed a 3D face model-only method that represents multiple global histograms of local normal patterns from multiple normal components and binary encoding scales. These features are on.3D facial geometry, thus achieving high accuracy. 3D scanner remains a limited application of expression recognition in real life because the sensor is not cheap and has slow acquisition speed.

On the contrary, some new 3D sensors such as the Kinect sensor have low cost but gain high speed. Some researchers proposed a new method using low quality 3D data (RGB-D) such as the approach of surface curvature measurements [7,16]. Billy Y.L. Li et al. [7] proposed shape and texture features for face recognition and in the experiment, fusing depth and texture images to improves the performance more considerably than both depth image-only method and color image-only method. Also, [16] proposed a mean curvature based on low quality of 3D face data. Moreover, the accuracy of fusion of 3D and 2D method is higher than that of 3D-only method and 2D only-method. Introduced as a proposal of Duc Fehr et al. [4], several features using covariance descriptors provide a low dimensional representation of the data for object recognition. On the other hand, representation of local feature based on covariance matrix, which has succeeded in the case of not only discrimination ability, but also facial expression recognition, is offered from [6,14].

3 Our Approach

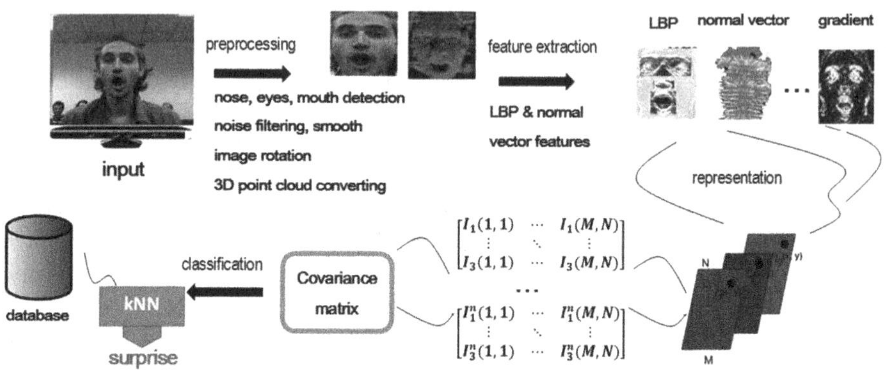

Fig. 1. Conceptual diagram of the proposed system

3.1 Pre-processing

We apply an adequate filter in order to normalize intensity of color image and approximate surface of point cloud, which affects the accuracy of the system due to both noise, holes in depth images and highly contrasted intensity of pixels in color images. We make algorithms for point cloud preprocessing with the aim of removing outliers known as an abnormal point near to hair, ear..., filling holes as well as robust smooth surfaces. To extract many vision information of images, holes existing on depth images are firstly filled by applying morphological closing. Besides, the histogram equalization can also be used to normalize the color images. After that, the face images are aligned at the same position. Then, some sub-regions of eyes, nose, and mouth have cropped the color face image to provide the most crucial facial expression feature. Finally, the depth images are converted into 3D point clouds (Fig. 1).

3.2 Feature Description

Local Binary Pattern on Color Image. The LBP operator is designed for texture descriptor and also, has been proved to be highly discriminative. It was introduced by Ojala et al. [13]. Each pixel in LBP image could be constructed in the way to threshold its neighborhood in the original image and then having the result as a binary number. By doing this, the pixel of LBP image is presented a decimal number, which is converting its the binary number. Formally, the LBP operator takes the form:

$$LBP_{8,1}(x_c, y_c) = \sum_{i=0}^{7} s(I_i - I_c) * 2^i \tag{1}$$

Where I_c, I_i are the grey level values at c, i. In this case, i is the label of parts around the center pixel location (x_c, y_c) and $s(x) = \begin{cases} 1 & \text{if} \quad x \geq 0 \\ 0 & \text{if} \quad x < 0. \end{cases}$

By doing this, the LBP operator become the technique of simplicity. To enhance the texture descriptor, we present some extensions of LBP operator including using neighborhoods of different sizes extended LBP and multi scale LBPs.

Using Neighborhoods of Different Sizes Extended LBP. [6] proposed $LBCM_{8,2}$, $LBCM_{16,2}$ to use neighborhoods of different sizes for each pixel in the whole facial image aim to recognize facial expression. Additionally, in their experiment, the LBP method has acquired the recognition precision rate higher than other methods. In this paper, we present the extended LBP operator to use 8 neighborhoods of size as 2, to capture dominant features from visually salient facial region including eyes, nose, and mouth. Notation $LBP_{8,2}$ is symbolic for our approach.

Multi-scale LBPs. Multi-scale LBPs considers the quantity of the neighborhood and the scale coefficient, which affect more the captured detail at various scales. Furthermore, it has expected to outperform other standard LBP operator because it encodes the micro structures of the facial expression but also provides a more extensive description than the standard LBP operator. By doing this, it enhances the discriminating ability for strong suitability in facial expression recognition. The approach can increase the computational cost of each LBP image. Otherwise, we propose a different approach to multi scale analysis using LBPs, which is defined as $LBP_1^{ms}(8, R)$ with R = 1..8. And $LBP_2^{ms}(8, R)$, LBP (8, R) with R=1, 2 and LBP(16,R) with R=2, 3, 4.

Normal Estimation on Point Clouds. In fact, the normal description which brings good results in processing 3D face model [8] is also successful in exploring in terms of 3D point cloud with many noises e.g. [15]. As a result, it has been expected to not only improve the discriminating power but capture more informative geometric information as well. Furthermore, all points in point cloud have already exacted the normal vector such variations in 3D point cloud [15]. To set up surface normal in a point cloud has the pseudo code known as Algorithm 1.

As a result, it has been expected to not only improve the discriminating power and robustness in facial expression but also capture more informative geometric information as well. Furthermore, all points in point cloud have already exacted the normal vector such variations in 3D point cloud.

Algorithm 1. Setting up surface normal in a point cloud

```
Input: 3D point cloud
Output: Normal vector description
procedure EstimateSurfaceNormal
    P // 3D point cloud
    k = 20 // number of nearest neighbors of point
    for p in P do
        Create the surface sampled around p: P_p = {p_1, p_2, .., p_k}
        Estimate the approximated normal vector n to P_p: n = f(P_p)
    end for
    Select the correct scale of all normal vector n
end procedure
```

3.3 Covariance Descriptor

The covariance descriptor was proposed by Duc Fehr et al. [4] for object detection and classification. Furthermore, Yanwei Pang et al. [14] also proposed face recognition using the covariance descriptor. We propose calculating covariance matrix of the given point cloud include RGB-XYZ channels, which have been constructing the data representation and denotes $F(x,y) = \phi(I, D, x, y) = z_i$.

For texture classification, we define the mapping function F(x,y)=(I,x,y) in color image as:

$$F_3 = \left[x \; y \; I \; I_x \; I_y \; \frac{I_x}{I_y} \; LBP(8,2) \right]. \tag{2}$$

Otherwise, for shape classification, we define the mapping function $F(x,y) = \phi(D, x, y)$ in point cloud as:

$$F_{12} = [x \; y \; n_x \; n_y \; n_z] \tag{3}$$

Finally, we propose to combine color and depth image the mapping function $F(x,y) = \phi(I, D, x, y)$ as

$$F_{Fusion} = \left[x \; y \; I \; I_x \; I_y \; \frac{I_x}{I_y} \; LBP(8,2) \; n_x \; n_y \; n_z \right]. \tag{4}$$

Förstner and Moonen [5] claimed that covariance matrices do not adhere to the Euclidean geometry, but span a Riemannian space with an associated Riemannian distance metric. Therefore, we present the distance function for calculating the difference between classes. One metric that can be used, the geodesic distance, has been introduced by [5,14] and is defined as follows:

$$\rho(C^G, C^P) = \sum_{i=1}^{12} \rho(C_i^G, C_i^P) - max[\rho(C_i^G, C_i^P)], \tag{5}$$

where $\rho(C_1, C_2) = \sqrt{\sum_{i=1}^{d} ln^2 \lambda_i(C_1, C_2)}$; C_i^G, C_i^P are sub-regions, located in the same regions such as eyes, nose and mouth regions. $\{\lambda_i(C_1, C_2) | i = 1, 2, .., d\}$ are generalized eigenvalues of C_1 and C_2 in $\lambda_1 * C_1 * u_1 = C_2 * u_2$.

4 Experiment

4.1 Database

In our experiments, we have used the face Warehouse database which was introduced by Cao et al. in [2]. Recall that Face Warehouse database, the facial expression database, is a sort of RGBD data based on Kinect camera. In addition, on the color face database, we have created 3 subsets, including the original subset, the distorting for dark and light of the original subset.

4.2 Experimental Results

For our experimental system, 4 main steps were used to recognize the facial expression, including preprocessing, feature extraction, feature representation and classification. In addition, to perform the preprocessing for both color and depth images, the first step is that we approach the STASM library automatically proposed by Tim Cootes [3] to display the landmarks of a face in a color image, i.e. eyes, nose, mouth, thus cropped 3 regions from the face image providing crucial facial expression information for expression recognition. Above all, it is linked with reducing the number of features. The extracted face image is accounted for the large size of features and then it is represented feature to adopt the formula of $d^2 * n^2 * 3$ where, d is a quantity of kinds of feature combined with our method, n denotes the number of sub-regions cropped from eyes, nose, mouth regions. We have done two trends of conducting an experiment included in early fusion and late fusion of technique of 2D and 3D processing. But the result shows that the early fusion of 2D and 3D images achieve much higher than what observed on the later fusion. In our proposed method, the number of features on face color image, depth image and fusion of color and depth image in the datasets are 588, 300, 1400, respectively. Finally, in order to classify facial exprcssion, it is based on the k-nearest neighbor. We perform a number of experiments to choose the parameter of k that is appropriate for accurate performance. To put it another way, the range value of k nearest neighbor is 1–20, thus the maximum accuracy is achieved whenever k = 9 to 14 in comparison to the others. All experiments are approached with 3-fold, except for the Table 3 which approached with leave-one-out cross-validation.

Experiment 1: Normalized Illumination. We show several types of features and also compare the accuracy of these types of feature. This allows us to find the feature descriptor which improves the facial expression recognition. All of the proposed 2D-only methods, F_4 feature achieves 84.13 % average recognition rate to be highest accuracy as shown in Table 1. In addition, F_4 feature consists of 588 dimensions. Classification rates achieved by F_1,F_2,F_3 are 81.75 %, 80.16 %, 81.75 %, respectively. Although, the number of feature of F_2 is 2352 features more than the number feature of F_3 account for 1252 but its classification rate is less rate than F_3. In the same way, the number of feature of F_1 (192 features) is less in comparison with F_3 (1452 features), but the accuracy rates are in contrast to the quantity of number of features. Comparing on classification rate on F_4, it shows that the size of feature vector is middle size range but its features is effectively improvement. Maximum curvature values: P_{max}, Minimum curvature values: P_{min}, Mean curvature:H, Gaussian curvature values:K, Shape index value: S. $F_1 = [x\ y\ I\ LBP(8,2)]$, $F_2 = \left[x\ y\ I\ I_x\ I_y\ \frac{I_x}{I_y}\ LBP_1^{ms}\right]$, $F_3 = \left[x\ y\ I\ I_x\ I_y\ \frac{I_x}{I_y}\ LBP_2^{ms}\right]$, $F_4 = \left[x\ y\ I\ I_x\ I_y\ \frac{I_x}{I_y}\ LBP(8,2)\right]$, $F_6 = [x\ y\ S]$, $F_5 = [H\ K\ S\ P_{min}\ P_{max}]$, $F_7 = [x\ y\ n_x\ n_y\ n_z\ P_{min}\ P_{max}]$, $F_8 = [x\ y\ n_x\ n_y\ n_z\ K]$, $F_9 = [n_x\ n_y\ n_z]$, $F_{10} = [x\ y\ n_x\ n_y\ n_z\ H]$, $F_{11} = [x\ y\ n_x\ n_y\ n_z\ S]$, $F_{12} = [x\ y\ n_x\ n_y\ n_z]$.

Also, Table 3 shows the several results of many 3D point cloud method, including $F_{5..11}$ on the corresponding range of 47.62 % to 69.05 %. It combines shape index, maximum curvature, minimum curvature, mean curvature, Gaussian curvature values with normal vector and location. Besides, the important result is the fact that F_{12} achieves 72.22 % on the average and has 300 features, which has a higher accuracy than the curvature features in Table 2. To put another way, the number of features of F_{12} is less size than most of the feature of $F_{5..11}$. Here, both F_{12} feature and other curvature features are known as the 3D point cloud-only method.

Table 1. Recognition rate (%) for original of FaceWarehouse dataset with color images-only under good illumination conditions). The 2D features utilize the LBP feature.

Features	F_1	F_2	F_3	F_4
Accuracy	80.95	80.16	81.75	84.13

Table 2. Recognition rate (%) for original of FaceWarehouse dataset with depth images-only under good illumination conditions

Features	F_5	F_6	F_7	F_8	F_9	F_{10}	F_{11}	F_{12}
Accuracy	47.62	62.70	66.67	68.25	62.70	62.70	69.05	72.22

Table 3. Recognition rate (%) for recognition facial expression in 3-Fold and leave-one-out cross-validation.

Datasets	Accuracy	
	3-fold	leave-one-out cross-validation
RGB dataset	84.13	87.3
Depth dataset	72.22	75.40
RGBD dataset	78.57	80.16

Table 4. Recognition rate (%) for recognition facial expression in 3-Fold and leave-one-out cross-validation

Datasets	Accuracy	
	Low illumination	High illumination
RGB dataset	53.97	74.98
Depth dataset	72.22	72.22
RGBD dataset	73.81	76.57

Experiment 2: Low, High Illumination. In these experiments, the color images are only distorted the light channel in order to closely resemble the real conditions, whereas the depth images are not distorted. As before, we do this experiment by fusing the color image and depth image in order to show the improvement of recognition performance in the real conditions when using our approach.

The most important result is the fact that, as shown in Table 4, the low and high illumination increase 19.84 % and 1.59 % average recognition rates, respectively, in the case of the fusion of color image and depth image method in comparison with the color image-only method. Indeed, this shows the fact that when using our recognition proposal, the fusion of color images and depth images gives a better result than the infusions under unexpected illumination, and namely the result approached the color image only method. Hence, it brings the performance of results what it is serious challenges in previous time.

Table 5. Recognition rate (%) for performance comparison of different methods with depth images- only method

Approach	Method	No. Express	Accuracy(%)
Mao [12]	AUs, FPPs, FPs	4	<72
Proposed method	**Normal descriptor, kNN**	**6**	**72.22**

Comparative Study. Table 5 compares two results approached different methods based on depth images of Facewarehouse database. The criteria for comparison of performing consist of the number of expressions, the recognition rate. Although the amount of emotion is different, experiments show that the result of our proposed approach is higher than its Mao [12]. Thus, it figures out our potential actually approach compete in the facial expression recognition.

5 Conclusion and Future Works

In summary, we have presented a novel descriptor and framework for automatic facial expression recognition in real conditions based on the fusion of both color image and depth image. In addition, the data representation is relied on the covariance matrix which depends on the number of features without the size of the region. We constructed the covariance matrices with LBP in color image and normal vector in facial point cloud. Therefore, our final descriptor brings both texture feature and shape feature contributing to the increase in the discrimination of emotional facial expressions. In specific, we indicated that the fusion of depth with color images method improves the accuracy in comparison with the depth image-only and color image only performances under low and high illumination condition. It is true that under bad illumination condition, the accuracy of the depth image-only method is not affected, whereas the accuracy of color image-only method is decreasing. The fusion of depth in cooperation with color images which has the less feature recognition in real conditions (low and high illumination), has the mutual support to increase the precision. This implies that this method is better than the color image-only method or depth image-only method.

We leave as future research the study of experiments which are based on an extension of our proposed framework in the challenge of head pose. Furthermore, the proposed framework used in this paper will be investigated for a sequence of images.

Acknowledgments. This research is funded by Vietnam National University Ho Chi Minh City (VNU-HCM) under grant number B2014-18-02.

References

1. Berretti, S., Amor, B.B., Daoudi, M., Del Bimbo, A.: 3D facial expression recognition using sift descriptors of automatically detected keypoints. Int. J. Vis. Comput. **27**(11), 1021–1036 (2011)
2. Cao, C., Weng, Y., Zhou, S., Tong, Y., Zhou, K.: Facewarehouse: a 3D facial expression database for visual computing. IEEE Trans. J. Vis. Comput. Graph. **20**(3), 413–425 (2014)
3. Tim, C.: Model-based methods in analysis of biomedical images. In: Baldock, R., Graham, J. (eds.) Image Processing and Analysis, pp. 223-247. Oxford University Press (2000)
4. Fehr, D.A.: Covariance based point cloud descriptors for object detection and classification. University of Minnosita, August 2013
5. Förstner, W., Moonen, B.: A metric for covariance matrices. In: Grafarend, E.W., Krumm, F.W., Schwarze, V.S. (eds.) Journal of Geodesy-The Challenge of the 3rd Millennium, pp. 299–309. Springer, Heidelberg (2003)
6. Guo, S., Ruan, Q.: Facial expression recognition using local binary covariance matrices. In: Proceedings of 4th IET International Conference on Wireless, Mobile and Multimedia Networks (ICWMMN 2011) (2011)
7. Li, B.Y., Mian, A.S., Liu, W., Krishna, A.: Using kinect for face recognition under varying poses, expressions, illumination and disguise. In: 2013 IEEE Workshop on Proceedings of Applications of Computer Vision (WACV), pp. 186–192. IEEE (2013)
8. Li, H., Chen, L., Huang, D., Wang, Y., Morvan, J.: 3D facial expression recognition via multiple kernel learning of multi-scale local normal patterns. In: 2012 21st International Conference on Proceedings of Pattern Recognition (ICPR), pp. 2577–2580. IEEE (2012)

9. Li, H., Morvan, J.-M., Chen, L.: 3D facial expression recognition based on histograms of surface differential quantities. In: Blanc-Talon, J., Kleihorst, R., Philips, W., Popescu, D., Scheunders, P. (eds.) ACIVS 2011. LNCS, vol. 6915, pp. 483–494. Springer, Heidelberg (2011)
10. Liu, M., Shan, S., Wang, R., Chen, X.: Learning expressionlets on spatio-temporal manifold for dynamic facial expression recognition. In: IEEE Conference on Proceedings of Computer Vision and Pattern Recognition, pp. 1749–1756. IEEE (2014)
11. Malawski, F., Kwolek, B., Sako, S.: Using kinect for facial expression recognition under varying poses and illumination. In: Ślęzak, D., Schaefer, G., Vuong, S.T., Kim, Y.-S. (eds.) AMT 2014. LNCS, vol. 8610, pp. 395–406. Springer, Heidelberg (2014)
12. Mao, Q., Pan, X., Zhan, Y., Shen, X.: Using kinect for real-time emotion recognition via facial expressions. J. Front. Inf. Tech. Electr. Eng. **16**, 272–282 (2015)
13. Ojala, T., Pietikainen, M., Maenpaa, T.: Multiresolution gray-scale and rotation invariant texture classification with local binary patterns. IEEE Trans. J. Pattern Anal. Mach. Intell. **24**(7), 971–987 (2002)
14. Pang, Y., Yuan, Y., Li, X.: Gabor-based region covariance matrices for face recognition. J. IEEE Trans. Circuits Syst. Video Technol. **18**(7), 989–993 (2008)
15. Rusu, R.B., Cousins, S.: 3D is here: point cloud library (PCL). In: IEEE International Conference on Proceedings of Robotics and Automation (ICRA), Shanghai, China, 9–13 May 2011, pp. 1–4. IEEE (2011)
16. Savran, A., Gur, R., Verma, R.: Automatic detection of emotion valence on faces using consumer depth cameras. In: 2013 IEEE International Conference on Proceedings of Computer Vision Workshops (ICCVW), pp. 75–82. IEEE (2013)
17. Shbib, R., Zhou, S.: Facial expression analysis using active shape model. J. Signal Process. Image Process. Pattern Recogn. **8**(1), 9–22 (2015)

Zero-Velocity Detectors for Orientation Estimation Problem

Agnieszka Szczesna[1(✉)], Przemysław Pruszowski[1], Andrzej Polański[1], Damian Peszor[1], and Konrad Wojciechowski[2]

[1] Institute of Informatics, The Silesian University of Technology, Gliwice, Poland
{Agnieszka.Szczesna,Przemyslaw.Pruszowski,Andrzej.Polanski, Damian.Peszor}@polsl.pl

[2] The Polish-Japanese Academy of Information Technology, Warsaw, Poland
Konrad.Wojciechowski@polsl.pl

Abstract. The paper concerns the application of the zero-velocity updates in orientation estimation filter for a inertial motion capture system. The evaluation of three commonly used detectors together with quaternion complementary filter to orientation estimation based on indications of gyroscope, magnetometer and accelerometer sensors is presented.

Keywords: Quaternion complementary filter · Orientation estimation · Zero-velocity update · Inertial motion capture

1 Introduction

Human motion analysis is receiving increasing attention from researchers, which can be motivated by a wide spectrum of applications. The research results described in this article can be used in the *Inertial Motion Capture System* for out-door acquisition of human motion [1]. Such full-body human measurement system is based on inertial sensors (IMU) attached to a costume, skeleton model and fusion algorithms for sensor orientation estimation. The mapping of estimated orientations over time to specific segments on body model (skeleton) which is composed of rigid bodies, allows to capture of subject motion. The orientations of the sensors are estimated by fusing a measurements of gyroscope (ω), an accelerometer (a), a magnetometer (m) and reference values such as Earth's gravity vector (g) and magnetic field vector (mg). With this knowledge and the structure of skeleton, the overall pose can be tracked [2].

The article concerns applications of the zero-velocity detectors in basic orientation estimation methods based on indications of IMU sensors. In this case the general and simple complementary filter was chosen as primary estimation algorithm. The evaluation was done on the basis of the average error of estimated orientations. Reference are the data from the optical mocap system.

In this paper we are not proposing a new algorithm for the estimation of the IMU sensor orientation. Instead we propose an evaluation of possibility to

N.T. Nguyen et al. (Eds.): ACIIDS 2016, Part II, LNAI 9622, pp. 388–396, 2016.
DOI: 10.1007/978-3-662-49390-8_38

application of zero-velocity update in orientation estimation filter, with a simple experiment. The following zero-velocity detectors were implemented and tested: the acceleration-moving variance detector, the acceleration-magnitude detector, and the angular rate energy detector. These techniques are commonly used in inertial pedestrian, personal navigation and gait analysis systems [3–5]. During normal walking cycles, a foot touches the ground almost periodically and stays stationary for a short time, which is called the zero-velocity interval. We tested these methods in general configuration not specially adapted to particular movement (such as human walking or running) or placement of sensor (for example on a shoe). Our purpose is apply detection of zero-velocity intervals in more general movement and used this information to improvement the estimated orientation.

2 Complementary Quaternion Filter

We are focusing only on algorithms based on a quaternion representation of body orientation. Orientation is coded by quaternion to improve computational efficiency and avoid singularities. The filters are based on the well-known correlation between the angular velocity ω and the quaternion derivative:

$$\dot{\boldsymbol{q}} = \frac{1}{2}\boldsymbol{q} \otimes \boldsymbol{\omega}, \tag{1}$$

where q is the orientation quaternion and $\otimes$ denotes quaternion multiplication.

In Eq. (1) the orientation is determined by integrating the output signal from the gyroscope (ω), so the accuracy of determining the angle depends on the sensor stability of zero. The integration accumulates the noise over time and turns noise into the drift, which yields unacceptable results and large error values.

To instantaneous orientation determination with respect to a reference frame, it is common to use a set of vectors measured in a body-fixed coordinate frame, using sensors such as magnetometers and accelerometers. This is possible for a stationary sensor in an environment free of magnetic anomalies. The task of resolving the attitude by comparing these vectors is known as Wahba's problem [6–8]. In more dynamic applications, high frequency angular rate information can be combined in a complementary manner with accelerometer and magnetometer data through the use of a sensor fusion algorithms such as complementary [9] or Kalman filters [10–14].

The complementary orientation estimator is modification of the motion Eq. (1) by using the non-linear feedback based on the instantaneous estimate of relative orientation:

$$q_{inst} = q_{inst}(a_k, m_k) \tag{2}$$

where a_k and m_k are output k sample from accelerometer and magnetometer.

The quaternion form dynamic equation for rotation estimate, with non-linear feedback, has the following form:

$$\frac{d}{dt}q(t) = \frac{1}{2}q(t) \otimes (q_\omega + k_p\tilde{\omega}), \tag{3}$$

where q_ω denotes the pure quaternion from gyroscope measurement ω_k, k_p is the feedback gain and $\tilde{\omega}$ is the non-linear term corresponding to the error between filter estimate q and the instantaneous estimate of the orientation quaternion $q_{inst}(a_k, m_k)$. The $\otimes$ is quaternion multiplication.

Computational implementation of the non-linear, complementary filter involves discrete approximation of (3):

$$q_{k+1} = [I + \frac{\Delta t}{2} M_R(q_\omega + k_p\tilde{\omega})]q_k, \tag{4}$$

where I is the 4×4 identity matrix, $M_R(q_\omega + k_p\tilde{\omega})$ stands for matrix representation of quaternion right multiplication by pure quaternion $q_\omega + k_p\tilde{\omega}$ and Δt is the sampling interval. Iterations of (4) are additionally augmented by normalising the unit quaternion q_k in each iteration. The initial condition for (4) is $q_0 = q_{inst(a_0, m_0)}$.

3 Zero-Velocity Updates

Due the nature of problem, zero velocity updates allows bounding the error growth in application uses the IMU sensors measurements. The updates are made when the IMU sensor is detected as not moving, and can be classified in to two groups: hard and soft [4]. In soft update, after detecting zero velocity the system resets the error state and then fed back to correct estimation filter. When orientations or inertial data are set to initial values, the updates are known as hard.

Whether hard or soft zero-velocity updates are used, the identification of the time intervals when the inertial sensors are stationary is required. Accordingly, a range of detectors has been proposed for detecting the zero-velocity time interval.

The zero-velocity detection checks if the IMU is moving or not, during a time interval consisting of N observations between the time samples n and $n+N-1$. This can be done by testing the detector function value:

$$f_D({y_k}_{k=n}^{n+N-1}), \tag{5}$$

where $y_k = \begin{bmatrix} \boldsymbol{\omega}_k \\ \boldsymbol{a}_k \\ \boldsymbol{m}_k \end{bmatrix}$ is measurement vector. If $f_D < \gamma$, where γ is detection threshold, the sensor is stationary, otherwise the sensor is moving.

The thee detectors proposed in the literature [4, 15], were tested with complementary quaternion filter:

- Acceleration-Moving Variance Detector (AMV)

$$f_D({y_k}_{k=n}^{n+N-1}) = \frac{1}{\sigma_a^2 N} \sum_{k=n}^{n+N-1} |a_k - \bar{a_n}|^2 \tag{6}$$

– Acceleration-Magnitude Detector (AM)

$$f_D({y_k}_{k=n}^{n+N-1}) = \frac{1}{\sigma_a^2 N} \sum_{k=n}^{n+N-1} (|a_k| - |g|)^2 \quad (7)$$

– Angular Rate Energy Detector (ARE)

$$f_D({y_k}_{k=n}^{n+N-1}) = \frac{1}{\sigma_\omega^2 N} \sum_{k=n}^{n+N-1} |\omega_k|^2 \quad (8)$$

Where σ_a and σ_ω are the variances of the accelerometer and gyroscope measurement noise. The $\bar{a_k}$ is a simple mean of N samples. The N is the window size which should be short and comparable to the duration of the shortest interval when the sensor is moving.

Two possibility applications of detecting zero-velocity internals for complementary filter are proposed. In first the feedback gain k_p (in (3)) value is changing which lead to gated version of complementary filter. When the sensor is stationary the higher weights k_p for q_{inst} based on accelerometer and magnetometer measurements are used. Otherwise the k_p is lower to increase the influence of gyroscope measurements on orientation estimation.

Second approach reduces the effect of sensor drift by hard zero-velocity updates. This technique has been successfully used in gait analysis and pedestrian navigation studies, where stationary stance periods are detected and used to reset the angular rate to zero. This resets the yaw angle estimation value. Then the greatest influence on the estimating orientation has measurements from accelerometer and magnetometer.

4 Experiments

In order to evaluate the zero-velocity detectors, a 3-segment pendulum has been build. As reference, data from the optical system of motion capture (Vicon system) is used. A similar experiment configuration was used in [16].

The pendulum consists three segments connected by movable joints. Each segment has an IMU sensor build at the *Silesian University of Technology, Department of Automatic Control and Robotics* [17]. On the pendulum markers for optical system were also attached, marked as IMU1, IMU2, IMU3 and R1, R2, W1, W2, W3, W4, W5, and W6 (Fig. 1). Twenty three capture recordings with eight different scenarios (each scenario repeated 3 times) were carried out using the Vicon Nexsus system with a frequency of 100 Hz. The IMU sensors also worked with such a frequency. A length of the recordings is from 9600 to 19840 samples. The recorded movement is characterized by high values of acceleration amplitudes of about 20 m/s^2.

In order to provide an informative comparison of orientation data streams (from inertial and optical system) the data must be normalized. Streams are recorded with different reference frames and measured according to separate

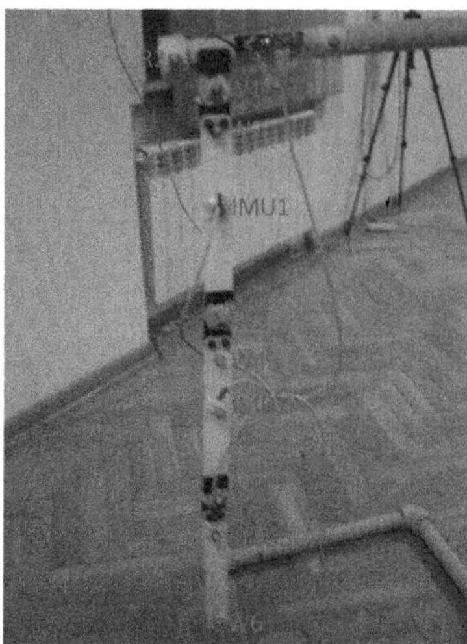

Fig. 1. The pendulum

timers with the same frequency. Normalization is divided into two steps: time synchronization and transforming orientations to the same reference frame.

The described techniques were evaluated on the basis of the average difference, over the experiment time horizon, between true (form optical system) and estimated orientation of the each segment [18]. We use the geodesic distance between two quaternions - filter estimate $\hat{q}$ and the true rotation q, on the hypersphere S^3:

$$DI = 2 * arccos(|\hat{q} * q|) \tag{9}$$

5 Results and Discussion

In the zero-velocity algorithms, some threshold values (both for detector function and window size) must be adjusted. These threshold values could differ significantly for different movement patterns. For very fast changing movements the threshold value should be larger to properly detect short zero-velocity intervals. On the other side, for slower movement we could detect wrong zero-velocity intervals. Figure 2 presents the value of AMV detector for first pendulum segment in dynamic scenario. Black line presents the detection of zero-velocity intervals (ZVI = 0). When ZVI = 1 the sensor is moving. Results are for windows size $N = 100$ samples and threshold value equals 0.05 % of AVM amplitude detector value ($\gamma = 1.5170e + 007$).

In Table 1 the results count of detected non zero-velocity intervals in dynamic test for each segments for different window size are presented. The true number of non zero-velocity intervals is 9. The best results are for ARE detector based on magnitude of angular rate. When the window size ($N < 100$) is too small the detection is performed not properly. This indicates that the gyroscope output

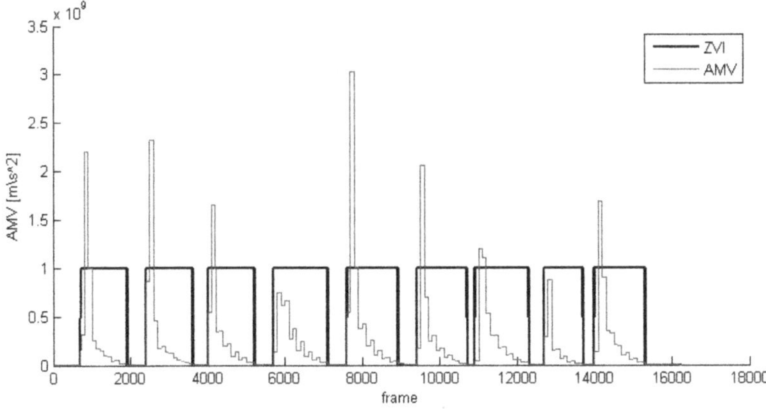

Fig. 2. Value of AMV detector and determined zero-velocity intervals (ZVI = 0)

hold the most reliable information for zero-velocity detection under the experimental conditions stated in this paper. Also for detectors AM and AMV we obtained good results. The variance of measurement noise are $\sigma_a = 0.0001$ and $\sigma_\omega = 0.005$.

Table 1. Count of detected non ZVI (intervals when sensor is moving)

Segment/ZVI detector	AMV	AMV	AMV	AM	AM	AM	ARE	ARE	ARE
N (window size)	10	50	100	10	50	100	10	50	100
S1	36	25	9	38	14	9	133	10	9
S2	47	18	10	97	12	9	117	9	9
S3	52	24	9	108	9	9	119	13	9

The improvement in orientation estimation by complementary quaternion filter with zero-velocity detector as a gate condition (NCF_ZVU_G) and additionally with hard update (NCF_ZVU_G_H) was evaluated in comparison to general complementary filter (NCF). In Fig. 4 the average errors are presented. The ARE detector with $N = 100$ was used. For gated version of NCF for detected ZVI $k_p = 2$, otherwise $k_p = 0.2$. Error is calculated by (9) and represent average angle between estimated orientation and orientation from reference system. The best results are obtained for NCF_ZVU_G_H filter (Euler angles are presented in Fig. 3). The hard ZVU causes attenuation of yaw angle, but a better fit for roll and pitch angles, so the overall error is lower.

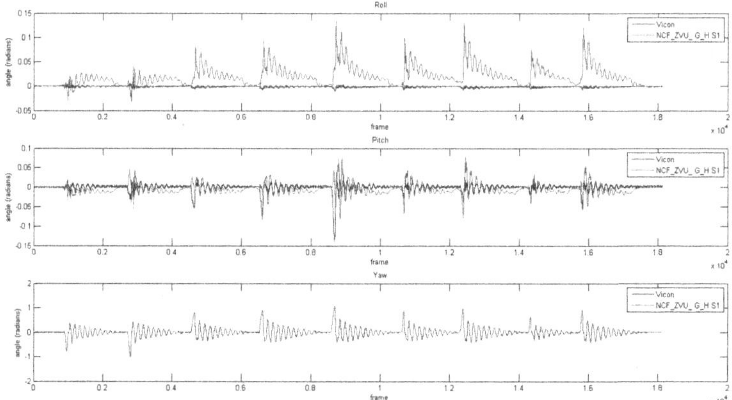

Fig. 3. Result Euler angles estimated by NCF_ZVU_G_H (green line) and reference data (blue line) for segment 1 of the pendulum (Color figure online)

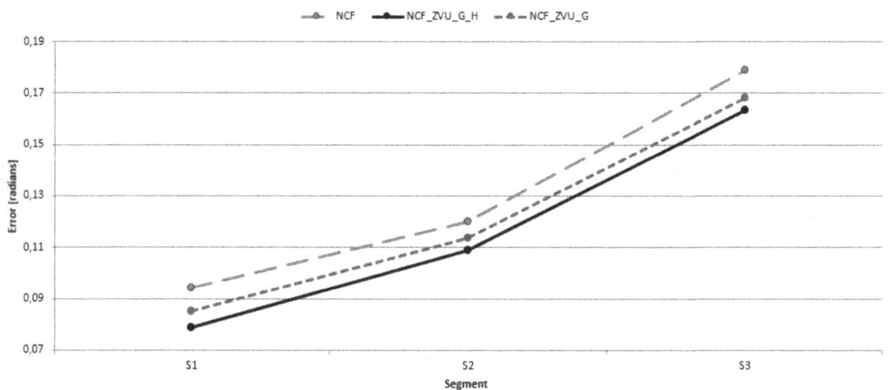

Fig. 4. Average error between orientation estimated by NCF, NCF_ZVU_G_H, NCF_ZVU_G filters and orientation from optical motion capture system

6 Summary

This article concerns the possibility of use the zero-velocity detectors for orientation estimation problem. To show influence to errors the very simple quaternion complementary filter was used as basic filter. Next the two additional elements were added: gate based on detected zero-velocity interval and hard zero-velocity update. Presented approach can be easily adapted to the specifics gaits and used for human motion analysis systems.

Acknowledgments. This work was supported by POIR/01.02.00-24-2 "System of autonomous landing of an UAV in unknown terrain conditions on the basis of visual data" and partly performed using the infrastructure supported by POIG.02.03.01-24-

099/13 grant: "GCONiI - Upper-Silesian Center for Scientific Computation". Data was captured in Human Motion Laboratory of Polish-Japanese Academy of Information Technology (http://hm.pjwstk.edu.pl/en/).

References

1. Kulbacki, M., Koteras, R., Szczesna, A., Daniec, K., Bieda, R., Słupik, J., Segen, J., Nawrat, A., Polański, A., Wojciechowski, K.: Scalable, wearable, unobtrusive sensor network for multimodal human monitoring with distributed control. In: Lacković, I., Vasic, D. (eds.) 6th MBEC 2014, pp. 914–917. Springer, Heidelberg (2015)
2. Sabatini, A.: Estimating three-dimensional orientation of human body parts by inertial/magnetic sensing. Sensors **11**(2), 1489–1525 (2011)
3. Park, S.K., Suh, Y.S.: A zero velocity detection algorithm using inertial sensors for pedestrian navigation systems. Sensors **10**, 9163–9178 (2010). Molecular Diversity Preservation International
4. Skog, I., Händel, P., Nilsson, J.-O., Rantakokko, J.: Zero-velocity detection - an algorithm evaluation. IEEE Trans. Biomed. Eng. **57**, 2657–2666 (2010). IEEE
5. Foxlin, E.: Pedestrian tracking with shoe-mounted inertial sensors. IEEE Comput. Graph. Appl. **25**, 38–46 (2005)
6. Wahba, G.: A least squares estimate of satellite attitude. SIAM Rev. **7**(3), 409–409 (1965)
7. Shuster, M.D.: The generalized Wahba problem. J. Astronaut. Sci. **54**(2), 245–259 (2006)
8. Shuster, M.D., Oh, S.D.: Three-axis attitude determination from vector observations. J. Guidance Control Dyn. **4**(1), 70–77 (1981)
9. Mahony, R., Hamel, T., Pflimlin, J.-M.: Nonlinear complementary filters on the special orthogonal group. IEEE Trans. Autom. Control **53**(5), 1203–1218 (2008)
10. Yun, X., Bachmann, E.: Design, implementation, and experimental results of a quaternion-based Kalman filter for human body motion tracking. IEEE Trans. Robot. **22**(6), 1216–1227 (2006)
11. Yun, X., Aparicio, C., Bachmann, E., McGhee, R.: Implementation and experimental results of a quaternion-based Kalman filter for human body motion tracking. In: Proceedings of Robotics and Automation, pp. 317–322 (2005)
12. Sabatini, A.: Quaternion-based extended Kalman filter for determining orientation by inertial and magnetic sensing. IEEE Trans. Biomed. Eng. **53**(7), 1346–1356 (2006)
13. Sabatini, A.: Kalman-filter-based orientation determination using inertial/magnetic sensors: observability analysis and performance evaluation. Sensors **11**(10), 9182–9206 (2011)
14. Słupik, J., Szczesna, A., Polański, A.: Novel lightweight quaternion filter for determining orientation based on indications of gyroscope, magnetometer and accelerometer. In: Chmielewski, L.J., Kozera, R., Shin, B.-S., Wojciechowski, K. (eds.) ICCVG 2014. LNCS, vol. 8671, pp. 586–593. Springer, Heidelberg (2014)
15. Skog, I., Nilsson, J.-O., Händel, P.: Evaluation of zero-velocity detectors for foot-mounted inertial navigation systems. In: 2010 International Conference on Indoor Positioning and Indoor Navigation (IPIN), pp. 1–6 (2010)

16. Szczesna, A., Pruszowski, P., Słupik, J., Peszor, D., Polański, A.: Evaluation of improvement in orientation estimation through the use of the linear acceleration estimation in the body model. In: Gruca, A., Brachman, A., Kozielski, S., Czachórski, T. (eds.) Man-Machine Interactions 4, pp. 377–387. Springer International Publishing, Heidelberg (2016)
17. Jedrasiak, K., Daniec, K., Nawrat, A.: The low cost micro inertial measurement unit. In: Proceedings of Industrial Electronics and Applications (ICIEA), pp. 403–408. IEEE (2013)
18. Gramkow, C.: On averaging rotations. J. Math. Imaging Vis. **15**(1–2), 7–16 (2001)

Analysis of Human Motion Data Using Recurrence Plots and Recurrence Quantification Measures

Henryk Josiński[2(✉)], Agnieszka Michalczuk[1], Romualda Mucha[3], Adam Świtoński[2], Agnieszka Szczęsna[2], and Konrad Wojciechowski[1]

[1] Polish-Japanese Academy of Information Technology, Koszykowa 86, 02-008 Warszawa, Poland
{amichalczuk,kwojciechowski}@pjwstk.edu.pl

[2] Institute of Informatics, Silesian University of Technology, Akademicka 16, 44-100 Gliwice, Poland
{Henryk.Josinski,Adam.Switonski,Agnieszka.Szczesna}@polsl.pl

[3] Medical University of Silesia, Batorego 15, 41-902 Bytom, Poland
romam28@wp.pl

Abstract. The authors present exemplary application of recurrence plots and recurrence quantification analysis for the purpose of exploration of experimental time series describing the movements of the hip joint during a few selected assisted rehabilitation exercises maintaining mobility of the hip in case of coxarthrosis. Time series were extracted from motion sequences which were recorded in the Human Motion Laboratory (HML) of the Polish-Japanese Academy of Information Technology in Bytom, Poland by means of the Vicon Motion Kinematics Acquisition and Analysis System. Additionally, some features of a recurrence plot were presented on the basis of the Rabinovich-Fabrikant system.

Keywords: Recurrence plot · Recurrence quantification analysis · Time series · Human motion analysis

1 Introduction

Recurrence plots (RP) introduced by Eckmann *et al.* [1] and recurrence quantification analysis (RQA) developed by Zbilut and Webber [2] have become important tools for analysis of dynamical systems. Recurrence plots have found application, among others, in evaluation of the postural instability in PD patients [3] (PD stands for Parkinson's disease) as well as in analysis of the heart rate variability [4–6], first of all with purpose of cardiac arrhythmia detection. The goal of this paper is to present their application to exploration of experimental time series describing the movements of the hip joint during a few selected assisted rehabilitation exercises maintaining mobility of the hip in case of coxarthrosis.

A dynamical system is any system that evolves in time [7]. The state of a dynamical system at a given instant of time can be represented by a point in the

N.T. Nguyen et al. (Eds.): ACIIDS 2016, Part II, LNAI 9622, pp. 397–406, 2016.
DOI: 10.1007/978-3-662-49390-8_39

phase space spanned by the state variables of the system. Successive points form a trajectory. The term "recurrence of states" refers to the situation in which after some time the trajectory returns close to a previously visited location. As a fundamental property of dissipative dynamical systems, recurrence of states is typical for nonlinear or chaotic systems. The system will return to close proximity of the former state even though in the meantime small perturbations cause exponential divergence of the states [8].

Characteristic invariants of a particular system can be determined by analyzing the time course of one of its variables. On the basis of Takens' embedding theorem the phase space can be reconstructed from a time series of observables using the method of delays [9]. Reconstruction consists in viewing a time series $x_k = x(k\tau_s)$, $k = 1, \ldots, Z$ in a Euclidean space $\mathbb{R}^m$, where m is the embedding dimension and τ_s is the sampling time [10]. Each m-dimensional embedding vector is formed as $\mathbf{x}_l = \left[x_l, x_{l+\tau}, x_{l+2\tau}, \ldots, x_{l+(m-1)\tau}\right]^T$, where τ is the delay time. Elements of the embedding vector $\mathbf{x}_l$ determine coordinates of the point X_l on the trajectory in the reconstructed phase space. Consequently, length of the trajectory built on the basis of Z-point time series is equal to $N = Z - (m-1)\tau$.

2 Recurrence Plots and Recurrence Quantification Analysis

Recurrence plots constitute a method of visualization of the recurrence of states in a phase space by means of binary symmetric $N \times N$ matrix in which element R_{ij} (a *recurrence point*) is defined as follows:

$$R_{ij} = \begin{cases} 1, & \|X_i - X_j\| \leq \epsilon \\ 0, & \|X_i - X_j\| > \epsilon \end{cases} \tag{1}$$

where $\|\cdot\|$ denotes a norm (e.g. the Euclidean norm) and the parameter ϵ, $\epsilon \geq 0$ is called the *recurrence threshold*. In other words, a value of 1 is assigned to the element R_{ij} whenever a point X_i on the trajectory is close enough to another point X_j [11]. Thus, a trajectory from m-dimensional phase space can be analyzed by means of two-dimensional representation of its recurrences.

As mentioned earlier, construction of a RP for experimental data requires some preparatory steps. First, two parameters of the phase space trajectory reconstruction – time delay τ and embedding dimension m – are estimated, e.g. by means of the following methods: the Average Mutual Information (τ) [7] and the False Nearest Neighbors (m) [12]. Consequently, the trajectory can be reconstructed giving the opportunity to construct the aforementioned binary matrix which is a basis for the RP, bearing in mind the proper estimation of the recurrence threshold ϵ. Several alternative "rules of thumb" – collated in [13], [14] – have been suggested for the choice of the ϵ: a value which should not exceed 10 % of the mean or the maximum phase space diameter, or a value that guarantees a density of recurrence points (which is one of the basic measures of the recurrence quantification analysis) of approximately 1 %.

The process of the RP construction will be presented on the example of the Rabinovich-Fabrikant model (R-F) which is described by the following set of nonlinear first-order differential equations (state equations) [15]:

$$\begin{cases} \dot{x} = y(z - 1 + x^2) + \gamma x \\ \dot{y} = x(3z + 1 - x^2) + \gamma y \\ \dot{z} = -2z(\alpha + xy) \end{cases} \tag{2}$$

Parameters α and γ are constants that control the evolution of the system – for $\alpha = 1.1$ and $\gamma = 0.87$ the system is chaotic (confirmed by the positive value of the largest Lyapunov exponent ≈ 0.2 computed using the Jacobian matrix constructed on the basis of state equations), but for $\alpha = 0.14$ and $\gamma = 0.1$ it tends to a stable cyclic orbit [16] (the largest Lyapunov exponent ≈ 0). For this reason the set of equations with initial conditions $x(0) = -1$, $y(0) = 0$, $z(0) = 0.5$ was solved numerically by the *ode45* MATLAB function in time span of [0, 100] for both aforementioned pairs of model parameters. The output step size was equal to 0.1. Time series for the x state variable was basis for the phase space reconstruction, parameters of which were as follows: (a) chaotic case: $m = 3$, $\tau = 7$, (b) stable orbit case: $m = 3$, $\tau = 5$. Time series, reconstructed phase space trajectories and recurrence plots for both stable and chaotic behavior of the Rabinovich-Fabrikant system are presented in Fig. 1. The recurrence threshold $\epsilon = 0.11$ was determined by the rule according to which the ϵ value should not exceed 10 % of the maximum phase space diameter. For the comparison purposes ϵ remained unchanged for both numerical experiments fulfilling the aforementioned condition for both phase spaces.

Diagonal lines in Fig. 1c denote that the evolution of trajectory to a stable orbit is similar at different time instants (excluding the initial fragment). Furthermore, equal distance between lines confirms the tendency of the time series to periodicity. On the other hand, the structures in Fig. 1f result from sudden and unexpected changes of system dynamics. Lack of vertical lines in both RPs reveals absence of states that change slowly or do not change at all.

Authors of the RQA proposed a set of measures quantifying features of the RPs. *Recurrence rate* (RR) is a density of recurrence points in an RP:

$$RR = \frac{1}{N^2} \sum_{i,j=1}^{N} R_{ij} \tag{3}$$

Determinism (DET) is a quota of recurrence points that form diagonal lines:

$$DET = \frac{\sum_{l=l_{min}}^{l_{max}} l \cdot P(l)}{\sum_{i,j=1}^{N} R_{ij}} \tag{4}$$

where $P(l)$ is the histogram of the lengths l of diagonal lines, l_{min} denotes the minimal length of such lines ($l_{min} = 2$ adjacent points) and l_{max} is the length of the longest diagonal line, excluding the main diagonal which is called *line of identity* (LoI).

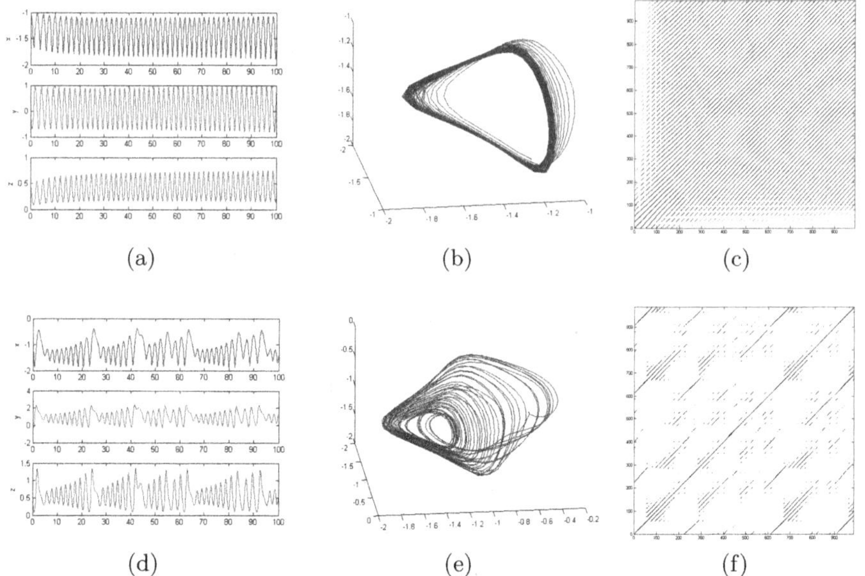

Fig. 1. Stable (a–c) and chaotic (d–f) behavior of the Rabinovich-Fabrikant system: (a, d) time series, (b, e) reconstructed phase space trajectories, (c, f) recurrence plots.

The average length of diagonal lines (LEN) is also calculated on the basis of the histogram $P(l)$:

$$LEN = \frac{\sum_{l=l_{min}}^{l_{max}} l \cdot P(l)}{\sum_{l=l_{min}}^{l_{max}} P(l)} \tag{5}$$

(denominator represents total number of diagonals). LEN should be interpreted as estimation of the average time interval during which two segments of the trajectory are close to each other. In case of small perturbations of chaotic dynamical systems LEN can suggest exponential divergence of the states.

Shannon entropy of the frequency distribution of the diagonal lines lengths (ENTR) is described by the following formula known from the theory of information:

$$ENTR = -\sum_{l=l_{min}}^{l_{max}} p(l) \cdot \ln p(l) \tag{6}$$

with frequency distribution $p(l)$ defined on the basis of the histogram $P(l)$ as follows:

$$p(l) = \frac{P(l)}{\sum_{l=l_{min}}^{l_{max}} P(l)} \tag{7}$$

ENTR reaches its maximum for the uniform frequency distribution. Consequently, higher ENTR values should refer to periodic behaviour of the system, lower values result from chaotic behaviour and values close to 0 are typical for noise.

Characterisation of the diagonal lines is supplemented by the *divergence* (DIV) which is defined as reciprocal of the l_{max}.

The values of RQA measures for both considered variants of the Rabinovich-Fabrikant system are included in Table 1. All computations were performed in the MATLAB environment. Recurrence plots along with the RQA measures were prepared using the CRP Toolbox [17].

Table 1. The values of RQA measures for both considered variants of the R-F system

	RR	DET	LEN	ENTR	l_{max}
Stable R-F system	0.07	0.97	12.08	2.30	753
Chaotic R-F system	0.01	0.95	7.55	2.36	125

Diagonal lines are, on average, longer in case of the stable R-F system. As far as the DET measure is concerned, the slight difference between variants results from the domination of diagonal lines on both RPs. In "stable" case initial fragment of the trajectory determined the uneven frequency distribution and, consequently, lower ENTR value.

It is worthwhile to mention that some measures refer to vertical (or horizontal) lines determining a quota of recurrence points that form vertical lines (*laminarity*; LAM), average length of vertical lines (*trapping time*; TT), and length of the longest vertical line (v_{max}).

3 Experimental Results

For the purpose of time series analysis, two subjects (described as B211 and B238) performed a few selected assisted rehabilitation exercises under supervision of the physiotherapist. B238 impersonated a patient with coxarthrosis (a degenerative osteoarthritis of the hip joint) which is a common disease of the hip joint in adults. B211 behaved naturally as a healthy person. The course of exercises was recorded in the Human Motion Laboratory (HML) of the Polish-Japanese Academy of Information Technology (http://hm.pjwstk.edu.pl) by means of the Vicon Motion Kinematics Acquisition and Analysis System equipped with 10 NIR cameras registering movements of a subject wearing a suit with attached markers (the *motion capture* process) with a frequency of 100 Hz. Positions of the markers in consecutive time instants constitute basis for reconstruction of their 3D coordinates. Recorded time series of length of several dozen seconds represent angles of hip joints' movements in sagittal plane which divides body into left/right parts. For the calibration purposes each exercise started and ended with a T-pose but the corresponding fragments were removed from the time series in the preprocessing phase. Simultaneous video recording provided the referential information.

Figure 2 presents both subjects performing the following exercises with a bar (the "healthy" B211 is always in the background): (a) alternate raising of knees, (b) crouches, and (c) alternate abduction of legs. The right body part of a "patient" is assumed to be affected by the disease.

(a) raising of knees (b) crouches (c) abduction of legs

Fig. 2. Rehabilitation exercises.

Figure 3 consists of time series representing movements of left and right hip joints of both subjects in the sagittal plane (left column; blue color is assigned to left body part, red – to right one) and trajectories reconstructed on basis of the time series of the B211 (middle column) and the B238 (right column) for particular rehabilitation exercises: alternate raising of knees (a–c), crouches (d–f), and alternate abduction of legs (g–i).

Figure 4 includes recurrence plots for both subjects – B211 (a–f) and B238 (g–l). Each row represents one side of the body – left (a–c, g–i) and right (d–f, j–l). Plots in each column relate to one of the rehabilitation exercises: raising of knees (a,d,g,j), crouches (b,e,h,k), and abduction of legs (c,f,i,l). Signatures for particular exercises are A, B, C, respectively. Symbols L, R stand for left and right hip, respectively.

For the comparison purposes value of recurrence threshold for all RPs should guarantee equal density of recurrence points of approximately 1 % (in practice, $RR = 0.0097$).

In case of the exercise A regular character of B211's time series (Fig. 3a) was confirmed by the domination of diagonal line segments in both corresponding RPs (Fig. 4a, d). The similarity degree between B238's RPs is smaller than in case of B211. Besides, the vertical (and horizontal) structures reveal presence of states that change rather slowly (e.g. in Fig. 4j the right hip joint of B238 stands still when the left leg raises three times in a row).

As far as the exercise B is concerned, the B211's time series in Fig. 3d are almost identical, so the similarity degree between corresponding RPs (Fig. 4b, e) is high. This time, dynamic changes of states result in lack of vertical (and horizontal) segments.

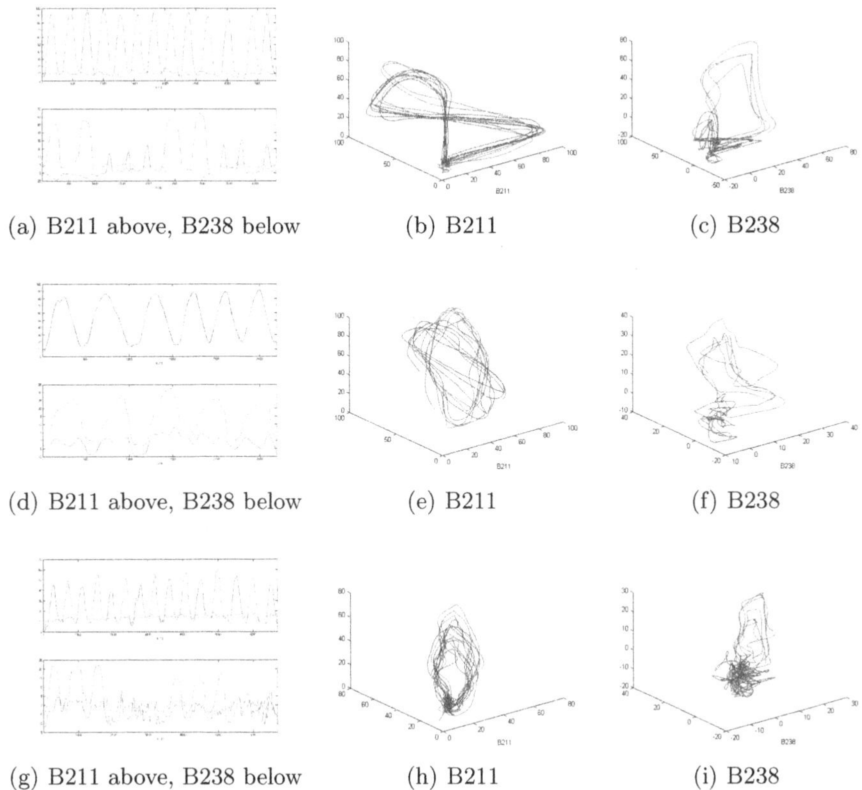

(a) B211 above, B238 below (b) B211 (c) B238

(d) B211 above, B238 below (e) B211 (f) B238

(g) B211 above, B238 below (h) B211 (i) B238

Fig. 3. Time series and reconstructed trajectories: (a–c) – alternate raising of knees, (d–f) – crouches, (g–i) – alternate abduction of legs (Color figure online).

The courses of time series from the exercise C, presented in Fig. 3g, are for both subjects the least regular from among all considered exercises. As a consequence, diagonals lines almost disappeared from the RPs.

Table 2 includes the values of selected RQA measures for both subjects and all 3 exercises at fixed value of $RR \approx 1\,\%$. Values of other measures did not show any significant differences between subjects within given exercise.

The exercise B requires simultaneous work of both body parts which in case of the healthy subject B211 results in almost identical behaviour of both hip joints. Consequently, the longest diagonal lines appear in Fig. 4b and e and the values of the LEN measure from Table 2 substantiate this observation. Generally, direct comparison of values of particular measures included in Table 2 between both subjects within given exercise and body part confirms through greater values of measures related to diagonals – LEN and ENTR – and, at the same time, lower values of the TT measure referring to vertical structures (excluding the C exercise in latter case) greater regularity of B211's time series. However,

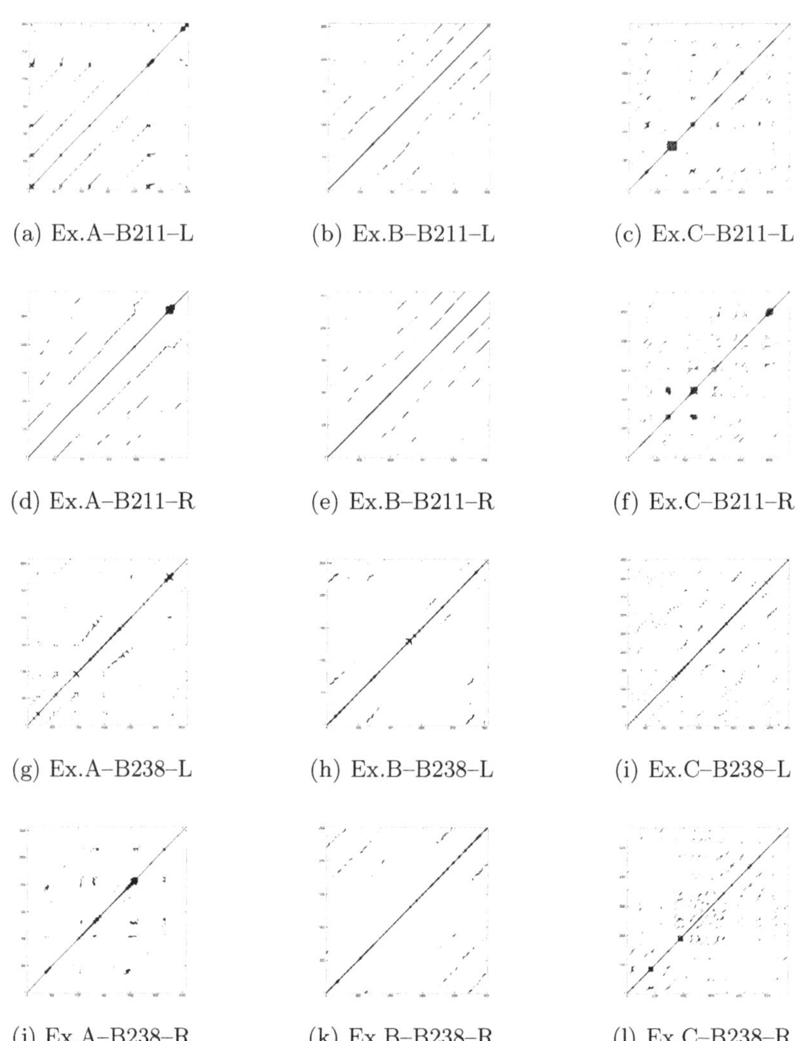

(a) Ex.A–B211–L (b) Ex.B–B211–L (c) Ex.C–B211–L

(d) Ex.A–B211–R (e) Ex.B–B211–R (f) Ex.C–B211–R

(g) Ex.A–B238–L (h) Ex.B–B238–L (i) Ex.C–B238–L

(j) Ex.A–B238–R (k) Ex.B–B238–R (l) Ex.C–B238–R

Fig. 4. Recurrence plots: (a–f) – B211, (g–l) – B238.

Table 2. Values of RQA measures for both subjects and all 3 exercises at fixed RR

	LEN (L)	LEN (R)	ENTR (L)	ENTR (R)	TT (L)	TT (R)
Ex.A–B211	27.46	38.50	4.01	4.12	8.22	8.48
Ex.A–B238	21.05	27.10	3.51	3.84	11.25	16.95
Ex.B–B211	53.98	66.88	4.10	4.18	7.80	7.79
Ex.B–B238	33.16	31.50	3.84	3.88	9.74	9.38
Ex.C–B211	29.84	25.38	3.95	3.79	16.77	16.17
Ex.C–B238	22.36	24.12	3.46	3.72	14.52	14.45

differences between compared values as well as in corresponding RPs are smaller than expected, probably due to too low value of the recurrence threshold.

4 Conclusion

The present paper includes only exemplary recurrence plots for human motion data, and the recurrence quantification analysis of their features is merely fragmentary. Nevertheless, it might be said that both tools seem to be very useful for the purpose of motion data analysis, especially together with other methods intended for examination of dynamical invariants of a system, e.g. relationship between lengths of diagonal lines from the RP and the largest Lyapunov exponent should be taken into consideration. Selected RQA measures could serve as one of tools for quality assessment of completion of rehabilitation exercises based on motion data which, after replacing the motion capture suit by IMU (inertial measurement units) sensors, could be helpful in virtual rehabilitation aimed for remote automatic control of exercises based on comparison of recording of patient's completion of given exercise with a pattern recording.

Besides, RQA measures could extend set of features useful for identification purposes. Finally, formulation of an universal rule for determination of optimal recurrence threshold constitutes one of the challenges of great importance for quality of recurrence quantification analysis.

Acknowledgements. This work has been supported by the following projects: UOD-DEM-1-183/001 "Intelligent video analysis system for behavior and event recognition in surveillance networks" from the National Centre for Research and Development and BK-263/2015 "Modern problems of graphics, video and computer simulation".

References

1. Eckmann, J.-P., Kamphorst, S.O., Ruelle, D.: Recurrence plot of dynamical systems. Europhys. Lett. **4**(9), 973–977 (1987)
2. Webber Jr., C.L., Zbilut, J.P.: Dynamical assessment of physiological systems and states using recurrence plot strategies. J. Appl. Physiol. **76**(2), 965–973 (1994)
3. Schmit, J.M., Riley, M.A., Dalvi, A., Sahay, A., Shear, P.K., Shockley, K.D., Pun, R.Y.: Deterministic center of pressure patterns characterize postural instability in Parkinson's disease. Exp. Brain Res. **168**(3), 357–367 (2006)
4. Marwan, N., Wessel, N., Meyerfeldt, U., Schirdewan, A., Kurths, J.: Recurrence plot based measures of complexity and its application to heart rate variability data. Phys. Rev. E **66**(2), 026702 (2002)
5. Kurths, J., Marwan, N., Wessel, N.: Recurrence plot based measures of complexity to predict life-threatening cardiac arrhythmias. In: Proceedings of the 16th European Conference on Circuits Theory and Design (ECCTD03), Kraków (2003)
6. Javorka, M., Trunkvalterova, Z., Tonhajzerova, I., Lazarova, Z., Javorkova, J., Javorka, K.: Recurrences in heart rate dynamics are changed in patients with diabetes mellitus. Clin. Physiol. Funct. Imaging **28**(5), 326–331 (2008)
7. Henry, B., Lovell, N., Camacho, F.: Nonlinear dynamics time series analysis. In: Akay, M. (ed.) Nonlinear Biomedical Signal Processing: Dynamic Analysis and Modeling, vol. 2, pp. 1–39. Wiley Online Library, published online (2012)

8. Marwan, N., Kurths, J.: Cross Recurrence Plots and Their Applications. Mathematical Physics Research at the Cutting Edge. Nova Science Publishers Inc., New York (2004)
9. Takens, F.: Detecting strange attractor in turbulence. Lect. Notes Math. **898**, 366–381 (1981)
10. Kugiumtzis, D.: State space reconstruction parameters in the analysis of chaotic time series - the role of the time window length. Phys. D: Nonlinear Phenom. **95**(1), 13–28 (1996)
11. Zbilut, J.P., Webber Jr., C.: Recurrence quantification analysis. In: Wiley Encyclopedia of Biomedical Engineering, Wiley (2006). doi:10.1002/9780471740360.edb1355
12. Kennel, M.B., Brown, R., Abarbanel, H.D.I.: Determining embedding dimension for phase-space reconstruction using a geometrical construction. Phys. Rev. A **45**(6), 3403–3411 (1992)
13. Schinkel, S., Dimigen, O., Marwan, N.: Selection of recurrence threshold for signal detection. Eur. Phys. J. Spec. Top. **164**, 45–53. Springer (2008)
14. Marwan, N.: How to avoid potential pitfalls in recurrence plot based data analysis. Int. J. Bifurcat. Chaos, **21**(4), 1003–1017. World Scientific Publishing Company (2011)
15. Sprott, J.C.: Chaos and Time-Series Analysis. Oxford University Press, Oxford (2003)
16. Danca, M.-F., Romera, M.: Algorithm for control and anticontrol of chaos in continuous-time dynamical systems. Dyn. Continuous Discrete Impulsive Syst. Ser. B: Appl. Algorithms **15**, 155–164 (2008)
17. Marwan, N., Romano, M.C., Thiel, M., Kurths, J.: Recurrence plots for the analysis of complex systems. Phys. Rep. **438**(5–6), 237–329 (2007)

Cellular Nuclei Differentiation Evaluated by Automated Analysis of CLSM Images

Julita Kulbacka[1(✉)], Marek Kulbacki[3], Jakub Segen[3], Grzegorz Chodaczek[4], Magda Dubinska-Magiera[2], and Jolanta Saczko[1]

[1] Department of Medical Biochemistry, Medical University, Chalubinskiego 10, 50-368 Wroclaw, Poland
julita.kulbacka@umed.wroc.pl

[2] Department of General Zoology, Zoological Institute, University of Wroclaw, Sienkiewicza 21, 50-335 Wroclaw, Poland

[3] Polish-Japanese Academy of Information Technology, Koszykowa 86, 02-008 Warszawa, Poland

[4] Wroclaw Research Centre EIT+, Laboratory of Confocal Microscopy, Stablowicka 147, Wroclaw, Poland

Abstract. Nuclear morphology abnormalities in cells are often the symptom of the cell death. However, minor disturbances in nuclear shape, such as a slight blebbing or deformation, may indicate characteristic medical disorders or therapeutic effect. The analysis of microscopic images requires time consuming observations and meticulous analysis that often are encumbered with human mistake. In the present work, image analysis of nuclei as a method of morphological verification is presented. The automated analysis of numerous cellular nuclei images may be a useful tool for any biological or analytical laboratory.

Keywords: Nucleus · Fluorescent image analysis

1 Introduction

The nucleus is one of the most important organelles, which is responsible for the transfer of genetic information and cell division. Unfortunately, very little is known about how it is formed, what determines its shape and its size. Several diseases show characteristic abnormal nuclear architecture, thus morphology of nuclei is the major interest of cytopathologists. Morphological modifications of nuclei can be demonstrated by light microscopy in routine staining, by confocal microscopy (DAPI staining, propidium iodide and trypan blue), and by electron microscope, which can show nucleus ultrastructure. The morphological changes are often connected with cellular dysfunctions and methods enabling objective morphology assessment can help us to understand and describe pathology of nucleus, diagnose various disorders and design new therapies. The verification of morphological changes of nuclei can determine possible cellular defects such as

N.T. Nguyen et al. (Eds.): ACIIDS 2016, Part II, LNAI 9622, pp. 407–416, 2016.
DOI: 10.1007/978-3-662-49390-8_40

alterations of nuclear matrix, membrane abnormalities and chromatin reorganization [1,2]. The perturbed nucleus morphology, frequently found also in cancer cells, presents many characteristic microscopic changes which can be observed as modification of nuclear size, shape, chromatin pattern, nuclear envelope and perinuclear space [3]. In particular, nuclei discrimination may be useful in cancer but also in normal cells analysis, during development, regeneration, reprogramming or even after therapeutic action. Malignant nuclei are usually characterized by variation of size which is commonly known as pleomorphism. The conventional theories cannot explain this phenomenon and, currently, the mutator phenotype due to genetic instability has been suggested to be its cause. The genetic mutations lead to formation of various subclones of cancer cells with morphological changes including modification of nuclear shape and size [3–6]. The determination of nucleus pleomorphism is essential for assessment of the degree of tumor differentiation and the level of histological malignancy of tumor [7]. The estimation of this parameter can be performed using basic research methods such as histochemical staining. The frequent morphological alterations of nucleus involve nuclear shape and margin abnormality [8,9]. The assessment of morphological changes of nucleus is very important to discriminate between different types of cell death such as apoptosis, necrosis or autophagy. For instance, fragmentation of nuclei with condensation and margination of chromatin is frequently observed during apoptotic pathway [10]. In contrast to apoptosis, during necrosis the late disintegration of nucleus is observed while chromatin condensation is rare [11–13]. Moreover, pyknotic and fragmented nuclei are not a common element in necrotic cell death. Increased participation of nuclear morphological modification and dysfunction is also considered in neurodegenerative diseases, which are characterized by the loss of structural and functional integrity of the central nervous system. In many cases, the etiology is unknown and abnormal aggregation of protein in the cytoplasm and nucleus of brain cells is found. The complete knowledge of morphological and functional organization of the nucleus in health and under pathological conditions can help in understanding the development and progression of various diseases such as malignant transformation leading to cancer. Importantly, the assessment of morphological modifications of nuclei plays a significant role in pathological diagnosis and prognosis of different disorders. The objective of this study was to identify characteristics that can be computed from images and aid in automatic differentiation between normal and abnormal nuclei. First, image analysis was conducted utilizing standard functionality of Fiji image analysis software [21], with a focus on nucleus shape features [22]. It was found that determined mean value of circularity parameter showed a statistically significant difference between the two groups. To find additional discriminative characteristics, more detailed analysis of shape contour was performed using custom methods. The results revealed strong differentiating potential of features related to local contour curvature, achieving recognition rate over 90 % in a classification test.

2 Materials and Methods

Cell Culture. The studies were performed on three cell lines: LoVo (human colon adenocarcinoma), MCF-7/WT (human breast adenocarcinoma) and CHO-K1 (normal hamster ovarian fibroblasts). Cells were harvested in complete culture medium containing 10 % fetal bovine serum (FBS, HyClone, Poland) and antibiotics (streptomycin/penicillin; Sigma). The cell lines were cultured in plastic flasks 25 or 75 cm^2 (Nunc, Denmark), which were stored at 37°C and 5 % CO_2 in an incubator (SteriCult, ThermoScientific, Alab, Poland). For the experiments, the cells were detached by trypsinization (Trypsin 0.025 %; Sigma) and neutralized by cell culture medium.

Confocal Laser Scanning Microscopy (CLSM) for the Nuclei Evaluation. Cells subject to various experimental conditions leading to nuclear damage were grown on coverslips for 24 h, then fixed with 4 % paraformaldehyde (PFA) in PBS. DNA was stained with DAPI (4,6-diamidino-2-phenylindole; 0.2 μg/ml). Cell were mounted in fluorescence mounting medium (DAKO). For imaging, FluoView FV1000 confocal laser scanning microscope (Olympus) was used. The images were recorded by employing a Plan-Apochromat 60x oil-immersion objective.

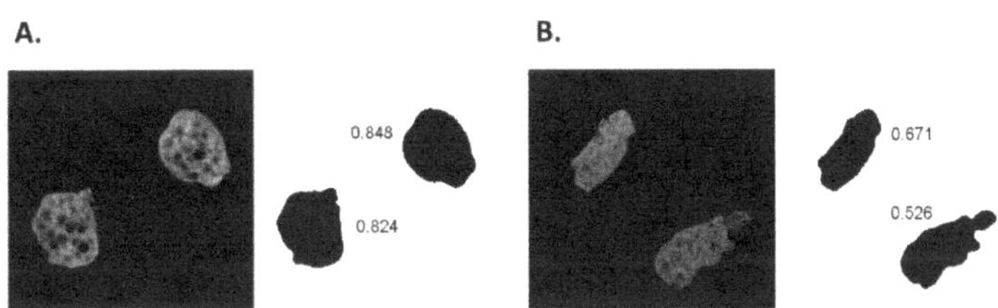

Fig. 1. The analysis of (**A**) normal and (**B**) abnormal nuclei by circularity values assessment

2.1 Nuclei Image Analysis

2.1.1 Morphological Image Analysis of Normal and Abnormal Nuclei in Cells

The image analysis was performed using standard operations in the Fiji software. Images showing cell nuclei were cleaned of background noise with *Subtract Background* function (radius 50), which also equalizes remaining background signal across images. Next, images were segmented using *Threshold* function and Huang algorithm that rendered nuclei area creating binary images as a result. Holes in nuclei were filled with *FillHoles* command and remaining small artifacts were removed with *Medianfilter* (radius 5). Morphology of nuclei was quantified with *AnalyzeParticles* function and *Circularity* parameter (Shape

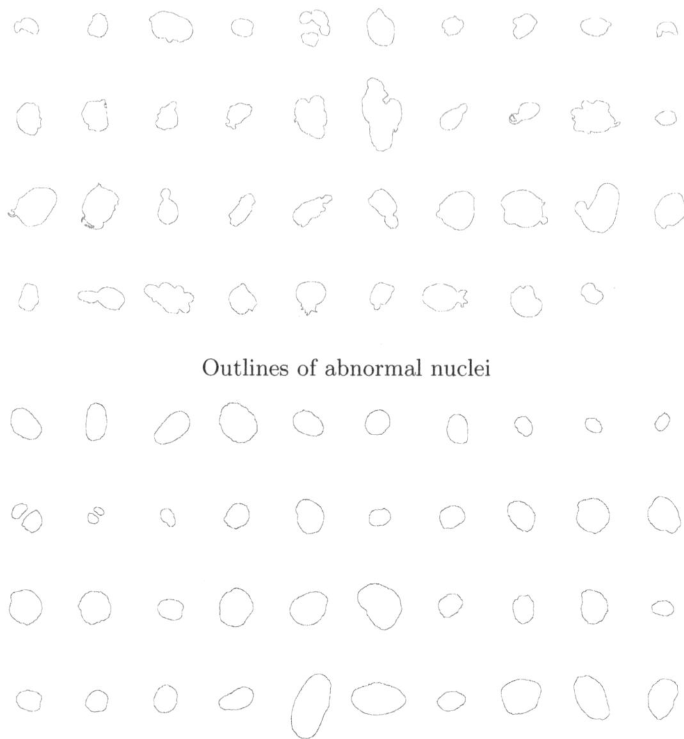

Outlines of abnormal nuclei

Outlines of normal nuclei

Fig. 2. Outlines of nuclei

descriptors module), which measures roundness of nucleus according to formula: $4\pi * area/perimeter^2$. A circularity value of 1.0 indicates a perfect circle. Clustered or dividing were excluded from the analysis. An exemplary analysis of normal and abnormal nuclei is presented in the Fig. 1.

2.1.2 Contour Based Shape Analysis

The first part of the analysis, similarly to Sect. 2.1.1 uses standard functions of the Fiji software [21,22], in order to segment out nuclei shapes and compute their contours. The second part concentrates on contour analysis.

Segmentation. Automatic approach has been used to calculate contours of nuclei. First, each image has been converted to binary image by automatic thresholding using Otsu method. The Otsu's threshold clustering algorithm searches for a threshold that minimizes the intra-class variance, defined as a weighted sum of variances of the two classes [23]. Holes in nuclei were filled with *FillHoles* command and small artifacts were removed with *Medianfilter* (radius = 2). *AnalyzeParticles* plugin in segmented image with option

$Show = Outlines$ displays a contour of the detected objects (Fig. 2). Selected particles were transformed into vector of X,Y coordinates of contour using command $SaveAs- > XYCoordinates$.

Contour Analysis. The outer contour of a 2-D shape contains complete information of the shape without holes. Measurements and estimation of shape properties, shape classification and shape transformations can be conveniently and efficiently performed through the analysis of contour [18]. A contour of a digital shape can be considered a discrete representation of a continuous curve. A useful characterization of a continuous planar curve is its curvature function $K(s)$, where s is the arc length parameter. Curvature is invariant under translation of the shape and under rotation up to a circular shift of $K(s)$. These invariances benefit shape classification based on curvature. For a discrete contour, estimating the continuous curvature is not simple, as it may involve curve fitting which depends on a chosen functional form and it is affected by noise. Instead, one can use a contour function which behaves similarly to the continuous curvature and has the same invariance properties. One such function is called k-curvature [19]. The k-curvature value for a contour point indexed p is measured as the angle between two chords, one from the point p to the point $p + k$, the other from p to $p - k$, as shown in Fig. 3. The shape features for classification of images of nuclei are computed from k-curvature functions of the contour for four values of $k, 2, 4, 7, 10$. Three features are calculated for each value of k: minimum k-curvature, maximum k-curvature, and the number of critical points. A critical point was defined here as a point of a contour where the k-curvature has a local extremum (maximum or minimum), and the absolute value of k-curvature is greater than a given threshold, set at 0.7 radians. Figure 4 shows examples of these features for two normal cases, and Fig. 5 for two abnormal, using the value $k = 2$. For each case, the contour and the k-curvature plot are shown. Detected critical points are marked on the contour. Figure 7 shows scatter plots for pairs of selected features. To assess the discriminating power of these features a training and classification experiment was performed using an SVM classifier with RBF kernel [20], selecting one pair of features at a time. The dataset was divided at random into two subsets: a training set composed of 70 % of instances and a test set with 30 % of instances. The selected RBF kernel type was $\exp(-\gamma * |u - v|^2)$ and a tune function was used to do a grid search over the supplied parameter ranges (cost, gamma) using the training set. In order to build an SVM model the best parameter values were selected with the tune() function. The resulting SVM model was used on the test set to predict classes. The results are summarized in Fig. 8.

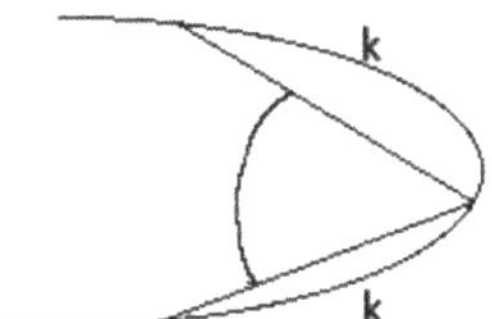

Fig. 3. k-curvature measurement

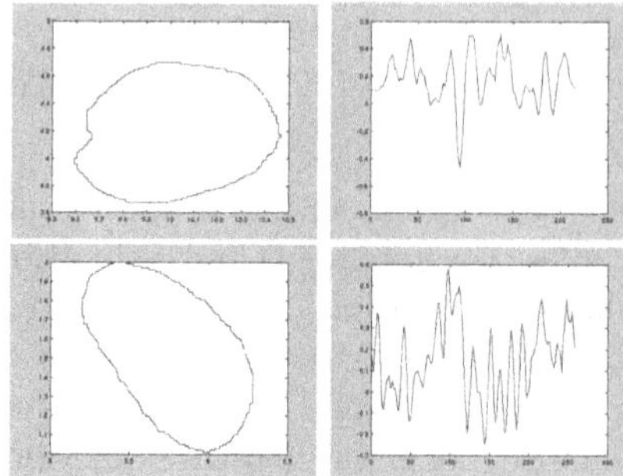

Fig. 4. Contour and curvature plot for normal nuclei

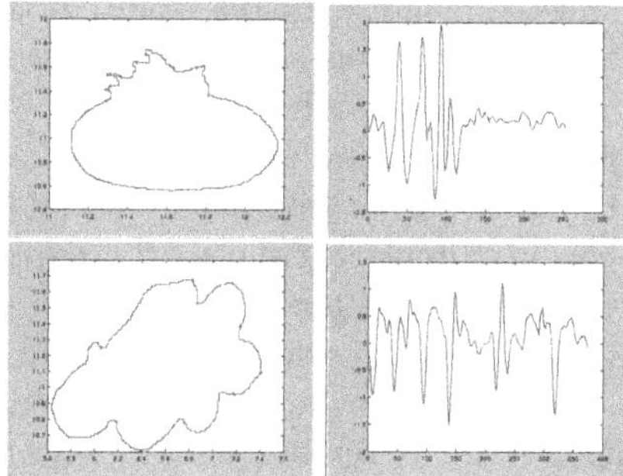

Fig. 5. Contour and curvature plot for abnormal nuclei

3 Results and Discussion

The changes of nucleus shape in neoplastic cells play the main role in the assessment of tumor malignancy. These changes concern its surface, volume, the nucleus/cytoplasm ratio, shape and density, as well as structure and homogeneity. The ultrastructural characteristic is related to nucleus segmentation, invaginations, changes in chromatin, such as heterochromatin reduction, increase of interchromatin and perichromatin granules, increase of nuclear membrane pores, formation of inclusions, etc. The analysis of nuclear changes and abnormalities can be used for interpretation of the disease induced changes or presence of different cell clones and genetic anomalies associated with these changes. The results obtained in the first study using the Fiji's *Analyze Particles* function [21] aimed at analysis of normal and abnormal nuclei are presented in Fig. 6. The Mann-Whitney test was used for statistical evaluation of the results. Schöchlin et al. [14] also used ImageJ software for nuclear analysis. Authors classified circularity values into four shape categories (spindled, elongated, oval, round) and demonstrated statistically significant differences in the spindled and round categories.

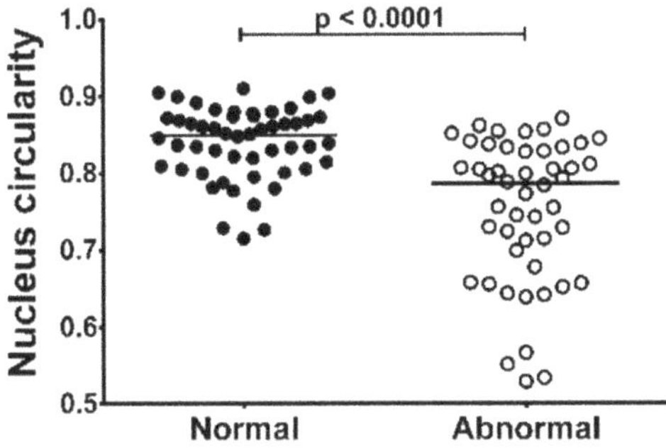

Fig. 6. The Mann-Whitney test for circularity analysis

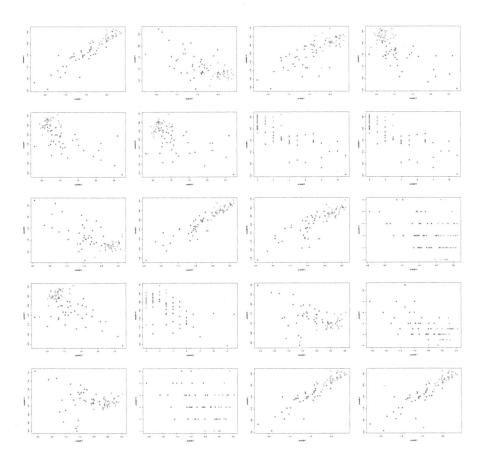

Fig. 7. Scatter plots for pairs of features that achieve accuracy > 91 % in SVM classifier test

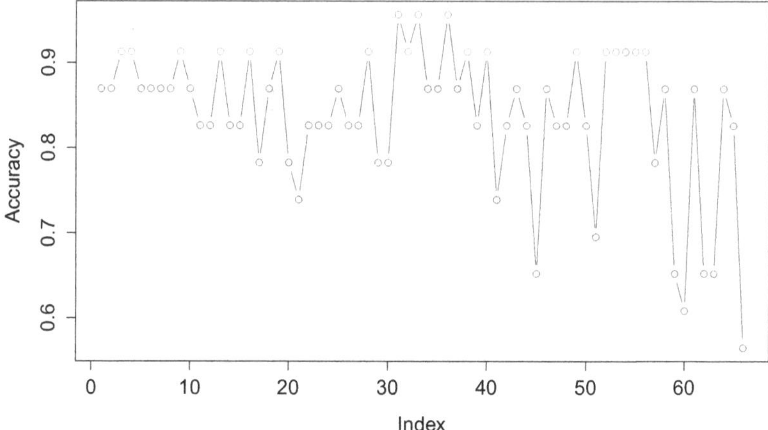

Fig. 8. The results of classification tests on all pairs of features. *Accuracy* >= 91 % green, *Accuracy* < 91 % red (Color figure online)

They concluded that DM (desmoplastic melanoma) contained more spindled nuclei than SM (spindle cell melanoma) and SM contained more round nuclei. Driscoll et al. [15] used an automated, quantitative method to study distributions of blebbing in large cell population. Similarly to our study, fluorescent images were also used for analysis. The following parameters were measured: area, perimeter and curvature of nucleus. In each set of treated cells, the curvature of all nuclei was visualized in a single plot. The obtained plots enabled quick assessment of the severity of blebbing in the population of nuclei. Authors efficiently used this method for examination of treated fibroblasts [15]. Other authors used a custom written program in MATLAB to provide 3D-rendering and automated analysis of confocal images in order to obtain quantitative measures of each cell, and the locations of its cytoskeleton and nucleus relative to the gel surface. They indicated that quantitative image-processing approach can reveal differences in the structure and internal organization between indenting, metastatic cells and benign cells [16]. Agley et al. [17] reported a straightforward method using immunofluorescent staining and the commercially available imaging program Adobe Photoshop, which allows to gather an objective and precise information concerning irregularly shaped cells. Authors applied this measurement technique to the analysis and quantification of the surface area of individual myotubes across a broad size range, measured nuclear areas, and quantified staining intensity within individual nuclei [17]. The results of contour based shape analysis (Sect. 2.1.2) characterize the behavior of k-curvature with respect to the type of a nucleus. The scatter plots of features based on k-curvature Fig. 7 show significant differentiating potential for some pairs of features, for example the pair in row 3, column 1 [Min k-curvature for k = 4; Max k-curvature for = 4] or the pair in row 2, column 1 [Max k-curvature for k = 2, Min k-curvature for k = 7]. The results of classification tests using the

SVM classifier on all pairs of features are summarized with a plot in Fig. 8. It shows the highest recognition accuracy values: 95.7 %, 95.65 %, 95.62 % for the feature pairs [Min k-curvature for k = 4; Max k-curvature for k = 4], [Min k-curvature for k = 4; Min k-curvature for k = 7] and [Min k-curvature for k = 4; Min k-curvature for k = 10]. The above results demonstrate the potential of local features of contour such as k-curvature to discriminate between normal and abnormal shapes of nuclei and to achieve recognition accuracy rates above 90 %. These results will be verified and refined when a larger dataset of images is available. A larger dataset will also make possible more complete classification experiments, with larger feature vectors that would include both curvature-like features and global features such as circularity and additionally features based on holes and occlusions, with the aim to reach classification rates around and above 98 %.

Acknowledgments. This work was supported by the National Science Centre in Poland grant no. 2014/14/E/NZ6/00365 (G.C.) and partially by grant of Wroclaw Medical University No. PBmn-131 and PBmn-132, and UMED sponsorship account.

References

1. Fulda, S., Gorman, A.M., Hori, O., Samali, A.: Cellular stress responses: cell survival and death. J. Cell Biol. **12**, 23–46 (2010)
2. Nafe, R., Franz, K., Shlote, W., Shneider, B.: Morphology of tumor cell nuclei is significantly related with survival time of patients with glioblastomas. Clin. Cancer Res. **11**(6), 2141–2148 (2005)
3. Misteli, T.: Beyond the sequence: cellular organization of genome function. Cell **128**, 787–800 (2007)
4. Dey, P.: Cancer nucleus: morphology and beyond. Dignostic Cytopathol. **38**(5), 1–23 (2010)
5. Bignold, L.P.: The mutator phenotype theory of carcinogenesis and the complex histopathology of tumours: support for the theory from the independent occurrence of nuclear abnormality, loss of specialization and invasiveness among occasional neoplastic lesions. Cell Mol. Life Sci. **60**, 883–891 (2003)
6. Smith, T.F., Waterman, M.S.: Identification of common molecular subsequences. J. Mol. Biol. **147**, 195–197 (1981)
7. Sapierzyski, R.: Rokowanie u pacjentw onkologicznych. Badania morfologiczne. ycie weterynaryjne **84**, 723–726 (2009)
8. Jogai, S., Jassar, A., Adisena, A., Dey, P.: Fine needle aspiration cytology of thyroid lesions. Acta Cytol. **49**, 483–488 (2005)
9. Rajesh, L., Dey, P., Joshi, K.: Fine needle aspiration cytology of lobular carcinoma: comparison with other breast lesions. Acta Cytol. **47**, 177–182 (2003)
10. Krysko, D.V., Vanden, B.T., D'Herde, K., Vandenabeele, P.: Apoptosis and necrosis: detection, discrimination and phagocytosis. Methods **44**(3), 205–221 (2008)
11. Elmore, S.: Apoptosis: a review of rogrammed cell death. Toxicol Pathol. **35**(4), 495–516 (2007)
12. Webster, M., Witkin, K.L., Fix, O.C.: Sizing up the nucleus: nuclear shape, size and nuclear-envelope assembly. J. Cell Sci. **122**, 1477–1486 (2009)

13. Ziegler, U., Groscurth, P.: Morphological features of cell death. News Physiol. Sci. **19**, 124–128 (2004)
14. Schöchlin, M., Weissinger, S.E., Brandes, A.R., Herrmann, M., Mller, P., Lennerz, J.K.: A nuclear circularity-based classifier for diagnostic distinction of desmoplastic from spindle cell melanoma in digitized histological images. J. Pathol. Inform. **5**, 40 (2014)
15. Driscoll, M.K., Albanese, J.L., Xiong, Z.M., Mailman, M., Losert, W., Cao, K.: Automated image analysis of nuclear shape: What can we learn from a prematurely aged cell? Aging **4**(2), 119–132 (2012)
16. Dvir, L., Nissim, R., Alvarez-Elizondo, M.B., Weihs, D.: Quantitative measures to reveal coordinated cytoskeleton-nucleus reorganization during in vitro invasion of cancer cells. New J. Phys. **17**, 043010 (2015)
17. Agley, C.C., Velloso, C.P., Lazarus, N.R., Harridge, S.D.R.: An image analysis method for the precise selection and quantitation of fluorescently labeled cellular constituents: application to the measurement of human muscle cells in culture. J. Histochem. Cytochem. **60**(6), 428–438 (2012)
18. Costa, L.F., Cesar Jr., R.M.: Shape Analysis and Classification: Theory and Practice. CRC Press, Boca Raton (2009)
19. Freeman, H., Davis, L.: A corner-finding algorithm for chain-coded curves. IEEE Trans. Comp. **C–26**, 297–303 (1977)
20. Shawe-Taylor, J., Cristianini, N.: Support Vector Machines and Other Kernel-Based Learning Methods. Cambridge University Press, New York (2000)
21. Schindelin, J., Arganda-Carreras, I., Frise, E., Kaynig, V., Longair, M., Pietzsch, T., Preibisch, S., Rueden, C., Saalfeld, S., Schmid, B., Tinevez, J.Y., White, D.J., Hartenstein, V., Eliceiri, K., Tomancak, P., Cardona, A.: Fiji: an open-source platform for biological-image analysis. Nat. Methods **9**(7), 676–682 (2012)
22. Wagner, T., Lipinski, H.G.: IJBlob: An ImageJ library for connected component analysis and shape analysis. J. Open Res. Softw. **1**(6), 68 (2013)
23. Otsu, N.: A threshold selection method from gray-level histograms. IEEE Trans. Sys. Man. Cyber. **9**, 62–66 (1979)

Selected Space-Time Based Methods for Action Recognition

Sławomir Wojciechowski[1], Marek Kulbacki[1(✉)], Jakub Segen[1], Rafał Wyciślok[1], Artur Bąk[1], Kamil Wereszczyński[1,2], and Konrad Wojciechowski[1]

[1] Polish-Japanese Academy of Information Technology, Koszykowa 86, 02-008 Warszawa, Poland
mk@pjwstk.edu.pl
[2] Institute of Informatics, Silesian University of Technology, Akademicka 16, 44-100 Gliwice, Poland

Abstract. A survey on very recent and efficient space-time methods for action recognition is presented. We select the methods with highest accuracy achieved on the challenging datasets such as: HMDB51, UCF101 and Hollywood2. This research focuses on two main space-time based approaches, namely the hand-crafted and deep learning features. We intuitively explain the selected pipelines and review good practices used in state-of-the-art methods including the best descriptors, encoding methods, deep architectures and classifiers. The best methods were chosen and some of them were explained in more details. Furthermore, we conclude how to improve the methods in speed as well as in accuracy and propose directions for further work.

Keywords: Action recognition · Robust space-time methods · Survey

1 Introduction

Action recognition still remains a very challenging task and the demands are continuously growing together with a big amount of videos recorded and uploaded every day. It could be employed in such areas as surveillance, human-computer interaction and could help to control almost each industrial process. Methods and algorithms have been getting better over the years and the problem moved from recognizing actions in constrained videos to more challenging datasets such as HMDB51 [30], UCF101 [32] and Hollywood2 [31]. To the best of our knowledge, there is still a lack of a good review of best performing methods in the last three years and we provide it here. The remaining of the paper is organized as follows. We shortly review other surveys in the successive chapter. Chapter 3 describes the popular pipelines. In Chap. 4 we choose the methods based on their results obtained on the aforementioned datasets. A comparison followed by the explanation of the most interesting ones is provided. We conclude our work in Chap. 5.

N.T. Nguyen et al. (Eds.): ACIIDS 2016, Part II, LNAI 9622, pp. 417–426, 2016.
DOI: 10.1007/978-3-662-49390-8_41

2 Related Work

The existing reviews on action recognition mostly try to capture the whole domain and summarize a big amount of methods [19,20]. They take a broad historical view about action recognition domain and could still serve as a good guideline to improve some interesting research directions. Another study of human action recognition approaches [21] provides a review of best methods up to 2012. Our goal is to explain and conclude the best up-to-date space-time based methods, which is only one out of many ways to recognize an activity in video.

3 Common Space-Time Approaches

The space-time based methods treat a video as a space-time cuboid and extract comparable representations which can serve as an input to a classifier. In this section, the pipelines for the two most popular approaches are given. We pay a lot of attention to the algorithms and tools utilized by methods described in the subsequent chapter.

3.1 Bag of Words Approach

The Fig. 1 presents a single layer pipeline for bag of words (BOW) methods. The presented pipeline is divided into phases. At the very beggining the *videos* from the datasets are taken. They could have annotations, such as bounding boxes marking humans. The next phase is optional and here the input videos are *preprocessed* by e.g. rescaling, flipping, cropping [3] or MPEG compression [25]. *Feature detector* decides which areas in the video serve as volumes for computing features. The most popular are space-time interest points (STIP) [26], which provide sparse representation, and dense trajectories (DT) [13]. Some methods do not exploit feature detectors. For instance, dense sampling [22] or random sampling [28] do not detect regions for feature extraction thus saving computational time. A big amount of evaluations claim dense sampling methods outperform STIP [22]. A *feature descriptor* captures important information for representing an action within these subvolumes. Amongst popular, histogram of oriented gradients (HOG) describes appearance features while histogram of optical flow (HOF) as well as motion boundary histogram (MBH) are based on optical flow (OF) and represent motion [11]. Both the information are captured by the lately introduced fast gradient boundary histograms (GBH) descriptor [7]. It avoids computing OF and measures the dynamics of image gradients which results in removing background and encoding the motion of human shapes [7]. However, actions have varying speed and skipping different frames detects features at different frequencies which together with mixing them later turned out to enhance the final representation [5]. Subsequently, in order to reduce the features dimensionality and decorrelate them *principal component analysis (PCA)* is widely used, which has a big impact on results [8]. In the next step a codebook is generated. *The PCA model and codebook* are often trained on a small subset of descriptors, e.g. as in [2,24]. The popular hard-assignment k-means method introduces a new feature space which is fully divided into fixed-size parts with codewords as centers. Further, the similar descriptors are gathered together during *feature encoding* phase. A good imagination of this representation is a k-cell histogram capturing the frequencies of occurrence of different descriptors in a video. Fisher vector (FV) is another popular encoding approach. Intuitively, FV remembers the mean and variance of the features in a codeword. It is obtained directly from the expected-maximization algorithm adjusting the parameters of Gaussian mixture model (GMM) [23]. It naturally exploits the GMM for the codebook generation. For a fair comparison of different videos, the same fixed-size codebook is generated for each movie when training and testing. The dimension of FV is 2DK, where D denotes the descriptor size and K is the number of utilized gaussians. Once the encoding is finished the final descriptors are *normalized* and fed into a *classifier*. BOW Histogram descriptor mostly works [24] and achieves the best results [9] with the RBF-χ^2 kernel support vector machine (SVM), whereas FV is usually fed to the linear SVM.

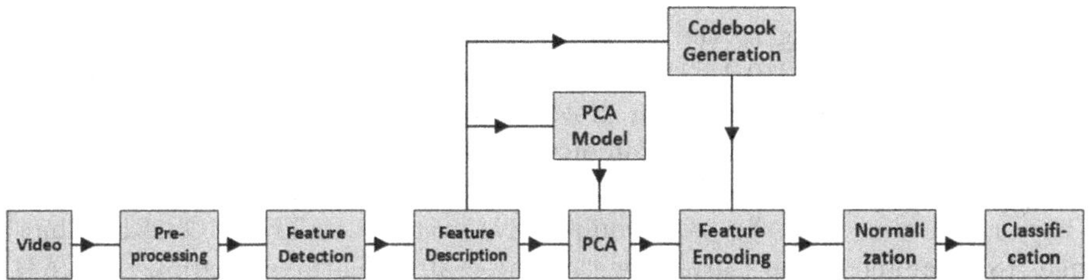

Fig. 1. Bag of words approach pipeline

Fusion Methods. The current systems often incorporate multiple descriptors. There are three main concatenation approaches, namely: descriptor level, representation (kernel) level and score level fusion [8]. Moreover, hybrid representation is similar to the kernel fusion but combines different BOW methods which could have different encoders [8].

Problems and Evaluations. It is worth highlighting that GMM with 256 gaussians and k-means with the codebook size of 8000 is a good balance between accuracy and computational time [8]. For the popular improved dense trajectories (iDT) [11] with PCA and representation level fusion it provides about 100 k dimensional representation for FV and about 25 k dimensional representation for BOW Histogram. However, the recent evaluation [9] shows that BOW histogram works at least 3 times slower than FV for that configuration. Besides, enlarging the codebook size four times is still inferior to the accuracy obtained by FV with 64 gaussians on HMDB51 dataset [9]. Moreover, linear SVM is faster than other kernel SVM [9]. This motivates the popularity of FV encoding connected with linear SVM. It should be noticed that during the feature encoding phase the information about the descriptor location in the space-time volume is lost, which was also observed by [28]. Recently, methods try to deal with this problem as described in the next chapter. Choosing the parts that fit together is of great importance for the final results [8]. For the comparison of feature detectors and descriptors please refer to the Wang's evaluation [22]. A lot of research on almost each stage of our pipeline has been made in the Peng's work [8]. We recommend this paper for designing an efficient and properly adjusted system. A fast implementation of the popular descriptors, evaluation of different optical flow methods and a trade-off between speed and accuracy is provided by the Uijlings' paper [23]. Idea for speeding up the motion flow based descriptors which is a major bottleneck of current methods was proposed by Kantorov [25]. The tremendous amount of algorithms, parameters and implementations prevents comparing even two similar looking methods. For that reason we find analyzing the results obtained on challenging datasets a fair evaluation.

3.2 Deep Learning Methods

Deep learning methods involve a multilayered architecture for capturing abstraction in data. According to the Table 1 only methods exploiting convolutional neural network (CNN) [10,14,15] are able to compete with current methods on HMDB51 and UCF101. The major drawbacks of these methods are the computational time and a need of huge amount of training data [10,29]. The Wang's work [14] explains how to deal with the training problems in very deep networks. It is worth noting that the most competitive systems at THUMOS2015 Challenge [18] combined deep learning with hand-crafted features or applied VLAD (simplified FV) for encoding deep learning features as proposed in [16].

4 The Selected Methods

Here, we present the current best methods for action recognition. They are ordered by their accuracy obtained on HMDB51. We follow the results reported by authors. In some cases, they are obtained by combining with other descriptors or methods. Therefore, we selected also some other interesting approaches which could have a big impact on the performance or the remaining methods, even if they alone report worse accuracy. It is worth noting that eight [1–6,8,9] out of ten best methods utilize DT or iDT and there are only three methods [1,3,10] involving CNN. Nevertheless, recently a lot of work has been made to improve CNN for action recognition [14–16]. These ideas were yet not tested on HMDB51 but the results obtained on UCF101 [14,15], and THUMOS Challenge 2015 [17] are similar or even better than the best hand-crafted methods. The BOW and deep learning approaches are often complementary and their combination give the best results [15,18]. In the following sections the selected approaches are explained in more details.

Table 1. Comparison of different methods

Method	HMDB51	UCF101	Hollywood2
15 k object categories [1]	71.3	88.5	66.6
Stacked Fisher vectors [2]	66.79	-	-
Trajectory-pooled deep-conv. descriptors [3]	65.9	91.5	-
Motion part regularization [4]	65.5	-	66.7
Multi-skip feature stacking [5]	65.1	89.1	68.0
Modeling video evolution [6]	63.7	-	73.7
Gradient boundary histogram [7]	62.0	86.6	-
Comprehensive study – hybrid Peng's system [8]	61.1	87.9	-
A robust and efficient INRIA method [9]	60.1	86.0	66.8
Two stream convolutional networks [10]	59.4	88.0	-
Improved dense trajectories [11]	57.2	-	64.3
Genetic programming approach [12]	48.4	-	46.8
Dense trajectories [13]	-	-	58.3
Very deep two-stream convnets [14]	-	91.4	-
3D convolutional networks [15]	-	90.4	-

4.1 Dense Trajectory Based Methods

At the beginning we describe the baseline methods. Then, other selected methods are explained. They utilize iDT but often their ideas can be adopted by other methods. For more details please refer to the original papers. They usually give an evaluation of different parameters, whereas we present only the settings for the best reported result. In the following methods GMM with 256 gaussians creates a codebook, FV encodes the features and the linear SVM was employed for classification, unless stated otherwise.

Dense Trajectories [13]. In this method, for each frame an OF for driving the trajectory is computed. Similarly to other feature detectors a space time volume for computing features has to be detected. Here, it can be imagined as a 32 by 32 spatial patch driven by a trajectory, for each of eight spatial scales separately. Frame by frame the length of the volume is verified and the tracking process is stopped for a trajectory extending 15 frames. Simultaneously, a 5 by 5 spatial area is checked and in the case of founding no points belonging to the trajectory volume, a new one is extracted. Trajectories in homogeneous areas are deleted. Inside the trajectory volumes HOG, HOF, MBHx and MBHy are computed in a 2 by 2 by 3 box. The main ideas of the method are depicted in Fig. 2. There are eight quantization possibilities for HOG, MBHx, MBHy and nine for HOF which gives 96 and 108-dimensional vectors for the whole volume, respectively. They are normalized by l2 norm. Besides, a trajectory shape descriptor is extracted. Each descriptor (trajshape, HOG, HOF and MBH which incorporates MBHx

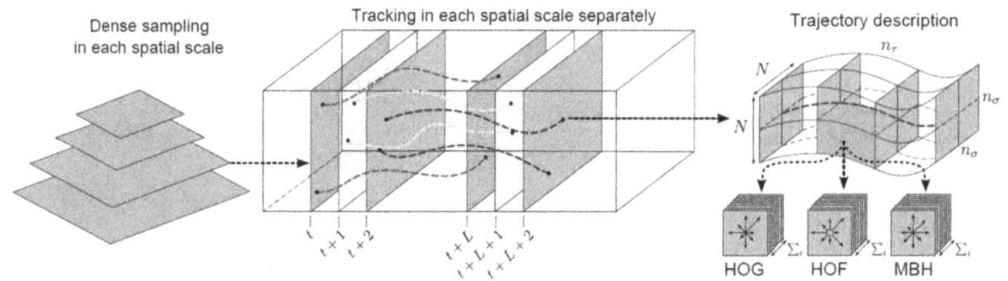

Fig. 2. Dense Trajectories approach visualisation [13]

and MBHy) is encoded by a BOW histogram and has a separate codebook with 4000 visual words. These four representations are concatenated and fed into SVM with χ^2 kernel.

Improved Dense Trajectories [11]. Here, the main advantage over the previous approach is camera motion compensation. For that, they estimated the homography by RANSAC algorithm. It involves matches both from speed up robust features descriptors and dense OF. Moreover, a human detector providing boundig boxes as annotations brings further improvements. This can be done manually or automatically. It enhances the camera motion removal by deleting areas containing human motion. The camera motion is deleted from OF field which influences both MBH and HOF descriptors. Furthermore, trajectories similar to the estimated motion are removed. Here, the descriptors are normalized by RootSIFT approach instead of l2 norm. Subsequently PCA is applied. For the final representation after FV power and l2 normalization is performed.

Multi-skip Feature Stacking [5]. This method simply extracts iDT features at different time scales by skipping frames. A great improvement was also achieved by taking advantage of descriptors' space-time normalized location information provided at the end of the descriptors. Next, the extracted at different time scales the same kind descriptors augmented with location information are mixed together. Their dimension is reduced by PCA. The FVs representing four kinds of descriptors are concatenated at representation level fusion as in iDT but they are normalized twice, one time before and one time after concatenation, which brings a light enhancement.

Comprehensive Study – Hybrid Peng's System [8]. In this method, a hybrid system for action recognition was proposed. In contrary to other methods composed by many descriptors here representations not only from different low-level descriptors, but also from different BOW methods including two encoding schemes, namely FV and a variant of super vector coding (SVC-k) are concatenated. They chose HOG, HOF and MBH with the same settings as in iDT. Next, PCA and whitening are applied to reduce the dimensionality by a factor of two. For FV the GMM with 512 gaussians is trained, while k-means with k = 512 generates the codebook for SVC-k. For the final representations power normalization followed by intra l2-normalization is performed.

A Robust and Efficient INRIA Method [9]. The method combines the iDT with spatial pyramid and spatial Fisher vector (SFV) in the same manner as proposed Oneata [24], trying to preserve the spatio-temporal location of features in the video. However, the Oneata's method utilizes an old-fashioned traditional DT with MBH and SIFT descriptors. Instead, here HOG, HOF and MBH from iDT are used. The descriptors dimensionality is two times lower by employing PCA. They are fused at representation-level. Spatial pyramid divides the space-time volume into subvolumes in spatial or temporal domain. For instance, in the case of dividing the temporal domain into three subvolumes and spatial domain into two volumes there are five subvolumes for which five different FVs are performed and concatenated with the whole video representation giving the final representation. Spatial Fisher vectors add at the end of the representation the mean and variance of 3D feature location in the space-time volume. It must be mentioned that SFV enlarges the final representation slightly, whereas spatial pyramid in this configuration would enlarge it six times. However, this is usually solved by summing up their kernels. Both the approaches enhance the accuracy by about 2 % on HMDB51.

Stacked Fisher Vectors [2]. This method takes the iDT as input and provides a two-layered encoding structure. Therefore, it captures more abstract information and can be treated as a mid-level approach though discriminative action parts are not extracted. Firstly, they extract a 396 dimensional descriptor, which is a concatenation of derived from the iDT method HOG, HOF and MBH descriptors in sampled subvolumes. The dimension is further reduced to 200 by PCA and whitening. The amount of subvolumes per video ranges from 600 to 6000 depending on the dataset. Subsequently, FV for each subvolume is performed producing a high-dimensional representation. Hence, max-margin reduction is utilized to deliver drastically lower 400-dimensional representation reduced again by PCA and whitening to 200. These FV representations of volumes are encoded by second FV. Stacked FVs perform very well with actions without much variations. It turned out that one layered, traditional FV slightly outperforms the stacked FVs, but they are complementary and the combination of the two representations gives the best results.

15 k Object Categories [1]. Here, objects are added to the video representation. Their own FVs obtained on descriptors from iDT, fused with stacked FVs and further enriched by object representation achieved 71.3 % accuracy on HMDB51. The other results were obtained without stacked FVs.

4.2 Random Sampling Methods

GBH method [7] takes advantage of random sampling method [28], which in turn involves local part model (LPM) [27]. Random sampling takes randomly 10,000 features points from very dense sampling grid. The utilized LPM is composed of one root part at half resolution and eight overlapping local parts in the surrounding. This model tries to capture the spatio-temporal organization of the different parts. For each part the descriptors are computed separately and then concatenated, which results in nine times richer representation. GBH method extends this approach by introducing the GBH descriptor, combined for the best result with MBH. Moreover, the FV (k = 128) preceded by PCA is employed. In addition, here the root and local parts channels are encoded separately and pooled later together also with other descriptors channels at representation level. This method takes only 11,46 % out of all dense sampling points on HMDB51. The amount of features per frame on HMDB51 is similar to DT.

4.3 CNN Based Methods

Two-Stream ConvNets [10] treats the spatial and temporal information from video separately. Both streams are built on a CNN architecture, which consists of five convolutional layers and three fully-connected layers. For the first, still images from ImageNet challenge dataset [33] were employed for pre-learning and only the last layer was trained on the target action recognition dataset. The later channel utilizes stacked multiple-frame OF field thus providing motion feature before the method indeed starts. They stack 10 subsequent frames for an input and compute a displacement vector fields between them, separately in x and y direction providing 20 input channels. For temporal net, multi-task learning on both HMDB51 and UCF101 datasets was performed. The obtained softmax scores were further combined by linear SVM resulting in best reported accuracy. It is worth remarking that spatial nets alone are notably inferior to temporal nets. It occurs that motion information have a big influence on the results. They argue that the popular hand-crafted motion features are generalized by their temporal representation. Due to a large amount of parameters and methods to be set here please refer to the original paper for more details.

5 Conclusion

We provided a discussion of space-time based methods for action recognition. Their results indicate a large amount of work to be done to achieve an accurate solution of this problem. The easiest way to improve the ranked methods is to leverage the good practices and ideas proposed by others. For instance, approaches using HOG, HOF and other descriptors based on consecutive frames should apply the idea of frame multi-skipping which could speed up the method or bring better performance. In the face of the lately emerged, well-performing descriptors such as GBH or obtained by the recently introduced genetic programming approach [12] there is a place for evaluation of different descriptors combinations and their influence on particular classes. This is, however, limited by different tools and technologies utilized by research teams. A big work for the whole researcher community would be an open source system that divides the different steps of BOW approach and is easy to fill with other methods. The spatial pyramid, SFV, LPM and location added at the end of descriptors are a step further into alleviating the location and spatio-temporal structure information lost during encoding. Nonetheless, elaborating an encoding method that efficiently preserves these information remains an interesting research direction. The fast improvement of deep learning methods in recent years and a big interest on this topic suggest that this trend will be continued. However, further enhancements concerns probably more complex architectures and a bigger amount of training data. The better progress in these areas the more computational time will be required. This limits the methods for common use. Due to a large number of approaches a more elaborately designed classification system taking into account representations and scores of different approaches seems to be an important research area.

Acknowledgments. This work has been supported by the National Centre for Research and Development (project UOD-DEM-1-183/001 "Intelligent video analysis system for behavior and event recognition in surveillance networks").

References

1. Jain, M., Gemert, J.C., Snoek, C.G.M.: What do 15,000 object categories tell us about classifying and localizing actions? In: CVPR, pp. 46–55 (2015)
2. Peng, X., Zou, C.Q., Qiao, Y., Peng, Q.: Action recognition with stacked fisher vectors. In: Fleet, D., Pajdla, T., Schiele, B., Tuytelaars, T. (eds.) ECCV 2014, Part V. LNCS, vol. 8693, pp. 581–595. Springer, Heidelberg (2014)
3. Wang, L., Qiao, Y., Tang, X.: Action recognition with trajectory-pooled deep-convolutional descriptors. In: CVPR, pp. 4305–4314 (2015)
4. Ni, B., Moulin, P., Yang, X., Yan, S.: Motion part regularization: improving action recognition via trajectory group selection. In: CVPR, pp. 3698–3706 (2015)
5. Lan, Z., Lin, X., Li, X., Hauptmann, A.G., Raj, B.: Beyond gaussian pyramid: multi-skip feature stacking for action recognition. In: CVPR, pp. 204–212 (2015)
6. Fernando, B., Gavves, E., Oramas, J., Ghodrati, A., Tuytelaars, T.: Modeling video evolution for action recognition. In: CVPR, pp. 5378–5387 (2015)
7. Shi, F., Laganiere, R., Petriu, E.: Gradient boundary histograms for action recognition. In: WACV, pp. 1107–1114 (2015)
8. Peng, X., Wang, L., Wang, X., Qiao, Y.: Bag of visual words and fusion methods for action recognition: comprehensive study and good practice. In: arXiv preprint arxiv:1405.4506 [cs.CV] (2014)
9. Wang, H., Oneata, D., Verbeek, J., Schmid, C.: A robust and efficient video representation for action recognition. In: arXiv preprint arxiv:1504.05524 [cs.CV] (2015)
10. Simonyan, K., Zisserman, A.: Two-stream convolutional networks for action recognition in videos. In: NIPS, pp. 568–576 (2014)
11. Wang, H., Schmid, C.: Action recognition with improved trajectories. In: ICCV, pp. 3551–3558 (2013)
12. Liu, L., Shao, L., Li, X., Lu, K.: Learning spatio-temporal representations for action recognition: a genetic programming approach. IEEE Trans. Cybern. **46**, 158–170 (2015)
13. Wang, H., Kläser, A., Schmid, C., Liu, C.-L.: Action recognition by dense trajectories. In: CVPR, pp. 3169–3176 (2011)
14. Wang, L., Xiong, Y., Wang, Z., Qiao, Y.: Towards good practices for very deep two-stream ConvNets. In: arXiv preprint arxiv:1507.02159 [cs.CV] (2015)

15. Tran, D., Bourdev, L., Fergus, R., Torresani, L., Paluri, M.: Learning spatiotemporal features with 3D convolutional networks. In: arXiv preprint arxiv:1412.0767 [cs.CV] (2015)
16. Xu, Z., Yang, Y., Hauptmann, A.G.: A discriminative CNN video representation for event detection. In: CVPR, pp. 1798–1807 (2015)
17. Xu, Z., Zhu, L., Yang, Y., Hauptmann, A.G.: UTS-CMU at THUMOS2015. In: THUMOS challenge 2015 (2015)
18. Gorban, A., Idrees, H., Jiang, Y.-G., Roshan Zamir, A., Laptev, I., Shah, M., Sukthankar, R.: THUMOS challenge: action recognition with large number of classes (2015). http://www.thumos.info/
19. Aggarwal, J.K., Ryoo, M.S.: Human activity analysis: a review. ACM Comput. Surv. **43**(3), 16 (2011)
20. Ke, S.-R., Thuc, H.L.U., Lee, Y.-J., Hwang, J.-N., Yoo, J.-H., Choi, K.-H.: A review on video-based human activity recognition. Computers **2**(2), 88–131 (2013)
21. Cheng, G., Wan, Y., Saudagar, A.N., Namuduri, K., Buckles, B.P.: Advances in human action recognition: a survey. In: arXiv preprint arxiv:1501.05964 [cs.CV] (2015)
22. Wang, H., Ullah, M.M., Kläser, A., Laptev, I., Schmid, C.: Evaluation of local spatio-temporal features for action recognition. In: BMVC, pp. 124.1–124.11 (2009)
23. Uijlings, J., Duta, I.C., Sangineto, E., Sebe, N.: Video classification with densely extracted HOG/HOF/MBH features: an evaluation of the accuracy/computational efficiency trade-off. IJMIR **4**(1), 33–44 (2014)
24. Oneata, D., Verbeek, J., Schmid, C.: Action and event recognition with fisher vectors on a compact feature set. In: ICCV, pp. 1817–1824 (2013)
25. Kantorov, V., Laptev, I.: Efficient feature extraction, encoding and classification for action recognition. In: CVPR, pp. 2593–2600 (2014)
26. Laptev, I., Lindeberg, T.: Space-time interest points. In: ICCV, pp. 432–439 (2003)
27. Shi, F., Petriu, E.M., Cordeiro, A.: Human action recognition from local part model. In: Proceedings of the IEEE International Haptic Audio Visual Environments and Games (HAVE) Workshop, pp. 35–38 (2011)
28. Shi, F., Petriu, E., Laganiere, R.: Sampling strategies for real-time action recognition. In: CVPR, pp. 2595–2602 (2013)
29. Karpathy, A., Toderici, G., Shetty, S., Leung, T., Sukthankar, T., Fei-Fei, L.: Large-scale video classification with convolutional neural networks. In: CVPR, pp. 1725–1732 (2014)
30. Kuehne, H., Jhuang, H., Garrote, E., Poggio, T., Serre, T.: HMDB: a large video database for human motion recognition. In: ICCV, pp. 1550–5499 (2011)
31. Marszalek, M., Laptev, I., Schmid, C.: Actions in context. In: CVPR, pp. 2929–2936 (2009)
32. Soomro, K., Zamir, A.R., Shah, M.: UCF101: a dataset of 101 human actions classes from videos in the wild. In: arXiv preprint arxiv:1212.0402 [cs.CV] (2012)
33. Berg, A., Deng, J., Fei-Fei, L.: Large scale visual recognition challenge (ILSVRC) (2010). http://www.image-net.org/challenges/LSVRC/2010

Recent Developments in Tracking Objects in a Video Sequence

Michał Staniszewski[1,2], Mateusz Kloszczyk[1], Jakub Segen[1], Kamil Wereszczyński[1,2], Aldona Drabik[1], and Marek Kulbacki[1(✉)]

[1] Polish-Japanese Academy of Information Technology, Koszykowa 86, 02-008 Warszawa, Poland
mk@pjwstk.edu.pl

[2] Institute of Informatics, Silesian University of Technology, Akademicka 16, 44-100 Gliwice, Poland

Abstract. Methods of tracking of multiple objects or people in video sequences have applications in many fields such as surveillance, art, transport or biology. This, over four decades old area is still very active, with multiple new contributions presented every year. Tracking methods must solve intricate problems, for example occlusion of many objects, crowded scenes, illumination of different places and motion of camera. This paper presents a brief survey of recent developments in video tracking based methods, focused mainly on the last three years. The surveyed methods are divided into two groups: tracking by detection, which includes methods that solve the problem of time-linking objects detected in all video frames, and tracking by correlation, containing methods that follow a selected object using cross correlation. The reviewed methods are collected in a table that lists for each method the benchmark datasets used for its evaluation, implementation environment, and whether it can track single or multiple objects.

Keywords: Object tracking · Computer vision · Statistical analysis · Video signal processing

1 Introduction

In the last four decades many approaches were designed in the field of tracking of multiple objects, which can be applied in surveillance, detecting human behavior and many other aspects of computer vision. There are many problems in this area that should be taken into consideration. In real tracking problems people may be occluded by other persons or objects. Video sequences may be influenced by different kind of light and illumination which lead to many changes in colors. Finally camera can move and the same person or object should be still traced. The development of tracking approaches from last decade led to creation of many methods having different type of input and output. Current methods were tested also on different data sets which is not reliable for larger evaluation. That problem was introduced in [3] and as a result authors collected and shared

N.T. Nguyen et al. (Eds.): ACIIDS 2016, Part II, LNAI 9622, pp. 427–436, 2016.
DOI: 10.1007/978-3-662-49390-8_42

publicly available data benchmarks in one place with important annotations consisting of many different factors affecting tracking performance.

The tracking algorithms may be divided into few parts, which can be formulated in many different ways. Representation model of object can be described in the manner of sparse representation presented and improved in [11]. Additionally tracking algorithms may use visual features such as histograms of oriented gradients [5] or color [4] and Haar-like features [6]. In order to discriminate the target many learning methods have been applied such as Support Vector Machines (SVM) with its modifications [7] or boosting methods [8]. In order to find the target localization deterministic [4] and stochastic [8] methods have been used. To deal with appearance changes the object model has to be updated which can be done effectively by proposed algorithms of online mixture model [9] or online boosting [6]. Surrounding context is also important in discriminating objects in occlusion which has been presented in [10] (Fig. 1).

Fig. 1. Possible problems that may be present during procedure of tracking. On the left side objects traveling in one group, on the right side many crossing tracks with possibility of occlusions.

Contributions of that paper into tracking problem rely on comparison of methods from last three years which have shared source code on project's web page (that gives opportunity to real comparison). Authors presented mainly methods that solve the most challenging problem which is tracking of multiple objects. The paper consists also of methods dealing only with single object but using correlation filters. In comparison to other current surveys that paper is much more comprehensive [1,2] because it describes public benchmarks and much more methods and parameters of evaluation.

The paper is organized in following way - in the second section authors describe current state-of-the-art in the groups of tracking by detection and by use of correlation filters. In the third and fourth sections authors present available data benchmarks containing video sequence and possible parameters used in order to mark investigated methods.

2 Current Methods

The section is built from descriptions of few methods which were divided into two groups. The first one contains methods based on tracking by detection while the second one methods using correlation filters. All of them have available source code which can be downloaded and tested. Authors added also few comments for each method in the order of advantages and disadvantages.

2.1 Tracking by Detection

Group of methods located in the term of tracking by detection has one common feature. The possible approach contains of application of discriminative appearance learning. Such methods contain online learning by use of appearance prediction of given object in the next frame of video. Such prediction can be obtained thanks to algorithms of SVM [7], random forest methods or boosting models [8]. A number of these methods link detections in different frames by solving an optimization problem defined on a graph. It can be solved either locally by application of hungarian algorithm where the problem is approximated or globally where instead of few frames, methods use batch of frames getting more global overview. Additionally methods of tracking by detection can have two categories - batch and online. Batch methods use in the analysis detections of all frames within batch and connect short parts (tracklets) in longer parts (tracks). Such approach requires much bigger computational power. On the other hand online methods can be applied for many issues in the real time by sequencing building of trajectories in connection frame by frame.

Tracking Multiple People Online and in Real Time [12]. The method of Online tracking of multiple objects uses a multi-stage cascade combined with a sliding temporal window. In fact input of given source code consists of videos and set of detections which can be described by its appearance feature, position in time and estimated velocity that object moves. In order to describe appearance of a person in algorithm an HSV color histogram is applied, but algorithm can be extended with many other descriptors without any modification of that method. For each pair of observations (coming from person detection) algorithm measures the evidence of being or not identical. The value of evidence is computed basing on data and calculated as a measure of correlation, which can be a number coming from the set (−inf, [−1, 1], +inf). If the pair of observations evidence is indicated by positive value while negative shows evidence against and zero is connected to indifference. Infinite (negative and positive) corresponds to hard constraints. The scheme of algorithm consists of two simple phases built from two stages each. In the first part partial detections are connected to form of short tracklets (in short time) basing on appearance and space-time affinities. In the second part method operates on entire temporal window which results in whole trajectories. **Advantages** Approach similar to GMCP [13], multi person tracking is done jointly for all identities (in contrast to sequential GMCP), result depends mainly on number of observations (not on length of sequence), implementation consists of separate parts connected to detections and tracklets. **Disadvantages** Formulation of correlation matrix used as a distance and similarity comperator causes time complexity.

The Way they Move: Tracking Multiple Targets with Similar Appearance [14]. The algorithm is computationally efficient and is dedicated for tracking of multi-object by detection that can overcome four main problems: camera movement, appearance similarity of many targets, lack of data due to being out of the field or occlusion, crossing trajectories. The method takes as input a set of short tracklets that have different lengths and no appearance information. The main problem is related with connection of tracklets that belong to similar trajectories. In this paper authors propose set of methods using dynamics of motion that can handle many objects described by long tracks with possibility of occlusion and lack of data. In the first part the problem is formulated as a generalized linear assignment (GLA) [15] of short tracklets which are used to built longer trajectories by use of similarity of motion which needs efficient algorithms in order to estimate such similarities. The algorithm does not require any prior assumption connected to starting and ending nodes or length of trajectories. Estimation of dynamics similarity can be done thanks to two algorithms proposed by authors. The first one, presented in paper, alternating direction method of multipliers (ADMM) [16] uses different optimization which has lower memory requirements and relies on solving a relaxation and assumes choice of parameters responsible for noise penalty and estimation of the rank basing on singular values. In second case authors propose new algorithm basing on iterative Hankel Total Least Squares (IHTLS) [17] which is connected in finding the rank of noisy data formulated in Hankel matrices. That algorithm was contributed by authors to clean noisy matrices and estimate rank. **Advantages** Method allows tracks to start and terminate anywhere in position and time, algorithm can operate at the tracklet level, but also all the data on a tracklet rather than a selected portion, does not assume any priors for the target motion. **Disadvantages** Algorithm should be compared with state-of-the-art tracking method basing on currently available benchmarks in order to check its reliability.

GMMCP Tracker: Globally Optimal Generalized Maximum Multi Clique Problem for Multiple Object Tracking [18]. The method of tracking is formulated as a data association of Generalized Maximum Multi Clique problem (GMMCP). Input for the algorithm is based on finding detection in each frame described in [31] to get the detection hypothesis in each frame. Long tracks are built from shorter tracklets in two layer framework. GMMCP starts with low level tracklets which are limited to a maximum of 10 frames due to elimination of tracklets shorter than 5 frames. In the next step low level tracklets in segments create the input which will form the first layer of framework returning mid level tracklets, which are used later in forming the final trajectories. In comparison to different tracklets, algorithm takes into consideration appearance affinity, using for each node color histogram. The histogram is calculated for each detection and the representation of the nodes are computed by the median appearance. In order to find the appearance affinity of two tracklets algorithm calculates the histogram intersection. Additionally in tracking method motion model is taken in the form of constant velocity. **Advantages** Contribution to GMCP [13] method as a new graph theoretic problem, efficient occlusion handling, improvement in comparison to other methods, performance close to real time. **Disadvantages** The time complexity increases with the number of objects.

Joint Tracking and Segmentation of Multiple Targets [19]**.** The algorithm is dedicated for the problem of multi-target tracking that can exploit low level image information and connect each pixel to an investigated object or treats it as background. The result of the method in the form of video segmentation can work in real world videos. The algorithm takes as an input video sequence and starts with assigning a unique ID not only for each target detection but also to specially defined superpixel in videos. Thanks to that idea the method can be used in recovery of trajectories in long occlusion due to existence of superpixels even when the detections are missed. Authors used multi-label conditional random field (CRF) in order to model the problem, which is stated as finding hypotheses of long trajectories that best describe the data basing on the low-level information. **Advantages** Improvement by 0,1 on average of missing recalls, while reduction of the number of ID switches. **Disadvantages** Needs improvement in order to work in time close to real time.

Tracking Multiple High-Density Homogeneous Targets [20]**.** The method was created to deal with detection and tracking multiple objects which are dense and homogeneous. Objects detection is made by a technique of gradients and usage of isocontour for intensity maps. The method of tracking recursively connects detection by use of graph in temporal window, which is solved by greedy algorithm. Problem of detection was solved by automatic finding of objects thanks to local maximum. Additionally detector uses isocontour on intensity maps, which contributes connection with objects. Tracking works on short tracklets and by use of greedy algorithm validates results backward in temporal window. Authors added possibility of online reduction of false tracks in connection step. The whole process consists of connection of detection in time, forecasting detection in particular frames and reduction of false detections. At the beginning short tracks are generated by Hungarian algorithm from those detections having the biggest affinities. Long trajectories are built from short tracks with probability of connection. **Advantages** Detection is background independent and does not require learning of model features such as color and texture. Good results for detections and tracking for prepared datasets. **Disadvantages** Complex algorithm with good results however not so competetive. For public benchmarks observable changes in ID switch.

Robust Online Multi-object Tracking Based on Tracklet Confidence and Online Discriminative Appearance Learning [21]**.** The main goal of the method was to create algorithm that will track multiple objects basing on tracklet confidence, which is evaluated on the base of detection and continuity. In the next step detections are connected into tracklets by use of Online discriminative appearance learning getting difference in objects. In order to face with occlusion across whole video sequence method uses tracklet confidence and the whole problem is solved depending on value of confidence - reliable tracklets with high confidence are connected locally while low tracklets with low confidence come back into remaining and not finished tracklets and detections. The crucial moment in presented method relies on connecting tracklets globally and locally in order to form one object. For that reason author proposed application of algorithm of Online discriminative appearance learning which is used for updating model according to results of tracking and online training of collecting data in order to distinguish occurring new objects. Such assumption can be satisfied by Incremental Linear Discriminant Analysis method (ILDA) [22], which helps in distinguishing many objects and updating model. **Advantages** Algorithm simple, intuitive and well tested. Application of tracklet confidence and discriminative online learning. **Disadvantages** Matlab implementation needs to be optimized, large time complexity.

2.2 Tracking by Correlation

In last years tracking has been enriched by methods of correlation filters. In comparison to tracking by detection methods, the results of application of correlation filters lead to less computational load even on hundreds of frames. The main advantage of that group lies in the lack of necessity in iterating over all objects and also in application of Fourier domain.

Accurate Scale Estimation for Robust Visual Tracking [23]**.** In contrast to other method that algorithm strongly takes into consideration robust scale estimation for visual tracking. The method start with videos formulated as raw pixels or Histograms of Oriented Gradients (HOG) [28] and uses learning discriminative correlation filters working on a representation of scale pyramid. The improvement of scale searching bases on learning different filters for translation and scale estimation. The method contributes the discriminative correlation filters first time applied in the MOSSE tracker [24]. One of the advantages relies on incorporating idea of scale estimation on other tracking frameworks. **Advantages** Method is more than 2.5 times faster than Struck [25], 25 times faster than Adaptive Structural Local Sparse Appearance (ASLA) [26] and 250 times faster than Sparsity Collaborative Model (SCM) [27] in median frame per second (FPS). **Disadvantages** Results are presented for single moving object and needs modification in order to serve for multiple objects.

Fast Visual Tracking via Dense Spatio-Temporal Context Learning [29]**.** The presented robust algorithm uses the dense spatio-temporal context for visual tracking and takes video sequence as an input. In the first part of algorithm a model of spatial context is evaluated between the object and its local background by means of spatial correlations by solving deconvolution problem. In the next frame the learned spatial context is used in order to update the model. Building trajectories is made by formulating a confidence map as a convolution problem. The best possible object position can be computed by getting maximum of the confidence map and adapting a novel scale estimation scheme which results in final track. **Advantages** The proposed algorithm is simple and fast that needs only 4 computation of Fast Fourier Transforms (FFT) at 350 FPS in MATLAB. Method uses explicit scale update **Disadvantages** Method was tested mainly on data prepared by authors.

High-Speed Tracking with Kernelized Correlation Filters [30]**.** The method uses a correlation filter in tracking problem for videos built from raw pixels or HOG descriptors [28]. The presented tool can operate on thousands of objects with different relative translations without a need of iterating on them. That solution works in the Fourier domain where some learning algorithms can operate even if new samples are added for specific model. Correlation filters use a feature that the convolution of two samples (dot-product at different translations) is equal to an element-wise product performed in the Fourier domain. Thanks to that advantage and application of the Fourier domain, the desired results of linear classifier can overcome problems for many translations and image shifts. **Advantages** Taking advantage of all 4 cores of a desktop computer, method take less than 2 min to process all 50 videos (29,000 frames). **Disadvantages** Lack of comparison to available benchmarks (Table 1).

Table 1. Summary of all described methods with respect to benchmarks applied for evaluation and statement of used code. Additionally possibility of method to deal either with single object or multi objects.

Method	Benchmarks	Code	Object
Tracking multiple people online in real time [12]	Towncenter [33], pets2009 [32], Parkinglot [13]	Matlab	Multi object
The way they move: tracking multiple targets [14]	Own data [14], TUD [34]	Matlab	Multi object
GMMCP tracker [18]	TUD [34], Parkinglot 2 [18]	Matlab	Multi object
Joint tracking and segmentation of multiple targets [19]	TUD campus [34]	Matlab	Multi object
Tracking multiple high-density homogeneous targets [20]	ETH [35], pets [32], TUD [34]	Matlab	Multi object
Multi-object tracking based on tracklet confidence [21]	ETH [35]	Matlab	Multi object
Accurate scale estimation for robust visual tracking [23]	Own data [23]	Matlab	Single object
Fast visual tracking via dense spatio context learning [29]	Own data [29]	Matlab	Single object
High-speed tracking with kernelized correlation filters [30]	Own data [30]	Matlab	Single object

3 Available Data Benchmarks

There are many currently used video benchmarks that are tested in order to compare presented methods. Thanks to such dataset already published methods can be compared in terms of similar aspects. All of benchmarks are built from many videos or set of pictures that are presenting different scenes with partial occlusion and other tracking problem. Some of them consist of ground truth table and detections that can be used as a reference. The most popular videos are stored in benchmarks:pets2009 [32], towncenter [33], Parkinglot and Parkinglot 2 [13,18], TUD (Crossing, Campus, Stadtmitte) [34], ETH (Bahnhof, Jelmoli, Sunny Day) [35], videos from own dataset and youtube containg moving objects, sport scenes etc. [14].

4 Methods of Evaluation

Authors of previous works used many different parameters which are used for comparison purposes. Most of them are based on a specially prepared ground truth table which consists of hand made detections. The most popular parameters are listed below.

- increase the number of false positive detections by adding random detections into the set or increase the number of false negative detections by removing correct detections from the set [14],
- number of ID switches when trajectories are crossing [14],
- Score as a relation of area of common part of detected bounding box and ground truth to their sum [14],
- distance precision calculated as the average Euclidean distance between the estimated centre location of the target and the ground-truth [23],
- centre location error computed as the relative number of frames in the sequence where the location error is smaller than a certain threshold [23],
- overlap precision defined as the percentage of frames where the bounding box overlap surpasses a threshold [23],
- multiple Object Tracking Accuracy (MOTA) combines the number of false positives, false negatives and identity switches over all frame indices [12].

5 Conclusion

New methods of tracking multiple objects are being developed at an even pace and the field remains competitive. The present survey of recent work in the area shows that current methods can be very effective in a wide range of environments and imaging conditions. However, there is still much room for improvement, especially in cases of multiple occlusions and where there are many objects in close proximity. Also, the speed of execution needed for real time tracking remains a challenge for the computationally more demanding methods. The continuing work will be devoted to a precise performance comparison among the current methods, by testing them on a unified benchmarking dataset and using the same configuration parameters.

Acknowledgments. This work has been supported by the National Centre for Research and Development (project UOD-DEM-1-183/001 "Intelligent video analysis system for behavior and event recognition in surveillance networks").

References

1. Li, X., Hu, W., Shen, C., Zhang, Z., Dick, A., van den Hengel, A.: A Survey of Appearance Models in Visual Object Tracking (2013). CoRR abs/1303.4803
2. Chu, D.M., Cucchiara, R., Calderara, S., Dehghan, A., Shah, M.: Visual tracking: an experimental survey. Pat. An. Mach. Intel. **36**, 1442–1468 (2013)
3. Wu, Y., Lim, J., Yang, M.-H.: Online Object Tracking: A Benchmark CVpPR 2013, pp. 2411–2418 (2013). http://visual-tracking.net
4. Comaniciu, D., Ramesh, V., Meer, P.: Kernel-based object tracking. PAMI **25**(5), 564–577 (2003)
5. Dalal, N., Triggs, B.: Histograms of oriented gradients for human detection. In: CVPR (2005)
6. Grabner, H., Grabner, M., Bischof, H.: Real-time tracking via on-line boosting. In: BMVC (2006)
7. Avidan, S.: Support vector tracking. PAMI **26**(8), 1064–1072 (2004)
8. Babenko, B., Yang, M.-H., Belongie, S.: Visual tracking with online multiple instance learning. In: CVPR (2009)
9. Jepson, A.D., Fleet, D.J., El-Maraghi, T.F.: Robust online appearance models for visual tracking. PAMI **25**(10), 1296–1311 (2003)
10. Santner, J., Leistner, C., Saffari, A., Pock, T., Bischof, H.: PROST: parallel robust online simple tracking. In: CVPR (2010)
11. Mei, X., Ling, H.: Robust visual tracking using L1 minimization. In: ICCV (2009)
12. Ristani, E., Tomasi, C.: Tracking multiple people online and in real time. In: 12th Asian Conference on Computer Vision, pp. 444–459 (2014). https://www.cs.duke.edu/ristani/bip_tracker.html
13. Zamir, A.R., Dehghan, A., Shah, M.: GMCP-Tracker: global multi-object tracking using generalized minimum clique graphs. In: Proceedings of the 12th European Conference on Computer Vision, pp. 343–356 (2012). http://crcv.ucf.edu/projects/GMCP-Tracker/
14. Dicle, C., Camps, O., Sznaier, M.: The Way They Move: Tracking Multiple Targets with Similar Appearance Computer Vision (ICCV) (2013). https://bitbucket.org/cdicle/smot
15. Rossand, G., Soland, R.: A branch and bound algorithm for the generalized assignment problem. Math. Program. **8**(1), 91–103 (1975)
16. Ayazoglu, M., Sznaier, M., Camps, O.: Fast algorithms for structured robust prin cipal component analysis. In: CVPR, pp. 1704–1711 (2012)
17. Park, H., Zhang, L., Rosen, J.: Low rank approximation of a hankel matrix by structured total least norm. BIT Numer. Math. **39**(4), 757–779 (1999)
18. Dehghan, A., Assari, S., Shah, M.: GMMCP tracker: globally optimal generalized maximum multi clique problem for multiple object tracking. In: The IEEE Conference on Computer Vision and Pattern Recognition (CVPR), pp. 4091–4099 (2015). http://crcv.ucf.edu/projects/GMMCP-Tracker/
19. Milan, A., Leal-Taixe, L., Schindler, K., Reid, I.: Joint tracking and segmentation of multiple targets CVPR (2015). https://bitbucket.org/amilan/segtracking
20. Poiesi, F., Cavallaro, A.: Tracking multiple high-density homogeneous targets. IEEE Trans. Circ. Syst. Video Technol. **25**, 623–637 (2015). http://www.eecs.qmul.ac.uk/andrea/thdt.html
21. Bae, S.-H., Yoon, K.-J.: Robust online multi-object tracking based on tracklet confidence and online discriminative appearance learning CVPR, pp. 1218–1225 (2014). https://cvl.gist.ac.kr/project/cmot.html

22. Kim, T.-K., Stenger, B., Kittler, J., Cipolla, R.: Incremental linear discriminant analysis using sufficient spanning sets and its applications. IJCV **91**(2), 216–232 (2011)
23. Danelljan, M., Hager, G., Shahbaz, K., F., Felsberg, M.: Accurate scale estimation for robust visual tracking. In: Proceedings of the British Machine Vision Conference (2014). http://www.cvl.isy.liu.se/en/research/objrec/visualtracking/scalvistrack/index.html
24. Bolme, D.S., Beveridge, J.R., Draper, B.A., Lui, Y.M.: Visual object tracking using adaptive correlation filters. In: Computer Vision and Pattern Recognition (2010)
25. Hare, S., Saffari, A., Torr, P.: Struck: structured output tracking with kernels. In: Computer Vision and Pattern Recognition (2011)
26. Jia, X., Lu, H., Yang, M.-H.: Visual tracking via adaptive structural local sparse appearance model. In: Computer Vision and Pattern Recognition (2012)
27. Zhong, W., Lu, H., Yang, M.-H.: Robust object tracking via sparsity based collaborative model. In: Computer Vision and Pattern Recognition (2012)
28. Dalal, N., Triggs, B.: Histograms of oriented gradients for human detection. In: Proceedings of IEEE Conference Computer Vision and Pattern Recognition, pp. 886–893 (2005)
29. Zhang, K., Zhang, L., Liu, Q., Zhang, D., Yang, M.H.: Fast visual tracking via dense spatio-temporal context learning. In: 13th European Conference, Zurich, pp. 127–141 (2014). http://www4.comp.polyu.edu.hk/cslzhang/STC/STC.htm
30. Henriques, J.F., Caseiro, R., Martins, P., Batista J.: High-Speed Tracking with Kernelized Correlation Filters, CoRR (2014). abs/1404.7584http://home.isr.uc.pt/henriques/circulant/
31. Felzenszwalb, P., Girshick, R., McAllester, B., Ramanan, D.: Object detection with discriminatively trained part based models. IEEE Trans. Pattern Anal. Mach. Intell. **32**(9), 1627–1645 (2010)
32. Ferryman, J.: Proceedings (pets 2009). Eleventh IEEE International Workshop on Performance Evaluation of Tracking and Surveillance (2009)
33. Benfold, B., Reid, I.: Stable multi-target tracking in real-time surveillance video. In: Computer Vision and Pattern Recognition (2011)
34. Andriluka, M., Roth, S., Schiele, B.: People-tracking-bydetectionandpeople-detection-by-tracking. In: Proceedings of IEEE Conference Computer Vision and Pattern Recognition, pp. 1–8 (2008)
35. Ess, A., Leibe, B., Schindler, K., Van Gool, L.: A mobile vision system for robust multi-person tracking. In: Computer Vision and Pattern Recognition (2008)

Recent Developments on 2D Pose Estimation From Monocular Images

Artur Bąk[1], Marek Kulbacki[1](✉), Jakub Segen[1], Dawid Świątkowski[1], and Kamil Wereszczyński[1,2]

[1] Polish-Japanese Academy of Information Technology, Koszykowa 86, 02-008 Warszawa, Poland
mk@pja.edu.pl

[2] Institute of Informatics, Silesian University of Technology, Akademicka 16, 44-100 Gliwice, Poland

Abstract. Human pose estimation from monocular images is one of the most significant aspects of modern computer vision tasks and its application demand is still increasing in such areas as automatic images indexing or human activity recognition from video. Among many approaches applied in these areas the one based on pose estimation gives, beyond all doubts, one of the most powerful representation of human on the picture in sense of sparsity and semantics. In this paper we provide a detailed survey of the most efficient methods in 2D pose estimation domain as well as the test results of selected methods on the LSP dataset, which is commonly used by state-of-the-art works.

Keywords: Human pose estimation · PSM · Poselet

1 Introduction

To make a simple and general definition of human pose in computer vision one can describe it as a geometric configuration of its particular parts on the image. It may have a form of body parts locations, the angles between adjacent parts or both of them. Regardless of its exact form it gives a strong semantic description of human activity caught on the image. Additionally, such sparse representation is very compact as opposed to dense representations where human figure description requires huge amount of information [14]. This makes the pose based representation computationally efficient, easy to interpret and well suited to many applications like automatic image indexing or video activity recognition.

All these advantages are true when the estimated pose is accurate, which is not easy due to body parts occlusion, brightness variation, illumination and high degree of freedom for human body configuration. Good pose estimation algorithm should overcome these problems efficiently.

The most popular approach in pose estimation domain is that based on pictorial structure model (PSM) [1], which is commonly used in the recent methods [2–5]. All body parts in PSM are modeled as a tree-structured graph with

N.T. Nguyen et al. (Eds.): ACIIDS 2016, Part II, LNAI 9622, pp. 437–446, 2016.
DOI: 10.1007/978-3-662-49390-8_43

flexible connections between them using spring-like model, which allows to detect the human pose in very efficient and flexible way.

Another commonly used branch of pose estimation consists of methods based on *poselet* concept, that is, the patches of image representing bigger body parts than single joints or segments between them. This method was originally proposed in [6] and used in some later methods [7,8]. There are also the hybrid approaches that merge the PSM and poselet methods in order to increase the pose estimation quality [8].

To optimize the pose estimation task and decrease its complexity there exists a special group of methods which cover only the pose of upper body [19–21] but their applications are limited only to domains that do not need the whole human body representation. Another enhancement is to use a video context from adjacent frames to enhance the pose detection on the single frame [22,23], which also causes a limitation to domains using the video instead of single images.

In the last years the deep learning methods and neural networks [24–27] are more and more popular in pose estimation. They achieve even better precision than classical methods but they are usually worse with regard to computational efficiency and interpretation of algorithm workflow in case of problems.

Consecutive chapters describe the approaches based on PSM and poselets as we found them to be a compromise between precision and computational efficiency as well as readiness to use them in practice. We also provide the empirical results achieved by these methods on the *Leeds Sports Poses* (LSP) dataset [18].

Related Work. There were a few works created in the last few years that provided different summaries of the human pose estimation domain as well as empirical results of selected methods. Among the most recent works, Andriluka *et al.* [9] summarized a recent state-of-the-art analysis in 2D human pose estimation domain as well as introduced a new dataset for methods comparison. They considered relatively few methods and rather focused on exploring different aspects that are most problematic in current pose estimation methods. Liu *et al.* [10] made a summary of latest technical progress with discussion of current limitations and perspective for the future.

The main goal of this paper is to collect the state-of-the-art methods that we found most interesting in sense of precision and efficiency as well as highlight the practical difference between them rather than considering the current capacities and perspectives of human pose estimation domain.

2 Pictorial Structure Model

The pictorial structure model (PSM) is relatively old approach in object recognition domain introduced by Fischler and Elschlager more than 40 years ago [15]. Nevertheless, it is still eagerly selected method for human pose estimation.

The object being recognized is modeled as a tree where nodes represent the object parts and edges represent their physical connections, which is used in optimization task to find the best matching of the tree with given image.

The PSM concept can be easily applied for human body figure, where nodes represent the physical body parts as head, torso or upper arm and edges represent their physical connections. Following one of the first applications of PSM in human pose estimation [1], let us assume that the object consists of n parts and it can be expressed on given image as a spatial configuration of its parts in form of locations $L = (l_1, ..., l_n)$. Additionally the relations between particular parts are modeled by graph $G = (V, E)$, where V represents n parts $(v_1, ..., v_n)$ and E represents the physical connections between the parts.

For given image I, the detection task is to find the optimal configuration of object $L^* = (l_1^*, ..., l_n^*)$ by optimizing the objective function F over different locations L. The general form of objective function F is following:

$$F(I, L) = \sum_{i=1}^{n} \phi(I, l_i) + \sum_{ij \in E} \varphi(l_i, l_j) \tag{1}$$

The appearance term $\phi(I, l_i)$ measures the cost of visual matching of part v_i at location l_i on image I. The pairwise term $\varphi(l_i, l_j)$ measures the deformation cost for two adjacent parts v_i and v_j at given locations (l_i, l_j).

The appearance term can be intuitively understood as a visual part detector while the pairwise term as a kinematic tree, that all possible configurations of parts must be compatible with. The spring-like model used in such kinematic tree may handle slight anatomical differences between humans and minimizes the influence of small variations in camera viewpoint. This flexibility of model mitigates significantly the important problems of monocular camera usage like parts foreshortening effect. The conceptual illustration of PSM model for two same poses of two anatomically different persons is presented in the Fig. 1.

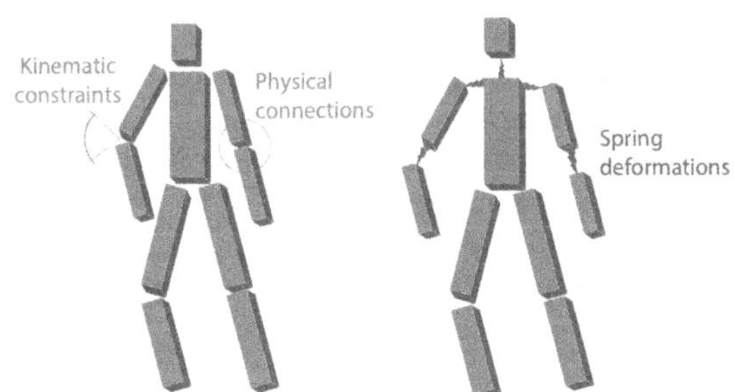

Fig. 1. PSM instances of the same pose for two anatomically different persons

Though modeling human figure in this way seems to be very simple and generic, the main problem with this approach is that optimizing the objective function efficiently is hard due to big number of model parameters. Additionally the local optima can be found instead of single global one, which may lead to highly unexpected results in case of pose estimation. Another aspect not easy to handle is the part detector invariant to colors and textures variability. Selected methods addressing these problems are presented in following sections.

2.1 Flexible Mixture Model

Yang and Ramanan in [2] proposed the flexible mixture of parts as an extension to original PSM model and this became the classical approach in current human pose estimation domain. Their method still achieves the state-of-the-art results and is the most willingly cited and compared method quoted in state-of-the-art works during the last few years. The main difference from standard PSM is that each part of body is treated as a mixture of templates, which quantizes the possible part orientations into limited and discrete number of classes - called *types*. The types for particular body part are obtained by clustering the different part orientations over the training set of images. Another significant difference is the use of color-invariant histogram of gradients (HOG) descriptor [17], which encodes the visual properties of body parts.

Assuming that $t = (t_1, ..., t_n)$ represents the types of all body parts in particular pose configuration, the Eq. (1) is replaced by following ones:

$$F(I, L, t) = S(t) + \sum_{i=1}^{n} \phi(I, l_i) + \sum_{ij \in E} \varphi(l_i, l_j) \tag{2}$$

$$S(t) = \sum_{i=1}^{n} b_i^{t_i} + \sum_{ij \in E} b_{ij}^{t_i, t_j} \tag{3}$$

The term $S(t)$ represents the types compatibility, where $b_i^{t_i}$ measures how typical is the type t_i for part v_i and $b_{ij}^{t_i, t_j}$ measures how typical is the co-occurrence of two types t_i and t_j for adjacent parts v_i and v_j. The appearance term $\phi(I, l_i)$ represents the visual part detector trained by latent support vector machines (SVM) formulated in [11] on training images with use of HOG descriptors extracted from part location l_i of image I. The pairwise term $\varphi(l_i, l_j)$ represents the standard deformation cost for adjacent parts. Finally, the optimization of the objective function is efficiently done by dynamic programming.

The use of part type concept significantly decreases the complexity of model parameter space while still keeping the possibility to represent the exponential number of possible poses. The use of HOG descriptor makes the part detector more efficient and less sensitive to color variabilities between different images.

2.2 Combined Parts

Wang and Li in [3] proposed an extension to standard PSM by introducing the additional node types to the tree. One of these additional node types is called *combined part* and may consist of a few single primitive parts e.g. the whole lower body consisting of four elements: left upper leg, left lower leg, right upper leg and right lower leg. The combined parts may represent not only the adjacent parts but also the parts that are not physically connected to each other. In this way one can encode the semantic relation among any single parts like the fact that two arms are usually symmetric. To connect the separate parts in the tree the *latent node* type is introduced, which plays the role of their virtual parent.

Similarly to flexible mixture model [2], the different visual categories of combined parts are extracted by clustering performed across the training set with use of geometric relations. Also the HOG descriptor and latent SVM classifier is used to train the visual detector of part. To make the final pose detection the similar objective function to (2) is used but the term $S(t)$ considers the visual categories of combined parts instead of types representing orientations.

Use of combined parts and visual categories significantly reduces the search space for objective function optimization and gives better precision than [2].

2.3 Hierarchical Model

An extension proposed by Tian *et al.* in [4] introduces the hierarchical model for kinematic tree with latent nodes representing the parents of adjacent body parts. This gives two kinds of nodes in the tree: (1) the *leaf node* for primitive body parts like head, and (2) the *latent node* for a parent of adjacent primitive body parts like whole arm consisting of upper arm and lower arm.

For all primitive body parts as well as for latent nodes the *types* of part are defined in similar way as in [2]. As a difference to simple orientation in [2], the type may indicate only a few typical configurations like upright or side-way straight mode and it is not only limited to orientations e.g. type can indicate open hand or close hand. The part types are generated by clustering over the training set by considering the parent-child spatial relationships as features.

The important difference between latent nodes here and combined parts of previous method [3] is that only the leaf nodes have the HOG detector used in the appearance term of objective function. The latent nodes are used only for keeping the compatibility of parent and child types.

Assuming that P represents the indexes of all leaf nodes of the tree and $c(j)$ defines the indexes of children of latent parent v_j, the objective function used for final pose detection has following form:

$$F(I,L) = \sum_{i \in P} \phi(I, l_i) + \sum_{j \notin P} \sum_{i \in c(j)} \Phi(v_i, v_j) \tag{4}$$

where unary term $\phi(I, l_i)$ measures the matching of primitive part at location l_i with use of HOG features and pairwise term $\Phi(v_i, v_j)$ captures the compatibility between the primitive part v_i and its latent parent v_j.

Similarly to [2,3], this extension makes a significant optimization in number of possible poses for optimization task. Additionally it allows to handle the high-order spatial relationships between body parts.

2.4 Inaccurate Annotations

The ground-truth annotation provided with the training set of images is a crucial factor for the efficiency that particular method can achieve. However obtaining the precise and reliable annotations is not always an easy and cheap task, especially for large datasets with untypical poses. The work of Johnson and Everingham [5] is focused on this aspect and proposes the method able to work with missing or inaccurate annotations.

They introduced the annotation error model integrated with PSM objective function, where two kinds of error are modeled: (1) displacement error in comparison to ground-truth annotation and (2) structural error made by an annotator like marking the left hand instead of the right one. Both error types are modeled using probability distributions estimated by comparison of the causal annotations delivered by freelancers and the reliable annotations delivered by experts.

All annotations used for PSM model training are treated as they would be inaccurate and are automatically corrected in expectation-maximization (EM) scheme [28] with use of the objective function and estimated probabilities encoded by error models.

The method [5] introduces also other significant extensions to standard PSM model. The first one is replacing the single PSM model with mixture of PSM-s obtained by clustering the pose space over the training set. The second extension is the use of a mixture of linear classifiers for appearance model, which gives the similar accuracy as non-linear classifier simultaneously keeping the computational efficiency of linear one.

3 Poselets

Apart from methods using PSM model in human pose estimation, there is also an approach based on *poselet* concept, where poselet defines the particular configuration of single body part or group of many parts visible on some patch of the image. Originally the poselet concept used for human pose estimation was proposed by Bourdev and Malik in [6]. In the first step they perform an automatic poselets selection from annotated images by clustering the annotated joints into the groups, where a consistent spatial configuration is the group membership criteria. Then a poselet detector is trained inside each group using linear SVM classifier and HOG descriptor. The final pose detection on the test image relies on the classification of each pixel as a center of particular poslets. The positive output from given pixel for some poselet defines the potential localizations of joints that are encoded by this poselet. Each such potential joint location can be treated as a *vote*. After collection of all votes from all positively classified poselets we can define the special map of votes across the image for given joint. The location that collected the most votes is treated as a real joint location. Such detection procedure for whole image can be efficiently executed using a hough transform with max-margin framework [12]. Figure 2 presents the examples of poselet training images and poselet detection result.

Original poselet approach presented in [6], though appealing one, is generally not so efficient as the recent state-of-the-art methods based on PSM. Nevertheless, the poselets concept is willingly used as an enhancement of other methods that are described in following sections.

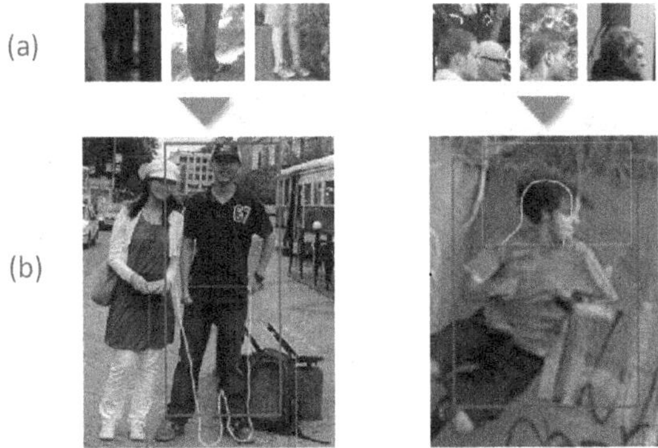

Fig. 2. Example of poselets generated by [6]: (a) training images, (b) poselet detection

3.1 Deformable Part Models with Poselets

Gkioxar *et al.* in [7] proposed the human pose estimation method based on deformable part models (DPM) [11] with use of poselets as appearance detector for each part. The DPM model is very similar to the PSM one but its goal is to detect the parts of the object as simple bounding boxes while the PSM detects the exact geometric configuration of all body parts in form of localizations and orientations. The DPM treats the physical body parts more implicitly and uses rather some general components containing many single parts, which can be even shared between many components. In the method [7] the trained poselets are used as such components as opposed to standard DPM approach which uses the simple HOG filters. The final physical body parts can be identified on the test image directly from poselets detection using the fact the joints layout is known for each poselet.

3.2 Mixing the Poselets and PSM

Pishchulin *et al.* in [8] uses the PSM approach enhanced by poselet detectors. Conceptually we can imagine that the kinematic tree of the standard PSM is common for entire pose space. It means in practice that such generic kinematic tree favors the pose configuration in a standing and rest position. This seems to be not the best solution as a diversity of poses in modern dataset is very high. Additionally the generic kinematic tree captures only adjacent parts dependencies that must be known a-priori. In [8] the poselet conditioned kinematic tree is introduced. In the first step they define a few groups of poses by clustering the training set of images with use of relative positions of body parts. Then the standard poselet method is used for early detection of body part locations and rotations. This allows for quick classification of obtained potential pose to appropriate group of poses and in this way the most fitting kinematic tree is used for final detection by optimizing the PSM objective function.

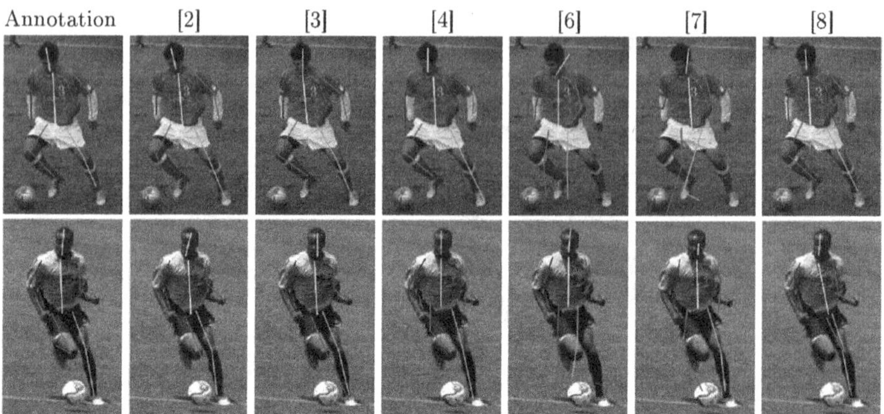

Fig. 3. Part of the visual results of selected methods on LSP dataset [18]: the relatively easy pose in the top row and more challenging one in the bottom row

Such approach gives more flexible model, which can achieve better precision than standard PSM model due to decreasing the space of possible kinematic constraints on body parts that must be considered. Another advantage is that it captures also the non-adjacent part dependencies implicitly as the poselet considers many body parts together regardless of their physical connections.

4 Empirical Evaluation

We performed some experiments to compare the selected methods described in this paper. For our comparison we selected only the methods for which the source code was publicly available. The same test images set and the same evaluation criterion were used for all methods. As a test set we use 1000 images from LSP dataset. As evaluation criterion for performance comparison we use the same method as in [13], namely the part is treated as correctly detected if both of its end points lie in less distance from ground-truth locations than 50 % of the part length. For all methods we used the pre-learned models, which were trained in original way suggested by their authors in appropriate works.

Table 1. Performance of selected methods on the LSP dataset [18] in percentages

Method	Evaluation (%)
Flexible mixture model [2]	64
Combined parts [3]	66
Hierarchical model [4]	59
Basic poselets [6]	48
DPM with poselets [7]	59
Mixture of PSM and poselets [8]	81

As was shown in Table 1, the methods based on PSM achieves better performance on LSP dataset than methods based only on poselets and DPM. The best result was achieved by mixture of PSM and poselets [8]. Figure 3 shows some visual results from our experiments.

5 Conclusion

We summarized and compared the selected state-of-the-art methods from 2D human pose estimation domain that we found most practical in sense of precision, efficiency and clarity of concept. Another important factor we took into account is the availability of source code or relatively easy possibility to implement them basing on available description. Our experiments shown that these methods are not perfect yet but may be good enough already for many practical application like action recognition where their imperfections can be minimized by enhancements based on video context [22,23]. We focused rather on simple message how these methods work conceptually and what are the major differences between them rather than analyzing the possible improvements of the current state of the art. For discussion on the current problems and perspectives in pose estimation domain one may refer to other surveys raising these aspects [9,10].

Acknowledgements. This work has been supported by the National Centre for Research and Development (project UOD-DEM-1-183/001 "Intelligent video analysis system for behavior and event recognition in surveillance networks").

References

1. Felzenszwalb, P., Huttenlocher, D.: Pictorial structures for object recognition. IJCV **61**(1), 55–79 (2005)
2. Yang, Y., Ramanan, D.: Articulated pose estimation with flexible mixtures-of parts. In: CVPR, pp. 1385–1392 (2011)
3. Wang, F., Li, Y.: Beyond physical connections: tree models in human pose estimation. In: CVPR, pp. 596–603 (2013)
4. Tian, Y., Zitnick, C.L., Narasimhan, S.G.: Exploring the spatial hierarchy of mixture models for human pose estimation. In: Fitzgibbon, A., Lazebnik, S., Perona, P., Sato, Y., Schmid, C. (eds.) ECCV 2012, Part V. LNCS, vol. 7576, pp. 256–269. Springer, Heidelberg (2012)
5. Johnson, S., Everingham, M.: Learning effective human pose estimation from inaccurate annotation. In: CVPR, pp. 1465–1472 (2011)
6. Bourdev, L., Malik., J.: Poselets: body part detectors trained using 3D human pose annotations. In: ICCV, pp. 1365–1372 (2009)
7. Gkioxari, G., Hariharan, B., Girshick, R., Malik, J.: Using k-poselets for detecting people and localizing their keypoints. In: CVPR, pp. 3582–3589 (2014)
8. Pishchulin, L., Andriluka., M., Gehler, P., Schiele, B.: Poselet conditioned pictorial structures. In: CVPR, pp. 588–595 (2013)

9. Andriluka, M., Pishchulin, L., Gehler, P., Schiele, B.: 2D human pose estimation: new benchmark and state of the art analysis. In: CVPR, pp. 3686–3693 (2014)
10. Liu, Z., Zhu, J., Bu, J., Chen, C.: A survey of human pose estimation: the body parts parsing based methods. JVCI **32**, 10–19 (2015)
11. Felzenszwalb, P., Girshick, R., McAllester, D., Ramanan, D.: Object detection with discriminatively trained part based models. PAMI **32**(9), 1627–1645 (2010)
12. Maji, S., Malik, J.: Object detection using a max-margin hough tranform. In: CVPR, pp. 1038–1045 (2009)
13. Ferrari, V., Marin-Jimenez, M., Zisserman, A.: Progressive search space reduction for human pose estimation. In: CVPR, pp. 1–8 (2008)
14. Wang, H., Klaser, A., Schmid, C., Liu, C.-L.: Action recognition by dense trajectories. In: CVPR, pp. 3169–3176 (2011)
15. Fischler, M., Elschlager, R.: The representation and matching of pictorial structures. IEEE Trans. Comput. **22**(1), 67–92 (1973)
16. Eichner, M., Ferrari, V.: Better appearance models for pictorial structures. In: BMVC, pp. 1–11 (2009)
17. Dalal, N., Triggs, B.: Histograms of oriented gradients for human detection. In: CVPR, vol. 1, pp. 886–893 (2005)
18. Johnson, S., Everingham, M.: Clustered pose and nonlinear appearance models for human pose estimation. In: BMVC, pp. 1–11 (2010)
19. Sapp, B., Taskar, B.: MODEC: multimodal decomposable models for human pose estimation. In: CVPR, pp. 3674–3681 (2013)
20. Gkioxari, G., Arbelaez, P., Bourdev, L., Malik, J.: Articulated pose estimation using discriminative armlet classifiers. In: CVPR, pp. 3342–3349 (2013)
21. Cherian, A., Mairal, J., Alahari, K., Schmid, C.: Mixing body-part sequences for human pose estimation. In: CVPR, pp. 2361–2368 (2014)
22. Wang, C., Wang, Y., Yuille, A.: An approach to pose-based action recognition. In: CVPR, pp. 915–922 (2013)
23. Nie, B., Xiong, C., Zhu, S.-C.: Joint action recognition and pose estimation from video. In: CVPR, pp. 1293–1301 (2015)
24. Fan, X., Zheng, K., Lin, Y.: Combining local appearance and holistic view: dual-source deep neural networks for human pose estimation. In: CVPR, pp. 1347–1355 (2015)
25. Ouyang, W., Chu, X., Wang, X.: Multi-source deep learning for human pose estimation. In: CVPR, pp. 2337–2344 (2014)
26. Toshev, A., Szegedy, C.: DeepPose: human pose estimation via deep neural networks. In: CVPR, pp. 1653–1660 (2014)
27. Tompson, J., Jain, A., LeCun, Y., Bregler, C.: Joint training of a convolutional network and a graphical model for human pose estimation. In: NIPS, pp. 1799–1807 (2014)
28. Dempster, A., Laird, N., Rubin, D.: Maximum likelihood from incomplete data via the EM algorithm. J. Roy. Stat. Soc. **39**(1), 1–38 (1977)

Video Editor for Annotating Human Actions and Object Trajectories

Marek Kulbacki[1(✉)], Kamil Wereszczyński[1,2], Jakub Segen[1], Michał Sachajko[1], and Artur Bąk[1]

[1] Polish-Japanese Academy of Information Technology, Koszykowa 86, 02-008 Warszawa, Poland
mk@pja.edu.pl
[2] Institute of Informatics, Silesian University of Technology, Akademicka 16, 44-100 Gliwice, Poland

Abstract. A system for managing, annotating and editing video sequences is a necessary tool in research on recognition of human actions and tracking people or objects. In addition annotation process is complex and expensive, so some people try to use crowdsourced marketplace based tools to make this process cost effective. Such a tool, video editor for annotating human actions and object trajectories -**VATRAC**, is presented. It enables flexible viewing video sequences under selected configuration of annotation layers, adding and editing of annotations for actions and trajectories of the entire objects or selected parts of the objects. Video sequences can be queried according to a variety of criteria and preferences for example searching for subsequences annotated with the action class.

Keywords: Annotation · Tracking · Video processing

1 Introduction

Annotating, editing and managing video sequences has a significant influence on preparation of image processing procedures [1] especially on training and evaluation phases in video analysis algorithms [2–4]. The form and structure of annotations depend on target system application. In projects connected with development of IVA (Intelligent Video Analysis) system we have focused on four essential for video analysis areas: object tracking, object identification, pose recognition and action recognition.

The video acquisition environment for obtaining training and testing data has a marked effect on data quality. The use of video data acquired from real-world, in comparison to video data from a laboratory, where an actor follows a screenplay, results in a more realistic and diverse scenarios. In general, examples created in a laboratory do not well represent real world situations. Therefore in **VATRAC** we assume more complex scenarios:

1. One scene can contain several actors.

N.T. Nguyen et al. (Eds.): ACIIDS 2016, Part II, LNAI 9622, pp. 447–457, 2016.
DOI: 10.1007/978-3-662-49390-8_44

2. One actor can be involved in many actions simultaneously.
3. Actions arranged in spatiotemporal hierarchical structure create a *behavior*.
4. An annotation of one action contains rectangular area adapted to the size of objects in each frame.

Annotations in video streams in the form of metadata are used for target system training and evaluation. Methods for motion analysis that have been used in our target system include in particular: tracking methods, nonskeleton based for action recognition methods, poselets, tracklets and manifolds based methods, as well as few others not categorized. The following annotation functionalities have been provided within the editor:

1. Annotations of actors and actions - for the purposes of a general and detailed level of action recognition methods and object identification.
2. Annotations of track paths (trajectories) of whole objects and objects parts (like arms, legs, head, torso) - for the purposes of object tracking and pose recognition.

We have been conducting research on actions analysis within the SAVA project. Researchers working on the project use video datasets created on the base of video streams from urban surveillance camera systems. These systems have been installed in two cities in Poland: one around Bytom's marketplace (VMASS [5]) and the second one on Gliwice streets (VMASS 2). Researchers also use popular reference video dataset like: Videoweb [6], UT [7], Weitzmann [8], KTH [9], UCF family [10–13], HMDB [14], HOHA [15] or Hollywood 2 [16]. Therefore, the crucial functionality for transformation from external to internal format of video streams and annotations as well as exchange data between different video datasets has been created. Several tools for data annotation have been developed in a form of stand alone applications [3,17,19] or web-based tools [2,18] or cloud base semi-automatic solutions [4]. Most of them are non-commercial tools, custom made to deliver required functionalities [3,4,17,19]. We aimed to develop the tool, that gives more flexibility in techniques intended for data annotation as well as for different reference datasets interoperability on the semantic, syntactic, and lexical level and this is the major novelty in comparison with existing tools.

In this paper we describe each type of used annotation, formulate the purpose of annotation creation, functionality and the architecture of an existing system and results of the annotation process.

2 Domain Problem

Our IVA system extensively uses machine learning algorithms, so video sequences are necessary for training and evaluation. In the first step, the system must be trained using a training set – a set of annotated video sequences. Then the trained system is evaluated on the set disjoint with a training set. Using well-known prediction quality measures like ROC Curve [20] (Receiver Operating

Characteristic Curve) system is evaluated. Then procedure goes back to the first step but in another configuration and with different parameters of a detector. That's why we need a quite big set of annotated video sequences. The details and form of particular annotations depend on a task goal: e.g. an action recognition system demands action annotation, whereas an object tracking system requires object trajectory annotations.

2.1 Annotation of Human Action

Human action annotation consists of two parts: a set of actions and actor qualifiers; time segments concerning appropriate qualifiers. Exemplary annotations are presented on Figs. 1 and 3.

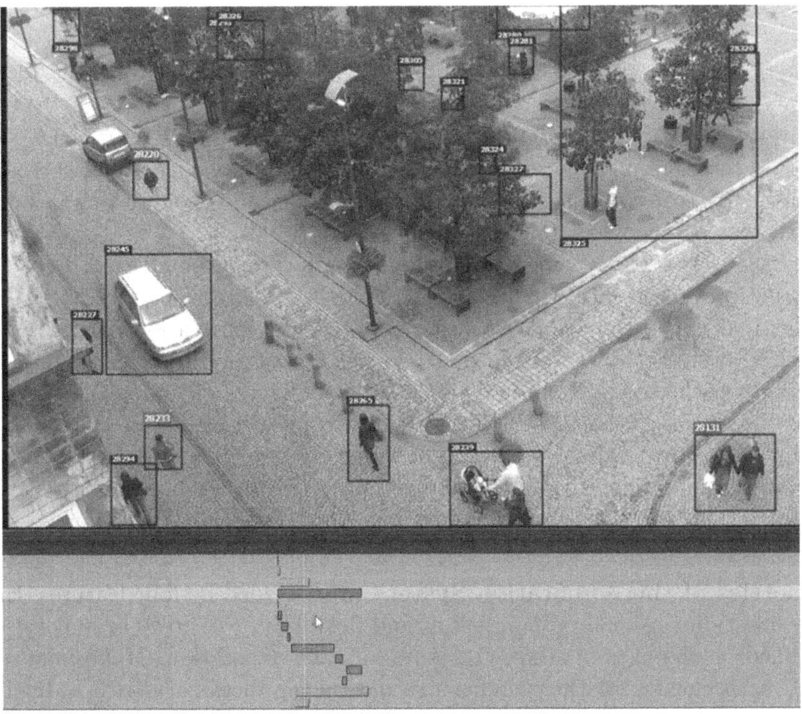

Fig. 1. Example of action annotation.

Let $Q = \{q_1, q_2, ..., q_n\}$ be qualifier dictionary, $T(q_i) = \{t_1, t_2, ..., t_K\}$ be tag dictionary for qualifier q_i with dimension K (exemplary part of tag dictionary is shown on Fig. 2). Let's consider a pair: $\alpha_R = (q_i \in Q, t_j \in T(q_i))$ and time segment $\tau(\alpha_R)$ which is continuous, and an action described by α is persistent in a whole segment for the same object R (which can change its shape and position

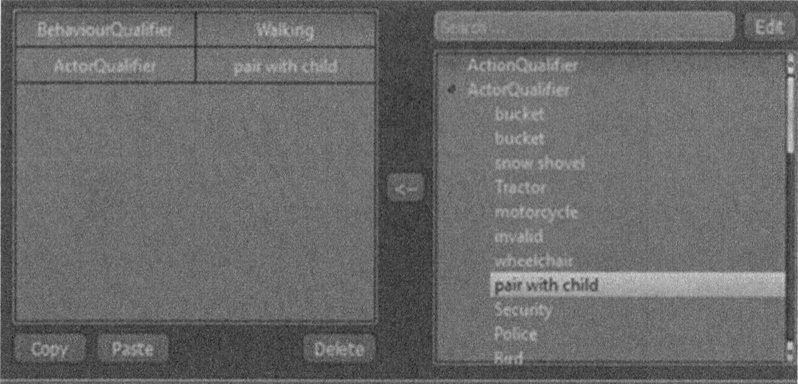

Fig. 2. Fragment of dictionary for actor qualifier

in time). The pair $(\alpha_R, \tau(\alpha_R))$ we call the Annotation Part for object R. Here, one actor or group of actors will be considered as an *object*.

Let's consider two annotation parts: $a = (\alpha_R, \tau(\alpha_R))$ and $b = (\beta_G, \tau(\beta_G))$. If β_G could be regarded as part of α_R on the base of a knowledge about surroundings and $\tau(\beta_G) \subset \tau(\alpha_R)$ then a will be regarded for parent of b which is a child of a.

In this case a tree composed of annotation parts can be created. Each tree construction process begins with finding unique annotation part r that has no parent; it is called root annotation part. Then child annotation parts are subordinated to root. This procedure is repeated iteratively until leaf annotation parts are reached. The structure in the form of a tree of annotation parts is called *Structural annotation.* Structural annotation can consist of many actions and involve many actors. The depth of structural annotation is not assumed - it follows from action structure, more complex action will have greater depth; simpler actions will have smaller depth.

Properties of the Annotation System are specified by four features: Hierarchy, multidimensionality, parallelism and flexibility. *Hierarchy* of annotations is defined on the base of a tree structure for each annotation. *Multidimensionality* - each element of the qualifier can have a tag dictionary with a different dimension. *Parallelism* is an assumption specifying that annotations:

1. Can overlap one another in time and in spatial perspective.
2. Annotation parts can overlap in time perspective.
3. One annotation part can belong to different annotations.
4. One region of interest can belong to different annotation parts.
5. One region of interest can be divided into separate annotation parts, which could be overlapped in time.

Flexibility means that there are no limits for sizes of qualifier and tag dictionaries and for a depth of annotation; these dictionaries can be modified anytime.

Fig. 3. Exemplary annotation of action composed of two qualifier-tags: ActionQualifier:Ride, ActorQualifier:Car.

2.2 Annotation of Object Trajectories

Object tracking methods require information about object trajectory to evaluate object position in each frame and a quality of an object re-identification (see Fig. 4). Position of object is defined by position given by one specific point in object: $p_k(O) = (x, y)$ within a frame k. This point could be defined geometrically (e.g. as the middle of object bounding box) or semantically, which is more comprehensive for humans (e.g. as in the middle of a torso). The sequence of object positions in consecutive frames creates object trajectory: $Tr_k^m(O) = \{p_k, p_{k+1}, ..., p_m\}$.

Some of methods use information of object parts (like hand, leg). That's why for trajectory annotation there is a hierarchical structure introduced as well. Let's assume that object O consists of l object parts: $O = \{Op_1, Op_2, ..., Op_l\}$. In that case

$$Tr_k^m(O = \{Op_1, Op_2, ..., Op_l\}) = \{Tr_k^m(O), Tr_k^m(Op_1), Tr_k^m(Op_2), ..., Tr_k^m(Op_l)\}.$$

It means that the same size trajectory is created for object and for each object part. All these trajectories together are treated as one trajectory annotation for object O. Still, information about object part trajectories could be used independently.

3 Video Editor

Based on the background described above the application authors created video editor for annotating of human actions and object trajectories **VATRAC**. It is a stand-alone application with user-friendly interface dedicated for users not

Fig. 4. Example of trajectory annotation.

familiar with annotation process. The annotation user interface was created as the result of conclusions taken from deep review of existing video annotation tools and the prototype evaluation. **VATRAC** is based on custom timeline interface, where sections of video streams are laid out in sequence for playback and edition.

3.1 Main Functionalities

Postprocessing Video Sequences including : forward and reverse playback with changing the speed, access to each frame separately and zooming the scene, trimming, cutting, splicing and arranging video streams across the timeline.

Dictionary allows the creation and modification of qualifier dictionary and tag dictionaries for each qualifier. It allows creation of new information layer for tag dictionaries. Tags are created in relationship many-to-many from one layer to the second one. In this way, knowledge enclosed in qualifier-tag system can be easy adapted to different standards of annotation.

Annotation Manager organizes annotations with information about all their hierarchical structure. Operations: add new annotation by creation from scratch; delete an annotation; add new annotation part to the existing nodes of annotations using automatically detected region of interest considered by object; modify and remove annotation parts; add, modify and remove a qualifier - tag information to annotation parts.

Trajectory Manager enables operations: add object and object parts of trajectories by picking out the first point of it; expand existing trajectories by new points in new frames; modify trajectories by changing the position of points belonging to it, remove trajectory.

Mapping Annotation to Trajectories enables split of information currently connected with region of interest to separate object existing in this area.

Searching introduces filtering and exporting functionalities, that allow finding sets of short clips concerning given qualifier-tag with user-defined size and exporting it to a sequence of images (JPG, PNG) or compressed video frames using well-known formats: AVI and MPEG4. There are two available modes of data filtering:

1. In output folder there are stored maximum n video clips responding for given query.
2. Output folder is divided in two subfolders: *positive*, containing n clips responding for given query and *negative* containing clips responding to query opposite to given one. This approach allows easy create positive and negative samples for training and evaluation process.

An example of sequence exported to set of JPEG files is shown on Fig. 5.

Dataset Interoperability. We use our own video data sets (VMASS, VMASS2) as well as existing reference data sets listed in the section Introduction. Each of them is different on the semantic, syntactic, and lexical level. Each has different data format and structure of annotations. We developed unified API for interoperability purposes. This API enables set of unified I/O operations and query mechanism required by **VATRAC** editor and at the same time keeps all datasets untouched in their original form.

3.2 Brief Architecture Review

The **VATRAC** application is being developed in C++ with elements of the standard C++14. The major external libraries include Qt (v5), boost (v1.58) and openCV (v3.0). **VATRAC** uses libraries developed in our lab for: camera control and stream retrieval, video sequences management, qualifier-tag dictionaries and annotation system. It's dedicated on x64 platform and is available for windows and Linux. User interface has been implemented using MVC (Model View Controller) software architectural pattern. The **Model** represents reusable application part, **View** it's screen presentation and the **Controller** defines the way the user interface reacts the user input. The **Model** is composed of our libraries:

1. *Acqlib* - implementation of annotation structure and search engine.

Fig. 5. Example of exporting sequence to the set of JPG images.

2. *CameraControl* - video management on the level of a frame, sequence and stream,
3. *Utils* - qualifier-tag dictionaries system.
4. *SequenceConverterLib* - datasets interoperability (export and I/O unification).

The **Control** encapsulates the response mechanism implementing mainly *CameraControl* library in a part of multi-thread video data acquisition system API. Current **View** is being implemented with Qt library and its Widget tupt.

Used Design Patterns and Idioms. Conceptual and implementation phases of **VATRAC** were based on object and generic programming paradigms. Design patterns used during **VATRAC** development are listed in Table 1.

4 Results and Discussion

The **VATRAC** - a multi-layered annotation tool have been developed to help authors improve the tasks of annotating video sequences with metadata required for surveillance-based research problems. It is also able to provide a great value to researchers in other disciplines like forensics, medicine, psychology or computer vision, where additional and accurate information correlated with video streams is required. One can create save, update and download customized lexicons The tool also enables a few of implemented operations for semiautomatic video processing to be used with all listed by authors external reference datasets. Future developments are directed to put annotation process into the video acquisition and processing pipeline. We started works on automation of segmentation process, background foreground separation, and tracking process. It will help

Table 1. Design patterns and generic idioms used for **VATRAC** implementation. DP-design pattern, I-idiom

Name	Kind	Libraries	Description
Singleton	DP	CameraControl	Support for camera operations in existing system
	I	acqlib	Creating video data buffer shared by many processes during acquisition
	I	sequenceConverterLib	Accessing dataset library
Builder	DP	Camera Control	Creation of camera handler object consisting of many compounds
Facade	DP	CameraControl	Utilizing library functionality
Decorator	DP	acqlib	Creating different boundaries around region of interests

Table 2. Properties of VMASS datasets

Effective recordings in hours	>10 000
Frame count [millions]	1200
Recordings quantity	>28 TB
Annotations	
Events	53 000
Behaviors	4600
Event types	ca. 150
Actor types	ca. 30
Scene types	ca. 50

to create next tool more effective, more accurate and make all the annotation process faster. With the currently available functionality using VALTRAC tool authors indexed huge video datasets (VMASS and VMASS 2) with annotations characterized in Table 2. Annotations have been made manually by three workers and have been used for development of practical methods:

1. Creation human motion path from surveillance data.
2. Non skeleton based method for clustering segments and paths of human motion.
3. Object tracking methods.
4. Action recognition methods.

For evaluation of above methods authors used all listed reference datasets as well as own created: VAMSS and VMASS2 datasets.

VMASS datasets have more variety and distinguish from existing datasets with the huge number of annotations addressed to action recognition, the environment of recordings, structure and number of annotations. These differences are shown on Table 3:

Table 3. Comparison of VMASS and listed reference datasets

	KTH	Weitzmann	HOHA1	TREC-VID	VIRAT	VMASS
Real environment	no	no	partly	partly	partly	yes
Resolution	160 × 120	180 × 144	540 × 240	720 × 576	1920 × 1080	1920 × 1080
Frames per second	25	50	24	N/A	N/A	18–27
event type count(e) and action type count (a)	6	10	8	10	23	150(e) 400(a)
scene types	N/A	N/A	many	5	17	>50
event count [thousand]	0.6	0.09	0.68	10	23	ca. 7.500
annotation type	1/ob	1/ob	1/ob	1/ob	n/ob	HPFM
annotation count	600	90	680	10.000	23.000	53.000

Acknowledgments. This work has been supported by the National Centre for Research and Development (project UOD-DEM-1-183/001 "Intelligent video analysis system for behavior and event recognition in surveillance networks").

References

1. Smith, J.R., Lugeon, B.: A visual annotation tool for multimedia content description. In: Proceedings of the SPIE Photonics East, Internet Multimedia Management Systems (2000)
2. Russell, B.C., Torralba, A., Murphy, K.P., Freeman, W.T.: LabelMe: a database and web-based tool for image annotation. Int. J. Comput. Vis. **77**, 157–173 (2008)
3. Korč, F., Schneider, D.: Annotation Tool. Technical report TR-IGG-P-2007-01, University of Bonn, Department of Photogrammetry (2007)
4. Grundmann, M., Kwatra, V., Han, M., Essa, I.: Efficient hierarchical graph based video segmentation. In: IEEE CVPR (2010)
5. Kulbacki, M., Segen, J., Wereszczyński, K., Gudyś, A.: VMASS: massive dataset of multi-camera video for learning, classification and recognition of human actions. In: Nguyen, N.T., Attachoo, B., Trawiński, B., Somboonviwat, K. (eds.) ACIIDS 2014, Part II. LNCS, vol. 8398, pp. 565–574. Springer, Heidelberg (2014)
6. Zhang, S., Staudt, E., Faltemier, T., Roy-Chowdhury, A.: A camera network tracking (CamNeT) dataset and performance baseline. In: IEEE Winter Conference on Applications of Computer Vision, Waikoloa Beach, Hawaii, January 2015
7. Chen, C.-C., Ryoo, M.S., Aggarwal, J.K.: UT-Tower dataset: aerial view activity classification challenge (2010). http://cvrc.ece.utexas.edu/SDHA2010/Aerial_View_Activity.html
8. Gorelick, L., Blank, M., Shechtman, E., Irani, M., Basri, R.: Actions as space-time shapes. Trans. Pattern Anal. Mach. Intell. **29**(12), 2247–2253 (2007)
9. SchÃijldt, C., Laptev, I., Caputo, B.: Recognizing human actions: a local SVM approach. In: Proceedings of the ICPR, pp. 32–36 (2004)
10. Soomro, K., Zamir, A.R., Shah, M.: UCF101: a dataset of 101 human action classes from videos in the wild. In: CRCV-TR-12-01, November 2012
11. Reddy, K.K., Shah, M.: Recognizing 50 human action categories of web videos. Mach. Vis. Appl. J. (MVAP) **24**, 971–981 (2012)

12. Liu, J., Luo, J., Shah, M.: Recognizing realistic actions from videos "in the Wild". In: IEEE International Conference on Computer Vision and Pattern Recognition (CVPR) (2009)
13. Rodriguez, M.D., Ahmed, J., Shah, M.: Action MACH: a spatio-temporal maximum average correlation height filter for action recognition. In: Computer Vision and Pattern Recognition (2008)
14. Jain, M., Jegou, H., Bouthemy, P.: Better exploiting motion for better action recognition. In: CVPR (2013)
15. Laptev, I., Marszałek, M., Schmid, C., Rozenfeld, B.: Learning realistic human actions from movies. In: IEEE Conference on Computer Vision & Pattern Recognition (2008)
16. Marszałek, M., Laptev, I., Schmid, C.: Actions in context. In: IEEE Conference on Computer Vision & Pattern Recognition (2009)
17. Kipp, M.: ANVIL - a generic annotation tool for multimodal dialogue. In: Proceedings of the 7th European Conference on Speech Communication and Technology (Eurospeech), pp. 1367–1370 (2001)
18. Vondrick, C., Patterson, D., Ramanan, D.: Efficiently scaling up crowdsourced video annotation. Int. J. Comput. Vis. (IJCV) **101**, 184–204 (2012)
19. Hailpern, J.: VCode and VData: Illustrating a new Framework for Supporting the Video Annotation Workflow. Google engEDU: Tech Talks, Mountain View, CA, 21 June 2008
20. Swets, J.A.: Signal Detection Theory and ROC Analysis in Psychology and Diagnostics : Collected Papers. Lawrence Erlbaum Associates, Mahwah, NJ (1996)

Learning Articulated Models of Joint Anatomy from Utrasound Images

Jakub Segen[1], Kamil Wereszczyński[1,2], Marek Kulbacki[1(✉)], Artur Bąk[1], and Marzena Wojciechowska[1]

[1] Polish-Japanese Academy of Information Technology, Koszykowa 86, 02-008 Warszawa, Poland
mk@pja.edu.pl

[2] Institute of Informatics, Silesian University of Technology, Akademicka 16, 44-100 Gliwice, Poland

Abstract. Parts of a joint anatomy, such as bones or the joint center can be robustly identified in an ultrasound image with the help of an articulated or structural model. Such a model is a structure of parts that represent the bones and skin as polygonal chains and the join as a point, where the parts remain within specified geometric relations. The parts are identified by registration or a match of a structural description derived from the ultrasound image with the articulated model. To account for anatomical differences between the subjects, a library of joint models must be constructed, each model representing a class of joints, where all models together cover the range of possible anatomies. A new method of unsupervised learning is proposed for constructing the library of joint models by clustering structural descriptions computed from image annotations. The clustering method uses an inter-model distance measure defined as a minimum of the objective function that measures a discrepancy between structural descriptions. The objective function is minimized through a search for a best match between two structural descriptions. The method presentation is illustrated with the results of its application to ultrasound images of finger joints.

1 Introduction

One of many medical applications of ultrasound imaging is focused on detection, assessment and monitoring of synovitis, an inflammation of synovial membrane often associated with rheumatoid arthritis [1,2]. While the examination and analysis of ultrasound images directed towards synovitis assessment is currently performed manually by specialists, there is a need for automating this process to decrease its cost and to reduce the discrepancy in human scoring. A project named Medusa [3], conducted in Poland and Norway has as its objective construction of a synovitis estimator, that will process unannotated, and unlabeled ultrasound images of joints, to assess for each the presence and a degree of synovitis. The first stage of such processing is the identification of anatomical elements in the image, namely the skin, bones and the joint, by comparing an

N.T. Nguyen et al. (Eds.): ACIIDS 2016, Part II, LNAI 9622, pp. 458–466, 2016.
DOI: 10.1007/978-3-662-49390-8_45

ultrasound image of a joint with structural descriptions of articulated models that are stored in a dataset called the Class Model Library (CML). As a result of the comparison a model is selected which gives the best registration score, or the best match. Using the best matching model, the skin, bones and joint in the target image can be identified according to the mapping provided by the result of the match operation. The matching between models from CML and the ultrasound image is computed using the recently proposed method for image registration using structural descriptions [4]. An example of a structural description, overlaid on an ultrasound image as yellow polygonal chains, is shown in Fig. 1. The focus of this paper is the construction of the CML library. A new method of unsupervised learning, based on clustering structural descriptions, uses as a distance measure for the clustering the minimum of the objective function that results from the method of registration of structural descriptions described in [4]. While the authors don't know of any work which is closely related to the proposed learning method, the approach to modeling, recognition and learning by parts in computer vision, represented by [5–8] is similar in spirit.

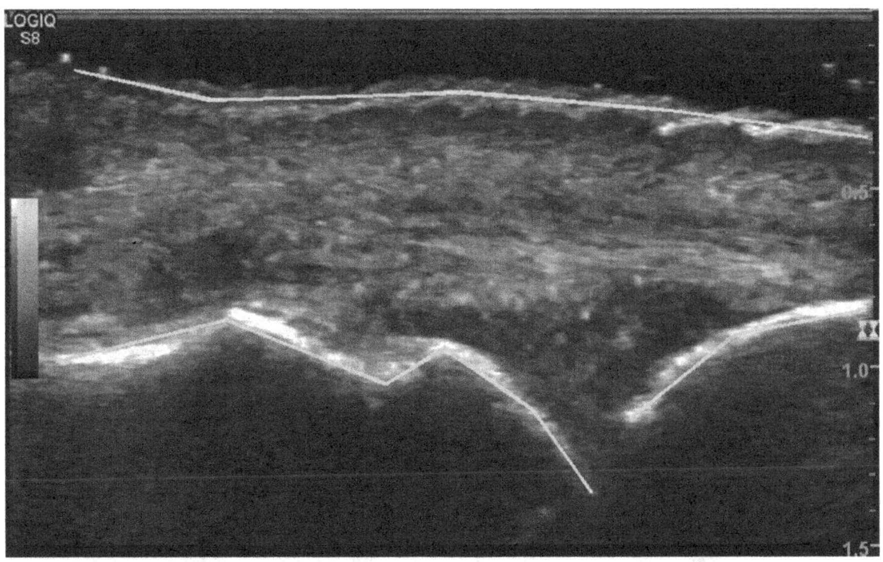

Fig. 1. The yellow polygonal chains that mark the skin and the bones form a structural description of the ultrasound image (Color figure online)

The CML is built during the learning phase, using as training data the annotations from a training set of ultrasound images of joints. The learning phase is done once, after the acquisition and annotation of ultrasound training images are completed. The registration method that compares the annotated images expects a structural description where the skin and bone parts are represented as a polygonal chain (piecewise linear curve), that is a chain of connected line segments. However, the skin and bone elements in the annotated ultrasound images are drawn using smooth curves. To obtain a structural description form required by the registration method, the annotation curves need to be approximated with polygonal chains, or linearized. The curve linearization method is described in Sects. 2 and 3 describes the learning method for constructing a CML.

2 Curve Linearization

The linearizing approximation of a smooth curve is done through a simple recursion, based on Douglas-Peuker algorithm [9]. In the first step a curve to be linearized is compared to a line drawn from the curve's first point to the last, and a point of the curve that is farthest from the line is found. If its distance is smaller than a threshold value, the line is the approximation result, otherwise the curve is divided at the farthest point, and the same operation is applied to each of the two segments, terminating when none of the curve segments needs to be divided. This process is illustrated in Figs. 2 and 3 shows the examples of linearization results for four images.

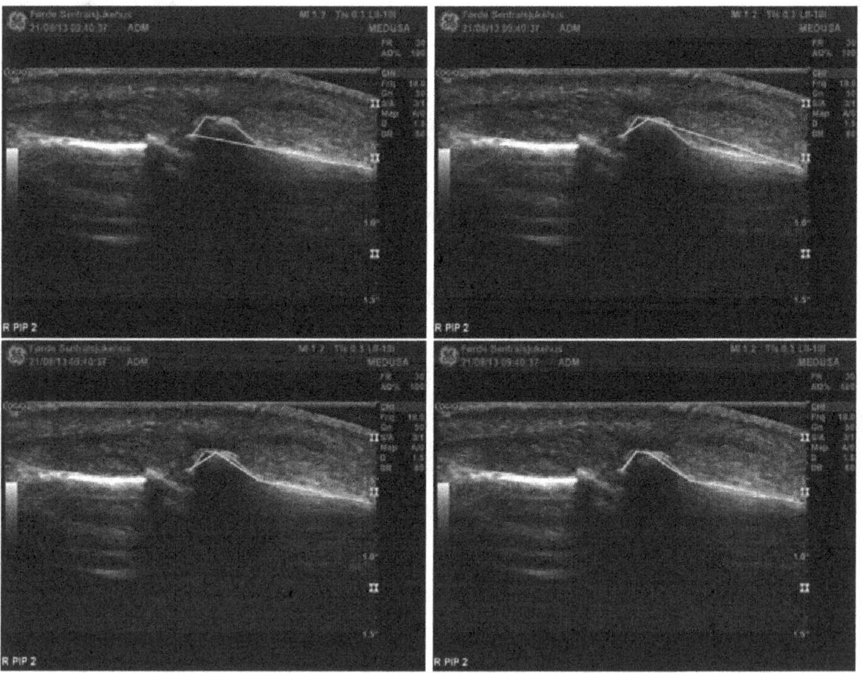

Fig. 2. Linearization: steps 1,2,3,4

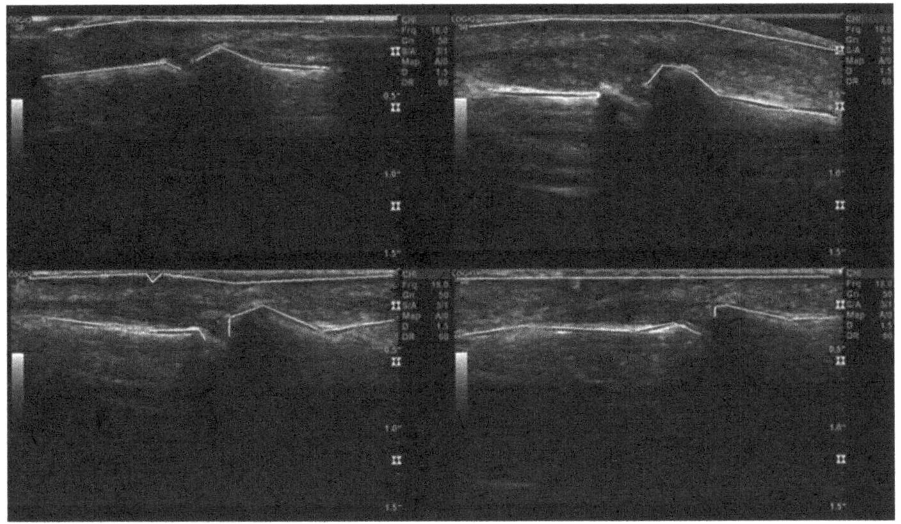

Fig. 3. Examples of linearization of annotation curves. Original curve - blue, polygonal chain - yellow (Color figure online)

3 Learning Class Models

The method for registering structural descriptions, described in [1] uses an objective function (cost) to guide a search for a best correspondence between two structural descriptions, X the target and R the reference, represented as sets of planar features. This search allows the reference R to be transformed by a rigid, planar transformation T, and it aims at finding a mapping Map that maps nodes of X to the nodes of R, and a transformation T, that to minimize the cost value:

$$Q(X, R, T, Map) = \sum_{i=1}^{n} d^2(X_i, T(R_{Map(i)})) + mC_R \tag{1}$$

where n and m are the numbers of nodes in the target and the reference structures, respectively, C_R is a regularization coefficient and $d^2(x, r)$ is the squared distance between the nodes x and r. The node distance d^2 is computed from the sum of a vector v_p that projects a midpoint of r onto x and a vector V_s, which is the minimal translation of the center of r in parallel to x, that makes the length of the intersection of x with a projection r_p, of r onto the line extending x, equal to the minimum length of r_p and x, as it was described in [1].

$$d^2(x, r) = (v_p + v_s)^t(v_p + v_s) \tag{2}$$

The minimal cost value achieved in the process of registration of two feature sets, X and R, is given in [1], in Eq. (2):

$$Q^{Opt}(X, R) = Min_{Map,T} Q(X, R, T, Map) \tag{3}$$

The value $Q^{Opt}(X, R)$ is used here as a squared distance measure between the sets X and R, that is

$$d^2(X, R) = Q^{Opt}(X, R) = Min_{Map,T} Q(X, R, T, Map) \tag{4}$$

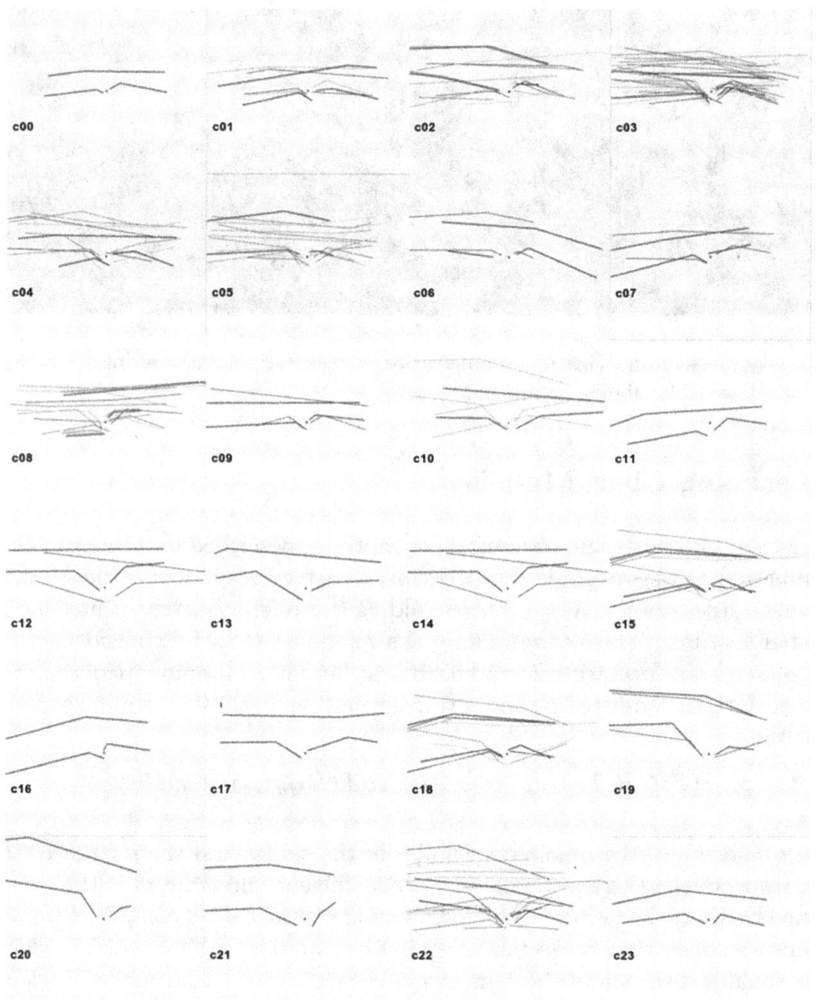

Fig. 4. Example of structural description clustering. Cluster centers are drawn in red (Color figure online).

The proposed method for learning class models, uses this distance measure for comparing and clustering structural descriptions obtained from annotated training images. A cluster of structural descriptions is a set $C = C(1), C(2), ...C(k)$

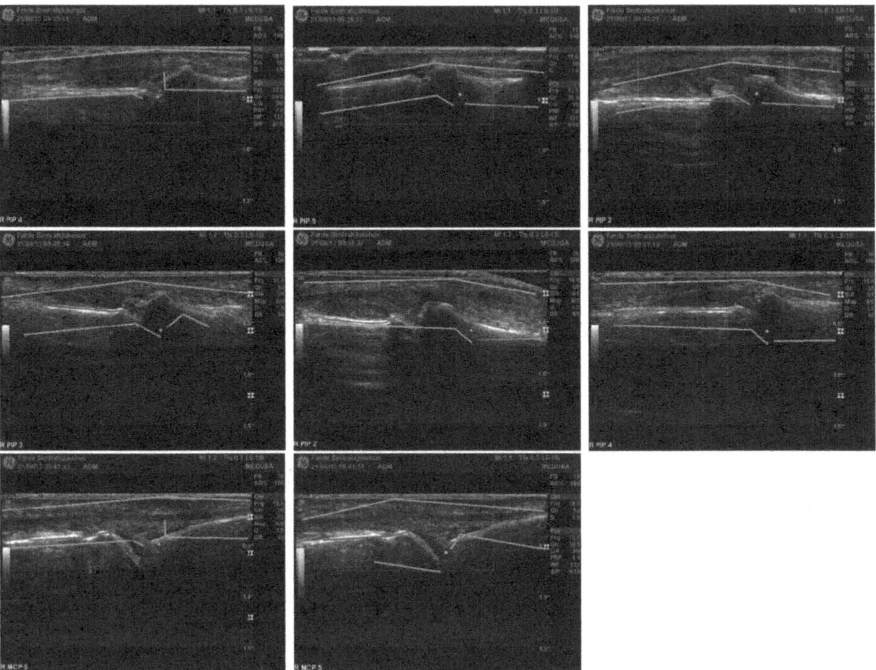

Fig. 5. Examples of image registration using CML. Each target image is shown with its annotation structure (red), and with a transformed center of the nearest matching cluster from CML (green) (Color figure online).

where $C(1), C(2), ...$ are structural descriptions, which are the cluster members. A cluster center C^m is defined as

$$C^m = C(argmin_i \sum_{j=1}^{k} d^2(C(i), C(j))) \tag{5}$$

Each of the clusters resulting from this process defines a distinct class. A class model is the central element, or center of a cluster. Cluster *center* is defined as a cluster member for which the sum of squared distances to the remaining cluster members is minimal. A distance between a structural description X and a cluster C, $D(X, C)$ is defined as the distance of this structural description from the cluster center C^m.

$$D(X, C) = d^2(X, C^m) \tag{6}$$

The set of all class models (cluster centers) is the Class Model Library. To identify the joint, skin and bones in an unlabeled image, the structural description of the image is compared with models in Class Model Library to find the nearest model and its mapping (Map) function, which identifies the features of the unlabeled structure. The following steps describe the proposed method for clustering structural descriptions. The structural descriptions are referred to as samples.

1. Create an empty cluster, include in it the first sample, making this sample the cluster center.
2. If all samples have been processed, terminate. Otherwise, take the next sample, compare it (register) with the center of each cluster, and select the nearest cluster.
3. If the distance of the sample to the nearest cluster is greater than a threshold T, create a new cluster and include in it the sample, making the sample the cluster center. Otherwise, add the new sample to its nearest cluster and recompute the center for this cluster.
4. Go to Step 2.

Example 1. Clustering. For illustration, the first 24 of the 25 clusters, that resulted from the application of this method to a set of 85 annotated ultrasound images are shown in Fig. 4. Each sub-image of Fig. 4 shows one cluster. Each member of a cluster is shown using a different color. The same color is applied to all the features of a cluster member. The red color is used to draw the central element of each cluster.

Example 2. Image Registration. In a pilot image registration experiment, structural descriptions derived from images are matched with clusters from the CML. For each image, the structural description of the center of the nearest matching cluster is transformed according to T given by Eq. 3, and overlaid on the image. Figure 5 shows images with annotations (red) and for each image an overlaid best matching cluster center transformed by T (green). One can see, that orientation and position along the y-axis between the target and the reference structural descriptions are close, but in several cases there is a significant difference in the position along the x-axis. However, the position of the joint has not been used in the CML construction or in the image registration, and the results are expected to improve when it is included.

4 Conclusions

A Class Model Library is a set of models of structural descriptions of ultrasound images of joints, where a structural description represents skin and bones as polygonal chains. A Class Model Library is constructed from a set of ultrasound images with annotations which mark skin and bones in the images. Its purpose is to serve as a reference set for identifying skin, bones and a joint in an unannotated ultrasound image. An approach to constructing a Class Model Library has been presented. Its first part to obtain structural descriptions from the smooth curves that mark skin and bone features in the annotated ultrasound images, using polygonal chain approximation. The second part is a novel unsupervised learning of class models by clustering the structural descriptions, where the minimal cost value of registration of structural descriptions is used as a measure of distance between structural descriptions and between clusters and structural descriptions. In continuation of this work a joint will be included in the structural description as one more feature, which should increase the registration accuracy, and a joint detector [10, 11] will be used to find joint candidate location in an image. The presented approach is quite general. It is not limited to images of a joint and it should be useful for learning structural models of other anatomical structures, and it should work with other imaging modalities than ultrasound.

Acknowledgments. The research leading to these results has received funding from the Polish-Norwegian Research Programme operated by the National Centre for Research and Development under the Norwegian Financial Mechanism 2009–2014 in the frame of Project Contract No. Pol-Nor/204256/16/2013.

References

1. Zufferey, P., Tamborrini, G., Gabay, C., Krebs, A., Kyburz, D., Beat, M., Moser, U., Villiger, P.M., So, A., Ziswiler, H.R.: Recommendations for the use of ultrasound in rheumatoid arthritis: literature review and SONAR score experience. Swiss Med. Wkly. **143**, w13861 (2013)
2. Vlad, V., Berghea, F., Libianu, S., Balanescu, A., Bojinca, V., Constantinescu, C., Abobului, M., Predeteanu, D., Ionescu, R.: Ultrasound in rheumatoid arthritis - volar versus dorsal synovitis evaluation and scoring. BMC Musculoskelet. Disord. **12**, 124 (2011)
3. Automated Assessment of Joint Synovitis Activity from Medical Ultrasound and-Power Doppler Examinations using Image Processing and Machine Learning Methods. http://eeagrants.org/project-portal/project/PL12-0015
4. Segen, J., Kulbacki, M., Wereszczyński, K.: Registration of ultrasound images for automated assessment of synovitis activity. In: Nguyen, N.T., Trawiński, B., Kosala, R. (eds.) ACIIDS 2015. LNCS, vol. 9012, pp. 307–316. Springer, Heidelberg (2015)
5. Biederman, I.: Recognition-by-components: a theory of human image understanding. Psychol. Rev. **94**(2), 115–147 (1987)
6. Yang, Y., Ramanan, D.: Articulated pose estimation using flexible mixtures of parts. In: Computer Vision and Pattern Recognition (CVPR) Colorado Springs, Colorado, June 2011
7. Felzenszwalb, P., Girshick, R., McAllester, D., Ramanan, D.: Object detection with discriminatively trained part based models. IEEE Trans. Pattern Anal. Mach. Intell. **32**(9), 1627–1645 (2010)
8. Segen, J.: Graph clustering and model learning by data compression. In: Proceedings of the Seventh International Conference on Machine Learning, Austin, Texas, USA, 21–23 June 1990
9. Douglas, D., Peucker, T.: Algorithms for the reduction of the number of points required for represent a digitzed line or its caricature. Can. Cartographer **10**(2), 112–122 (1973)

10. Wereszczyński, K., Segen, J., Kulbacki, M., Mielnik, P., Fojcik, M., Wojciechowski, K.: Identifying a joint in medical ultrasound images using trained classifiers. In: Chmielewski, L.J., Kozera, R., Shin, B.-S., Wojciechowski, K. (eds.) ICCVG 2014. LNCS, vol. 8671, pp. 626–635. Springer, Heidelberg (2014)
11. Wereszczyński, K., Segen, J., Kulbacki, M., Wojciechowski, K., Mielnik, P., Fojcik, M.: Optimization of joint detector for ultrasound images using mixtures of image feature descriptors. In: Nguyen, N.T., Trawiński, B., Kosala, R. (eds.) ACIIDS 2015. LNCS, vol. 9012, pp. 277–286. Springer, Heidelberg (2015)

Facial Reconstruction on the Basis of Video Surveillance System for the Purpose of Suspect Identification

Damian Pęszor[1,2(✉)], Michał Staniszewski[2], and Marzena Wojciechowska[1]

[1] Polish-Japanese Academy of Information Technology, Koszykowa 86, 02-008 Warsaw, Poland
{damian.peszor,mwojciechowska}@pja.edu.pl

[2] Institute of Informatics, Silesian University of Technology, Akademicka 16, 44-100 Gliwice, Poland
michal.staniszewski@polsl.pl

Abstract. Growing importance and commonness of video surveillance systems brings new possibilities in the area of crime suspect identification. While suspects can be recognized on video recordings, it is often a difficult task, because in most cases parts of suspect's face are occluded. Even if there are multiple cameras, and the recordings are long enough to expose entirety of suspect's face, it is challenging for an observer to accumulate information from different cameras and frames. We propose to solve this problem by reconstructing a three-dimensional mesh that could be presented to an observer, so he could identify suspect based on accumulated information rather than fragmented one, while choosing any angle of observation. Our approach is based on extraction of anthropological features, so that even with imperfect recordings, the most important features in terms of facial recognition are preserved, while those not registered might be supplemented with generic facial surface.

Keywords: Facial reconstruction · Suspect identification · Facial composite · Surveillance

1 Introduction

The advent of video-based surveillance systems provided new tools for forensic investigators. In the past, when surveillance systems were few and far between, facial composites were constructed on the basis of eyewitness' memory with the help of trained professional artist. Because of the stressful nature of the experience, imperfection of perception and memory as well as inability to unambiguously convey information about human appearance, the margin of error using such approach is relatively high. Accessibility of recordings obtained through video-based surveillance systems made it possible for an eyewitness to recognize the suspect on a recording which provides the investigators with more reliable data. However, such recognition is not as certain as one might think. Even

N.T. Nguyen et al. (Eds.): ACIIDS 2016, Part II, LNAI 9622, pp. 467–476, 2016.
DOI: 10.1007/978-3-662-49390-8_46

though eyewitness does not have to verbalise his recollections and might not be consciously aware of his memories, he still might not recognize the suspect. Some of the reasons that the chances for successful recognition are lowered, are related to the fact, that eyewitness experience recordings in specific way, such that:

- Recordings from different cameras are viewed sequentially, which means that data from different recordings are processed separately. The effect of synergy is therefore greatly diminished, as information from one camera is not perceived while eyewitness watches another recording.
- Each surveillance camera is installed in a fixed point of space and records the scene from fixed perspective. This means, that eyewitness is not able to change perspective to the one that better corresponds with his observations and therefore his certainty is reduced.
- In most cases, suspect will not stand still while being recorded. Facial features will be visible clearly for several frames, but then suspect will change his pose, or occlusion will occur. Since eyewitness does not see all recorded features at once, but only partial image of face at any given time, his certainty will be reduced.

To resolve these problems, we employ a part of a method originally designed for the purpose of facial animation. This method, designed to retarget mimicry obtained from a video-based performance capture to different facial structure involves reconstruction of facial mesh from multiple recordings on the basis of positions of fiducial points characteristic from anthropometric point of view. Quality of reconstruction depends on the number of cameras which recorded the suspect, their positions, parameters and movement of suspect and occlusions which makes it hard to estimate the error. This solution is not intended to be an automatic facial recognition software, which would require much more reliable data and could be used mostly in case of identification of previously apprehended criminals. Rather than that, the authors intend to assist eyewitness in suspect identification at the stage of facial composite creation, when the data is present, but it is too distributed for eyewitness to properly analyse. With our approach, following statements hold:

- Data from different recordings are combined, so that eyewitness perception benefits from all of them at once.
- Eyewitness is able to see suspect without partial occlusions, as entire face is reconstructed. Even in case of part of face not being present on any recording, it is reconstructed from generic model, which eliminates eyewitness' confusion.
- Eyewitness is able to see suspect's face from different perspective and is able to use one that corresponds best to his experience, therefore being able to perceive features that he remembers.

2 Materials and Methods

2.1 Facial Area Localization

To develop and test suggested approach, a number of recordings was taken in different environments. Apart from recordings specifically taken for the purpose of facial reconstruction, VMASS video sequences dataset [1] was used to test proposed approach with real-life data. Since proper reconstruction requires taking into account camera parameters in order to be able to reconstruct each pixel of an image into a three-dimensional point, a calibration of each camera is required. In case of video surveillance systems, calibration itself is an issue that has to be recognized, due to two factors - working infrastructure and cameras' specificity. One cannot assume particulars places, in which suspect will be present, and therefore there is no possibility of establishing infrastructure that will be properly calibrated. Investigators have to use existing, working infrastructure. Since most of such systems are not designed for processing of video recordings, one cannot expect that there will be any markers that will faciliate calibration, therefore calibration procedure has to employ data available on the video at the time of the recording without any previous preparation.

While some of cameras might be stable enough that their movement is negligible, one cannot assume that this is the case. Tremors related with heavy machinery operating, cars on a street, or even human steps in case of loosely installed camera might have significant influence on calibration process. Also, cameras might change their perspective based on program or human intervention. For example, change in tilt will not only result in small tremors induced by tilting mechanism, but will also change the observed volume which will result in a need for new calibration.

To deal with such issues, as first step of suggested approach, we employ a navigation algorithm previously described in [2]. It is used to calibrate cameras based on features present in the background image of the recording. By using calibration procedure described in [3], intrinsic parameters of cameras are obtained in order to be used later to adjust projection of feature points to their correspondences on the captured images.

Data used in further face detection will be in first stage applied for tracking of multiple objects in views from many video cameras. The first problem present in such situation will be identification of particular object/person and application of obtained information across all views. Objects will be therefore identified in single view, however the problem of reidentification in another camera's view is still present. Here, extremely important will be solution for problems of change in illumination, possibility of people moving in one view and the place between two views of cameras where object can be not visible.

One of the current methods that is suited for the problem at hand was presented during Multi-Camera Object Tracking (MCT) Challenge Zurich, 12th September 2014 in conjunction with ECCV 2014. The method formed in [4] extends the basic tracking idea. The general concepts of the method relies on searching for affinities between tracks basing on features and spatial temporal context. Algorithm consists of four main parts: online sample collection, discriminative appearance learning, relative appearance context learning and track association.

Online sample collection - basing on tracks computed by multi-target tracking by on-line learned discriminative appearance models and spatial context the presented method obtain trajectories describing each object. It may be assumed

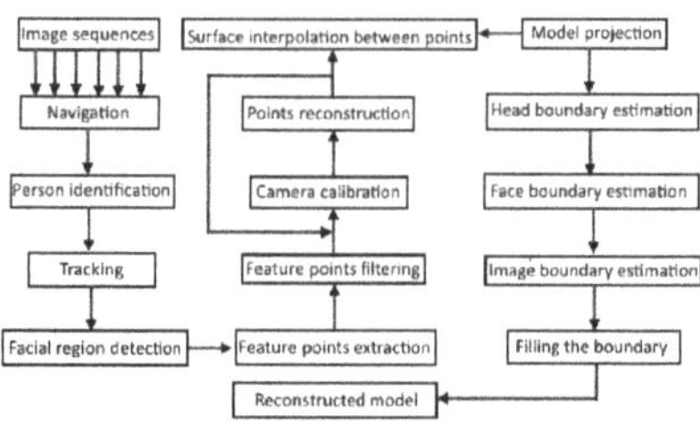

Fig. 1. Scheme of proposed approach

Fig. 2. Sample captured frame with tracking

that one object may occur only ones in one frame. In that part the presented method collects tracks and assigns them as positive or negative depending on possibility of fitting object inside one camera and between many.

Discriminative appearance learning - that part of algorithm relies on creation of strong feature model that will be able to distinguish very similar objects. In order to calculate the cost of affinity, this method applies the following pre-processing steps. Firstly color normalization has to be introduced which will compensate differences in colors across many cameras. On such processed image algorithm applies feature descriptors as a histograms of RGB colors or HOG [5], in which affinities are computed basing on correlation coefficients. Features are then used in order to discriminate objects across many cameras.

Relative appearance context learning - method introduces scheme for dealing with many objects moving in one group which is usually difficult to discriminate. The solution lies in identification of groups having similar velocities and distance.

Track association - last part of algorithm which connects short tracks in long trajectories related to particular object. Problem is formulated as construction of cost matrix solved by Hungarian algorithm.

The result of tracking algorithm which will be visible in the form of bounding box will be used in detection of faces. Tracking method gives reliability that faces in bounding boxes in different camera views will be related to particular object. One of the possible solution was described in [6] which detects parts of objects. Problem of object detection is solved by mixtures of multiscale deformable part models. System relies on methods for discriminative training of classifiers that make use of latent information. It also relies on efficient methods for matching deformable models to images.

2.2 Reconstruction Using Feature Points

Once facial region is obtained, facial feature points can be extracted from the images. Few issues have to be considered in selection of correct approach. First, since goal of the method is to reconstruct anthropometrically correct facial model on the basis of many frames from different cameras, the amount of extracted points is important, the higher the amount of points that the method can extract, the better. Second, surveillance cameras in most cases are installed above the

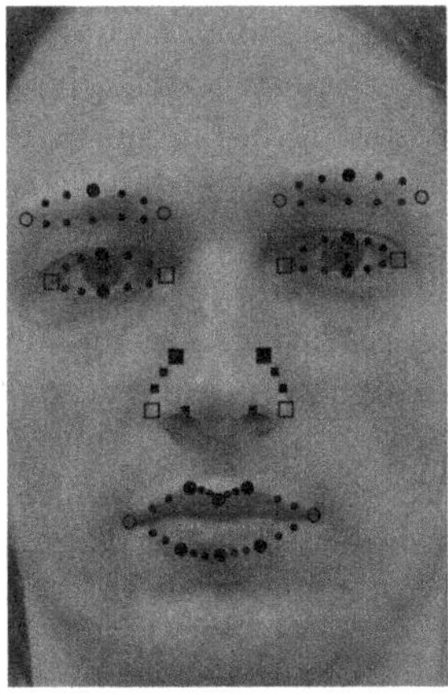

Fig. 3. Structural set (*rectangles*) and expressive set (*circles*) of contour points (*small figures*) and fiducial points (*big figures*); both global (*empty figures*) and local (*filled figures*).

height of human face, which is a first source of rotation from frontal pose. Second source is that one cannot expect suspect to look in the direction of camera, so the face will mostly be recorded with horizontal rotation as well. The method used for extraction of feature points has to be possibly robust, so that rotations of facial pose will not render most of frames useless. Third, the method has to be able to handle some occlusions, for example - with one eye occluded it has to be able to correctly establish the position of other feature points. This allows to use frames which do not have every point found, which is a commonplace in case of surveillance cameras - those points will be reconstructed using different frames. Fourth issue, is that the method should be able to use data from different frames. In many cases either occlusions or rotations will prove difficult to find feature points, an information from previous frame can be used to decide where to look for a given feature point, and therefore minimize the possibility of finding different feature instead. This aspect is also important due to facial expressions which can significantly influence the process of extracting feature points.

A method that proved to meet requirements of approach presented in this paper and therefore was selected is based on multi-state hierarchical shape model as described in [7]. This method is able to extract 26 fiducial points and many contour points between them (authors use 56 contour points, and though it can be easily changed, it proved to be a reasonable number for our approach). Since in proposed approach, the data from different frames is accumulated, there is a possibility, that some of those frames might contain different facial expressions. Using all feature points in further steps would therefore introduce errors related to differences between facial images. To mitigate the problem, all facial feature and feature contour points are divided into two sets: structural and expressive. Structural set contains points which are only slightly affected by facial expressions, those will be used in primary reconstruction. Expressive set contains points which are greatly affected by expressions, those might be filtered out so that same facial expression will be reconstructed using different frames. Membership of each point is presented in Fig. 3. Further part of proposed approach was inspired by facial reconstruction algorithm presented in [8], although there are notable changes in described method, due to the differences in data between controlled environment (as in [8]) and environment that is under surveillance. Using automatic feature detection rather than manual one (as in mentioned article) significantly decreases the amount of time needed for proper reconstruction. Although the mentioned approach proposes using *Downhill Simplex* algorithm [9] to minimize residual error value and thus find appropriate camera's intrinsic parameters, it is not suitable for presented application. This is due to the following:

- Since feature points are selected in an automatic way, their position at this point is not necessarily as precise as in case of manual selection and some of them might be inappropriate, which would heavily influence the residual error value.
- Only small portion of recorded image represents suspect's face, it is therefore not a good representation of entire image and thus camera's properties.
- Intrinsic parameters can be calculated on the basis of entire recorded image, which is much more precise.

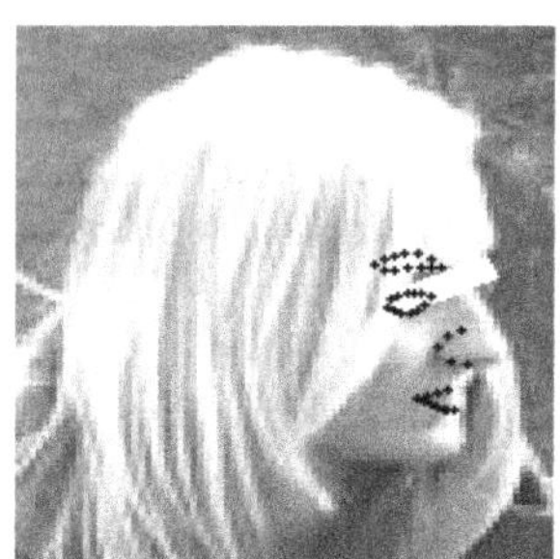
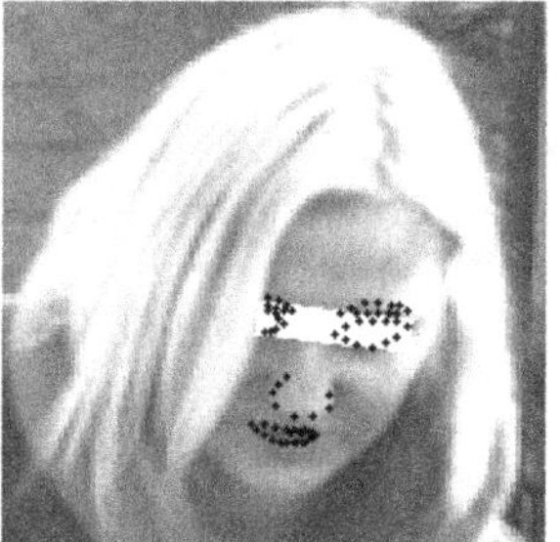
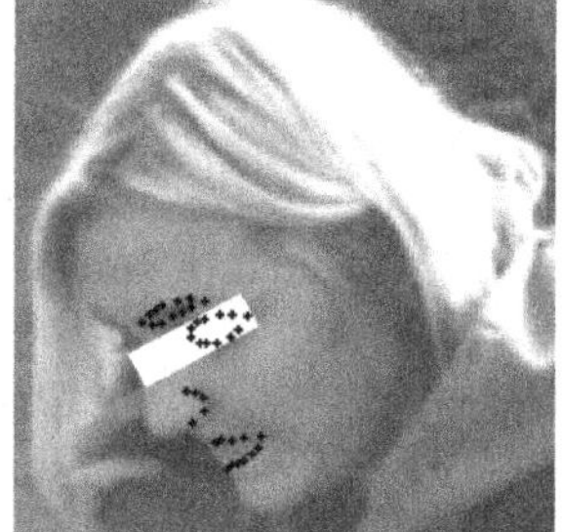

Fig. 4. Detected feature points

With known intrinsic parameters, the distance between found feature points that belong to the structural set and projection of their correspondences on generic model, is iteratively minimized using POSIT [10]. This step might be considered as calibration, and therefore be confusing since calibration was already performed. The difference is, that in this step, the orientation and position of camera around the face is found. For each image in the filtered set, different position and orientation of camera is found, even if in fact it is the same camera. This is due to the fact, that the position of face in three-dimensional space changed, which is modeled as different position of camera in relation to face.

Once orientation and position of the camera regarding to the face is obtained, the position of feature points is reconstructed by adaptive symmetry as in [8], which changes generic feature points' positions to ones more related to the shape of reconstructed face. Since their position is still not perfect, calibration/reconstruction process is repeated until convergence. Once reconstruction of feature points from structural set is complete and the position and orientation data is calibrated, the feature points belonging to expressive set are reconstructed. There are two aspect that need to be considered. First, there is a chance, that automatic feature points selection will yield wrong results. Second, some of expressive points might be under influence of facial expressions (which is why those are not considered in calibration), which will reduce the effectiveness of reconstruction. Therefore, not every image in the filtered set has to be used to reconstruct every point. For each point in expressive set for each image, we calculate sum of distances to every point in reconstructed structural set. Only those expressive points in an image which have a sum of distances different by not more than standard deviation of all the sums of this expressive point in all images are considered a valid source of reconstruction. Having reconstructed all feature points from both structural and expressive sets, radial basis function in the form of $\sigma(r) = r$ is used to interpolate the changes in feature points and modify the surface between them.

2.3 Correcting Reconstructed Mesh Using Boundaries

Using radial basis function will properly reconstruct most of anthropometrically important features. However, due to the fact that it is not guaranteed that all feature points will be recognized, as well as the fact that recognized feature points might be far from each other, it is necessary to correct the estimation. Otherwise, it would be possible that interpolation from neighbouring feature points would result in creation of unwanted features in minimas of sum of feature points' influence. The method proposed by [8] is used here as well. The edges of the mesh are projected onto the image using calibration data creating the *projected boundary*. The desired boundary of the model should coincide with the boundary of face/head on the image, so the *image boundary* has to be found.

In case in which the calibration is correct, *projected boundary* is similar to the real *image boundary*, which makes it suitable for use as an initial clue in finding of *image boundary*. In other cases, due to following logical operations, the method will not yield any boundary, and therefore will not negatively affect the overall reconstruction. To obtain *image boundary*, the method presented in [11] is used with some modifications.

Both head boundary and face boundary is extracted using mentioned method. Since the projected model consist of entire head rather than only part of it, the logical disjunction of both boundaries is used. Next, the conjunction of obtained boundaries and the *projected boundary* broadened by half of intercanthal width as obtained from found feature points. The selection of intercanthal width is based on few reasons; the stability of this distance around different populations (see [12]), the fact that inner eye corners are feature points which are obtained with most accuracy, and the fact that in tests it proved to be usable without any further scaling. The conjunction removes estimates of boundaries

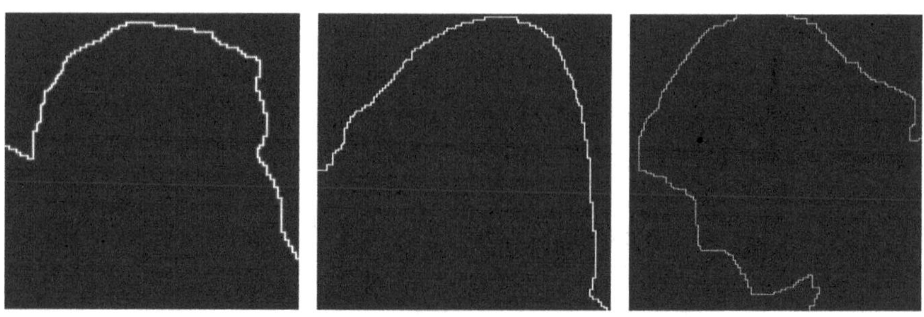

Fig. 5. Detected head boundaries

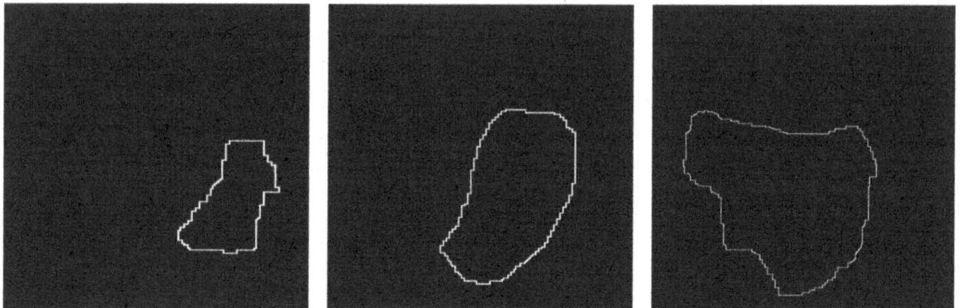

Fig. 6. Detected face boundaries

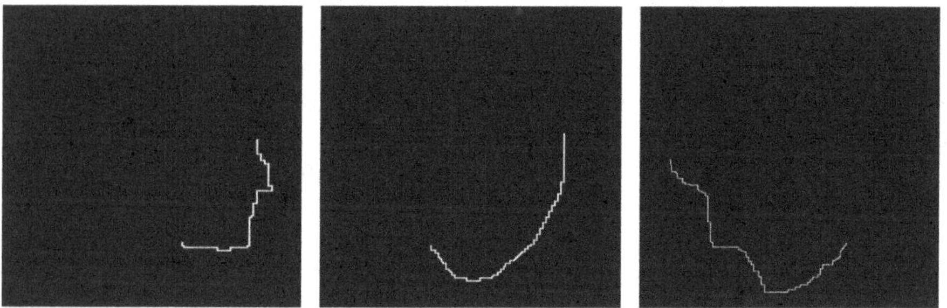

Fig. 7. Fragments of boundaries used for reconstruction

that are based on different features than those presented in projected model, e.g. line of hair above the forehead, hair on the side of face, shoulders and neck. Since the disjunction of head and face boundary was used, for each branch inside of conjunction, the decision is made to use the one with smallest average distance to *projected boundary*, thus obtaining *image boundary*. With both boundaries found, the aforementioned method [8] is used; maxima of distances between boundaries are found and used to find RBF interpolation that will transform mesh so that *projected boundary* coincides with *image boundary* thus reconstructing parts of 3D mesh that are not reconstructed well using found feature points.

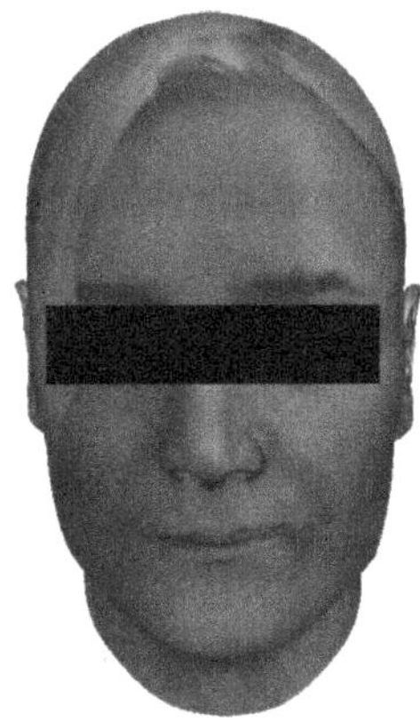

Fig. 8. Reconstructed model

3 Conclusion

The approach presented in this paper is a composite solution for reconstruction of facial mesh from video-based surveillance systems' data. Presented solution produces satisfactory results in case of videos, where face is fully visible from different angles, however it still requires further work in case of low quality data (as in Figs. 4, 5, 6 and 7), from which the presented reconstructed model (Fig. 8) was built. Further work will be focused mostly on obtaining additional feature points through boundaries of facial features. Still though, results are promising enough to consider this approach worth further work so that it could be used in assisting eyewitnesses in suspect identification.

Acknowledgments. This work has been supported by the National Centre for Research and Development (project UOD-DEM-1-183/001 "Intelligent video analysis system for behavior and event recognition in surveillance networks")

References

1. Kulbacki, M., Segen, J., Wereszczyński, K., Gudyś, A.: VMASS: massive dataset of multi-camera video for learning, classification and recognition of human actions. In: Nguyen, N.T., Attachoo, B., Trawiński, B., Somboonviwat, K. (eds.) ACIIDS 2014, Part II. LNCS, vol. 8398, pp. 565–574. Springer, Heidelberg (2014)
2. Gudyś, A., Wereszczyński, K., Segen, J., Kulbacki, M., Drabik, A.: Camera calibration and navigation in networks of rotating cameras. In: Nguyen, N.T., Trawiński, B., Kosala, R. (eds.) ACIIDS 2015. LNCS, vol. 9012, pp. 237–247. Springer, Heidelberg (2015)
3. Hartley, R.I.: Self-calibration from multiple views with a rotating camera. In: Eklundh, J.-O. (ed.) ECCV 1994. LNCS, vol. 800, pp. 471–478. Springer, Heidelberg (1994)

4. Cai, Y., Medioni, G.: Exploring context information for inter-camera multiple target tracking. In: Applications of Computer Vision (WACV) 2014, pp. 761–768. IEEE (2014)
5. Dalal, N., Triggs, B.: Histograms of oriented gradients for human detection. In: Proceedings of IEEE Conference Computer Vision and Pattern Recognition, pp. 886–893. IEEE (2005)
6. Felzenszwalb, P., Girshick, R., McAllester, D., Ramanan, D.: Object detection with discriminatively trained part based models. Pattern Anal. Mach. Intell. **32**(9), 1627–1645 (2010)
7. Tong, Y., Wang, Y., Zhu, Z., Qiang, J.: Robust facial feature tracking under varying face pose and facial expression. Pattern Recogn. **40**(11), 3195–3208 (2007)
8. Roussel, R., Gagalowicz, A.: Realistic face reconstruction from uncalibrated images. In: VMV 2004, pp. 141–149. Aka GmbH (2004)
9. Nelder, J.A., Mead, R.: A simplex method for function minimization. Comput. J. **7**(4), 308–313 (1965)
10. DeMenthon, D.F., Davis, L.S.: Model-based object pose in 25 lines of code. In: Sandini, G. (ed.) ECCV 1992. LNCS, vol. 588, pp. 335–345. Springer, Heidelberg (1992)
11. Shih, F.Y., Chuang, C.-F.: Automatic extraction of head and face boundaries and facial features. Inf. Sci. **158**, 117–130 (2004)
12. Farkas, L.G., Katic, M.J., Forrest, C.R.: International anthropometric study of facial morphology in various ethnic groups/races. J. Craniofac. Surg. **16**(4), 615–646 (2005)

Efficient Motion Magnification

Mariusz Domżał, Dawid Sobel, Jan Kwiatkowski,
Karol Jędrasiak(✉), and Aleksander Nawrat

Institute of Automation, Silesian University of Technology, Gliwice, Poland
karol.jedrasiak@polsl.pl

Abstract. Video motion magnification (VMM) allows to amplify hardly visible changes in the video input sequence. It is possible to amplify breathing motion of hospital patients or oscillations of some mechanical element without need to connect measuring devices. On the other hand modified VMM algorithm can be used to extinguish motion before processing with another method. Unfortunately, algorithm in its full, but slowest version uses new pyramids instead laplacian ones and requires remarkable computational power in each stage of processing and that problem is solved by GPU-based implementation of basic operations using CUDA technology and scheduling them by CPU. That approach lets to run all computations on GPU. Additionally in comparison with previous studies temporal filter was changed to butterworth one. Although testing hardware was only able to run low resolution in real time, there are no doubts that better GPU would be able to run 640×480 resolution in real time.

Keywords: Image processing · GPU-based · Motion magnification

1 Introduction

Since the invention of first computers there have always been need for more computational power. Over the years CPUs have became faster and faster, but for many purposes it is not enough. Right now it is possible to run calculations also on GPU, which offers a huge step in computation power in comparison to CPU. For example currently fastest CPU Inter Core i7-5960X offers up to 384 GFLOPS while cheaper GPU nVidia GeForce GTX 980 Ti provide over 5500 GFLOPS. That ratio remains true also in lower segments of market. Unfortunately, GPU architecture allows to unleash that power only for massive amount of parallel calculations. While CPUs offers up to 8 cores, typical GPU consist of hundreds or even thousands of cores (mentioned GPU is supplied with 2816 cores). Image processing and vision algorithms mostly can perform on GPU well, as each image can be treated as set of pixels or groups of pixels and computations can be executed simultaneously over multiple groups.

Video motion magnification method implemented in this work is based on [1]. It yields effects similar to [2], but do not require as much resources and allows to amplify subtle changes in surrounding environment, so that people can easily notice them. Probably it could be used also in preprocessing to remove small motions from video sequence for other algorithms. Additionally, In [1] was shown that laplacian pyramid based

N.T. Nguyen et al. (Eds.): ACIIDS 2016, Part II, LNAI 9622, pp. 477–486, 2016.
DOI: 10.1007/978-3-662-49390-8_47

version can run in real time on CPU and here it is shown that GPU allows to run version using new pyramids.

In further part, GPU programming techniques are shortly introduced. The oldest approach to run calculations on GPU involved utilization of shading languages (GLSL or HLSL) through standard graphic pipeline. That approach offered high performance, but was uncomfortable to use. Later appeared 2 technologies namely: OpenCL and CUDA. The first one permit implementation of application on most GPUs (it is not limited to products of one manufacturer) and CPUs in alike manner, but offers worse performance. On the other hand CUDA is much faster, but is supported only by products of nVidia company. Relatively narrow range of supported products is compensated by access to low level device function (for example different kinds of memory).

For purpose of this work here is short description of CUDA programming model (based on [8]). All threads called also kernels run in grid further called block and blocks run in a grid too. These grids supports up to 3D layout. For instance block of size (16,2) launched in (4,4) grid would result in (16*2)*(4*4) threads. Additionally, used types of memory are described below:

- global memory – slowest, but largest, offers speedup when nearby threads (in grid) accesses nearby addresses
- constant memory – part of global memory, cannot be modified from inside thread, broadcasts value to all requesting threads (if any number of threads requests the same value in the same time, they receive it with cost of 1 memory read)
- shared memory – fast, onchip memory, shared across threads in the same block, accessible only from inside of device
- local memory – in most cases uses registers (fast), but if become too big part will be stored in global memory.

2 Algorithm

Implemented VMM method consist of few steps (in order): spatial decomposition, phase calculation, temporal and spatial filtering, phase amplification and spatial reconstruction. Below each step is described in more detail.

Firstly image is spatially decomposed into laplacian like pyramid as it was proposed in [1] Each band is constructed by using 4 times McClellan transform recursively on input band so that they form set, then that set is combined into one image with use of Chebysheve polynomials corresponding to highpass filter. Input for next level construction is created also by combining mentioned set with Chebysheve polynomials, but corresponding to lowpass filter and then downsampled. More details are shown in Fig. 1.

Image reconstruction (shown in Fig. 2) in case of that structure is very similar, as this pyramid is self-inverting. Input band becomes upsampled and lowpassed with use of McClellan transform and Chebysheve polynomials and added to highpassed in the same way upper level of pyramid. Result of these operations becomes input for next iteration.

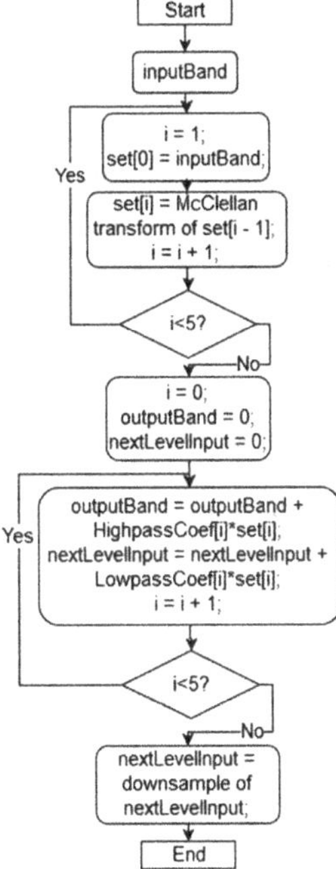

Fig. 1. Flowchart of pyramid band construction

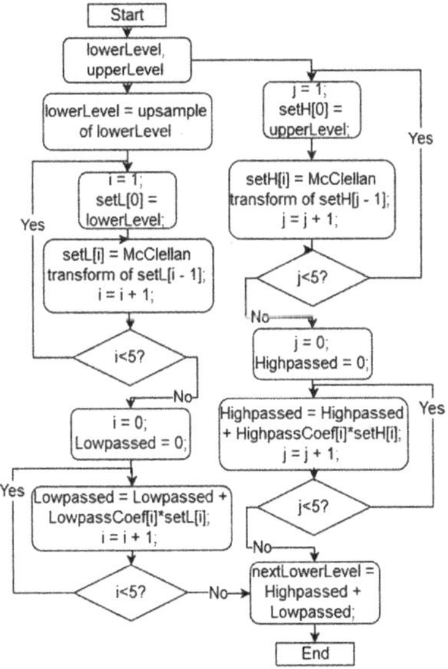

Fig. 2. Flowchart of pyramid based image reconstruction

For comparison in case of laplacian pyramid band construction is done by firstly blurring input and then subtracting that from the input (blurred image is then downsampled and become input for next level construction). Reconstruction consist of upsampling input, then blurring it and adding corresponding pyramid level to it. So it is clear that new pyramid will significantly slow down processing.

After decomposition follows phase calculation. This step is done as shown in [1]. Firstly input band I is filtered with [0.5 0 −0.5] and $[0.5\ 0\ -0.5]^T$ kernels and that form responses R_1 and R_2 and can be converted to spherical coordinates:

$$\begin{aligned} I &= Acos(\phi), \\ R_1 &= Asin(\phi)\,cos(\theta), \\ R_2 &= Asin(\phi)\,sin(\theta). \end{aligned} \tag{1}$$

Analytical solution of above equation yields local amplitude A, local phase φ and local orientation θ.

In next step phase is temporally and spatially filtered, but it can not be done straight because of ambiguity between (φ, θ) and (- φ, θ + π). As it is proposed in [1] filtered are quantities:

$$\begin{aligned} q_1 &= \phi sin(\theta), \\ q_2 &= \phi cos(\theta). \end{aligned} \tag{2}$$

Those values are resistant to abrupt changes of phase sign and are filtered with butterworth bandpass filter and blurred with Gaussian filter.

After that phase amplification takes place. Initially filtered quantities q_1 and q_2 are recombined to obtain filtered phase φ_f as it is shown below.

$$\phi_f = cos(\theta)\left(q_2 * K_\rho\right) + sin(\theta)\left(q_1 * K_\rho\right), \tag{3}$$

$K\rho$ stands for Gaussian kernel convolution with standard deviation ρ and * for convolution. We resigned from amplitude weighting shown in [1].

According to [1] phase is amplified with following equation:

$$O = Icos\left(\alpha\phi_f\right) + Asin(\phi)\,sin\left(\alpha\phi_f\right), \tag{4}$$

where O stands for output band and α for amplification coefficient. Finally frame is reconstructed as it was described earlier.

3 Implementation

Prepared implementation involves openCV library [9] which allows to acquire, save or display frames of video sequence and [10] for calculation of butterworth filter coefficients. Remaining operations run on GPU by using CUDA to implement basic operations: downsampling, upsampling, 1D convolution in horizontal and vertical direction, addition, subtraction, phase calculation, phase magnification and temporal filtering.

Following part describes those operations in more detail and next show how they are combined together to form another operations.

Each thread in case of up and downsampling connects 2×2 pixel area in upper band with one pixel in lower band. This way memory accesses are saved during upsampling – imagine that upsampling is done this way: thread connects one pixel in lower band to one pixel in upper band. In this scenario each pixel in lower band is accessed 4 times, when proposed approach make only 1 access to get the same result.

Convolution is the most often used elementary operation in this algorithm, so a lot of effort was put into it to make it fast. Proposed solution involves using constant memory to store kernel and shared memory to speedup memory access to image (as convolution is bandwidth bound). In first part of processing all data is loaded into shared memory of threads block (with required neighborhood) as it is shown in Fig. 3 - assuming mask length is N and radius $R = [(N\text{-}1)/2]$ then each thread load value with offset of $-R$ and then last $N\text{-}1$ in block loads also value with offset of R. In case of boundaries of image value of last pixel is cloned. Usage of constant memory for kernel also reduces amount of memory reads, as all threads ask for these values in the same time. This way it is possible to minimize amount of accesses to global memory, exact number k can be calculated from equation:

$$k = \frac{XY}{B}(B + N - 1) + N, \tag{5}$$

where B stands for size of block in corresponding dimension, X for image width and Y for height. Afterwards operations of multiplication and addition are performed in unrolled loop and then result is stored. For comparison easiest approach based on direct access to global memory for pixel and mask values would result in:

$$k = 2NXY \tag{6}$$

In case of 640×480 gray image and mask size 5 Eq. (5) yields 345605 when $B = 32$ and (6) 3072000.

Remaining operations are performed in one pixel of input image, or images (for example addition, subtraction) to one of output and do not use complicated memory access patterns.

Phase calculation and temporal filtering are done as one operation, as it allows to avoid part of global memory reads/writes – data remains stored in registers. Calculations are done by use of analytical solutions of Eq. (1), then quantities (2) are created, temporally filtered and stored in global memory with other metadata (if global synchronization from inside of thread would be possible phase magnification and spatial filtering would join to mentioned pair).

Temporal filtering is achieved by bandpass butterworth filter consisting of biquad sections, which replaced 2-order IIR filter - comparison of their characteristics are shown in Fig. 4. Each biquad section has 4 banks for previous values 2 for input values and 2 for output and during processing on GPU only banks with older data are overridden and structure of data is maintained by swap of addresses done by CPU.

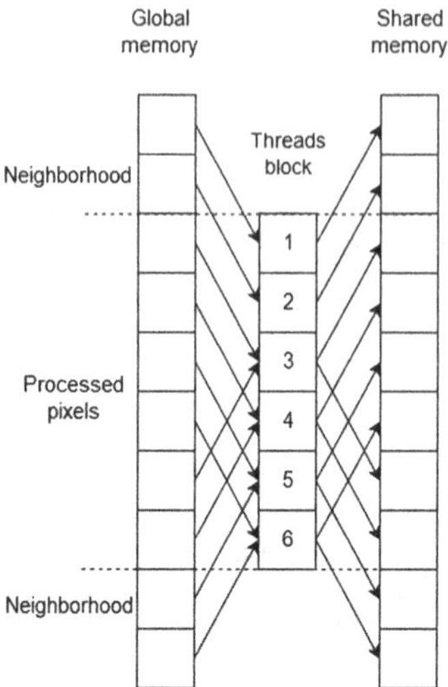

Fig. 3. Pattern of loading data to shared memory for mask size of 5

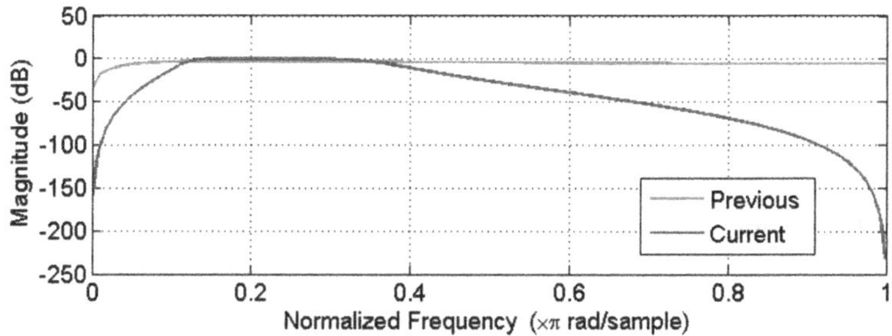

Fig. 4. Comparison of previously used filter with current

Phase amplification involves recombining filtered quantities (2) to obtain phase (3) and then amplify it with (4).

Spatial filtering is done by 2 convolutions with 1D gaussian mask – one in horizontal and second in vertical direction.

The last not explained so far part is McClellan transform. It can be seen as convolution with [4]:

$$\begin{bmatrix} 0.125 & 0.25 & 0.125 \\ 0.25 & -0.5 & 0.25 \\ 0.125 & 0.25 & 0.125 \end{bmatrix} \tag{7}$$

Kernel and is not separable in that form. Fortunately, it is possible to obtain it with following operations:

Fig. 5. Comparison of frame fragment of input (up) and result (down).

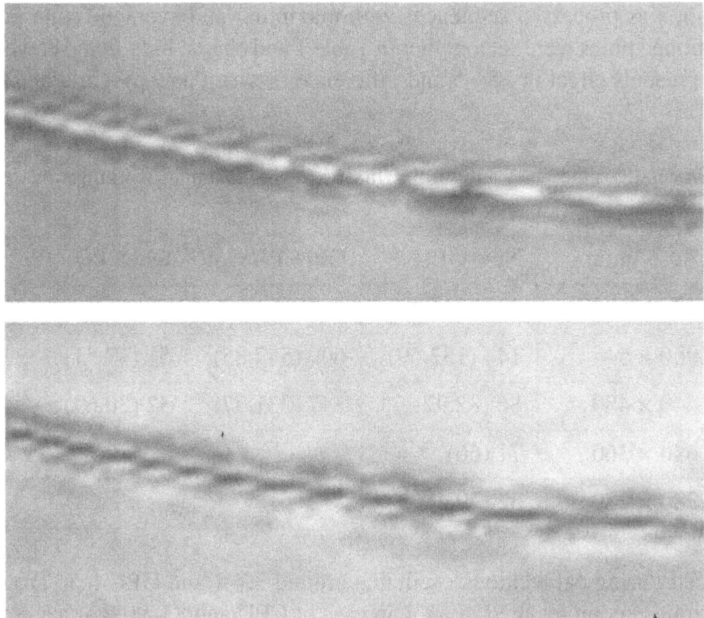

Fig. 6. Difference of output between laplacian (up) and new pyramid (down).

$$\begin{bmatrix} 0.25\sqrt{2} & 0.5\sqrt{2} & 0.25\sqrt{2} \end{bmatrix} * \begin{bmatrix} 0.25\sqrt{2} \\ 0.5\sqrt{2} \\ 0.25\sqrt{2} \end{bmatrix} - \begin{bmatrix} 0 & 0 & 0 \\ 0 & 1 & 0 \\ 0 & 0 & 0 \end{bmatrix} \quad (8)$$

so vertical and horizontal 1D convolution followed by subtraction of input allows to get McClellan transform of input.

4 Results

Results was obtained on laptop supplied with Inter Core i7-3537U running at 2.9 GHz during calculations and nVidia GeForce GT 740 m. They are compared with previous version of algorithm (based on laplace pyramid and much simpler temporal filter) running on GPU and CPU. Temporal filter was 8-order Butterworth bandpass filter. Frames fragments shown is this work comes from [7] - drum vibrations captured with high fps camera and chosen fragments show part of its membrane. Effects of the processing should be particularly visible around edges undergoing small motions, while homogenous areas should be left almost untouched. Input to program was video sequences (with varying resolution) and set of parameters (amplification coefficient, band start, band stop, number of pyramid levels to built), which were constant during tests. Output was processed sequences with maximum and average frame processing time. Mentioned times were set together in Table 1 and charts: Figs. 7 and 8, while frame fragments presents effect in Fig. 5 and difference between new pyramid and laplacian one - Fig. 6.

Table 1. Processing time per frame in ms, first value is maximum working time, second (in brackets) is average.

Resolution	New GPU	Old CPU	Old GPU
1280 × 720	227 (224.44)	–	–
960 × 544	144 (132.69)	609 (547.85)	51 (47.51)
640 × 480	86 (82.97)	347 (331.77)	32 (30.69)
640 × 360	71 (66)	–	–
280 × 280	32 (29.4)	–	–

CPU load during calculations oscillated around 33 % and GPU load around 80 % while previous version resulted in 33 % in case of CPU and 85–90 % in case of GPU.

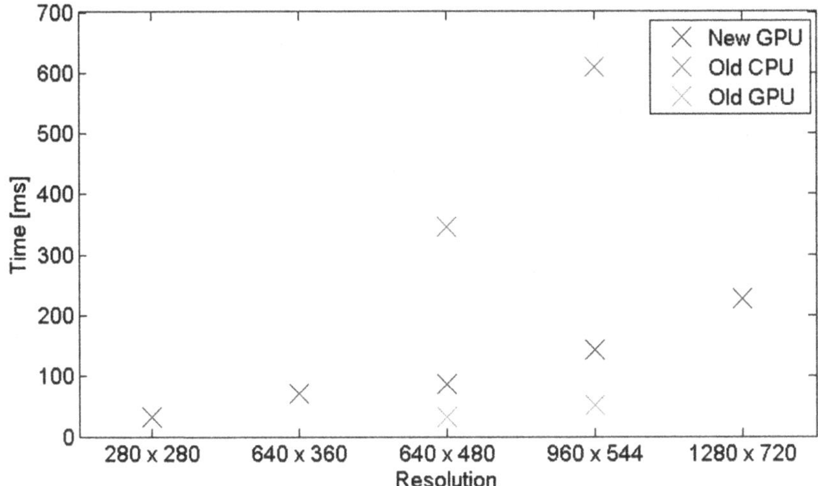

Fig. 7. Chart of maximum frame processing time in dependence of resolution.

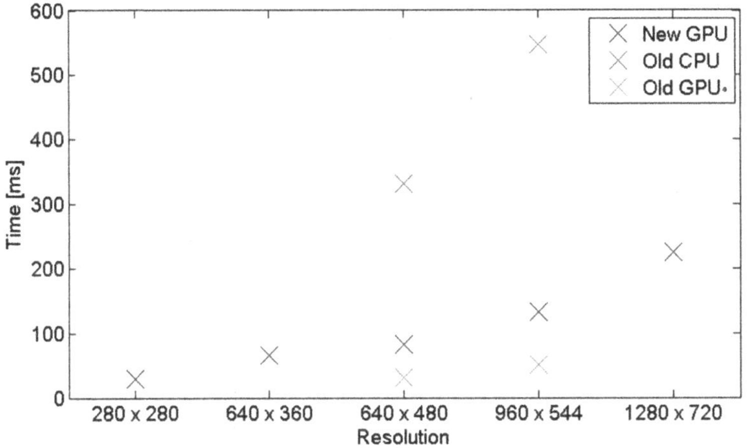

Fig. 8. Chart of average frame working time in dependence of resolution.

5 Conclusions

Presented algorithm is much more complex in computation than one used in earlier studies. This observation is supported by obtained results, working time with new pyramids was about 2.5 times longer than one with laplacian run on GPU, but still remains 4 times faster than laplacian pyramid on CPU. There is also difference in GPU load, new algorithm causes lower load, but that shows there is still room for further optimizations.

It is likely that presented implementation would run in real time on desktop supplied with one of new high end GPUs (nVidia GeForce GTX 960 or better).

Image quality is increased by utilization of new pyramid. Figure 6 shows that new pyramid allows to obtain motion magnified frame, which is much less distorted.

Shown implementation can be used as preprocessing technique that extinguishes unnecessary motion before Eulerian video magnification [3] with ideal filter or other methods take place.

Future works will probably include further improvement to presented implementation or GPU-based implementation of other algorithms for instance [5] or [6]. Particularly interesting would be trial to prepare real time GPU-based implementation of [2] as it as shown in [1] is about 4 times slower than implemented method.

Acknowledgement. This work has been supported by National Centre for Research and Development as a project ID: DOB-BIO6/11/90/2014, Virtual Simulator of Protective Measures of Government Protection Bureau.

References

1. Wadhwa, N., Rubinstein, M., Durand, F., Freeman, W.T.: Riesz pyramids for fast phase-based video magnification. In: IEEE International Conference on Computational Photography (ICCP) (2014)
2. Wadhwa, N., Rubinstein, M., Durand, F., Freeman, W.T.: Phase-based video motion processing. ACM Trans. Graph. **32**(4), 80:1–80:10 (2013)
3. Wu, H.-Y., Rubinstein, M., Shih, E., Guttag, J., Durand, F., Freeman, W.: Eulerian video magnification for revealing subtle changes in the world. ACM Trans. Graph. **31**(4), 65:1–65:8 (2012)
4. Lim, J.S.: Two-Dimensional Signal and Image Processing. Prentice Hall, Inc., Englewood Cliffs (1990)
5. Ryt, A., Sobel, D., Kwiatkowski, J., Domzal, M., Jedrasiak, K., Nawrat, A.: Real-time laser point tracking. In: Chmielewski, L.J., Kozera, R., Shin, B.-S., Wojciechowski, K. (eds.) ICCVG 2014. LNCS, vol. 8671, pp. 542–551. Springer, Heidelberg (2014). ISBN (Print) 978-3-319-11330-2, ISBN (Online) 978-3-319-11331-9
6. Nawrat, A., Jedrasiak, K.: SETh System Spation-Temporal Object Tracking Using Combined Color And Motion Features, Electrical And Computer Engineering Series, pp. 67–72 (2009)
7. Internet source, accessed August 2015. http://people.csail.mit.edu/nwadhwa/riesz-pyramid/videos/Drum.mp4
8. Internet source. http://docs.nvidia.com/cuda/. Accessed June 2015
9. Internet source. http://opencv.org/. Accessed June 2015
10. Internet source. https://github.com/ruohoruotsi/Butterworth-Filter-Design. Accessed August 2015

Real Time Thermogram Enhancement by FPGA-Based Contrast Stretching

Jan Kwiatkowski, Krzysztof Daniec, Karol Jędrasiak, Dawid Sobel, Mariusz Domżał, and Aleksander Nawrat(✉)

Institute of Automatic Control, Silesian University of Technology, Gliwice, Poland
{jkwiatkowski,aleksander.nawrat}@polsl.pl

Abstract. Image enhancement is a challenging problem in the world of digital technology. Currently, producers of video cameras, are providing devices working with increasingly higher frequencies and resolutions of acquired video frames. However, that creates a need of development of methods used in image enhancement. Furthermore, real time processing of a significant amount of data require utilization of parallel processing. The paper presents the results of research on the problem of real time histogram stretching. Proposed solution utilizes parallel processing, supported by FPGA (Field Programmable Gate Array). The problem of histogram stretching is closely associated with thermal imaging, in which, in most of the cases, the range of measured infrared radiation, in the observed scene, is generally narrow, in relation to total possible range of radiation measured by IRFPA (Infrared Focal Plane Array), which results in narrow histogram and thus low contrast of the acquired image. The proposed solution can be used in every branch of contemporary industry, wherever video cameras are used. Real time operating hardware implementation, eliminates a need of further post-processing, and may be considered as useful and eligible in many modern applications.

Keywords: Thermal imaging · Real time video processing · Histogram stretching · Normalization · FPGA · Hardware acceleration

1 Introduction

Nowadays, the problem of image processing is becoming more and more challenging task. Producers of cameras are trying to confront and fulfill users' demands, in terms of image quality, resolution and frequency of data acquisition. Larger amount of data per second means, that larger amount of data has to be preprocessed with utilization of FPA (Focal Plane Array) driver. Moreover, fast, computer-independent algorithms for video preprocessing are eligible in implementations, in which computer carries out extremely advanced image analysis algorithms, such as dense motion estimation [7], object extraction and tracking [8, 10, 11] or advanced filtering [9]. With this solution, the need of burdening of the computer with tasks, such as contrast enhancement or noise reduction, is generally eliminated. This article discusses the problem of contrast enhancement and the method taken under consideration is contrast stretching, called

N.T. Nguyen et al. (Eds.): ACIIDS 2016, Part II, LNAI 9622, pp. 487–496, 2016.
DOI: 10.1007/978-3-662-49390-8_48

also histogram stretching. This procedure aims to increase distinguishability of details, through increasing of the spread of gray levels in captured image.

A characteristic feature of the proposed solution is working in real time, determined by the speed of readout of data, captured by infrared sensors, which are a parts of IRFPA (InfraRed Focal Plane Array). This means that, taking into account the limitations of used RAM (Random Access Memory), computing the parameters required for the correction, and the correction of value of new pixel, are completed within one cycle of the PCLK (Pixel CLocK).

1.1 Contrast Stretching

Methods used in histogram enhancement can be divided into two groups: non-adaptive algorithms, in which parameters are independent of the observed scene, and adaptive, in which method parameters vary over time, depending on the received data. In addition, adaptive methods can be divided according to the size of the test environment of particular pixel, which is taken into account, when determining the parameters of conversion functions. Because of that, adaptive methods can be divided into local, where a close environment of the corrected pixel is taken into account, and global, in which the whole picture is taken into account. Local methods are characterized by strong exposure of details. However, utilization of local method may result in distorted image. The proposed solution is based on adaptive and global approach, depending on the observed scene and taking into account the whole picture during determination of the parameters of conversion function. The algorithm of histogram stretching, which is taken into consideration, is well-known in the scientific literature and is shown in Eq. (1) [1–5]. Sample histogram before and after stretching is presented in Fig. 1.

$$\mathrm{f(i)} = (i - I_1)\frac{I_{max}}{I_2 - I_1} \tag{1}$$

where: I_{max} – maximum gray level, I_1, I_2 – gray levels adequate to percentiles 5 % and 95 % of cumulative histogram, i – number of subsequent pixel.

During the testing process, an additional adaptive GHE (Global Histogram Equalization) method was used for comparison purposes. The method uses appropriately converted cumulative histogram, as a way of conversion of gray levels. The histogram is quantized to integer values, and converted in such a way that the maximum histogram value is equal to the maximum possible value of gray level. Every value of cumulative histogram is converted proportionally. In other words, cumulative histogram is normalized and then every value is multiplied by maximum possible gray level, and rounded to integer value. The converted cumulative histogram is used as a lookup table (LUT), and the conversion is done by searching for the gray level corresponding to the input gray level and using it as a replacement of input value.

1.2 Definition of the Problem

The problem of histogram stretching appeared during working over the implementation of IRFPA (Infrared Focal Plane Array) driver, which is a part of high resolution

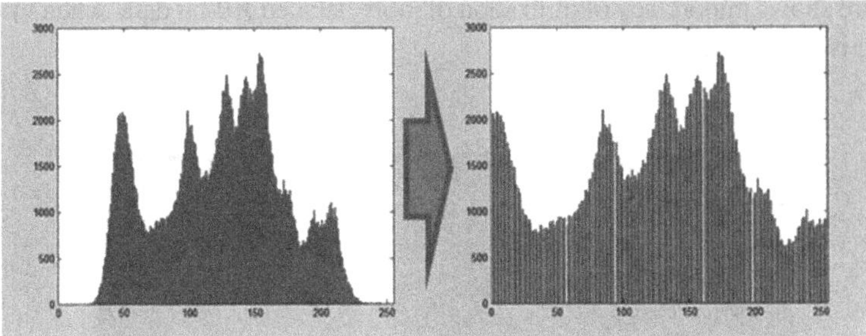

Fig. 1. Original (left) and stretched (right) histogram.

infrared camera (Fig. 2), and is closely linked with infrared imaging. Analog to digital Converter, through which voltage is read from infrared sensors included in FPA, has a resolution of 10 bit. In the current implementation, to preserve data integrity, two LSB of 10-bit words standing as pixel values, are omitted, and because of that, pixel values are represented by 8-bit words, which are sent to a computer, where histogram stretching is performed. However, the range of measured infrared radiation, in the observed scene, is generally narrow, in relation to total possible range of radiation measured by IRFPA, which results in narrow histogram and thus low contrast of the acquired image. For this reason, the loss of 2 LSB of data depth causes a significant loss of information contained in the image, which affects the quality of resulting enhanced image. That created a need for a rapid algorithm for histogram stretching, which would be working with more detailed 10-bit data, using FPGA. In addition, to avoid loss of data integrity, the conversion must be done in real time, determined by rate of appearance of further information, which are successive pixels gray level values, read from IRFPA.

Sample image acquired by the camera is presented in Fig. 3, section A, while the same image, postprocessed with utilization of a computer, is presented in Fig. 3, section B.

Fig. 2. High resolution IR camera.

The picture shows importancy of utilization of more detailed, 10-bit data, when histogram stretching is performed.

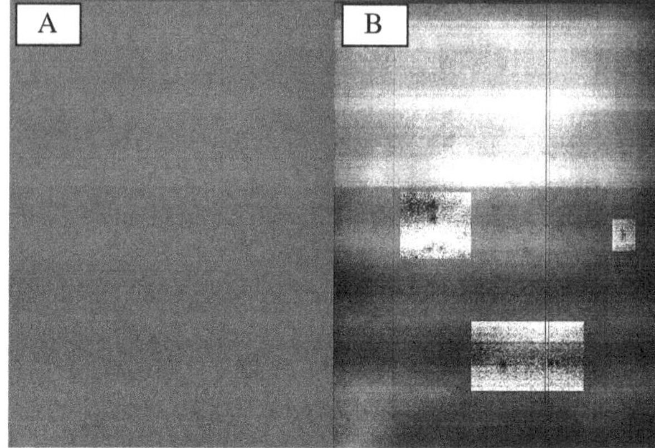

Fig. 3. Sample frame acquired by IR camera, before (A) and after (B) postprocessing with utilization of a computer, highlighted regions are two people and a car, which are completely unseen before contrast enhancement.

2 Proposed Solution

The most significant problem, that emerged during implementation of the described method, is meeting the requirement regarding real-time working implementation. What is worth mentioning, there exists a necessity of immediate access to I1 and I2 parameters, calculated from the preceding frame. Moreover, the parameters has to be computed within time between readout of last pixel from previous frame, and readout of the first pixel of a new frame, so the new information could be corrected. Furthermore, the method of histogram stretching, described by Eq. (1), had to be modified, because of the usually multi-cycle-performed implementation of a division operation, and a complexity of multiplication operation, which are present in Eq. (1). All these reasons created a requirement of proposing a new, modified solution, which enables a possibility of single-cycle correction of incoming data. Each complex mathematical operation is described in the following sections.

2.1 Real-Time Performed Multiplication

The problem of multiplication is possible to be performed in real time with utilization of parallel processing, supported by FPGA. In accordance with the technique illustrated in Fig. 4, mathematical operation of multiplication of two words can be realized as a sum of words.

```
   1011
 x  110
   0000     1011*110 = 1011*0 +
  1011      + 10110*1 + 101100*1
+ 1011 .
 111010
```

Fig. 4. Binary multiplication.

2.2 Real-Time Performed Division

The most problematic of the necessary mathematical operations turned out to be division. Division is generally performed in multiple clock cycles, until required accuracy is achieved. However, there are two particular cases, in which division is possible to be conducted within a single clock cycle. The first of proposed solutions assumes, that table of known logarithms of every possible gray level is available during computation. Then, according to Eq. (2):

$$\log\left(\frac{a}{b}\right) = \log(a) - \log(b) \tag{2}$$

an argument, for which the value of the logarithm is closest to the difference presented in Eq. (2), can be determined within one clock cycle.

However, the first proposed method requires access to large amounts data, the size of which depends on the required accuracy and number of possible gray levels. In the light of the need of collection of data regarding histogram, which is to define the parameters I1 and I2, the concept of using of logarithm tables is proven to be impossible to be implemented. Therefore, realization of the project required utilization of the second method of single-cycle division. The method includes the approximation of result of division present in Eq. (1), by computation of the average of the results of dividing a dividend by two numbers, being the power of 2, which are closest to divisor, if the divisor is not a power of 2. In the case of binary words, a division, when divisor is a power of 2, is becoming a simple operation of bit shift. In addition, an average of two binary words can be computed within a single clock cycle by right-directed, single digit shift of the word formed from the sum of the two binary words. This method of performing division guarantees achieving a result within a single PCLK cycle, however, due to non-linearity of division operation, it is almost always encumbered with an error. This error, however, is negligible in relation to the obtained results, as shown in test results section.

2.3 An Iterative Method of Determination of I1 and I2 Parameters

Another problem that emerged during the implementation of the histogram stretching is computation of I1 and I2 parameters. I1 and I2 are the values of gray levels, for which, on cumulative histogram, a number of 5 and 95 percent of all pixels in image, is exceeded. In typical approach, determination of these thresholds would require determination of histogram, and following, determination of cumulative histogram, used to determine I1 and I2. That approach, however, would create a need of storing a lot of irrelevant data, because, as for the cumulative histogram, only values regarding I1 and I2 are important. From the point of view of the FPGA, which is highly ineffective in terms of memory storage, operations on large blocks of data should be avoided as much as possible. Therefore, data collection was limited to the determination of the histogram, stored in the external RAM (Random Access Memory), while for the determination of the values of I1 and I2, an iterative method is proposed. The algorithm is performed as follows:

1. Before reading of the first pixel of a new frame, a three counters, Bank0, Bank1 and Bank2 are reset to 0, while levels I1 and I2 are respectively reset to 0 and MAX_GRAY_LEVEL.
2. All newly emerging pixel values are incrementing one of three counters: Bank0 when the value is less than or equal to I1, Bank2, when the value is greater than I2 and all values other than described are incrementing Bank1. In parallel, the histogram is created iteratively.
3. If the number of 90 % of all pixels of full frame is exceeded by Bank1 counter, a shift operation is performed:
   (a) When Bank0 is greater than or equal to Bank2, I2 is decremented by 1, and the value Bank2 is incremented by the current value of histogram of gray level corresponding to the previous value I2, while Bank1 is decremented by the same value.
   (b) When Bank0 is lesser than Bank2, I1 is incremented by 1, and the value Bank0 is incremented by the current value of histogram of gray level corresponding to the previous value I1, while Bank1 is decremented by the same value.

According to the algorithm, after the reading of whole frame, I1 and I2 should be adequate to percentiles 5 and 95 of cumulative histogram.

3 Test Results

The testing process was based on the two cases. The first case presents a situation, where contrast of test image was similar to contrast of image acquired by infrared camera (narrow histogram). The second test involves an example, illustrating the negative impact of loss of information, occurring when the number of possible gray levels in the image is reduced.

3.1 Comparison of Described Global Methods for Histogram Stretching

The test is meant to show efficiency of described method, in enhancing low contrast image, A histogram of test image is narrow, the number of gray levels in the image is

Table 1. Description of Figs. 5 and 6 sections.

Section of figure	Method description
A	Histogram Stretching, classical approach
B	original image
C	Global Histogram Equalization
D	slope of the line (1) approximated as a result of single-cycled division, where divisor is the closest power of 2, which is lesser than I2-I1
E	Proposed solution, Slope of the line (1) approximated as an average of slopes from D and F
F	slope of the line (1) approximated as a result of single-cycled division, where divisor is the closest power of 2, which is greater than I2-I1

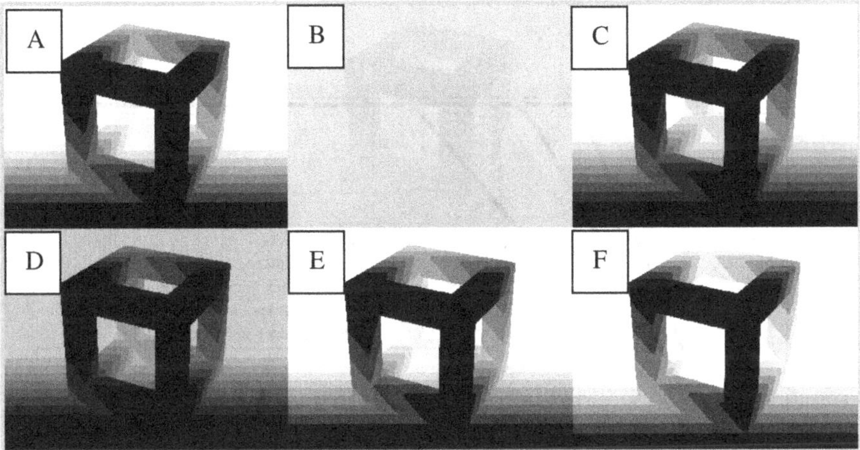

Fig. 5. Comparison of results acquired by utilization described approaches, sections of the image are described in Table 1.

small in relation to all possible gray levels. Description of methods presented in Figs. 5 and 6 is included in Table 1:

The test confirmed validity of assumption, that the results obtained using simplified division operation, are visually similar to the results, obtained by the classic approach. The histograms for each of sections A, C and E, are showing small differences, but all of the methods have proven to be effective in image enhancement. Resulting image is characterized by a lot better contrast than the input image. Furthermore, proposed method is designed for FPGA, to work in real time, which is unattainable for the classical approach. That feature determines supremacy of proposed method (Fig. 7).

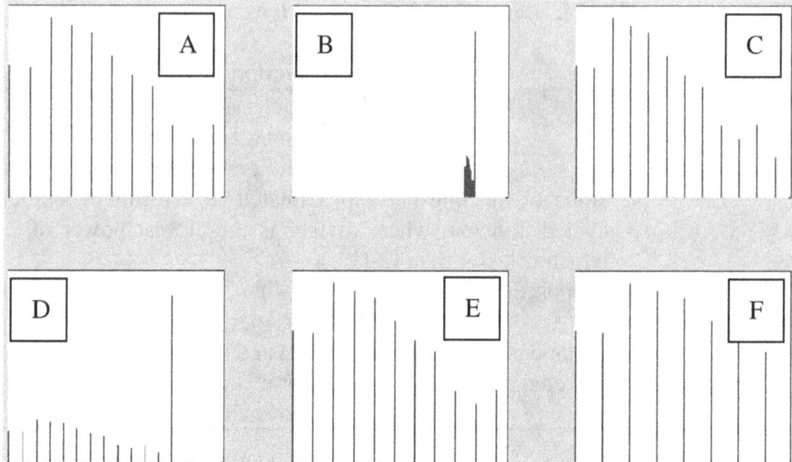

Fig. 6. Comparison of histograms acquired from histogram stretching with utilization of described approaches, sections of the image are described in Table 1.

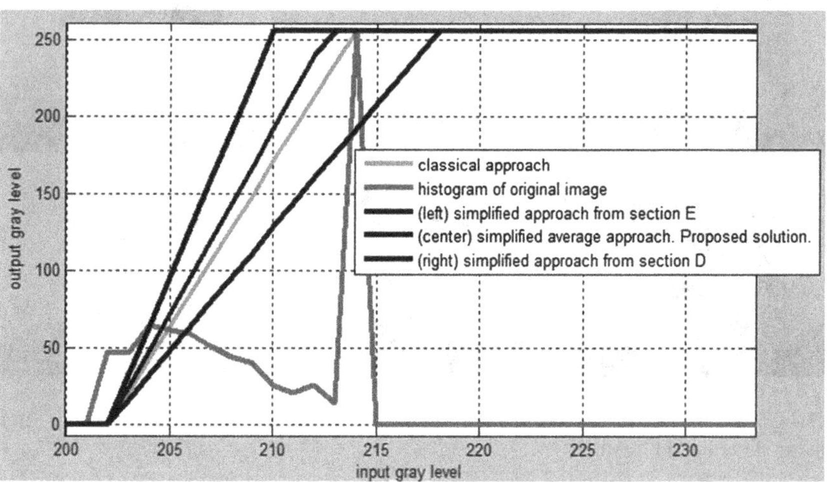

Fig. 7. Comparison of LUT tables, achieved by implementation of each mentioned approach.

3.2 An Influence of Decreasing the Numer of Possible Gray Levels

Second test is designed to show the impact of decreasing the number of possible gray levels, and to observe its influence on the outcome of proposed solution. The test purpose is to show the importance of preserving as much detailed data as possible, especially in enhancement of low contrasted video, captured by thermal imaging camera (Fig. 8).

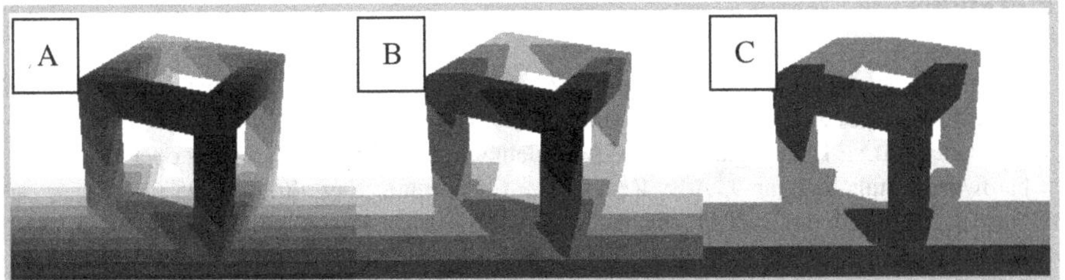

Fig. 8. Comparison of achieved results with source image from section B of Fig. 3, where number of possible gray levels is equal to 256 (A), 128 (B) and 64 (C).

Tests showing the impact of reduced data depth has shown, that for the described case of thermal imaging and narrow histogram, the depth of data, which is used to perform histogram stretching, is extremely important. Fourfold reduction of the amount of achievable gray levels, results in a significant loss of quality of performance of proposed method, as well as classical approaches. Furthermore, the test outcome confirms the validity of assumption, that histogram stretching should be performed with utilization of FPGA, before number of possible gray levels is reduced.

4 Summary

The problem of histogram stretching is widely described in scientific literature [1–5], and methods used are implemented using both computer and FPGA. However, there is no known solution, which tries to cope with a problem of real-time working application, which utilizes parallel processing, with utilization of FPGA. The proposed solution has not only proven its ability to achieve real-time performance, determined by the rate of occurring of new information, but also is highly effective in improvement of image quality.

The solution can be found as useful in many branches of contemporary industry, both civil and military. In particular, it can be used wherever there exists a need of implementation of acquisition and enhancement of an image quality, which should not or must not be performed by computer or other video receiver [6].

In summary, the resulting solution is effective and fast, moreover, its flexibility and versatility allows it to be used it in any FPA driver, as solution is independent of the source of the data processed.

Acknowledgement. This work has been supported by National Centre for Research and Development as a project ID: DOB-BIO6/11/90/2014, Virtual Simulator of Protective Measures of Government Protection Bureau.

References

1. Bittibssi, T.M., Salama, G.I., Mehaseb, Y.Z., Henawy, A.E.: Image enhancement algorithms using FPGA. Int. J. Comput. Sci. Commun. Netw. **2**(4), 536–542 (2012)
2. Warkari, M.D.S., Kshirsagar, D.: FPGA implementation of point processing operation using hardware simulation. Int. J. Adv. Res. Comput. Commun. Eng. **4**(4), 91–95 (2015)
3. Nirmala, S.O., Dongale, T.D., Kamat, R.K.: Review on image enhancement techniques: FPGA implementation perspective. Int. J. Electron. Commun. Comput. Technol. (IJECCT) **2**, 270–274 (2012)
4. Jane, P.G., Narkhede, N.P.: Image enhancement algorithm implemented on reconfigurable hardware. Int. J. Comput. Appl. International Conference on Quality Up-gradation in Engineering, Science and Technology (ICQUEST-2014)
5. Dakre, K.A., Pusdekar, P.N.: Image enhancement using hardware co-simulation for biomedical applications. Int. J. Recent Innovation Trends Comput. Commun. **3**(2), 869–877 (2015)
6. Kwiatkowski, J., Sobel, D., Jędrasiak, K., Nawrat, A.: FPGA based omnidirectional video acquisition device (OVAD). In: Recent Advances in Electrical Engineering and Computer Science
7. Kwiatkowski, J., Sobel, D., Ryt, A., Domżał, M., Jędrasiak, K., Nawrat, A.: Real time dense motion estimation using FPGA based omnidirectional video acquisition device. In: Nawrat, A., Jędrasiak, K. (eds.) Innovative Simulation Systems, pp. 87–108. Springer International Publishing, Switzerland (2015)
8. Wilk, P., Targiel, T., Sobel, D., Kwiatkowski, J., Jędrasiak, K., Nawrat, A.: The concept of an active thermal camouflage for friend-foe identification system. In: Nawrat, A., Jędrasiak, K. (eds.) Innovative Simulation Systems, pp. 67–76. Springer International Publishing, Switzerland (2015)
9. Sobel, D., Kwiatkowski, J., Ryt, A., Domżał, M., Jędrasiak, K., Janik, Ł., Nawrat, A.: The system for augmented reality motion measurements visualization. In: Nawrat, A., Jędrasiak, K. (eds.) Innovative Simulation Systems, pp. 189–200. Springer International Publishing, Switzerland (2015)
10. Domżał, M., Ryt, A., Sobel, D., Kwiatkowski, J., Jędrasiak, K., Nawrat, A.: GPU-based parameters estimation for anisotropic diffusion. In: Nawrat, A., Jędrasiak, K. (eds.) Innovative Simulation Systems, vol. 33, pp. 109–117. Springer International Publishing, Switzerland (2015)
11. Ryt, A., Sobel, D., Kwiatkowski, J., Domzal, M., Jedrasiak, K., Nawrat, A.: Real-time laser point tracking. In: Chmielewski, L.J., Kozera, R., Shin, B.-S., Wojciechowski, K. (eds.) ICCVG 2014. LNCS, vol. 8671, pp. 542–551. Springer, Heidelberg (2014)
12. Ryt, A., Sobel, D., Kwiatkowski, J., Domżał, M., Jędrasiak, K., Nawrat, A.: Real-time multiple laser points tracking. In: Nawrat, A., Jędrasiak, K. (eds.) Innovative Simulation Systems, vol. 33, pp. 201–213. Springer International Publishing, Switzerland (2015)

Real World Applications in Engineering and Technology

Dijkstra-Based Selection for Parallel Multi-lanes Map-Matching and an Actual Path Tagging System

Mick Chang-Heng Lin[1(✉)], Fu-Ming Huang[2], Po-Ching Liu[1], Yu-Hsiang Huang[1], and You-shan Chung[1]

[1] Institute of Information Science, Academia Sinica, Taipei, Taiwan
{mikelin, pliu, yuhsiang, yschung}@iis.sinica.edu.tw
[2] Research Center for Information Technology Innovation, Academia Sinica, Taipei, Taiwan
angushuang@iis.sinica.edu.tw

Abstract. Map-matching between a road network and a raw GPS trajectory must be done in order to analyze the urban traffic computing. A weight-based map-matching algorithm has proposed some important features to solve this problem, such as perpendicular distance between a raw GPS point and a road segment, bearing difference and connectivity. However, the connectivity of a map-matching problem becomes complex when the raw trajectory traveled a parallel multi-lanes road network segments, even humans will have difficulty selecting the correct road segment. To solve this problem, a dijkstra-based selection map-matching (DBSMM) algorithm is asserted by us. Candidate segment set formation, dijkstra-based selection and a friendly driver tagging system are presented in this paper. With the driver-tagged actual paths of our tagging system, it is possible to evaluate the DBSMM algorithm. Therefore, the precise map-matched network traffic data can be the basis for more further traffic researches.

Keywords: Map-matching · Shortest path selection · Ground truth · Tagging system · User-friendly · Parallel Multi-lanes · Elevated road

1 Introduction

The development of positioning techniques triggers various ubiquitous and location-based applications. Systems track users' location to provide corresponding services and users want to know the locations to make their life more convenient and smart. Furthermore, Luxen and Vetter [1] developed a hand-held application working with OpenStreetMap to provide actual-time and exact shortest path computation on continental sized networks with millions of street segments in addition to routes dragging and round-trip planning service. Many drivers often hope that their chosen route is the best. Kawak Kim Liu Nath and Iftode [2] proposed a DoppelDriver system to solve this problem by determining actual times of arrival from participatory users contributed data on the road networks. In depth, they explored the potential benefits of ex-post feedback of travel time and how snapshots of travel time comparisons can be

N.T. Nguyen et al. (Eds.): ACIIDS 2016, Part II, LNAI 9622, pp. 499–508, 2016.
DOI: 10.1007/978-3-662-49390-8_49

used to support strategic route decision making. In addition to accurate route routing, Saremi, Fatemieh, Ahmadi, Wang, Abdelzaher, Ganti, Liu, Hu, Li and Su [3] presented a GreenGPS to help drivers to find the most fuel-efficient routes customized for their vehicles between arbitrary end-points; and Yanagi, Yamamoto and Takahashi [4] presented a mobile voice navigation system to guide users to destination by using user map annotations, landmark information, and automatic navigation sentences generation.

Map-matching means that matching a raw GPS trajectory to roads on a digital map. Many location-based applications use GPS devices to record users' position. However, for saving the communication and energy cost, users often set their GPS report in a low frequency. In addition, the civilian GPS modules often have low accuracy and sampling error is common in various application scenarios. The occurrence of the collected low sampling rate trajectories also leads to critical challenges to existing map matching algorithms. Therefore, the paper is devoted to the study of map-matching problem and ground truth collection mechanism.

The existed weight-based map-matching algorithm [5] performs well in the general roads. However, while the GPS trajectory traveled in a parallel multi-lanes road, the algorithm failed to select a correct road segment. Hence, we propose a dijkstra-based selection map-matching algorithm to solve the parallel multi-lanes roads map-matching problem.

In general, this paper proposed two contributions.

1.1 Multi-lanes Map-Matching Algorithm

In urban computing, people care most about the traffic information of the main roads, like highways, freeways, expressways and trunks. The dijkstra-based selection map-matching algorithm converts the inaccurate raw GPS data to the correct lane of a multi-lanes road. Therefore, users can obtain the correct information, like whether this GPS data is on the elevated road. They can use the correct information to do further researches or to inform people the historical traffic status.

1.2 Driver Tagging System for Actual Paths

A human tagged path as the ground truth for a GPS trajectory is an ubiquitous method [6]. However, the tagging task is complicated when the road network is complex [7]. Thus, we constructed a driver tagging system which lets drivers to drag the predicted points of the predicted path to the actual roads they passed. A ground truth created by this system contributes not only the comparing basis to a map-matching problem but also the correction to OpenStreetMap. Figure 1 shows a part of the tagging system.

The outline of this paper is as follows. Section 2 reviews the related work of map-matching, shortest path selection and traffic time estimation. Section 3 formulates the map matching problem and term definition. The Dijkstra-based map-matching algorithm is presented in Sect. 4. The experiment results are described in Sect. 5. Section 6 concludes the paper and discusses the future work.

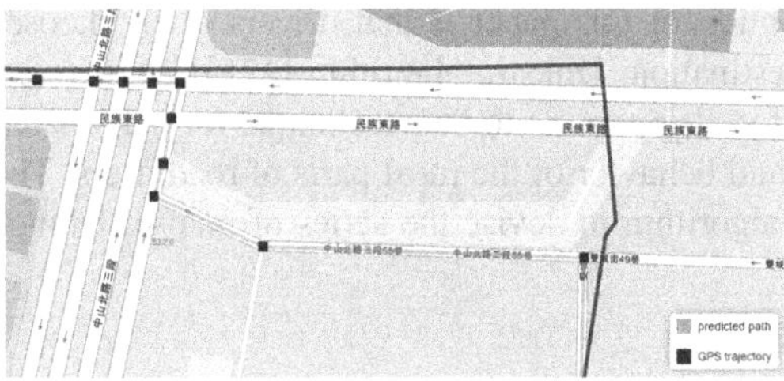

Fig. 1. Shows our tagging system which allows the driver to drag the green square points to the actual roads he/she traveled, http://www.plash.tw/~mikelin/antrack/vote.html

2 Related Work

Map matching focuses on measuring the collected location data to the road network in order to infer the driver's actual path. In Newson's research [8], the authors presented a principled map matching algorithm based on Hidden Markov Model. The solution accounted for measurement noise and the layout of the road network and utilized the feature of time-stamped sequence of latitude and longitude pairs to find the most likely road route. When the sampling rate is low or the data collected is sparse, the arc-skipping problem will be critical [9]. The author proposed a topologically based matching procedure to solve this problem, and tested the solution with low quality GPS data to measure the robustness. In this era, the GPS applications are popular and have very good sensor calibration and sensor fusion technologies. However, inaccuracies in the positioning sensors are often inevitable. Ochieng, Quddus and Noland [10] proposed an improved probabilistic map matching algorithm to reconcile inaccurate locational data with inaccurate digital road network data. Based on the model, they integrated the error positioning sensors sources, the historical trajectory of the vehicle, topological information on the road network and the heading and speed information of the vehicle. Then they successfully identified the correct link on which the vehicle is travelling. On the side, to solve the problem of low frequency GPS data, some researchers developed a new weight-based shortest path and vehicle trajectory aided map-matching algorithm [4]. They derived the shortest path between two points with the well-known A* search algorithm. And they considered link connectivity and turn restrictions at junctions.

However, the collected GPS data often have the circumstance of amount disparity. The freeway or industrial district roads may be shown by thousands of trajectories, but residential roads may only have small amount of trajectories. In Biagioni's research [11], the authors proposed an extensible map inference pipeline to mitigate GPS error, admit less-frequently traveled roads, and scale to large datasets. From this work, they can automatically inferring maps from large collections of opportunistically collected GPS trips. In Yuan's research [6], the authors considered the spatial and temporal information of a GPS trajectory and devised a voting-based strategy to model the weighted mutual influences between GPS points.

The basic assumption of this paper is that drivers often choose the shortest path from the origin to destination. Dijkstra algorithm [12] performs very well in shortest path selection. And it is also popular to be a technique for map-matching, because it fit in with the thinking and behavior of the most parts of road users. Therefore, this paper utilized the Dijkstra algorithm to devise the series of map-matching solution.

3 Preliminary

3.1 Problem Definition

The map-matching problem is to identify a correct road segments combination, which can construct a predicted path P for a raw GPS trajectory T in a given road network G (V,E). A weight-based map-matching method [4] is mainly for a low frequency GPS data, like 0.2 Hz or 5 s interval, 0.033 Hz or 30 s interval and 0.0166 Hz or 60 s interval. It advocated four different weights to each of the candidate road segment to attain the goal of choosing the correct candidate road segment. First, they set the distance weight by measuring the perpendicular distance between the GPS raw point and the candidate road segment. Second, they compare the bearing difference for weight of bearing. Third, they determine the weight of the shortest-path distance between the two consecutive candidate road segments to quantify the connectivity relationship. Fourth, they concern the weight for the vehicle trajectory heading difference. However, the total weight between the two consecutive candidate road segments would be the same when the vehicle traveled in a parallel multi-lanes road. The Fig. 1 shows the baffled situation for the weight-based map-matching method, which could generate an unexpected result. Both candidate road segments are qualified in Fig. 2(a) for the weight-based map-matching method. Here, we proposed a Dijkstra-based selection map-matching algorithm, which could identify a correct road segments combination in a parallel multi-lanes road. Therefore, we can acquire the more reasonable result revealed in Fig. 2(b).

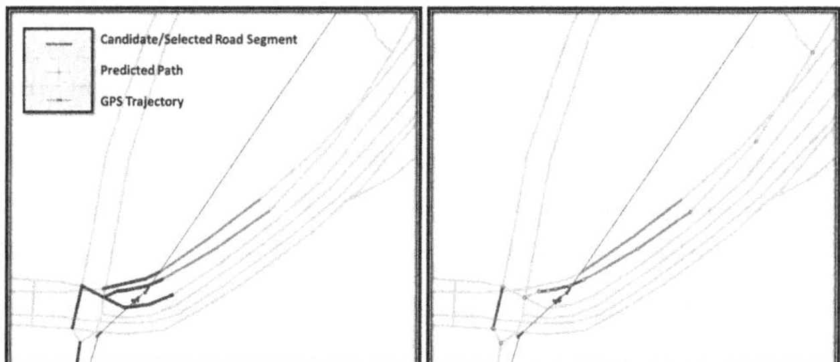

Fig. 2. (a) shows that the two parallel road segments is challenging for the decision making of a weight-based map-matching algorithm and 2(b) shows the selected road segments and predicted path of the dijkstra-based selection map-matching algorithm

3.2 Term Definition

GPS Trajectory. AGPS trajectory T is a point sequence $p_1 \rightarrow p_2 \rightarrow \cdots \rightarrow p_n$ ordered by time ascending of the GPS points. Each GPS point p_i includes p_i.lat, p_i.lng, p_i.time, which are its latitude, longitude and time.

Road Network (RN). Aroad network is a directed graph G(V,E), where E is a set of edges meaning the road segments. V is a set of vertices meaning the nodes of each edges. Each edge e is constituted of a starting point *e.start*, an ending point *e.end*, a road segment length *e.cost* and a road segment bearing *e.bearing*. The bearing is north-based and is measured clockwise: North = 0; East = PI/2; South = PI; West = 3PI/2.

Predicted Path. Given two vertices V_i, V_j in graph G(V, E). We can generate a path which is set of edges, $e_1 \rightarrow e_2 \rightarrow \cdots \rightarrow e_n$. V_i is e_1.start and V_j is e_n.end.

4 Dijkstra-Based Map-Matching Algorithm

4.1 Candidate Road Segments (CRS) Formation

Near Field Points Cleaning. We apply a similar near field points cleaning to Mapbox map-matching tidy algorithm [13]. We set a separate radius ε between p_{i-1} and p_i. Once the p_i escapes the radius ε, we start to form its Candidate Road Segments (CRS).

Heading Coherence Checking. A actual path should be coherent in a traveling process. Especially, in our data, the GPS point update frequency is 30 s a point. The sudden change of the vehicle's heading is abnormal. Figure 3(a) shows an example of this abnormal occurrence. Therefore, we design a 3-point-window to detect and delete the sudden change occurrence. Further details can be referred in Algorithm 1.

Radius Bound and Bearing Bound. For a GPS trajectory $p_1 \rightarrow p_2 \rightarrow \cdots \rightarrow p_n$, we generate a set of candidate road segments (CRS) $\{c_1^1, c_1^2, \cdots, c_1^{k1}, c_2^1, \cdots, c_{n-1}^{k-1}, c_n^1, c_n^2, \cdots, c_n^{kn}\}$. In Fig. 3(b), there is a nearby searching with a circle center p_i and a radius bound. Because of this nearby searching, we can obtain a group of nearby road segments from the road network G(V,E). Furthermore, by considering the bearing feature, every candidate set for a point p_i can be filtered. We compare the bearing between $p_{i.bearing}$ and $e_{.bearing}$.

$$p_i.bearing = \angle(p_{i-1} \rightarrow p_i)$$

$$\cos(p_i.bearing - e.bearing) > \frac{3}{5}$$

Therefore, only within-radius and within-bearing segments can be remained in the CRS.

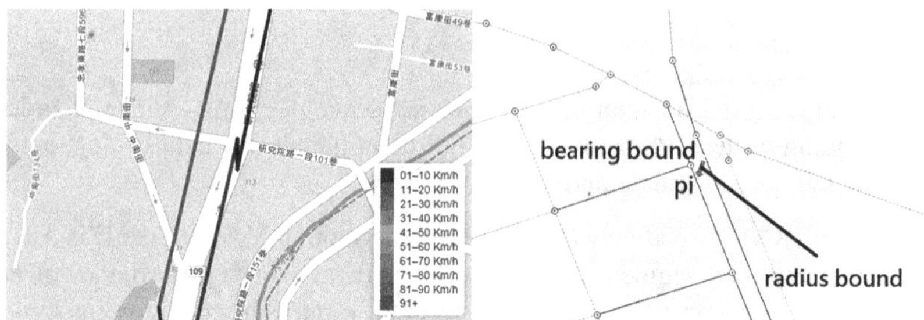

Fig. 3. (a) shows sudden heading change in a raw GPS trajectory and 3(b) shows the within-radius and within-bearing segments, the solid lines, which are remained in the CRS of p_i, p_i is the center of nearby searching circle and $p_{i.bearing}$ is calculated by the bearing from $p_{i-1} \rightarrow p_i$

Algorithm 1. Candidate Road Segments Formation Algorithm

```
Input:
        Directed Road Network G(V,E),  GPS trajectory T: p1→p2→···→pn ,
        radius bound r, bearing bound α,
Output:
        Candidate-road-segment set CRS {c^1_1, c^2_1,···, c^k1_1, c^1_2, c^2_2,···, c^k-1_n-1, c^1_n,
        c^2_n, ···, c^kn_n}  for this T

        for each p_i ∈ T do :
                if (|p_i-1, p_i|> ε)
                        generate  CRS_i ← {e| |e, p_i | ≤ r, e ∈ E } ;
        for  i=1 to n :
                if ( i ==1 ) ∠(p_1→p_2) →∠(p_0→p_1);
                for each m ∈ CRS_i in ascending order of |e, p_i |
                        if (|c^m_i.bearing- ∠(p_i-1→p_i)| > α )
                                delete c^m_i from CRS_i ;
        for i=2 to n-1
                k=0;
                for each q ∈ {-1, 1}
                        if (|c^min_i+q.bearing-  c^min_i.bearing| > 150°) k-- ;
                if ( k == -2 ) delete CRS_i ;
        return CRS;
```

4.2 Dijkstra-Based Selection

The weight-based map-matching method [4] considered the connectivity of each candidate road segments. They use A* shortest path algorithm to connect each candidate road segments. Based on their concept, we use dijkstra shortest path algorithm to connect each CRS_i. Furthermore, we consider a overall connectivity situation of the GPS trajectory. It can be remodeled as a new dijkstra shortest path problem. We define a directed graph D(S,C), Fig. 4. S are the vertices of the graph D(S,C). We copied the CRS_i to S_i. C are the edges. The cost of each edge $C_i^{j,q}$ is the cost of dijkstra shortest path from CRS_i^j to CRS_{i+1}^q on RN graph G(V,E). We define the remodeled dijkstra shortest path problem as a Dijkstra-based selection which generates a lowest distance cost combination of S_i^j. For the shortest path to S_{i+1}^j, denoted $d[S_{i+1}^j]$, formula below indicates that the shortest distance to S_{i+1}^j is either the previously known distance to S_{i+1}^j, or the result of going from S_1 to some vertex S_i^j and then directly from S_i^j to S_{i+1}^j. Finally, we save the combination of S_i^j into selected road segments (SRS) in order to perform the final connection of the predicted path.

$$d\left[S_{i+1}^j\right] = min\left(d\left[S_{i=1}^j\right], d\left[S_i^j\right] + C_i^{j,q}\right)$$

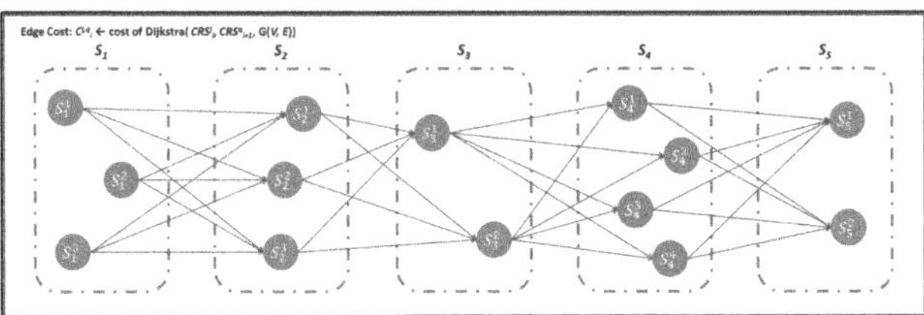

Fig. 4. Shows the concept of D(S,C)

Algorithm 2. Dijkstra-based Selection Algorithm

Input:
Candidate-road-segment set CRS from Candidate Road Segments Formation Algorithm, Directed Road Network G(V,E),

Output:
Selected sequential road segments SRS $\{c^{s1}{}_1, c^{s2}{}_2, \cdots, c^{sn}{}_n\}$

Define Graph D(S, C), S are vertices and C are edges ;
for each CRS_i and CRS_{i+1} ∈ CRS, ∀j∈CRS_i,∀q∈CRS_{i+1} do :
  $S^j_i \leftarrow CRS^j_i$;
  $C^{j,q}_i \leftarrow$ cost of Dijkstra(CRS^j_i, CRS^q_{i+1}, G(V, E)) ;
SRS←vertices of Dijkstra(S_1, S_n, D(S, C));
return SRS;

4.3 Dijkstra-Based Interpolation

The sampling rate of a raw GPS trajectory is discrete so there must be missing segments for a raw GPS trajectory. Intuitively, we can connect the separated selected road segments (SRS) by utilizing dijkstra shortest path algorithm to fill up the missing segments [14].

Algorithm 3. Dijkstra-based interpolation Algorithm

Input:

Selected-road-segment set SRS from Dijkstra-based selection Algorithm, Directed Road Network G(V,E), GPS trajectory T: $p_1 \rightarrow p_2 \rightarrow \cdots \rightarrow p_n$

Output:

Predicted Path M : $m^1{}_1 \rightarrow m^2{}_1 \rightarrow \cdots \rightarrow m^{k1}{}_1 \rightarrow m^1{}_2 \rightarrow \cdots \rightarrow m^{kn-1}{}_{n-1} \rightarrow m^1{}_n \rightarrow \cdots \rightarrow m^{kn}{}_n$

for each $p_i \in T$ do :
    $m^1{}_i \leftarrow$ projection(p_i ,SRS_i);
    if(intersect(SRS_i, SRS_{i+1}))
        $m^2{}_i \leftarrow$ intersection(SRS_i, SRS_{i+1});
    else
    for each $v_q \in$ vertices of Dijkstra(SRS_i, SRS_{i+1}, G(V,E))
        $m^{q+2}{}_i \leftarrow v_q$;
return M;

5 Experiment

5.1 Respondents' Confidence of 20-Predicted-Path Questionnaire

We randomly select 20 predicted paths to question 54 participants in this questionnaire. The participants are consisted of some university students, the research assistants in Academia Sinca and the employees of Taipei City Government Traffic transportation department. We apply the Likert Scale to represent the degree of respondents. The average mean of 20-predicted-path is 3.72. In the questionnaire, respondents expressed a high level of confidence, with scores of > 3 out of a maximum of 5. The questionnaire interface: http://www.plash.tw/~mikelin/antrack/vote_demo.html.

5.2 Comparison with Mapbox Map-Matching

The dijkstra selection map-matching approach complements the deficiency of Mapbox map-matching [15]. Mapbox map-matching was released at March 10, 2015. It implements the method from Paul Newson and John Krumm [8]. They solved the Hidden-Markov-Models problem by using viterbi algorithm [16]. It's a well crafted implementation at the most of time. However, sometimes, the output from Mapbox map-matching doesn't fill up a path. Figure 5(a)(b) demonstrates the occurrence. Therefore, we can contribute the dijkstra-based selection algorithm to Mapbox map-matching to complement their deficiency.

5.3 Comparison with Driver-Tagged Actual Paths

A driver-tagged actual path is the most accurate solution to evaluate the map-matching algorithm. Luckily, we found 30 actual paths from the actual drivers. However, the tagging task is relatively challenging because of the complexity of the road network in Taipei. Therefore, we support them a friendly tagging system to complete this task. We hope that this tagging system can be used on other's map-matching works. In the long run, people can refine the map-matching algorithm because of the driver-tagged actual paths. Figure 5(c) shows the accuracy of 30 predicted paths.

$$\text{Accuracy} = 1 - (\text{corrected points from drivers}/\text{total points of the predicted path})$$

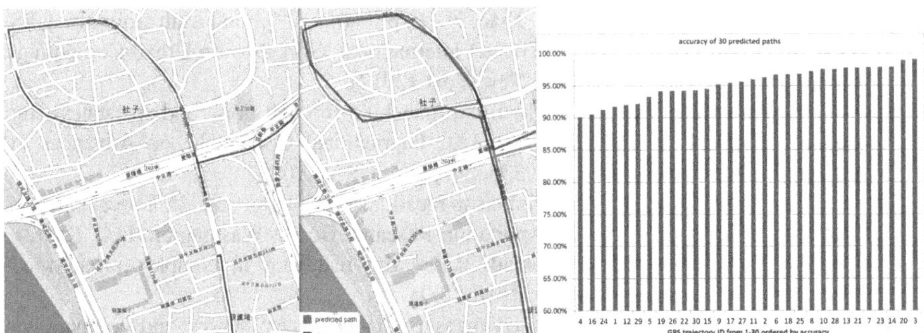

Fig. 5. (a) is the output of Mapbox map-matching, 5(b) the results of dijkstra-based selection map-matching, 5(c) shows the accuracy of 30 predicted paths

6 Conclusion

The dijkstra-based selection map-matching (DBSMM) algorithm finds the most possible path for a raw GPS trajectory, even in a parallel multi-lanes road. This map-matching work will help researchers who are interested in urban traffic computing. Due to the page limit, the further discussion of accuracy, complexity and running time of DBSMM will be presented in the journal version.

Acknowledgement. This project was partly supported by the Ministry of Science and Technology of Taiwan under grant NSC101-2221-E-001-021-MY3 and NjiSC100-2219-E-001-002.

References

1. Luxen, D., Vetter, C.: Actual-time routing with OpenStreetMap data. In: Proceedings of the 19th ACM SIGSPATIAL International Conference on Advances in Geographic Information Systems, GIS 2011, NY, USA, pp. 513–516 (2011)

2. Kawak, D., Kim, D., Liu, R., Nath, B., Iftode, L.: DoppelDriver: counterfactual actual travel times for alternative routes. In: 2015 IEEE International Conference on Pervasive Computing and Communications (PerCom), pp. 178–185 (2015)
3. Saremi, F., Fatemieh, O., Ahmadi, H., Wang, H., Abdelzaher, T., Ganti, R., Liu, H., Hu, S., Li, S., Su, L.: Experiences with GreenGPS – fuel-efficient navigation using participatory sensing. In: IEEE Transactions on Mobile Computing, vol. 99 (2015)
4. Quddus, M., Washington, S.: Shortest path and vehicle trajectory aided map-matching for low frequency GPS data. Transp. Res. Part C Emerg. Technol. **55**, 328–339 (2015)
5. Yanagi, T, Yamamoto, D.; Takahashi, N.: Development of mobile voice navigation system using user-based mobile maps annotations. In: 2015 IEEE/ACIS 14th International Conference on Computer and Information Science (ICIS), Las Vegas, NV, pp. 373–378 (2015)
6. Yuan, J., Zheng, Y., Zhang, C., Xie, X, Sun, G.Z.: An interactive-voting based map matching algorithm. In: 2010 Eleventh International Conference on Mobile Data Management (MDM), pp. 43–52. IEEE, Kansas City (2010)
7. Mao, H., Luo, W., Tan, H., Ni, L.M., Xiao, N.: Exploration of ground truth from raw GPS data. In: Proceedings of the ACM SIGKDD International Workshop on Urban Computing, UrbComp 2012, pp. 118–125. ACM, New York (2012)
8. Newson, P., Krumm, J.: Hidden markov map matching through noise and sparseness. In: Proceedings of the 17th ACM SIGSPATIAL International Conference on Advances in Geographic Information Systems, pp. 336–343, NY, USA. (2009)
9. Greenfeld, J.S.: Matching GPS observations to locations on a digital map. In: Proceedings of the 81st Annual Meeting of the Transportation Research Board, Washington, D.C. (2002)
10. Ochieng, W.Y., Quddus, M.A., Noland, R.B.: Map-matching in complex urban road networks. Braz. J. Cartography **55**(2), 1–18 (2004)
11. Biagioni, J., Eriksson, J.: Inferring road maps from gps traces: survey and comparative evaluation. In: Transportation Research Board 91st Annual Meeting (2012)
12. Dijkstra's algorithm. https://en.wikipedia.org/wiki/Dijkstra%27s_algorithm
13. Tidy algorithm. https://github.com/mapbox/geojson-tidy
14. Lou, Y., Zhang, C., Zheng, Y., Xie, X., Wang, W., Huang, Y.: Map-matching for low-sampling-rate GPS trajectories. In: Proceedings of the 17th ACM SIGSPATIAL International Conference on Advances in Geographic Information Systems, GIS 2009, pp. 352–361. ACM, New York (2009)
15. Mapbox map-matching. https://www.mapbox.com/blog/map-matching/
16. Viterbi algorithm. https://en.wikipedia.org/wiki/Viterbi_algorithm

PBX Autoresponder System for Information Lookup of Pupil Records

Nguyen Hong Quang(✉), Trinh Van Loan, and Bui Duy Chien

Hanoi University of Science and Technology, 1 Dai Co Viet Road, Hanoi, Vietnam
{quangnh,loantv}@soict.hust.edu.vn, dtxthanhha@gmail.com

Abstract. This paper presents the integration of a Vietnamese recognition module into the digital PBX Asterisk and construction of a PBX autoresponder system for information lookup of pupil records. This system allows direct interaction with the user entirely in Vietnamese, as it receives requests from users and then responds in Vietnamese. An experiment has been performed to test Vietnamese recognition using a set of commands containing from 1 to 6 syllables. The acoustic unit selection method was tested. The test results showed that use of acoustic units with diphones and rhyming syllables gives better recognition results than use of units with single phonemes. The scores of the Vietnamese recognition system in the command mode were 78.52 % for one-syllable commands and 74.71 % for six-syllable commands.

Keywords: Asterisk autoresponder system · Vietnamese speech recognition · Acoustic unit · Telco services

1 Introduction

A combination of computer and telephone networks have been extensively developed and the deployment of automatic information service significantly reduces labor costs for providers of telecommunications services by communicating using voice via the telephone network. Autoresponder applications are very popular in business, health and education services; these automated systems also allow users to search information, check their account, and look up information about the weather, the traffic situation, etc. In the education sector, the development of education has led to the need to exchange information between families and schools, but the information search method according to traditional methods is limited, as it is time-consuming and inflexible, for example. Thus there is an urgent need for autoresponder telephony system for pupil information lookup.

The PBX (private branch exchange) system is a device that allows a connection between subscriber s and provides various information services to subscribers [3]. A switchboard provides transmission lines to transmit information simultaneously in two directions. The first generation of switchboards used operators performing the basic functions of the PBX as the connection between the subscriber and the reply service. Then the semi-automatic electromechanical

N.T. Nguyen et al. (Eds.): ACIIDS 2016, Part II, LNAI 9622, pp. 509–518, 2016.
DOI: 10.1007/978-3-662-49390-8_50

switchboard appeared, followed by the digital switchboard. A commonly used digital switchboard that is entirely based on software is Asterisk PBX [1].

In Vietnam, Vietnamese recognition is growing quite strongly. Although there have been many studies on Vietnamese recognition and some success achieved [5–8], the creation of products applying them is still limited. This article describes methods of applying Vietnamese recognition for schoolchildren and their applications on the Asterisk PBX system. This system was built with the goal of providing online information services by telephone. In this system, a Vietnamese recognition module is integrated with Asterisk PBX. Users can search information by dialing the call center and following the instructions. The information management department of the school is able to manage pupil information and updates each day through the administration system interface. Then the PBX system will give the pupil information quickly and accurately through voice commands, communicating directly with the users.

The next section of the paper will provide an overview of the system and the method for integrating the Vietnamese recognition module into the PBX system. Section 3 describes scenarios for providing services. Section 4 presents the system of Vietnamese recognition, and Sect. 5 describes the test experiment of commands in Vietnamese. Finally, Sect. 6 summarizes the results and provides directions for further development.

2 General Diagram of System

2.1 Digital Telephone Switchboard Asterisk

Asterisk is a software implementation of the functions of internal telephone exchange (PBX), that allows the extension phone to make calls to other phones

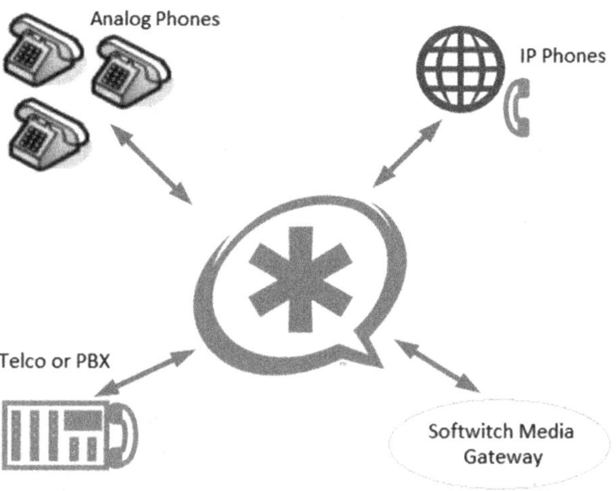

Fig. 1. General diagram of Asterisk [2]

and connect with other phone systems, including regular public switched telephone network (PSTN) and Voice over IP (VoIP) [2]. Asterisk has the full features of a commercial PBX, including voicemail, conferencing, and interactive voice. Asterisk supports many protocols, such as SIP and H.323 VoIP, acting as a connectivity station between IP phones and PSTN [3]. The name Asterisk is derived from the symbol *, which is used in Unix and Linux to express any option [1]. Asterisk is an open source software switching system that was written in the C language and runs on the Linux operating system. Asterisk is also a set of tools for voice applications and a full-function server for handling calls [4].

Figure 1 shows the communication ability of the Asterisk PBX: communicating with an ordinary analog phone, Voice IP telephony equipment, PSTN, and other VoIP providers.

2.2 PBX Autoresponder System

A block diagram of the system is described in Fig. 2. The execution steps of the system are as follows:

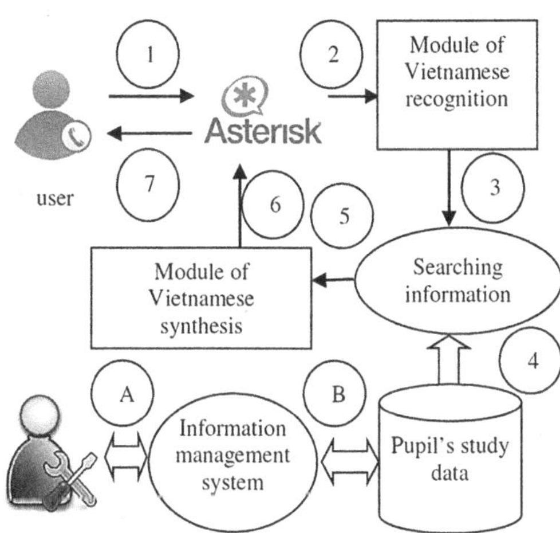

Fig. 2. Diagram of autoresponder system

- Step 1: After dialing the PBX, users send a request by "speaking an order", one of the corresponding commands to a scripts that users need to search for information. A voice command will be sent to the Asterisk server system by the server of the voice service provider (Telco). We have asked Telco to provide a phone number for this service.
- Step 2: Audio data received by the Asterisk PBX is encrypted in GSM (Global System for Mobile) format, so the sound is converted to 16-bit WAV PCM (Pulse-code modulation) format with a GSM decoding algorithm. Then the

sound comes through the automatic recognition system to identify the request of the user.
- Step 3: This request is sent to the module of Vietnamese recognition, which will be described in detail in Part IV. The recognition process gives the results as text, which constitutes the user requirement.
- Step 4 and Step 5: The user requirement is processed by an information query module. This module performs a search of information in the database of pupil records. These data are managed and updated by an administrator through an information management system (step A and step B in Fig. 2). The result returned by this module is information in text format.
- Step 6: The information resulting in text will be converted into digital audio format (WAV PCM 16 bit) by a Vietnamese synthesis module. This module is described in Sect. 3.2.
- Step 7: PBX Asterisk encodes the audio results in GSM and switches to the Telco server to send them to the user.

All of the steps described above have been implemented and integrated into the control program for the operation of Asterisk PBX (written in the script of Asterisk). The above steps are executed repeatedly when the user still needs to look up information. The support of the operator can be added into this scenario in case the user is not satisfied with the automated interactive system, such as when the statement of the recognition system is inaccurate or the user wants to request more special information.

3 System Scenario for Providing Information

3.1 Scenario

The system was tested using the service for looking up pupils in Thanh Ha Continuing Education Center, Hai Duong, Vietnam. Three scripts were built:

- Set of commands 1: zero, one, two, three, four, five. These commands are used to test the operability of the system.
- Set of commands 2: this set concerns the subjects: math, physics, chemistry, literature, history, geography, informatics, English, and music. This script allows users to look up information by subject.
- Set of commands 3: oral test grade, 15 min test grade, 45 min test grade, semester grades, final grades of the first semester, final grades of the second semester. This set allows users to look up detailed information about each grade of a course.

3.2 Vietnamese Synthesis Module

The amount of information provided by this system only represents a limited information field, so to synthesize the Vietnamese, we chose a method of concatenation using the available recorded words. The recorded content serving speech synthesis includes fixed notices Table 1, command sets, and numeric values describing the grades of subjects.

Table 1. Sentences recorded to serve speech synthesis by concatenating words

No.	Statements
1	Enter name of subject
2	Enter grade
3	Subjects mark is

4 Vietnamese Recognition System

4.1 Speech Recognition

The need for equipment (machinery) that can recognize and understand speech has become very necessary. This was predicted by human beings as well as scientists and research projects on speech recognition over the last century. Up to now, we have gained much advancement in building and developing important Vietnamese speech recognition systems.

Speech recognition is a pattern recognition process with the aim of classifying input in the form of a speech signal into a pattern that has been studied and stored in memory. Models are acoustic units; they can be words or phonemes. If these models are immutable and unchangeable, the recognition process will simply compare speech data to models studied and stored in memory. The basic difficulties of speech recognition change with the duration of speech, and there are major discrepancies in the speech of different speakers, speed, contexts, and environment.

A general speech recognition system includes 2 parts: training and recognition. Training is a process in which the system uses data to train the statistical models of acoustic units. Recognition is the process of determining which word is read based on a trained lexical set.

The state-of-the-art system for automatic recognition of Vietnamese speech uses a hidden Markov model for presenting acoustic units [6–8]. Thus in this study, we also use the same method, and training data are used to train the HMM model based on the Baum-Welch algorithm. Specifically, the forward-backward algorithm is used to measure the probability of command data being recorded in each HMM model of each command [6]. The command being recognized is the one that has the largest probability.

4.2 Mel Frequency Cepstral Coefficients

In the process of speech training and recognition, first, each speech unit is presented by an array of parameter vectors. The most common parameter set is Mel frequency cepstral coefficients (MFCC). The algorithm for calculating the parameters is shown in Fig. 3.

In Fig. 3, the speech signals are framed with the length of 0.1 s each, and the offset of the frame is 0.01 s. Then, pre-processing will be performed for each speech frame using Eq. 1.

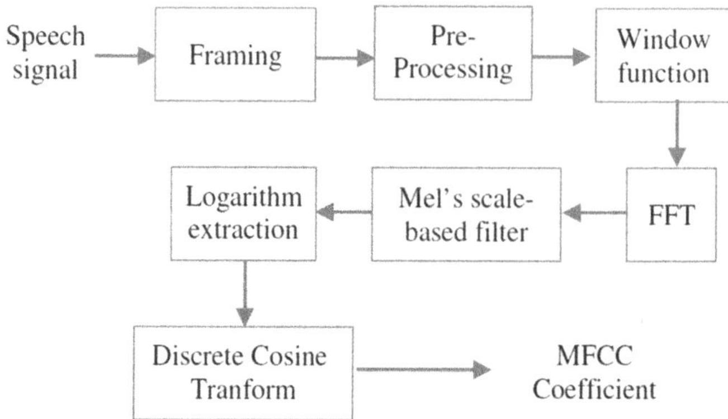

Fig. 3. Diagram of the MFCC coefficient extraction algorithm

$$y(n) = x(n) - x(n-1) \tag{1}$$

The speech signal after pre-processing will be sent to the Hamming window function (In Eq. 2, N is the number of samples of each speech frame).

$$w(n) = 0.54 - 0.46cos\left(\frac{2\pi n}{N-1}\right), 0 \leq n \leq N-1 \tag{2}$$

Then, a fast Fourier transform (FFT) will be conducted on the signal that has already been through the window function. The signal spectrum will be processed through the mel scale filter. The number of the filter is the number of the required MFCC coefficients. Finally, applying the logarithm to the filter output and applying a discrete cosine transform will determine the MFCC coefficients.

5 Speech Recognition Experiment

5.1 Vietnamese Database

The speech recording process was carried out in a computer room at Thanh Ha Continuing Education Center. The studio was far away from the pupils study area and has good soundproofing. The data were recorded with the sampling frequency of 16000 Hz, 16 bits per sample in mono.

The database was recorded by 6 speakers, all of whom were from Northern Vietnam. They recorded 3 sets of commands [Sect. 3.1]. Each person recorded each command in a set 50 times. They spoke all of the commands in the corpus in order for each recording.

This database was divided into 2 sets: the training set had a database of 4 speakers, and the test set had a database of the other two. The two methods of choosing data to train the acoustic model were as follows:

- Method 1 was an acoustic model of a set of commands that depended solely on the speech recording data in that set. When we did a speech recognition experiment on corpus 1, we used only the recorded speech on the corpus 1 in the training set, a total of 200 files. We did the same for corpus 2 and 3 so that each corpus would have a different acoustic model.
- Method 2 was only one acoustic model trained for all of the recording files in the training set. We used all of the recording files of the speakers in the training set to train the acoustic model. The number of files amounted to 600.

5.2 Choosing Acoustic Units for Vietnamese

The Vietnamese syllable structure is described in Table 2.

Table 2. Vietnamese syllable structure

Vietnamese syllable			
Initial (22 consonants)	Final		
	Medial (1 semi vowel)	Nucleus (13 vowels/3 diphthong)	Ending (2 semivowels/6 consonants)

Vietnamese is a tonal language with six tones. For every syllable, there is one and only one tone. The tone is very important in determining the meaning of a word. If two similar monosyllabic words have different tones, their meaning is different.

Table 3. Methods for selecting an acoustic unit for the vietnamese recognition module

Method	Description	Example with word "BA'N" (Sell)
AC1	Single phonemes without tone	/b/ + /a/ + /n/
AC2	Single phonemes with tone	/b/ + /a'/ + /n/
AC3	Double phonemes without tone	/ba/ + /an/
AC4	Double phonemes with tone	/ba/ + /a'n/
AC5	Consonant + rhyme without tone	/b/ + /an/
AC6	Consonant + rhyme with tone	/b/ + /a'n/

There are some methods for choosing acoustic units, and they are described in Table 3. AC1, AC3, and AC5 do not use tonal information. We still used these methods for the experiment because there was no command in the command set with the same phonemes but different tones. AC1 and AC2 only used single phonemes, while AC3 and AC4 used double phonemes. AC5 and AC6 used consonants and rhyming syllables.

The number of phonemes in each set is illustrated in Table 4. We can see that set 1 has the fewest commands (5 commands), and all commands are single syllables. Set 2 has the most commands (9 commands), but like in set 1, all commands are single syllables. Therefore, set 2 has more acoustic units than set 1. On the contrary, though set 3 does not have the most commands (7 commands), all commands include many syllables (2–6 syllables), so set 3 has the most acoustic units. Data selection method 2 uses all of the acoustic command sets, so this method has the most acoustic units in Table 4.

Table 4. Statistics of acoustic units for command sets with the relevant selection method

Method for selecting the acoustic unit	Data selection method 1			Data selection method 2
	Set 1	Set 2	Set 3	
AC1	10	10	20	27
AC2	10	17	23	36
AC3	12	20	25	53
AC4	12	20	25	51
AC5	11	17	20	41
AC6	11	17	20	39

5.3 Experiment of Command Set Recognition

We dealt with context-independent acoustic models for Vietnamese, so a hidden Markov model was used for each Vietnamese phoneme. We chose a 5-state left-right topology with 3 emitting states; the entry and exit states of an HMM are non-emitting. Each emitting state consists of 8 Gaussian mixtures. The feature vectors are extracted every 10 ms and contain 39 dimensions of 13 MFCC, plus their first and second derivatives. This acoustic model was trained with all the data in the training corpus by using the expectation-maximization (EM) algorithm.

The result of data selection method 1 and 2 is shown in Table 5. These results are determined by the percentage of commands recognized correctly.

In Table 5, some results are the same, such as the recognition results for AC1 and AC2 in set 1 and those for AC3 and AC4 in set 2. It can be explained that the phonemes are different but the number of phonemes is similar [Table 4].

Table 5 indicates that with the experiments to select the acoustic unit following method 1, command sets that have more commands will have lower recognition results. For example, set 2 has a lower percentage of correct recognition than set 1.

Table 5. Recognition scores following the acoustic selection method with data selection method 1

Method for selecting the acoustic unit	Data selection method 1			Data selection method 2		
	Set 1	Set 2	Set 3	Set 1	Set 2	Set 3
AC1	63.76	41,11	34.43	47.65	31.82	31.86
AC2	63.76	43.03	34.57	62.42	48.89	41.57
AC3	78.52	51.92	56.43	64.93	56.57	55.29
AC4	78.52	51.92	56.43	71.31	52.42	59.43
AC5	77.35	52.67	57.43	67.28	54.75	68.29
AC6	77.35	52.67	57.43	70.30	54.65	74.71

Table 5 shows that although data selection method 2 used more data to train the phonemes (using all data of 3 sets Sect. 5.1), the score for recognition was lower than that of method 1. For example, for AC1 in set 1, the score for recognition of method 1 was 63.76 %, while that of method 2 was 47.65 %. It could be explained by the fact that method 2 needed to be trained on more acoustic units than method 1 (Table 3). However, in some situations, method 2 gave better results: the best recognition results were those of set 3 (74.71 % compared with 57.43 % of method 1) and set 2 (56.57 % compared with 52.67 % of method 1). This reflects that in general, when there are many commands and the commands have many syllables, the large number of training data will be a decisive factor in increasing the recognition score of the system.

The results in Table 5 also show that acoustic unit selection methods AC3, AC4, AC5, and AC6 have better recognition scores than methods AC1 and AC2. This means that using acoustic units with double syllables and rhyming syllable gives better recognition than using acoustic units with single syllables.

6 Conclusion and Orientation for Next Research

This paper presents a method for integrating Vietnamese speech recognition into the digital telephone switchboard Asterisk and then building an autoresponder system to help pupils search for information. Experiments to recognize Vietnamese were executed with a command set including commands from 1 to 6 syllables. Two acoustic unit selection methods were tested. The results show that using acoustic units with double syllables and rhyming syllables gives better recognition than one with single syllables. The results of Vietnamese recognition in the form of a command set were 78.52 % for 1-syllable instructions and 74.71 % for 6-syllable instruction. In the future, we will continue to expand the research on other basic features of speech, such as fundamental frequency, and integrate it into the speech recognition system. An orientation for future research is to build a Vietnamese synthesis system with unlimited vocabulary.

References

1. Gomillion, D., Dempster, B.: Building Telephony Systems With Asterisk. PACKT Publisher, Birmingham (2006)
2. Van Meggelen, J., Madsen, L., Bryant, R.: AsteriskTM: The Definitive Guide. Oreilly Publisher, Sebastopol (2011)
3. Van Meggelen, J., Madsen, L., Smith, J.: Asterisk - The Future of Telephony. OReilly Media, Sebastopol (2007)
4. Rabiner, L.R.: A tutorial on hidden markov models and selected applications in speech recognition. Proc. IEEE **77**(2), 257–286 (1999)
5. Diep, D.T.T., Quang, N.H., Van Loan, T., Hung, P.N.: Text-dependent speaker recognition for Vietnamese. In: Proceedings of the 2013 Fifth International Conference of SoftComputing and Pattern Recognition (SoCPaR 2013), Hanoi, Vietnam, 15–18 December 2013 (2013)
6. Quan, V.U., et al.: Advances in acoustic modeling for Vietnamese LVCSR. In: 2009 International Conference on Asian Language Processing (IALP 2009), Singapore, 7–9 December 2009 (2009)
7. Le, V.B., Tran, D.D., Castelli, E., Besacier, L., Serignat, J.-F.: Spoken and written language resources for Vietnamese. In: LREC, Lisbon, Portugal, 26–28 May 2004 (2004)
8. Quang, N.H., Van Loan, T., Dat, L.T.: Automatic speech recognition for Vietnamese using HTK system. In: IEEE RIVF International Conference on Computing and Communication Technologies (RIVF 2010), Hanoi, Vietnam, 01–04 November 2010 (2010)

Possibilities for Development and Use of 3D Applications on the Android Platform

Tomas Marek[1], Ondrej Krejcar[1(✉)], and Ali Selamat[1,2]

[1] Faculty of Informatics and Management, Center for Basic and Applied Research, University of Hradec Kralove, Rokitanskeho 62, 500 03 Hradec Kralove, Czech Republic
Tomas.Marek.5@uhk.cz, ondrej@krejcar.org

[2] Faculty of Computing, MaGIC-X (Media and Game Innovation Centre of Excellence) UTM-IRDA Digital Media Centre, Universiti Teknologi Malaysia, 81310 Johor Bahru, Johor, Malaysia
aselamat@utm.my

Abstract. Computer graphics in combination with mobile devices finds many applications in the fields of entertainment, education or displaying data. The amount of information that can be shown to the user in real time depends mostly on the speed of visualisation process and thus on the optimization of the graphic chain during application development. For the development of optimized solution, the developers need to be familiar with the possibilities and differences of the developing platform. An appropriate way can be the implementation of graphics engine and sample scenes that show basic indicators of a quality level of current smart devices. This paper focuses on the problem of selecting a proper graphics engine for selected visualization task as well as the optimization of several most used techniques for visualisation of different types of 3D effects. Quality of the implementation is finally evaluated on the developed testing engine which provides relevant data in the sense of Frame per Second (FPS) based on the visualisation speed. The second aspect of quality is evaluated by the visual correctness of four selected scenes such as water level, volumetric light scattering, fur simulation and forest.

Keywords: Mobile device · Engine · Computer graphics · Optimization

1 Introduction

A higher penetration rate of desktop or portable PCs is going to 100 % limit not only in Europe, but in the World context [ref]. There is no financial problem to own any PC. There is an increasing computational power of the devices that enable us to develop the application environment in which it is possible to find tools for solving many tasks. These devices are also constantly available and ready for use. Due to the higher performance of Smart Devices, there is an increasing demand for smarter applications in various fields [4, 17, 19–24].

One possibility to develop a smart application [1] by using computer graphics with a special interface known as Open Graphics Library for Embedded Systems (OpenGL ES)

N.T. Nguyen et al. (Eds.): ACIIDS 2016, Part II, LNAI 9622, pp. 519–529, 2016.
DOI: 10.1007/978-3-662-49390-8_51

for Android and a new Metal API (application programming interface) that has been introduced with the Apple IOS8 [15]. The range of applications where it can be applied is very wide. For example, the effects for recording video and photos as depth of field effect [2] or 3D talking avatar [3]. Significantly most profitable and most popular categories we can find across applications are games. However, 2D and 3D graphics are popularly being used in other areas of education or transportation. Generally graphics can be used for professional programs modules, viewing data or augment reality [12]. Even a cheap mobile device may display large medical data by volumetric rendering [5]. With this trend there are new mobile versions of the large graphics engines such as Unity [ref], Unreal engine [ref] or CryENGINE [ref]. However, use of such special engines is not simple, thus the application as well as a developers studio arises with question "how to start creating a graphical content?".

Many available engines that can be obtained freely from the web have several disadvantages that need to be considered. Firstly, the overall quality and speed of updates are not available [ref]. Also the developer has no control over the engine and must rely that the errors will be corrected in time. Some modules of used engines can also be written inappropriately for specific solutions that we need. The big advantage is verified cross-platform solution of such engine. Another problem is however the learning time needed to understand the engine. Engines come with their own development tools and their full program code understanding may take as long as creating of a new engine. The use of smaller custom solutions is therefore better in many cases [22–24].

2 Problem Definition

Generally, 2D and 3D engines are tools for effective work with graphics for the purposes of displaying a content to the users [13, 14, 16]. The engines may also have other tools, such as particle system or scene graph. This article outlines the essential parts of the engine for rendering and resource handling.

Resources represent data loaded and processed by engine. These are the object models, their textures, sounds, but also shaders, animations, scripts and more. The models are the elements of a real world or a designer mind and are created in special modelling tools such as Blender or Autodesk 3ds Max. Models can also be called a mesh data and are the sum of all the information needed to render the scene geometry. For transmission of models between programs and engines there are different kinds of formats, for example OBJ [6], COLLADA [7], 3DS [8] and others.

Important parts of the program are the shaders. These are short programs that affect the course of rendering. They are divided into vertex and fragment shaders [9] according to their focus. All the rendered data pass through them. Vertices of the rendered objects are transformed by the vertex shader. The data for these operations can be supplied to the shader while being renderedusing uniform variables. In contrast to a uniform, which contains the same data for a model, the model properties (position, normal, colour etc. for each vertex) are transmitted via the attribute variables. This involves matrices required for the proper rotation and deformation of the scene. The result of the vertex shader is further processed and applied in the fragment shader, which is used for conducting operations at the level of individual pixels. The transferred data between the shaders are referred to as varying variables.

Displaying a graphical content is almost not possible without using textures. Texture is a 2D image, generally defining a colour using the RGB (colour mixing method), which is mapped to the surface of the body. Textures are used for a wide variety of effects, such as to define a roughened surface, for storing shadow or highlight, for glitter or directly specular reflections. Only rarely it is possible to map a texture to an object in the ratio of one to one. In other cases OpenGL uses the settings for texture filtering, which can define a method of obtaining the data from the texture. Various settings have different results and different impacts on the rendering speed.

Optimization of the data means minimizing their size or their preliminary preparation to simplify subsequent rendering. The haders together with the complex models and textures are the main engine load. With an increasing complexity of the scene it may contain more complex calculations for physics, collisions, animation and more. Rendering can be described as the transfer of data to the graphics card and their passage through the graphic chain, which includes the shaders. To reduce the complexity it is worthwhile to consider over data types, the use of compressed textures, texture atlases, optimization model data, the method of rendering etc. Rendering means processing the data while sending them on the graphics card. It is a use of glDrawArrays call for not indexed data and glDrawElements for indexed data [10]. Indexing the data leads to a reduction of duplications and mostly to increase rendering speed. By a parameter mode you can set the type of graphics primitives to be rendered. The aim is to speed up rendering and reduce the data size. Graphics primitives are points, lines, triangles, special kinds of connected triangles (TRIANGLE_STRIP, TRIANGLE_FAN) and others. Before calling the renderer it is required to load and send the data to the graphics card. For various scenes there are different ways of transmitting the data. The basic one is vertex arrays object (VAO). Then the data are stored in RAM memory. A better solution then is to use vertex buffer object (VBO) and store the data in video memory.

3 New Solution

Android devices that work with 2D and 3D graphics use programming interface (API) OpenGL ES (Open Graphics Embedded Systems Libraryfor), which is supported directly by processors. This interface is used for basic processing of vector graphics and is available in several versions. The most widespread OpenGL ES 2.0 is supported from Android 2.2 Froyo.

Future engine will use a modular structure. The basic packages will display geometry, load textures and shaders. OpenGL is used in Android through GLSurfaceView class. This view shows the final render of OpenGL thread and is assigned to the main activity that caters to run the application and provides a user input.

OpenGL thread is represented by renderer class that uses the other modules. The result of drawing is stored in the graphics card's framebuffer (image output device). The objects are composed out of graphics primitives, which are the basis for rendering by OpenGL and are set when calling glDraw. To define the basic shape it is needed to set the vertices where each vertex is a point in space. Each vertex is passed with its properties and connections of these points, which define the entire model [13].

The results of the vertex shader, which runs for each vertex, are final vertex coordinates after the transformation and eventual varying variables. The next step, except for automatic translation for the following processes, is creation of the graphics primitives and cropping them to the view volume. The following step is rasterization (converting to 2D image), starting fragment shader and saving the value to the framebuffer. In case the triangle is partially offscreen, some of its vertices are removed and to maintain the shape new points are connected to the edge of the view volume. When processing fragment (color, depth, varying data) it is sometimes necessary to re-read the previous record from the framebuffer. For example, while dealing with transparency.

On the provided base it can be implemented variety of graphic scenes and effects that help to test the capabilities of current devices and the embedded engine. Any such scene uses the base class renderer and helps to extend the engine with new features. Many effects are created mainly by using shaders. In addition to the basic classes for the work with geometry and the data it is also implemented a package for loading models. These are the vertices organized into specific shapes and optionally containing additional information (texture, normal vectors etc.). Selected scenes can be used for testing the devices and the engine.

3.1 Water Level Scene

The first realized scene is terrain with trees models and a water level. The main idea connected with the water surface is to use only four peaks forming the base area, normal mapping (change the behaviour of light on the surface of the object and to create a small bumps), to create waves and a texture carrying reflection of the environment. The scene is shown in Fig. 1a. To create the reflection it is necessary to render the geometry once more and use the resulting image in the second rendering.

Fig. 1. (a) Water level scene (left) and (b) Volumetric light scattering scene (right)

3.2 Volumetric Light Scattering Scene

This effect is based on a post processing. The finished scene is further adjusted to the level of a 2D image. Volumetric light scattering effect is formed by light scattering in the environment with a high content of dust particles, water vapor etc. The effect is realized by the accumulation of the light intensity towards the source and requires a large amount of texture reading. To create separate beams of light it is necessary to add one pass forming special texture. It is therefore needed to render the scene, another pass is required for preparing special texture and ultimately the final pass creates the effect shown in Fig. 1b.

3.3 Fur Simulation Scene

Fur simulation is a simple effect based on geometry repetition. By repeating there are formed layers, which in conjunction with a semi-transparent noise texture create the effect of fibre, as shown in (Fig. 2a).

Fig. 2. (a) Fur simulation scene (left) and (b) Forest scene with Bounding Volumes (right)

3.4 Forest Scene

The last scene forms a landscape where there are 99 trees located. This composition primarily tests the number of peaks that can be used on mobile devices. Moreover the scene tests also a version of implementation of a frustum culling, which is used to identify the objects outside the viewing frustum defined by the view matrix. These objects are not rendered, thus they reduce power requirements. The sample scene contains a landscape, which forms the basis for drawing objects. The landscape is composed of 4096 vertices and is wrapped with a skybox. This part of the scene is not affected by cropping using the viewing volume. The influence of the applied method is tested on a tree model consisting of 1648 vertices. A figure (Fig. 2b) shows the forest scene and the bounding volumes for each tree as simplified models for frustum culling.

4 Implementation

For working with graphics it is needed OpenGL ES and a special class GLSurfaceView for its initialization in Android. This class provides display configuration, rendering in special thread and helps to manage the life cycle of an activity [9, 18]. Rendering is then performed on the part of display called surface or viewport. This view is subsequently allocated to an activity as the current content view [11]. With GLSurfaceView it is also possible to control the type of rendering (setRenderMode) and to set the continuous rendering or rendering on request. Now it's necessary to create a class that will cater the actual drawing to OpenGL surface view. This class is implemented by GLSurfaceView.Renderer and contains three basic methods:

OnSurfaceCreated – this method is called when you create a surface. Here it is possible to load the data and to create buffers.

OnDrawFrame – called whenever it is necessary to render something. Here it is a call glDraw.

OnSurfaceChanged – called for the surface formation and whenever the image is resized, for example, when you rotate the display.

The basic activity with GLSurfaceView and renderer form the basis for the future engine. Another element is to provide the communication between the user thread (activity) and OpenGL thread in the background [18]. The UI thread can directly work with renderer methods and transmit the information about the user's input [13]. Such basic information is a finger move across the screen, which can be determined by OnTouchListener in the UI thread, or GLSurfaceView. But if it is necessary to provide, for example, displaying FPS (redraw speed - Frame Per Second), it is a communication from a background thread to the main thread and you must use a Handler (communication between the Android fibers). FPS value can be easily displayed in the activity label or in a connected view, where it is also possible to define additional UI elements. After securing the communication with the main UI thread basis it is finally ready for the implementation of classes and packages that will form the actual functionality of the engine. These are utility classes for working with buffers and matrices, classes caring for animation, shaders, data loading and any future effects. Conceptual design is shown in Fig. 3.

Application main activity is stored in the main package. This activity is responsible for the main menu, which will contain a list of all the activities planned for each scene and ensures that they run when tapped.

To display each start of the scene link it is used ListView, which renders scrolling list of the items. Each item will include a preview scene image, the main description and any additional notes to the sample and is representative of a class rowItem (Fig. 4).

These data are passed to the ListView by CustomListAdapter class, which is an extension of BaseAdapter. CustomListAdapter then takes care of acquisition and allocation of appropriate data for each row in the ListView. The rows of the ListView then refer to the activity of GlSurfaceView. All these activities are stored in the package opengl_ukazky. Renderer package then contains the classes for rendering assigned to each of the activities from opengl_ukazky. Packages skeletalAnimations and clothSimulation will contain the necessary operation for the planned

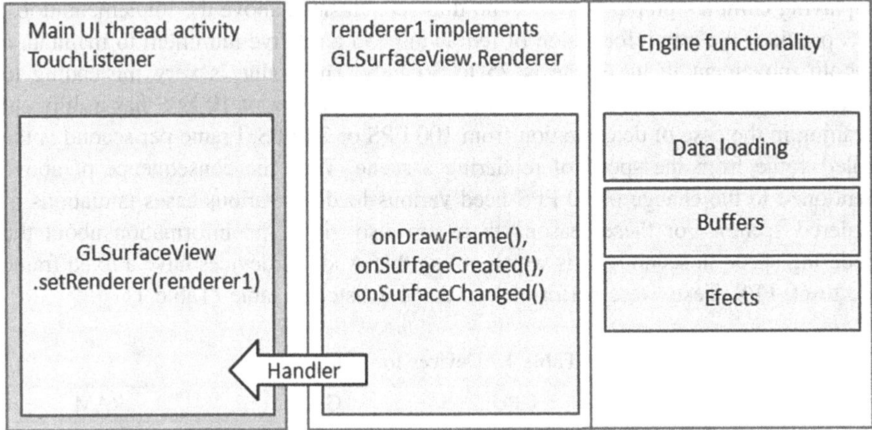

Fig. 3. Engine scheme

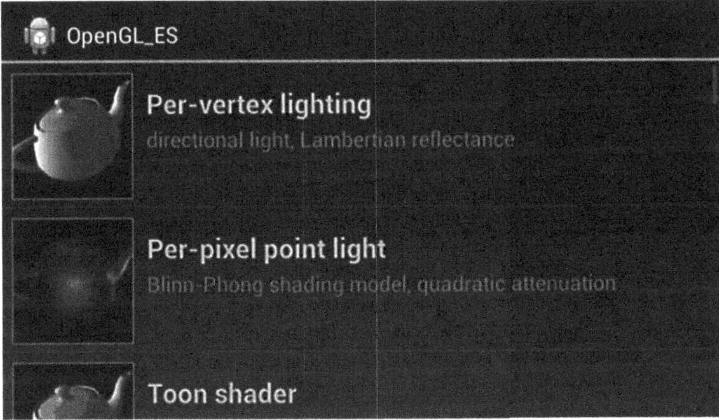

Fig. 4. Main activity with list view

implementation of these effects. The package "environment" is prepared for securing the data for effects showing scene elements, such as the preparation of the underlying flat surface. Within the package "model" it will be implemented a class to represent the geometry objects, which will be used by the base class for creating buffers, shaders and textures as a preparation of "utils" package. Last package "transforms" contains utility classes for mathematical computations with matrices, vectors and vertices.

5 Testing of Developed Application

The selected scenes were tested on the main performance indicator, which is the rendering speed. Rendering speed indicator is displayed in each scene using unit FPS (frame per second). The goal of testing is to assess the capabilities of current devices in displaying complex graphics scenes and draw conclusions above the implementations. FPS provides the rate information of redrawing and a relative minimum to maintain a smooth movement in the scene is 25 to 30 FPS. This value is very misleading to measure the actual performance. For example, deterioration of 10 FPS has a different meaning in the case of deterioration from 100 FPS or 20 FPS. Frame per second is the scaled value from the speed of rendering a scene. With the consequence of above mentioned to the change of 10 FPS need various loads in various cases (situations of rendered scene). For these reasons there are also given the information about the rendering speed in seconds. It is worth noting that Android devices have a fixed frame rate to 60 FPS. Tests were performed on devices listed in table (Table 1).

Table 1. Devices for testing

Device	SoC	CPU	GPU	RAM
Samsung S4	Qualcomm Snapdragon 600 APQ8064T	Krait 300, 1900 MHz, Cores: 4	Qualcomm Adreno 320, 400 MHz, Cores: 4	2 GB, 600 MHz
Samsung S3	Samsung Exynos 4 Quad 4412	ARM Cortex-A9, 1400 MHz, Cores: 4	ARM Mali-400 MP4, 440 MHz, Cores: 4	1 GB, 400 MHz
Huawei y300	Qualcomm Snapdragon S4 Play MSM8225	ARM Cortex-A5, 1000 MHz, Cores: 2	Qualcomm Adreno 203	512 MB
Zopo zp990	MediaTek MT6592	ARM Cortex-A7, 1700 MHz, Cores: 8	ARM Mali-450 MP4, 700 MHz, Cores: 4	2 GB, 666 MHz

The test results are listed in table (Table 2) for all the devices in units of the FPS and the speed of one rendering in seconds. Each value is the average of a thousand individual results specific to redraw in the landscape mode for one scene.

The overall performance is obviously significantly lower than a PC or newer consoles. The ability to show detail scenes that represent model complexity varies between performance and price classes of devices. The number of vertices that can be used without problems ranges in tens of thousands. To improve the visual impression it can be used a variety of effects, such as normal mapping. Still the amount of geometry is relatively low for rendering complete detailed scenes and it is necessary to use many compromises and optimizations. However, this problem dramatically decreases today and the following years will bring processors capable of rendering hundreds thousands and more vertices. For per-pixel effects applies that complex shaders used in an areas occupying most of the screen are highly ineffective. Effects using transparency are at a comparable level. The water level in the results maintained in most cases above the 30 FPS. The performance varies considerably according to the area of the screen that the

Table 2. The results of testing

Device	Water level [FPS]/[s]	Volumetric light scattering [FPS]/[s]	Fur simulation [FPS]/[s]	Forest scene [FPS]/[s]	Forest scene with frustrum culling [FPS]/[s]
Samsung S4	58/0,017	15/0,067	59/0,017	25/0,04	57/0,018
Samsung S3	44/0,023	22/0,045	59/0,017	35/0,028	50/0,02
Huawei y300	26/0,038	13/0,077	23/0,043	12/0,083	21/0,048
Zopo zp990	42/0,024	16/0,063	59/0,017	29/0,034	43/0,023

water just occupies. As a big problem it was revealed a multiple reading from textures (volumetric light scattering) which is often used in more complex effects (such as image blur or shadow).

6 Conclusions

The paper describes a development of a graphical engine primarily suitable for testing of quality and speed of visualisation processes of modern Smart devices. Five different scenes have been tested on the created engine, while the whole process of engine creation is a source of experience and can be the basis for any graphics applications using OpenGL ES on the Android platform. At the same time several capabilities of modern mobile devices were tested. An appropriate use of the identified properties can create advanced graphics applications.

Mobile applications compared to desktop versions provide a significant opportunity to achieve profits due to a faster development, lesser content and easier publishing [22–24].

Acknowledgement. This work and the contribution were supported by project "SP-2016 - Smart Solutions for Ubiquitous Computing Environments" Faculty of Informatics and Management, University of Hradec Kralove, Czech Republic. The Universiti Teknologi Malaysia (UTM) and Ministry of Higher Education (MOHE) Malaysia, under research grant FRGS 4F550 and GUP 02G31, are hereby acknowledged for some of the facilities utilized during the course of this research work.

References

1. Behan, M., Krejcar, O.: Modern smart device-based concept of sensoric networks. EURASIP J. Wirel. Commun. Networking **155**(1), 2013 (2013). doi:10.1186/1687-1499-2013-155

2. Wang, Q.S., Yu, Z., Rasmussen, C., Yu, J.Y.: Stereo vision-based depth of field rendering on a mobile device. J. Electron. Imaging **23**(2), 023009 (2014)
3. Lin, H.J., Jia. J., Wu, X.J., Cai, L.H.: Stereo talking android: an interactive, multimodal and real-time talking avatar application on mobile phones. In: Signal and Information Processing Association Annual Summit and Conference (APSIPA), 2013 Asia-Pacific, Kaohsiung, Taiwan, 29 October 2013-1 November 2013, pp. 1–4 (2013). doi:10.1109/APSIPA.2013.6694211
4. Krejcar, O., Jirka, J., Janckulik, D.: Use of mobile phone as intelligent sensor for sound input analysis and sleep state detection. Sensors. **11**(6), 6037–6055 (2011)
5. Hachaj, T.: Real time exploration and management of large medical volumetric datasets on small mobile devices-evaluation of remote volume rendering approach. Int. J. Inf. Manag. **34**, 336–343 (2014)
6. Wavefront OBJ File Format Summary. FileFormat.info (2014). http://www.fileformat.info/format/wavefrontobj/egff.htm. Cited 18 December 2014
7. Khronos Group. Collada (2014). https://collada.org/. Cited 18 April 2014
8. Horák, J., Růžička, J., Novák, J., Ardielli, J., Szturcová, D.: Influence of the number and pattern of geometrical entities in the image upon PNG format image size. In: Pan, J.-S., Chen, S.-M., Nguyen, N.T. (eds.) ACIIDS 2012, Part II. LNCS, vol. 7197, pp. 448–457. Springer, Heidelberg (2012)
9. Brothaler, K.: OpenGL ES 2 for Android: A Quick-Start Guide. The Pragmatic Programmers, Raleigh (2013)
10. Khronos Group. OpenGL ES Common Profile Specification Version 2.0.25 (2010). https://www.khronos.org/registry/gles/specs/2.0/es_full_spec_2.0.25.pdf. Cited 18 April 2015
11. Article 11 - Beginner's Guide to Android Animation/Graphics. CodeProject 2014. http://www.codeproject.com/Articles/822412/Article-Beginners-Guide-to-Android-Animation-Gr. Cited 28 January 2015
12. Bilek, O., Krejcar, O.: Development of augmented reality application on android OS. In: Donnellan, B., Helfert, M., Kenneally, J., VanderMeer, D., Rothenberger, M., Winter, R. (eds.) DESRIST 2015. LNCS, vol. 9073, pp. 488–495. Springer, Heidelberg (2015)
13. Behan, M., Krejcar, O.: Adaptive graphical user interface solution for modern user devices. In: Pan, J.-S., Chen, S.-M., Nguyen, N.T. (eds.) ACIIDS 2012, Part II. LNCS, vol. 7197, pp. 411–420. Springer, Heidelberg (2012)
14. Marek, T., Krejcar, O.: Optimization of 3D rendering in mobile devices. In: Younas, M., Awan, I., Mecella, M. (eds.) MobiWIS 2015. LNCS, vol. 9228, pp. 37–48. Springer, Heidelberg (2015). doi:10.1007/978-3-319-23144-0_4
15. IOS8. Apple Inc. Apple. https://www.apple.com/cz/ios/. Cited 18 April 2015
16. Marek, T., Krejcar, O.: Optimization of 3D rendering by simplification of complicated scene for mobile clients of web systems. In: Núñez, M., Nguyen, N.T., Camacho, D., Trawiński, B. (eds.) ICCCI 2015. LNCS, vol. 9330, pp. 3–12. Springer, Heidelberg (2015). doi:10.1007/978-3-319-24306-1_1
17. Benikovsky, J., Brida, P., Machaj, J.: Proposal of user adaptive modular localization system for ubiquitous positioning. In: Pan, J.-S., Chen, S.-M., Nguyen, N.T. (eds.) ACIIDS 2012, Part II. LNCS, vol. 7197, pp. 391–400. Springer, Heidelberg (2012)
18. Krejcar, O.: Threading possibilities of smart devices platforms for future user adaptive systems. In: Pan, J.-S., Chen, S.-M., Nguyen, N.T. (eds.) ACIIDS 2012, Part II. LNCS, vol. 7197, pp. 458–467. Springer, Heidelberg (2012)
19. Kasik, V., Penhaker, M., Novák, V., Bridzik, R., Krawiec, J.: User interactive biomedical data web services application. In: Yonazi, J.J., Sedoyeka, E., Ariwa, E., El-Qawasmeh, E. (eds.) ICeND 2011. CCIS, vol. 171, pp. 223–237. Springer, Heidelberg (2011)

20. Penhaker, M., Kasik, V., Snasel, V.: Biomedical distributed signal processing and analysis. In: Saeed, K., Chaki, R., Cortesi, A., Wierzchoń, S. (eds.) CISIM 2013. LNCS, vol. 8104, pp. 88–95. Springer, Heidelberg (2013)
21. Machacek, Z., Slaby, R., Hercik, R., Koziorek, J.: Advanced system for consumption meters with recognition of video camera signal. Elektron. Ir Elektrotechnika. **18**(10), 57–60 (2012)
22. Chlouba, T., Cimler, R., Tomaskova, H.: Synthesizing mobile communication. In: World Conference on Educational Technology Researches (WCETR), Procedia Social and Behavioral Sciences, vol. 28 (2011). doi:10.1016/j.sbspro.2011.11.067
23. Simkova, M., Tomaskova, H., Nemcova, Z.: Mobile education in tools. In: Cyprus International Conference on Educational Research (CY-ICER), Procedia Social and Behavioral Sciences, vol. 47, pp, 10–13 (2012). doi:10.1016/j.sbspro.2012.06.604
24. Tomaskova, H., Nemcova, Z., Simkova, M.: Usage of virtual communication in university environment. In: World Conference on Educational Technology Researches (WCETR), Procedia Social and Behavioral Sciences, vol. 28 (2011). doi:10.1016/j.sbspro.2011.11.068

Ontology-Based Software Development

Ensuring the Correctness of Business Workflows at the Syntactic Level: An Ontological Approach

Thi-Hoa-Hue Nguyen[1,2(✉)] and Nhan Le-Thanh[2]

[1] Information Technology Faculty, Vietnam-Korea Friendship Information Technology College, Da Nang, Vietnam
huenth@gmail.com
[2] WIMMICS - The I3S Laboratory - CNRS - INRIA, Nice Sophia Antipolis University, Sophia Antipolis, France
Nhan.LE-THANH@unice.fr

Abstract. High quality business workflow definitions play an important role in the organization. An incorrectly defined workflow may lead to unexpected results. Therefore, each business workflow definition should be carefully analyzed before it is put into use. In this paper, we propose an ontological approach which is suitable for ensuring the syntactic correctness of business workflows. In details, to represent CPNs with OWL DL, we first introduce the CPN Ontology. Then, we define axioms, which are added to the CPN Ontology to provide automated support for establishing the correctness of business workflows. Finally, by relying on the CORESE semantic engine, SPARQL queries are implemented to detect shortcomings in concrete workflows. To the best of our knowledge, this is a novel approach for the representation and verification of business workflows based on ontologies.

Keywords: Business workflow · Correctness · OWL DL · SPARQL · Verification

1 Introduction

The current tendency in e-business has resulted in more complex business processes. However, the specification of a real-world business process is generally manual and is thus vulnerable to human error. An incorrectly designed workflow may lead to failed workflow processes, execution errors or not meet the requirements of customers, etc. In fact, existing techniques applied to check the correctness of workflows are particularly used in commercial business workflow systems. Most of them assume that a workflow is correct if it complies with "the constraints on data and control flow during execution" [1]. Whether the workflow is in conformity with the design requirements is neither specified

T.-H.-H. Nguyen—This work was done as part of a collaboration between Nice Sophia Antipolis University and Da Nang University.

N.T. Nguyen et al. (Eds.): ACIIDS 2016, Part II, LNAI 9622, pp. 533–543, 2016.
DOI: 10.1007/978-3-662-49390-8_52

nor proved. There is thus a great need for developing a thorough and rigorous method that automatically supports workflow designers to ensure workflows being well-formed.

In this study, we extend our previous work [2] in designing well-formed CPNs-based business workflow templates (CBWTs) and checking their correctness. The approach is based on Knowledge Engineering, Coloured Petri Nets (CPNs) and Semantic Web technologies which provide semantically rich business process definitions and automated support for CBWTs verification. Our contributions are:

- Presenting a classification of syntactic constraints in modelling business processes and creating their related axioms using Description Logic (DL) in order to support workflow designers;
- Showing the SPARQL [3] query language is able to verify workflow templates.

The rest of this paper is structured as follows: In Sect. 2, we briefly introduce our CPN Ontology as a representation Coloured Petri Nets (CPNs) with OWL DL. We then present syntactic constraints and create their related axioms added to the CPN Ontology to support designers in establishing well-formed workflow templates in Sect. 3. In Sect. 4, we introduce the SPARQL query language used to verify CBWTs at the syntactic level. In Sect. 5 we give related work. Finally, Sect. 6 concludes the paper with an outlook on the future research.

2 Modelling Business Processes with Coloured Petri Nets - the CPN Ontology

On one hand, Coloured Petri Nets (CPNs) [4] have been developed into a full-fledged language for the design, specification, simulation, validation and implementation of large-scale software systems. CPN is a well-proven language which is suitable for modelling workflows or work processes. Therefore, CPNs are chosen as the workflow language in our work to transform a business process into a control flow-based business workflow template. However, it is difficult to interoperate, share and reuse business processes modelled with CPNs, i.e., business workflows, because of the lack of semantic representation of CPN components [2].

On the other hand, an ontology with its components, which provides machine-readable definitions of concepts, can represent semantically rich workflow definitions. Once workflow definitions are stored as semantically enriched workflow templates, developers can easily build their appropriate software systems from these templates. Therefore, in this section, we shortly introduce the CPN Ontology, which is first proposed in [2] as a representation of CPNs with OWL Description Logic (OWL DL). The main purpose is to facilitate business processes modelled with CPNs to be easily shared and reused.

In order to develop the CPN Ontology, we translate each element of CPNs into a corresponding OWL concept. The core concepts of the CPN ontology is depicted in Fig. 1. The ontology is described based on DL syntax and the axioms supported by OWL.

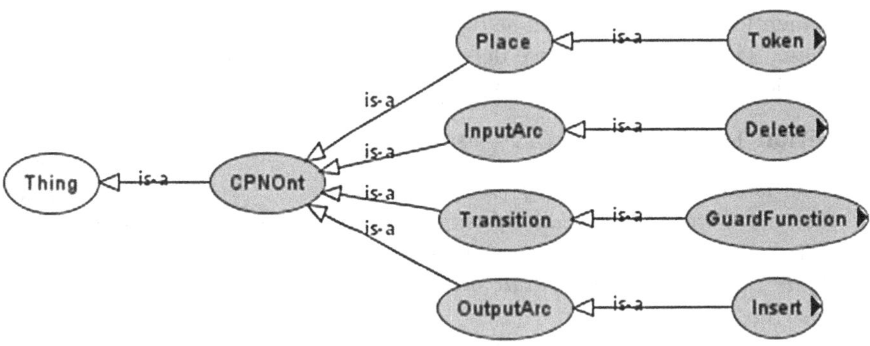

Fig. 1. CPN Ontology (excerpt)

In the CPN Ontology, we define the concept **CPNOnt** for all possible business processes modelled with CPNs. We define the concept **Place** and the concept **Transition** to represent all places and transitions of a process model, respectively. In order to represent all directed arcs from places to transitions and all directed arcs from transitions to places, we define the concept **InputArc** and the concept **OutputArc**, respectively. In our case, one place contains no more than one token at one time, therefore, the concept **Token** is defined for all tokens inside places. To express all transition expressions, the concept **GuardFunction** is defined. Transitions consist of control and activity nodes. We define the concept **CtrlNode** for occurrence condition in the former nodes and the concept **ActNode** for occurrence activity in the latter nodes. The concepts **Delete** and **Insert** are defined for all expressions in input arcs and output arcs, respectively. In order to express all attributes of individuals, we define the concept **Attribute**. And for all subsets of $I_1 \times I_2 \times \ldots \times I_n$ where I_i is a set of individuals, the concept **Value** is defined.

Properties between the concepts in the CPN Ontology are also indicated. For example, a class **Transition** has two properties **connectsPlace** and **hasGuardFunction**. Consequently, the concept **Transition** can be glossed as "The class **Transition** is defined as the intersection of: (i) any class having at least one property **connectsPlace** whose value is equal to the class **Place** and; (ii) any class having one property **hasGuardFunction** whose value is restricted to the class **GuardFunction**" [2].

3 Taxonomy of Constraints in Modelling Business Processes

To provide automated support for workflow designers in establishing the correctness of ontology-based workflow representations, in this Section we introduce a set of syntactic constraints. The constraints are categorized into two groups. Axioms related to the constraints are also defined using a DL as $\mathcal{SHOIN}(\mathcal{D})$ to complete the CPN Ontology.

As mentioned earlier, we aim at representing the correct CBWTs in a knowledge base. Therefore, at first, we define the soundness property that is used as the criterion to check the correctness of workflow processes at the syntactic level.

Definition 1. (Sound). *A CPN-based process model, PM, is sound iff:*

(i) PM is connected and well-formed;
(ii) For every state M_j reachable from state Start M_0, there also exists another firing sequence starting from state M_j to state End M_e;
(iii) State End M_e is the only state which is reachable from state Start M_0 with one token in place e;
(iv) There is no deadlock, no infinite cycle and no missing synchronization in PM.

3.1 Syntactic Constraints Related to the Definition of Process Model

- **Constraints related to places.**

  Constraint 1. For every place $p \in P$, p connects and/or is connected with transitions via arcs.

  We create the axiom corresponding to Constraint 1 as follows:

  $hasPlace^{-}.CPNOnt \sqcap \neg(\exists connectsTrans.hasTrans^{-}.CPNOnt \sqcup \exists connectsPlace^{-}.hasTrans^{-}.CPNOnt) \sqsubseteq \bot$

  Constraint 2. There is one and only one start point in a process model.

  We create the axiom corresponding to Constraint 2 as follows:

  $CPNOnt \sqcap \neg(= 1\ hasPlace.(connectsTrans.hasGuardFunction.has Activity.\ ActNoce \sqcap \neg(\exists\ connectsPlace^{-}.hasTrans^{-}.CPNOnt))) \sqsubseteq \bot$

  Constraint 3. There is one and only one end point in a process model.

  We create the axiom corresponding to Constraint 3 as follows:

  $CPNOnt \sqcap \neg(= 1\ hasPlace.(connectsPlace^{-}.hasGuardFunction.has Activity.\ ActNode \sqcap \neg(\exists\ connectsTrans.hasTrans^{-}.CPNOnt))) \sqsubseteq \bot$

  Constraint 4. A place has no more than one leaving arc. If a place is connected to a transition, there exists only one directed arc from the place to the transition.

  We create the axiom corresponding to Constraint 4 as follows:

$Place \sqcap \neg(\leq 1\ hasPlace^{-}.InputArc) \sqsubseteq \bot$

Constraint 5. A place has no more than one entering arc. If a transition is connected to a place, there exists only one directed arc from the transition to the place.

We create the axioms corresponding to Constraint 5 as follows:

$Place \sqcap \neg(\leq 1\ connectsPlace^{-}.(=1 hasTrans^{-}.OutputArc)) \sqsubseteq \bot$

Constraint 6. There are no pairs of activity nodes connected via a place.

We create the axiom corresponding to Constraint 6 as follows:

$Place \sqcap \exists connectsTrans.hasGuardFunction.hasActivity.ActNode \sqcap \exists connectsPlace^{-}.hasGuardFunction.hasActivity.ActNode \sqsubseteq \bot$

Constraint 7. There are no pairs of control nodes connected via a place.

We create the axiom corresponding to Constraint 7 as follows:

$Place \sqcap \exists connectsTrans.hasGuardFunction.hasControl.CtrlNode \sqcap \exists connectsPlace^{-}.hasGuardFunction.hasControl.CtrlNode \sqsubseteq \bot$

- **Constraints related to transitions.**

Constraint 8. A transition is on the path from the start point to the end point of a process model.

If a transition has no input place, it will never be enabled;
If a transition has no output place, it will not lead to the end.

Consequently, each transition in a workflow must have at least one entering arc and at least one leaving arc.

We create the axiom corresponding to Constraint 8 as follows:

$Transition \sqsubseteq\ \geq 1\ connectsPlace.Place \sqcap \geq 1\ connectsTrans^{-}.Place$

Constraint 9. An activity node has only one entering arc and one leaving arc.

We create the axiom corresponding to the Constraint 9 as follows:

$hasGuardFunction.hasActivity.ActNode \sqsubseteq\ = 1\ connectsPlace.Place \sqcap = 1\ connectsTrans^{-}.Place$

Constraint 10. A control node does not have both multi-leaving arcs and multi-entering arcs.

We create the axiom corresponding to the Constraint 10 as follows:

$$\geq 2\ connectsPlace.Place \sqcap \ \geq 2\ connectsTrans^{-}.Place \sqcap \\ hasGuardFunction.hasControl.CtrlNode \ \sqsubseteq \ \bot$$

- **Constraints related to directed arcs.**

Constraint 11. Directed arcs connect places to transitions or vice versa.

We create the axioms corresponding to the Constraint 11 as follows:

$$hasPlace^{-}.InputArc \ \equiv \ connectsTrans.hasTrans^{-}.CPNOnt \\ hasTrans^{-}.OutputArc \ \equiv \ connectsPlace.hasPlace^{-}.CPNOnt$$

3.2 Syntactic Constraints Related to Uses of Control Nodes

A poorly designed workflow due to improper uses of control nodes can result in deadlock, infinite cycle or missing synchronization. However, these errors can be detected when designing a workflow template and therefore, we can get rid of them. To do that, we next introduce Constraint 12 and the symptoms related to deadlock, infinite cycle or missing synchronization.

Constraint 12. There is no deadlock, no infinite cycle and no missing synchronization.

- **Deadlock:** A deadlock is a situation in which a process instance falls into a stalemate such that no more activity can be enabled to execute. Figure 2 shows three simple deadlock simulations.
- **Infinite cycle:** An infinite cycle is derived from structural errors where some activities are repeatedly executed indefinitely. A simple infinite simulation is depicted in Fig. 3(a).
- **Missing synchronization:** Missing synchronization is a situation in which the mismatch between the building blocks leads to neither deadlock nor infinite cycle, but results in unplanned executions. Figure 3(b) shows a simple simulation of missing synchronization.

We next create the axioms related to the control nodes (one of two types of transitions), including $And-split$, $And-join$, $Xor-split$ and $Xor-join$, used to detect deadlock, infinite cycle or missing synchronization.

- **And-split** is connected to at least two output places. Every output place contains one token. We create the axiom corresponding to *And-split* as follows:

$$AndSplit \ \sqsubseteq \ Transition \sqcap connectsPlace.hasMarking.Token \sqcap \\ connectsTrans^{-}.hasMarking.Token \ \sqcap \ hasGuardFunction.hasControl. \\ CtrlNode \sqcap \ =1\ connectsTrans^{-}.Place \ \sqcap \ \geq 2\ connectsPlace.Place$$

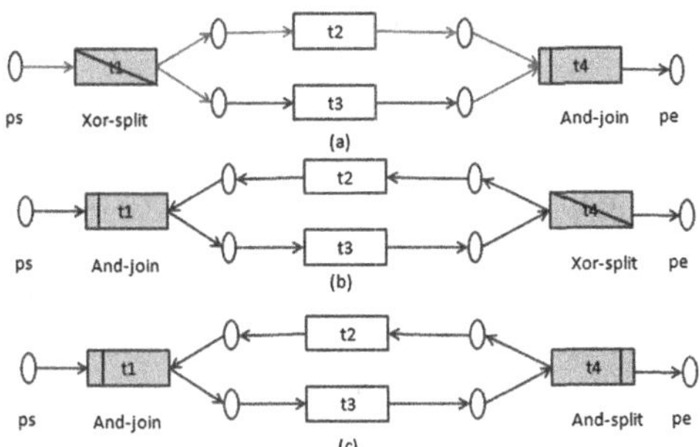

Fig. 2. Deadlock simulations

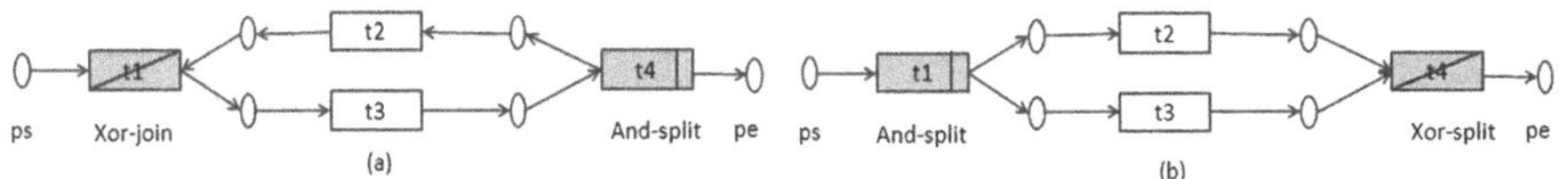

Fig. 3. Infinite cycle simulation

- **And-join**: There are at least two input places connected to *And-join*. In order to activate *And-join*, every input place has to contain one token. We create the axiom corresponding to *And-join* as follows:

$AndJoin \sqsubseteq Transition \sqcap connectsPlace.hasMarking.Place \sqcap connectsTrans^{-}.hasMarking.Token \sqcap hasGuardFunction.hasControl.CtrlNode \sqcap \geq 2connectsTrans^{-}.Place \sqcap =1\ connectsPlace.Place$

- **Xor-split** is connected to at least two output places. Unlike *And-split*, at any time, one and only one output place of *Xor-join* can contain a token. We create the axiom corresponding to *Xor-split* as follows:

$XorSplit \sqsubseteq Transition \sqcap \neg AndSplit \sqcap hasGuardFunction.hasControl.CtrlNode \sqcap =1\ connectsTrans^{-}.Place \sqcup \geq 2\ connectsPlace.Place \sqcup connectsTrans^{-}.hasMarking.Token$

- **Xor-join**: There are at least two input places connected to *Xor-join*. Unlike *And-join*, *Xor-join* is activated if one and only one input place contains a token. We create the axiom corresponding to *Xor-join* as follows:

$XorJoin \sqsubseteq Transition \sqcap \neg AndJoin \sqcap connectsPlace.hasMarking.Token \sqcap \geq 2\ connectsTrans^{-}.Place. \sqcap hasGuardFunction.hasControl.CtrlNode \sqcap =1\ connectsPlace.Place$

3.3 A Wrong Workflow Example

An example of a wrongly designed business process modelled with CPNs is illustrated in Fig. 4. The air ticket agent first requires a customer to provide some information related to the flights that he or she wants to book, including name(s), depart, destination, date and class. It then looks for the requested ticket(s) on its partner websites. For simplicity, we assume that two websites are utilized. The obtained results, which may consist of no results, some results or time out, are then evaluated in order to make a decision.

As shown in Fig. 4, the example model contains syntactic errors. There are three end points, i.e., *Timeout*, *End*2 and *End*3. Besides, the combination of a Xor-split (the transition $t2$ - $Preparetolookforaflight$) and an And-join (the transition $t5$ - *Collectresults*) causes a deadlock. Assuming that the place $Requestverified$ contains a token that makes the transition $t2$ to be enabled. If the transition Xor-split $t2$ fires, it consumes the token from its input place $Requestverified$ and then produces one token for only one of its output places. Consequently, either $t3$ or $t4$ may be activated. Since only one of the two transitions $t3$ and $t3$ can fire, not all input places of the transition And-join $t5$ can get its token. As a result, a deadlock occurs because the transition $t6$ will never be enabled to fire.

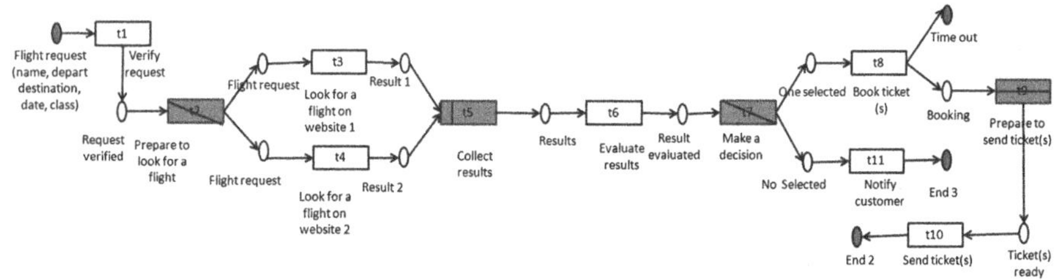

Fig. 4. A wrongly designed workflow model for the airline booking process

We have introduced the CPN ontology represented in OWL DL and axioms which are defined to support designers in verifying CPNs-based process models. It is necessary to note that, to develop or modify CBWTs (i.e., CPN models), manipulation operations [2], such as inserting new elements, deleting existing elements, etc., on business process models are required. Therefore, at design time, workflow templates stored in RDF format need to be verified before they are put into use. In the next Section, we present the SPARQL query language used to detect shortcomings in workflow templates represented in RDF syntax.

4 Using SPARQL to Verify Workflow

The CORESE [5], a semantic search engine, developed for answering SPAQRL queries asked against an RDF knowledge base, is used in our work. We choose the SPAQRL query language because: (i) It is an RDF query language; (ii) It is a W3C Recommendation and is widely accepted in the Semantic Web and also AT community; (iii) Its syntax is quite simple which allows for a query to include triple patterns, conjunctions, disjunctions and optional patterns; and (iv) It can be used with any modelling language.

In order to verify a workflow template, SPARQL verification queries are created based on the syntactic constraints. Two query forms are used in our work, including ASK and SELECT. According to [3], SELECT query is used to extract values, which are all, or a subset of the variables bound in a query pattern match, from a SPARQL endpoint. The variables that contain the return values are listed after a SELECT keyword. In the WHERE clause, "one or more graph patterns can be specified to describe the desired result" [6]. ASK query is used to return a boolean indicating whether a query pattern matches or not.

```
<?xml version="1.0"?>
<sparql xmlns='http://www.w3.org/2005/sparql-results#'>
  ...
  <result>
    <binding name='xorsplit'>
      <uri>http://WFTemplate/AirlineBooking#t2</uri>
    </binding>
    <binding name='andjoin'>
      <uri>http://WFTemplate/AirlineBooking#t5</uri>
    </binding>
  </result>
  ...
</sparql>
```

Fig. 5. Checking deadlocks caused of the two control nodes $Xor-split$ and $And-join$

The following query[1], for example, is used to check whether there exist errors related to improper uses of control nodes or not. This query is used to detect if there are any deadlocks caused by the combination of pairs of control nodes, Xor-split and And-join.

```
SELECT distinct ?xorsplit ?andjoin
WHERE {
?xorsplit  rdf:type h:Xor-split
?andjoin rdf:type h:And-join
?t1 h:hasGuardFunction/h:hasActivity _:b1
?t2 h:hasGuardFunction/h:hasActivity _:b2
```

[1] The prefix is assumed as: $PREFIXh$:<http://www.semanticweb.org/CPNWF#>.

```
?xorsplit h:connectsPlace/h:connectsTrans ?t1
?xorsplit h:connectsPlace/h:connectsTrans ?t2
?t1 h:connectsPlace/h:connectsTrans ?andjoin
?t2 h:connectsPlace/h:connectsTrans ?andjoin
FILTER(?t1!=?t2)}
```

As a result of the execution of each SPARQL query created based on the syntactic constraints, we obtain an XML file which results in nodes consisting of required information (e.g., the name) and causes shortcomings. For example, Fig. 5 shows the result of the execution of the above query applied to check whether the workflow, depicted in Fig. 4, contains deadlocks or not.

The query presented above does not only demonstrate that we can use the SPARQL query language to check the syntactic correctness of workflow processes, but also the useful of terminology provided by the CPN Ontology, such as *Xor-split* and *hasGuardFunction*.

5 Related Work

Today, the problem of ensuring the correctness of process models have been paid attention in various researches. However, researchers mainly focused on checking the compliance of models concerning aspects of the syntax and formal semantics. To process modelling, there exist some formal criteria, such as "soundness", "completeness", "well-structureness". These criteria are used to examine anomalies, e.g., deadlock, livelock, missing synchronization and dangling references. There are some methods have been proposed to verify workflow models, such as Petri Nets-based [7,8], logic-based [9,10], graph reduction-based [11] methods. However, most of them check the conformance of a workflow process based on the principle that if the constraints on data and control flow are met during execution, the workflow is correct.

In fact, the ontology-based approach for modelling business process is not a new idea [6]. In order to support (semi-)automatic system collaboration, some works, such as [12,13], made efforts to build business workflow ontologies. Machine-readable definitions of concepts and interpretable format, therefore, are provided via these ontologies. However, they do not mention the issues relating to a taxonomy of constraints and also the verification of workflows at the syntactic level.

In our work, the Web Ontology Language is used to develop the CPN Ontology for representing business processes modelled with CPNs. Our ontological approach enables the formulation of constraints added to the CPN Ontology to ensure the soundness of workflow patterns. The constraints are then applied to concrete CBWTs using an RDF engine in order to automatically verify workflow processes at the syntactic level.

6 Conclusion

In this paper, we introduce an ontological approach to support designers in verifying workflow processes. We shortly present the CPN Ontology, a representation of CPNs and OWL DL, which is defined to take advantage of powerful reasoning systems. Then, we describe two groups of constraints that ensure the soundness of well-formed workflow processes. We concentrate on defining axioms corresponding to the syntactic constraints and introduce some axioms involving the use of control nodes.

To verify concrete CBWTs, which are represented in RDF format, we specify the syntactic errors and errors related to improper uses of control nodes as SPARQL queries. By relying on the CORESE semantic engine, we show that the SPARQL query language is usable to workflow verification.

We know that checking workflow templates at build-time is not enough to ensure the proper execution of a workflow template. The ability to check the correctness of workflow execution is also needed. In our future work, a run-time environment is going to develop for workflow verification.

References

1. Lu, S., Bernstein, A.J., Lewis, P.M.: Automatic workflow verification and generation. Theor. Comput. Sci. **353**, 71–92 (2006)
2. Nguyen, T.-H.-H., Le-Thanh, N.: An Ontology-Enabled Approach for Modelling Business Processes. In: Kozielski, S., Mrozek, D., Kasprowski, P., Małysiak-Mrozek, B. (eds.) BDAS 2014. CCIS, vol. 424, pp. 139–147. Springer, Heidelberg (2014)
3. W3C: Sparql 1.1 query language. W3C Recommendation (2013). http://www.w3.org/TR/sparql11-query/
4. Kristensen, L.M., Christensen, S., Jensen, K.: The practitioner's guide to coloured petri nets. STTT **2**, 98–132 (1998)
5. Corby, O., et al.: Corese/kgram. https://wimmics.inria.fr/corese
6. Nguyen, T.H.H., Le-Thanh, N.: Ensuring the semantic correctness of workflow processes: an ontological approach. In: Nalepa, G.J., Baumeister, J. (eds.) Proceedings of 10th Workshop on Knowledge Engineering and Software Engineering (KESE10) Co-located with 21st European Conference on Artificial Intelligence (ECAI 2014), vol. 1289. CEUR Workshop Proceedings, Prague (2014)
7. van der Aalst, W.M.P.: The application of petri nets to workflow management. J. Circ. Syst. Comput. **8**, 21–66 (1998)
8. Verbeek, H., Basten, T., van der Aalst, W.: Diagnosing workflow processes using woflan. Comput. J. **44**, 246–279 (1999)
9. Bi, H.H., Zhao, J.L.: Applying propositional logic to workflow verification. Inf. Technol. Manage. **5**, 293–318 (2004)
10. Wainer, J.: Logic representation of processes in work activity coordination. In: Proceedings of the 2000 ACM Symposium on Applied Computing, SAC 2000, vol. 1, pp. 203–209. ACM, New York (2000)
11. Sadiq, W., Orlowska, M.E.: Analyzing process models using graph reduction techniques. Inf. Syst. **25**, 117–134 (2000)
12. Koschmider, A., Oberweis, A.: Ontology based business process description. In: EMOI-INTEROP, pp. 321–333. Springer (2005)
13. Sebastian, A., Tudorache, T., Noy, N.F., Musen, M.A.: Customizable workflow support for collaborative ontology development. In: 4th International Workshop on Semantic Web Enabled Software Engineering (SWESE) at ISWC 2008 (2008)

Semantic Integration via Enterprise Service Bus in Virtual Organization Breeding Environments

Wilcilene Maria Kowal Schratzenstaller[1(✉)], Fabiano Baldo[1], and Ricardo José Rabelo[2]

[1] Santa Catarina State University (UDESC), Joinville, SC, Brazil
wilcilenekowal@gmail.com.br, baldo@joinville.udesc.br
[2] Federal University of Santa Catarina (UFSC), Florianópolis, SC, Brazil
ricardo.rabelo@ufsc.br

Abstract. To attend demands faster and with lower cost, small and medium sized enterprises (SME) act in a collaboratively way in organization breeding environments through resource sharing and integration of systems. This integration is easier by the use of service oriented architecture and enterprise service bus (ESBs). However is still complex due the heterogeneity of existing data. To solve this compatibility problem among data, the ESBs could be enabled to mediate divergences with semantic support. But how to add semantic support into the ESBs to facilitate data mapping of SME in order to integrate its systems? This paper proposes the creation of an ontology as a method of semantic data analysis with the objective of solving incompatibilities among data that should be exchanged between SME. To validate the proposal some metrics are applied, they evaluate the coverage of the search terms, the wealth of knowledge representation and ontology detail level, as well as the prototype of semantic mapper developed was executed and analyzed.

Keywords: Matching process · Ontology · Semantic mapper · Virtual breeding environment

1 Introduction and Justification

As a way of ascending competitiveness, obtained both by time reduction and cost of production, small and medium sized enterprises (SME) unite in strategic alliances as the collaborative networked organizations (CNO) [1].

CNO is a broad concept of collaborative network where Virtual Organization Breeding Environment (VBE), Virtual Enterprise (VE) and Virtual Organization (VO) are the most common forms of manifestations. The VBE is a long time alliance with the objective of offering the necessary ways to SME preparation to work in a collaborative form with the objective of answering business opportunities in an agile way. After SME are prepared, they can join alliances that are dedicated to realize collaborative business. These alliances usually are established as VEs or VOs. A VE is a temporary and dynamic strategic alliance among SME that work in a collaborative

N.T. Nguyen et al. (Eds.): ACIIDS 2016, Part II, LNAI 9622, pp. 544–553, 2016.
DOI: 10.1007/978-3-662-49390-8_53

way to reach a common objective and earn profit [2]. A VO has a similar objective, however, without the requirement of earning profit.

To enable SME to act in a collaborative way, its information and communication systems must be integrated. The punctual integration between systems itself is a complex task due the heterogeneity and semantic missing about the data used by them. When this integration demands to happen on large scale, as in the case of VEs and VOs, it becomes harder, mainly at the creation phase. Apart from the integration scale, due the dynamics of VEs/VOs, where enterprises can enter or leave from the business at any moment, the integration becomes even a greater challenge.

Among the existing approaches used to integrate system, one that stands out most is the service oriented architecture (SOA). One of the benefits of SOA is that it provides a standard language to specify interfaces that facilitate the integration and reuse of software components [3]. To provide security, reliability, transaction management and process orchestration to the services, technologies such as enterprise service bus (ESB) are used. Although the advantages gained of the combination among services and ESB, the way that the ESBs enable the integration among services included in the business process is by using manually configured adapters [4]. The integration made in manual mode decreases the process agility, since a human interaction is necessary and this cannot be available to integration in the necessary moment, besides it suffer variation in time according to the tacit knowledge from the ones involved in the moment of required fields identification.

These problems can be solved by using automatic ways of integration. However, to provide the ESB these characteristics, is necessary to incorporate on it the necessary knowledge to perform the integration. This knowledge can be represented by an ontology. The process automation can be done through a mechanism that queries the ontology that performs the necessary mappings to integrate the services in the business process. Sward and Whitacre [5] highlight the lack of semantic support in the integration via ESB.

Given the benefits that the use of semantics can provide for services integration, this paper approaches the following research problem: How to add semantics into the ESB system to enable it of automating the service integration process in order to facilitate and accelerate the establishing VEs and VOs?

As scientific contribution, this paper aims to propose a method of semantic support to enable map data available by individual SME through its web services via ESB to integrate its systems in a flexible and automatic way. This approach assumes that the BPMN describes the orchestration of participant authors (SME), through the calling of pre-defined services and proposes a method for services integration by reference to BPMN. Thus, the syntactic and semantic incompatibilities that exist between the service defined in BPMN, and which are implemented effectively by EMS are solved.

It is assumed as hypothesis that the semantic support may be added by creating an ontology to represent the concepts and metadata provided by the SME participants of VBE.

The methodological procedure begins with the bibliographical review about CNOs and their manifestations in VBE, VE and VO, as well as data matching processes, systems integration and process management. Moreover, it is also reviewed the involved technologies, such as, SOA and ESB. Subsequently, is made a systematic literature review of papers that addressed the incorporation of semantics as a way of system integration with focus in the use of ontologies. About ontologies are also revised some evaluation methods and metrics. In parallel to bibliographical review, it is designed and built a simulation environment and a generic business process for the VBE context, where both are used as the basis for ontology creation. To evaluate the proposal are performed experiments on the prototype implemented and collected some metrics on the specified ontology.

The sequence of the paper is structured as follows. Section 2 presents the literature review and the related works. Section 3 presents the proposed solution, including the ontology and method specifications. Section 4 presents the evaluation made. Finally, Sect. 5 presents the conclusion and future work.

2 Literature Review

For SME work collaboratively in VBE their systems must be integrated. In order to facilitate this integration SOA architecture allows SME make available its information building software as services. One of the benefits of this approach is that services can be accessed with standard communication protocols known by other participants. In order to facilitate the invocation and coordinated execution of services contained in business process emerged the ESBs tools [3, 6]. They also provide security, reliability and transaction management. However, they do not provide automated methods for metadata matching and, consequently, for systems integration. To perform integration of services within ESBs it is necessary the intervention of human agents.

To provide autonomy for ESBs in automatic identification of metadata and in posterior integration of services, it is necessary to incorporate all the semantics involved in the process. Ontologies are one of the most appropriate ways to represent, understand and retrieve knowledge, so that both humans and machines can understand it [7].

Ontologies can be represented by means of text, tables, charts, or specific languages. Some of the representation languages are: RDF (Resource Description Framework) which has good flexibility of representation and is equivalent to a semantic network; OIL (Ontology Inference Layer) which is derived from RDF and expands its representative capacity and, finally, the OWL (web ontology language) which is the W3C recommendation for representing ontologies to web [8]. The differences between languages are mainly at expressiveness and computational property that each one offer [9].

The use of ontologies allows adding relationships between terms and rules to formalize, describe and organize knowledge. Thus, compared to other approaches of knowledge representation, ontology enables better definition of semantics, formalization and readability by both humans and computers [9]. These way of knowledge representation allows them, for example, to automatic identify synonyms stored under different designations. This benefit is crucial for SME participating in VBE. They can have their exchanged information represented in an ontology using a set of concepts and possible relationships in a way that computers become able to understand the information contained therein and thus perform the mapping process.

The use of ontologies as a tool for automatic semantic mediation has been explored in several scientific works. Among which those dealing more specifically semantic integration services are presented below.

Wang, Wang and Wang (2010) propose the creation of an ontology for each participant with subsequent mapping to a public ontology.

Tramontin Junior, Rabelo and Hanachi (2010) present a model based on the use of ontologies to represent the context of each user in a RCO.

Santos (2011) establishes a similar model proposed by Wang, Wong and Wang (2010), but from the same enterprise systems.

Li, Xie and Tang (2014) approach the creation of ontologies (task-level services and product features) per participant and interconnection through inference rules (business rules).

Tu Zacharewicz and Chen (2014) proposed the creation of an ontology for each system that will be integrated that should be sent with the message.

The aforementioned works mainly suggest the creation of an ontology for each SME belonged to the VBE that are merged or mapped when used for integration. Spite of their benefits, they can not be applied in the VBE scenario due to the dynamics imposed to the collaboration within VO and VE. So, VOs/VEs need agility in integration processes, because as longer as the design phase takes, the more time the SME are waiting the distribution and start of production activities. Since the start date of production directly impacts on the delivery date, if the setup of communication takes more time than expected, the subsequent release of the participants to work in other demands is delayed. So, the time of carrying out this activity need to be reduced.

This work aims to create a single ontology for the VBE, where metadata of the SME involved are added at the ontology only once for each one when they arrive at the VBE, in the setup phase. Thus, when the establishment of communication is needed, the semantics of the parameters are already known and the matching is done automatically.

3 Proposed Solution

The proposed solution for systems integration of the SME involved in VEs/VOs include semantic annotations using an ontology without the need to change the interfaces of the services provided. This enables the use of services for other purposes in addition to cooperation in OVS. The solutions are composed of two elements:

1. The ontology that represent the services contained in the business processes used to manage the creation, operation, evolution and dissolution of VEs/VOs. Besides the services and their respective input and output parameters, ontology also stores instances of these services provided by the interfaces of each SME belonging to VBE. The purpose of this ontology is to support the process of matching services and thus enable the integration of SME contained in the VBE. The population of ontology with instances of the services provided by an enterprise must be held before it is selected for participation in a VE/VO. The ontology population is simplified because only occurs only in leaf nodes. In order that the differences between the structures of XML and ontology does not constitute a problem;

2. The integration method that performs the matching of the service invoked by the client (a SME participating of VE/VO) to a service implemented by the server (another SME participating of VE/VO), including their respective parameters. This matching is done by querying the ontology that defines the names and parameters of the services involved in the creation, operation, evolution and dissolution of VEs/VOs. After finding the corresponding required parameter names, the subsequent matching of them is made and the service is invoked. Thus, ontology performs the role of a thesaurus where the synonymous of each reference service, and its respective parameter, is consulted for further mapping.

Figure 1 shows an example of the operation of the proposed integration method. It can be comprehended as follows. (1) SME1 (sender) sends a request to the SME2 (receiver) via ESB, by invoking the respective reference service specified in the ontology; (2) The ESB invokes the Mapper requesting it to returns the corresponding call to SME2 service implementation, which was requested by the SME1. (3) The mapper queries the ontology to find out how the service requested by SME1 was implemented by SME2, and is responsible for identifying the required parameters and map them as specified in the ontology. (4) The mapper returns to ESB data properly mapped to the invocation of the service in SME2. (5) Finally, the ESB performs the SME2 service invocation, after that, the SME2 service response is mapped by the mapper to the result expected by SME1.

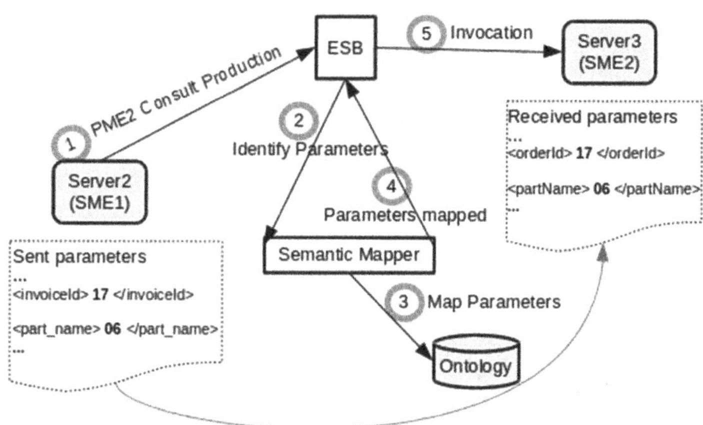

Fig. 1. Communication process

The proposed ontology is developed in owl on prótegé as presented in Fig. 2. The Thing class, root of ontology, contains two subclasses, one that represent reference services extracted from the business processes used to manage VEs/VOs [15] and another that represents the respective objects used as parameters of these services and their synonyms. In addition, each service has a relationship "HasObject" that relates it to its input and output parameters, as well as each object has "ObjectSynonym" relationship with their respective synonyms.

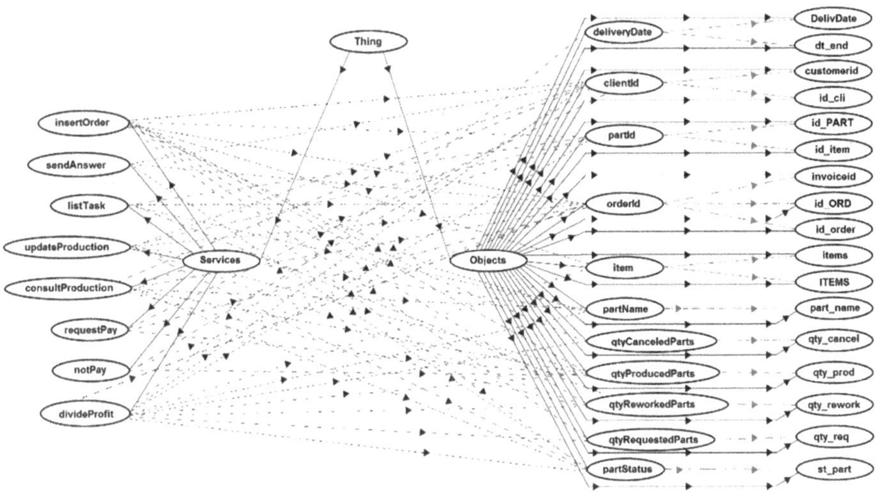

Fig. 2. Proposed ontology

The population of ontology is a critical component in the success of the integration process, because if it does not cover the whole set of services used by companies belonged to the VBE, the matching of services and their parameters can not be performed. Therefore, whenever a new SME joins the VBE leaf nodes of the ontology must be updated, otherwise the communication is stopped if the parameter sent in the invocation is not found in the ontology.

The prototype implementation was developed using JAVA language with queries SPARQL on the ontology. This prototype can be invoked by ESB tools on the access points where occurs message exchange between services provided by different SMEs.

To perform the service integration a mapper prototype has been developed. The developed prototype can be coupled as a plug-in to an ESB open source tools. After being connected to the ESB, during design phase, the semantic mapper can be added as a component into the business process whenever integration is required.

In details, the mapper intercepts the SOAP envelops exchanged between sender and receiver, analyses the message payload that is specified in XML, queries the tags (metadata) into the ontology, finds the corresponding synonymous and creates the XML message expected by the receiver as payload of the SOAP message. This process is done transparently during the business process execution.

4 Results Evaluation

To evaluate the proposal it has been built an environment composed of 5 virtual machines where one of them is the VBE and the other four representing the SME participants. The VBE server hosts the suite of computational tools necessary for the service integration. That is, it contains the ESB, ontology and knowledge base, and the semantic mapper. The other 4 servers, which represent the SME that are part of VBE, contain the services of each SME participants.

Eight services were simulated with 32 different forms of invocation performed by the four SME of the environment. The data used for representing the data contained in the SMEs databases, so, already added to the ontology when they were accepted into the VBE. Thus, it can perform the method in a way that each simulated SMEs invoke the services available in the environment with automatic mapping between the used parameters that are already in the ontology. Figure 3 shows part of the result obtained with an invocation of the proposed method.

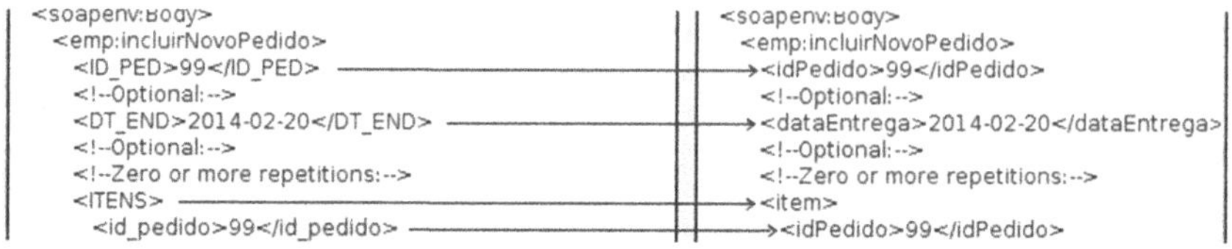

```
<soapenv:Body>
  <emp:incluirNovoPedido>
    <ID_PED>99</ID_PED>
    <!--Optional:-->
    <DT_END>2014-02-20</DT_END>
    <!--Optional:-->
    <!--Zero or more repetitions:-->
    <ITENS>
      <id_pedido>99</id_pedido>
```

```
<soapenv:Body>
  <emp:incluirNovoPedido>
    <idPedido>99</idPedido>
    <!--Optional:-->
    <dataEntrega>2014-02-20</dataEntrega>
    <!--Optional:-->
    <!--Zero or more repetitions:-->
    <item>
      <idPedido>99</idPedido>
```

Fig. 3. Identification of synonyms parameters

For each of the fields identified in the XML document sent by the sender, the semantic mapper identifies the contained parameters and searches the corresponding parameters in the ontology, and then performs the matching and delivery of the document to the receiver. In Fig. 3 it can be observed that some sent parameters are equivalent to the expected, while for others it is necessary to find the correct match of them on the receiver. All parameters are searched in the ontology, because only after searching the mapper can know that the sent parameter is equivalent to the expected and there is no need for replacement by another denomination.

The evaluation of the ontology was based on the application of some metrics proposed by Bouiadjra and Benslimane [16]. Among the metrics suggested by them were applied the ones to evaluate the degree of search terms coverage, richness in knowledge representation and ontology of level of detail,

The terms cover (Ct) measures the ratio of class identified in the ontology based on the search term. For example, if the search term is "deliveryDate" and there is a class with this name in the ontology, it must be returned. In the evaluations Ct value was 100 %, whereas for all the searched parameters corresponding classes were found. This is because whenever a new SME integrates VBE, its services and parameters are registered in the ontology. If the term searched has not been registered, and the communication between companies can not be established. So the Ct of 100 % is critical to the success of integration.

The richness of relations (rR) indicates the diversity of relationships in the ontology. The value obtained for rR was 0.63, meaning 63 % of relationships are not hierarchical. The ontology is considered rich in terms of relationships. As closer as rR is to 1, as richer it is [17]. The richness of attributes (rA) evaluates the amount of information in instances. The value of rA was 1.68, that is, each class has on average 1.68 non-hierarchical relationships with other classes. This makes it rich with respect to the attributes. It is considered that the larger the number of attributes, the greater the knowledge that transmits the ontology [18].

The overall level of detail metric (dNG) reflects how well the knowledge is clustered between classes and subclasses. The value obtained for dNG was 0.95. Values close to zero indicate a horizontal ontology while values close to 1 indicate a vertical ontology [17]. Vertical ontologies represent details of a particular domain [17, 19]. As the proposed ontology aims to represent a domain in details, the value of 0.95 indicates that it is well detailed. The metric specific level of detail (dNE) indicates the average level at which the search terms are. The value obtained for dNE was 3. Searches on ontologies that run less than 5 levels are considered fast. So, 3 levels means the search in the proposed ontology is not complex.

The results obtained in evaluation metrics indicate that the proposed ontology needs to have full coverage (100 %) on the search terms to be considered robust. Based on experiments, it was observed that it presents this characteristic, that is, has the ability to return synonyms parameters correctly. Besides that, it might be considered richness in terms of knowledge representation in both relations and attributes. That is because there are enough relationships and attributes to enable the mapper to perform the queries and find the necessary classes. It may also be observed that it has sufficient level of detail to classify the knowledge represented and to enable the mapper go through a few levels when searching for a specific term. This enables searches to find results in an agile way.

The results obtained with experiments performed with the prototype implementation indicate that the search for synonyms in an ontology allows the application of the concept of semantic search to ESBs. To make this possible, the developed module (ontology + mapper) must be coupled to the ESB and its component should be added into the process during the designing phase whenever integration between services of different SME need to performed. Analyzing the results obtained by evaluating the prototype, it can see that the proposed method is able to offer greater flexibility in system integration of SME involved in VBEs.

5 Conclusion

In order to reduce cost and time SME come together in strategic alliances strongly grounded in collaboration and sharing of resources, risk and benefits. Among the manifestations of existing alliances, stand out VBEs, which serve as means for creating VEs and VOs. However, for SME to effectively collaborate within these alliances their systems must be integrated. The SOA architecture with support from ESB tools is the most suitable to play this role. However, while the ESB promotes safe and coordinated invocation for system integration, it does not have support for automating systems integration of the SME involved in VEs/VOs.

Therefore, this paper aims to propose a method to incorporate semantic to ESB in order to automate the identification process of involved data, and consequently accelerate the process of SME systems integration involved in VBE. To make this possible an ontology has been specified, which served as a source of knowledge to help in identifying similar data and map the parameters required for system integration of the SME members of the VEs/VOs.

In order to validate the proposed solution, a prototype was implemented, that automates the search in ontology and the identification of data for the integration mapping, and on it were applied some metrics. The metrics assessed the coverage of the search terms, the richness of knowledge representation and ontology detail level. The results indicate that the ontology presents 100 % coverage of the searched terms, so all the expected results returned correctly when existing in the ontology. They also indicate that the overall richness of ontology is 2.31, so, ontology is considered rich in terms of both relationship and attributes. The ontology presents a detail level of 3.95 which classifies it as vertical (created for detailed representation of a particular area) and the search gets results on average in the third level of the ontology.

Based on the evaluation results, it can see that the proposed solution is able to offer greater flexibility and agility in the integration of SME systems involved in VBEs. This is critical to the success of VOs and VEs because the shorter the time of matching and creation, faster it comes into operation and consequently earlier deliver the product to the customer, this way attending one of its main goals that is the agility and quickly.

Although research has reached the goals initially, it was observed in the course of development that other characteristics can be explored in order to improve the proposal, they are: Identification and treatment of homonyms; Analysis of the values contained in the XML with subsequent correction of the differences in existing measures or formatting and; Problem of excessive or too many parameters that need to be completed.

References

1. Missikoff, M., Smith, F., Taglino, F.: Ontology building and maintenance in collaborative virtual environments. Concurrency Comput. Pract. Experience **27**(11), 2796–2817 (2015)
2. Camarinha-Matos, L. M.; Afsarmanesh, H.: The virtual enterprise concept. In: IFIP Working Conference on Infrastructures for Virtual Enterprises (1999)
3. Hewit, E.: SOA Implementation Recipes, Tips, and Techniques: Java SOA Cookbook. O'Reilly Media, Inc., Sebastopol (2009)
4. Chen, L.: Integrating cloud computing services using enterprise service bus (ESB). Bus. Manage. Res. (2012). doi:10.5430/bmr.v1n1p26
5. Sward, R.E., Whitacre, K.J.: A multi-language service-oriented architecture using an enterprise service bus. In: Proceedings of the ACM Annual International Conference on SIGA (2008)
6. Li, Q., et al.: Business processes oriented heterogeneous systems integration platform for networked enterprises. Comput. Ind. **61**(2), 127–144 (2010)
7. Bem, R.M. Coelho, C.C.: Instrumentos de Representação do Conhecimento para práticas de Gestão do Conhecimento: taxonomias, tesauros e ontologias. InCID: Revista de Ciência da Informação e Documentação, vol. 4, Brazil (2013)
8. Abel, M., Fiorini, S.R.: Uma revisão da engenharia do conhecimento: Evolução, paadigmas e aplicações. Int. J. Knowl. Eng. Manag. **2**, 1–35 (2013)
9. Nazario, D.C., Danta, M., Todesco, J.L.: Representação de Conhecimento de Contexto e Qualidade de Contexto. Jornada Iberoamericana de Ingeniería del Software e Ingeniería del Conocimiento (2012)
10. Wang, X., Wong, T.N., Wang, G.: Service-oriented architecture for ontologies supporting multi-agent system negotiations in virtual enterprise. J. Intell. Manuf. **23**(4), 1331–1349 (2010)

11. Tramontin, Jr., R.J., Rabelo, R.J., Hanachi, C.: Customising knowledge search in collaborative networked organisations through context-based query expansion. Prod. Plan. Control., **21**(2), 229-246 (2010)
12. Santos, K.C.P.: Utilização de ontologias de referência como abordagem para interoperabilidade entre sistemas de informação utilizados ao longo do ciclo de vida dos produtos. Dissertação (Engenharia Mecânica e de Materiais) Universidade Tecnológica do Paraná, Curitiba (2011)
13. Li, C., Xie, T., Tang, Y.: GMVN oriented S-BOX knowledge expression and reasoning framework. J. Intell. Manuf. **25**(5), 993–1011 (2012). doi:10.1007/s10845-012-0722-x
14. Tu, Z., Zacharewicz, G., Chen, D.: A federated approach to develop enterprise interoperability. J. Intell. Manuf. **27**(1), 11–31 (2014). doi:10.1007/s10845-013-0868-1
15. Baldo, F., Rabelo, R.J.: Uma Abordagem Estruturada para Implementação de Ambientes de Criação de Empresas Virtuais. Anais SIMPOI. (2010)
16. Bouiadjra, A.B., Benslimane, S.: FOEval: full ontology evaluation: model and perspectives. In: 7th International Conference on IEEE Natural Language Processing and Knowledge Engineering (NLP-KE), Tokushima (2011)
17. Sicilia, M.A., et al.: Empirical findings on ontology metrics. Expert Syst. Appl. **39**(8), 6706–6711 (2012)
18. Moura, H. et al.: Developing a ubiquitous tourist guide. In: ACM WebMedia 2013. Salvador, Brazil (2013)
19. David, P.B., et al.: Análise da Afetividade em Fóruns Virtuais: Construção de uma Ontologia de Domínio. Nuevas Ideas en Informática Educativa TISE (2014)

Conceptual Modeling Using Knowledge of Domain Ontology

Hnatkowska Bogumiła(✉), Huzar Zbigniew, Tuzinkiewicz Lech, and Dubielewicz Iwona

Wroclaw University of Technology, Wyb. Wyspiańskiego 27, 50-370 Wrocław, Poland
{Bogumila.Hnatkowska,Zbigniew.Huzar,Lech.Tuzinkiewicz, Iwona.Dubielewicz}@pwr.edu.pl

Abstract. Conceptual modeling play an important role in software development. In initial phase of conceptual modeling domain models are constructed. They should be consistent with the reality they intend to present. Existing ontologies can be perceived as a valuable source for domain knowledge. The paper presents a method supporting development of domain models on the base of knowledge represented by a domain ontology. The main benefit of the method is that the developed domain model preserves consistency with the domain ontology in the context of the given domain problem. The proposed method is illustrated by an example.

Keywords: Ontology · Knowledge acquisition · Conceptual modeling

1 Introduction

In the software systems development the quality of domain models strongly affects the quality of the final products. The domain model provides basic knowledge for a particular domain of discourse that is necessary for problem domain understanding and its presentation. It is very important the model is externally consistent with a domain it intends to represent. Otherwise, the risk, that a software system the domain model is part of, will not meet the customer needs, is very high. The need for a domain model construction results from a problem stated in the application domain. Its construction is based on the knowledge from a given application domain. Most often it is the knowledge delivered by domain experts but, in the last few year, domain ontologies are considered as a valuable source of this knowledge. Domain ontologies may also be used as a base for validation of already elaborated domain models.

By definition, ontologies represent an objective and agreed viewpoint of the universal domain of discourse, which is independent of possible applications, while domain models are concentrated on specific problems on the background of the domain of discourse. For the purpose of domain modeling only a fragment from a given ontology, that is relevant for problem domain solution, should be extracted. How to select the proper fragment of a related ontology is the main problem discussed in the paper.

N.T. Nguyen et al. (Eds.): ACIIDS 2016, Part II, LNAI 9622, pp. 554–564, 2016.
DOI: 10.1007/978-3-662-49390-8_54

The paper presents background for designing of a tool called SUME (SUMO-UML Modeling Environment), supporting developers in the business modeling. The possibility of using a tool supporting class diagrams creation is clearly conducive to improving the quality of the created model and, in addition, is useful for verification of a previously elaborated domain model with respect to the ontology.

The proposed approach to that problem is presented assuming that the domain knowledge is represented by SUMO ontology, and that the elaborated domain models are represented in the UML.

The Suggested Upper Merged Ontology (SUMO) has been chosen because of some interesting properties. SUMO is a high level ontology developed under auspices of IEEE, and was the base for elaboration of many other middle level and low level (domain) ontologies [9, 10]. SUMO and a large set of extended ontologies are available under GNU General Public License. SUMO defines formally a hierarchy of classes and related rules and relationships, using the SUO-KIF language. In the context of domain modeling, a particular useful feature is that the notions of SUMO have not only formal definitions but also are mapped to the WordNet lexicon [8]. A mapping from WordNet synsets to SUMO has also been defined.

The Unified Modeling Language (UML) belongs to the most popular and the most general modeling languages used in software development. With UML we can represent both static and dynamic aspects of domain of discourse. In further, we restrict ourselves to the its static aspect.

The structure of the paper is as follows. Related works are described in Sect. 2. Formal outline of the approach is contained in Sect. 3, while some pragmatic remarks about it in Sect. 4. An illustrative example of approach application is presented in Sect. 5. The last Sect. 6 concludes the paper.

2 Related Works

The idea to use ontologies in software development is not new. The examples of current works are [2, 12]. They were also some works focused on the construction of software tools to support business modeling, e.g. [1].

Ontology and software development are interwoven in many ways. Particularly important is their interrelation with the early phases of software development, i.e. business modeling, requirements specification, and analysis. It is because the involved software engineers usually are not domain experts, and different understanding of the domain concepts may lead to an ambiguous and incomplete specification. So it seems reasonable that in these early software development phases, the existing ontologies are treated as a 'codified' knowledge and some pieces of it is to be transferred to the application domain being considered. This trend fits the ontology usage approach proposed in [6] The author contrasts a conceptual modeling as practiced in software systems projects with the ontology approach. According to the author's opinion the approach called Ontology-based Software Engineering (OBSE) when applied would allow to get better models and more re-usable components and to decrease costs in the software development projects. OBSE might become an attractive software engineering paradigm. Our work follows this paradigm.

During software development ontologies are used also in other ways. For example, in [11] ontologies are considered as:

- Ontologies of a Domain: understood as a representation (partial) knowledge of a particular domain within Software Engineering and Technology, and
- Ontologies as Software Artifacts: ontologies are used as some kind of artifacts during a software engineering processes: at development time and other processes (i.e. management and organization processes), or during the run-time (e.g. as Architectural Artifacts when ontologies are parts of the software system architecture and as Information Resources when ontologies are used as resource by the system at its run time).

The approach presented in this paper concentrates on the ontologies of domain.

Basing on two dimensions taking for comparison: the role of ontologies (at run time vs. at development time) and the kind of knowledge (domain vs. infrastructure) the authors of [6] distinguish four types of ontology usage. Combining these dimensions they define following usage types: Ontology Driven Development (ODD), Ontology Enabled Development (OED), Ontology Based Architecture (OBA), and Ontology Enabled Architectures (OEA).

In our case, we are especially interested in ODD approach, where ontology supports developers in their tasks. More precisely we would like to use ontology in the process of requirements engineering where a glossary is drawn up from requirements. We feel that it is especially important for improving the accuracy of the glossary when software engineers are not domain experts.

This paper is a continuation of previous authors' works. In [5] an identification of basic problems relating possible mapping between SUMO-like ontologies and domain models expressed in UML was given. An approach to the systematic approach of developing a CIM (Computational Independent Model) model on the basis of a selected domain ontology was outlined in [3]. This paper is a refinement of the method of business modeling based of SUMO ontologies outlined in [4]. In the paper [7] a translator from SUMO sub-ontology to UML was described.

3 Formal Outline of the Approach

Construction of a programming tool supporting software developers in the phase of domain modeling could be based on the approach presented below. The approach is based on two assumptions.

The first assumption says that we have acquired a set of initial domain notions (IDN) which may be both classes and relationships. This initial set of domain notions consists of key notions pulled out from the informal description of system domain.

There are many forms of such description, for example, plain text, business rules, and system vision documents. Regardless of which format of description we have considered we are interested in acquisition of a set of the domain key notions, where each key notion is represented by a phrase retrieved from a description.

In any description there are different phrases for consideration, nouns or noun phrases, and other phrases, especially verb phrases, which present relationships among nouns. As this paper is continuation of work [4], therefore only nouns and noun phrases are taken into account further in this paper. We retrieve them during analysis of (system) vision document [13] containing a concise description of user needs and expected system features which are supposed to fulfill them. Additionally we assume that these system features are expressed as a set of user stories and each user story is presented as statement in a standardized form called "role-feature-reason":

As a < type of user >, I want < some goal > so that < some reason > .

Such an approach is justified because information about relationships among nouns, which usually is contained in verb phrases, will be drawn up from the domain ontology. So such a set of key notion phrases is assumed to represent a set of initial domain notions (IDN) where some nouns or noun phrases may represent not only classes but also relationships.

The second assumption says that there exists a complete and consistent domain ontology, which covers all notions from the set of initial notions IDN and is defined in suitable form.

Considering form of ontology presentation we assume that its definitions include at least the following components: sets and their elements, classes and their instances, relationships among classes and/or instances, and a set of assertions that comprise the overall theory on the ontology. In general, relations may have any other relations as their arguments. In the paper we assume that only sets or classes are allowed to be arguments of these relationships. There is also a special kind of relationship – a taxonomy (subsumption) relationship. It can be described as a classification of entities from the general to the specific ones, or as an abstract definition of a hierarchy of concepts. In software engineering this relation is called a generalization. Other relationships that are not taxonomic, are called associations.

Basing on these assumptions, we aim to define such a sub-ontology, or, more precisely, an excerpt of the given ontology, which contains the set of all elements of the ontology, which are equivalent to the elements from the set of domain initial notions, denoted $EQ(IDN)$, and, additionally, the set of all other ontology elements, denoted $AD(EQ(IDN))$, that are related with the elements from the set $EQ(IDN)$. The elements of $EQ(IDN)$ may be domains (sets), classes or relations.

Formally, EQ is a mapping of the signature:

$$EQ{:}IDN \rightarrow ONT$$

where ONT is a set of all ontology notions. The mapping is defined by a domain expert, which should take into account a possibility that some of domain initial notions may be synonyms or may have more than one meaning.

Similarly, AD may be treated as a relation of the signature:

$$AD \subseteq EQ(IDN) \times ONT$$

where $< x, y > \in AD$ means that:

- if x is a class and $y \in SUP(x)$ which, by definition, is a set of all classes that are direct or indirect super-classes of the class x;
- if x is a class and $y \in ASS(x)$ which, by definition, is a set of relationships in which x is a domain; ASS(x) includes also relationships inherited by the class x;
- Remark. It is assumed that the set of relationships $ASS(x)$ for the class x contains also the relation determining the set of attributes of this class.
- if x is a relation and $y \in CLS(x)$ which, by definition, is a set of classes that are domains of the relation x.

Therefore, the extract of the given ontology generated by the set *IDN* is defined by:

$$ONT\,(IDN) = EQ(IDN) \cup AD(EQ\,(IDN))$$

where

$$AD(EQ\,(IDN)) = \bigcup\nolimits_{x \in EQ(IDN)} (SUP\,(x) \cup ASS\,(x) \cup CLS\,(x))$$

The set $EQ\,(IDN)$ contains elements that are not questionable relevant to the domain problem. Now, we have to decide which elements from the set $AD(EQ(IDN))$ are also relevant to the domain problem. So, we need to define a subset of chosen elements $CAD(EQ(IDN))$, i.e.

$$CAD\,(EQ\,(IDN)) \subsetneq AD\,(EQ\,(IDN))\,.$$

There are two practical problems related to such a selection of elements of the set $CAD(EQ(IDN))$. The first one: there is no possibility to decide about the selection fully automatically, therefore the selection should be made by a domain expert. The second one: an analysis and validation done by the domain expert of elements of the set $ONT(IDN)$ may be costly and time-consuming. Therefore, we propose a method, which allows a domain expert for a selection of the relevant fragment.

The subsequent steps of building $AD(EQ(IDN))$ should follow the rules:

1. The initial set of recommended notions for consideration (by an expert) $RN: = EQ(IDN)$.
2. If $r \in EQ(IDN)$ then $RN := RN \cup CLS(r)$
3. If $r \in \bigcup_{x \in EQ(IDN)} ASS(x)$, where r represents a relation, which all domains are classes from $EQ(IDN)$ then $RN: = RN \cup \{r\}$.
4. If $r \in \bigcup_{x \in EQ(IDN)} ASS(x)$, where r represents a relation, which some domains are from $EQ(IDN)$ and classes $d'_1, \ldots, d'_k \in \bigcup_{x \in EQ(IDS)} SUP(x)$, which are directed or undirected parents (super-classes) of $d_1, \ldots, d_k \in EQ(IDN)$, then $RN := RN \cup \bigcup_{i=1}^{k} chain(d_i, d'_i)$, where $chain\left(d, d'\right)$ is a set of classes $\{d_0, \ldots, d_n\}$ such that d_j is a direct parent of d_{j-1} for $j = 1, \ldots, n$ and $d = d_0, d_n = d'$.

It appears that the set of recommended notions *RN* contain all elements that are potentially relevant to the domain problem. Now, set *RN* should be analyzed by a domain expert with respect to its completeness or to its excess in the context of a given domain problem. If the set *RN* is not complete it means that the set of initial domain notions does not describe the problem, and, in consequence, the informal description of the application domain should be extended. The set *RN* may be redundant; in such a case the domain expert should reformulate the informal description of the application domain by removing unnecessary notions.

In the analysis, the expert may use axioms of the ontology. In our presentation, we skip a discussion of how to use the axioms, especially how to derive constraints relating to the elements from *RN*.

The set of recommended notions *RN* is the base for a UML class diagram definition. This class diagram is considered as a model representing static aspect of the application domain. Construction of the diagram may be done automatically. The main components of UML class diagrams contain classes, associations and generalizations. Therefore, the further consideration is limited to theses main components.

4 Pragmatic Treatment of the Approach

The proposed approach of creating a domain model using a domain ontology consists of the following five stages: (1) preparation, (2) extraction, (3) construction, (4) refactorization, (5) validation.

Below, a brief description of these stages is given, especially the problems that may occur are presented. Some of these problems are addressed by the proposed approach, which means that they may be solved or at least weakened. In the description these problems are written in italics.

The preparation stage consists of three steps.

1. In the first step a scope and vision of the system is discussed and agreed with stakeholders and a set of user stories is defined. In general, the potential problems that may occur here are:
   - ambiguous understanding and interpretation of the problem domain,
   - incorrect description of the scope of problem domain,
   - *incomplete or excess features expressed in user stories.*
2. In the next step, basing on the user stories of the future system, a glossary of initial domain notions (*IDN*) is created. Within this step the following activities are performed:
   - extraction of noun phrases (it is possible to perform this process automatically using software tools),
   - interpretation of noun phrases (semantic definition),
   - identification of synonyms (grouping nouns phrases that have the same meaning),
   - definition of a glossary of initial domain notions *IDN*.

   Note that the last three activities should be performed by domain expert. Potential problems that may occur are:

- *ambiguous or obscure interpretation of noun phrases,*
- *noun phrases are outside the domain glossary or scope of the problem domain.*

3. In the last step, a set of recommended terms of ontology $RN := EQ(IDN)$ is created by the expert who retrieves relevant concepts from ontology for each group of synonyms in the glossary. When there is no term in the ontology which is identical to any element in the set of synonyms, this ontology term should be included to the set of synonyms, what may result in notion standardization. Potential problems that may occur are:
   - *there is no relevant notions in chosen ontology,*
   - lack of required additional knowledge and skills for interpreting notions represented in ontology.

The model extraction and construction stages are realized automatically using software tools (SUME) implemented by the authors.

The model refactorization stage is realized by business analyst in collaboration with business expert. The aim of this stage is to make a model more readable, e.g. by representing some relations as class' attributes. The main potential problem that may occur is communication between business analyst and business expert resulting from their different experience and skills.

The model validation stage requires the participation of a domain expert who evaluates the created model in the context of user needs and specified scope of the domain problem. Negative validation reveals a necessity of user stories improvement, which entails vision improvement and repetition (next iteration) of the entire process.

In the proposed approach, the domain expert engagement concentrates on two activities. Within the first activity, the expert analyzes noun phrases extracted from the set of user stories, brings their interpretation, and confronts them with concepts found in the domain ontology based on interpretation of the terms (their semantics). Within the second activity, the expert cooperates with the business analyst during model refactorization and validation stages.

5 Illustrating Example

This chapter presents a short example of proposed method. The example is limited to the three first stages: preparation, extraction and model construction.

Within the first, preparation stage, we present execution of the second and third steps which aims in finding the mappings between noun phrases, being a part of user stories, and their representation in selected domain ontology ($EQ(IDN)$).

Let assume that we have 2 user stories, given below, in which noun phrases are written in italics.

1. As a *potential customer*, I want to see information about *hotel*, *hotel rooms*, *rooms' amenities* and *prices* so that I can decide whether to become a *customer*.
2. As a *potential customer*, I want to check availability of selected room (*room availability*) in a given *reservation period* so that to be able to decide if to make *reservation* or not.

The resulting *IDN* set is defined below:

IDN = {potential customer, hotel, hotel room, room amenity, price, customer, reservation, reservation period}

In Table 1 mappings of *IDN* to ontology notions (merge.kif, mid-level-ontology.kif, hotel.kif, dining.kig) is presented.

Table 1. Results of activity 1

Vision term (input)	SUMO term (output)	Type	SUMO documentation
Potential customer	Potential customer	Rel.	"(potentialCustomer ?CUST ?AGENT) means that it is a possibility for ?CUST to participate in a financial transaction with ?AGENT in exchange for goods or services"
Hotel	Hotel building	Class	"A residential building which provides tempo-rary accommodations to guests in exchange for money"
Hotel room	Hotel room	Class	"hotel room refers to a room that is part of a hotel (building) that serves as a temporary residence for travelers"
Room amenity	Room amenity	Rel.	"(room amenity ?ROOM ?PHYS) means that traveler accommo-dation provides physical ?PHYS in hotel unit ?ROOM"
Room price	price	Rel.	"(price ?Obj ?Money ?Agent) means that ?Agent pays the amount of money ?Money for ?Obj"
Customer	customer	Rel.	"A very general relation that exists whenever there is a financial tran-saction between the two agents such that the first is the destination of the financial transaction and the second is the agent"
Reservation Period	Reservation Start Reservation End	Rel. Rel.	(reservationStart ?TIME ?RESERVE) means that the use of a resource or consumption of a service which is the object of ?RESERVE starts at ?TIME")
Reserva tion	Hotel reservation	Class	"hotel reservation refers to a reservation specifi-cally for traveler acco-mmoda-tion stays"

The second, extraction phase was performed automatically by SUME. The results of its steps are presented in subsequent paragraphs.

Step 1 The initial set of recommended notions for consideration of an expert *RN* : = *EQ*(*IDN*) = {potentialCustomer, HotelBuilding, HotelRoom, roomAmenity, price, customer, reservationStart, reservationEnd, HotelReservation}

Step 2 *RN* is extended with classes introduced by relations *RN* = {*Physical, Physical, Agent, Reservation, CognitiveAgent, CurrencyMeasure, TimePoint, *HotelUnit}∪{potentialCustomer, HotelBuilding, HotelRoom, roomAmenity, price, customer, reservationStart, reservationEnd, HotelReservation}
Note: *Type means a power set of a given Type, i.e. that type represents all subclasses of Type.

Step 3 *RN* is extended with relationships that connect elements in *EQ*(*IDN*). Nothing was added. *RN* remained the same

Step 4. *RN* is extended with relationships that connect elements in *EQ*(*IDN*) and their parents (direct or indirect) *RN* = {reservedRoom, HotelUnit}∪{*Physical, Physical, Agent, Reservation, CognitiveAgent, CurrencyMeasure, TimePoint, *HotelUnit, potentialCustomer, HotelBuilding, HotelRoom, roomAmenity, price, customer, reservationStart, reservationEnd, HotelReservation]

The third, model construction phase was also performed automatically. The resulting UML model is presented as a class diagram shown in Fig. 1.

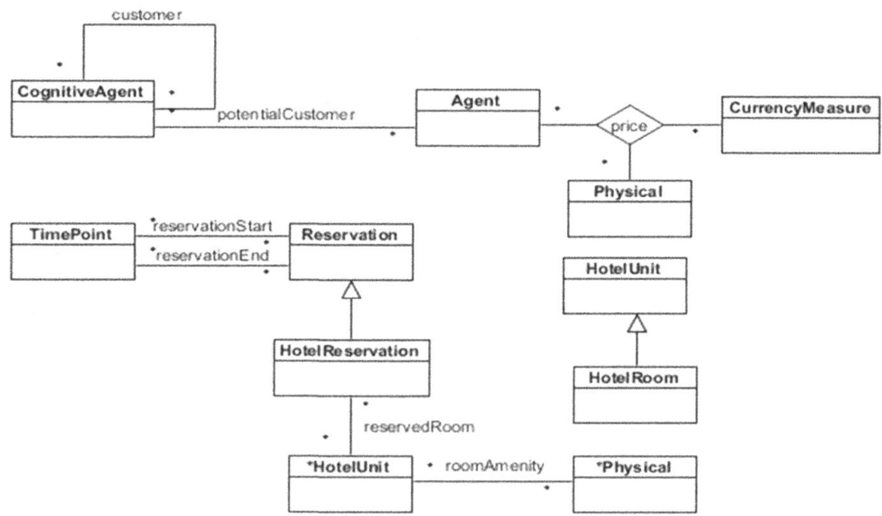

Fig. 1. The manually rewritten domain model (class diagram)

6 Conclusions

The aim of the paper is to present an approach to extraction such knowledge from an ontology which is relevant to the informal description of the described domain problem. The approach consists of five stages from which two stages could be performed automatically.

The main benefit of the approach is that the developed domain model preserves consistency with the domain ontology. A side positive effect is a possibility of standardization of domain notions. Moreover, the approach has positive influence on the quality of system vision within requirement specification. Revealed errors in the vision document may be removed in subsequent iteration of the approach.

Application of the approach is supported by the SUME tool performing extraction and construction stages. The part relating construction stage was described in [7]. Additionally, an open problem is concerned with elaboration of tools supporting other stages.

Future works are concentrated on experiments aiming at assessment of effectiveness of the proposed approach. However, at this stage of research our experience tells us that these works should be preceded by careful examination of existing domain ontologies. In our experiments we discovered many discrepancies of SUMO ontology, especially incompleteness and inconsistencies [7].

References

1. Benevides, A.B., Guizzardi, G.: A model-based tool for conceptual modeling and domain ontology engineering in OntoUML. In: Filipe, J., Cordeiro, J. (eds.) Enterprise Information Systems. LNBIP, vol. 24, pp. 528–538. Springer, Heidelberg (2009)
2. Jing, D., Yang, H., Tian, Y.: Abstraction based domain ontology extraction for idea creation. In: 13th International Conference on Quality Software, pp.341–348. IEEE (2013)
3. Dubielewicz, I., Hnatkowska, B., Huzar, Z., Tuzinkiewicz, L.: Domain modeling in the context of ontology. Found. Comput. Decis. Sci. **40**(1), 3–15 (2015)
4. Dubielewicz, I., Hnatkowska, B., Huzar, Z., Tuzinkiewicz, L.: Development of domain model based on SUMO ontology. In: Zamojski, W., et al. (eds.) Proceedings of the Tenth International Conference on Dependability and Complex Systems DepCoS-RELCOMEX, pp. 163–173. Springer, Switzerland (2015)
5. Dubielewicz, I., Hnatkowska, B., Huzar, Z., Tuzinkiewicz, L.: Problems of SUMO-Like ontology usage in domain modelling. In: Nguyen, N.T., Attachoo, B., Trawiński, B., Somboonviwat, K. (eds.) ACIIDS 2014, Part I. LNCS, vol. 8397, pp. 352–363. Springer, Heidelberg (2014)
6. Hesse W.: Ontologies in the Software Engineering process, Fachbereich Mathematik und Informatik, Univ. Marburg. http://ceur-ws.org/Vol-141/paper1.pdf
7. Hnatkowska, B.: Towards automatic SUMO to UML translation. In: Kościuczenko, P., Śmiałek, M. (eds.) From Requirements to Software, Research and Practice, pp. 87–100. Polskie Towarzystwo Informatyczne, Katowice (2015)
8. Niles, I., Pease, A.:. Linking lexicons and ontologies: mapping WordNet to the suggested upper merged ontology. In: Proceedings of the IEEE International Conference on Information and Knowledge Engineering, pp. 412–416 (2003)

9. Niles, I., Pease, A.: Toward a standard upper ontology. In: Chris Welty, C.H., Smith, B. (eds.) Proceedings of the 2nd International Conference on Formal Ontology in Information Systems (FOIS-2001), pp. 2–9 (2001)
10. Pease, A.: Ontology: A Practical Guide. Articulate Software Press, Angwin (2011)
11. Ruiz, F., Hilera, J.: Using ontologies in software engineering and technology. In: Calero, C., Ruiz, F., Piattini, M. (eds.) Ontologies for Software Engineering and Software Technology, pp. 49–102. Springer, Heidelberg (2006)
12. Si-Said Cherfi, S., Ayad, S., Comyn-Wattiau, I.: Improving business process model quality using domain ontologies. J. Data Semant. **2**, 75–87 (2013). Springer
13. Wikipedia. Vision document – wikipedia, the free encyclopedia (2015). Accessed 20 September 2015

Intelligent and Context Systems

Using Context-Aware Environment for Elderly Abuse Prevention

Maciej Huk(✉)

Department of Computer Science, Wroclaw University of Technology, Wroclaw, Poland
maciej.huk@pwr.edu.pl

Abstract. Numerous national, international and private institutions indicate that in the field of elderly care there is an important, growing and unsolved problem of elderly abuse – also in certified nursing homes. Intelligent environments and contextual processing can be used as a set of tools for elderly abuse prevention and detection, to decrease possibility of a set of abusive behaviors such as neglect and psychological abuse. Thanks to the context of care giving facilities, intelligent sensors can act not only as a help for caregivers to monitor the patient, but can also provide a method to monitor caregivers work in private homes as well as in long term nursing and residential care facilities. The aim of this text is to present which elements of context-aware intelligent environments such as high-level contextual analysis of data, continuous monitoring and placement of main data processing outside of the care giving area are needed to build systems that can help to prevent elderly abuse. Text describes also how low-level contextual processing can enhance functionality of such systems by extending sensors battery life as well as decreasing usage of radio channels.

Keywords: Context analysis · Elderly care · Intelligent sensors · Work monitoring · Abuse prevention · Neural networks · Context-aware environments

1 Introduction

World Health Organization and other numerous national, international and private institutions indicate that in the field of elderly care there is an important, growing and unsolved problem of elderly abuse [1–4]. There are various types of elder abuse, such as: physical, psychological, sexual, financial, neglect etc. Reports show that many of them can be observed frequently in certified nursing homes and other long-term care facilities worldwide. The abuse against elderly who get professional long-term care in domestic settings is also present. Studies of elderly abuse show that in certified nursing homes [5, 6]:

- 44 % of interviewed residents had said they had been abused,
- 95 % said they had been neglected or seen another resident be neglected,
- 50 % of nursing homes staff admitted to mistreating older patients in the last year (e.g. physical violence, mental abuse, neglect – mostly neglect (75 %)).

N.T. Nguyen et al. (Eds.): ACIIDS 2016, Part II, LNAI 9622, pp. 567–574, 2016.
DOI: 10.1007/978-3-662-49390-8_55

The situation in Europe is similar. Nearly 40 % of professional care staff respondents in Hanover in 2006 reported at least one incident of abuse or neglect of an older in-home care recipient within the last 12 months. Again – psychological abuse/verbal aggression (21 %) and neglect (19 %) were most common.

Presented numbers of the elderly abuse cases indicate that there is substantial and actual need to implement countermeasures that will be able to decrease the scale of that problem. This is especially relevant to aging societies, e.g.: Germany, Japan, etc. Thus developing automated, persistent elderly abuse detection systems based on intelligent environments and contextual processing can be of great value both for the societies as a whole and, what is more important, for millions of individuals worldwide.

2 Current Response to the Elderly Abuse Problem

Currently government authorities and institutions dedicated to respond to and prevent further maltreatment concentrate on public and professional awareness campaigns, screening of potential victims and abusers as well as on caregiver support interventions (e.g. stress management, respite care). There are also efforts to organize caregiver training on dementia, mandatory reporting of maltreatment to authorities, audits and visitations by social workers and certifications programs for personnel and nursing homes. Additionally special funds are allocated for providing safe-houses, emergency shelters and adult protective services (systems to place complaints, court-appointed individuals that manage financial assets and property of an adult unable to do it, advocate help etc.) [7]. However, despite such actions being needed, and having been carried on for at least 5 years, the current size of the problem shows that these measures are not sufficient. This is not a surprise, when e.g. in Germany even after the increase of frequency of the quality audits, each certified long–term care facility is inspected only once per year [8]. Those observations are consistent with the fact that US CDC also informs about "absence of proven prevention strategies" against elder abuse [9].

On the other hand there are currently only a few types of products available that target "elderly abuse prevention":

- hidden cameras and microphones (to observe the care giving personnel),
- patient positioning systems (e.g. for wandering-off detection; neglect prevention),
- fall monitors (for neglect prevention),
- wireless "call for help" buttons with events logging (neglect prevention),
- systems that support screening for criminal history of long-term caregivers [10].

Unfortunately the above mentioned products despite their advantages, from the point of view of care-taking person and his/her family, have significant imperfections:

- hidden cameras and microphones, generate a lot of data which has to be regularly analyzed. Regular physical access to the device can be a problem if its hidden functions have to be kept in secret for a long time.
- systems for screening care-giving personnel data do not detect abuse cases,

- fall monitors, patient positioning systems and "call for help" devices are targeted to prevent neglect in cases usually not caused by care-giving persons, and often do not monitor if the help was really given.

At the same time, scientists involved in the research on intelligent environments have developed various solutions for assisting caregivers in better realization of such tasks as elderly position and fall detection as well as their behavior and health classification (activity levels, anxiety, etc.) [11]. Others have proposed various specialized intelligent sensors to help in realization of elderly monitoring tasks (bed sensors, boots sensors, movement sensors, etc.). Few discuss safety of personal information of caretakers that is collected during their monitoring [12]. And to the author's best knowledge there are no propositions of systems that are specialized in monitoring caregivers work quality and/or built for elderly abuse prevention.

3 Context-Aware Automated System for Abuse Prevention

Analysis of the elderly abuse problem, current state of intelligent environments research and of available technologies shows that already it should be possible to build effective, automated system for prevention against selected types of elderly abuse. With use of machine learning models, intelligent sensors can detect abusive behaviors (elements of psychological abuse and neglect) and their short and long-time effects. The narrow context of elderly abuse allows gathering important information about care-giving environment just with a small set of sensors such as detectors of doors/windows opening and presence/movement. Additional sensors of the noise level, temperature, smell or behaviors can give more possibilities. In effect, the ability to immediately report and respond to events as well as documented, long-term care quality evaluation should decrease the number of abusive behaviors of caregivers.

The general architecture of the proposed automated system for abuse prevention (ASAP) is presented in the Fig. 1. It consists of the three main blocks: (1.) intelligent care environment equipped with suitable sensors and data transmission devices, (2.) distributed data processing system equipped with adequate knowledge base, reasoning and anomaly detection possibilities used for personal health, activity and care quality modeling and (3.) sensitive and selective reporting subsystem used for immediate alarming relatives or authorities about cases of abuse, as well as long-term care and caretaker health quality.

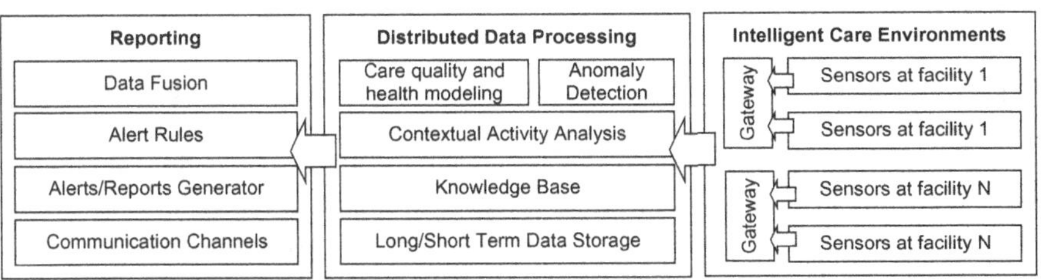

Fig. 1. Architecture of the proposed automated system for abuse prevention

An important element of the ASAP solution is the placement of the data storage and processing system outside of the care-giving facilities. This makes it not prone to unwanted influences both from caregivers and caretakers on the already collected data and on its processing or reporting. Potential switching off of the sensors (e.g. not changing batteries for time) or prolonged malfunction of the communication devices can be used as indication of decreased care quality (lower care transparency) and as information for authorities about the need for on-site inspection in given care facility.

Due to the expected high volume of the data collected from the sensors, the storage subsystem should have at least three different parts. Thus one part of it keeps processed and aggregated data and models of caretakers, caregivers and monitored facilities as well as other effects of automatic reasoning and information about events and alerts generated by the system. The second part temporarily stores all collected data for processing and reasoning, and removes it after processing. The last part is devoted for permanent storing of portions of original data selected by the reasoning subsystem as related with abusive behaviors or other crucial events. These data can be further used for long-term reasoning, analysis by humans and as the evidence for judicial processes.

Having the data from many sensors the contextual analysis of collected observables can be made with use of machine learning techniques to detect cases of elderly abuse when e.g.:

- the door and movement sensor do not notice activities for a long time (e.g. patient was left alone locked, possibly no water and food is given at normal hours),
- the door sensor will note rapid door opening or closing (possible psychological abuse as the care giving person is nervous, or in a hurry/overworked),
- the window sensor will not detect window opening for very long time (possible neglect: lack of ventilation),
- the window sensor is opened for very long time, especially in winter (neglect: lack of supervision; too low temperature is sometimes used as a punishment).

ASAP knowledge base and reasoning rules should be built and maintained to maximize system sensitivity and to limit possibility of overuse of its functionality by caretakers. But even if single events of similar types do not necessary mean abuse, their high frequency for prolonged time should be noted as significant and reported.

4 Low-Level Contextual Data Processing for Abuse-Aware Environment

The previous section presents how the top-level contextual information processing can be used within elderly abuse prevention system for the care-giving environment. But the contextual data processing can be beneficial for abuse prevention also when performed at the lowest levels of ASAP system information aggregation. It can be embedded within sensors to extend such important parameter as sensors battery life time. This parameter can greatly influence the value of the ASAP system, where uninterrupted monitoring of the environment and of the sensor itself are crucial. Additionally, in wireless sensor nodes context-aware processing can additionally reduce radio bandwidth usage thanks to better channel status prediction. This is also important for ASAP systems in care-giving environments where hundreds of sensors can operate at the same time.

The ASAP environment can include various sensor types, both wired and wireless, but in most cases the latter are used due to their mobility and low costs of installation. In turn, wireless sensor node often includes many individual sensors such as accelerometers, thermometers, GPS positioning, ultrasonic or video cameras, as well as communication subsystems. All those elements require energy for their functioning and limiting their use when not needed is still an actual problem. One of the solutions that can be used is incorporating into intelligent sensors contextual neural networks due to their ability to select from which sensors data needs to be collected in relation to the environment state [13] and when to perform communication, especially in the cognitive radio networks setups [14, 15].

After training contextual neural network can decide that in the given situation data from some sensor is not needed for proper system functioning, and the sensor can be turned off till the time when it will be needed. This allows not only for energy savings but also increases further signals analysis correctness as neglecting not important inputs decreases the amount of the noisy data which has to be processed. In some cases such solution can even allow the system to work properly when less important sensors are not working correctly. What is interesting, those properties can be achieved with the use of feed-forward neural networks architectures analogical to MLP and trained with generalized error back-propagation algorithm [13, 16].

As intelligent sensors rarely include more than twenty basic sensors and signals from the care environment should not be chaotic, the size of the problem of training the neural network controlling the work of basic sensors can be regarded as comparable to the mid-size benchmark problems from the UCI ML repository. Then, how effective such control mechanism can be observed e.g. on a well known example of Votes problem dataset. Depending on the neural network configuration, the achieved average neural network inputs activity *nia* (average relation of the number of neural network inputs used during input vector processing to the number of neural network inputs) can be as low as 20 % of the full inputs activity, without noticeable decrease of the neural network outputs generation accuracy, both for training and testing data (see Fig. 2, result after [13]). It can be also seen, that thanks to the contextual processing, some of the neural network inputs can be not used at all (see *niu* - the percentage of inputs used for all test vectors processing). Such results suggest that in the case of physical sensors contextual control of basic sensors usage can give energy savings as high as 80 %. The actual savings will depend on the construction of basic sensors, their energy needs, and the parameters of the environment in which they will be used. But in the case of ASAP system energy savings even much lower than 80 % can be worth using low–level contextual processing, as this decreases chances to miss the abusive event due to not working sensor, as well as lowers costs and time of sensors maintenance for care-giving facilities.

At the same time, communication is one of the most energy-intensive tasks while collecting information and determining context from sensors embedded in the intelligent environments. Within ASAP systems in professional care giving facilities the number of wireless sensors can be high (more than a few hundred) and the communication

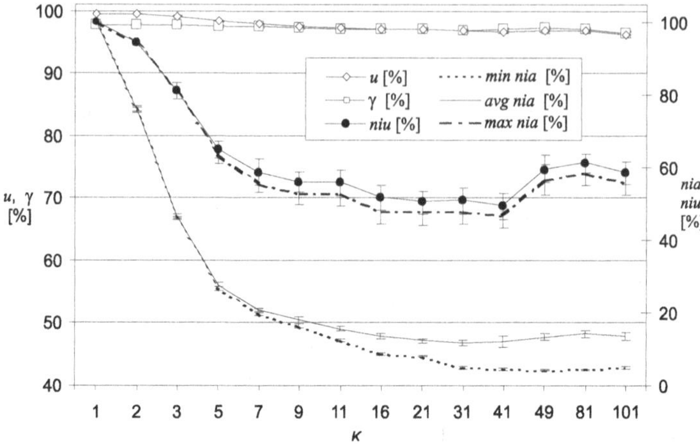

Fig. 2. The contextual Sigma-if network input activity *nia*, the number of network input used *niu*, and the classification accuracy of training *u* and test data γ for the UCI Votes problem versus the assumed number of hidden neuron input connections groups K ($K = 1$ means MLP, network: 48 inputs, 2 hidden neurons, 2 outputs). Lines between points used for clarity. [13]

conflicts can occur forcing devices to retransmit data what would be a waste of the energy and communication channel bandwidth. On the other side, sensing if the channel is free takes time and again costs energy. In such case contextual neural network classifier can be trained to predict the communication channel status in the next time slot based on the slot status history. The variable length of input vector of the contextual neural network causes that for some input data the predictor can give prediction with shortened status history, decreasing sensing time and preserving energy. Thus contextual data processing can also increase the average channel status prediction probability, additionally widening of the communication channel and increasing sensor energy savings.

Figure 3 depicts results of the cognitive radio communication simulation in which Knuth method was used to generate Poisson distribution of the primary user traffic of 50 % channel traffic intensity with inter arrival times from 10 to 20 slots and the length of the training and testing data were 1000 and 9000 slots respectively. After the initial experiments the Sigma-if neural network was constructed with 60 inputs, 15 hidden neurons and two outputs. Network was trained with the learning rate 0.001, maximal number of epochs 200, aggregation threshold $\varphi^* = 0.6$ and number of inputs groups K from the set {1, 2, 3, 5, 7, 10, 20, 30, 40, 60}. Measured values were calculated as averages from 10-fold cross validation repeated 10 times.

The observed average probability of proper prediction of the communication channel status was higher for the Sigma-if network than for the MLP. However, the highest generalization γ and fastest computation was achieved for $K = 5$ and 7 respectively. For $K = 7$ the reduction of outputs computation time τ is connected with almost 32 % decrease of the internal network activity, and the average decrease of neural network inputs use was from 0 to 5.2 %. This was the result of not requesting up to 8 time slots from the network input vector, and not processing of six times more inputs by the individual hidden neurons. This was possible as the inputs excluded from processing by the

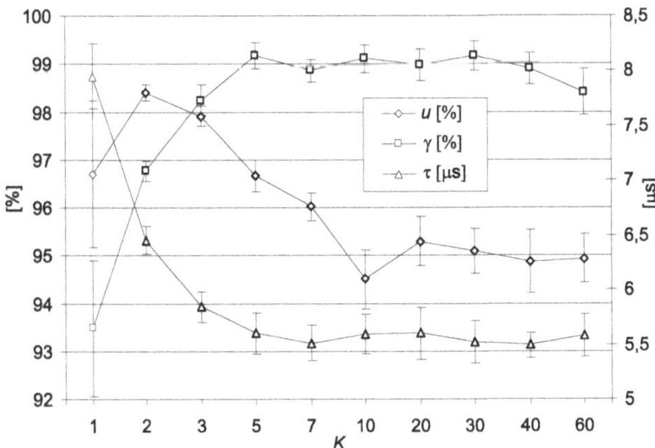

Fig. 3. Average communication channel status prediction probability by the contextual Sigma-if neural network for training u and testing data γ with average output generation time τ for various numbers of inputs groups K ($K = 1$ means MLP). Lines between points used for clarity.

hidden neurons were not only the ones at the edges of input vector. One can also observe that the averaged probability of correct prediction for the training data u decreases with increase of K, what can be interpreted as the decrease of overfitting effect typical for MLP neural networks.

5 Conclusion

Elderly abuse is a growing problem in certified nursing homes and other long-term care facilities worldwide. As the response for this situation the architecture of automated system for abuse prevention was proposed and discussed. In a basic version it can be built with existing technologies and can help to decrease the number of elderly abuse cases of neglect and psychical maltreatment in nursing homes and other long-term care facilities as well as ease to authorities and relatives of caretakers to monitor their health, care quality level and react when needed.

On the above background, it was presented that a contextual information processing can play important role within ASAP system, both during abuse detection as well as during low-level signals accumulation by the intelligent sensors. This is because the specific context of care giving facilities allows for specification of rules usable for detection of common elderly abuse situations, and contextual usage of sensors and optimizing wireless sensors communication with selective attention can increase sensors battery operation time, ASAP system reliability and decrease its maintenance costs.

References

1. Sethi, D., Wood, S., Mitis, F., Bellis, M., Penhale, B., Marmolejo, I., Lowenstein, A., Manthorpe, G., Kärki, F.: European Report on Preventing Elder Maltreatment. World Health Organization (2011)

2. Georgantzi, N.: Elder Abuse and Neglect in the European Union. UN Open-ended Working Group on Ageing, AGE Platform Europe (2012)
3. Elder Abuse: Definitions. US Centers for Disease Control & Prevention, National Center for Injury Prevention (2015)
4. Choi, N.G., Mayer, J.: Elder abuse, neglect, and exploitation: risk factors and prevention strategies. J. Gerontological Soc. Work **33**, 5–25 (2000)
5. Abuse of Residents of Long Term Care Facilities Fact Sheet. National Center on Elder Abuse, University of California (2012)
6. Post, L., Page, C., Conner, T., Prokhorov, A., Fang, Y., Biroscak, B.: Elder abuse in long-term care: types, patterns, and risk factors. Res. Aging **32**(3), 323–348 (2012)
7. Teaster, P., Dugar, T., Mendiondo, M., Abner, E., Cecil, K., Otto, J.: The Survey of State Adult Protective Services: Abuse of Adults 60 Years of Age and Older. National Center on Elder Abuse (2006)
8. McDaid, D.: Measuring the quality of long-term care. Eurohealth **16**(2), 1–44 (2010)
9. Report on Elder Maltreatment: Prevention Strategies. US Centers for Disease Control and Prevention, Division of Violence Prevention, Atlanta, USA (2014)
10. Abujarad, F., Swierenga, S.J., Dennis, T.A., Post, L.A.: Rap backs: continuous workforce monitoring to improve patient safety in long-term care. In: Marcus, A. (ed.) DUXU 2013, Part III. LNCS, vol. 8014, pp. 3–9. Springer, Heidelberg (2013)
11. Hossain, M.A., Ahmed, D.T.: A human caregiver support system in elderly monitoring facility. In: IEEE International Conference on Multimedia and Expo Workshops, pp. 435–440. IEEE (2012)
12. Chen, C.L., Yang, T.T., Leu, F.Y.: A secure authentication scheme of health care based on the cloud. In: Eighth International Conference on Innovative Mobile and Internet Services in Ubiquitous Computing, pp. 499–504 (2014)
13. Huk, M.: Learning distributed selective attention strategies with the Sigma-if neural network. In: Akbar, M., Hussain, D. (eds.) Advances in Computer Science and IT, pp. 209–232. InTech, Vukovar (2009)
14. Roy, N., Misra, A., Julien, C., Das, S.K., Biswas, J: An energy-efficient quality adaptive framework for multi-modal sensor context recognition. In: IEEE International Conference (2011)
15. Xing, X., Jing, T., Cheng, W., Huo, Y., Cheng, X.: Spectrum prediction in cognitive radio networks, next generation cognitive cellular networks. IEEE Wirel. Commun. **20**(2), 90–96 (2013)
16. Huk, M.: Backpropagation generalized delta rule for the selective attention sigma-if artificial neural network. Int. J. Appl. Math. Comput. Sci. **22**(2), 449–459 (2012)

Post-search Ambiguous Query Classification Method Based on Contextual and Temporal Information

Shahid Kamal, Roliana Ibrahim(✉), and Imran Ghani

Faculty of Computing, Universiti Teknologi Malaysia (UTM), 81310 Skudai, Johor, Malaysia
skamaltipu@gmail.com, {roliana,imran}@utm.my

Abstract. Web search involves user queries to process and then in response provide information. Commonly, the provided information results much irrelevant information which need to be filtered according to the user needs. Queries submitted to search engines are by nature ambiguous. The ambiguous queries constitute a significant fraction of such instances and pose real challenges to the web search. It has also created an interest for the researchers to deal with search by considering the context along with temporal perspective. Furthermore, contextual as well as temporal information retrieval has been a topic of excessive interest in recent years. The purpose is to enhance the effectiveness of retrieved information in documents and queries. This paper presents a new method PsAQCM of classifying the ambiguous queries based on the post-search results by applying content similarity approach. Java-based prototype is developed to derive the contextual and temporal information from the web results based on the 220, 44, and 114 ambiguous queries of GISQC_DS, AMBIENT and MORESQUE dataset separately. Our proposed method attained 51 %, 82 % and 78 % independently, improved results in terms of query ambiguity resolution. In future work, we intend to develop a small scale search engine which will enable us to carry out a full text analysis in order to improve the search performance in case of ambiguous queries.

Keywords: Ambiguous · Context · Temporal information · Disambiguation · Web search

1 Introduction

The fundamental research objectives in the field of web search are known to be the disambiguation of the searched queries, information to be captured according to the needs and then to enhance the performance. Suppose, there is a collection of documents, and a person (user) articulates a question (query) to which the answer is a set of documents that are appropriate according to his/her requirements. Then he/she will analyze all the retrieved documents keeping hold of the relevant documents and disposing of irrelevant ones. This solution is perceptibly unviable because, a user will not have either enough time to read all the retrieved documents or it may be substantially impossible.

The fast growth of the information world and the internet saturation into all fields has increased the importance of the problem of information search. Web search has supported the individuals to use their ability, in order to explore and then acquire the

N.T. Nguyen et al. (Eds.): ACIIDS 2016, Part II, LNAI 9622, pp. 575–583, 2016.
DOI: 10.1007/978-3-662-49390-8_56

knowledge. The organizations and individuals today produce and share the information in vast quantity with the advent of the Internet. It has become impossible to discover useful information without the assistance of an Information Systems (IR) because of this rapid growth of available information. In pursuance of communication being made with an IR system, the user must express his/her needs in the form of the queries. However, due to the lack of domain knowledge and the limitation of the natural languages such as synonymy and polysemy, many system users cannot formulate their needs into effective queries [1].

In response of user queries, the search engines give thousands of web pages for a query, most of these are irrelevant. The main reason behind getting a bulk of irrelevant information is the unclear contents of the user query. This situation is called ambiguity and it mainly creates a confusion regarding the understanding of the query and its results. The dissatisfying relevance in web search is considered as one of the vital problems, caused by the ambiguity in search queries. As query terms are commonly short, containing one to three terms only [2], so these are ascribed to be naturally ambiguous because of polysemy i.e. coexistence of many possible meanings for a word or phrase.

The ambiguity can be resolved by putting on knowledge of the available domain and then applying refinement process over the query with the addition of different features representing some entity e.g., place or time. During the precision-oriented search, the search process is enhanced by adding some features related to time and context to find the better results. This led to an introduction of temporal search [3–5] and contextual search [6–8] as well. Hence in this study, we intend to propose a Post-search Ambiguous query Classification Method (PsAQCM) for user queries to disambiguate them in order to improve the search results using the contextual information i.e., location and temporal information i.e., year.

The rest of the paper is organized as follow; Sect. 2 describes the previous work being done in the field of query classification that led to the disambiguation. Section 3 gives an overview of our proposed method and then results are elaborated in detail in Sect. 4. At the end of the paper, concluding remarks and some prominent future directions are described in the Sect. 5.

2 Related Work

Optimized web search is aimed at producing high-quality relevant information using the search engines over the internet. However, as the web grows in size and complexity, achieving the goals of relevant and accurate information based on user needs to be become more challenging. For this reason, web search often needs the user input queries for retrieving the information. Moreover, these user queries are ambiguous in nature [9]; often relinquish the performance in terms of accuracy of the retrieved information. Further, leading to ambiguous query identification, which is an arduous and complicated task. In order to develop a basic understanding of the problem background, if we give a search query “Cultural Show” in well-known search engine Google, it produces 324 million results in 0.39 s. Hence, retrieving much irrelevant information, that is because of the unclear query being given.

Disambiguating the search not only help in getting much relevant information but also positively contribute to increasing the performance of the search engine [10]. In this context, the disambiguation techniques such as [8–11] support by identification of the ambiguous queries were introduced. Ricardo, et al. highlighted the disambiguation of text queries with respect to temporal feature time in terms of a year [3]. The approach was based on clustering the search results based on the temporal feature. However, their work solely relies on temporal features thereby compromising the accuracy of the results.

The temporal features refer to the characteristics like timestamps, associated with the documents, retrieved in a response of user queries. For example, email and news, that can be placed with timeline i.e., sent date, time, information in the headers of the email. Traditional IR systems do not exploit temporal information even though there is enough temporal information exists with the retrieved documents [12]. The idea of the context has been used long ago in the multiple computer science applications. It is addressed in user modeling, adaptive hypermedia, and information retrieval [13–15].

The notion context can be defined as, any information that is used to characterize the surrounding application i.e., user, objects, interactions, and have an association with itself [15]. Further, the query disambiguation problem is investigated by [8] using contextual information, However, our work is differentiated with the authors in terms of using combined contextual and textual information being present in the search results produced in a response of the queries rather than using separately. Despite several existing approaches, the efforts need furtherance with respect to the specific case in which the contextual and temporal information is exploited using the query disambiguation.

3 Proposed Work

In order to carry out this research, the methodology is based on a hybrid approach, i.e. a combination of exploratory research and experimentation. The research work focuses on the web queries ambiguity identification and disambiguation approaches and their correctness for web information retrieval. In order to disambiguate the web queries to find relevant search results according to user intentions, we must first identify the nature of the query whether it is ambiguous (unclear) or not.

Due to a continuous increase in a size of the web and resulting issues related to web information retrieval, an efficient retrieval of information based on user queries is needed. To deal with effective information retrieval, in this section we present our method to deal with ambiguous queries in order to make them clear for accurate information retrieval.

The input query can be explicit (combination of text and location) or implicit (just text). In this component of the proposed approach, we deal with the former one where the availability of contextual information is helpful in refining the search results. We denote the input query by q and consider a result set n of 10 entries against each query of three different datasets namely; GISQC_DS dataset and AMBIENT dataset and MORESQUE dataset separately. In order to investigate the queries for different categories to be processed, we use two features mainly named as contextual that represents the location and temporal representing the year information being retrieved in the search results (see Fig. 1).

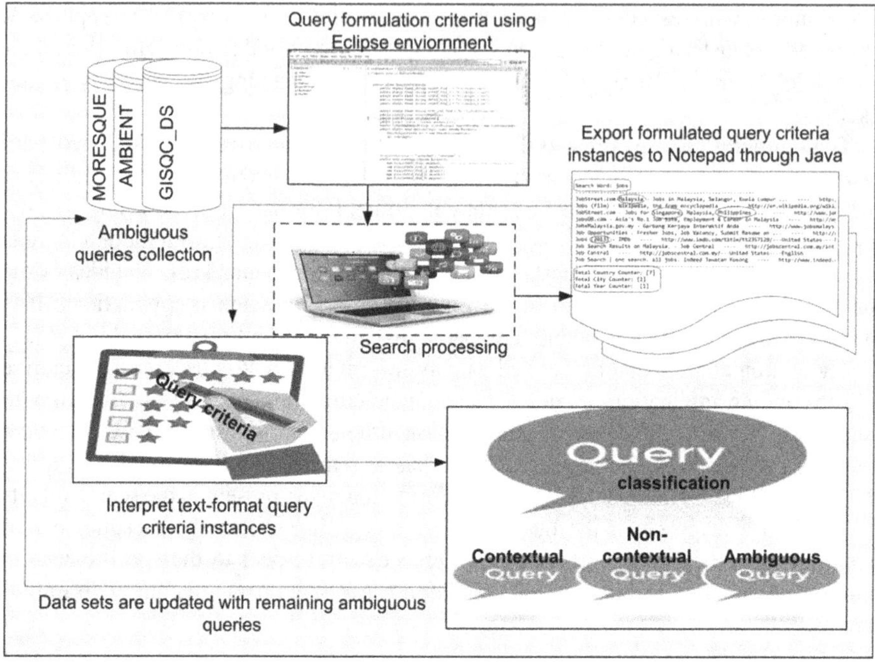

Fig. 1. Work diagram of the proposed method

Based on the processing performed over the search results, we categorize the queries into three different categories namely; contextual, non-contextual and ambiguous. Contextual queries are the queries which are delicate to location i.e. country or city name are recorded in the search results. Non-contextual queries are said to be non-contextual if the corresponding results include some temporal features i.e. year. Ambiguous queries are those where no contextual information (country or city) and temporal features (year) are returned in the search results after query execution.

The Disambiguation Approach (DA) is implemented as middleware between users and Google search engine. For the implementation of DA, we used the Eclipse IDE on top of a 64-bit system with 8 GB RAM and 2.5 GHz Core i5 processor.

4 Experiments and Results

To evaluate our approach, we used three datasets namely: GISQC_DS, AMBIENT (Ambiguous Entries) and MORESQUE (MORE Sense-tagged QUEries). The Google Insights for Search Query Classification dataset (GISQC_DS) [16] consists of 540 queries extracted from Google Insights for Search. Out of these 540 queries, 220 queries are categorized as ambiguous that we utilized to make them unclear.

We have also used another dataset AMBIENT [17] designed for evaluating subtopic information retrieval. It consists of 44 topics, each with a set of subtopics and a list of 100 ranked documents. The topics were selected from the list of ambiguous Wikipedia entries; i.e., those with "disambiguation" in the title. We have additionally utilized another dataset MORESQUE [18] intended for assessing data retrieval. It comprises of 114 queries. It is designed as a complement for AMBIENT. The Table 1 given below presents the number of ambiguous queries that are being refined by using our proposed method.

Table 1. Number of queries identified as ambiguous in different datasets

Sr.	Dataset	Number of ambiguous queries
1	GISQC_DS	220
2	AMBIENT	44
3	MORESQUE	114

As our main concern is to address the ambiguity of input queries, therefore, we focused and processed only the 220, 44 and 114 ambiguous queries of the data sets described in Table 1 independently. As part of our disambiguation approach, we further classified the ambiguous queries being identified by Generic Temporal Evaluation (GTE) as ambiguous, clear and broad solely based on temporal feature i.e., a year into three categories namely: contextual, non-contextual and ambiguous.

We carried out our approach by using the results retrieved in a response of the queries. Upon retrieval, we analyzed 10 search results against each query and looked into each result to find contextual, non-contextual information. Figure 2 below presents the analysis process in which we used three counters to indicate searched information separately.

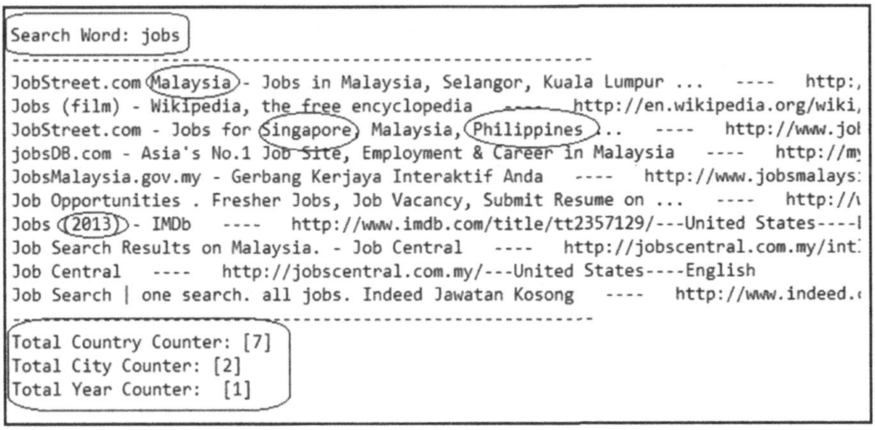

```
Search Word: jobs
---------------------------------------------------
JobStreet.com Malaysia - Jobs in Malaysia, Selangor, Kuala Lumpur ...   ----   http:/
Jobs (film) - Wikipedia, the free encyclopedia   ----   http://en.wikipedia.org/wiki/
JobStreet.com - Jobs for Singapore, Malaysia, Philippines ...   ----   http://www.jol
jobsDB.com - Asia's No.1 Job Site, Employment & Career in Malaysia   ----   http://my
JobsMalaysia.gov.my - Gerbang Kerjaya Interaktif Anda   ----   http://www.jobsmalays
Job Opportunities . Fresher Jobs, Job Vacancy, Submit Resume on ...   ----   http://
Jobs (2013) - IMDb   ----   http://www.imdb.com/title/tt2357129/---United States----
Job Search Results on Malaysia. - Job Central   ----   http://jobscentral.com.my/int
Job Central   ----   http://jobscentral.com.my/---United States----English
Job Search | one search. all jobs. Indeed Jawatan Kosong   ----   http://www.indeed.
---------------------------------------------------
Total Country Counter: [7]
Total City Counter: [2]
Total Year Counter:  [1]
```

Fig. 2. Results analysis process

The contextual information refers to the results that are in the form of web snippets; a short description of the retrieved result contains information about any place. While non-contextual information focuses on the web snippets containing information having time information in the form of year. The results which do not contain either location or time information are categorized as ambiguous queries, which are to be analyzed in future using some other parameters like date, month combined with places.

Below in Table 2, the total number of ambiguous queries are 220, 44 and 114 in the datasets namely: GISQC_DS, AMBIENT, and MORESQUE respectively, that are further processed according to the proposed approach. These queries are categorized into three different types namely; contextual, non-contextual and ambiguous. The contextual queries are in number 62, hence categorized as 28 % of the entire 220 ambiguous queries. The second category i.e. non-contextual queries are counted as 51 and, therefore, concluded as 23 % of the ambiguous queries. Furthermore, 107 queries are being leftovers as ambiguous i.e. not satisfying the defined criteria to be categorized in the case of GISQC_DS dataset to be processed in future.

Table 2. Query classification results with different datasets

Processed queries/DATASET	GISQC_DS	AMBIENT	MORESQUE
Ambiguous queries	220	44	114
Contextual queries	62	23	21
Non-contextual queries	51	13	68
Clear queries after processing	113	36	89
Leftovers ambiguous queries	107	8	25
Performance achieved	**51 %**	**82 %**	**78 %**

We further classified all of the ambiguous queries from the AMBIENT dataset into three categories namely: contextual, non-contextual, and ambiguous as shown in Table 2 based on criteria defined in our proposed approach. The contextual queries are being identified in number 23, non-contextual in number 13 and ambiguous as 8 in number, hence representing 52 %, 30 %, and 18 % of the aggregate respectively while, eight (8) queries are leftovers ambiguous queries.

Utilizing our query arrangement approach, we further characterized the majority of the ambiguous queries from the MORESQUE dataset into three classifications in particular: contextual, non-contextual and ambiguous as indicate in Table 2 in view of criteria characterized in our proposed methodology. The contextual queries are being recognized in number 21, non-contextual in number 68 and an ambiguous as 25 in number, subsequently speaking to 18 %, 60 % and 22 % of the total separately. While twenty-five (25) queries are leftovers ambiguous queries to be processed in future. The Fig. 3 shows the number of queries being processed and then classified bu our proposed method PsAQCM shown for the thorough understanding.

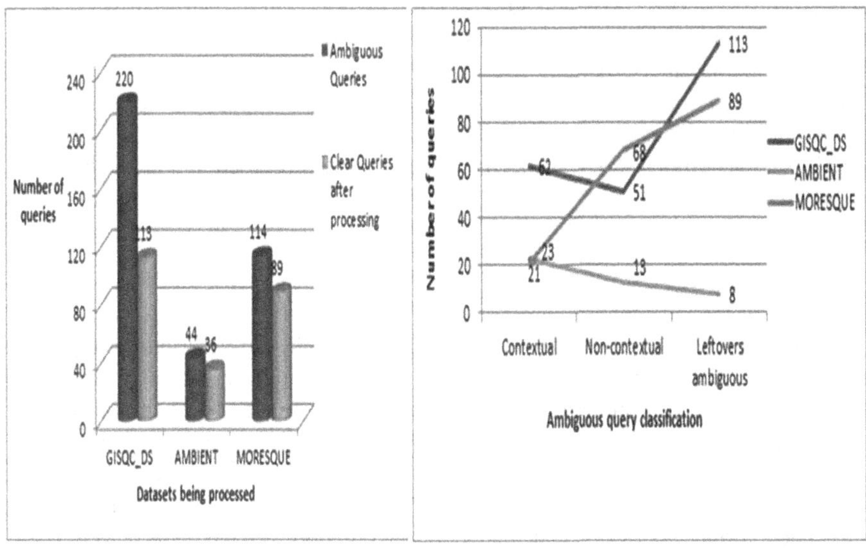

Fig. 3. Queries being processed and classified by PsAQCM (Color figure online)

Figure 3 shows performance of our approach regarding the query classification criteria defined in the approach. The blue-colored bar shows the outcomes of our proposed method PsAQCM while, red-colored bar is to show the leftover ambiguous queries, that are to be processed in future.

Furtherance in exemplification about Fig. 4, there we used two legends namely PsAQCM performance and leftover ambiguous queries in blue-colored and red-colored

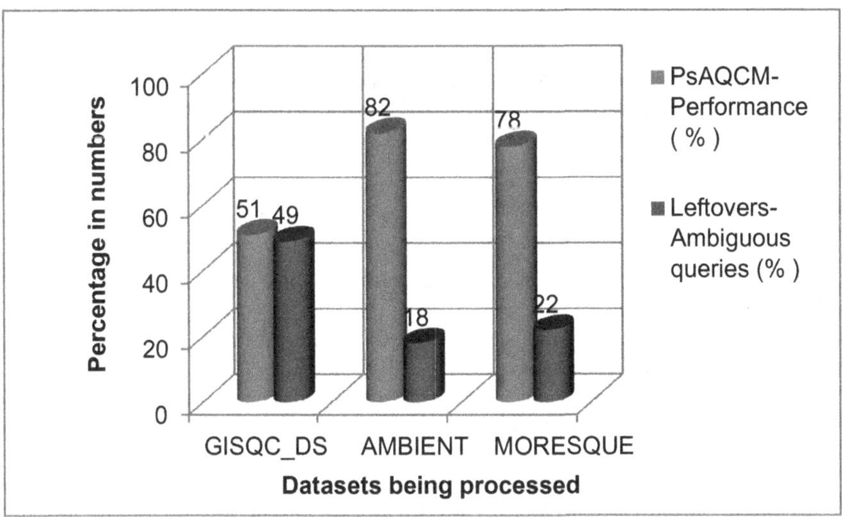

Fig. 4. Results come out from different datasets (Color figure online)

bars respectively. The x-axis legend represents the different datasets that we used for the experimentation while the y-axis is representing the percentage that have been represented as numbers. Among the results it has been observed that our method PsAQCM significantly contributed to making the ambiguous queries as clear, while the leftover ambiguous queries i.e., 49 %, 18 %, and 22 % in the datasets GISQC_DS-DS, AMBIENT and MORESQUE respectively need to be processed and analyzed by the researchers.

5 Conclusion

Ambiguous queries that are given for searching particular information often retrieve irrelevant information that the users don't want to read or analyze. However, as the web grows in the size and complexity, achieving goals of relevant and accurate information based on user needs, becoming more challenging. In this paper, we used a large proportion of ambiguous queries i.e. 378 from three different datasets to make them clear for the purpose of having accurate information. The research used the standard datasets independently and has made clear the existing ambiguous queries, which were previously declared as unclear. By making use of both temporal features as well as keeping track of location found in the web snippets, it addresses the inherent ambiguities in input queries and filters out irrelevant results in response to user's search over the web. Our proposed PsAQCM method has revealed an improved performance and attained 51 %, 82 %, and 78 % of clear queries using three datasets separately, which have been declared as ambiguous in the past. We are also profound to test our method by using diverse datasets to verify its robustness. In addition, we also plan to develop a small scale search engine as part of future work, which will enable us to carry out a full-text analysis using the contextual information in the search queries.

Acknowledgement. We would like to thank the Universiti Teknologi Malaysia and the Malaysia Ministry of Higher Education (MOHE) Research University Grant Scheme (Vot No. Q.J130000.2528.05H84) and also (Vot No: 4F315) for the facilities as well as support to conduct this research study. Moreover, we would also like to say thanks to the Higher Education Commission of Pakistan and the Gomal University Dera Ismail Khan, Pakistan.

References

1. Zhang, H., Adviser-Jacob, E.: Query enhancement with topic detection and disambiguation for robust retrieval (2013)
2. Roul, R.K., Sahay, S.K.: An effective information retrieval for ambiguous query. arXiv preprint (2012). arXiv:1204.1406
3. Campos, R., et al.: Disambiguating implicit temporal queries by clustering top relevant dates in web snippets. In: 2012 IEEE/WIC/ACM International Conferences on Web Intelligence and Intelligent Agent Technology (WI-IAT). IEEE (2012)
4. Drew, T., Wolfe, J.M.: Hybrid search in the temporal domain: monitoring an RSVP stream for multiple targets held in memory. J. Vis. **12**(9), 1276 (2012)

5. Lan, R., et al.: Temporal search and replace: an interactive tool for the analysis of temporal event sequences. HCIL, University of Maryland, College Park, Maryland, Technical report HCIL-2013-TBD (2013)
6. Kraft, R., et al.: Searching with context. In: Proceedings of the 15th International Conference on World Wide Web. ACM (2006)
7. Mizzaro, S., Vassena, L.: A social approach to context-aware retrieval. World Wide Web **14**(4), 377–405 (2011)
8. Anastasiu, D.C., et al.: A novel two-box search paradigm for query disambiguation. World Wide Web **16**(1), 1–29 (2013)
9. Song, R., et al.: Identification of ambiguous queries in web search. Inf. Process. Manag. **45**(2), 216–229 (2009)
10. Bunescu, R.C., Pasca, M.: Using encyclopedic knowledge for named entity disambiguation. In: EACL (2006)
11. Mihalcea, R., Csomai, A.: Wikify!: linking documents to encyclopedic knowledge. In: Proceedings of the Sixteenth ACM Conference on Information and Knowledge Management. ACM (2007)
12. Jones, R., Diaz, F.: Temporal profiles of queries. ACM Trans. Inf. Syst. (TOIS) **25**(3), 14 (2007)
13. Salton, G., Buckley, C.: Term-weighting approaches in automatic text retrieval. Inf. process. Manag. **24**(5), 513–523 (1988)
14. Singhal, A.: Modern information retrieval: a brief overview. IEEE Data Eng. Bull. **24**(4), 35–43 (2001)
15. Tamine-Lechani, L., Boughanem, M., Daoud, M.: Evaluation of contextual information retrieval effectiveness: overview of issues and research. Knowl. Inf. Syst. **24**(1), 1–34 (2010)
16. Campos, R.: Google Insights for Search Query Classification Dataset (GISQC_DS) (2011)
17. Carpineto, C., Romano, G.: Ambient dataset (2008). http://credo.fub.it/ambient/
18. Navigli, R., Crisafulli, G.: Inducing word senses to improve web search result clustering. In: Proceedings of the 2010 Conference on Empirical Methods in Natural Language Processing (EMNLP 2010). MIT Stata Center, Massachusetts (2010)

A Context-Aware Implicit Feedback Approach for Online Shopping Recommender Systems

Luu Nguyen Anh-Thu, Huu-Hoa Nguyen, and Nguyen Thai-Nghe(✉)

College of Information and Communication Technology,
Can Tho University, 3-2 Street, Can Tho City, Vietnam
{lnathu,nhhoa,ntnghe}@ctu.edu.vn

Abstract. Recommender Systems are widely used in many areas such as entertainment, education, science, especially e-commerce. Integrating recommender system techniques to online shopping systems to recommend suitable products to users is really useful and necessary. In this work, we propose an approach for building an online shopping recommender system using implicit feedback from the users. For building the system, first we propose a method to collect the implicit feedback from the users. Then, we propose an ensemble method which combine several extended matrix factorization models which are specialized for those implicit feedback data. Next, we analyze, design, and implement an online system to integrate the aforementioned recommendation techniques. After having the system, we collect the feedback from the real users to validate the proposed approach. Results show that this approach is feasible and can be applied for the real systems.

Keywords: Recommender systems · Product recommendation · Implicit feedback · Matrix factorization

1 Introduction

Recommender Systems (RS) are successfully used in practice to help users dealing with problem of information overloading [2, 7]. Using RS, the users can choose relevant items (products, services,...) based on their feedback in the past. These feedback could be **explicit**, e.g., user's rating (1 star to 5 stars; like (1)/dislike (0), etc.) or the feedback could be **implicit** which means that the users need not to provide their evaluation. The implicit feedback could be the number of clicks, quantity of purchased products, the time that the user has browsed/viewed products, etc. [2].

There are many companies/organizations using RS to recommend services, products and necessary information to the users such as Amazon.com: recommending the products that customers may be interested; Youtube.com: recommending video clips; Netflix.com: recommending movies, etc. These systems collect the feedbacks from their customers, e.g., from 1 star (dislike) to 5 stars (like the most) on Amazon, Netflix, Movielens,...; like/dislike on Youtube; etc. [5].

N.T. Nguyen et al. (Eds.): ACIIDS 2016, Part II, LNAI 9622, pp. 584–593, 2016.
DOI: 10.1007/978-3-662-49390-8_57

Using explicit feedback, the systems can easily determine the level of user favorite on the products, thereby predicting the next products that users might like to recommend them. However, explicitly collecting the feeback from the user is not convenient because the users are not always pleased to leave their feedback, so the system should determine what the user needs through their implicit feedback [1,4].

In this work, we propose an approach for building online shopping recommender systems using implicit feedback from the users (such as the number of user clicks on products and the quantity of purchased products). First, we propose methods to collect and exploit the potential implicit feedback from the users. Then, we propose an ensemble of models which are specialized for implicit feedback data. Next, the whole system is built to integrate the recommendation models. After having a complete system, we collect data from the real users to evaluate the effectiveness of the proposed approach. Experimental results show that the proposed approach can recommend suitable products to the users.

2 Recommender Systems (RS)

Usually, recommender systems have three objects: user, item (e.g., movie, paper, song, product, etc. In this work, we consider item as a product), and the user feedback on that product (e.g., rating, number of clicks, etc.). Let $\mathbf{U}$ be a set of users and u be a particular user ($u \in \mathbf{U}$). Let $\mathbf{I}$ be a set of items and i be a particular item ($i \in \mathbf{I}$). Let $\mathbf{R}$ ($\mathbf{R} \subset \mathbb{R}$) be a set of feedback values which are used to estimate the preferences of the users, and $r_{ui} \in \mathbf{R}$ is a feedback by user u on item i. As aforementioned, this work is concerned about how to determine r_{ui} implicitly.

In the next sections, we introduce the state-of-the-art methods in RS which can be used to predict the user feedback.

2.1 Matrix Factorization (MF)

Matrix factorization is the task of approximating a matrix $\mathbf{X}$ by the product of two smaller matrices $\mathbf{W}$ and $\mathbf{H}$ such that $\mathbf{X}$ can be re-constructed from these two smaller matrices [6], i.e. $\mathbf{X} \approx \mathbf{W}\mathbf{H}^T$.

An illustration of matrix decomposition is presented in Fig. 1.

In the context of recommender systems, the matrix $\mathbf{X}$ is the partially observed ratings matrix; $\mathbf{W} \in \mathbb{R}^{|\mathbf{U}| \times K}$ is a matrix where each row u is a vector containing the K latent factors ($K << |\mathbf{U}|, K << |\mathbf{I}|$) describing the user u and $\mathbf{H} \in \mathbb{R}^{|\mathbf{I}| \times K}$ is a matrix where each row i is a vector containing the K factors describing the item i.

Let w_{uk} and h_{ik} be the elements of $\mathbf{W}$ and $\mathbf{H}$ ($\mathbf{w}$ and $\mathbf{h}$ are their vectors, respectively), respectively, then the feedback/preference (e.g., rating) given by a user u to an item i is predicted by:

$$\hat{r}_{ui} = \mathbf{w} \cdot \mathbf{h}^T = \sum_{k=1}^{K} w_{uk} h_{ik} \tag{1}$$

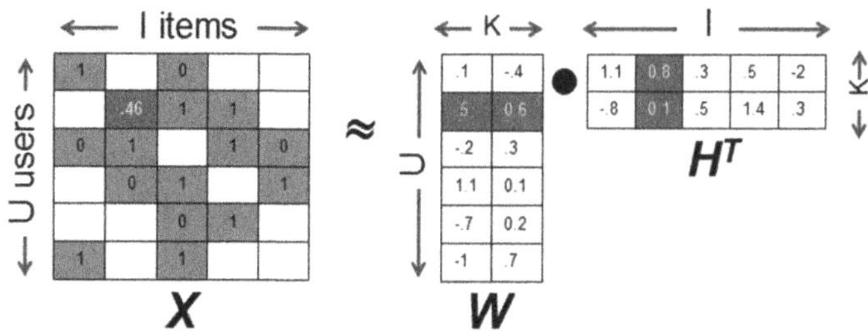

Fig. 1. An illustration of matrix factorization

where **W** and **H** are the latent matrices (model parameters) and can be learned by optimizing an objective function, e.g.,

$$\mathcal{O}^{MF} = (r_{ui} - \hat{r}_{ui})^2 + \lambda \left(||\mathbf{W}||_F^2 + ||\mathbf{H}||_F^2\right) = (r_{ui} - \mathbf{w}_u \cdot \mathbf{h}_i^T)^2 + \lambda \left(||\mathbf{W}||_F^2 + ||\mathbf{H}||_F^2\right) \quad (2)$$

where λ is a regularization term ($0 \leq \lambda < 1$) which is used to prevent overfitting and $||.||^2$ is a Frobenius norm.

After the training phase, there are the two optimal latent factors **W** and **H**, the remaining task is straightforward. The rating of user u in for a given item i is predicted easily by Eq. 1. For more details, please refer to [6,10].

2.2 SVD++

SVD++ (extended from Singular Value Decomposition) [5] is a variant of matrix factorization which can be used for implicit feedback data. SVD++ incorporates the user and item biases to the standard MF. The feedback/preference of user u on item i is predicted by

$$\hat{r}_{ui} = \mu + \mathbf{b}_u + \mathbf{b}_i + \left(\mathbf{w}_u + \frac{1}{\sqrt{|N(u)|}} \sum_{j \in N(u)} \mathbf{y}_j\right) \cdot \mathbf{h}_i^T \quad (3)$$

where μ is a global average, $\mathbf{b}_u$ and $\mathbf{b}_i$ are user and item biases, respectively. $N(u)$ is a set of implicit feedback for users u and $\mathbf{y}_j$ is a latent factor vector that represents for implicit feedback. For details about this method, please refer to paper [5].

3 Proposed Method

In this section, first, we introduce methods for gathering implicit feedback from users, then we propose an ensemble models of recommender system techniques for those implicit feedback data.

3.1 Methods for Collecting Implicit Feedback from the Users

Usually, existing online shopping systems collect the feedback from users by using explicit schema, which means that the users should explicitly provide their preferences, such as 1 to 5 stars, or like/dislike, etc. However, these methods could be inconvenient since most of the users rarely leave their feedback. Therefore, we propose to collect the user feedback by using implicit schema, e.g., the quantity of purchased products (if available) and the number of user's clicks on the products. For the feedback in term of quantity of purchased products, it is easy to collect after a shopping transaction. However, for the other one (number of clicks/views), it is more difficult to get them. When the user logins to the system, the system will automatically update his/her feedback to database. However, if the user does not login to the system, the system will automatically create a temporary account for this user to record his/her feedback. This case has three contexts:

- **Context 1 (existing user)**: If the user has finished his session and signed into the system, the number of his views/clicks are updated to database and the system removes information from the temporary account.
- **Context 2 (new user)**: If the user create a new account, the feedback information from the temporary account will be mapped to this new account.
- **Context 3**: If the user leaves the system without registration, we store his/her IP address together with his/her feedback information. These information will be utilized to recommend to the new users by using **item average** [10] approach.

3.2 Model Combination

Recommender systems usually use one value of the user's feedback (e.g., rating value). In this study, to take advantage of the feedback information in order to increase the reliability of the models, we used two implicit feedback information, which is the quantity of purchased products and the number of user clicks on the products. Please note that, in case the user who has not previously purchased the products, we cast this information as the number of user views/clicks. Based on SVD++ [5], we present two models in the following.

In the first model, the objective function that need to be optimized for getting the parameters is presented as

$$\mathcal{O}^1 = \left(r_{ui} - \mu - \mathbf{b}_u - \mathbf{b}_i - (\mathbf{w}_u + \frac{1}{\sqrt{|Q(u)|}} \sum_{j \in Q(u)} \mathbf{y}_j) \cdot \mathbf{h}_i^T \right)^2 + \lambda \cdot \left(||\mathbf{b}_u||_F^2 + ||\mathbf{b}_i||_F^2 + ||\mathbf{W}||_F^2 + ||\mathbf{H}||_F^2 + ||\mathbf{y}_j||_F^2 \right) \quad (4)$$

This objective function is extended from the standard matrix factorization in Eq. (2), where $Q(u)$ is the set of user implicit feedback which is the quantity of purchased products. The feedback for user u on item i is predicted by

procedure SVDPlusPlus-SGD($\mathcal{D}^{train}$: training data; N: Implicit feedback data; λ_1, λ_2, λ_3, λ_4, λ_5: Regularization term; β_1, β_2, β_3, β_4, β_5, β_6: Learning rate; K: #Latent factors; a stopping criterion)
    $\mu \leftarrow \frac{\sum_{r_{ui} \in \mathcal{D}^{train}} r_{ui}}{|\mathcal{D}^{train}|}$ ▷ global average
    for each user u **do**
        $\mathbf{b}_u \leftarrow \frac{\sum_i (r_{ui} - \mu)}{|\mathcal{D}_u^{train}|}$ ▷ user bias
    end for
    for each item i **do**
        $\mathbf{b}_i \leftarrow \frac{\sum_u (r_{ui} - \mu)}{|\mathcal{D}_i^{train}|}$ ▷ item bias
    end for
    $W \leftarrow \mathcal{N}(0, \sigma^2)$ ▷ Draw from normal distribution
    $H \leftarrow \mathcal{N}(0, \sigma^2)$
    $\mathbf{y} \leftarrow \mathcal{N}(0, \sigma^2)$
    while (a stopping criterion is NOT met) **do**
        Draw randomly (u, i, r_{ui}) from $\mathcal{D}^{train}$
        $\hat{r}_{ui} \leftarrow \mu + \mathbf{b}_u + \mathbf{b}_i + \left(\mathbf{w}_u + \frac{1}{\sqrt{|N(u)|}} \sum_{j \in N(u)} \mathbf{y}_j\right) \cdot \mathbf{h}_i^T$
        $e_{ui} = r_{ui} - \hat{r}_{ui}$
        $\mu \leftarrow \mu + \beta_1 \cdot e_{ui}$
        $\mathbf{b}_u \leftarrow \mathbf{b}_u + \beta_2 \cdot (e_{ui} - \lambda_1 \cdot \mathbf{b}_u)$
        $\mathbf{b}_i \leftarrow \mathbf{b}_i + \beta_3 \cdot (e_{ui} - \lambda_2 \cdot \mathbf{b}_i)$
        $\mathbf{w}_u \leftarrow \mathbf{w}_u + \beta_4 \cdot (e_{ui} \cdot \mathbf{h}_i - \lambda_3 \cdot \mathbf{w}_u)$
        $\mathbf{h}_i \leftarrow \mathbf{h}_i + \beta_5 \cdot (e_{ui} \cdot \left(\mathbf{w}_u + \frac{1}{\sqrt{N(u)}} \sum_{j \in N(u)} \mathbf{y}_i\right) - \lambda_4 \cdot \mathbf{h}_i)$
        for each $j \in N(u)$ **do**
            $\mathbf{y}_j \leftarrow \mathbf{y}_j + \beta_6 \cdot \left(e_{ui} \cdot \frac{1}{\sqrt{N(u)}} \mathbf{h}_i - \lambda_5 \cdot \mathbf{y}_j\right)$
        end for
    end while
    return $\{W, H, \mathbf{y}, \mathbf{b}_u, \mathbf{b}_i, \mu\}$
end procedure

Fig. 2. Learning a SVDPlusPlus using Stochastic Gradient Descent

$$\hat{r}_{1ui} = \mu + \mathbf{b}_u + \mathbf{b}_i + \left(\mathbf{w}_u + \frac{1}{\sqrt{|Q(u)|}} \sum_{j \in Q(u)} \mathbf{y}_j\right) \cdot \mathbf{h}_i^T \tag{5}$$

In the second model, the objective function that need to be optimized for getting the parameters is presented as

$$\mathcal{O}^2 = \left(r_{ui} - \mu - \mathbf{b}_u - \mathbf{b}_i - (\mathbf{w}_u + \frac{1}{\sqrt{|C(u)|}} \sum_{j \in C(u)} \mathbf{y}_j) \cdot \mathbf{h}_i^T\right)^2 + \lambda \cdot \left(||\mathbf{b}_u||_F^2 + ||\mathbf{b}_i||_F^2 + ||\mathbf{W}||_F^2 + ||\mathbf{H}||_F^2 + ||\mathbf{y}_j||_F^2\right) \tag{6}$$

where $C(u)$ is the set of user implicit feedback which is the number of clicks on the product. Then, the feedback for user u on item i is predicted by

$$\hat{r}_{2ui} = \mu + \mathbf{b}_u + \mathbf{b}_i + \left(\mathbf{w}_u + \frac{1}{\sqrt{|C(u)|}} \sum_{j \in C(u)} \mathbf{y}_j \right) \cdot \mathbf{h}_i^T \tag{7}$$

Finally, we use an ensemble model [3] to integrate two above models.

$$\hat{r}_{ui} = (\hat{r}_{1ui} + \hat{r}_{2ui}/\Theta)/2 \tag{8}$$

In facts, once a user has purchased the products, this implicitly indicates that he prefers/needs those products. Thus we propose that the weight for users' purchased products should be higher than the weight for clicked/viewed products without purchasing. Moreover, in any e-commerce website, the number of user clicks/views is enormous while the number of purchases are smaller. To reduce these disparities, we introduce a hyperparameter Θ as in Eq. (8). This hyperparameter should be determined in advance, e.g., using cross-validation approach (in our system, after evaluating on available data sets, we set $\Theta = 4$).

Details of an implementation for the SVD++ is presented in Fig. 2. Here we use Stochastic Gradient Descent for learning process. The regularization and learning rate are separately used for each parameter.

4 Experimental Results

We have applied the proposed approach for building a product (IT products) recommender system. After hosting the system, we collect the real feedback from the users, then evaluating the performance of the system.

4.1 Typical Screenshots

Figure 3 presents the main interface of the proposed product recommendation based on user implicit feedback. Using the past information, the system can automatically recommend the products to the user as presented in the left column of Fig. 3.

4.2 Model Evaluation

Evaluation on Real Data Set: We have collected the feedback from 186 real users, 174 products and 637 feedback (including of the number of purchased products and the number of user clicks on the products).

Table 1 presents results of Mean Absolute Error (MAE) when using single model (as described in Eqs. 4 and 6) and an ensemble model (Eq. 8). Experimental results clearly show that by using a combination of two feedback from users, the MAE is significantly improved.

Comparison with Other Baselines: Figure 4 shows the comparison results of the proposed approach which other baselines such as Global Average, User Average, and Item Average (please refer to [6, 7, 9] for these baselines). From the results, we can see that using implicit feedback approach also gets improvement.

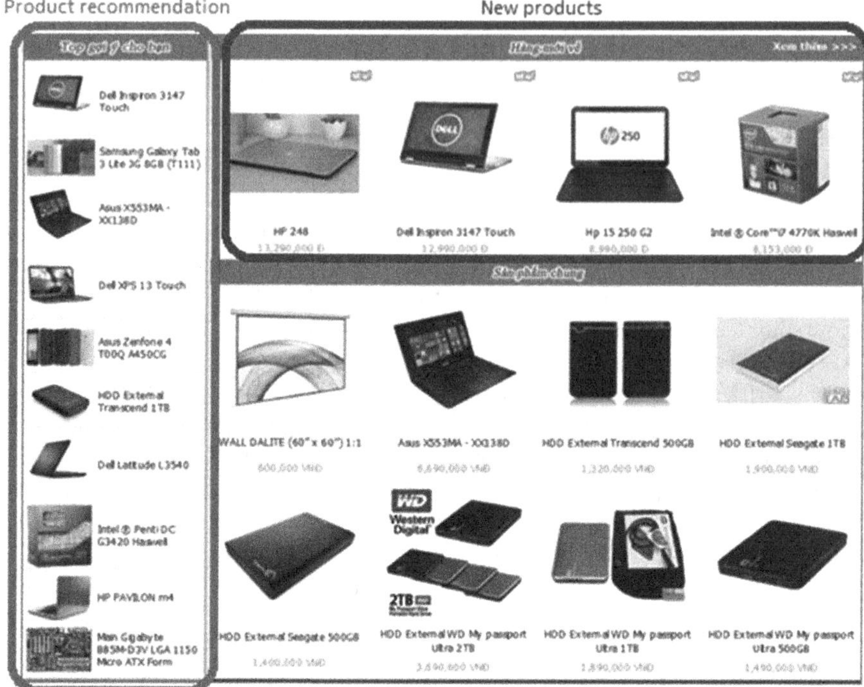

Fig. 3. Main interface of the product recommender system using implicit feedback

Table 1. MAE on single models ($\hat{r}_{1ui}, \hat{r}_{2ui}$) vs. combination model ($\hat{r}_{ui}$)

Test No.	$\hat{r}_{1ui}$	$\hat{r}_{2ui}$	$\hat{r}_{ui}$
1	0.4306	0.3533	0.1848
2	0.4399	0.3365	0.1922
3	0.4373	0.3466	0.1873
4	0.4284	0.3416	0.1917
5	0.4349	0.3424	0.1942
6	0.4386	0.3431	0.1883
7	0.4317	0.3553	0.1864
8	0.4317	0.3519	0.1837
9	0.4365	0.3486	0.1829
10	0.4470	0.3453	0.1869
Average	**0.4087**	**0.3464**	0.1878

Evaluation on the Effectiveness of the System: Besides evaluating the system using standard measure (e.g., MAE) as presented above, we have also evaluated the effectiveness of the recommendation to understand whether the

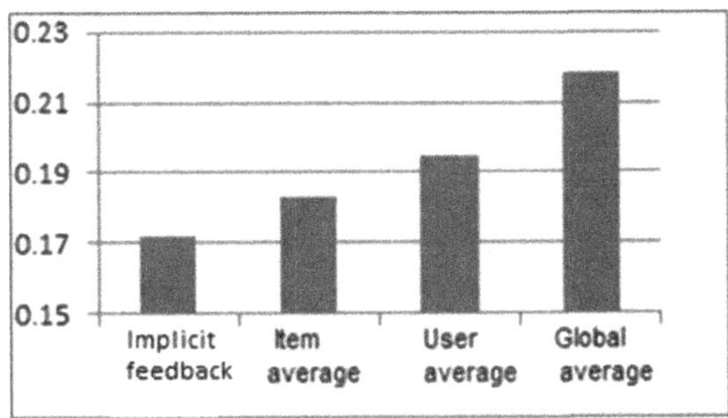

Fig. 4. Comparison with other baselines on a real data set using MAE

recommendation results are appropriate for the users. The recommendations are considered as suitable to the users if they select the products from the list of recommendation products [8]. This evaluation process is described in the following steps:

- Create a train and a test data set: For each user, we select 70 % of data for training and 30 % for testing.
- Build the models on the training set.
- Predict all items which are not in the train set for each user. Get Top-K (e.g., K = 15) products with the highest predicted values to check whether these predicted results are in the test set. If yes, this prediction is considered as suitable for the user.
- Repeat for all users in the test set.

For easy in visualization, we randomly selected 5 users (user 21 has 43 feedback; user 22 has 34 feedback; user 46 has 11 feedback; user 48 has 7 feedback; and user 56 has 6 feedback) to display the prediction results as described in Table 2. Please note that the evaluation process is done for 186 users as presented above and this visualization is just a small part of the data set.

We perform the experiments to check whether Top-15 recommendation products are in the test set for those 5 users. In Table 2, each double-column is a user including the number of products and product IDs which are correctly recommended to that user. For example, in the first test (first row in this table), their are 2 products which are correctly recommended to user 21 (product IDs: 134, 164), one for user 22, and one for user 46.

From these experimental results, we found that the accuracy of the recommendation is sensitive to the "frequent members" who have more interactions with the system than other "infrequent members". For example, users 21, 22, and 46 are frequent users (visited the system several times) so the recommendations for them are better than that of the infrequent users (users 48 and 56) since the system has not enough information to learn.

Table 2. Example of recommendation accuracy for 5 random users

Test No.	User 21		User 22		User 46		User 48		User 56	
	#Item	ItemID	#Item	ItemID	#Item	ItemID	#Item	ItemID	#Item	ItemID
1	**2**	**134, 164**	1	38	1	130	0		0	
2	2	134, 144	2	38, 70	1	130	1	33	0	
3	3	134,144, 164	22	38	1	30	0		0	
4	2	134,164	2	35, 38	1	130	0		0	
5	1	164	0		1	130	0		0	
6	1	164	1	38	1	130	0		0	
7	2	134, 164	1	158	2	130, 105	1	33	0	
8	1	134	0		1	130	1	33	0	
9	2	134, 164	1	38	1	130	0		0	
10	0		1	38	1	130	0		0	
AVG	90%		80%		100%		30%		0%	

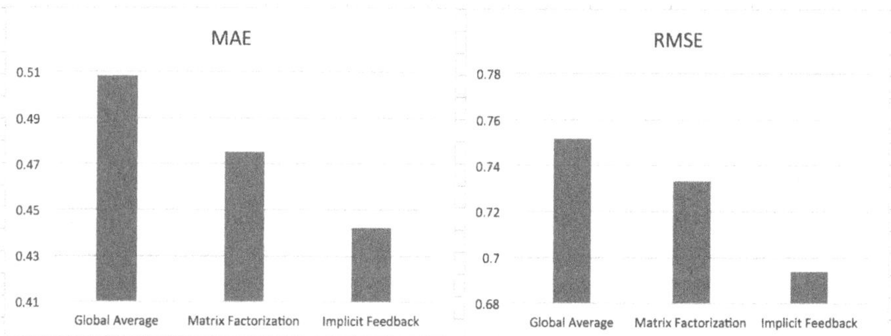

Fig. 5. Comparison results on a benchmark data set

Evaluation on Public Data Set: We have also compared the proposed approach, which is using implicit feedback, with other methods on a benchmark data set which is available at www.recsyswiki.com. The data set is called **D12 Ta-Feng** which is collected from a shopping system including 15,447 users, 1,780 items, and 178,216 ratings. Experimental results on the Mean Absolute Error (MAE) and Root Mean Squared Error (RMSE) are presented in Fig. 5. In these experiments, the proposed approach is also compared with the state-of-the-art matrix factorization [6]. Results show that using implicit feedback approach has lower error than other methods.

5 Conclusion

This study proposes an approach for building an online shopping recommender system using implicit feedback from the users. For building the system, first we propose a method to collect the users implicit feedback. Then, we propose an ensemble method which combine several extended matrix factorization models which are specialized for those implicit feedback. Next, we analyze, design, and implement an online system to integrate the aforementioned recommendation techniques. After having the system, we collect the feedback from the real users to validate the proposed approach. Results show that this approach is feasible and can be applied for the real systems.

References

1. Bauer, J., Nanopoulos, A.: Recommender systems based on quantitative implicit customer feedback. Decis. Support Syst. **68**, 77–88 (2014)
2. Bobadilla, J., Ortega, F., Hernando, A., GutiéRrez, A.: Recommender systems survey. Know.-Based Syst. **46**, 109–132 (2013)
3. Dietterich, T.G.: Ensemble methods in machine learning. In: Kittler, J., Roli, F. (eds.) MCS 2000. LNCS, vol. 1857, pp. 1–15. Springer, Heidelberg (2000)
4. Hu, Y., Koren, Y., Volinsky, C.: Collaborative filtering for implicit feedback datasets. In: Proceedings of the 2008 Eighth IEEE International Conference on Data Mining, pp. 263–272. IEEE Computer Society (2008)
5. Koren, Y.: Factorization meets the neighborhood: a multifaceted collaborative filtering model. In: Proceedings of the 14th ACM SIGKDD International Conference on Knowledge Discovery and Data Mining, KDD 2008, pp. 426–434. ACM, New York (2008)
6. Koren, Y., Bell, R., Volinsky, C.: Matrix factorization techniques for recommender systems. IEEE Comput. Soc. Press **42**(8), 30–37 (2009)
7. Ricci, F., Rokach, L., Shapira, B., Kantor, P.B. (eds.): Recommender Systems Handbook. Springer, USA (2011)
8. Shani, G., Gunawardana, A.: Evaluating recommendation systems. In: Ricci, F., Rokach, L., Shapira, B., Kantor, P.B. (eds.) Recommender Systems Handbook, pp. 257–297. Springer, USA (2011)
9. Thai-Nghe, N., Drumond, L., Horváth, T., Krohn-Grimberghe, A., Nanopoulos, A., Schmidt-Thieme, L.: Factorization techniques for predicting student performance. In: Santos, O.C., Boticario, J.G. (eds.) Educational Recommender Systems and Technologies: Practices and Challenges (ERSAT 2011). IGI Global (2011)
10. Thai-Nghe, N., Schmidt-Thieme, L.: Factorization forecasting approach for user modeling. J. Comput. Sci. Cybern. **31**(2), 133–148 (2015)

Improving Efficiency of Sentence Boundary Detection by Feature Selection

Thi-Nga Ho[1,2](✉), Tze Yuang Chong[1], Van Hai Do[3], Van Tung Pham[1,2], and Eng Siong Chng[1,2]

[1] School of Computer Engineering, Nanyang Technological University, 50 Nanyang Avenue, Singapore, Singapore
ngaht@ntu.edu.sg

[2] Temasek Laboratories@NTU, Nanyang Technological University, 50 Nanyang Drive, Singapore, Singapore

[3] Advanced Digital Sciences Center, 1 Fusionopolis Way, Singapore, Singapore

Abstract. The goal of sentence boundary detection (SBD) is to predict the presence/absence of sentence boundary in an unstructured word sequence, where there is no punctuation presented. In this paper, we propose a feature selection approach to obtain more effective features used for the SBD classifier. Specifically, the observed words are considered its correlation with the sentence boundary based on the pointwise mutual information before being used as the feature of the classifier. By using the linear chain CRF model to predict sentence boundaries of a text sequence, the experimental results on a part of the English Gigaword 2nd Edition corpus show that the proposed method helps to reduce the number of model parameters up to 44.87 % while maintaining a comparable F1-score to the original model.

Keywords: Sentence boundary detection · Feature selection · Structure event detection · Rich transcription

1 Introduction

Sentence boundaries are essential information in many text processing applications, such as natural language processing and machine translation. However, they are absent from several text sources. For example, pure speech recognition systems produce running text from speech signals without any punctuation. Hence, the Sentence Boundary Detection (SBD) task has received significant attentions from the research community in recent years [2,4,5,8,9].

To improve SBD performance, many types of input feature have been investigated, e.g. word identity, n-gram, Part-Of-Speech (POS) tags [1,2]. These features are typically modelled by frameworks such as Hidden Markov Model (HMM) [2], Maximum Entropy (MaxEnt) [4], Conditional Random Fields (CRF) [5,8,9]. Among these, CRF is a commonly used model, especially in dealing with text features [3,5,6]. The experiments in [8,9] showed that by including more

N.T. Nguyen et al. (Eds.): ACIIDS 2016, Part II, LNAI 9622, pp. 594–603, 2016.
DOI: 10.1007/978-3-662-49390-8_58

features, higher accuracy could be obtained. However, using more input features without selection may lead to redundancy and noise which results in increasing computational cost and required storage, and may also reduce system accuracy.

In this paper, we propose a feature selection approach to select a useful input feature set for the SBD task. The objective is to significantly reduce the input feature size while maintaining system accuracy. In the proposed approach, a feature selection module, named as trigger model, is employed to pre-process the input features. Only useful features are selected and further processed, redundant or noisy features are removed. In this study, we examine word identities as the original feature set. The useful words after feature selection, called triggers, are used as the input for the SBD model. To select triggers, pointwise mutual information (PMI) measurement [10], which reflects the correlation between word identities and sentence boundaries, is employed as a selection criterion. Subsequently, a linear-chain CRFs is used to predict sentence boundaries.

The rest of this paper is organized as follows. Section 2 presents related works to our proposed approach. Section 3 describes the proposed framework in details. Section 4 describes experimental setup. Section 5 shows the experimental results and discussion. Finally, Sect. 6 concludes our work and proposes several potential future directions.

2 Related Works

Feature selection is an essential process in many machine learning tasks [12]. Various feature selection strategies have been proposed, e.g. trying with several subset of features, placing a penalty to features based on their usefulness, measuring the relevance of features to the target labels [12]. For CRF models [15], the feature selection is much more important since they require global optimization for the whole input sequence. McCallum [13] introduced a feature induction algorithm for linear-chain CRFs. In this algorithm, feature conjunction is iteratively constructed to reduce the parameter count as well as increase the role of important features into the model. The work showed impressive results on increasing system accuracy while reducing the number of used features for name entity recognition and noun phase segmentation tasks.

In this work, we propose to use trigger model, the approach of capturing the distance correlations between features and target events, as a feature selection module for our SBD system. In the past, the trigger model was investigated in various tasks such as language modelling (LM) and spoken language understanding (SLU). In the LM task, Rosenfeld [11] incorporated a traditional language modelling method, named N-grams model, with a trigger model to form an interpolated model. The experimental results reported that the model perplexity is improved significantly. In the SLU task, Jeong et al. [14] used triggers to predict flight information. They concluded that, thanks to the capacity of capturing long distance correlation, triggers help to improve the performance when being adhoc to the existing local features. Triggers were also explored in the SBD task. Gavalda et al. [7] used only 30 triggers and POS tags to predict clause boundaries and achieved a 85.2 % F1-score on a small conversational corpus, which contains only 1669 clause boundaries in total after pre-processing. Although triggers have been investigated, to the best of our knowledge, we have not seen any research on using trigger model as a stand alone feature selection method for the SBD task.

3 Feature Selection for Sentence Boundary Detection

In this section, we present the proposed SBD framework in detail. The framework overview is described in Sect. 3.1. Trigger model and CRF model are presented in Sects. 3.2 and 3.3, respectively.

3.1 Framework Overview

Figure 1 illustrates the proposed framework for the SBD task. The input text sequence is first selected by a trigger model. Then the selected input, called as the triggers, are used as the input to the CRF model. The output sequence is the original input text with added sentence boundaries.

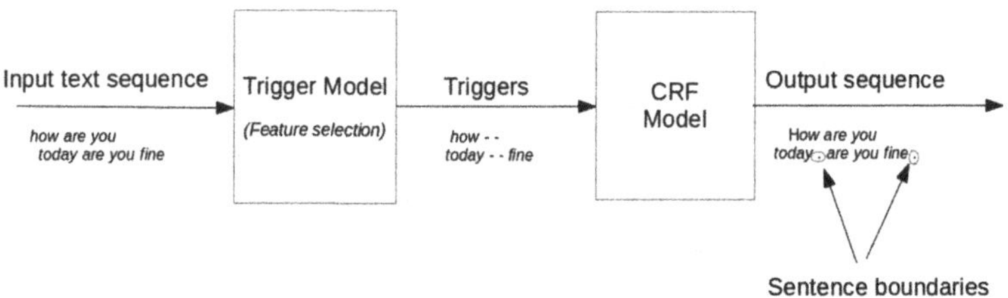

Fig. 1. The proposed SBD framework.

3.2 Trigger Model

We propose to use a feature selection module, called trigger model, to pre-process the input feature before passing them to a classifier for sentence boundary detection. In this module, we employ pointwise mutual information (PMI) [10] as the feature selection criterion.

Pointwise Mutual Information. The essential idea of the proposed feature selection is to employ a specific measurement as selection criterion to remove unnecessary features. Such measurement could be obtained using pointwise mutual information (PMI). The PMI measures the possibility of a pair of event x and event y occurring together in a data set. The PMI is defined as:

$$PMI(x,y) = \log \frac{p(x,y)}{p(x)p(y)} = \log \frac{p(x|y)}{p(x)} \quad (1)$$

where $p(x, y)$ is the joint distribution of x and y, $p(x)$ and $p(y)$ are marginal distribution, $p(x|y)$ is the conditional distribution of x given y. In this work, x refers to a word identity in the training set, y refers to the sentence boundary.

Trigger Selection. Trigger model is used to select useful features, i.e. triggers as shown in Fig. 1. To capture the "word - sentence boundary" relationship, for each sentence boundary, we examine the correlations between the boundary and all of its word neighbours. These word neighbours are limited within a context window. Specifically, with a context window size of n, the examined window includes $n/2$ words in the left context, and $n/2$ words in the right context of the boundary. Within the window, the PMI value between each word position and the boundary is calculated. Only words which have the PMI value higher than a pre-defined threshold are selected as triggers.

Figure 2 shows the trigger selection process for the context window size of 10 and PMI-threshold is 0.8. Two words, *said* and *addition*, which have PMIs, i.e. 0.937 and 1.240, higher than the threshold, hence they are considered as triggers. There is no trigger at other positions within the context window in this example. At each relative position to a boundary, the selected triggers form a trigger set. Therefore, within the context window size of n, n trigger sets are formed. The accurate pre-defined PMI threshold and window size, n need to be estimated on the given dataset for the best system performance.

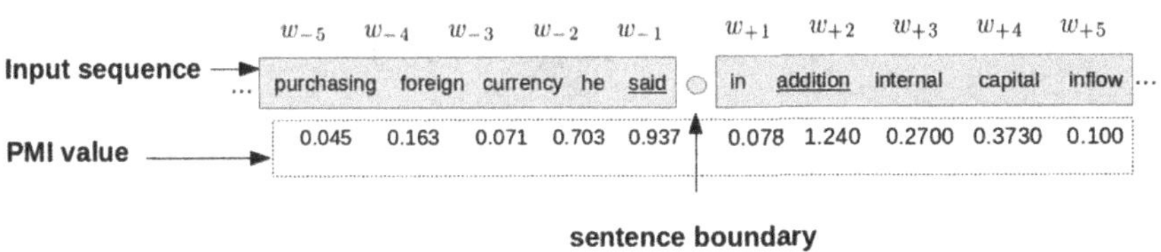

Fig. 2. An example of triggers selection in a context window of 10.

3.3 CRF Model

Triggers selected from the original words by the trigger model are passed to a linear-chain CRF to detect sentence boundaries. Linear-chain CRFs [15,16] are widely used for the sequence labelling tasks [12], and therefore the SBD tasks [5,8]. The Linear-chain CRFs equation could be defined as follows:

$$P_\lambda(y|x) = \frac{1}{Z(x)} \exp\left(\sum_m \lambda_m f_m(x_i, y_i)\right) \tag{2}$$

Where f_m are the binary feature functions, which equal to 1 when given conditions is satisfied, otherwise equal to 0. The λ_m are parameters estimated from training data, m is the number of observations and $Z(x)$ is the normalization factor to ensure the conditional probabilities over all possible outcomes sum up to 1.

For each input sequence, CRFs create a set of feature functions. During the learning process, different values are assigned to each feature function to weight the importance of that function to the formed model. In general, the number of feature functions, equivalent to the number of parameters that need to be estimated, proportionally increases with the number of unique input features. In a simple case, when only independent features are used as the input, the total number of parameters is:

$$X = NL \tag{3}$$

where N is number of unique input features and L is number of classes which is 2 in the SBD[1]. Therefore, the more number of unique features are used, the more parameters need to be estimated.

4 Experimental Setup

4.1 Evaluation Metrics

Following previous studies on the SBD task [5,8,9], we use standard F-measure metrics to evaluate our experimental results. The precision, recall and F-measure are defined as follows:

$$Precision = TP/(TP + FP) \tag{4}$$

$$Recall = TP/(TP + FN) \tag{5}$$

$$F - measure = \frac{(1 + \beta^2) * Precision * Recall}{\beta^2 * Recall + Precision} \tag{6}$$

Where *TP* (True Positive), *FN* (False Negative) and *FP* (False Positive) - are calculated based on the confusion matrix as in Table 1. Note that the total number of word boundaries is *TP* + *FN* + *FP* + *TN* and the total number of true sentence boundaries is *TP* + *FN*.

In this paper, β is set to 1 and F-measure becomes F1-score.

Table 1. The confusion matrix between the system output and reference.

	System True	System False
Reference True	*TP*	*FN*
Reference False	*FP*	*TN*

[1] The formula is explained in the document of this CRF++ Toolkit: https://taku910.github.io/crfpp/.

4.2 Data Description

To evaluate the proposed system, we use a subset of English Gigaword 2nd Edition (LDC2005T12)[2] released by the Linguistic Data Consortium (LDC), from The University of Pennsylvania, for our experiments. The whole corpus was archived from several news publishers, e.g. Agence France-Presses, English Service, the New York Times, Newswire Service in several years. In our experiments, only the news crawled from Agence France-Presses, English Service in 1995, is used. Detailed information of the used corpus is shown in Table 2.

Table 2. The dataset for experiments.

	# words	# sentence boundaries
Train	31.12 M	1.17 M
Development	3.81 M	144 k
Test	3.91 M	146 k

The three data subsets, i.e. train, development and test sets, are separately employed for different purposes. At first, the train set is used for training the model. Then, the development set is used for tuning parameters, e.g. choosing length of context window size and selecting an appropriate PMI threshold for each position in a context window. Finally, the model and the chosen PMI thresholds are tested on test set to see the efficiency.

5 Results and Discussion

To evaluate the effectiveness of the proposed feature selection, two major experiments are conducted. The first experiment evaluates the usefulness of feature selection at single context positions. Specifically, the feature selection is only applied on one context position at a time. The second experiment examines the feature selection at multiple context positions, i.e. feature selections are done simultaneously on various positions within the context window. The baseline and the two previously mentioned experiments are presented and discussed in the following sections.

5.1 Baseline

A baseline result, where all observed features are used to train the CRF, is used to compare with the results of our proposed approach. The baseline result is selected from the best F1-score achieved by varying the context window size n. The original words within the chosen context window are treated as independent

[2] https://catalog.ldc.upenn.edu/LDC2005T12.

features to the CRF model. Figure 3 shows the F1-scores on the development set for various context window size, from 2 to 12. It can be seen that when n is small, increasing n results in a significant improvement of system performance. However, this improvement is saturated when n reaches to 10. Thus, for simplification, we chose the performance at $n = 10$ as the baseline for further experiments. The F1-score for the baseline is 58.00 %. The corresponding F1-score on the test set is 62.37 %.

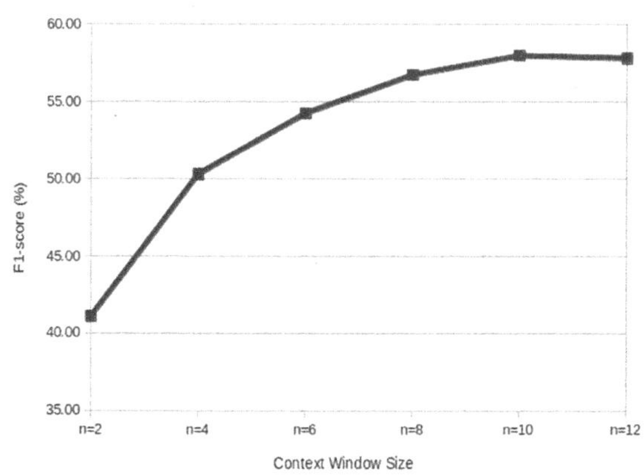

Fig. 3. Context window size varies from 2 to 12.

5.2 Results of Feature Selection at Single Context Positions

This experiment shows the SBD performance when applying feature selection at each single context positions within the context window size of 10, as described in Sect. 3.2. Specifically, within the context window of 10, the trigger selection is applied on only one specific position while the other positions keep it original words. Figure 4 presents the experimental results on the development set of the feature selection at the left and right contexts. The x-axis represents the PMI threshold value, while the y-axis represents F1-score. The "baseline" label in the x-axis refers to the baseline in Sect. 5.1, i.e. without feature selection. From the figure, it can be seen that the system performance is improved at all positions when feature selection is applied. Another important observation is that the closer the triggers are to sentence boundaries, the higher impact to the SBD performance. Specifically, the F1-scores for feature selection at position L1 and R1, where L and R denote the left and right respectively while the number 1 denote the context position, drop dramatically when the PMI threshold values increase, while the F1-scores of feature selection at position L5, R5 are almost stable.

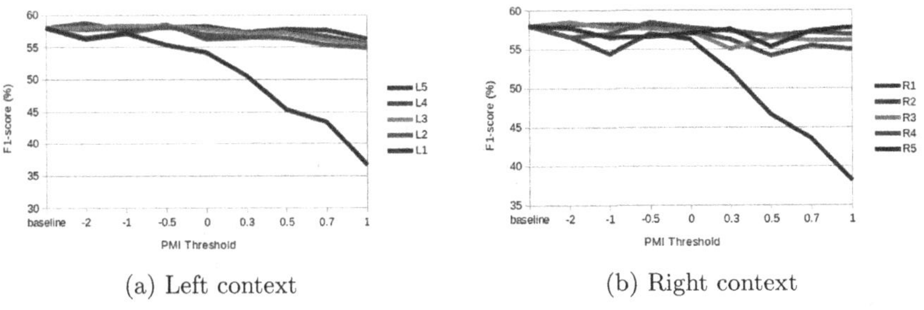

(a) Left context (b) Right context

Fig. 4. Feature selection at single context positions within context window size of 10.

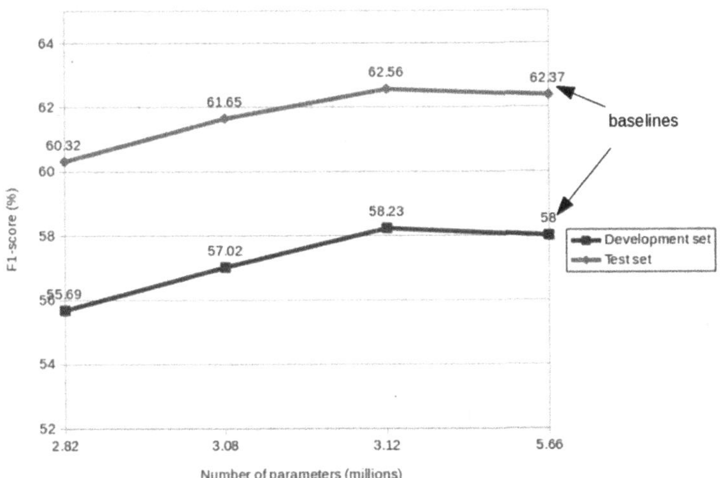

Fig. 5. The F1-score and number of model parameters obtained on the development set using the feature selection at L543R345 when varying the PMI threshold.

Table 3. The F1-score and number of model parameters on test set obtained using feature selection at L543R345.

Feature selection	F1-score(%)	# of parameters
Baseline	62.37	5.66 M
L543R345	62.56	3.12 M

5.3 Results of Feature Selection at Multiple Context Positions

This experiment evaluates the effectiveness of feature selection at multiple context positions. As showed in Sect. 5.2, the positions near sentence boundaries, i.e. L1, L2 and R1, R2, are important and very sensitive with the feature selection. Figure 4 shows that at positions L1, L2, R1 and R2, the F1-scores drop significantly when the higher PMI thresholds increase, meaning that the less triggers are selected. Thus, in this experiment, feature selection is only applied on the combination of context positions L5, L4, L3, R3, R4, R5, named as L543R345.

Since the optimal PMI thresholds for feature selection at single context positions might not be optimal when applying for multiple context positions, a new threshold needs to be examined in this experiment. Figure 5 shows the trade-off between F1-score and number of model parameters obtained on the development set and test set when varying the PMI threshold. Note that, the reducing in number of parameters reflects the increasing of global thresholds. The first point, where the number of parameters is 5.66 M, is the baseline. When applying the feature selection, a lot of words which never occur within the context window of 10, are not considered as triggers. This leads to a significant decrease in number of parameters, from 5.66 M to 3.12 M, while the F1-score increases. At the other points, where less triggers are used, although the numbers of parameters decrease, it costs a slight reduction in F1-scores.

The experimental results on test set, which correspond to the optimal PMI threshold on the development set, are presented in Table 3. We achieve 0.2 % absolute improvement of F1-score with a significant reduction of 44.87 % number of parameters. It can be concluded that the proposed feature selection is effective for multiple context positions.

6 Conclusion

In this work, we proposed a feature selection approach using the trigger model. The original word identities are preprocessed by the trigger model using the pointwise mutual information as the feature selection criterion. Experimental results on a subset of Gigaword dataset showed that, when feature selection is applied at single context positions, our approach achieves better SBD performance at all positions as compared to the original features. When feature selection is performed at multiple context positions, a dramatic reduction of number of features is achieved while the system performance is slightly improved. This work only examines feature selection on the independent feature, i.e. word identity. Whether the trigger model works on correlated features is still an open question. Another perspective we would like to explore in the future is how the trigger selection could work when multiple features, such as POS Tags or signal-based features, are included.

Acknowledgements. This work is supported by DSO funded project MAISON DSOCL14045.

References

1. Christensen, H., Gotoh, Y., Renals, S.: Punctuation annotation using statistical prosody models. In: Proceedings of ISCA Workshop on Prosody in Speech Recognition and Understanding, Red Bank, NJ, USA (2001)

2. Shriberg, E., Stolcke, A., Hakkani-Tür, D., Tür, G.: Prosody-based automatic segmentation of speech into sentences and topics. Speech Commun. **32**(1–2), 127–154 (2000)
3. Liu, Y., Stolcke, A., Shriberg, E., Harper, M.: Comparing and combining generative and posterior probability models: Some advances in sentence boundary detection in speech. In: Proceedings of EMNLP, Barcelona, Spain (2004)
4. Huang, J., Zweig, G.: Maximum Entropy model for punctuation annotation from speech. In: Proceedings of INTERSPEECH, Denver, Colorado, USA (2002)
5. Liu, Y., Stolcke, A., Shriberg, E., Hillard, D., Ostendorf, M., Harper, M.: Enrich speech recognition with automatic detection of sentence boundaries and disfluencies. IEEE Trans. Audio Speech Lang. Process. **14**(5), 1426–1540 (2006)
6. Kolář, J.: Automatic segmentation speech into sentence-like units. Ph.D. thesis, University of West Bohemia in Pilsen, Pilsen, Czech (2008)
7. Gavalda, M., Zechner, K., Aist, G.: High performance segmentation of spontaneous speech using part of speech and trigger word information. In: Proceedings of 5th Conference on Applied Natural Language Processing, Washington D.C., USA (1997)
8. Xu, C., Xie, L., Huang, G., Xiao, X., Chng, E.S., Li, H.: A deep neural network approach for sentence boundary detection in Broadcast News. In: Proceedings of INTERSPEECH 2014, Singapore (2014)
9. Huang, G.P., Xu, C., Xiao, X., Xie, L., Chng, E.S., Li, H.: Multi-view features in a DNN-CRF model for improved sentence unit detection on English Broadcast News. In: APSIPA 2014, Cambodia (2014)
10. Bouma, G.: Normalized (pointwise) mutual information in collocation extraction. In: Proceedings of the Biennial GSCL Conference 2009 (2009)
11. Rosenfeld, R.: A maximum entropy approach to adaptive statistical language modelling. Comput. Speech Lang. **10**(3), 187–228 (1996)
12. Dietterich, T.G.: Machine learning for sequential data: a review. In: Caelli, T.M., Amin, A., Duin, R.P.W., Kamel, M.S., de Ridder, D. (eds.) SPR 2002 and SSPR 2002. LNCS, vol. 2396, pp. 15–30. Springer, Heidelberg (2002)
13. McCallum, A.: Efficiently inducing features of conditional random fields. In: Proceedings of the 19th Conference on Uncertainty in Artificial Intelligence, pp. 403–410. Morgan Kaufmann Publishers Inc. (2002)
14. Jeong, M., Gary, G.L.: Practical use of non-local features for statistical spoken language understanding. Comput. Speech Lang. **22**(2), 148–170 (2008)
15. Lafferty, J., McCallum, A., Pereira, F.: Conditional random fields: probabilistic models for segmenting and labelling sequence data. In: Proceedings of ICML, pp. 282–28 (2001)
16. Sha, F., Pereira, F.: Shallow parsing with conditional random fields. In: Proceedings of HLT/NAACL (2003)

Modelling and Optimization Techniques in Information Systems, Database Systems and Industrial Systems

Pointing Error Effects on Performance of Amplify-and-Forward Relaying MIMO/FSO Systems Using SC-QAM Signals Over Log-Normal Atmospheric Turbulence Channels

Duong Huu Ai[1(✉)], Ha Duyen Trung[2], and Do Trong Tuan[2]

[1] VietNam Korea Friendship Information Technology College, Hoa Quy Ward, Ngu Hanh Son District, Da Nang, Vietnam
aidh@viethanit.edu.vn

[2] School of Electronics and Telecommunications, Hanoi University of Science and Technology, 405/C9, no. 1, Dai Co Viet, Hai Ba Trung District, Hanoi, Vietnam
{trung.haduyen, tuan.dotrong}@hust.edu.vn

Abstract. In this work, we analyze pointing error effects on performance of Amplify-and-Forward (AF) relaying multiple-input multiple-output (MIMO) free space optical (FSO) communication system employing subcarrier quadrature amplitude modulation (SC-QAM) signal over log-normal distributed atmospheric turbulence channels. We study the pointing error effect by taking into account the influence of beam-width, aperture size and jitter variance on the average symbol error rate (ASER), which is derived in closed-form expressions of MIMO/FSO and SISO/FSO systems. In addition, the number of relaying stations is taken into account in the statistical model of the combined channel including atmospheric loss, atmospheric turbulence and pointing error. The numerical results show that by combining AF relaying stations and MIMO/FSO configurations, the link length can be extended due to the transmitted power is reduced accordingly to the amplifier gain. Moreover, performance of AF relaying MIMO/FSO systems is better than that of AF relaying SISO/FSO systems at the same link length.

Keywords: AF · Atmospheric turbulence · ASER · FSO · QAM · Pointing error

1 Introduction

Free-space optics (FSO) is known as a line-of-sight (LOS) green communication technology, which can be used for a variety of applications ranging from high date-rate links of inter-building connections within a campus, high quality video surveillance and monitoring of a city, back-haul for next wireless mobile networks, disaster recovery links and ground to satellites [1]. The FSO's special characteristics are unlimited bandwidth, licensing-free requirements, high security, reduced interference,

N.T. Nguyen et al. (Eds.): ACIIDS 2016, Part II, LNAI 9622, pp. 607–619, 2016.
DOI: 10.1007/978-3-662-49390-8_59

cost-effectiveness and simplicity of communication system design and deployment [2]. However, there are numerous challenging issues when deploying FSO systems including the harmful effects of scattering, absorption, turbulence and the presence of pointing errors caused by misalignment between transmitter and receiver. They therefore severely impair the system performance in the link error probability [3–8]. To mitigate the impact of turbulence, multi-hop relaying FSO systems have been proposed as a promising solution to extend the transmission links and the turbulence-induced fading. Recently, performance of multi-hop relaying FSO systems over atmospheric turbulence channels has been studied in [9–11]. Moreover, recent studies have shown that, similar to wireless communications, the effect of turbulence fading with pointing errors can be significantly relaxed by using multiple-input multiple-output (MIMO) technique with multiple lasers at the transmitter and multiple photo-detectors at the receiver [12–18].

Previous works focus on MIMO FSO systems using On-Off keying (OOK) with intensity modulation and direct detection (IM/DD) and Pulse-position modulation (PPM) techniques. OOK is its simplicity and low-cost, widely used for FSO systems. However, OOK modulation needs an adaptive threshold to achieve its optimal performance, while PPM has poor bandwidth efficiency. To overcome the limitation of OOK and PPM modulation, FSO systems using sub-carrier (SC) intensity modulation schemes, such as sub-carrier quadrature amplitude modulation (SC-QAM), have been studied as the alternative modulation scheme for FSO systems [19–25]. All above studies, however, to the best of our knowledge, the pointing error effects on performance of AF relaying MIMO FSO using SC-QAM signals over log-normal atmospheric turbulence channels has not been clarified.

In this work, the ASER expressions of AF MIMO FSO systems in the log-normal atmospheric turbulence channel are analytically obtained taking into account the influence of pointing errors represented by beam-width, aperture size and jitter variance. The SC-QAM scheme is adopted for the performance analysis. Moreover, the number of relaying stations is included in the statistical model of the combined channel together with atmospheric loss, atmospheric turbulence and pointing error.

The remainder of the paper is organized into 6 sections: Sect. 2 introduces the system models, Sect. 3 discusses the atmospheric turbulence model of AF MIMO/FSO/SC-QAM systems with pointing error. Section 4 is devoted to ASER derivation of AF MIMO/FSO links. Section 5 presents the numerical results and discussion. The conclusion is reported in Sect. 6.

2 System Models

2.1 The AF Relaying SISO/FSO System Model

We start by investigating a typical serial multi-hop FSO system depicted in Fig. 1, which operates over independent and not identically distributed fading channels. The source node S can transmit data to the destination node D via multiple LOS free-space links arranged in an end-to-end configuration such that the source S communicates with the destination D through c relay stations R_1, R_2, . . ., R_{c-1}, R_c.

S → R_1 → R_2 → .... → R_{c-1} → R_c → D

Fig. 1. An illustration of a serial multi-hop relaying SISO/FSO system

It is assumed that all relay nodes concurrently receive and transmit signals in the same frequency band. In the Fig. 2, at the source node (Fig. 2 (a)), input data is first modulated into SC-QAM symbols at a subcarrier frequency f_c. The electrical SC-QAM signal at the output of electrical QAM modulator can be written as

$$e(t) = s_\mathrm{I}(t)cos(2\pi f_c t) - s_\mathrm{Q}(t)sin(2\pi f_c t). \tag{1}$$

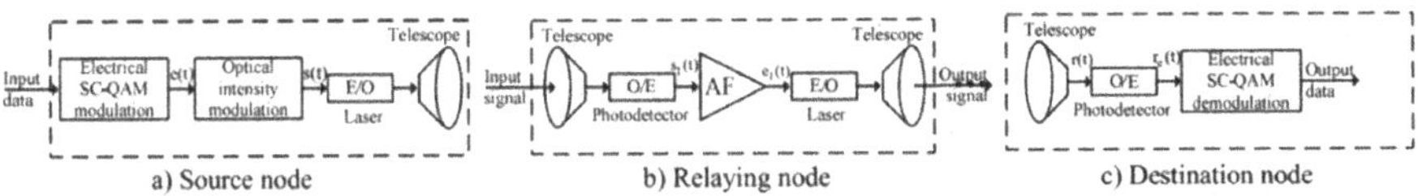

Fig. 2. A SISO/FSO system model: the source node, relay node and destination node

In Eq. (1), $s_\mathrm{I}(t) = \sum_{i=-\infty}^{i=+\infty} a_i(t)g(t - iT_s)$ and $s_Q(t) = \sum_{j=-\infty}^{j=+\infty} b_j(t)g(t - jT_s)$ are the in-phase signal and the quadrature signal, respectively. $a_i(t)$, $b_j(t)$ are the in-phase amplitude and the quadrature information amplitude of the transmitted data symbol, respectively, $g(t)$ denotes the shaping pulse and T_s is the symbol interval. The QAM signal is then used to modulate the intensity of an electrical-to-optical (E/O) laser before pointing laser beam through a telescope of the transmitter to the relaying node, the transmitted signal can be expressed as

$$s(t) = P_s\{1 + \kappa[s_I(t)\cos(2\pi f_c t - s_Q(t)\sin(2\pi f_c t)]\}, \tag{2}$$

where P_s denotes the average transmitted optical power per symbol at each hop and κ $(0 \leq \kappa \leq 1)$ is the modulation index. At the first relay node, the received optical signal can be written as

$$s_1(t) = XP_s\{1 + \kappa[s_I(t)\cos(2\pi f_c t - s_Q(t)\sin(2\pi f_c t)]\}. \tag{3}$$

In Eq. (3), X is the signal scintillation caused by log-normal atmospheric turbulence and pointing error. At each relay node, an AF module is used for signal amplification as illustrated in Fig. 2(b). The electrical signal output of the AF module at the first relay node will be

$$e_1(t) = \Re P_1 X P_s \kappa e(t) + \upsilon_1(t), \tag{4}$$

where $\Re$ is the photodiode (PD) responsivity, P_1 is the amplification power of the first AF module, $\upsilon_1(t)$ is the receiver noise that can be modeled as an additive white Gaussian noise (AWGN) process with power spectral density N_0.

Repeating such steps above through the number of relay stations, c, the electrical signal at the PD's output of the destination node can be obtained as

$$r_e(t) = P_s e(t)\left[\prod\nolimits_{i=0}^{c} X_{i+1}(t)\Re^{2i+1}P_i\right] + \sum\nolimits_{i=0}^{c} \upsilon_i(t), \tag{5}$$

where $\sum_{i=0}^{c} \upsilon_i(t)$ is the total receiver noise. X_{i+1} and P_i are the stationary random process for the turbulence channel and the average transmitted optical power per symbol at the ith relay station, respectively.

2.2 The AF Relaying MIMO/FSO System Model

Next, we consider a general AF relaying $M \times N$ MIMO/FSO system using SC-QAM signals with M-lasers pointing toward an N-aperture receiver as depicted. The source node, relaying node and destination node diagrams are presented in Fig. 3. The channel model of MIMO/FSO systems can be expressed by $M \times N$ matrix, which are denoted by $X = [\mathbf{X}_{mn}]_{m,n=1}^{M,N}$. The electrical signal at the input of QAM demodulator of the destination node can be expressed as

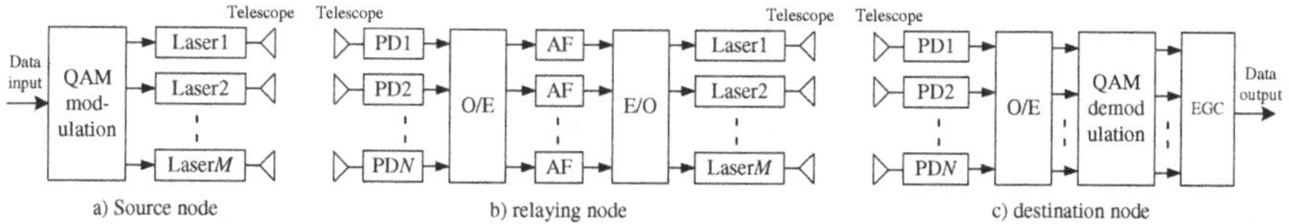

Fig. 3. An AF relaying MIMO/FSO system model

$$r_e(t) = P_s e(t)\left[\sum_{m=1}^{M}\sum_{n=1}^{N}\prod_{i=0}^{c} (X_{i+1})_{mn}\Re^{2i+1}P_i\right] + \sum_{i=0}^{c}\left(\sum_{m=1}^{M}\sum_{n=1}^{N} v_{mn}\right)_i (t), \tag{6}$$

where X_{mn} denotes the stationary random process of the turbulence channel from the mth laser to the nth PD. In this system model, we use an equal gain combining (EGC) scheme at the destination node to the estimate the received signal from sub-channels, the instantaneous electrical SNR will be

$$\gamma = \left(\sum_{m=1}^{M}\sum_{n=1}^{N}\prod_{i=0}^{c}\left(\sqrt{\gamma_{i_{mn}}}\right)\right)^2, \tag{7}$$

where $\gamma_{i_{mn}}$ is the random variable (r.v.) defined as the instantaneous electrical SNR component of the sub-channel from the mth laser to the nth PD, it can be described by

$$\gamma = \left(\frac{1}{MN}\kappa\Re^{2i+1}P_s\prod_{i=1}^{c}X_{i+1}P_i\right)^2/N_0 = \overline{\text{SNR}}\left(\prod_{i=0}^{c}X_{i+1}\right)^2 = \bar{\gamma}\left(\prod_{i=0}^{c}X_{i+1}\right)^2. \quad (8)$$

3 Log-Normal Atmospheric Turbulence with Pointing Error

In Eqs. (7) and (8), X represents the channel state, which models the optical intensity fluctuations caused by atmospheric loss X_l, atmospheric turbulence induced fading X_a and pointing error X_p. They can be described as

$$X = X_l X_a X_p. \quad (9)$$

3.1 Atmospheric Loss

Firstly, atmospheric loss X_l is no randomness and a deterministic component. Therefore, it acts as a fixed scaling factor over a long time period and modeling in [2] as

$$X_l = e^{-\sigma \mathrm{L}}, \quad (10)$$

where σ denotes a attenuation coefficient and L is the link length.

3.2 Log-Normal Atmospheric Turbulence

Secondly, the most widely used model for the case of weak atmospheric turbulence regime is the log-normal distribution that has been validated by studies [2, 3]. The pdf of the irradiance intensity of log-normal channel is given as

$$f_{X_a}(X_a) = \frac{1}{X_a\sigma_I\sqrt{2\pi}}\exp\left(-\frac{[\ln(X_a)+0.5\sigma_I^2]^2}{2\sigma_I^2}\right), \quad (11)$$

where $\sigma_I^2 = (\exp(\omega_1+\omega_2)-1)$ is the log intensity, ω_1 and ω_2 are defined as

$$\omega_1 = \frac{0.49\sigma_2^2}{\left(1+0.18d^2+0.56\sigma_2^{12/5}\right)^{7/6}} \quad \text{and} \quad \omega_2 = \frac{0.51\sigma_2^2(1+0.69\sigma_2^{12/5})^{-5/6}}{1+0.9d^2+0.62\sigma_2^{12/5}}. \quad (12)$$

In Eq. (12), $d = \sqrt{kD^2/4L}$, $k = 2\pi/\lambda$ is the wave number, λ is the wavelength, D is the receiver aperture diameter, and σ_2 is the Rytov variance, which is defined as

$$\sigma_2 = 0.492C_n^2k^{7/6}L^{11/6}, \quad (13)$$

where C_n^2 is the refractive-index structure parameter.

The pdf of (c + 1) turbulence channels, X^{c+1}, for AF relaying MIMO/FSO systems will be

$$f_{\mathbf{X}_{mn}}\left(X^{c+1}\right) = \frac{1}{(c+1)X^{c+1}\sigma_I\sqrt{2\pi}}\exp\left(-\frac{\left[\ln(X)+0.5\sigma_I^2\right]^2}{2\sigma_I^2}\right). \tag{14}$$

3.3 Pointing Error

Thirdly, a statistical model of pointing error induced fading channel is developed in [7, 8], the pdf of X_p is given as [7]

$$f_{X_p}(X_p) = X_p^{\xi^2-1}\left(\xi^2 / A_0^{\xi^2}\right), \qquad 0 \le X_p \le A_0, \tag{15}$$

where $A_0 = [\mathrm{erf}(v)]^2$ is the fraction of the collected power at radial distance 0, the parameter v is given by $v = \sqrt{\pi}r/(\sqrt{2}\omega_z)$ with r and ω_z respectively denote the aperture radius and the beam waist at the distance z, and $\xi = \omega_{zeq}/2\sigma_s$, where the equivalent beam radius can be represented by [7]

$$\omega_{zeq} = \omega_z\left(\sqrt{\pi}\mathrm{erf}(v)/2v\exp(-v^2)\right)^{1/2}. \tag{16}$$

In Eq. (16), $\omega_z = W_0\left(1+\varepsilon(\lambda L/\pi W_0^2)^2\right)^{1/2}$, where W_0 is the transmitter beam waist radius at z = 0, $\varepsilon = (1+2W_0^2)/\rho_0^2$ and $\rho_0 = (0.55C_n^2k^2L)^{-3/5}$ is the coherence length [8].

3.4 Combined Channel Model

Finally, we derive a completed statistical model of the channel considering the combined effect of atmospheric lost, atmospheric turbulence and pointing error. The unconditional pdf of the combined channel state is expressed as [8]

$$f_X(X) = \int f_{X|X_a}(X|X_a)f_{X_a}(X_a)dX_a, \tag{17}$$

where $f_{X|X_a}(X|X_a)$ denotes the conditional probability of given atmospheric turbulence state, which can be defined by [8]

$$f_{X|X_a}(X|X_a) = \frac{1}{X_aX_l}f_{X_p}\left(\frac{X}{X_aX_l}\right). \tag{18}$$

As a result, we can derive the unconditional pdf of X can be derived as

$$f_X(X) = \frac{\xi^2}{(c+1)(A_0 X_l)^{\xi^2}} X^{\xi^2 - 1} \int_{(X/X_l A_0)}^{\infty} \frac{1}{X_a^{\xi^2 + c + 1} \sigma_l \sqrt{2\pi}} \exp\left\{ \frac{[\ln(X_a) + 0.5\sigma_l^2]^2}{2\sigma_l^2} \right\} dX_a. \tag{19}$$

To simplify, we can let $t = (\ln(X_a) + a)/(\sqrt{2}\sigma_l)$, as the result Eq. (19) can be obtain in the closed-form expression as

$$f_X(X) = \frac{\xi^2}{(c+1)(A_0 X_l)^{\xi^2}} X^{\xi^2 - 1} \frac{1}{2} e^b \times \mathrm{erfc}\left(\frac{\ln(X/X_l A_0) + a}{\sqrt{2}\sigma_l} \right), \tag{20}$$

where $a = 0.5\sigma_l^2 + \sigma_l^2(\xi^2 + c)$ and $b = \sigma_l^2(\xi^2 + c)\{1 + (\xi^2 + c)\}/2$.

4 Average Symbol Error Rate

Now we can derive the average symbol error rate of the MIMO/FSO systems using SC-QAM signals under the effect of atmospheric turbulence and pointing error. The system's ASER, $\overline{P}_{se}^{MIMO}$, can be generally expressed as

$$\overline{P}_{se}^{MIMO} = \int_{\Gamma} P_e(\gamma) \times f_\Gamma(\Gamma) d\Gamma, \tag{21}$$

where $P_e(\gamma)$ is the conditional error probability (CEP), $\Gamma = \{\Gamma_{nm},\ n = 1, \ldots, N,\ m = 1, \ldots, M\}$ is the matrix of the MIMO FSO channels. When using SC-QAM signals for modulating the data symbol, the CEP can be given as [21]

$$P_e(\gamma) = 2q(M_I)Q(A_I\sqrt{\gamma}) + 2q(M_Q)Q(A_Q\sqrt{\gamma}) - 4q(M_I)q(M_Q)Q(A_I\sqrt{\gamma})Q(A_Q\sqrt{\gamma}), \tag{22}$$

where M_I and M_Q are in-phase and quadrature signal amplitudes, respectively; $q(x) = 1 - x^{-1}$, $Q(\mathrm{x})$ is the Gaussian Q-function, A_I and A_Q are defined as

$$A_I = \left(\frac{6}{\left[(M_I^2 - 1) + r^2(M_Q^2 - 1)\right]} \right)^{1/2} \text{ and } A_Q = \left(\frac{6r^2}{\left[(M_I^2 - 1) + r^2(M_Q^2 - 1)\right]} \right)^{1/2}, \tag{23}$$

where $r = d_Q/d_I$ is the quadrature to in-phase decision distance ratio.

Let us assume that MIMO/FSO sub-channels' turbulence processes are uncorrelated, independent and identically distributed (iid). According to Eqs. (8) and (20), we obtain the pdfs of AF relaying MIMO/FSO systems over log-normal channel as

$$f_{\Gamma_{mn}}\left(\gamma_{mn}^{\frac{c+1}{2}}\right) = \frac{\xi^2}{2(c+1)(A_0X_l)^{\xi^2}} \frac{\gamma_{mn}^{0.5\xi^2-1}}{\gamma_{mn}^{0.5\xi^2}} \frac{1}{\sqrt{\pi}} e^b \times \mathrm{erfc}\left(\frac{0.5\ln\left(\frac{\gamma_{mn}}{X_l^2A_0^2\bar{\gamma}_{mn}}\right)+a}{\sqrt{2}\sigma_l}\right). \tag{24}$$

Substituting Eqs. (22) and (24) into Eq. (21), we obtain the systems' ASER as

$$\begin{aligned}\overline{P}_{se}^{MIMO} &= 2q(M_I)\int_\Gamma Q(A_I\sqrt{\gamma})f_\Gamma(\Gamma)d\Gamma + 2q(M_Q)\int_\Gamma Q(A_Q\sqrt{\gamma})f_\Gamma(\Gamma)d\Gamma \\ &- 4q(M_I)q(M_Q)\int_\Gamma Q(A_I\sqrt{\gamma})Q(A_Q\sqrt{\gamma})f_\Gamma(\Gamma)d\Gamma.\end{aligned} \tag{25}$$

5 Numerical Results and Discussion

As mentioned above, using the derived expressions, Eqs. (24) and (25), we present numerical results of ASER performance of the MIMO/FSO systems. The ASER can be calculated via multi-dimensional numerical integration with the help of the Matlab™ software. Systems' parameters are provided in Table 1.

Table 1. Systems' parameters

Parameter	Symbol	Value
Laser wavelength	λ	1550 nm
Photodetector responsivity	$\Re$	1 A/W
Modulation index	κ	1
Total noise variance	N_0	10^{-7}A/Hz
In-phase, Quadrature signal amplitudes	M_I, M_Q	8, 4
The number of relay stations	c	0, 1, 2
Index of refraction structure	C_n^2	$10^{-15}\text{m}^{-2/3}$

In Figs. 4 and 5, the system's ASER is presented as a function of transmitter beam waist radius for various values of relay stations. Figure 4 illustrates the ASER against W_0 for various values of the pointing error displacement standard deviation $\sigma_s =$ 0.18 m, 0.02 m and 0.22 m. Whereas, Fig. 5 shows the ASER versus W_0 for various values of aperture radius $r = 0.074$ m, 0.075 m and 0.076 m. From these figures it can be seen that for a given condition including specific values of the number relay stations, aperture radius and average SNR, the best ASER performance can be obtained at a specific value of W_0, ranging from 0.02 to 0.025 m. This optimal value of W_0 is the optimal transmitter beam waist radius, the more the value of transmitter beam waist radius comes close to the optimal one, the lower the value of system's ASER achieved.

Figures 6 and 7 illustrate the ASER performance versus the pointing error displacement standard deviation of the different AF relaying MIMO/FSO systems. More specifically, we compare the ASER performance of 2 × 2 and 4 × 4 MIMO/FSO configurations with SISO/FSO system, for various values of transmitter beam waist and

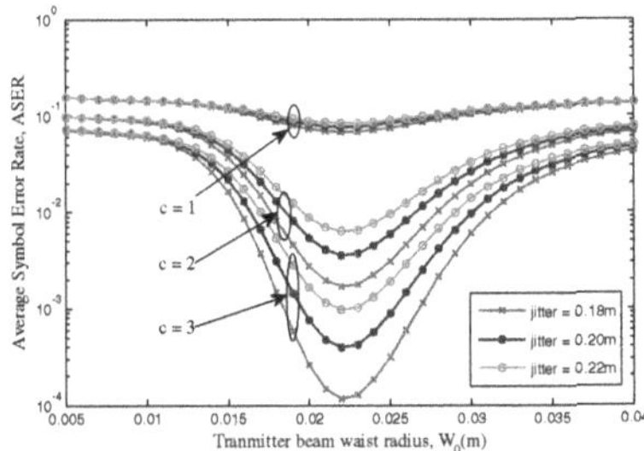

Fig. 4. ASER versus transmitter beam waist radius under various values of σ_s.

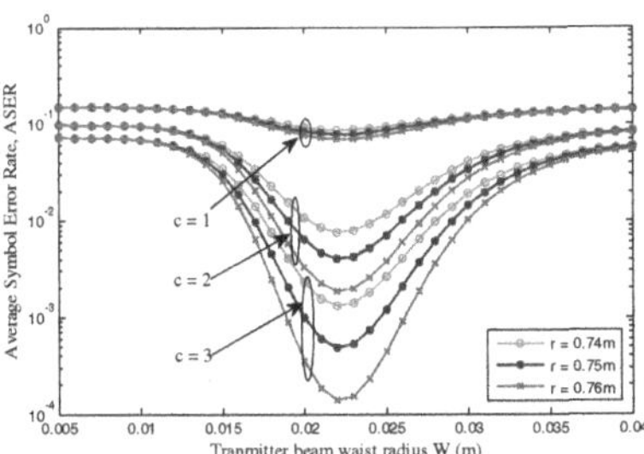

Fig. 5. ASER versus transmitter beam waist radius under various values of aperture radius with $\sigma_s = 0.35$ m.

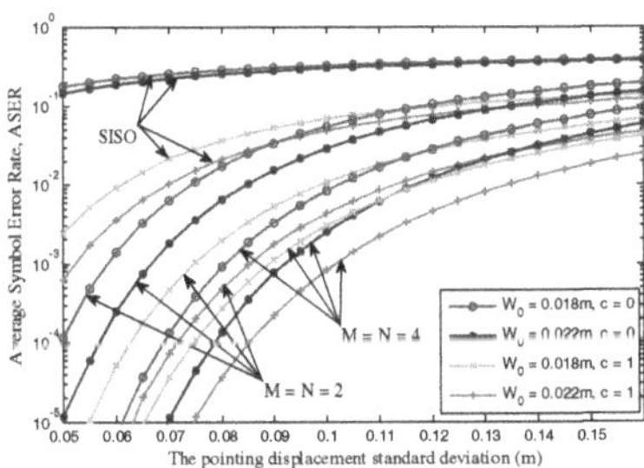

Fig. 6. ASER versus the pointing error displacement standard deviation under various values of transmitter beam waist radius, $r = 0.055$ m, and SNR = 25 dB.

aperture radius, the link distance $L = 1000$ m and SNR = 25 dB. It is also noted that the amplifier gain at each relay station is set by 3.5 dB. The system's ASER is significantly decreases when the pointing error displacement standard deviation decreases with the same MIMO/FSO configuration and the number of relay station. The impact of the aperture radius and the transmitter beam waist radius on the system's performance is more significant in low σ_s values than in high σ_s values. We thus can more easily adjust the system performance when the pointing error displacement standard deviation is low

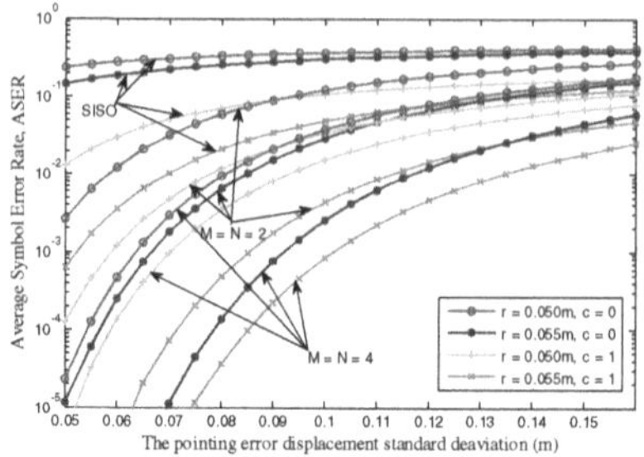

Fig. 7. ASER versus the pointing error displacement standard deviation under various values of aperture radius, $W_0 = 0.022$ m, SNR = 25 dB.

by appropriately changing the values of the aperture radius and the transmitter beam waist radius.

Figures 8 and 9 illustrate the ASER performance versus the aperture radius under various the numbers of relay stations and MIMO/FSO configurations. As the result, the system's ASER is significantly decreases when the values of aperture radius and number relay stations increase. It can be found that, in the low-value region when aperture radius increases, system's ASER is not much change. However, when aperture

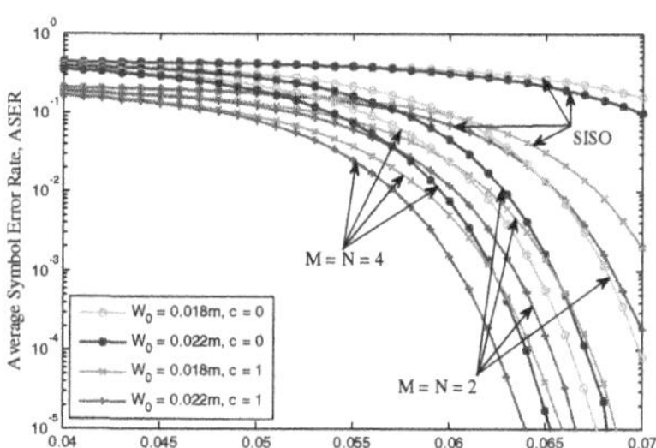

Fig. 8. ASER versus the aperture radius under various values of transmitter beam waist radius with $\sigma_s = 0.16$ m, SNR = 25 dB.

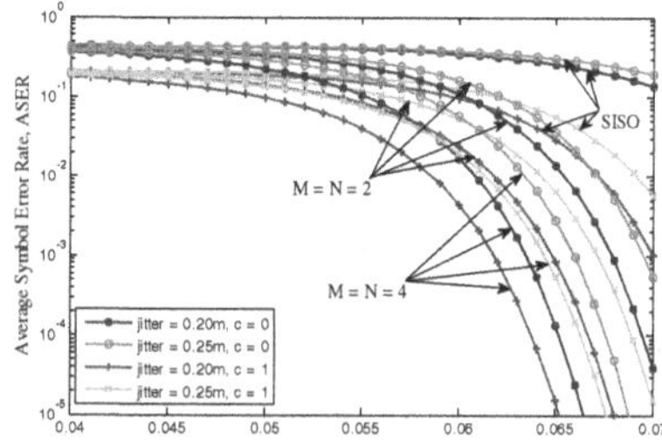

Fig. 9. ASER versus the aperture radius under various values of σ_s, $W_0 = 0.022$ m, SNR = 25 dB.

radius exceeded the threshold value ASER plummeted when aperture radius increases. More clearly, we could not control the pointing error displacement standard deviation but we could control the aperture radius, number relay stations, the transmitter beam waist radius and MIMO/FSO systems.

In Fig. 10, the system's ASER is presented as a function of average SNR under various values of the transmitter beam waist radius, the number relay station $c = 1$ and the aperture radius $r = 0.055\,\mathrm{m}$. Besides, comparison between 2×2 and 4×4 MIMO/FSO systems and SISO/FSO system, is performed. It can be observed from Fig. 10 that the ASER decreases with the increase of the SNR, number of relay stations and MIMO/FSO configuration. The best ASER performance is achieved when the optimal beam waist radius of 0.022 m is applied. It can be confirmed that simulation results are closed agreement with analytical results. Finally, Fig. 11 depicts the ASER performance as function of the transmission link distance L for various number of relay stations $c = 0$, $c = 1$, and $c = 2$. It can be seen from the figure that the ASER increases when the transmission link distance is longer. In addition, when using relay stations combined MIMO/FSO systems, ASER will get better performance.

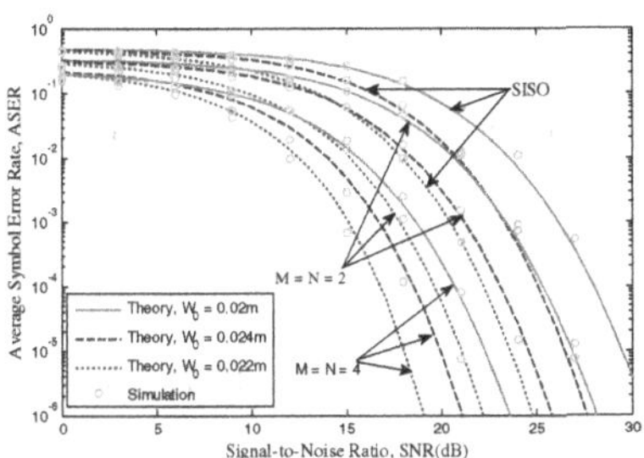

Fig. 10. ASER versus average SNR with pointing error under various values of W_0, $\sigma_s = 0.16\,\mathrm{m}$, the number of relay stations $c = 1$ and $r = 0.055\,\mathrm{m}$.

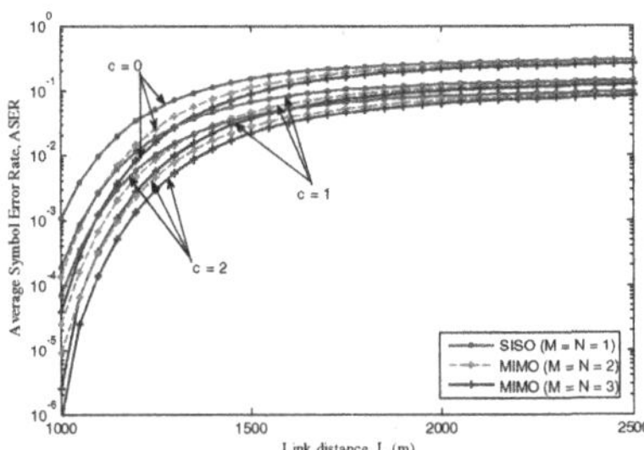

Fig. 11. The ASER versus transmission link distance under various number of relay stations with the aperture radius $r = 0.055\,\mathrm{m}, \sigma_s = 0.16\,\mathrm{m}$.

6 Conclusion

We have studied the AF relaying MIMO/FSO scheme using SC-QAM signals over log-normal atmospheric turbulence channels with pointing error. We have derived theoretical expressions for ASER performance of SISO and MIMO systems taking into account the number of AF relay stations, MIMO configurations, and the pointing error effect. The numerical results showed the impact of pointing error on the system's performance. By analyzing ASER performance, we can conclude that using proper values of aperture radius, transmitter beam waist radius, be partially surmounted pointing error and number relay stations combined with MIMO/FSO configurations could greatly benefit the performance of the such systems.

References

1. Kedar, D., Arnon, S.: Urban optical wireless communication networks: the main challenges and possible solutions. IEEE Commun. Mag. **42**, S2–S7 (2004)
2. Majumdar, A.K., Ricklin, J.C.: Free-Space Laser Communications: Principles and Advances. Springer, New York (2008)
3. Andrews, L., Phillips, R., Hopen, C.: Laser Beam Scintillation with Applications. SPIE Press, Bellingham (2001)
4. Garca-Zambrana, A., Castillo-Vzquez, C., Castillo-Vzquez, B.: Outage performance of MIMO FSO links over strong turbulence and misalignment fading channels. Opt. Express **19**, 13480–13496 (2011)
5. Lee, E., Ghassemlooy, Z., Ng, W.P., Uysal, M.: Performance analysis of free space optical links over turbulence and misalignment induced fading channels. In: Proceedings of IET International Symposium on Communication Systems, Networks and Digital Signal Processing (2012)
6. Ahmed, A., Hranilovic, S.: Outage capacity optimization for free-space optical links with pointing error. J. Lightw. Technol. **25**, 1702–1710 (2007)
7. Arnon, S.: Effects of atmospheric turbulence and building sway on optical wireless communication systems. Opt. Lett. **28**, 129–131 (2003)
8. Trung, H.D., Tuan, D.T., Anh, T.P.: Pointing error effects on performance of free-space optical communication systems using SC-QAM signals over atmospheric turbulence channels. AEU-Int. J. Elec. Commun. **68**, 869–876 (2014)
9. Safari, M., Uysal, M.: Relay-assisted free-space optical communication. IEEE Trans. Wirel. Commun. **7**, 5441–5449 (2008)
10. Tsiftsis, T.A., Sandalidis, H.G., Karagiannidis, G.K., Sagias, N.C.: Multihop free-space optical relaying communications over strong turbulence channels. In: IEEE International Conference on Communication, pp. 2755–2759 (2006)
11. Datsikas, C.K., et al.: Serial free-space optical relaying communications over Gamma-Gamma atmospheric turbu-lence channels. IEEE/OSA J. Optical Commun. Netw. **2**, 576–586 (2010)
12. Wilson, S.G., Brandt-Pearce, M., Cao, Q., Leveque, J.H.: Free-space optical MIMO transmission with Q-ary PPM. IEEE Trans. Commun. **53**, 1402–1412 (2005)
13. Takase, D., Ohtsuki, T.: Optical wireless MIMO communications (OMIMO). In: IEEE Global Telecommunication Conference (GLOBECOM 2004), pp. 928–932 (2014)

14. Takase, D., Ohtsuki, T.: Performance analysis of optical wireless MIMO with optical beat interference. In: IEEE Internatioanl Conference on Communication (ICC2005), pp. 954–958 (2005)
15. Takase, D., Ohtsuki, T.: Spatial multiplexing in optical wireless MIMO communications over indoor environment. Trans. IEICE **E89-B**, 1364–1371 (2006)
16. Takase, D., Ohtsuki, T.: Optical wireless MIMO (OMIMO) with backward spatial filter (BSF) in diffuse channels. In: IEEE International Conference on Communication (ICC2007), pp. 2462–2467 (2007)
17. Trung, H.D., Pham, T.A.: Performance analysis of MIMO/FSO systems using SC-QAM over atmospheric turbulence channels. IEICE Trans. Fund. Elec. Commun. Comput. Sci. **1**, 49–56 (2014)
18. Aminikashani, M., Kavehrad, M.: On the performance of MIMO FSO communications over double generalized gamma fading channels. In: IEEE ICC (2015)
19. Popoola, W., Chassemlooy, Z.: BPSK subcarrier intensity modulated free-space optical communications in atmospheric turbulence. J. Light. Tech. **27**, 967–973 (2009)
20. Trung, H.D., Bach, T.V., Anh, T.P.: Performance of free-space optical MIMO systems using SC-QAM over atmospheric turbulence channels. Proc. IEEE ICC **2013**, 3846–3850 (2013)
21. Peppas, K.P., Datsikas, C.K.: Average symbol error probability of general order rectangular QAM of optical wireless communication systems over atmospheric turbulence channels. J. Optical Commun. Netw. **2**, 102–110 (2010)
22. Prabu, K., Kumar, D.S., Malekian, R.: BER analysis of BPSK-SIM-based SISO and MIMO FSO systems in strong turbulence with pointing errors. Int. J. Light Electron Opt. **125**, 6413–6417 (2014)
23. Hassan, M.Z., Song, X., Cheng, J.: Subcarrier intensity modulated wireless optical communications with rectangular QAM. J. Opt. Commun. Netw. **4**, 522–532 (2012)
24. Trung, H.D., Tuan, D.T.: Performance of free-space optical communications using SC-QAM signals over strong atmospheric turbulence and pointing errors. In: Proceedings of IEEE ICCE14 pp. 42–47 (2014)
25. Trung, H.D., Tuan, D.T.: Performance of amplify-and-forward relaying MIMO free-space optical systems over weak atmospheric turbulence channels. In: Proceedings of IEEE NICS15, pp. 16–18 (2015)

Portfolio Optimization by Means of a χ-Armed Bandit Algorithm

Mahdi Moeini(✉), Oliver Wendt, and Linus Krumrey

Chair of Business Information Systems and Operations Research (BISOR), Technical University of Kaiserslautern, Postfach 3049, Erwin-Schrödinger-Str., 67653 Kaiserslautern, Germany
{mahdi.moeini,wendt}@wiwi.uni-kl.de, linus.krumrey@googlemail.com

Abstract. In this paper, we are interested in studying and solving the portfolio selection problem by means of a machine learning method. Particularly, we use a χ-armed bandit algorithm called Hierarchical Optimistic Optimization (HOO). HOO is an optimization approach that can be used for finding optima of box constrained nonlinear and nonconvex functions. Under some restrictions, such as locally Lipschitz condition, HOO can provide global solutions. Our idea consists in using HOO for solving some NP-hard variants of the portfolio selection problem. We test this approach on some data sets and report the results. In order to verify the quality of the solutions, we compare them with the best known solutions, provided by a derivative-free approach, called DIRECT. The preliminary numerical experiments give promising results.

Keywords: Portfolio selection problem · Machine learning · Multi-armed bandit · Derivative-free optimization

1 Introduction

Suppose that a certain amount of money (as capital) and a set of assets are given. The objective of the portfolio selection problem consists in finding an optimal way in order to diversify the capital among these assets. A portfolio is any combination of these assets [4]. Each portfolio carries an amount of return and is associated with a certain level of risk. Any portfolio that provides the minimum level of risk for a given amount of return is considered to be an optimal (or efficient) portfolio. The Mean-Variance (MV) model of Markowitz [8,9] is one of the classical models for finding the optimal portfolios [2–5,11]. In this model, the risk is defined as the variance of the returns. However, Markowitz's model does not consider typical practical constraints such as *buy-in threshold* [2,11,12] or *round-lot* constraints [3,5]. The buy-in threshold constraints prevent the investor from investing very small amounts in a single asset. The round-lot constraints require the invested fractions to be a multiple of an integer (i.e., the round-lot). These constraints are useful in practice, because most assets must be bought in integer quantities rather than arbitrary (real-valued) fractions.

N.T. Nguyen et al. (Eds.): ACIIDS 2016, Part II, LNAI 9622, pp. 620–629, 2016.
DOI: 10.1007/978-3-662-49390-8_60

Introducing these realistic constraints into the MV model produces NP-hard problems that are difficult to solve, even for moderate instances.

In this study, we formulate the portfolio selection problems under *buy-in threshold* and *round-lot* constraints in form of box-constrained nonconvex optimization problems. Then, we investigate a machine learning approach called *Hierarchical Optimistic Optimization (HOO)* algorithm for solving the resulting mathematical models. Machine learning methods and, in particular, bandit learning have already been used in portfolio optimization [14]. However, this is the first time that a machine learning approach, and in particular HOO, is used for solving portfolio selection problem under *buy-in threshold* and *round-lot* constraints. In order to verify the efficiency of HOO, we use the instances that have been already used in [2,3] and compare the results by the optimal values. According to our numerical results, HOO provides promising results.

We organize this paper as follows: In Sect. 2 the notations and the models are introduced and explained. Section 3 is devoted to presenting the HOO algorithm which is applied for solving the portfolio optimization problems. In Sect. 4 the numerical results of our experiments are presented. Finally, some conclusions are drawn in Sect. 5.

2 The Portfolio Selection Problem

Before going deeper into the optimization problems, we present the notations that we want to use in this paper. Let n be the number of available assets. We want to diversify our capital over these assets such that x_i shows the fraction of the total capital invested in asset i (for $i = 1, \dots, n$). The risk of each portfolio is defined by $V(\mathbf{x}) = \mathbf{x}^t Q \mathbf{x}$, where Q is the $n \times n$ variance-covariance matrix of historical returns and $\mathbf{x} = (x_1, \dots, x_n)^t$. We show the mean return of the assets by $\mathbf{r}$. The objective of the following model consists in finding the solution portfolio $\mathbf{x}$ such that the risk $V(\mathbf{x})$ is minimized and the expected return R is reached:

$$\min \{V(\mathbf{x}) : \mathbf{r}^t\mathbf{x} = R, \mathbf{e}^t\mathbf{x} = 1, x_i \geq 0 : i = 1, \dots, n\}, \tag{1}$$

where $\mathbf{e}$ is the n-dimensional vector which consists solely of ones. In this model, the solution must have a sum of one (i.e., the complete capital must be invested) and short-selling of assets is forbidden.

The model (1) is a convex quadratic programming problem that can be solved efficiently by using some classical methods or standard solvers, such as IBM CPLEX or Gurobi. As we explained in Sect. 1, this model does not take into account typical practical constrains such as *buy-in threshold* [2,11,12] or *round-lot* constrains [3,5]. After introducing these constraints in (1), the generalized model becomes nonconvex and very difficult to solve by classical approaches.

In the following sections, we present these realistic constraints. After introducing each constraint in (1), we formulate the resulting model in form of a nonlinear nonconvex programming problem. The new model will be used for solving the corresponding portfolio selection problem.

2.1 Portfolio Selection with Buy-In Threshold Constraints

The purpose of introducing the *buy-in threshold* constraints consists in preventing very small investments in each asset. For a given asset i, the buy-in threshold constraints are mathematically read as follows

$$x_i = 0 \text{ or } 0 < x_{min} \leq x_i \quad : \quad i \in \{1, \dots, n\}, \tag{2}$$

where x_{min} is the minimum permitted level of investment in each asset i. In order to reflect this constraint, we first define the penalty function $\omega(\mathbf{x})$ as follows

$$\forall i \in \{1, \dots, n\}: \quad \omega(x_i) = \begin{cases} \left(\frac{4x_i(x_i - x_{min})}{x_{min}^2}\right)^2 & : 0 < x_i < x_{min}, \\ 0 & : \text{otherwise.} \end{cases} \tag{3}$$

This function penalizes any violation of the threshold constraints. Then, by introducing positive penalty parameters η and μ, we transform (1) to minimization of the following function [3]

$$V_{threshold}(\mathbf{x}) = \mathbf{x}^t Q \mathbf{x} + \eta \left(\frac{\mathbf{r}^t \mathbf{x}}{R} - 1\right)^2 + \eta \left(\mathbf{e}^t \mathbf{x} - 1\right)^2 + \mu \sum_{i=1}^{n} \omega(x_i), \tag{4}$$

such that $0 \leq x_i \leq 1$ (for all $i \in \{1, \dots, n\}$).

2.2 Portfolio Selection with Round-Lot Constrains

In portfolio optimization, one of the well known discrete asset choice restrictions are the *round-lot constrains*. These constraints require integer values for invested amounts in each asset. More precisely, the invested fractions must be multiple of some values known as *lot size*. Due to discrete property of these constraints, including them into the classical MV model makes the new problem NP-hard and consequently difficult to solve by classical methods [3,5]. In order to overcome this issue, we suggest using a χ-armed bandit approach called *Hierarchical Optimistic Optimization (HOO)* algorithm. In this section, we present the way by which the round-lot constrains are included in the MV model. Then, we reformulate the modified model for the purpose of using HOO algorithm.

In order to present the round-lot constrains, we need some additional notations. Let M be the total amount of investment and p_i be the price of asset i. The total number of items that can be held in asset i should be equal to z_i:

$$z_i := \frac{M x_i}{p_i}, \tag{5}$$

where x_i is the fraction of M that is invested in asset i. Since z_i must be integer and multiple of a certain lot size, such as $l_i \in \mathbb{Z}_+$, one way for defining the round-lot constrains consists in using the following function:

$$\theta(x_i) = \frac{M x_i}{p_i} - \left\lfloor \frac{M x_i}{p_i} \right\rfloor, \tag{6}$$

where $\lfloor . \rfloor$ is used for defining the integer part of a real number. By using (6), the round-lot constrains are (mathematically) described as follows:

$$\phi(x_i) = \theta(x_i)(1 - \theta(x_i)) = 0 \quad : \quad i = 1, \ldots, n. \tag{7}$$

Due to the round-lot constrains, it may be impossible to actually invest the complete capital, hence the sum of all invested fractions may be less then one. Consequently, in place of the quadratic function, we need to use the modified risk measure $V(\mathbf{x}) = \frac{\mathbf{x}^t Q \mathbf{x}}{(\mathbf{e}^t \mathbf{x})^2}$ in the MV model [3,10]. Furthermore, we need to take into account the constraint $\mathbf{e}^t \mathbf{x} \leq 1$. For this purpose, by introducing penalty parameters η and μ, we can transform the MV model under round-lot constrains to the following minimization problem:

$$\min\ V_{rl}(\mathbf{x}) = \frac{\mathbf{x}^t Q \mathbf{x}}{(\mathbf{e}^t \mathbf{x})^2} + \eta \left(\frac{\mathbf{r}^t \mathbf{x}}{R} - 1 \right)^2 + \eta[\min(1 - \mathbf{e}^t \mathbf{x}, 0)]^2 + \mu \sum_{i=1}^{n} \phi(x_i)^2, \tag{8}$$

where η and μ are sufficiently large positive real numbers and the minimization is done over the hypercube $0 \leq x_i \leq 1$ (for all $i \in \{1, \ldots, n\}$).

3 Solution Methods

In this section, we present the methods that we use for solving the models (4) and (8) over the unit hypercube.

3.1 Hierarchical Optimistic Optimization

In literature, a stochastic bandit problem refers to a gambler who uses a slot machine to play sequentially with its arms (with initially unknown payoffs) in order to maximize his revenue [13]. Each arm has its own payoff and the gambler follows a strategy for choosing an arm. His strategy takes the rewards into account and gathers new information about the distribution of payoffs. Many studies have been done on bandits with finite and infinite number of arms [1]. One of the recent works in bandits concerns the χ-armed bandits where χ is a measurable space of arms [1]. Through this study, the Hierarchical Optimistic Optimization (HOO) is presented that estimate the mean-payoff function over χ. The HOO algorithm finds the global maxima of a function f within the unit hypercube by sampling f at strategically chosen points. To generate these points, HOO incrementally builds a binary tree T which estimates f accurately around the local maxima and less accurately everywhere else. A pleasant side-effect is that the shape of T resembles the shape of f.

HOO builds such a tree T as follows: Every node in T covers a certain part of the region in which f is explored. In fact, every parent node covers the region that is covered by both of its children. In the beginning, the root is the only node and covers the complete region. For every node N, HOO stores not only the region covered by the node but also: how often this node was traversed (from here onwards called c_N), the average function value reached in iterations where this node was traversed (denoted by $\bar{v}_N(i)$), and an optimistic estimate for the maximum function value within the region covered by this node (denoted by B_N and named the $B-value$ of node N). At iteration i (i.e., the ith round is played), the value of B_N is computed by

$$B_N(i) = \begin{cases} \infty & : \text{if } N \text{ is a leaf,} \\ \min\{U_N(i), \max(B_{N_L}(i), B_{N_R}(i))\} & : \text{otherwise,} \end{cases} \quad (9)$$

where N denotes a node of the tree with N_L as its left and N_R as its right child. According to (9), $U_N(i)$ is a key value in determining $B_N(i)$. Indeed, $U_N(i)$ is an estimate for the maximum value reachable at N, but it is less optimistic than B_N. More precisely, $U_N(i)$ is an initial estimate and is defined as follows

$$U_N(i) = \begin{cases} \infty & : \text{if } N \text{ is a leaf,} \\ \bar{v}_N(i) + \sqrt{\frac{2ln(i)}{c_N(i)}} + \nu\rho^h & : \text{otherwise,} \end{cases} \quad (10)$$

where $\bar{v}_N(i)$ is the average function values (rewards) obtained in all iterations so far. At iteration i, $c_N(i)$ shows in how many iterations the node N has been visited. Furthermore, $\nu > 0$ and $\rho \in (0, 1)$ are two parameters passed to HOO, and h is the depth of the node N in T (see [1] for more technical details).

Algorithm 1 illustrates pseudo-code for different steps of HOO. In fact, at each iteration, HOO works as follows: One leaf of the tree is selected by following the path from the root to the leaf where each node of the path has the maximum B-value, in comparison to its sibling. Ties can be broken arbitrarily. This selected leaf is then extended with two children and f is evaluated on the center of the region covered by the selected leaf. Afterwards, the average function value $\bar{v}_N(i)$ of each node along the path from the root to the selected leaf is updated. Finally, the complete tree is traversed in order to recalculate $U-$ and $B-$values.

Figure 1 shows an example. On this figure, the trees that HOO has constructed (after 1000 and 5000 iterations) are depicted. In Fig. 1, the nodes with children are drawn in red while leaves are shown in black. It is shown that after 1000 iterations the shape of the tree already resembles the shape of f . It is also visible that the area around the global maxima is quite densely sampled while the non-optimal parts of f are less sampled. These effects are even more accentuated after 5000 iterations.

Algorithm 1. A formal description of the HOO algorithm.

Parameters: Two real numbers $\nu > 0$, $\rho \in (0, 1)$.
Initialization: T consists only of the root which covers the whole search space i.e., the interval $(0, 1)$.

for $i = 0, 1, \ldots$ **do**
  node $N \leftarrow$ root
  while N is not a leaf **do**
    if $B_{N_R} > B_{N_L}$ **then**

    $N \leftarrow N_R$
  else if $B_{N_R} < B_{N_L}$ **then**
    $N \leftarrow N_L$
  else
    $N \leftarrow$ arbitrarily chosen child (e.g., randomly)
end while
$v = f(p)$ where p represents any point covered by node N
$N_L \leftarrow$ creation of the left child of N with $B_{N_L} = \infty$
$N_R \leftarrow$ creation of the right child of N with $B_{N_R} = \infty$
while node $N \neq$ **null do**
  $c_N \leftarrow c_N + 1$
  $\bar{v}_N = \left(1 - \frac{1}{c_N}\right) \bar{v}_N + \frac{v}{c_N}$
  $N \leftarrow$ parent node of N
end while
for all $N \in T$ in postorder **do**
  recompute U_N via (10)
  recompute B_N via (9)
end for all
end for

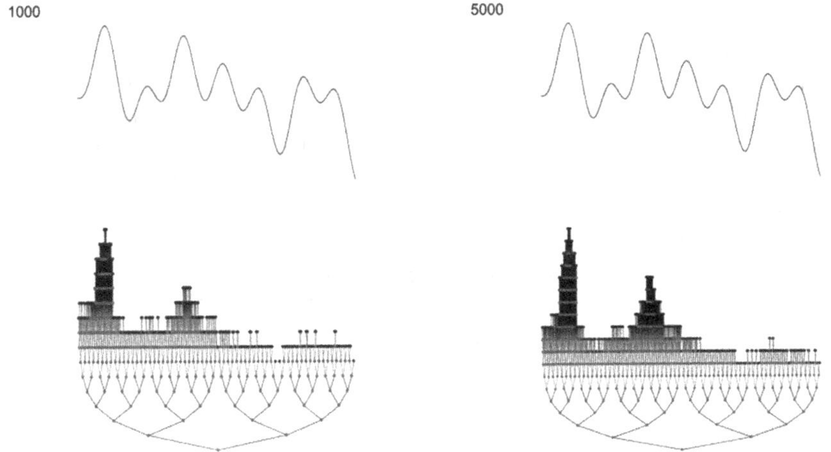

Fig. 1. A visualization of an HOO tree after 1000 and 5000 iterations for the function $f(x) = 0.7sin(20x)sin(11x)sin(14x)+1$. The parameters of HOO are set $\nu = 1, \rho = 0.5$.

3.2 Direct

In order to verify the efficiency of HOO and validate the correctness of its solutions, we solve (4) and (8) by a classical derivative-free approach that is well known under the name *DIRECT*. This method has a good rate of convergence toward high quality solutions [2,3,6]. Detailed information about DIRECT can be found in [2,3,6,7]; however, for the sake of completeness, we provide a brief description of the method for a minimization problem.

In a similar way to HOO, the method DIRECT divides the search space into smaller intervals and samples the midpoint of each interval. However, DIRECT uses a (computationally) cheap sampling procedure. Indeed, at each iteration of the algorithm, a set of intervals is selected and each of them is partitioned to three same-sized subintervals. Then, all of them are sampled at their midpoints. The selection of intervals is done in a way to explore the search space in the best possible way. For this purpose, a combination of local and global strategies is used. More precisely, DIRECT classifies the intervals according to their size, i.e., all intervals of same size are put in a same class. Then, from each class, the interval that has the smallest value at its midpoint is selected for further exploration [2,3,6,7]. The procedure can be stopped by setting a limit on the number of iterations.

4 Computational Experiments

In order to verify the efficiency of HOO, we carried out some numerical simulations. For this purpose, we used HOO and DIRECT for solving (4) and (8). In particular, we did our experiments by using instances that have been already used in [2,3]. These instances composed of 5 and 10 dimensional return vectors and variance-covariance matrices. We implemented the algorithms in *Java* and ran on a Samsung laptop running Windows 10 with a Intel Core i5-4200U processer clocked at 1.6 GHz and equipped with 4.00 GB of memory.

In our experiments, we set $\eta = 1000$ and $\mu = 1$ because these values provide the best results. In a similar way to [2,3], the amount of money to be invested (i.e., M) is 1000 and the price for assets $p_i = 1$ for all $i \in \{1, \ldots, n\}$. The buy-in threshold is set to $x_{min} = 0.05$. Furthermore, the HOO algorithm requires values for its two parameters: ν and ρ. Our default values are $\nu = 1$ and $\rho = 0.5$.

The functions (4) and (8) have been optimized over the 5 and 10 dimensional unit hypercube. In order to compare DIRECT versus HOO, we ran them under the same conditions and for the same amount of iterations.

The following tables present the results of our experiments. The first column (i.e., *problem*) shows which problem (i.e., (4) or (8)) is solved. The second column presents the algorithm used to produce the result in the corresponding row. The third column indicates the dimension of the instance. In these tables, *best value* shows the best value for the associated function (i.e., either (4) or (8)) while the column *risk* only the actual risk value, without the penalty terms. Just as a reminder, the risk was defined as $V(\mathbf{x}) = \mathbf{x}^t Q \mathbf{x}$ for (4) and as $V(\mathbf{x}) = \frac{\mathbf{x}^t Q \mathbf{x}}{(\mathbf{e}^t \mathbf{x})^2}$ for (8). In the last two columns, we present running time of the algorithms and number of function evaluations (i.e., *func. eval.*) at the sampled points.

Table 1 summarizes the results of algorithms DIRECT and HOO for solving (4) and (8) in a 5 dimensional setting. In these results, the maximum number

Table 1. Results of solving (4) and (8) in a 5 dimensional setting.

Problem	Algorithm	Dimension	Best value	Risk value	Func. eval	Time (second)
(4)	DIRECT	5	0.691	0.691	202008	2.547
(4)	HOO	5	1.241	1.189	5000	13.731
(8)	DIRECT	5	0.695	0.694	202942	2.593
(8)	HOO	5	1.296	1.186	5000	15.044

of iterations is limited to 5000. According to our observations, for this data set, there was no considerable improvement in higher number of iterations.

Table 2. Results of solving (4) and (8) in the 10 dimensional setting.

Algorithm	Problem	Iterations	Best value	Risk value	Func. eval	Time (second)
DIRECT	(4)	5000	0.260	0.260	236384	4.13
		10000	0.258	0.258	529852	14.297
		20000	0.242	0.242	1161868	28.148
		30000	0.242	0.242	1836630	54.071
HOO	(4)	5000	0.298	0.227	5000	15.037
		10000	0.250	0.229	10000	62.799
		20000	0.247	0.228	20000	267.357
		30000	0.243	0.229	30000	611.043
DIRECT	(8)	5000	0.174	0.142	185524	4.156
		10000	0.174	0.142	392014	9.328
		20000	0.119	0.116	827218	21.375
		30000	0.119	0.116	1265176	33.141
HOO	(8)	5000	0.322	0.186	5000	13.953
		10000	0.322	0.186	10000	62.107
		20000	0.294	0.185	20000	270.473
		30000	0.247	0.186	30000	630.929

The second set of our experiments consists in using a 10 dimensional instance and running DIRECT and HOO for solving (4) and (8). The selected values for the parameters are the same as the case of 5 dimensional data set. However, in 10 dimensional setting, we present the results of our experiments for different number of iterations. For this purpose, in Tables 2 and 3, there is a new column (i.e., *iterations*) that indicates the number of iterations by which DIRECT and HOO could provide the corresponding solutions.

Since the parameters of HOO has influence on the way of exploring the search space, we carried out a third set of experiments. More precisely, we studied the

Table 3. Changing the value of ν from 1 to 0.1 improves the results of HOO.

Algorithm	Iterations	Best value	Risk value	Func. eval	Time (second)
DIRECT	5000	0.260	0.260	236384	4.313
	10000	0.259	0.258	529852	14.297
	20000	0.242	0.242	1161868	28.148
	30000	0.242	0.242	1836630	54.071
HOO	5000	0.255	0.187	5000	15.016
	10000	0.211	0.188	10000	69.911
	20000	0.207	0.187	20000	280.316
	30000	0.207	0.187	30000	636.120

influence of parameters on performance of HOO. Table 3 shows the results of HOO for solving (4), where ν is set to 0.1. Indeed, we tested different values for ν and ρ; however, Table 3 presents the most interesting observations.

Comments on the Results

- According to the results described in Table 1, we see that DIRECT beats HOO in solving (4) and (8) when the dimension is 5. DIRECT finds better solutions in a shorter time. Our experiments show that letting the algorithms run longer does not change the results significantly. Both algorithms improve only by very small steps and HOO cannot outperform DIRECT.
- When the dimension of the instance is 10, as Table 2 shows, the results provided by HOO (for solving (4)) are quite similar to the results of DIRECT. We observe that the computational time of DIRECT is better than HOO; however, HOO performs a small number of function evaluations.
- According to Table 3, modifying the HOO parameter ν from $\nu = 1$ to $\nu = 0.1$ changes the results drastically. We observe that HOO can find better solutions than DIRECT. However, as it can be seen in Table 2, DIRECT outperforms HOO in solving (8). In this case, changing the parameters of HOO did not change its results significantly.

As a final comment, it is important to mention that the solutions provided by both methods violate *slightly* the buy-in threshold and round-lot constraints.

5 Conclusion

In this paper, we study the learning approach, *Hierarchical Optimistic Optimization (HOO)*, for solving Markowitz's Mean-Variance (MV) model under buy-in threshold and round-lot constraints. These problems are NP-hard and difficult to solve. By introducing some penalty terms, the extended variants of MV model can be transformed to optimization problems over unit hypercube. We show that HOO can find an acceptable solution in a short time; however, in some cases,

HOO is outperformed by the *DIRECT* algorithm. According to our observations, in higher dimensions, the HOO can be more competitive and gives better results than DIRECT. As a future work, we are interested in parallelizing HOO in order to improve its performance in solving large scale instances. Our research in this direction is in progress and the results will be reported in future.

Acknowledgements. The authors acknowledge the chair of Business Information Systems and Operations Research (BISOR) at the TU-Kaiserslautern (Germany) for the financial support, through the research program "CoVaCo".

References

1. Bubeck, S., Munos, R., Stoltz, G., Szepesvàri, C.: χ-Armed bandits. J. Mach. Learn. Res. **12**, 1655–1695 (2011)
2. Bartholomew-Biggs, M.C.: Nonlinear Optimization with Financial Applications, 1st edn. Kluwer Academic Publishers, Dordrecht (2005)
3. Bartholomew-Biggs, M.C., Kane, S.J.: A global optimization problem in portfolio selection. Comput. Manag. Sci. **6**, 329–345 (2009)
4. Fernández, A., Gómez, S.: Portfolio selection using neural networks. Comput. Oper. Res. **34**, 1177–1191 (2007)
5. Jobst, N., Horniman, M., Lucas, C., Mitra, G.: Computational aspects of alternative portfolio selection models in the presence of discrete asset choice constraints. Quant. Finance **1**, 1–13 (2001)
6. Jones, D.R., Perttunen, C.D., Stuckman, B.E.: Lipschitzian optimization without the lipschitz constant. J. Optim. Theor. Appl. **79**(1), 157–181 (1992)
7. Jones, D.R.: The DIRECT global optimization algorithm. In: Floudas, C.A., Pardolos, P.M. (eds.) Encyclopaedia of Optimization, pp. 431–440. Kluwer, Dordrecht (2001)
8. Markowitz, H.M.: Portfolio selection. J. Finance **7**(1), 77–91 (1952)
9. Markowitz, H.M.: Portfolio Selection. Wiley, New York (1959)
10. Mitchell, J.E., Braun, S.: Rebalancing an investment portfolio in the presence of convex transaction costs. Rensselaer Polytechnic Institute (2004)
11. Le Thi, H.A., Moeini, M.: Portfolio selection under buy-in threshold constraints using DC programming and DCA. In: International Conference on Service Systems and Service Management (IEEE/SSSM 2006), pp. 296–300 (2006)
12. Le Thi, H.A., Moeini, M.: Long-short portfolio optimization under cardinality constraints by difference of convex functions algorithm. J. Optim. Theor. Appl. **161**(1), 199–224 (2014)
13. Robbins, H.: Some aspects of the sequential design of experiments. Bull. Am. Math. Soc. **58**, 527–535 (1952)
14. Shen, W., Wang, J., Jiang, Y.-G., Zha, H.: Portfolio choices with orthogonal bandit learning. In: Proceedings of the Twenty-Fourth International Joint Conference on Artificial Intelligence (IJCAI 2015), pp. 974–980 (2015)

MSDN-TE: Multipath Based Traffic Engineering for SDN

Khoa Truong Dinh[1(✉)], Sławomir Kukliński[1,2], Wiktor Kujawa[1], and Michał Ulaski[1]

[1] Warsaw University of Technology, Warsaw, Poland
k.truongdinh@stud.elka.pw.edu.pl,
slawomir.kuklinski@tele.pw.edu.pl,
wkujawa@mion.elka.pw.edu.pl, michal.ulaski@gmail.com
[2] Orange Polska, Warsaw, Poland

Abstract. Software-Defined Networking is a new networking paradigm that has recently gained a lot of attention in the networking community. Its fundamental idea lies on the separation of the centralized control plane and the distributed forwarding plane. Due to global network view and the programmability, SDN is a better tool for solving traffic engineering problems than mechanisms, which exist in classical IP networks. In this paper a simple approach to SDN traffic engineering based on multipath forwarding is presented. The approach lies on dynamic selecting of the best path among several paths for forwarding of an incoming flow. The concept has been simulated using the OpenDaylight controller and Mininet simulator. Simulations have confirmed advantages of the proposed concept over the Shortest Path First (built-in the OpenDaylight controller) and Spanning Tree Protocol (Mininet simple forwarding techniques) approaches.

Keywords: SDN · Traffic engineering · Multipath · OpenDaylight

1 Introduction

In classical IP networks, IP packets are forwarded using the least (shortest) cost paths toward the destination. The shortest cost path algorithms (implemented as a part of IS-IS, CSPF or OSPF protocols) are usually based on simple, static metrics such as link weight or number of hops. The cost for the path is the sum of the weights of all the links belonged to this path. The available path bandwidth is typically not taken into account – even if taken, the time required to update or to gather the entire network state information is typically too long for fast network reaction. In case when the network has spare links, if a failure happens, the routers will compete for the best new path, which will likely be the shortest path. The moved traffic therefore will be forwarded through the new shortest path causing this path to become congested, whereas links on other paths remain under-utilized. This example shows the importance of traffic load balancing among multiple paths. It is worth to mention that the load balancing mechanism contributes to increased network efficiency (ability to handle more intensive traffic) and increased QoS for the best effort service by reduction of network congestion and packet loss rate.

N.T. Nguyen et al. (Eds.): ACIIDS 2016, Part II, LNAI 9622, pp. 630–639, 2016.
DOI: 10.1007/978-3-662-49390-8_61

There are already existing some approaches dealing with Traffic Engineering problem in the networks such as MPLS-TE, RSVP-TE, ECMP, etc. RSVP-TE extends RSVP with many features (such as LSP set up, Fast Reroute, GMPLS extension) in order to provide traffic engineering mechanisms [2]. MPLS-TE control plane is seen to be complex and makes dynamic operation difficult and expensive. Another traffic engineering solution, ECMP (Equal Cost Multipath) has some limitations, it cannot be used to split traffic between parallel paths with different capacities [3].

The creation of Software Defined Networking (SDN) that is based on the OpenFlow [7] protocol, used for controlling flows forwarding has recently gained enormous interest. SDN [8] decouples the programmable control plane from network forwarding plane. SDN consists of the forwarding plane, the control plane and the application plane [9]. SDN approach comes with multiple benefits. It has been proved that such centralized approach provides much shorter time of paths setup than using existing, distributed routing protocols [1]. Moreover, by limiting the number of control plane devices and simplifying data plane device architecture, the complexity and cost of the network can be significantly reduced. In addition, the traffic engineering (TE) mechanisms don't have to be implemented in every network element. The programmability of SDN, easy interface to applications and flow oriented operations are also important in the context of TE and creation of application-aware networks. The centralized controller becomes the brain of the entire network, and with network global overview, it can efficiently decide where to forward flows in order to fulfil application (end-user) demands and at the same time to optimize the forwarding plane resource utilization. The following properties of SDN show that an efficient TE based can be implemented:

- The whole network topology is known to the SDN controller, therefore all existing paths and their actual load can be easily identified and more than a single route can be used for traffic forwarding between any source-destination pair.
- The controller, using OpenFlow, can change the configuration of Flow Switching Tables (FST) of all switches in real-time.

A hybrid attempt based on MPLS and SDN has been already proposed in [4], which uses the standard MPLS data plane and an OpenFlow based control plane. This is seen as one of the first step to use SDN to handle the problem of TE, despite the fact that only bandwidth reservation mechanism combined with the shortest path algorithm in this approach was proposed. In the B4 solution [10], Google deployed TE in a hybrid, commercial network that includes SDN switches and classical IP routers. It has been claimed that the B4 approach drives most of links (ca. 70 %) to near 100 % utilization, providing 14 % increase in average network throughput due to forwarding the data flows among multiple paths in comparison to previous, MPLS based solution. In B4 the splitting of the traffic applies to new coming flows only, for the on-going flows no actions are taken. In [11] it has been shown that SDN network significantly outperforms OSPF routing in terms of network throughput, delay and packet loss ratio.

The goal of this paper is to exploit and evaluate the properties of SDN and the ability to manipulate flows in real-time for TE. Section 1 (this section) introduces TE issues in IP networks and benefits of SDN in that context. Section 2 presents multipath routing and dynamic routing issues. The proposed MSDN-TE approach is described in Sect. 3. The simulation environment and results of MSDN-TE approach are included in Sect. 4. Finally, Sect. 5 summarizes the paper.

2 Traffic Engineering Overview

The main goal of TE is to use effectively network resources and to avoid congestion of links or paths, while serving multiple network users. In order to achieve a high level of users' satisfaction, all CBR (constant bitrate) flows should obtain sufficient average bitrate in a short-term whereas the TCP traffic can adapt to changing network conditions and if source or destination provides high capacity a TCP section can saturate an existing path. The traffic properties between two end-devices in most cases are asymmetrical. In case of the dynamic routing, the routing decision can be taken according to path's cost that is based on the actual path's metric. In order to achieve that goal, the state of all network paths has to be evaluated and the information about the overall network's load and other statistics need to be collected.

In case when multiple paths are found, there are two different possibilities of traffic forwarding:

- The best path can be selected and only this path is used for traffic forwarding whereas the second path serves as a backup and is used in case of network failure. Most IP networks TE nowadays are applying this approach.
- The traffic can be distributed among the existing paths. There are several possibilities of such distribution, e.g. per packet or per flow basis. The first approach provides higher granularity of traffic distribution for the sake of packets disordering, whereas the second one is simpler, but less flexible.

The main problems with dynamic routing lies on:

- Incorrect decision related to path swapping, which may lead to the ping-pong effect due to increased load of the switched paths. As a result, routing instability can be observed.
- Unpredictable traffic volume in the switched paths caused by intrinsic TCP mechanism that can grab all the available bitrate therefore leading to the new path congestion.
- In case of per packet splitting there is a need to reorder packet at the destination. Such operation has to be supported by appropriate receiver buffer used for restoring the proper order of packets, which introduces additional transmission delay.
- In case of per flow traffic splitting, the granularity of switched traffic is potentially low, and the knowledge about the flow properties can be beneficial for this process. However, not much information about incoming flow can be obtained until it actually occurs.

3 MSDN-TE Description

In this paper, the multipath traffic engineering approach for SDN (MSDN-TE) is proposed. This approach is based on the usage of multiple paths to forward flows taking into account the actual path's load for the forwarding decisions. The approach relies on computing k-paths available to forward flows between any Source-Destination pair (S-D) and to select the least loaded path to handle an incoming flow. The main goal of the MSDN-TE algorithm is to avoid congested points in the network, moreover more distributed traffic among the existing paths contributes to easier handling of faults.

For finding multiple paths between the S-D, the k-shortest paths i.e. the Eppstein algorithm [12] has been used. At present, for the sake of simplicity, only the load of links contributed to paths is taken into account as a parameter to evaluate the path's cost. The number of paths used (k-number) is dependent on the network size. In a small network used during experiments, all available loop-free paths were used. However, in big networks, the number of paths can be dependent on network topology and is expected to be in range 2–5.

The MSDN-TE procedure is following:

1. *When new flow is coming the existence of an appropriate path is checked*
   - *If k-path between S-D doesn't exist such k-path is created*
2. *The cost of all paths (in this version load only) in k-path set is calculated and the flow is assigned to the lowest cost path*
3. *Network overall statistics are read and compared with previous values*
4. *Users' resource consumption is recorded (aggregated user's throughput)*
5. *All links (and paths) metrics are updated (available bitrate, packet loss rate, cost)*
6. *Go to 3 (or 1 if a new flow is detected).*

3.1 MSDN-TE Implementation

The MSDN-TE approach has been implemented as an additional module (OSGi bundle) to the OpenDaylight controller. In the implementation MSDN-TE consists of three main functional blocks: MONFUN, ACTFUN and TE Algorithm. MONFUN is a set of functions responsible for collecting information about the network. This information is input to the TE Algorithm, an algorithm which is responsible for decisions regarding flow to path assignment. This process is supported by ACTFUN, a block responsible for taking certain actions based on decision of TE algorithm. The overview of the MSDN-TE implementation is depicted in Fig. 1.

Monitoring Functions (MONFUN) retrieve monitoring parameters (such as network topology, flow's statistics, link utilization, etc.) from the underlying network elements. The set of monitoring data includes: PLR (*packet_loss_ratio*), *link_load*, *packet_delay*. MONFUN also provides information about flows and their assignments to paths. The total number of active flows in network is retrieved with their *idle_timeout/hard_timeout* (if no packet has matched the rule for such period of time, since the flow was inserted, the switch removes the entry and sends *FLOW_REMOVED* message to controller). Besides, the number of flows per host is also obtained to analyse how much traffic each host generates in the network. The refresh rate of path metrics is in range 10–15 s and is dependent on the ability of the controller to collect this information from switches. The network topology update is also retrieved in such time period. Three synthetic network parameters are collected: the *Link_load is* an averaged load of each link over certain period of time; the *Average_network_load* refers to the averaged load of all network links; and the *total_network_dropped* is the number of all dropped network packets.

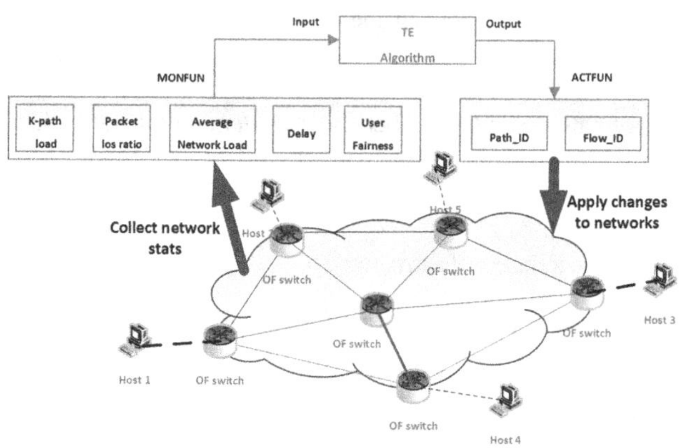

Fig. 1. Overview of MSDN-TE architecture

The monitoring of links is natively provided by OpenDaylight. On that basis there is possible to evaluate the path's quality and calculating quality parameters of all links constituting the path. The link parameters are *link_load, packet_loss_rate* and *delays*.

Actuating Functions (ACTFUN) are used to make changes to the network forwarding plane in a simple way. Controller using this functions can dynamically adjust the acting parameters *(path_ID and flow_ID)* to assign flows to the best available path. The *path_ID* is the path where flow is to be assigned and the *flow_ID* is the identification of the flow.

4 Simulation Environment and Results

In order to check the behavior of the proposed approach the testbed and testing scenarios have been created. The testbed was composed of:

- OpenFlow network emulator: Mininet (OpenVirtualSwitch with OF v1.0).
- OpenDayLight [13] Hydrogen version based SDN controller, enhanced by additional OSGi bundles. These new bundles implemented MONFUN, ACTFUN and the traffic engineering algorithm of MSDN-TE.
- I-DTG (Internet – Distributed Traffic Generator) used to generate network traffic and to monitor different network parameters (*jitter*, *packet_delay*) [14].

4.1 Testing Scenarios

Two kinds of benchmarks of different network topology types were used. The first one (Benchmark 1) is applied to simple network topologies (ring topology and small mesh network) in order to validate proper behaviour of the implemented functions. The second one (Benchmark 2) is using more realistic network topologies, i.e. network topologies taken from the Internet Topology Zoo database (AGIS and Abilene topology as shown in Figs. 5 and 6 respectively) [15]. In order to evaluate the efficiency of the MSDN-TE algorithm different testing scenarios have been performed for each benchmark.

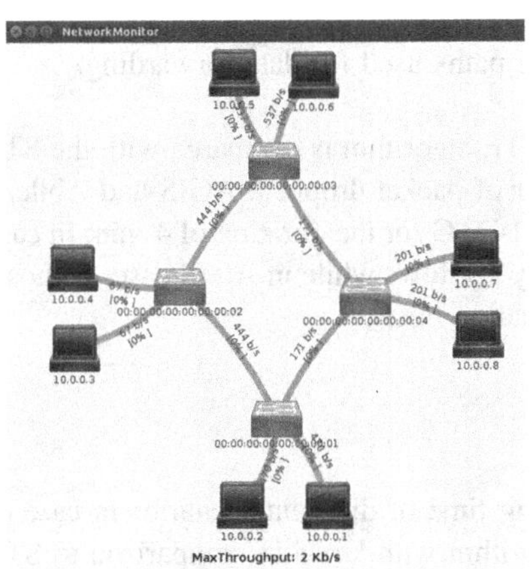

Fig. 2. Scenario 1: Ring topology

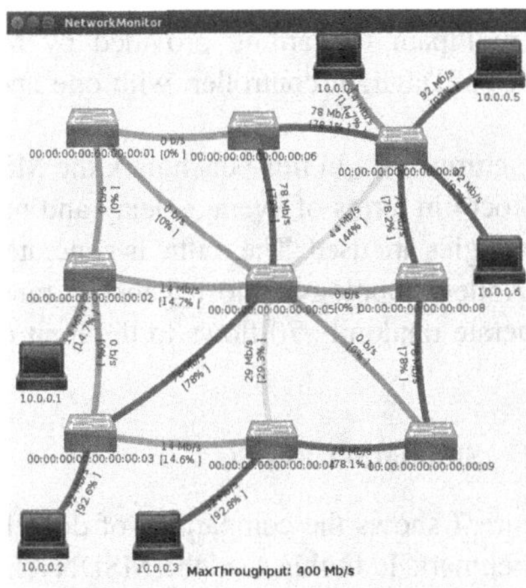

Fig. 3. Scenario 2: Traffic distribution in case of MSDN-TE (k = 5)

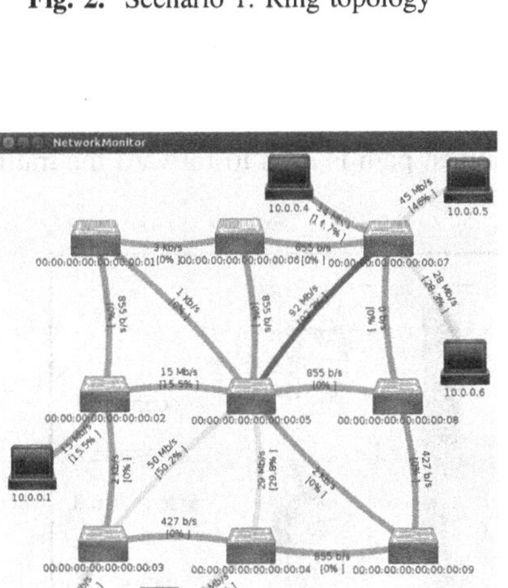

Fig. 4. Scenario 2: Traffic distribution in case of SPF

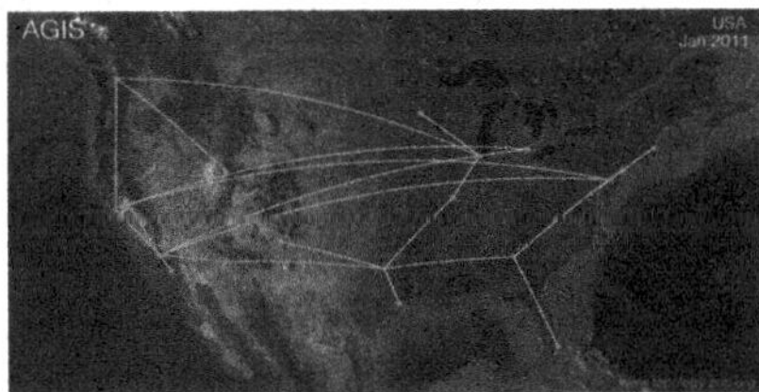

Fig. 5. AGIS topology

Fig. 6. Abilene topology

Benchmark 1. The description of testing scenarios is given below:

- Scenario 1: a ring network topology as shown in Fig. 2. Hosts are downloading a 100 Mb files from other hosts using the HTTP protocol.
- Scenario 2: mesh network (9 nodes, see Figs. 3 and 4), no background traffic provided. Hosts are downloading a 100 Mb files from other hosts using the HTTP protocol.
- Scenario 3: a modified Scenario 2 with the background traffic added.

Three data forwarding mechanisms were compared in Benchmark 1:

- the spanning tree protocol (STP) built-in natively in OpenVirtualSwitch;
- the shortest path first (SPF) protocol built-in OpenDaylight (ODL);
- multipath forwarding provided by MSDN-TE, implemented as an extension to OpenDaylight controller, with one and five paths used for data forwarding.

Benchmark 2. In this benchmark the MSDN-TE algorithm is compared with the STP protocol in terms of average delay and number of packet dropped. AGIS and Abilene topologies are used. The traffic is generated by I-DTG for the duration of 4 min. In case of Abilene topology, 7 hosts generate randomly 18 flows while in AGIS case, 25 hosts generate randomly 70 flows to different destination hosts.

4.2 Simulation Results

Figure 7 shows the comparison of downloading time of different scenarios in case of Benchmark 1. In this case the MSDN-TE algorithm with k = 5 in comparison to STP reduces significantly the downloading time: by more than 56 % in Scenario 1, about 33 % in Scenario 2 and more than 55 % in Scenario 3. In Scenario 1, due to the limited paths between every source-destination pair as a result of ring topology, the downloading times observed in case of SPF, MSDN-TE (k = 1) and MSDN-TE (k = 5) are similar. Even in Scenario 2, the MSDN-TE algorithm with k = 1 provides similar result in comparison with SPF, because only one shortest path is used to forward the traffic.

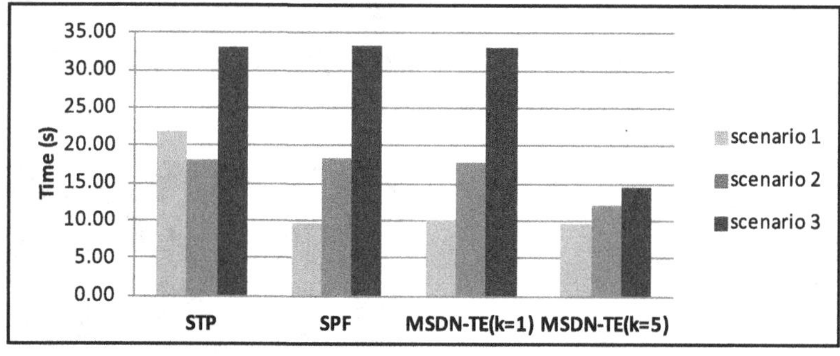

Fig. 7. Time of downloading files (using HTTP)

Figures 3 and 4 show that MSDN-TE is able to increase more than two times total network throughput in comparison to SPF (400 Mbps vs. 184 Mbps for mesh topology).

Figures 8 and 9 present the average network delay and the total number of dropped packets for STP and MSDN-TE algorithm in case of the Abilene topology whereas Figs. 10 and 11 present the same results but for the AGIS topology (Benchmark 2). For such networks, MSDN-TE reduces remarkably the average network delay and the number of dropped packets in comparison to STP. Table 1 shows that the overall delay was reduced by 35 % for the AGIS topology and 65 % for the Abilene topology; whereas the total number of dropped packets was reduced by 72.9 % in case of AGIS and more than 90 % in Abilene case.

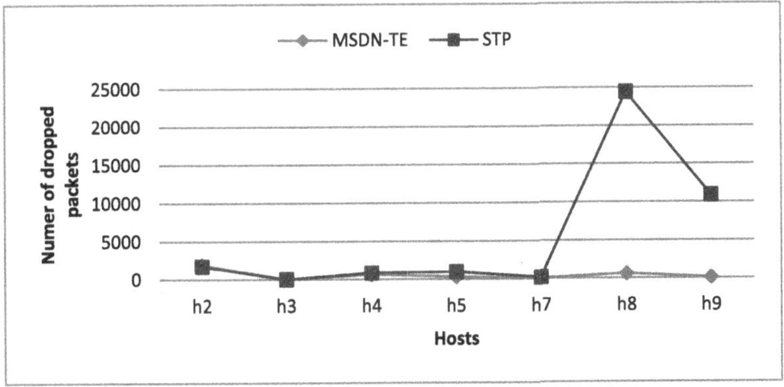

Fig. 8. Number of dropped packets in case of Abilene

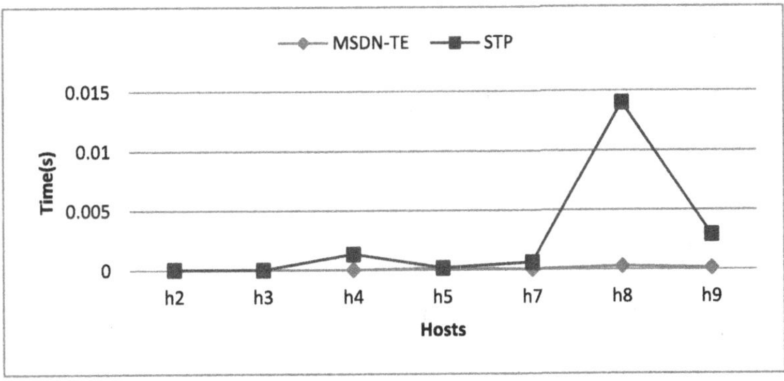

Fig. 9. Average network delay in case of Abilene

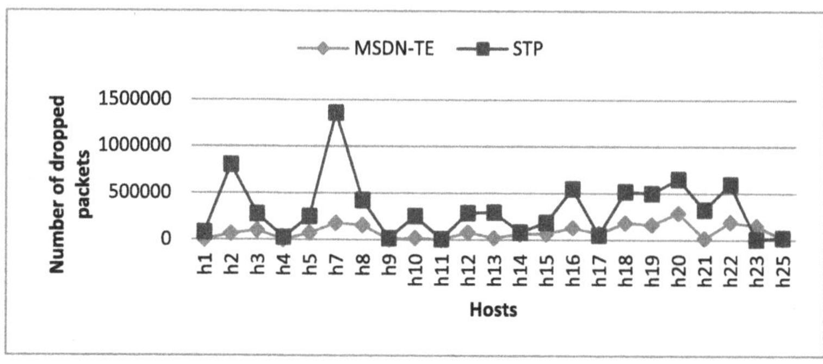

Fig. 10. Number of dropped packets in case of AGIS

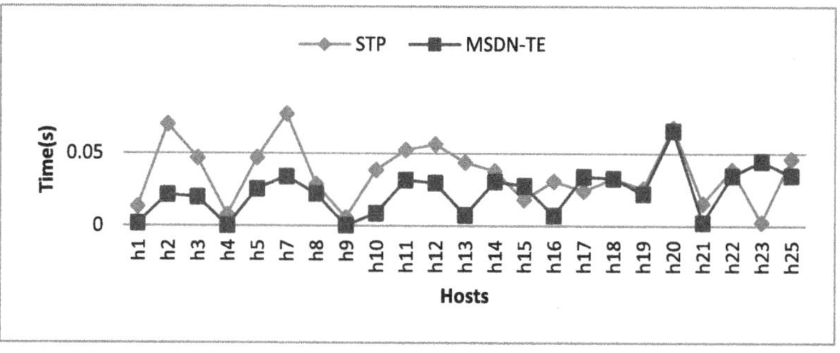

Fig. 11. Average network delay in case of AGIS

Table 1. Average delay and number of dropped packets comparison between STP and MSDN

Mechanisms / Parameters	STP		MSDN-TE (k=5)	
	Abilene	AGIS	Abilene	AGIS
Average network delay (s)	0.02	0.83	0.007 (65%)	0.54 (35%)
Number of dropped packets	39037	7571268	3542 (91%)	2047934 (72.9%)

5 Conclusions

In this paper a simple concept of multipath based forwarding approach for SDN networks has been described and evaluated. By taking advantage of SDN properties (global view of all links load), this approach, i.e. MSDN-TE, significantly outperforms simple traffic forwarding mechanisms based on STP and SPF in terms of packet dropped level, average network delay and the total network throughput. At present only the mechanism of initial assignment of flows to paths was implemented and assessed in MSDN-TE. However, during the flow assignment to path the flow's properties are not known yet. It is therefore desirable to add a mechanism to dynamically redirect the on-going flows to less loaded paths after their initial assignment. Such extensions to MSDN-TE has been already implemented and simulations are on-going.

Acknowledgement. This work has been partially conducted as part of the CoSDN (Cognitive Software Defined Networks) project, which is funded by FNR Luxembourg and NCBiR Poland.

References

1. Matsui, K., Kaneda, M., Matsuda, K.: Evaluation of a server-based traffic engineering architecture suitable for large-scale MPLS networks. In: 2010 8th Asia-Pacific Symposium on Information and Telecommunication Technologies (APSITT), pp. 1–6. IEEE, June 2010
2. Awduche, D., Berger, L., Gan, D., Li, T., Srinivasan, V., Swallow, G.: RSVP-TE: extensions to RSVP for LSP tunnels (No. RFC 3209) (2001)
3. Farrel, A., Ayyangar, A., Vasseur, J.P.: Inter-Domain MPLS and GMPLS Traffic Engineering–Resource Reservation Protocol-Traffic Engineering (RSVP-TE) Extensions (No. RFC 5151) (2008)
4. Das, S., Sharafat, A., Parulkar, G., McKeown, N.: MPLS with a simple OPEN control plane. In: Optical Fiber Communication Conference, p. OWP2. Optical Society of America, March 2011. [RL] Sutton, R.: Reinforcement learning: An introduction. MIT Press (1998)
5. Doria, A., Salim, J., Haas, R., Khosravi, H., Wang, W., Dong, L., Gopal, R., Halpern, J.: "Forwarding and Control Element Separation (ForCES) Forwarding Element Model," Internet Engineering Task Force (IETF) (2010)
6. Vasseur, J., Roux, J.: "Path Computation Element (PCE) Communication Protocol (PCEP)," Internet Engineering Task Force (IETF) (2009)
7. McKeown, N., Anderson, T., Balakrshman, H., Parulkar, G., Peterson, L., Rexford, J., Shenker, S., Tuner, J.: OpenFlow: enabling innovation in campus networks. Sigcomm Comput. Commun. **38**(2), 69–74 (2008)
8. McKeown, N.: Software-defined networking. INFOCOM Keynote Talk **17**(2), 30–32 (2009)
9. ONF, Software-Defined Networking: The New Norm for Networks (2012). https://www.opennetworking.org/images/stories/downloads/sdn-resources/white-papers/wp-sdn-newnorm.pdf
10. Jain, S., Kumar, A., Mandal, S., Ong, J., et al.: B4: experience with a globally deployed software defined WAN. In: SIGCOMM 2013, 12–16 August 2013, Hong Kong, China (2013)
11. Agarwal, S., Kodialam, M., Lakshman, T.: Traffic engineering in software defined networks. In: Proceedings of the 32nd IEEE International Conference on Computer Communications, INFOCOM 2013, pp. 2211–2219, April 2013
12. Eppstein, D.: Finding the k shortest paths. SIAM J. Comput. **28**, 652–673 (1999)
13. https://www.opendaylight.org/downloads
14. Botta, A., Dainotti, A., Pescapè, A.: A tool for the generation of realistic network workload for emerging networking scenarios. Comput. Netw. **56**(15), 3531–3547 (2012). Elsevier
15. Knight, S., Nguyen, H.X., Falkner, N., Bowden, R., Roughan, M.: The internet topology zoo. IEEE J. Sel. Areas Commun. **29**(9), 1765–1775 (2011)

DC Programming and DCA for Enhancing Physical Layer Security via Relay Beamforming Strategies

Tran Thi Thuy[1], Nguyen Nhu Tuan[2(✉)], Le Thi Hoai An[3], and Alain Gély[3]

[1] FPT University, Hoa Lac High Tech Park, Hanoi, Vietnam
thuytt@fpt.edu.vn
[2] Department of Cryptography,
Academy of Cryptography Technique, Hanoi, Vietnam
nguyennhutuan@bcy.gov.vn
[3] Laboratory of Theoretical and Applied Computer Science (LITA) UFR MIM,
University of Lorraine, Ile du Saulcy, 57045 Metz, France
{hoai-an.le-thi,alain.gely}@univ-lorraine.fr

Abstract. Beside of cryptography - the primary traditional methods for ensuring information security and confidentiality, the appearance of the physical layer security approach plays an important role for not only enabling the data transmission confidentially without relying on higher-layer encryption, but also enhancing confidentiality of the secret key distribution in cryptography itself. In this paper, we mention a technique used in physical layer security, this is relay beamforming strategy with Decode-and-Forward relaying technique. The optimization problems with the aim of maximizing the sum of secrecy rates subject to individual relay power constraints are formulated as a nonconvex problem. These problems can be rewritten as DC (difference of two convex functions) programs. We apply the standard DCA scheme to find solution to this problem. The main benefit of the proposed method is that it provides an optimal value for the system secrecy rate while the previous existing methods only give its upper bound.

Keywords: Decode-and-forward · Relay beamforming · Physical layer security · DC programming and DCA

1 Introduction and Related Works

Physical layer security approach is an alternative method, beside the cryptographic-based traditional method to ensure the confidentiality of data transmission in the wireless network. The key idea of physical layer security is to make it possible the swap of confidential messages over a wireless network system containing illegal eavesdroppers, without depending on encryption in higher layers. It exploits the inherent randomness property of noise and communication channels to restrict the quantity of information that might be eavesdropped by

N.T. Nguyen et al. (Eds.): ACIIDS 2016, Part II, LNAI 9622, pp. 640–650, 2016.
DOI: 10.1007/978-3-662-49390-8_62

fraudulent receivers. Physical layer security approach was pioneered by Wyner [29] in 1975. In this work, he proved that the system can achieve positive secrecy capacity if the eavesdropper has a worse channel than the receiver. After several early theoretical works performed by Wyner, the issue of secrecy at the physical layer has attracted researchers' attention only in the recent years. The recent researches have generalized Wyner's idea in various situations by taking advantage of user cooperation techniques such as cooperative jamming [6], cooperative beamforming and cooperative relaying with the popular decode-and-forward and amplify-and-forward schemes [3,8,18,27].

The requirement of channel conditions as the better quality of source-destination channel compared with that of source-eavesdropper channel causes difficulty for applying the traditional physical layer security methods. Some improvement was proposed in the recent researches to overcome these challenges, for example the employment of multiple antennas system such as multiple-input multiple-output (MIMO) [9,20], single-input multiple-output (SIMO) [21], multiple-input single-output (MISO) [25] systems. In spite of the advantages of employing multiple antennas, it is sometimes impossible to equip a terminal with multiple antennas because of its size and power limitation.

Another way to enhance traditional physical layer security is to cooperate with relays and then employ relay beamforming technique, especially when there is no existence of the direct link between the source and the destination. In terms of network architecture, various relay network designs have been considered. The simplest one is the model with a single source-destination pair and one relay studied in [7,19]. Later researchers consider relay network system with one source-destination but with multiple relays [10,26]. Recently, many works have extended to system with multiple source-destination pairs and multiple relays [2,5]. In terms of secrecy, there have been two principal kinds of cooperative relay system considered in the literature. The first one is based on untrusted relays in which the relay plays a role not only as a helper to enhance the communication between the source and the destination, but also as an unintended receiver from which the source message has to be hided. This system has been considered in several papers, e.g., [4,8]. The second one, which is considered in this paper, is relied on trusted relays where the message is not necessary to be kept secret from those devices but it has to be concealed from an external eavesdropper.

The central concepts linked to Physical layer security are achievable secrecy rate and secrecy capacity. An achievable secrecy rate is a rate at which information can be confidentially transmitted from source to its legal destination, and secrecy capacity is defined as the maximal achievable secrecy rate. It is evaluated by the maximum difference of mutual information. Constructing the suitable strategies to attain secrecy capacity actually leads us to solving an optimization problem that is often nonconvex and thus hard to solve. DC programming and DCA (DC Algorithms) are shown as a high-powered approach to deal with such problems. It is undoubtedly that almost all of the arduous nonconvex optimization problems in general and the ones related to physical layer security in practice are capable of being reformulated as DC programs. Furthermore, the prominence of DCA in comparison with other algorithms is shown in a lot of works [13,14,16,17]. In fact, DC programming and DCA is widely used by researchers in many fields as an advanced method for coping with nonconvex programming due to its effectiveness and efficiency [1,11,28].

In this paper, we consider a communication system where the source tries to transmit signal to the destination with the aid of relays employing beamforming technique so that the transmitted information is kept secret as much as possible from the eavesdropper. The perfect channel state information (CSI) is assumed to be available. The purpose of these problems is to determine the beamforming coefficients at relays in order to maximize the secrecy rate of this system subject to individual relay power constraints. This model was introduced in [31] and the existing method using the semidefinite relaxation (SDR) technique gave un upper bound for the system secrecy rate. We will first reformulate this problem as a DC program and then apply DCA to solve them. The proposed method provides an optimal value for the system secrecy rate while the existing method only gives its upper bound.

The remains of this paper is arranged as follows. Because the methodology used in this paper is DC programming and DCA, we present in Sect. 2 the outline of these theoretical and algorithmic tools. In Sect. 3, we describe the Decode-and-Forward relay beamforming design and the derived optimization problem of maximizing the secrecy rate as well as how to employ DC programming and DCA to deal with this problem. The experimental results and the analysis of its characters are presented in Sect. 4. Finally, Sect. 5 concludes the paper.

Notation: Let $()^T$, $()'$ and $(.)^*$ denote transpose, conjugate transpose and conjugate, respectively; $\text{diag}(X)$ denotes a vector of diagonal entries of matrix X; $\text{tr}(X)$ denotes the trace of matrix X, I_m is the identity matrix of size $m \times m$; $E\{.\}$ denotes expectation.

2 Brief Summary of DC Programming and DCA

DC Programming and DCA constitute the backbone of smooth/nonsmooth nonconvex programming and global optimization. They were introduced by Pham Dinh Tao in 1985 in their preliminary form and have been extensively developed by Le Thi Hoai An and Pham Dinh Tao since 1994 to become now classic and more and more popular. DCA is a continuous primal dual subgradient approach. It is based on local optimality and duality in DC programming in order to address standard DC programs which are of the form

$$\alpha = \inf\{f(x) := g(x) - h(x) : \; x \in \mathbb{R}^n\}, \quad (P_{dc})$$

with $g, h \in \Gamma_0(\mathbb{R}^n)$, which is a set of lower semi-continuous proper convex functions on $\mathbb{R}^n$. Such a function f is defined as a DC function, the difference $g - h$ is called DC decomposition of f, and the convex functions g and h are regarded as DC components of f. By means of the indicator function, all constrained DC programs with convex feasible set always can be converted into unconstrained DC ones. That is why it is sufficient to construct DCA theory for unconstrained DC programs.

Recall that, a subgradient of a convex function φ at x_0, denoted as $\partial\varphi(x_0)$, is defined by

$$\partial\varphi(x_0) := \{y \in \mathbb{R}^n : \varphi(x) \geq \varphi(x_0) + \langle x - x_0, y\rangle, \forall x \in \mathbb{R}^n\}$$

The conjugate function of φ, denoted by φ^*, is given by

$$\varphi^*(y) := \sup\{\langle x, y\rangle - \varphi(x) : x \in \mathbb{R}^n\}, \ \forall y \in \mathbb{R}^n.$$

DC duality associates the primal DC program P_{dc} with its dual D_{dc}, which is also a DC program with the same optimal value and defined by

$$\alpha = \inf\{h^*(y) - g^*(y) : \ y \in \mathbb{R}^n\}, \quad (D_{dc})$$

and studies the relation between primal and dual solution sets.

The principle idea of DCA is quite straightforward, that is, at the iteration k of DCA, the second convex component h is replaced by its linear approximate at x^k given by $\bar{h}(x, x^k) = h(x^k) + \langle y^k, x - x^k\rangle$, where $y^k \in \partial h(x^k)$, and it leads to solving the resulting convex program.

$$x^{k+1} \in \arg\min_{x \in \mathbb{R}^n} \{g(x) - \bar{h}(x, x^k)\}. \quad (P_k)$$

The computation of DCA is directly relied on DC components g and h but not the function f itself. Actually, one can offer infinitely many DC decomposition with respect to each DC function and those decompositions will produce various versions of DCA. The sophistication of DCA resides in the choice of an appropriate DC decomposition since the properties of DCA such as efficiency, convergence speed, robustness, global character of obtained solutions, and so on are all impacted by the respective DC decomposition. DCA is therefore a philosophy rather than an algorithm. For each problem one can propose a collection of DCA based algorithms. To the best of our knowledge, DCA is actually one of the seldom algorithms for nonsmooth nonconvex programming which permits to address large-scale DC programs. In fact, DCA was made use successfully to solve numerous diverse nonconvex optimization problems, which even gave global solutions. The success of employing DCA in many works indicates that DCA is more robust and more efficient than existing methods [22–24] and the list of reference in [12].

This is a DCA Generic Scheme:

- **Initialization.** Let $x^0 \in \mathbb{R}^n$ be an initial point. $0 \longleftarrow k$
- **Repeat.**
  Step 1. For each k, x^k is known, compute $y^k \in \partial h(x^k)$.
  Step 2. Find a solution x^{k+1} to the convex problem below

$$\min_{x \in \mathbb{R}^n} \{g(x) - h(x^k) - \langle x - x^k, y^k\rangle\}.$$

  Step 3. $k \leftarrow k + 1$.
- **Until** stopping criterion is satisfied.

The convergence features of DCA and its theoretical fundamental characters have been thoroughly presented in [15,22,23]. It is worth evoking here some typical important properties of DCA.
(i) DCA is a descent method without line search but with global convergence: the sequences $\{g(x^k) - h(x^k)\}$ and $\{h^*(y^k) - g^*(y^k)\}$ are decreasing.
(ii) If the optimal value α of DC program is finite and the infinite sequences $\{x^k\}$ and $\{y^k\}$ are bounded, then every limit point x^* (resp. y^*) of sequence $\{x^k\}$ (resp. $\{y^k\}$) is a critical point of $(g-h)$ (resp. (h^*-g^*)), i.e. $\partial g(x^*) \cap \partial h(x^*) \neq \emptyset$ (resp. $\partial h^*(y^*) \cap \partial g^*(x^*) \neq \emptyset$)
(iii) DCA has a linear convergence for DC programs.
(iv) DCA has a finite convergence for polyhedral DC programs.

3 Decode-and-Forward (DF) Relay Beamforming Design

3.1 Problem Formulation

We take into account a communication system including a source, a destination, an eavesdropper and M relays, as in Fig. 1. In this DF relay beamforming model, the signal s is transmitted by source S to relays with power $E(|s|^2) = P_s$. Afterwards, the message s is decoded at each relay R_m and then is normalized it as $s' = \frac{s}{\sqrt{P_s}}$. Subsequently, the normalized message is multiplied by a weight coefficient w_m to create a transmitted signal $r_m = w_m s'$. Each relay R_m has an output power computed by $E(\|r_m\|^2) = E(\|w_m s'\|^2) = |w_m|^2$. Let $[d_1, ..., d_M]$ and $[e_1, ..., e_M]$ denote the channel coefficients between the destination and the relays, and the eavesdropper and the relays, respectively. The received signals at the destination and the eavesdropper are given by

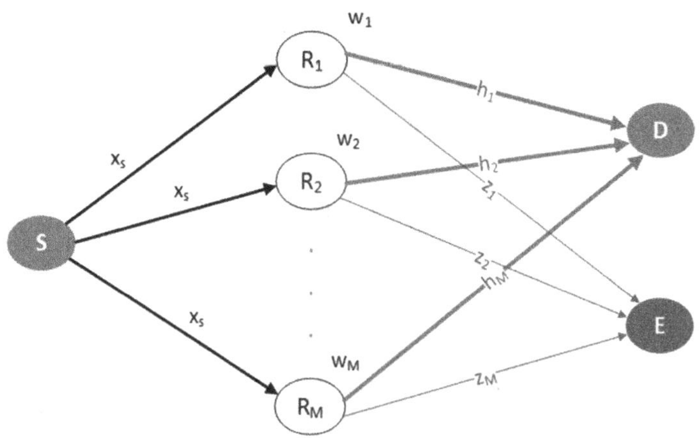

Fig. 1. Decode and Forward relay beamforming system

$$y_D = \sum_{m=1}^{M} d_m w_m s' + n_d,$$

$$y_E = \sum_{m=1}^{M} e_m w_m s' + n_e,$$

where n_d and n_e are the Gaussian background noise components at the destination D and the eavesdropper E, respectively, with zero mean and variance V_0.

The received SNR levels at the destination and the eavesdropper are given by

$$\Gamma_D = \frac{|\sum_{m=1}^{M} d_m w_m|^2}{V_0},$$

$$\Gamma_E = \frac{|\sum_{m=1}^{M} e_m w_m|^2}{V_0}.$$

Denote $d = [d_1^*, ..., d_M^*]^T$ and $e = [e_1^*, ..., e_M^*]^T$. The secrecy rate of this system is given by

$$\begin{aligned} R_s &= \log_2(1 + \Gamma_D) - \log_2(1 + \Gamma_E) \\ &= \log_2\left(\frac{V_0 + w'dd'w}{V_0 + w'ee'w}\right) = \log_2\left(\frac{V_0 + w'Dw}{V_0 + w'Ew}\right), \end{aligned}$$

where $D = dd'$ and $E = ee'$. We deal with the optimization problem of maximizing the secrecy rate with constraints on power at the relays as follows

$$\begin{aligned} &\max_w \ \log_2\left(\frac{V_0 + w'Dw}{V_0 + w'Ew}\right) \\ &\text{s.t.} \ |w_m|^2 \leq p_m, \ \forall m = 1, ..., M. \end{aligned} \tag{1}$$

Rather than solve the problem (1), we solve the following equivalent problem:

$$\begin{aligned} &\max_w \ \frac{V_0 + w'Dw}{V_0 + w'Ew} \\ &\text{s.t.} \ |w_m|^2 \leq p_m \ , \forall m = 1, ..., M. \end{aligned} \tag{2}$$

Recall that if the total relay power constraint equality is imposed, i.e., $\|w\|^2 = P_T$ then the problem (2) is completely solved using the generalized eigenvalue. In more detail, we have

$$\begin{aligned} &\max_{\|w\|^2 = P_T} \frac{V_0 + w'Dw}{V_0 + w'Ew} \\ = &\max_{\|w\|^2 = P_T} \frac{V_0 w' I_M w / P_T + w'Dw}{V_0 w' I w / P_T + w'Ew} \\ = &\max_{\|w\|^2 = P_T} \frac{w'(V_0/P_T.I_M + D)w}{w'(V_0/P_T.I_M + E)w} \\ &= \lambda_{max} \left(V_0/P_T.I_M + D, V_0/P_T.I_M + E\right), \end{aligned}$$

where $\lambda_{max}(A, B)$ is the largest generalized eigenvalue of the matrix pair (A, B). In case of total relay power constraint inequality, i.e., $\|w\|^2 \leq P_T$, if there exist at least a point satisfying this constraint such that $R_s \geq 0$, which is equivalent to $\frac{V_0 + w'Dw}{V_0 + w'Ew} \geq 1$, then it is not difficult to show that

$$\begin{aligned} &\max_{\|w\|^2 \leq P_T} \frac{V_0 + w'Dw}{V_0 + w'Ew} \\ &= \max_{\|w\|^2 = P_T} \frac{V_0 + w'Dw}{V_0 + w'Ew} \\ &= \lambda_{max}\left(V_0/P_T.I_M + D, V_0/P_T.I_M + E\right). \end{aligned}$$

Actually, we should choose the channel coefficients such that the above condition is satisfied. Otherwise, the secrecy rate of the system will be zero, which is insignificant.

When the individual relay power constraints are used, the problem (2) becomes more difficult to solve. A widely used approach to deal with this case is the semidefinite relaxation (SDR) technique. By denoting the square matrix X as $w.w'$, the problem (2) can be reformulated as follows:

$$\begin{aligned} \max_{X,t} \ & t \\ \text{s.t.} \ & \text{diag}(X) \leq p, \\ & X \geq 0, \\ & \text{rank}(X) = 1, \\ & \text{tr}((D - tE)X) \geq V_0(t-1), \end{aligned} \tag{3}$$

where $p = [p_1, ..., p_M]^T$. If the rank constraint is disregarded, the resulting problem is able to be efficiently solved by the interior point method with bisection algorithm [30]. However, because of ignoring the rank constraint, it only provides an upper bound of the secrecy rate but not its optimal value.

To get an optimal value for secrecy rate, we proposed in this paper a technique to equivalently transform the problem (2) into a standard DC program. By introducing the variable s the resulting problem is of convexly constrained fractional programming (the objective function is a ratio of quadratic function to linear function). The standard DCA scheme was applied to solve this program.

3.2 DC Formulation and DCA for DF Relay Beamforming Model (2)

The problem (2) is equivalently reformulated to the following one.

$$\begin{aligned} \max_{w,s} \ & \frac{V_0 + w'Dw}{s} \\ \text{s.t.} \ & |w_m|^2 \leq p_m, \ \forall m = 1, ..., M, \\ & V_0 + w'Ew \leq s. \end{aligned} \tag{4}$$

The above problem is actually a standard DC program of the form:

$$\begin{aligned} \min_{w,s} \quad & G(w,s) - H(w,s) \\ \text{s.t.} \quad & |w_m|^2 \leq p_m, \ \forall m = 1, ..., M, \\ & V_0 + w'Ew \leq s, \end{aligned} \tag{5}$$

where $G(w,s) = 0$ and $H(w,s) = \frac{V_0 + w'Ew}{s}$ are convex functions.

The linear approximation of $H(w,s)$ at the point (w^k, s^k) is given by

$$\bar{H}(s^k, w^k) = \frac{V_0 + (w^k)'Dw^k}{s^k} - \frac{V_0 + (w^k)'Dw^k}{(s^k)^2}(s - s^k) + \frac{2}{s^k}\text{Re}((w - w^k)'Dw^k).$$

Applying the DCA generic scheme for the DC program (4), we get its DCA scheme as follows.

The DCA Scheme for the Problem (5).

Initialization: choose randomly $x^0 = (s^0, w^0) \in (\mathbb{R}^+, \mathbb{C}^M)$ as an initial guess, set the value for tolerance ϵ, $k \leftarrow 0$.
Repeat
- Calculate $x^{k+1} = (s^{k+1}, w^{k+1})$ by solving the following convex subproblem

$$\begin{aligned} \min_{w,s} \quad & 0 - \bar{H}(w^k, s^k) \\ \text{s.t.} \quad & |w_m|^2 \leq p_m, \ \forall m = 1, ..., M, \\ & V_0 + w'Ew \leq s, \end{aligned}$$

- $k \leftarrow k + 1$,

Until $\left(\frac{\|x^{k+1} - x^k\|}{1 + \|x^k\|} < \epsilon \text{ or } \frac{|F(x^{k+1}) - F(x^k)|}{1 + |F(x^k)|} < \epsilon \right)$, where $F(x^k) = \frac{V_0 + (w^k)'Dw^k}{s^k}$

4 Numerical Results

We tested DCA on some generated datasets and compared it with the existing approach based on the semidefinite relaxation technique (SDR) that is described in the last paragraph of Sect. 3.1. Note that SDR only gives an upper bound for the secrecy rate because the rank constraint in the problem (3) is ignored.

In our experiment, all algorithms were implemented in the Matlab 2013, and performed on a laptop Intel core i7, 4 GB RAM. We stop all algorithms with the tolerance $\epsilon = 10^{-5}$. The channel coefficients $\{d_m\}$ and $\{e_m\}$ are assumed to be complex, circularly symmetric Gaussian random variables with zero mean and variances σ_d^2, and σ_e^2, respectively. The fixed parameters are $\sigma_d = 1, \sigma_e = 2$

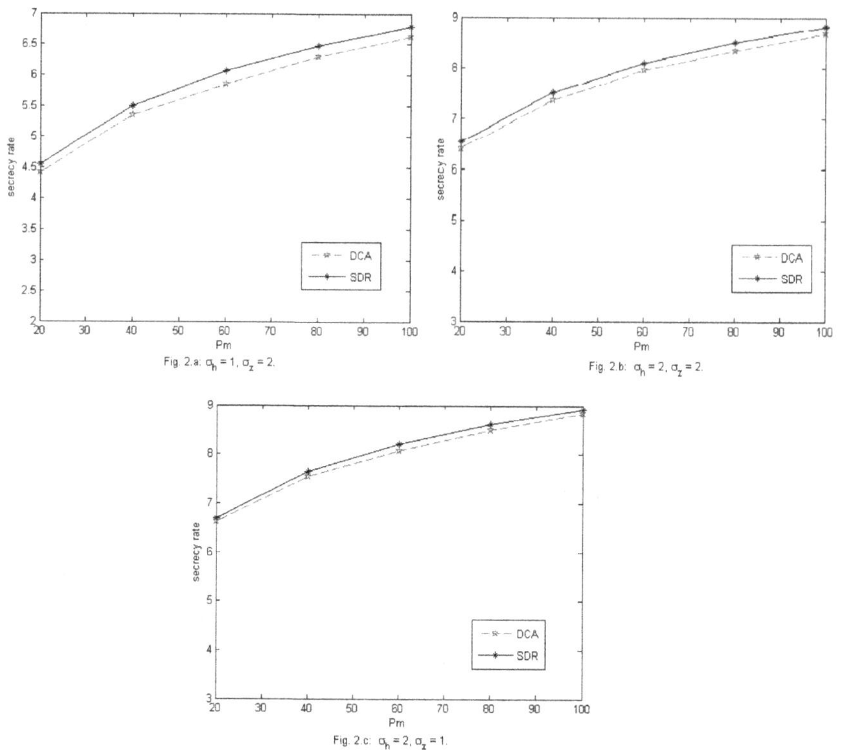

Fig. 2. DF secrecy rate vs. individual relay power constraints

for Fig. 2a; $\sigma_d = 2, \sigma_e = 2$ for Fig. 2b; $\sigma_d = 2, \sigma_e = 1$ for Fig. 2c and $M = 5, V_0 = 1$ for all figures. It is also assumed that the relays have equal power budgets, i.e., $p_m = \frac{P_T}{M}\ \forall m$. Data were randomly generated 100 times and these 100 generated datasets were shared for both DCA and SDR based algorithms. The average of secrecy rates obtained from both algorithms corresponding to these datasets were recorded.

In Fig. 2, we depict the value of secrecy rates achieved by SDR and DCA in the respective cases that the eavesdropper has a stronger channel ($\sigma_e > \sigma_d$, Fig. 2a), the channel conditions are the same for user and eavesdropper ($\sigma_e = \sigma_d$, Fig. 2b), and the eavesdropper has a weaker channel ($\sigma_e < \sigma_e$, Fig. 2c). It can be observed from these figures that DCA can achieve the security for data transmission even when the channel condition is better on the channel of eavesdropper (Fig. 2a). Moreover, the highest level of security obtained when the channel of the user is stronger than that of the eavesdropper (Fig. 2c). In general, there has been an upward trend in the value of secrecy rate achieved from both algorithms in all cases when the power budget of each relay increases. Besides, in all figures, DCA provides the optimal values of secrecy rate really close to its upper bound obtained by SDR. It means that the proposed algorithms DCA give the nearly global optimal value for the system secrecy rate.

5 Conclusions

In conclusions, the goal of this article is to introduce an efficient algorithm based on DCA to solve the optimization problem of maximizing the secrecy rate of the DF relay beamforming network system subject to individual relay power constraints. This problem is nonconvex and thus intractable. We reformulated it as a DC program and then employed DCA for solving it. While the existing method based on SDR technique or SOCP (second order cone programming) only gives an upper bound for the system secrecy rate, the proposed algorithm provides a feasible solution for the problem and gives an optimal value for the system secrecy rate. More interestingly, the experimental results show that the optimal values obtained by the proposed algorithm are nearly global. In the future, we plan to employ DC programming and DCA to investigate models in Amplify-and-Forward (AF) relay networks with multiple eavesdroppers and multiple users.

References

1. Al-Shatri, A., Weber, T.: Achieving the maximum sum rate using DC programming in cellular networks. IEEE Trans. Sig. Process **60**(3), 1331–1341 (2012)
2. Cheng, Y., Pesavento, M.: Joint optimization of source power allocation and distributed relay beamforming in multiuser peer-to-peer relay networks. IEEE Trans. Sig. Process. **60**(6), 2962–2973 (2012)
3. Dong, L., Han, Z., Petropulu, A., Poor, H.: Improving wireless physical layer security via cooperating relays. IEEE Trans. Sig. Process **58**(3), 1875–1888 (2010)
4. Ekrem, E., Ulukus, S.: Secrecy in cooperative relay broadcast channels. IEEE Trans. Inf. Theory **57**(1), 137–155 (2011)
5. Fazeli-Dehkordy, S., ShahbazPanahi, S., Gazor, S.: Multiple peer-to-peer communications using a network of relays. IEEE Trans. Signal Process. **57**(8), 3053–3062 (2009)
6. Goel, S., Negi, R.: Guaranteeing secrecy using artificial noise. IEEE Trans. Wireless Commun. **7**(6), 2180–2189 (2008)
7. Havary-Nassab, V., ShahbazPanahi, S., Grami, A.: Joint receive-transmit beamforming for multi-antenna relaying schemes. IEEE Trans. Signal Process. **58**(9), 4966–4972 (2010)
8. He, X., Yener, A.: Cooperative jamming: the tale of friendly interference for secrecy. In: Liu, R., Trappe, W. (eds.) Securing Wireless Communications at the Physical Layer, pp. 65–88. Spinger, Heidelberg (2010)
9. Hero, A.: Secure space-time communication. IEEE Trans. Inf. Theory **49**(12), 3235–3249 (2003)
10. Jing, Y., Jafarkhani, H.: Network beamforming using relays with perfect channel information. IEEE Trans. Inf. Theory **55**(6), 2499–2517 (2009)
11. Kha, H.H., Tuan, H.D., Nguyen, H.H.: Fast global optimal power allocation in wireless network by local DC programming. IEEE Trans. Wireless Commun. **11**(2), 510–512 (2012)
12. Le Thi, H.A.: DC Programming and DCA. http://www.lita.univ-lorraine.fr/~lethi/

13. Le Thi, H.A., Nguyen, M.C., Pham Dinh, T.: A DC Programming Approach for Finding Communities in Networks. Neural Comput. **26**(12), 2827–2854 (2014)
14. Le Thi, H.A., Nguyen, M.C., Pham Dinh, T.: Self-organizing maps by difference of convex functions optimization. Data Min. Knowl. Disc. **28**, 1336–1365 (2014)
15. Le Thi, H.A., Pham Dinh, T.: The DC (difference of convex functions) programming and DCA revisited with DC models of real world nonconvex optimization problems. Ann. Oper. Res. **133**, 23–46 (2005)
16. Le Thi, H.A., Vo, X.T., Le, H.M.: DC approximation approaches for sparse optimization. Eur. J. Oper. Res. **244**(1), 26–46 (2015)
17. Le Thi, H.A., Vo, X.T., Pham Dinh, T.: Feature selection for linear SVMs under uncertain data: Robust optimization based on difference of convex functions algorithms. Neural Netw. **59**, 36–50 (2014)
18. Li, J., Petropulu, A., Weber, S.: On cooperative relaying schemes for wireless physical layer security. IEEE Trans. Sig. Process **59**(10), 4985–4997 (2011)
19. Munoz-Medina, O., Vidal, J., Agustin, A.: Linear transceiver design in nonregenerative relays with channel state information. IEEE Trans. Sig. Process. **55**(6), 2593–2604 (2007)
20. Negi, R., Goel, S.: Secret communication using artificial noise. In: 2005 IEEE 62nd Vehicular Technology Conference, VTC-2005-Fall, vol. 3, pp. 1906–1910, September 2005
21. Parada, P., Blahut, R.: Secrecy capacity of simo and slow fading channels. In: Proceedings of the International Symposium on Information Theory, ISIT 2005, pp. 2152–2155, September 2005
22. Pham Dinh, T., Le Thi, H.A.: Convex analysis approach to DC programming: theory, algorithms and applications. Acta Mathematica Vietnamica **22**(1), 289–357 (1997)
23. Pham Dinh, T., Le Thi, H.A.: Optimization algorithms for solving the trustregion subproblem. SIAM J. Optim. **8**, 476–505 (1998)
24. Pham Dinh, T., Le Thi, H.A.: Recent advances in DC programming and DCA. In: Nguyen, N.-T., Le-Thi, H.A. (eds.) TCCI 2013. LNCS, vol. 8342, pp. 1–37. Springer, Heidelberg (2014)
25. Shafiee, S., Ulukus, S.: Achievable rates in gaussian miso channels with secrecy constraints. In: IEEE International Symposium on Information Theory, 2007, ISIT 2007, pp. 2466–2470, June 2007
26. ShahbazPanahi, S., Dong, M.: Achievable rate region under joint distributed beamforming and power allocation for two-way relay networks. IEEE Trans. Wireless Commun. **11**(11), 4026–4037 (2012)
27. Stanojev, I., Yener, A.: Improving secrecy rate via spectrum leasing for friendly jamming. IEEE Trans. Inf. Forensics Secur. **12**(1), 134–145 (2013)
28. Vucic, N., Schubert, M.: DC programming approach for resource allocation in wireless networks. In: IEEE, Proceedings of the 8th International Symposium on Modeling and Optimization in Mobile, Ad Hoc and Wireless Networks, pp. 380–386, May 2010
29. Wyner, A.D.: The wire-tap channel. Bell Sys. Tech. J. **54**, 1355–1387 (1975)
30. Zhang, J., Gursoy, M.: Collaborative relay beamforming for secrecy. In: 2010 IEEE International Conference on Communications (ICC), pp. 1–5, May 2010
31. Zhang, J., Gursoy, M.: Relay beamforming strategies for physical-layer security. In: 2010 44th Annual Conference on Information Sciences and Systems (CISS), pp. 1–6, March 2010

Bees and Pollens with Communication Strategy for Optimization

Tien-Szu Pan[1], Thi-Kien Dao[1], Trong-The Nguyen[1(✉)], Shu-Chuan Chu[2], and Jeng-Shyang Pan[3]

[1] Department of Electronics Engineering, National Kaohsiung University of Applied Sciences, Kaohsiung, Taiwan
tpan@cc.kuas.edu.tw, jvnkien@gmail.com, vnthe@hpu.edu.vn

[2] School of Computer Science, Engineering and Mathematics, Flinders University, Adelaide, Australia

[3] College of Information Sciences and Engineering, Fujian University of Technology, Fuzhou, China

Abstract. Due to interference phenomena among constrained dimensions of the multimodal optimization or complex constrained optimization problems, a local optimum is easily converged, rather than for the expected global optimum. The enhanced diversity agent in optimal algorithms is one of the solutions to this issue. This paper proposes a novel optimization algorithm, namely BPO, based on the communication of the bees in artificial bee colony optimization (ABC), with the pollen in flower pollination algorithm (FPA) to solve the multimodal optimization problems. A new communication strategy for Bees and Pollens is presented to explore and exploit the diversity of the algorithm. Six multimodal benchmark functions are used to verify the convergent behavior, the accuracy, and the speed of the proposed algorithm. Experimental results show that the proposed scheme increases the accuracy more than the original algorithms.

Keywords: Artificial bee colony optimization · Flower pollination algorithm · Multimodal optimization · Communication strategy

1 Introduction

The implementing empirical optimal complexity applications is often desirable to simultaneously locate multiple global and local optima of a given objective function [1]. Due to physical constraints in real-world problems, there are different optimal solutions in the search space and the best results of the obtained optimum cannot always be realized. Collaboration between two algorithms is to employ the strength points of the two types of algorithms. The strength point of the algorithms is considered motivation to merge them in parallel to overcome this issue. The parallel processing of them plays an important role for the efficient and effective computations of function optimizations [2–4]. The communications among agents while parallel processing would enhance the cooperating individuals, share the computation load, and increase the diversity optimizations. The idea of this paper is based on the communication

N.T. Nguyen et al. (Eds.): ACIIDS 2016, Part II, LNAI 9622, pp. 651–660, 2016.
DOI: 10.1007/978-3-662-49390-8_63

strategies in parallel processing for swarm intelligent algorithms. They can exchange information between the populations whenever the communication strategy is triggered. The existing methods of these related fields had been introduced including the parallel particle swarm optimization with communication strategies [5], Island-model genetic algorithm [6], and Parallel bat algorithm [7]. The parallelized structure of artificial agents in these algorithms proved that the accuracy and extends the global search capacity are more increases than the original structure.

The advantages of the meta-heuristics are considered throughout this paper consisting of Flower pollination algorithm (FPA) [8], and Artificial bee colony optimization (ABC) [9]. The real-life processes of the flower pollination such as self-pollination or cross-pollination are applied for the achievement of FPA. The cross-pollination can be considered as the global pollination because of the pollinators such as bees, bats, birds and flies can fly the long distances. However, the self-pollination is the fertilization of one flower, such as peach flowers, from the pollen of the same flower or different flowers of the same plant, thus they can be considered as the local search. The artificial bee colony optimization (ABC) [9] works are based on inspecting the behaviors of real honeybees on finding nectar and sharing the food information to the bees in their hive. In this algorithm, the employed bee, the onlooker, and the scout are defined for different roles in the optimization process. The employed bees stay on a food source. The spots in the solution space are represented by a food source, and the coordinate of them are provided for the onlookers in the hive for reference. The onlooker bee selects one of the food sources to gather the nectar whenever it receives the locations of the food sources. New food sources are discovered by moving of the scout into the solution space.

There are lots of advantages of these algorithms, and many applications have been solved successfully by them. However, these algorithms also have the disadvantages such as a premature convergence in the later search period and the accuracy of the optimal value which cannot meet the requirements sometimes. In this paper, the communication strategy and the concepts of the parallel processing are applied to develop a diversity enhanced optimization. In the proposed method, the several weaker individuals in FPA will be replaced with the better artificial agents from ABC algorithm after fixed iterations. On the contrary, the poorer agents of ABC will be replaced with the better pollens of FPA. The benefit of this strategy is to avoid the locally converged optimal in complex constrained optimization problems.

The rest of this paper is organized as follows. A brief review of FPA and ABC is given in Sect. 2. The analysis and designs for the proposed method is presented in Sect. 3. A series of experimental results and the comparison of the original algorithms, genetic algorithm (GA) and the proposed method are discussed in Sect. 4. Finally, the conclusion is summarized in Sect. 5.

2 Related Work

2.1 Flower Pollination Algorithm

Flower Pollination Algorithm (FPA) was developed by emulating the characteristic of the biological flower pollination in flowering plant [8]. The rules of the flow pollination process of flowering plant are applied in this algorithm as follows. (i) The global pollination processes are biotic and cross-pollination through which the pollen transports pollinators in a way that obeys Lévy flights. (ii) Local pollination is explored as abiotic and self-pollination. (iii) Reproduction probability is considered as flower constancy which is proportional to the resemblance of the two flowers in concerned. (iv) The switching probability $p \in [0, 1]$ can be used to control between the local and global pollination. Local pollination can have fraction p that is significant in the entire processes of the pollination because of physical proximity and wind. To simplify the proposed algorithm development, it was assumed that each plant has a single flower and each flower emit only a single pollen gamete. This means that a flower or pollen gamete is viewed as a solution x_i to a problem. FPA was designed with major stages as global and local pollination. To model the local pollination, both rule (ii) and rule (iii) can be represented as.

$$x_i^{t+1} = x_i^t + u(x_j^t - x_k^t) \tag{1}$$

where x_j^t and x_k^t are pollen from different flowers of the same plant species. u is drawn from a uniform distribution in [0, 1] and it is considered as a local random walk, if x_j^t and x_k^t comes from the same species or selected from the same population. In the global pollination, the pollens of the flowers are moved by pollinators e.g. insects and pollens can move for a long distance since the insects typically fly for a long range of distances. This process guarantees pollination and reproduction of the fittest solution represented as g^*. The flower constancy can be represented mathematically as:

$$x_i^{t+1} = x_i^t + \gamma \times L(\lambda) \times (x_i^t - g^*) \tag{2}$$

where x_i is solution vector at iteration t, γ is a scaling factor to control the step size. Lévy flight can be used to mimic the characteristic transporting of insects over a long distance with various distance steps, thus, $L > 0$ from a Lévy distribution.

$$L = \frac{\lambda \Gamma(\lambda) \times \sin(\frac{\pi\lambda}{2})}{\pi \times s^{i+\lambda}}, (s \gg s_0) \tag{3}$$

where $\Gamma(\lambda)$ is the standard gamma function, and this distribution is valid for large steps $s > 0$.

The switch probability or the proximity probability p can be effectively used likely in rule (4) to switch between common global pollination to intensive local pollination. The effectiveness of the PFA can be attributed to the following two reasons: (i) Insect pollinators can travel in long distances which enable the FPA to avoid local landscape to search in a very large space (explorations). (ii) The FPA ensures that similar species of the flowers are consistently chosen which guarantee fast convergence to the optimal solution (exploitation). To begin with, a naive value of p = 0.55 can be set as an initially value. A preliminary parametric showed that p = 0.8 might work better for most applications [8].

2.2 Artificial Bee Colony

The behaviors of real bees on sharing the food information to the other bees in the hive and finding nectar were imitated to develop the Artificial Bee Colony optimization (ABC) [9]. The base steps of ABC process are described as follows.

Initialization: The proportion of the populations is generated randomly into a solution space. So call the nectar amount is computed based on the fitness values of these candidates. The positions of solution for these candidates are called the employed bees. The nectar amount of them is taken into account by evaluating the occupied bees' fitness values. A selecting probability of the positions of solution is given as.

$$P_i = \frac{F(\theta_i)}{\sum_{k=1}^{S} F(\theta_k)} \tag{4}$$

Move the Onlookers: The probability of selecting a food source is calculated by Eq. (4), where P_i is the probability of selecting the *i-th* employed bee, θ_i denotes the position of the *i-th* employed bee, $F(\theta_i)$ denotes the fitness function, and S is represented as the number of employed bees. A food source is selected to move for every Onlooker bees by using the roulette wheel selection method and then the nectar amounts are determined based on this source. The Onlookers are transformed by Eq. (5), where $\varphi(.)$ is a series of random variable [−1,1], x_i denotes the position of the *i-th* Onlooker bee, t denotes the number of iteration, θ is the chosen employed bee randomly, and j represents the dimension of the solution.

$$x_{ij}(t+1) = \theta_{ij}(t) + \varphi(\theta_{ij}(t) - \theta_{kj}(t)) \tag{5}$$

Update the Best Food Source: The best position of solution and the finest fitness value which are found so far by the bees are recorded in memory.

Move the Scouts: If the employed bees have the fitness values that do not be improved by a "Limit" of the iterations continuously, the food sources will be abandoned, and these employed bees could become the scouts. Limit is a continuous predetermined number of iterations. The scouts are transformed by Eq. (6), where ε is a number of random variable and ε ∈ range from 0 to 1.

$$\theta_{ij} = \theta_{jmin} + \varepsilon \times (\theta_{jmax} - \theta_{jmin}) \tag{6}$$

Termination checking: The amount of the iterations whether satisfies the termination condition or not. If the stopping condition is satisfied, the program will be terminated the results are output and recorded; otherwise go back to *move the Onlookers*.

3 A Communication Strategy Bees and Pollens

The advantages of both FPA and ABC algorithms are strong robustness, fast convergence and high flexibility. They have been applied to solve successfully many problems in engineering, financial, and management fields [10, 11]. However, the disadvantage of them also exists such as the premature convergence in the later search period. This could make the accuracy of their optimal values are not to meet the requirements in sometimes [12, 13]. It could be easy to converge to a local optimum, if the swarm size is too small for searching solution based on its own best historical information. This issue could be overcome by applying the enhanced optimizations. Enhanced optimization can be implemented by constructing the communication between two algorithms. The exchange information among subpopulations can be figured out whenever the communication strategy triggers. The communication strategy for exchanging information between Bees and Pollen can be described as follows. The best bees in ABC could be copied to fly to other subpopulations in FPA replace the poorer pollens of them, and update the positions of all subpopulations in every period of exchanging time. The flow information of communicating the bees and pollens is employed with the communication strategy. In contrast, the finest artificial pollens among all the flowers of FPA's population would migrate to the weaker bees in ABC, replace them and update all positions for each population during every period exchanging time.

A parallel structure is made up in several groups by dividing the population into subpopulations. The diversity agents for the optimal method are built based on constructing of the parallel processing. The subpopulations are evolved into regular iterations independently. The advantages of each side of algorithms are taken into account by replacing the poorer individuals of them with the finest ones, and the benefit of cooperation between them is archived. During all iterations of the proposed method BPO, the exchanging period time of communication between FPA and ABC is set to R. The population size of BPO is set to N. The numbers of the population sizes of ABC and FPA are N_1 and N_2 be set to $N/2$ respectively. The top fitness k agents of in group with N_1 will be copied to the place of worst agents in group with N_2 for replacing the same number of the agents, where t is the current iteration, during running with $t \cap R \neq \emptyset$. The description of the proposed method can be summarized the basic steps as follows.

Step 1. **Initialization:** Population size of BPO is generated randomly by initializing the solutions of FPA and ABC. The number iteration of R is defined for executing the communication strategy. The N_1 and N_2 are the number of bees and pollens in solutions X_{ij}^t and S_{ij}^t for populations of ABC and FPA respectively, $i = 0, 1,.., N_{1,2}$ -1, $j = 0, 1,..D$. where D is dimension of solutions and t is current iteration with setting initializing to 1.

Step 2. **Evaluation:** The fitness function values of $f_2(X_{ij}^t), f_1(S_{ij}^t)$ are evaluated by both ABC and FPA in each iteration according to the fitness function. The evolvement of the populations is executed by both FPA and ABC.

Step 3. **Update:** The global pollination and local pollination of FPA are updated by using Eqs. (1–3) and the bees and food source positions of ABC are updated by using Eqs. (5) and (6). The best fitness value and their positions are memorized.

Step 4. **Communication Strategy:** The best pollens among all the flowers of FPA's population are copied with k the top fitness pollens in $N_{1,}$ migrate to other place of group in ABC population then replace the weaker bees in N_2, and update for each population in every R iterations. In contract, do the same with bees among all the individuals of ABC's population.

Step 5. **Termination:** Go to Step 2 if the predefined value of the function is not achieved or the maximum number of iterations has not been reached, otherwise, ending with minimum of the best value of the functions: Min(f(S^t), $f(X^t)$), and the best bee position among all the bees S^t or the best pollen among all the agents X^t. are recorded.

4 Experimental Results

The performance quality of the proposed algorithm of BPO is evaluated by using a set of multimodal benchmark functions [14, 15] to test the accuracy and the speed of it. The outcome values of the test functions in the experiments are averaged over 25 runs with different random seeds. All the optimizations for the test functions are to minimize the outcome.

The simulation results of the proposed method are compared with those obtained results of the previous algorithms as the ABC, FPA and Genetic algorithm (GA) [16], in terms of their performance of the accuracy and running speed. Let $S = \{s_1, s_2,\ldots, s_m\}$, $X = \{x_1, x_2,\ldots, x_m\}$, and $G = \{g_1, g_2,\ldots, g_m\}$ be the real value vectors of m-dimensional for ABC, FPA and GA respectively. The optimization goal is to minimize the outcome for all benchmarks. The outcome of the performed optimal for the test benchmark function is a minimizing problem. The population size for the methods of BPO, ABC, FPA and GA are set to 40 for all runs in the experiments. The setting parameters for ABC, FPA, and GA could be found in [8, 16, 17]. Table 1 lists the initial range, the dimension and total iterations for all test functions.

The parameters setting for BPO with ABC side is the initial with setting Limit to 10. The percentage of onlooker and employed bees are set to 50 %. The total

Table 1. The initial range and the total iteration of the multimodal benchmark functions

Multimodal test functions	Range	Dimension	Iteration
$F_1(x) = \sum_{i=1}^{n} \sin(x_i).(\sin(\frac{ix_i^2}{\pi}))^{2m},\ m = 10$	$0, \pi$	30	1000
$F_2(x) = \sum_{i=1}^{n} -x_i \sin(\sqrt{\lvert x_i \rvert})$	±500	30	1000
$F_3(x) = \sum_{i=1}^{n} [x_i^2 - 10\cos(2\pi x_i) + 10]$	±5.12	30	1000
$F_4(x) = [e^{-\sum_{i=1}^{n}(x_i/\beta)^{2m}} - 2e^{-\sum_{i=1}^{n} x_i^2}] \prod_{i=1}^{n} \cos^2 x_i,\ m = 5$	±20	30	1000
$F_5(x) = -\sum_{i=1}^{4} c_i \exp(-\sum_{j=1}^{6} a_{ij}(x_j - p_{ij})^2)$	0,10	4	1000
$F_6(x) = -\sum_{i=1}^{5} [(X - a_i)(X - a_i)^T + c_i]^{-1}$	0,10	4	1000

population size N_1 is set to $N/2$ as equal to 20 and the dimension of solution space d is set to 30, as in ref. [17]. Corresponding to the parameters setting with FPA side is the initial probability p is set to 0.55, λ is set to 1.5, the total population size N_2 set to 20 and the dimension d is set to 30, as in ref. [7]. Each benchmark function is tested with 1000 iterations per a run. The performance is evaluated in the average of the results from all runs. Comparing percentage is set to abs (BPO-original algorithm)*100/ (BPO).

Table 2 compares the performance quality for the multimodal optimization problems of three methods of ABC, FPA and the proposed BPO. Observed, the results of the proposed method on all of these cases of testing multimodal benchmark problems show that BPO method almost increases higher than those obtained from original methods of ABC and FPA. The maximum case obtained from BPO method increases higher than those obtained from the ABC and FPA methods are up to 44 % and 45 % respectively. However, the figure for the minimum cases is only the increase 08 % and 05 % for ABC and FPA respectively. Thus, in general the proposed algorithm BPO obtained the average cases of various tests multimodal optimization problems for the convergence, and accuracy increased more than those obtained from the ABC and FPA methods are 28 % and 30 % respectively.

Table 2. The quality performance evaluation comparison of ABC, FPA, and BPO for solving the multimodal optimization problems

Test function	Function values			Comparison performance	
	ABC	FPA	**BPO**	with ABC	with FPA
1	1.45E + 00	1.48E + 00	1.05E + 00	38 %	41 %
2	−4.87E + 03	−4.10E + 03	−6.09E + 03	20 %	33 %
3	1.65E + 02	1.73E + 02	1.29E + 02	28 %	34 %
4	1.60E − 03	1.60E − 03	1.10E − 03	44 %	45 %
5	−2.94E + 00	−3.04E + 00	−3.32E + 00	8 %	5 %
6	−7.15E + 00	−8.15E + 00	−9.72E + 00	26 %	16 %
Avge	**−7.86E + 02**	**−6.56E + 02**	**−9.95E + 02**	**28 %**	**30 %**

Figures 1, 2 and 3 show the experimental results for the first three multimodal benchmark functions over 25 runs output obtained from GA, ABC, FPA and proposed BPO methods with the same iteration of 1000.

The above figures show clearly that, all of the cases of testing functions for BPO have performance quality highest in terms of the accuracy and convergence.

Table 3 shows the performing quality and running time comparison of the proposed BPO with GA method for the multimodal optimization problems. The columns of comparison times and qualities are calculated as absolute of the obtained from BPO minus that obtained from GA then divided the obtained value of the BPO method. Clearly, the results of the proposed method on all of these cases of testing multimodal benchmark problems show that BPO method almost increases higher quality and

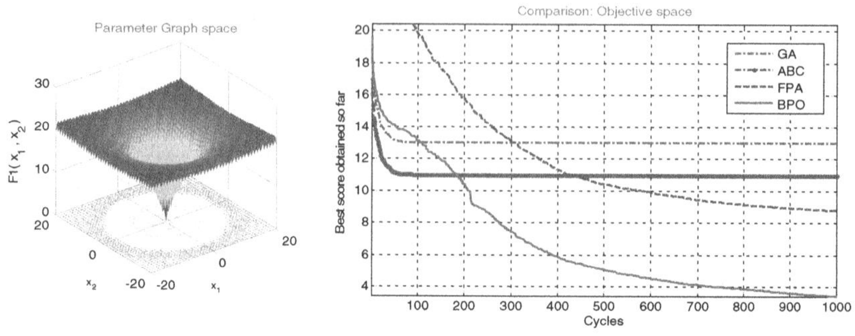

Fig. 1. The experimental results of function F1

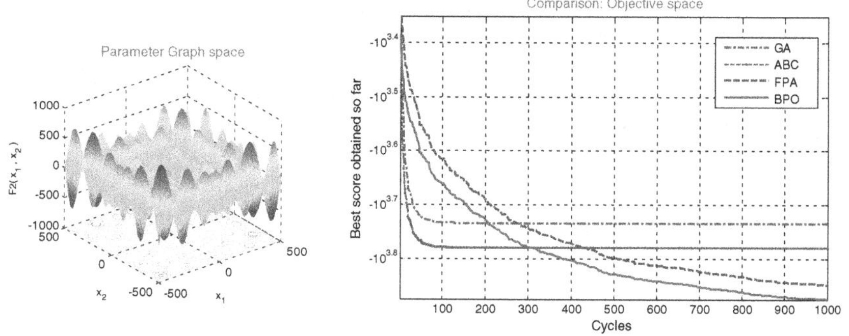

Fig. 2. The experimental results of function F2

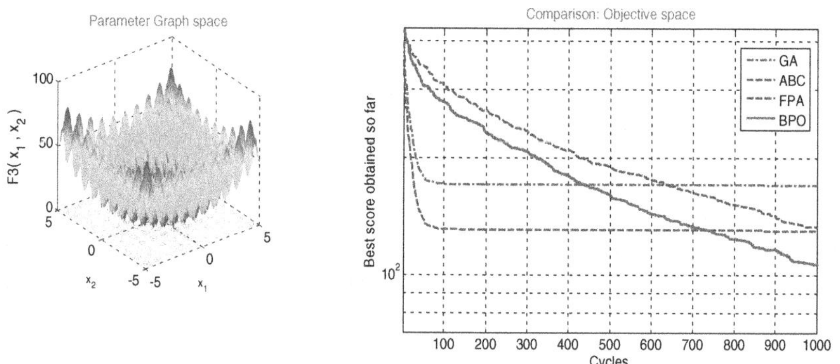

Fig. 3. The experimental results of function F3

shorter running time than those obtained from GA method. In general, the proposed algorithm obtained the average cases of various tests multimodal optimization problems for the convergence, and accuracy increased more than those obtained from the GA method is 37 %, and for the speed faster than that got from GA method is 2 %.

Table 3. The quality performance evaluation comparison of GA and BPO for solving the multimodal optimization problems

Test function	Consumption Time		Comp. times	Performances		Comp. qualities
	GA	BPO		GA	BPO	
1	2.4793	2.4371	2 %	1.91E + 00	1.25E + 00	51 %
2	0.897	0.8687	3 %	−5.27E + 03	−6.19E + 03	15 %
3	0.9873	0.9502	4 %	1.89E + 02	1.29E + 02	47 %
4	0.7776	0.7834	1 %	1.70E-03	1.10E-03	39 %
5	0.8918	0.9072	2 %	−2.24E + 00	−3.29E + 00	47 %
6	0.9893	0.9793	1 %	−8.05E + 00	−9.72E + 00	21 %
Avg	**1.1703**	**1.1543**	**2 %**	**−8.48E + 02**	**−1.01E + 03**	**37 %**

5 Conclusion

A novel proposed method for the multimodal optimization problems was presented in this paper with the enhanced optimization based on communication for Bees and Pollens, namely BPO. The making diversity agents to enhance optimization could take an important significance in the solutions for the issue of losing the global optimum in the optimal algorithms for the multimodal or complex constrained optimization problems. The proposed communication strategy is innovative because of the introduction of strength points of ABC and FPA in the cooperation of optimization algorithms. By this way, the poorer pollens in FPA could be replaced with new best bees from ABC after running the exchanging period. In contrast, the worst bees in ABC could be replaced with fresh finest pollens from FPA in every exchanging period. The performance of BPO algorithm is better than both original ABC and FPA, and GA in terms of convergence and accuracy. The numeric results from a set of various multimodal problems show that BPO for the maximum cases increase higher than those obtained from the ABC, FPA, and GA methods are up to 44 %, 45 %, and 51 % respectively. However, these figures for the minimum cases are only the increase 08 %, 05 %, and 15 %. Thus, in general the proposed algorithm obtained the average cases of various test problems for the convergence, and accuracy increased more than those by 28 %, 30 %, and 39 % from ABC, FPA, and GA respectively.

References

1. Qu, B.Y., Suganthan, P.N., Das, S.: A distance-based locally informed particle swarm model for multimodal optimization. IEEE Trans. Evol. Comput. **17**(3), 387–402 (2013)
2. Wolfe, M., Banerjee, U.: Data dependence and its application to parallel processing. Int. J. Parallel Program. **16**(2), 137–178 (1987)
3. Dao, T.-K., Pan, T.-S., Nguyen, T.-T., Chu, S.-C.: A compact articial bee colony optimization for topology control scheme in wireless sensor networks. J. Inf. Hiding Multimedia Sign. Proces. **6**(2), 297–310 (2015)

4. Pan, T.-S., Dao, T.-K., Nguyen, T.-T., Chu, S.-C.: Optimal base station locations in heterogeneous wireless sensor network based on hybrid particle swarm optimization with bat algorithm. J. Comput. **25**(4), 14–25 (2015)
5. Chu, S.C., Roddick, J.F., Pan, J.-S.: A parallel particle swarm optimization algorithm with communication strategies. J. Inf. Sci. Eng. **21**(4), 1–9 (2005)
6. Whitley, D., Rana, S., Heckendorn, R.B.: The island model genetic algorithm: on separability, population size and convergence. J. Comput. Inf. Technol. **7**(1305), 33–48 (1998)
7. Tsai, C.-F., Dao, T.-K., Yang, W.-J., Nguyen, T.-T., Pan, T.-S.: Parallelized bat algorithm with a communication strategy. In: Ali, M., Pan, J.-S., Chen, S.-M., Horng, M.-F. (eds.) IEA/AIE 2014, Part I. LNCS, vol. 8481, pp. 87–95. Springer, Heidelberg (2014)
8. Yang, X.-S.: Flower pollination algorithm for global optimization. In: Durand-Lose, J., Jonoska, N. (eds.) UCNC 2012. LNCS, vol. 7445, pp. 240–249. Springer, Heidelberg (2012)
9. Karaboga, D.: An idea based on honey bee swarm for numerical optimization. Technical report-TR06, Erciyes University, Engineering Faculty, Computer Engineering Department, vol. T2005 (2005)
10. Karaboga, D., Gorkemli, B., Ozturk, C., Karaboga, N.: A comprehensive survey: artificial bee colony (ABC) algorithm and applications. Artif. Intell. Rev. **42**(1), 21–57 (2014)
11. Wang, R., Zhou, Y.: Flower pollination algorithm with dimension by dimension improvement. Math. Probl. Eng. **2014**, 9 (2014)
12. TSai, P.-W., Pan, J.-S., Liao, B.-Y., Chu, S.-C.: Enhanced artificial bee colony optimization. Int. J. Innovative Comput. Inf. Control **5**(12), 5081–5092 (2009)
13. Wang, H., Sun, H., Li, C.H., Shahryar, R., Pan, J.-S.: Diversity enhanced particle swarm optimization with neighborhood search. Inf. Sci. **223**, 119–135 (2013). 20 Feb 2013
14. Yao, X., Liu, Y., Lin, G.: Evolutionary programming made faster. IEEE Trans. Evol. Comput. **3**(2), 82–102 (1999)
15. Suganthan, P.N., Hansen, N., Liang, J.J., Deb, K., Chen, Y.-P., Auger, A., Tiwari, S.: Problem definitions and evaluation criteria for the CEC 2005 special session on real-parameter optimization. KanGAL report, vol. 2005 (2005)
16. Whitley, D.: A genetic algorithm tutorial. Stat. Comput. **4**(2), 65–85 (1994)
17. Karaboga, D., Basturk, B.: On the performance of artificial bee colony (ABC) algorithm. Appl. Soft Comput. **8**(1), 687–697 (2008)

Online DC Optimization for Online Binary Linear Classification

Ho Vinh Thanh[1(✉)], Le Thi Hoai An[1], and Bui Dinh Chien[2]

[1] Laboratory of Theoretical and Applied Computer Science EA 3097, University of Lorraine, Ile du Saulcy, 57045 Metz, France
{vinh-thanh.ho,hoai-an.le-thi}@univ-lorraine.fr
[2] FPT University, Hoa Lac High Tech Park, Hanoi, Vietnam
chienbd@fpt.edu.vn

Abstract. This paper concerns online algorithms for online binary linear classification (OBLC) problems in Machine learning. In a sense of "online" classification, an instance sequence is given step by step and on each round, these problems consist in finding a linear classifier for predicting to which label a new instance belongs. In OBCL, the quality of predictions is assessed by a loss function, specifically 0–1 loss function. In fact, this loss function is nonconvex, nonsmooth and thus, such problems become intractable. In literature, Perceptron is a well-known online classification algorithm, in which one substitutes a surrogate convex loss function for the 0–1 loss function. In this paper, we investigate an efficient DC loss function which is a suitable approximation of the usual 0–1 loss function. Basing on Online DC (Difference of Convex functions) programming and Online DCA (DC Algorithms) [10], we develop an online classification algorithm. Numerical experiments on several test problems show the efficiency of our proposed algorithm with respect to Perceptron.

Keywords: Online DC optimization · Online DCA · Perceptron · Online binary classification · DC programming · DCA

1 Introduction

Online classification is an interesting topic in machine learning in general and in the area of online learning in particular [17]. While "offline" classification is the problem of determining to which label a new instance belongs, on the basis of whole instances whose label is known in advance, in online classification, the classifier is updated incrementally after the arrival of a new datapoint at each online round. In this paper, we study an online binary linear classification (OBLC) problem in which the label set is defined by $\mathcal{Y} = \{-1, 1\}$ and a linear classifier is considered. Moreover, OBLC is performed in a sequence of consecutive rounds, as shown in Fig. 1. Indeed, on each round t, after receiving an instance $x_t \in \mathcal{X}$ ($\mathcal{X} = \mathbb{R}^n$), the learner will find a linear classifier $w_t \in \mathcal{X}$ and predict the corresponding label $p_t = p_t(w_t) \in \mathcal{Y}$ defined by $p_t(w) = \mathsf{sign}(\langle w, x_t \rangle)$, where $\mathsf{sign} : \mathcal{X} \to \mathcal{Y}$ is a sign function. After that, the correct answer $y_t \in \mathcal{Y}$ is

N.T. Nguyen et al. (Eds.): ACIIDS 2016, Part II, LNAI 9622, pp. 661–670, 2016.
DOI: 10.1007/978-3-662-49390-8_64

revealed. In such a case, the quality of predictions is assessed by a loss function between the prediction p_t and the correct answer y_t, specifically the 0–1 loss function, ℓ_t, defined as

$$\ell_t(p_t, y_t) = \mathbb{1}_{\{p_t \neq y_t\}} = \mathbb{1}_{\{y_t\langle w_t, x_t\rangle \leq 0\}} = \begin{cases} 1 & \text{if } y_t\langle w_t, x_t\rangle \leq 0, \\ 0 & \text{otherwise.} \end{cases} \quad (1)$$

The main goal of OBLC is to make a sequence of predictions that minimizes the cumulative loss suffered along its run.

Online Binary Linear Classification

for $t = 1, 2, \ldots$
  receive an instance vector $x_t \in \mathcal{X}$
  predict a label $p_t = \mathsf{sign}(\langle w_t, x_t\rangle) \in \mathcal{Y}$
  receive a correct label $y_t \in \mathcal{Y}$
  suffer loss $\ell_t(p_t, y_t)$
  update the linear classifier w_{t+1} based on (x_t, y_t)

Fig. 1. A framework of online binary linear classification

In recent years, there are many online classification algorithms with the different predictions and loss functions such as the Approximate Maximal Margin Classification Algorithm [3], Relaxed Online Maximum Margin Algorithm [12], and the Margin Infused Relaxed Algorithm [2],... (see more details in [5]). In the context of OBLC, the classic Perceptron is known as the simplest algorithm [13,16,19]. In Perceptron, on round t, one receives a surrogate convex loss function, specifically the hinge-loss, which upper bounds the 0–1 loss function ℓ_t. This hinge-loss is defined as follows

$$\ell_{\text{hinge}}(h_w(x), y) = \max\{1 - y\langle w, x\rangle, 0\}. \quad (2)$$

In fact, Perceptron can be derived from the Online convex optimization framework. The paradigm of Online convex optimization is described in [18], where the linear classifier set $\mathcal{S}$ and the loss function are convex. In this context, there are two existing natural approaches to find the linear classifiers: Follow-the-Leader (FTL) [6] in which the leaner chooses the prediction based on the past rounds and the regularized form of Follow-the-Leader (FoReL) [17,18] that makes a stability for the predictions of online learning algorithms. From these approaches, many effective online convex algorithms were proposed such as Online Gradient Descent (with projections) (OGD) [20], Exponentiated Gradient (EG) [1,7,8], p-norm [4],... After all, based on OGD with the learning rule FoReL, one derives the classical Perceptron, described as Algorithm 1.

Recall that the OBLC problems usually consider the nonconvex 0–1 loss function. Such problems are intractable to solve. Instead of substituting the surrogate convex loss function as mentioned above, we take into consideration

Algorithm 1. Perceptron

Initialize: $w_1 = 0$.
for $t = 1, \ldots, T$ **do**
  Receive an instance vector $x_t \in X$ Predict a label: $p_t = \mathsf{sign}(\langle w_t, x_t \rangle)$
  Receive the correct label $y_t \in Y$ Suffer loss: $l_t(p_t, y_t) = \mathbb{1}_{\{p_t \neq y_t\}}(p_t, y_t)$ **if** $l_t(p_t, y_t) = 1$ **then**
    $w_{t+1} = w_t + y_t x_t$
  else
    $w_{t+1} = w_t$
  end
end

a DC loss function that is a better approximation of 0–1 loss function. In such a case, OBLC can be cast as an Online DC optimization problem for which an efficient optimization approach, namely Online DC programming and DCA [10], can be applied. DC programming and DCA were introduced by Pham Dinh Tao in a preliminary form in 1985 and extensively developed since 1994 by Le Thi Hoai An and Pham Dinh Tao (see [11,14,15] and the references therein). This work is motivated by the fact that (offline) DCA has been successfully applied to many (smooth or nonsmooth) large scale nonconvex programs in various domains of applied sciences, in particular in Machine Learning, for which they provide quite often a global solution and are proved to be more robust and efficient than the standard methods (see the list of references in [9]).

In this paper, our main contribution is fourfold. Firstly, we describe a new DC loss function for a particular OBLC problem. Secondly, we develop an OBLC algorithm, namely Online DCA, based on the generic Online DC programming and DCA [10]. Thirdly, we deal with the specific problem by combining Online DCA with the efficient online convex algorithms. Finally, in order to evaluate the efficiency of our proposed online algorithm, we conduct numerical experiments on several benchmark data in comparison with Perceptron.

The rest of the paper is organized as follows. In Sect. 2, we first present a short introduction of DC programming and DCA and then briefly show the elegant model of Online DC optimization and Online DCA scheme in OBLC. How to solve the particular classification problem by Online DCA is described in Sect. 3. Section 4 reports the numerical results on several test problems which is followed by some conclusions in Sect. 5.

2 Online DC optimization

2.1 Introduction of DC Programming and DCA

DC Programming and DCA constitute the backbone of smooth/nonsmooth nonconvex programming and global optimization. They address the problem of minimizing a function f which is a difference of convex functions on the whole space $\mathbb{R}^p$. Generally speaking, a DC program takes the form

$$\alpha = \inf\{f(x) := g(x) - h(x) : x \in \mathbb{R}^p\} \quad (P_{dc}) \tag{3}$$

where $g, h \in \Gamma_0(\mathbb{R}^p)$, the set contains all lower semicontinuous proper convex functions on $\mathbb{R}^p$. Such a function f is called DC function, and $g - h$, DC decomposition of f while g and h are DC components of f. The convex constraint $x \in C$ can be incorporated in the objective function of (P_{dc}) by adding the indicator function of C ($\chi_C(x) = 0$ if $x \in C$, $+\infty$ otherwise) to the first DC component g of the DC objective function f.

A standard DC program with a convex constraint C (a nonempty closed convex set in $\mathbb{R}^n$):

$$\alpha = \inf\{f(x) := g(x) - h(x) : x \in C\} \tag{4}$$

can be expressed in the form of (P_{dc}) by adding the indicator function of C ($\chi_C(x) = 0$ if $x \in C$, $+\infty$ otherwise) to the function g.

Recall that, for $\theta \in \Gamma_0(\mathbb{R}^p)$ and $x \in \text{dom}\,\theta := \{u \in \mathbb{R}^p | \theta(u) < +\infty\}$, the subdifferent of θ at x, denoted $\partial\theta(x)$, is defined as

$$\partial\theta(x) := \{y \in \mathbb{R}^p : \theta(u) \geq \theta(x) + \langle u - x, y\rangle, \forall u \in \mathbb{R}^p\} \tag{5}$$

which is a closed convex set in $\mathbb{R}^p$. It generalizes the concept of derivative in the sense that θ is differentiable at x if and only if $\partial\theta(x)$ is reduced to a singleton set, i.e. $\partial\theta(x) = \{\nabla\theta(x)\}$. Each $y \in \partial\theta(x)$ is called subgradient of θ at x.

The necessary local optimality condition for the primal DC program, (P_{dc}), is $\partial h(x^*) \subset \partial g(x^*)$. This condition is also sufficient for many important classes of DC programs [11,14]. A point that x^* satisfies the generalized Kuhn-Tucker condition $\partial g(x^*) \cap \partial h(x^*) \neq \emptyset$ is called a critical point of $g - h$.

Generic DCA scheme

Initialization: Let $x^0 \in \mathbb{R}^p$ be a best guess, $k := 0$.
Repeat
- Calculate $y^k \in \partial h(x^k)$
- Calculate $x^{k+1} \subset \text{argmin}\{g(x) - h(x^k) - \langle x - x^k, y^k\rangle :$ $x \in \mathbb{R}^p\} (P_k)$
- $k = k + 1$

Until convergence of $\{x^k\}$.

The main idea of DCA is quite simple: each iteration k of DCA approximates the concave part $-h$ by its affine majorization corresponding to taking $y \in \partial h(x^k)$ and minimizes the resulting convex function (that is equivalent to determining $x^{k+1} \in \partial g^*(y^k)$).

Convergence properties of the DCA and its theoretical basis are described in [11,14,15]. In the past years, DCA has been successfully applied in several works of various fields among them learning machines, financial optimization, supply chain management, etc. (see the list of references in [9]).

2.2 Online DC Optimization and Online DCA

As mentioned in Sect. 1, we further assume that the set of linear classifiers $\mathcal{S}$ ($\mathcal{S} = \mathcal{X}$) is convex. Under this assumption, Online DC optimization, referred in [10], is described as in Fig. 2. In such context, the DC objective function changes along the online process.

Online DC optimization

for $t = 1, 2, \ldots, T$
predict $w_t \in \mathcal{S}$
receive a DC function $f_t^{DC} : \mathcal{S} \to \mathbb{R}$ whose DC components are g_t and h_t
suffer $f_t^{DC}(w_t)$

Fig. 2. A framework of online DC optimization

In this paper, we consider FoReL with the Euclidean regularization function $R(w) = \frac{1}{2\eta}||w||_2^2$ as mentioned in Sect. 1. More precisely, on round t, we have

$$w_{t+1} \in \operatorname{argmin} \left\{ \sum_{i=1}^{t} f_i^{DC}(w) + R(w) : w \in \mathcal{S} \right\}, \ \forall t = 1, \ldots, T. \tag{6}$$

In general, (6) is a DC program. Hence, the main idea of DCA can be applied and we obtain one classification algorithm, namely Online DCA, shown in Fig. 3 and referred in [10].

Online DCA

Initialization: $w_1 \in \mathcal{S}$
for $t = 1, 2, \ldots, T$
receive a question $x_t \in \mathcal{X}$
predict an answer p_t based on w_t
receive the correct answer $y_t \in \mathcal{Y}$
update $w_{t+1} \in S$ by the following procedure:
compute $z_t \in \partial h_t(w_t)$
compute w_{t+1}, an optimal solution of the convex program

$$\min \left\{ \sum_{i=1}^{t} g_i(w) + R(w) - \sum_{i=1}^{t} \langle z_i, w \rangle : w \in \mathcal{S} \right\}.$$

Fig. 3. Generic online DCA scheme in online binary classification

It would be interesting to note that if the sequence of predictions $\{w_t\}_{t=1,\dots,T}$ is generated by Online DCA, then we obtain the following property:

$$\sum_{t=1}^{T} f_t^{DC}(w_t) \leq \sum_{t=1}^{T} f_t^{C}(w_t), \tag{7}$$

where $f_t^C(w) = g_t(w) - (h_t(w_t) + \langle z_t, w - w_t \rangle), t = 1, \dots, T,$ is a convex majorization of $f_t^{DC}(w)$. Despite the fact that we consider a convex majorization, $f_t^C(w)$, of $f_t^{DC}(w)$ as online convex optimization, these convex functions are updated dynamically along the online process. Thus, the power of online convex optimization can be exercised. That will be shown in the next section.

3 Online DCA for Online Binary Classification

In this section, we propose a DC loss function f_t^{DC}, a suitable approximation of ℓ_t. In case $t \notin \mathcal{M}$, i.e. the learner predicts true, we define $f_t^{DC}(w) = 0, \forall w \in \mathcal{S}$. For round $t \in \mathcal{M}$, we define the function f_t^{DC} satisfying the following condition

$$f_t^{DC}(w_t) \geq \ell_t(p_t, y_t). \tag{8}$$

This condition is necessary to find the best relative mistake bound of our algorithm. Indeed, on round $t \in \mathcal{M}$, for two parameters η_{1t}, η_{2t} $(\eta_{1t} > 0, \eta_{2t} \geq 0)$, we have

$$\begin{aligned} f_t^{DC} &: \mathbb{R}^n \to [0,1] \\ w &\mapsto f_t^{DC}(w) = \max\left\{0, \min\left\{\frac{\eta_{2t}}{\eta_{1t}}, \frac{-y_t\langle w, x_t\rangle}{\eta_{1t}}\right\}\right\}. \end{aligned} \tag{9}$$

If $\eta_{2t} = \eta_{1t}$, then it is evident that $0 \leq f_t^{DC}(w) \leq \ell_t(w) \leq 1$ and $f_t^{DC}(w) \to \ell_t(w)$ as $\eta_{1t} \to 0$, for any $w \in \mathbb{R}^n$. Moreover, DC components of f_t^{DC} are

$$g_t(w) = \max\left\{0, \frac{-y_t\langle w, x_t\rangle}{\eta_{1t}}\right\} \text{ and } h_t(w) = \max\left\{0, \frac{-\eta_{2t} - y_t\langle w, x_t\rangle}{\eta_{1t}}\right\}. \tag{10}$$

Specially, when $y_t\langle w_t, x_t\rangle = 0$, $\ell_t(p_t, y_t) = 1$ and $f_t^{DC}(w_t) = 0$. Thus, (8) does not hold. Therefore, we investigate another DC loss function, f_t^0, defined as:

$$f_t^0(w) = \max\left\{0, \min\left\{1, \frac{\eta_{1t} - y_t\langle w, x_t\rangle}{\eta_{1t}}\right\}\right\} \tag{11}$$

and the corresponding DC components are

$$g_t^0(w) = \max\left\{0, \frac{\eta_{1t} - y_t\langle w, x_t\rangle}{\eta_{1t}}\right\}, \ h_t^0(w) = \max\left\{0, \frac{-y_t\langle w, x_t\rangle}{\eta_{1t}}\right\}. \tag{12}$$

Instead of solving the convex program on each round, we apply the simplest algorithm OGD [20] in Online convex optimization with the corresponding convex loss functions $f_1^C, \dots, f_T^C$. More concretely, on round t, we will iteratively update the linear classifier as follows

$$w_{t+1} \leftarrow w_t + \mathbb{1}_{\{t \in \mathcal{M}\}} s_t, \tag{13}$$

where $s_t \in \partial g_t(w_t) - \partial h_t(w_t)$.

Compute $\partial g_t - \partial h_t$: By the definition of g_t and h_t, we have

$$\partial g_t(w) - \partial h_t(w) = \begin{cases} \{0\} & \text{if } y_t\langle w, x_t\rangle < -\eta_{2t} \text{ or } y_t\langle w, x_t\rangle > 0, \\ \left[\frac{-y_t x_t}{\eta_{1t}}, 0\right] & \text{if } y_t\langle w, x_t\rangle = -\eta_{2t} \text{ or } y_t\langle w, x_t\rangle = 0, \\ \left\{\frac{-y_t x_t}{\eta_{1t}}\right\} & \text{if } -\eta_{2t} < y_t\langle w, x_t\rangle < 0. \end{cases} \tag{14}$$

It is similar to the case of $y_t\langle w_t, x_t\rangle = 0$. To generally satisfy the condition $f_t^{DC}(w_t) \geq \ell_t(p_t, y_t)$ for all t, we choose the parameters $\eta_{2t} = -y_t\langle w_t, x_t\rangle$ and

$$\eta_{1t} = \begin{cases} \min\{\eta, -y_t\langle w_t, x_t\rangle\} & \text{if } y_t\langle w_t, x_t\rangle < 0, \\ \eta & \text{if } y_t\langle w_t, x_t\rangle = 0, \end{cases}$$

where $\eta > 0$.

From Online DCA scheme, DCA applied to OBLC is given by the following algorithm.

Algorithm 2. Online DCA for Online Binary Classification (Online DCA)

Initialize : $w_1 = 0 \in \mathbb{R}^n$
Parameter: $\eta > 0$
for $t = 1, 2, \ldots, T$ **do**
  receive an instance vector $x_t \in \mathcal{X}$ predict a label: $p_t = \mathsf{sign}(\langle w_t, x_t\rangle) \in \mathcal{Y}$
  receive the correct label $y_t \in \mathcal{Y}$ **if** $-\eta \leq y_t\langle w_t, x_t\rangle < 0$ **then**
    $w_{t+1} = w_t - \dfrac{x_t}{\langle w_t, x_t\rangle}$
  else if $y_t\langle w_t, x_t\rangle = 0$ or $y_t\langle w_t, x_t\rangle < -\eta$ **then**
    $w_{t+1} = w_t + \dfrac{y_t x_t}{\eta}$
  else
    $w_{t+1} = w_t$
  end
end

From (7) and (8), we have the relative mistake bound of Algorithm 2 defined as follows: for any $u \in \mathcal{S}$,

$$|\mathcal{M}| \leq \sum_{t\in\mathcal{M}} f_t^C(u) + U||u||_2\sqrt{\sum_{t\in\mathcal{M}} f_t^C(u) + U^2||u||_2^2}, \tag{15}$$

where $U = \max_t \frac{||x_t||_2}{\eta_{1t}}$ and $\mathcal{M}$ is the set of rounds on which Algorithm 2 makes a wrong prediction.

4 Numerical Experiments

In this section, we perform Online DCA for online binary classification. We compare our algorithm with the well-known Perceptron algorithm in terms of the number of prediction mistakes and the consuming time. Both algorithms were implemented in Visual C++ version 11.0 and run on a PC Intel(R) Xeon(R) CPU E5-2630, 2.60GHz of 32GB RAM. We tested on a variety of benchmark datasets from UCI Machine Learning Repository[1] and LIBSVM website[2]. All datasets used in our experiment are randomly chosen and shown in Table 1.

Table 1. Datasets used in our experiments

Data-set	Name	# instances	# features
D1	cod-rna	264923	8
D2	gisette	5601	5000
D3	svmguide1	5672	4
D4	mushrooms	6500	112
D5	magic04	15217	10
D6	a8a	26050	123

Table 2. Comparison between Online DCA and Perceptron

Dataset	Online DCA		Perceptron	
	No. mistakes (%)	CPU (s)	No. mistakes (%)	CPU (s)
D1	**47473.45** (17.920)	0.0102	47494.95 (17.928)	0.0093
D2	**649.05** (11.588)	0.0974	659.10 (11.768)	0.0765
D3	**1465.65** (25.840)	0.0001	1472.85 (25.967)	0.0003
D4	**105.65** (1.625)	0.0016	**105.65** (1.625)	0.0014
D5	**5564.20** (36.566)	0.0007	5572.35 (36.619)	0.0004
D6	5596.90 (21.485)	0.0097	**5443.40** (20.896)	0.0075

The number of rounds, T, is identical to the number of instance vectors. In order to choose a good value of η for Algorithm 2, we perform it by running over one random permutation of each dataset on the set $\{0.05, 0.10, \ldots, 4.95, 5.00\}$ and then take the value corresponding to the best results. After that, all algorithms were conducted over 20 runs of different random permutations for each dataset. The average results over these 20 runs of both algorithms were reported

[1] http://www.ics.uci.edu/~mlearn/MLRepository.html.
[2] http://www.csie.ntu.edu.tw/~cjlin/libsvmtools/.

in Table 2. We use CPU, No.Mis. and % to denote, respectively, CPU time in seconds, the number of mistakes of algorithms and the ratio of No. Mistakes to T.

The results in Table 2 show that Online DCA is more efficient than Perceptron. Indeed, in terms of the number of mistakes, Online DCA makes wrong predictions less than Perceptron's ones in 4/6 datasets. As for the CPU time, both algorithms are, in general, quite fast, less than 0.1 seconds in all cases. In addition, the CPU time of Online DCA is moderately comparable to that of Perceptron.

5 Conclusion

In this work, we have investigated Online DC optimization for the Online binary linear classification problem. We have proposed an effective, new DC loss function and described Online DCA based on the generic Online DC programming and DCA. Moreover, numerical experiments on various benchmark datasets have proved the efficiency of our proposed algorithm and Online DC optimization approach as well. In the future, we will apply Online DC programming and Online DCA to many different online classification problems in which they can be regarded as an Online DC optimization problem. Furthermore, we may be exploring more the efficient DC loss functions, the power of online convex optimization in a variety of problems for Online DC optimization in online learning.

Acknowledgements. This research is funded by Foundation for Science and Technology Development of Ton Duc Thang University (FOSTECT), website: http://fostect.tdt.edu.vn, under Grant FOSTECT.2015.BR.15.

References

1. Azoury, K., Warmuth, M.: Relative loss bounds for on-line density estimation with the exponential family of distributions. Mach. Learn. **43**(3), 211–246 (2001). http://dx.doi.org/10.1023/A3A1010896012157
2. Crammer, K., Singer, Y.: Ultraconservative online algorithms for multiclass problems. J. Mach. Learn. Res. **3**, 951–991 (2003). http://dx.doi.org/10.1162/jmlr.2003.3.4-5.951
3. Gentile, C.: A new approximate maximal margin classification algorithm. J. Mach. Learn. Res. **2**, 213–242 (2002). http://dl.acm.org/citation.cfm?id=944790.944811
4. Gentile, C.: The robustness of the p-norm algorithms. Mach. Learn. **53**(3), 265–299 (2003). http://dx.doi.org/10.1023/A:1026319107706
5. Hoi, S.C.H., Wang, J., Zhao, P.: Libol: a library for online learning algorithms. J. Mach. Learn. Res. **15**(1), 495–499 (2014). http://dl.acm.org/citation.cfm?id=2627435.2627450
6. Kalai, A., Vempala, S.: Efficient algorithms for online decision problems. J. Comput. Syst. Sci. **71**(3), 291–307 (2005). http://www.sciencedirect.com/science/article/pii/S0022000004001394

7. Kivinen, J., Warmuth, M.: Relative loss bounds for multidimensional regression problems. Mach. Learn. **45**(3), 301–329 (2001). http://dx.doi.org/10.1023/A:1017938623079
8. Kivinen, J., Warmuth, M.K.: Exponentiated gradient versus gradient descent for linear predictors. Inf. Comput. **132**(1), 1–63 (1997). http://www.sciencedirect.com/science/article/pii/S0890540196926127
9. Le Thi, H.A.: DC programming and DCA (2012). http://www.lita.univ-lorraine.fr/~lethi
10. Le Thi, H.A., Ho, V.T.: Online DC programming and DCA. LITA, Technical report, University of Lorraine, Metz, France (2015)
11. Le Thi, H.A., Pham Dinh, T.: The DC (difference of convex functions) programming and DCA revisited with DC models of real world nonconvex optimization problems. Ann. Oper. Res. **133**(1–4), 23–46 (2005)
12. Li, Y., Long, P.: The relaxed online maximum margin algorithm. Mach. Learn. **46**(1–3), 361–387 (2002). http://dx.doi.org/10.1023/A:1012435301888
13. Novikoff, A.B.: On convergence proofs for perceptrons. Proc. Symp. Math. Theor. Automata **12**, 615–622 (1963). http://citeseer.comp.nus.edu.sg/context/494822/0
14. Pham Dinh, T., Le Thi, H.A.: Convex analysis approach to d.c. programming: theory, algorithms and applications. Acta Mathematica Vietnamica **22**(1), 289–355 (1997)
15. Pham Dinh, T., Le Thi, H.A.: DC optimization algorithms for solving the trust region subproblem. SIAM J. Optim. **8**(2), 476–505 (1998)
16. Rosenblatt, F.: The perceptron: a probabilistic model for information storage and organization in the brain. Psychol. Rev. **65**(6), 386–408 (1958)
17. Shalev-Shwartz, S.: Online learning: theory, algorithms, and applications. Ph.D. thesis, The Hebrew University of Jerusalem (2007)
18. Shalev-Shwartz, S., Singer, Y.: A primal-dual perspective of online learning algorithms. Mach. Learn. **69**(2–3), 115–142 (2007). http://dx.doi.org/10.1007/s10994-007-5014-x
19. Van Der Malsburg, C.: Frank rosenblatt: principles of neurodynamics: perceptrons and the theory of brain mechanisms. In: Palm, G., Aertsen, A. (eds.) Brain Theory, pp. 245–248. Springer, Heidelberg (1986). http://dx.doi.org/10.1007/978-3-642-70911-1_20
20. Zinkevich, M.: Online convex programming and generalized infinitesimal gradient ascent. In: Fawcett, T., Mishra, N. (eds.) Proceedings of the 20th International Conference on Machine Learning (ICML-03), pp. 928–936 (2003). http://www.aaai.org/Papers/ICML/2003/ICML03-120.pdf

Robust Optimization for Clustering

Xuan Thanh Vo[1(✉)], Hoai An Le Thi[1], and Tao Pham Dinh[2]

[1] Laboratory of Theoretical and Applied Computer Science EA 3097, University of Lorraine, Ile de Saulcy, 57045 Metz, France
{xuan-thanh.vo,hoai-an.le-thi}@univ-lorraine.fr

[2] Laboratory of Mathematics, INSA - Rouen, University of Normandie, 76801 Saint-Etienne-du-Rouvray Cedex, France
pham@insa-rouen.fr

Abstract. In this paper, we investigate the robust optimization for the minimum sum-of squares clustering (MSSC) problem. Each data point is assumed to belong to a box-type uncertainty set. Following the robust optimization paradigm, we obtain a robust formulation that can be interpreted as a combination of MSSC and k-median clustering criteria. A DCA-based algorithm is developed to solve the resulting robust problem. Preliminary numerical results on real datasets show that the proposed robust optimization approach is superior than MSSC and k-median clustering approaches.

Keywords: Robust optimization · Clustering · DC programming · DCA

1 Introduction

Clustering is a powerful exploratory data mining technique and has many applications in various fields. It aims at dividing a given dataset into groups (clusters) of similar objects according to certain similarity metric measure. Approaches to clustering can be classified into two groups namely partitional and hierarchical. In this paper, we focus on the partitional approach with a specific description as follows. Given a dataset $\mathcal{X} = \{x_1, \ldots, x_m\}$ of m points in $\mathbb{R}^n$, a "distance" measure d defined on $\mathbb{R}^n \times \mathbb{R}^n$, and an integer $c\,(2 \leq c \leq m)$, the problem is to determine c centers (or centroids) v_i $(i = 1, \ldots, c)$ in $\mathbb{R}^n$ such that the sum of distances of each member of $\mathcal{X}$ to its closest center is minimized. This problem can be mathematically formulated as follows.

$$\min_{U,V} \left\{ \frac{1}{2} \sum_{k=1}^{m} \sum_{i=1}^{c} u_{i,k} d(x_k, v_i) : U \in \mathcal{M},\ V = (v_1, \ldots, v_c) \in \mathbb{R}^{n \times c} \right\} \tag{1}$$

where $U = (u_{i,k}) \in \mathbb{R}^{c \times m}$ is the matrix of memberships: $u_{i,k} = 1$ if the object x_k is assigned to cluster i^{th} and 0 otherwise,

$$\mathcal{M} = \{U = (u_{i,k}) \in \mathbb{R}^{c \times m} : \sum_{i=1}^{c} u_{i,k} = 1,\ u_{i,k} \in \{0, 1\},\ \forall i, k\}.$$

N.T. Nguyen et al. (Eds.): ACIIDS 2016, Part II, LNAI 9622, pp. 671–680, 2016.
DOI: 10.1007/978-3-662-49390-8_65

If d is the squared Euclidean distance, the problem (1) is known as the minimum sum-of squares clustering (MSSC) [16,19]. The k-means [20] is the most popular algorithm for this clustering formulation. A less popular clustering formulation is the use of the ℓ_1-norm distance metric (Manhattan distance) equipped with k-median algorithm [2]. In what follows, we will use the term *k-median clustering* to indicate the clustering using the ℓ_1-norm distance metric.

Although the k-means is the most commonly used algorithm for clustering (MSSC formulation), it is sensitive with respect to outliers and does not work well in a noisy environment and inaccuracy. Several variants of the k-means have been proposed to deal with this drawback and k-median is one of them. In this paper, we follow the robust optimization approach [1]: assume that the input data are inaccurate by arbitrary disturbances and minimize the worst possible objective value of (1) under such disturbances. We consider a *box-type* uncertainty for the MSSC clustering model and show that the resulting robust problem is a combination of the MSSC and the clustering using ℓ_1-norm distance metric. Our method for solving the robust problem is based on DC (Difference of Convex functions) programming and DCA (DC Algorithms) (see [15,18,21] and the references therein). Our work is motivated by the fact that DCA is a powerful method for many (smooth or nonsmooth) large-scale nonconvex programs in various domains of applied sciences, particularly in Machine Learning [5–14,16,17,19].

Robust optimization has been widely applied in Machine Learning, especially in classification (see [22] and the references therein). However, to our knowledge, this is the first time the use of robust optimization in clustering is addressed.

For beginning, let us describe the notations used in this paper. All vectors will be column vectors unless transposed to a row vector by a superscript T. For a vector $x = (x_1, \ldots, x_n) \in \mathbb{R}^n$, $|x|$ will denote a vector in $\mathbb{R}^n$ of absolute values of the components of x, that is $|x| = (|x_1|, \ldots, |x_n|)$. For vectors $x, y, z \in \mathbb{R}^n$, the inner product is $\langle x, y\rangle = \sum_{i=1}^n x_i y_i$, the notation $x \leq y$ means that $x_i \leq y_i$ for any $i = 1, \ldots, n$, the element–wise product $z = x \circ y$ means $z_i = x_i y_i, \forall i$, and the notation $\text{sgn}(z)$ denotes the vector of components $(\text{sgn}(z))_i = \text{sgn}(z_i) = 1$ if $z_i > 0$ and -1 otherwise.

The rest of this paper is organized as follows. The next section constructs the robust formulation of clustering problem under the box uncertainty. In Sect. 3 we briefly present DC programming and DCA for general DC programs. Section 4 shows how to apply DCA to solve our robust clustering problem. Finally, the numerical experiments are presented in Sects. 5 and 6 concludes the paper.

2 Robust Formulation for Clustering

We consider that the input data points x_k $(k = 1, \ldots, m)$ are subject to perturbations on each of their coordinates. These perturbations could be represented by additional vectors Δx_k $(k = 1, \ldots, m)$. We assume that Δx_k's vary in box-type uncertainty sets, i.e. $\Delta x_k \in B_k = \prod_{j=1}^n [-(w_k)_j, (w_k)_j]$, with $w_k \geq 0$ are known. Considering problem (1) with squared Euclidean distance and following the robust optimization paradigm, we aim to find the optimal solution in the worst case sense. That is, we need to solve the following min–max problem

$$\min_{U,V}\left\{\frac{1}{2}\sum_{k=1}^{m}\max_{\Delta x_k\in B_k}\left(\sum_{i=1}^{c}u_{i,k}\|(x_k+\Delta x_k)-v_i\|^2\right):U\in\mathcal{M},\,V\in\mathbb{R}^{n\times c}\right\}. \quad (2)$$

Using the condition $\sum_{i=1}^{c}u_{i,k}=1$, we have

$$\arg\max_{\Delta x_k\in B_k}\sum_{i=1}^{c}u_{i,k}\|(x_k+\Delta x_k)-v_i\|^2=$$

$$\arg\max_{\Delta x_k\in B_k}\left\{\|\Delta x_k\|^2-2\langle\Delta x_k,\sum_{i=1}^{c}u_{i,k}(v_i-x_k)\rangle\right\}=-\text{sgn}\Big(\sum_{i=1}^{c}u_{i,k}(v_i-x_k)\Big)\circ w_k.$$

Therefore,

$$\begin{aligned}
&\max_{\Delta x_k\in B_k}\sum_{i=1}^{c}u_{i,k}\|(x_k+\Delta x_k)-v_i\|^2\\
&=\sum_{i=1}^{c}u_{i,k}\|x_k-v_i\|^2+2\langle w_k,\Big|\sum_{i=1}^{c}u_{i,k}(v_i-x_k)\Big|\rangle+\|w_k\|^2\\
&=\sum_{i=1}^{c}u_{i,k}\left(\|x_k-v_i\|^2+2\langle w_k,|v_i-x_k|\rangle+\|w_k\|^2\right).
\end{aligned}$$

The last equality is by the conditions $\sum_{i=1}^{c}u_{i,k}=1$ and $u_{i,k}\in\{0,1\}$.

Then the robust problem (2) becomes

$$\min_{U,V}\left\{\frac{1}{2}\sum_{k=1}^{m}\sum_{i=1}^{c}u_{i,k}\left(\|x_k-v_i\|^2+2\langle w_k,|v_i-x_k|\rangle+\|w_k\|^2\right):U\in\mathcal{M},V\in\mathbb{R}^{n\times c}\right\}$$

This problem can be reformulated as a bilevel program as follows

$$\min\left\{F(V):=\frac{1}{2}\sum_{k=1}^{m}\min_{i=1,\dots,c}d_B(x_k,v_i):v_i\in\mathbb{R}^n,\ i=1,\dots,c\right\} \quad (3)$$

where $d_B(x_k,v_i)=\|x_k-v_i\|^2+2\langle w_k,|v_i-x_k|\rangle+\|w_k\|^2$.

In Sect. 4, we will develop an algorithm based on DC programming and DCA to solve the problem (3).

3 Outline of DC Programming and DCA

In this section we briefly present DC programming and DCA. For a complete study of this subject the reader is referred to [15,18,21] and the references therein.

A general DC program has the form

$$\alpha=\inf\{F(x):=G(x)-H(x)\,|\,x\in\mathbb{R}^n\}\quad(P_{dc}),$$

where G, H are lower semi-continuous proper convex functions on $\mathbb{R}^n$. Such a function F is called a DC function, and $G - H$ is a DC decomposition of F while G and H are the DC components of F. Note that, the closed convex constraint $x \in C$ can be incorporated in the objective function of (P_{dc}) by using the indicator function on C denoted by χ_C which is defined by $\chi_C(x) = 0$ if $x \in C$, and $+\infty$ otherwise.

For a convex function θ, the subdifferential of θ at $x_0 \in \text{dom}\theta := \{x \in \mathbb{R}^n : \theta(x_0) < +\infty\}$, denoted by $\partial\theta(x_0)$, is defined by

$$\partial\theta(x_0) := \{y \in \mathbb{R}^n : \theta(x) \geq \theta(x_0) + \langle x - x_0, y\rangle, \forall x \in \mathbb{R}^n\}.$$

A point x^* is called a *critical point* of $G - H$, or a generalized Karush-Kuhn-Tucker point (KKT) of (P_{dc})) if

$$\partial H(x^*) \cap \partial G(x^*) \neq \emptyset. \tag{4}$$

Based on local optimality conditions and duality in DC programming, the DCA consists in constructing two sequences $\{x^k\}$ and $\{y^k\}$ (candidates to be solutions of (P_{dc}) and its dual problem respectively). Each iteration k of DCA approximates the concave part $-H$ by its affine majorization (that corresponds to taking $y^k \in \partial H(x^k)$) and minimizes the resulting convex function.

Generic DCA Scheme
Initialization: Let $x^0 \in \mathbb{R}^n$ be an initial guess, $k \leftarrow 0$.
Repeat
- Calculate $y^k \in \partial H(x^k)$
- Calculate $x^{k+1} \in \arg\min\{G(x) - \langle x, y^k\rangle : x \in \mathbb{R}^n\} \quad (P_k)$
- $k \leftarrow k + 1$

Until convergence of $\{x^k\}$.

Convergences properties of DCA and its theoretical basic can be found in [15,18,21]. It is worth mentioning that

- DCA is a descent method (*without linesearch*): the sequence $\{G(x^k) - H(x^k)\}$ is decreasing.
- If $G(x^{k+1}) - H(x^{k+1}) = G(x^k) - H(x^k)$, then x^k is a critical point of $G - H$. In such a case, DCA terminates at k-th iteration.
- If the optimal value α of problem (P_{dc}) is finite and the infinite sequences $\{x^k\}$ and $\{y^k\}$ are bounded then every limit point x^* of the sequence $\{x^k\}$ is a critical point of $G - H$.
- DCA has a *linear convergence* for general DC programs.

A deeper insight into DCA has been described in [15]. For instant it is crucial to note the main feature of DCA: DCA is constructed from DC components and their conjugates but not the DC function f itself which has infinitely many DC decompositions, and there are as many DCA as there are DC decompositions. Such decompositions play a crucial role in determining the speed of convergence, stability, robustness, and globality of sought solutions. It is important to study various equivalent DC formulations of a DC problem. This flexibility of DC programming and DCA is of particular interest from both a theoretical and an algorithmic point of view.

4 Solving the Robust Problem (3)

Because of convexity of $d_B(x_k, v_i)$ w.r.t v_i, $F(V)$ has a DC decomposition $F(V) = G(V) - H(V)$ where DC components G, H are given by

$$G(V) = \frac{1}{2}\sum_{k=1}^{m}\sum_{i=1}^{c} d_B(x_k, v_i), \quad H(V) = \sum_{k=1}^{m} H_k(V), \tag{5}$$

with $H_k(V) = \max_{j=1,\ldots,c} \frac{1}{2}\sum_{i=1,i\neq j}^{c} d_B(x_k, v_i)$ for any $k = 1, \ldots, m$.

DCA for solving the problem (3) consists of constructing two sequence $\{V^{(t)}\}$ and $\{Y^{(t)}\}$ such that

- $Y^{(t)} \in \partial H(V^{(t)})$ for all $t = 0, 1, 2, \ldots$, and
- $V^{(t+1)}$ solves the convex sub-problem

$$\min\left\{\frac{1}{2}\sum_{k=1}^{m}\sum_{i=1}^{c} d_B(x_k, v_i) - \sum_{i=1}^{c}\langle y_i^{(t)}, v_i\rangle : v_i \in \mathbb{R}^n,\ i = 1, \ldots, c\right\}. \tag{6}$$

- *Calculation of* $Y \in \partial H(V)$*:* We have $\partial H(V) = \sum_{k=1}^{m} \partial H_k(V)$. Thus,

$$Y \in \partial H(V) \Leftrightarrow Y = \sum_{k=1}^{m} Y^{[k]} \text{ with } Y^{[k]} \in \partial H_k(V) \text{ for } k = 1, \ldots, m.$$

For $k = 1, \ldots, m$, we have $H_k(V) = \max_{j=1,\ldots,c} H_{k,j}(V)$, where

$$H_{k,j}(V) = \frac{1}{2}\sum_{i=1,i\neq j}^{c} d_B(x_k, v_i). \tag{7}$$

Denote by $I_k(V) = \arg\max_{j=1,\ldots,c} H_{k,j}(V) = \arg\min_{j=1,\ldots,c} d_B(x_k, v_j)$. Then a subgradient $Y^{[k]} \in \partial H_k(V)$ is given by

$$Y^{[k]} = Y^{[k,j(k)]} \in \partial H_{k,j(k)}(V),$$

where $j(k) \in I_k(V)$. Besides, a subgradient $Y^{[k,j]} \in \partial H_{k,j}(V)$ is given by

$$y_j^{[k,j]} = 0; \quad y_i^{[k,j]} = (v_i - x_k) + w_k \circ \operatorname{sgn}(v_i - x_k), \forall i \neq j. \tag{8}$$

Or, under matrix form,

$$Y^{[k,j]} = V - X^{[k]} + W^{[k]} \circ \operatorname{sgn}(V - X^{[k]}) - [(v_j - x_k) + w_k \circ \operatorname{sgn}(v_j - x_k)]e_j^T, \tag{9}$$

where e_j $(j = 1, \ldots, c)$ is the canonical basis of $\mathbb{R}^c$ and $X^{[k]}$ (resp. $W^{[k]}$) is the $n \times c$ matrix whose columns are all equal to x_k (resp. w_k).

Therefore, a $Y \in \partial H(V)$ is given by

$$Y = m(V-B)+\sum_{k=1}^{m} W^{[k]}\circ \mathrm{sgn}(V-X^{[k]})-\sum_{k=1}^{m}[(v_{j(k)}-x_k)+w_k\circ \mathrm{sgn}(v_{j(k)}-x_k)]e_{j(k)}^T, \quad (10)$$

where B is the $n \times c$ matrix whose columns are all equal to $\bar{x} = \frac{1}{m}\sum_{k=1}^{m} x_k$.

• *Solving the sub-problem:* Solving problem (6) boils down to solving c convex problems

$$\min\left\{\frac{1}{2}\sum_{k=1}^{m} d_B(x_k, v_i) - \langle y_i^{(t)}, v_i\rangle : v_i \in \mathbb{R}^n\right\}, \quad i = 1,\ldots,c. \quad (11)$$

Or equivalently,

$$\min\left\{\frac{1}{2}\left\|v_i - \left(\frac{y_i^{(t)}}{m} + \bar{x}\right)\right\|^2 + \sum_{k=1}^{m}\langle\frac{w_k}{m}, |v_i - x_k|\rangle : v_i \in \mathbb{R}^n\right\}, \quad i = 1,\ldots,c. \quad (12)$$

Solving these problems boils down to solving $c \times n$ univariate convex problems of the form (13) (cf. Appendix A).

For summary, DCA for solving the problem (3) is described as follows.

DCA-Box
Initialization: Let $\epsilon > 0$, $V^{(0)} \in \mathbb{R}^{n\times c}$, $t = 0$.
Repeat
- Compute $Y^{(t)} \in \partial H(V^{(t)})$ by using (10)
- Compute $V^{(t+1)}$ by using Algorithm 1 (cf. Appendix A) to compute, $\forall i = 1,\ldots,c;\ j = 1\ldots,n$,

$$v_{ij}^{(t+1)} = \arg\min\left\{\frac{1}{2}\left(v_i - \left(\frac{y_{ij}^{(t)}}{m} + \bar{x}_j\right)\right)^2 + \sum_{k=1}^{m}\frac{w_{kj}}{m}|v_{ij} - x_{kj}| : v_{ij} \in \mathbb{R}\right\},$$

- $t \leftarrow t+1$.
Until $\|V^{(t+1)} - V^{(t)}\| < \epsilon(\|V^{(t)}\|+1)$ or $|F(V^{(t+1)}) - F(V^{(t)})| < \epsilon(|F(V^{(t)})|+1)$.

5 Numerical Experiments

In this section, we carry out some experiments on real datasets to validate the proposed method. We will compare the robust clustering (3) (using Algorithm DCA-Box) with MSSC and k-median clustering. For k-median clustering, we use the algorithm developed in [2]. For MSSC, we use the DCA-based algorithm developed in [16] that has been proved to be better than the popular k-means algorithm [20].

For algorithm DCA-Box, the parameters w_k's are determined as follows

$$(w_k)_j = \left(\frac{1}{m}\sum_{l=1}^{m}((x_l)_j - \bar{x}_j)^2\right)^{1/2}, \quad \forall k = 1,\ldots,m; j = 1,\ldots,n.$$

Table 1. Real datasets used in experiments

Dataset	# samples	# features	# classes
Iris	150	4	3
Papillon	23	4	4
Satellite images	4435	36	6
Vote	435	16	2
Wine	178	13	3
Comp	3891	10	3
Glass	214	10	6
Ecoli	336	7	8

Table 2. Comparative results on real datasets

Dataset	Criterion	K-Median	MSSC	DCA-Box
Iris	Adj RI	0.717	0.730	**0.744**
	F-measure	0.884	0.891	**0.898**
Papillon	Adj RI	0.315	**0.517**	0.315
	F-measure	0.616	**0.797**	0.616
Satellite images	Adj RI	**0.546**	0.466	0.506
	F-measure	**0.724**	0.684	0.703
Vote	Adj RI	0.529	0.543	**0.584**
	F-measure	0.866	0.870	**0.884**
Wine	Adj RI	0.371	0.371	**0.375**
	F-measure	0.719	0.714	**0.720**
Comp	Adj RI	0.902	**0.928**	0.907
	F-measure	0.966	**0.975**	0.967
Glass	Adj RI	0.550	0.527	**0.568**
	F-measure	0.669	0.656	**0.676**
Ecoli	Adj RI	0.386	0.401	**0.437**
	F-measure	0.614	0.635	**0.689**

The quality of clustering solutions was evaluated by two criteria: Adjusted Rand index [3] and F-measure [4]. The higher these criteria, the better clustering performance.

We test on 8 real datasets taken from UCI[1] (Table 1). For each dataset, we run each algorithm 50 times from random initializations and select the solution corresponding to the smallest objective value. The results are presented in Table 2.

[1] http://archive.ics.uci.edu/ml/.

Observe from the experiment results, we see that DCA-Box outperforms k-median clustering and MSSC on five datasets. On the other three datasets, except for Papillon dataset, DCA-Box is better than either k-median or MSSC.

6 Conclusion

This work considered the robust optimization approach in the clustering problem. By assuming uncertainty of the input data in which box–type uncertainty sets are considered, we obtained a robust formulation from the clustering model using minimum sum-of-squares Euclidean distance. This robust formulation can be expressed as a combination of the cluster-compactness criterion using square Euclidean distance and the one using Manhattan distance. Thus, the new robust formulation can take advantage over minimum sum-of-squares clustering and k-median clustering methods. Based on DC programming and DCA, we have developed an efficient algorithm to solve the resulting optimization problem. Comparative numerical results also showed the efficiency and the superiority of the proposed robust optimization approach with respect to the standard MSSC and k-median clustering. In future works, we will consider other kinds of uncertainty sets and study the relation of robust optimization approach with other robust approaches in the clustering problem.

Acknowledgements. This research is funded by Foundation for Science and Technology Development of Ton Duc Thang University (FOSTECT), website: http://fostect.tdt.edu.vn, under Grant FOSTECT.2015.BR.15.

A Appendix

Given $a, a_i \in \mathbb{R}$, $b_i \in \mathbb{R}_{++}$ $(i = 1, \dots, m)$, consider the problem

$$\min \left\{ f(x) := \frac{1}{2}(x-a)^2 + \sum_{i=1}^{m} b_i |x - a_i| : x \in \mathbb{R} \right\}. \tag{13}$$

Assume that $\{a_i\}_{i=1}^m$ is in ascending order $a_1 \le a_2 \le \dots \le a_m$. Denote by $f^-(x)$ (resp. $f^+(x)$) the left (resp. right) derivative of f at x. We have

$$f^-(x) = x - a + \sum_{i=1}^{m} b_i \delta_i, \quad f^+(x) = x - a + \sum_{i=1}^{m} b_i \sigma_i, \tag{14}$$

where $\delta_i = -1$ if $x \le a_i$ and 1 otherwise, $\sigma_i = -1$ if $x < a_i$ and 1 otherwise $(\forall i = 1, \dots, m)$. For convenience, let $a_0 = -\infty$ and $a_{m+1} = +\infty$. Note that (13) is strongly convex, so the solution x^* is unique. We can find out the place where x^* is by using the following property.

Proposition 1. *Let* $\bar{a} = \arg\min \{\sum_{i=1}^m b_i |x - a_i| : x \in \mathbb{R}\}$, $l_b = \min(a, \bar{a})$, *and* $u_b = \max(a, \bar{a})$. *We have the following assertions:*

(i) $x^* \in [l_b, u_b]$.
(ii) *If* $f^-(a_i) > 0$, *then* $x^* \in (a_0, a_i)$. *If* $f^+(a_i) < 0$, *then* $x^* \in (a_i, a_{m+1})$. *If* $f^-(a_i) \leq 0$ *and* $f^+(a_i) \geq 0$, *then* $x^* = a_i$.

Proof. Given g is any finite convex function on $\mathbb{R}$. We have $\partial g(x) = [f^-(x), f^+(x)]$ for any $x \in \mathbb{R}$, and $\tilde{x} \in \arg\min_{x \in \mathbb{R}} g(x) \Leftrightarrow 0 \in \partial g(\tilde{x}) \Leftrightarrow f^-(\tilde{x}) \leq 0 \leq f^+(\tilde{x})$. Moreover, if $\tilde{x}$ is the unique optimum of g on $\mathbb{R}$, then for any $x \neq \tilde{x}$,

$$0 > g(\tilde{x}) - g(x) \geq y(\tilde{x} - x), \quad \forall y \in \partial g(x).$$

Therefore, $g^+(x) < 0$ if $x < \tilde{x}$, and $g^-(x) > 0$ if $x > \tilde{x}$. We have ii) is proved.

Let $f_1(x) = \frac{1}{2}(x-a)^2$ and $f_2(x) = \sum_{i=1}^m b_i|x - a_i|$. We have f_1 and f_2 are finite convex functions on $\mathbb{R}$, and a (resp. $\bar{a}$) is optimum of f_1 (resp. f_2) on $\mathbb{R}$. Without loss of generality we assume that $a \leq \bar{a}$. Then $f^+(a) = f_2^+(a) \leq 0$ and $f^-(\bar{a}) = f_1^-(\bar{a}) \geq 0$. This implies that $a \leq x^* \leq \bar{a}$. Thus, i) is proved.

Once we find out the interval where x^* is and f is differentiable, x^* is easily determined by solving the equation $f'(x) = 0$. The specific procedure for finding the solution x^* of problem (13) is given in Algorithm 1. It is clear that Algorithm 1 terminates after at most m steps.

Algorithm 1. Solving problem (13)

for $i = 1, \ldots, m$ **if** $l_b \leq a_i \leq u_b$ **do**
  Compute $f^-(a_i)$ using (14)
  if $f^-(a_i) > 0$ (then $x^* \in (a_{i-1}, a_i)$ and $f'(x^*) = x^* - a + \sum_{j \leq i-1} b_j - \sum_{j>i} b_j = 0$) **then**
    $x^* = a - \sum_{j \leq i-1} b_j + \sum_{j>i} b_j \Rightarrow$ STOP.
  end if
  Compute $f^+(a_i)$ using (14)
  if $f^+(a_i) \geq 0$ **then**
    $x^* = a_i \Rightarrow$ STOP.
  else
    if $i = m$ (then $x^* \in (a_m, a_{m+1})$ and $f'(x^*) = x^* - a + \sum_{j=1}^m b_j = 0$) **then**
      $x^* = a - \sum_{j=1}^n b_j \Rightarrow$ STOP.
    end if
  end if
end for

References

1. Ben-Tal, A., El Ghaoui, L., Nemirovski, A.: Robust Optimization. Princeton University Press, Princeton (2009)
2. Bradley, P.S., Mangasarian, O.L., Street, W.N.: Clustering via concave minimization. In: Mozer, M.C., Jordan, M.I., Petsche, T. (eds.) NIPS 9, pp. 368–374. MIT Press, Cambridge, MA (1997)

3. Hubert, L., Arabie, P.: Comparing partitions. J CLASSIF **2**, 193–218 (1985)
4. Gullo, F., Tagarelli, A.: Uncertain centroid based partitional clustering of uncertain data. Proc. VLDB Endowment (ACM) **5**(7), 610–621 (2012)
5. Le Thi, H.A., Le, H.M., Nguyen, V.V., Pham, D.T.: A DC programming approach for feature selection in support vector machines learning. J. Adv. Data Anal. Classif. **2**, 259–278 (2008)
6. Le Thi, H.A., Le, H.M., Pham, D.T.: Fuzzy clustering based on nonconvex optimisation approaches using difference of convex (DC) functions algorithms. J. Adv. Data Anal. Classif. **2**, 1–20 (2007)
7. Le Thi, H.A., Le, H.M., Pham, D.T., Huynh, V.N.: Binary classification via spherical separator by DC programming and DCA. J. Global Optim. **56**(4), 1393–1407 (2013)
8. An, L.T.H., Cuong, N.M.: Efficient algorithms for feature selection in multi-class support vector machine. In: Nguyen, N.T., van Do, T., Thi, H.A. (eds.) ICCSAMA 2013. SCI, vol. 479, pp. 41–52. Springer, Heidelberg (2013)
9. Le Thi, H.A., Vo, X.T., Pham, D.T.: Robust feature selection for SVMs under uncertain data. In: Perner, P. (ed.) ICDM 2013. LNCS, vol. 7987, pp. 151–165. Springer, Heidelberg (2013)
10. Le Thi, H.A., Le, H.M., Pham, D.T.: Feature selection in machine learning: an exact penalty approach using a difference of convex function algorithm. Machine-Learning (published online 04.07.14). doi:10.1007/s10994-014-5455-y
11. Le Thi, H.A., Le, H.M., Pham, D.T.: New and efficient DCA based algorithms for minimum sum-of-squares clustering. Pattern Recogn. **47**(1), 388–401 (2014)
12. Le Thi, H.A., Vo, X.T., Pham, D.T.: Feature selection for linear SVMs under uncertain data: robust optimization based on difference of convex functions algorithms. Neural Netw. **59**, 36–50 (2014)
13. Le Thi, H.A., Nguyen, M.C., Pham, D.T.: A DC programming approach for finding communities in networks. Neural Comput. **26**(12), 2827–2854 (2014)
14. Le Thi, H.A., Pham, D.T., Le, H.M., Vo, X.T.: DC approximation approaches for sparse optimization. Eur. J. Oper. Res. **244**(1), 26–46 (2015)
15. Le Thi, H.A., Pham, D.T.: The DC (difference of convex functions) Programming and DCA revisited with DC models of real world nonconvex optimization problems. Ann. Oper. Res. **133**, 23–46 (2005)
16. Le Thi, H.A., Belghiti, T., Pham, D.T.: A new efficient algorithm based on DC programming and DCA for clustering. J. Glob. Optim. **37**, 593–608 (2006)
17. Le Thi, H.A., Le, H.M., Pham, D.T.: Optimization based DC programming and DCA for hierarchical clustering. Eur. J. Oper. Res. **183**, 1067–1085 (2007)
18. Pham, D.T., Le Thi, H.A.: Convex analysis approach to DC programming: theory, algorithms and applications. Acta Math. Vietnamica **22**(1), 289–357 (1997)
19. Le Thi, H.A., Le, H.M., Pham, D.T.: New and efficient DCA based algorithms for minimum sum-of-squares clustering. Pattern Recogn. **47**(1), 388–401 (2014)
20. MacQueen, J.B.: Some methods for classification and analysis of multivariate observations. In: Proceedings of Berkeley Symposium on Mathematical Statistics and Probability, pp. 281–297 (1967)
21. Pham, D.T., Le Thi, H.A.: DC optimization algorithms for solving the trust region subproblem. SIAM J. Oppt. **8**, 476–505 (1998)
22. Xu, H., Caramanis, C., Mannor, S.: Robustness and regularization of support vector machines. J. Mach. Learn. Res. **10**, 1485–1510 (2009)

Using Valid Inequalities and Different Grids in LP-Based Heuristic for Packing Circular Objects

Igor Litvinchev(✉), Luis Infante, and Edith Lucero Ozuna Espinosa

Department of Mechanical and Electrical Engineering,
Nuevo Leon State University, Monterrey, Nuevo Leon, Mexico
igorlitvinchev@gmail.com

Abstract. Using a regular grid to approximate a container, packing is reduced to assigning objects to the nodes of the grid subject to non-overlapping constraints. The packing problem is then stated as a large scale linear 0–1 optimization problem. A problem of packing unequal circles in a fixed size rectangular container is considered. The circle is considered as a set of points that are all the same distance (not necessary Euclidean) from a given point. Different shapes, such as ellipses, rhombuses, rectangles, octagons, etc. can be treated similarly by simply changing the definition of the norm used to define the distance. Valid inequalities are used to strengthening the LP-relaxation. An LP-based heuristic is proposed. Numerical results on packing circles and octagons are presented to demonstrate the efficiency of the proposed approach.

Keywords: Packing · Integer programming · Large scale optimization · Heuristic

1 Introduction

Packing a set of items of known dimensions into one or more large objects or containers to minimize a certain objective (e.g., the unused part of the objects or waste) constitutes a family of natural combinatorial optimization problems applied in computer science, industrial engineering, logistics, manufacturing and production processes, material science and nanotechnology, medicine [2, 5–9, 11, 21, 22].

Packing problems for regular shapes (circles and rectangles) of objects and/or containers are well studied (see, e.g., a review [12] for circle packing). In circle packing problem the aim is to place a certain number of circles, each one with a fixed known radius inside a container. The circles must be totally placed in the container without overlapping. The shape of the container may vary from a circle, a square, a rectangular, etc.

Many variants of packing circular objects in the plane have been formulated as nonconvex (continuous) optimization problems with decision variables being coordinates of the centres. The nonconvexity is mainly provided by no overlapping conditions between circles. These conditions typically state that the Euclidean distance separating the centres of the circles is greater than a sum of their radii. The nonconvex problems can be tackled by available nonlinear programming (NLP) solvers; however

N.T. Nguyen et al. (Eds.): ACIIDS 2016, Part II, LNAI 9622, pp. 681–690, 2016.
DOI: 10.1007/978-3-662-49390-8_66

most NLP solvers fail to identify global optima. Thus, the nonconvex formulation of circular packing problem requires algorithms which mix local searches with heuristic procedures in order to widely explore the search space. It is impossible to give a detailed overview on the existing solution strategies and numerical results within the framework of a single short paper. We will refer the reader to review papers presenting the scope of techniques and applications for the circle packing problem (see, e.g., [1, 4, 17, 18, 20] and the references therein).

In this paper we study packing circular-like objects using a regular grid to approximate the container. The circular-like object is considered in a general sense, as a set of points that are all the same distance (not necessary Euclidean) from a given point. Thus different shapes, such as ellipses, rhombuses, rectangles, octagons can be treated the same way by simply changing the norm used to define the distance. The nodes of the grid are considered as potential positions for assigning centers of the circles. The packing problem is then stated as a large scale linear 0–1 optimization problem. Valid inequalities are used to strengthening the original formulation and improve LP-bounds. Reduced costs of the LP-solution are used to fix some variables in the original problem to get an approximate integer solution. Numerical results on packing circles and regular octagons are presented to demonstrate efficiency of the proposed approach. This work is a continuation of [16]. The rest of the paper is organized as follows. Section 2 presents the main constructions: basic binary model and valid inequalities. Heuristic LP-based algorithm and results of a numerical experiment are presented in Sects. 3, 4 while the last section concludes.

2 Basic Model

Suppose we have non-identical circles C_k of known radius R_k, $k \in K = \{1, 2, \ldots K\}$. Here we consider the circle as a set of points that are all the same distance R_k (not necessary Euclidean) from a given point. In what follows we will use the same notation C_k for the figure bounded by the circle, $C_k = \{z \in \mathbb{R}^2 : \|z - z_{0k}\| \leq R_k\}$, assuming that it is easy to understand from the context whether we mean the curve or the figure. Denote by S_k the area of C_k. Suppose that at most M_k circles C_k are available for packing and at least m_k of them have to be packed. Denote by $i \in I = \{1, 2. \ldots, n\}$ the node points of a regular grid covering the rectangular container. Let $F \subseteq I$ be the grid points lying on the boundary of the container. Denote by d_{ij} the distance (in the sense of norm used to define the circle) between points i and j of the grid. Define binary variables $x_i^k = 1$ if centre of a circle C_k is assigned to the point i; $x_i^k = 0$ otherwise.

Since the circle C_k assigned to the point i has to be no-overlapping with other circles packed, then we have $x_j^l = 0$ for $j \in I$, $l \in K$, such that $d_{ij} < R_k + R_l$. For fixed i, k let $N_{ik} = \{j, l : i \neq j,\ d_{ij} < R_k + R_l\}$. Let n_{ik} be the cardinality of $N_{ik} : n_{ik} = |N_{ik}|$. Then the problem of maximizing the area covered by the circles can be stated as follows:

$$\max \sum_{i \in I} \sum_{k \in K} S_k x_i^k \tag{1}$$

subject to

$$m_k \leq \sum_{i \in I} x_i^k \leq M_k, \quad k \in K, \tag{2}$$

$$\sum_{k \in K} x_i^k \leq 1, \quad i \in I \backslash F, \tag{3}$$

$$R_k x_i^k \leq \min_{j \in F} d_{ij}, \quad i \in I,\ k \in K, \tag{4}$$

$$x_i^k + x_j^l \leq 1, \text{ for } i \in I, k \in K, (j,l) \in N_{ik} \tag{5}$$

$$x_i^k \in \{0,1\}, \quad i \in I,\ k \in K \tag{6}$$

Constraints (2) ensure that the number of circles packed is between m_k and M_k; constraints (3) that at most one centre is assigned to any grid point; constraints (4) that the point i can not be a centre of the circle C_k if the distance from i to the boundary is less than R_k; pair-wise constraints (5) guarantee that there is no overlapping between the circles; constraints (6) represent the binary nature of variables. More details on the problem (1)–(6) and its properties one can find in [14–16]. In what follows we will assume for simplicity that for the case $K = 1$ the objective function (1) be $\sum_{i \in I} x_i$, the number of circles packed.

Note that all constructions proposed above remain valid for any norm used to define a circular-like object. In fact, changing the norm affects only the distance d_{ij} used in the definitions of the sets N_{ik} in the non-overlapping constraints (5). That is, by simple pre-processing we can use the basic model (1)–(6) for packing different geometrical objects of the same shape. It is important to note that the non-overlapping condition has the form $d_{ij} \geq R_k + R_l$ no matter which norm is used.

For example, a circular object in the maximum norm $\|z\|_\infty := \max_i\{|z_i|\}$ is represented by a square, taxicab norm $\|z\|_1 := \sum |z_i|$ yields a rhombus. In a similar way we may manage rectangles, ellipses, etc. Using a superposition of norms, we can consider more complex circular objects. For

$$\|z\| := \max_i\{|z_i|, \gamma \sum |z_i|\} \tag{7}$$

and a suitable $0.5 < \gamma < 1$ we get an octagon, an intersection of a square and a rhombus.

3 LP-Based Heuristic

We may expect that the linear programming (LP) relaxation of the problem (1)–(6), i.e., substituting $x_i^k \in \{0,1\}$ for $0 \leq x_i^k \leq 1$, provides a poor upper bound (LP bound) for the optimal objective. For example, for $K = 1$ and suitable M_k, m_k the point $x_i^k = 0.5$ for all $i \in I$ may be feasible to the relaxed problem with the corresponding objective growing linearly with respect to the number of grid points.

To tightening the LP-relaxation for (1)–(6), we consider valid inequalities aimed to ensure that no grid point is covered by two circles. Define matrix $\left[\alpha_{ij}^k\right]$ as follows. Let $\alpha_{ij}^k = 1$ for $d_{ij} < R_k$, $\alpha_{ij}^k = 0$ otherwise. By this definition, $\alpha_{ij}^k = 1$ if the circle C_k centred at i covers point j. The following constraints ensure that no points of the grid can be covered by two circles:

$$\sum_{k \in K} \sum_{j \in I} \alpha_{ij}^k x_j^k \leq 1, \quad i \in I. \tag{8}$$

We can treat (8) as a relaxed non-overlapping conditions and expect that refining the grid reduces overlapping. The valid inequalities (8) hold for any norm used to define the circular object. Numerical experiments presented in [14–16] demonstrate that aggregating valid inequalities (8) to the original problem (1)–(6) improves significantly the value of the corresponding LP-bound. Moreover, valid inequalities change the structure of the optimal LP-solution.

Let G be a set of the nodes of an original grid and FG be a set of the nodes of the refined grid constructed such that $G \subseteq FG$, i.e., all nodes of the original grid remain the nodes of the refined one. Let z_G and z_{FG} be the optimal values of the integer problems (1)–(6) or (1)–(6), (8) obtained for corresponding grids G, FG. Then we have

$$z_G \leq z_{FG} \leq LP_{FG}, \tag{9}$$

where LP_{FG} is the value of the LP-bound corresponding to the grid FG. Note that the problem (1)–(6), (8) still includes constraints (5). Here the first inequality holds since we may construct a feasible solution to the problem corresponding to FG by setting $x_i = 0$ for $i \in FG - G$ and leaving all other components equal to G- optimal solution. Thus we can construct LP-bounds for the original objective using grids different from the original one.

Figure 1 presents LP-solutions and corresponding LP bounds z_{LP} for the problem (1)–(6), (8) for different grids for the case of the container 20×20 and $R = 2.5$ with the objective to pack as many circles as possible. Here the scale of the gray corresponds to the value of the primal variables in the relaxed problem such that totally black corresponds to 1, while totally white to 0. We see that the structure of the LP-solution becomes closer to the optimal one when the number of grid points is increased.

Suppose that a relaxed problem for the grid FG is solved and corresponding reduced costs [19] are known. The heuristic algorithm below aims to reduce the number of variables in integer problem (1)–(6), (8) by fixing $x_i = 0$ for the nodes of G with sufficiently negative reduced costs. We think that the reduced cost is a tighter indicator than the value of the prime variable since it permits to compare variables equal to zero in LP-relaxation.

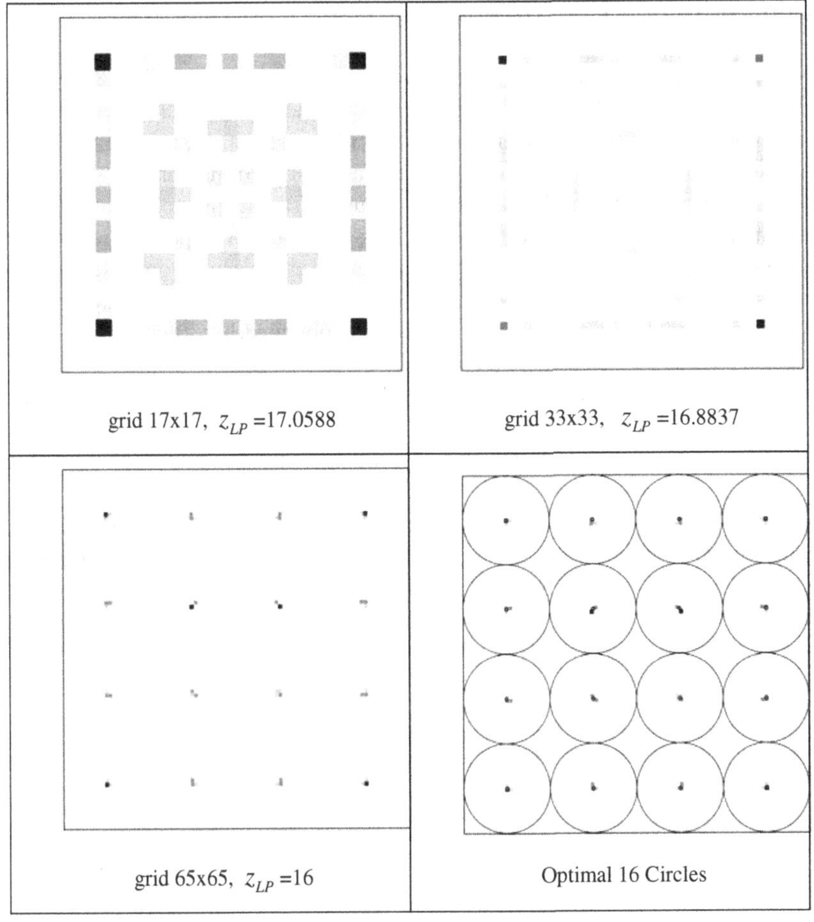

Fig. 1. LP-solutions for different grids.

3.1 LP-Based Heuristic

Step 1. For the original grid G define a refined grid FG, $G \subseteq FG$, and solve LP-relaxation for the grid FG. Denote by d_i, $i \in FG$ the corresponding reduced costs.

Step 2. Define the set of non-positive reduced costs, $FG_{-} = \{i \in FG : d_i \leq 0\}$.

Step 3. For $i \in FG_{-}$ define scaled reduced costs $\bar{d}_i \in [0, 1]$ as follows: $\bar{d}_i = |d_i| / (\max_{i \in FG_{-}} |d_i|)$

Step 4. For a fixed parameter $\delta \in (0, 1)$ define a set of "sufficiently" negative reduced costs: $FG_{\delta-} = \{i \in FG_{-} : \bar{d}_i \geq \delta\}$

Step 5. Solve the integer problem (1)–(6), (9) corresponding to the grid G fixing $x_i = 0$ for $i \in FG_{\delta-} \cap G$.

4 Computational Experiments

In this section we numerically compare LP-relaxations obtained for different grids and study the impact of introducing valid inequalities for the case of packing equal circular objects without the limits (2) for the availability of the objects. A rectangular uniform grid of size Δ along both sides of the container was used to form an initial grid. The test bed set of 9 instances from [10, Table 3] was used for packing maximal number of objects into a rectangular container of width 3 and height 6. All optimization problems were solved by the system CPLEX 12.6. The runs were executed on a desktop computer with CPU AMD FX 8350 8-core processor 4 GHz and 32 GB RAM.

First, we compare linear programming bounds obtained by different grids for circular objects defined by the Euclidian norm (circles). The LP-bound was calculated for the problem (1)–(4), (9), that is non-overlapping constraints (5) were substituted for valid inequalities (8). The following three grids were used: original grid of size Δ generated the same way as in [10, Table 3] with n_Δ nodes and two refined grids with $n_{\Delta/2}$ and $n_{\Delta/3}$ nodes obtained by reducing the original grid size to $\Delta/2$ and $\Delta/3$, respectively. The results of the numerical experiment are given in Table 1. Here the first four columns present instance number, radius R used to define the circular object, original size of the grid Δ and the optimal integer solution value z_I obtained for the grid Δ. The rest of the columns give the number of grid points (n), value of the corresponding LP-bound (z_{LP}) and CPU time T (in seconds).

Table 1. LP-bounds for circles

#	R	Δ	z_I	n_Δ	z_{LP}	T	$n_{\Delta/2}$	z_{LP}	T	$n_{\Delta/3}$	z_{LP}	T
1	0.5	0.125	**18**	697	**19**	0	2673	**18.06**	3	8017	**18.14**	23
2	0.625	0.078125	**10**	1403	**10**	1	5445	**10**	38	16333	**10**	390
3	0.5625	0.0625	**13**	2449	**14.07**	5	9577	**13.96**	130	28729	**13.7**	1500
4	0.375	0.09375	**32**	1425	**36.33**	0	5537	**34.54**	10	16609	**34.75**	88
5	0.3125	0.078125	**45**	2139	**53.4**	1	8357	**50.77**	23	25069	**50.77**	350
6	0.4375	0.546875	**21**	3666	**23.86**	5	14399	**24.01**	200	43195	**24.19**	3400
7	0.25	0.0625	**74**	3649	**90.98**	2	14337	**85.76**	180	43009	**85.24**	3100
8	0.275	0.06875	**61**	2880	**72**	2	11289	**67.78**	70	33865	**67.52**	1100
9	0.1875	0.046875	**140**	6897	**152.9**	35	27233	**151.8**	3401	81697	**151.8**	3591

Table 2 provides results obtained by the heuristic proposed in the previous section for $\delta = 0.1$ and the grid with $n_{\Delta/2}$ nodes used to form the relaxed problem. Here the first four columns present instance number, number of integer variables n_Δ corresponding to the original grid, optimal integer solution z_I and the corresponding CPU time. For all integer problems in Table 2 *mipgap* = 0 was set for running CPLEX. Computations to get integer solution z_I were interrupted after the computational time exceeded 12 h CPU time and the value in parenthesis gives the corresponding *mipgap*. More details on getting z_I one can find in [14–16]. The last three columns give the number of integer variables in the reduced problem $(n_{reduced})$, corresponding integer solution (z_H) and CPU time. For the heuristic the time limit was set to 1 h CPU.

Table 2. Heuristic solutions for circles

#	n_Δ	z_I	CPU	$n_{reduced}$	z_H	CPU
1	697	**18**	1	404	**18**	0
2	1403	**10**	41	718	**10**	4
3	2449	**13**	186	738	**13**	9
4	1425	**32**	4	790	**32**	1
5	2139	**45**	114	1372	**45**	28
6	3666	**21**	17654	1441	**21**	130
7	3649	**74** (5 %)	>12 h	2491	**74**	1400
8	2880	**61**	177	2105	**61**	24
9	6897	**140** (5 %)	>12 h	5649	**139** (1.3 %)	3600

Tables 3, 4 present LP-bounds and heuristic solutions obtained for packing regular octagons corresponding to $\gamma = 1/\sqrt{2}$ in (7). In both Tables CPU time T was limited to 1 h.

Table 3. LP-bounds for octagons

#	R	Δ	z_I	n_Δ	z_{LP}	T	$n_{\Delta/2}$	z_{LP}	T	$n_{\Delta/3}$	z_{LP}	T
1	0.5	0.125	**18**	697	**19**	0	2673	**18.00**	12	8017	**18.00**	20
2	0.625	0.078125	**9**	1403	**10**	1	5445	**9.524**	37	16333	**9.571**	450
3	0.5625	0.0625	**12**	2449	**14.07**	3	9577	**13.18**	140	28729	**13.00**	1700
4	0.375	0.09375	**26**	1425	**30.95**	0	5537	**30.94**	9	16609	**30.94**	80
5	0.3125	0.078125	**41**	2139	**53.40**	1	8357	**50.00**	19	25069	**49.79**	250
6	0.4375	0.546875	**20**	3666	**22.55**	5	14399	**22.75**	200	43195	**22.84**	2800
7	0.25	0.0625	**72**	3649	**90.98**	2	14337	**81.57**	71	43009	**83.99**	170
8	0.275	0.06875	**50**	2880	**59.01**	2	11289	**59.03**	67	33865	**59.21**	570
9	0.1875	0.046875	**106**	6897	**134.3**	25	27233	**134.0**	2800	81697	**134.0**	3487

Table 4. Heuristic solutions for octagons

#	n_Δ	z_I	CPU	$n_{reduced}$	z_H	CPU
1	697	**18**	1	371	**18**	0
2	1403	**9**	52	692	**9**	2
3	2449	**12**	202	752	**12**	5
4	1425	**26**	49	830	**26**	1
5	2139	**41**	6850	1383	**41**	77
6	3666	**20**	1430	1795	**20**	59
7	3649	**72**	22	2261	**72**	25
8	2880	**50**	20495	2254	**50** (4.8 %)	3600
9	6897	**106**	> 12 h.	5652	**106** (5.7 %)	3600

Figures 2, 3, 4 and 5 present optimal packing and grid points left after heuristic node-reduction based on reduced costs.

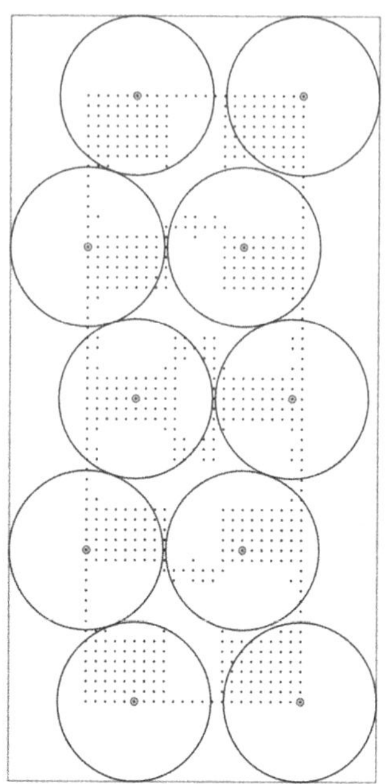

Fig. 2. Circles, instance 2

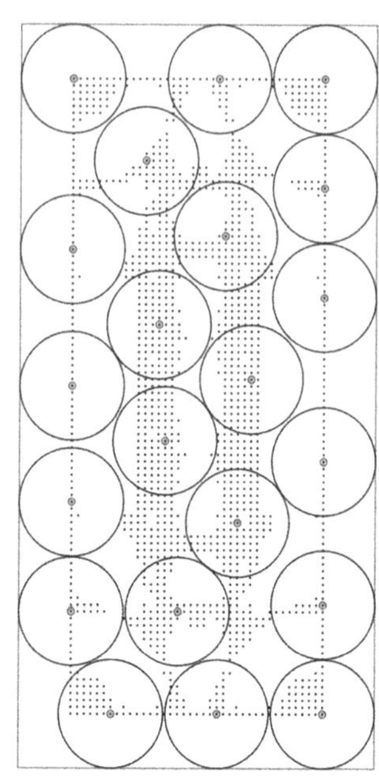

Fig. 3. Circles, instance 6

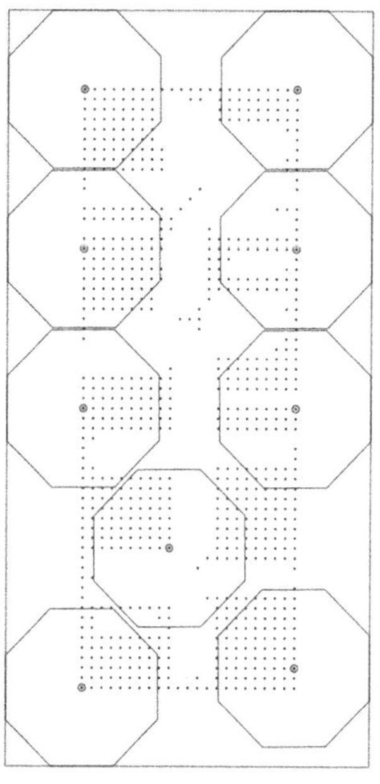

Fig. 4. Octagons, instance 2

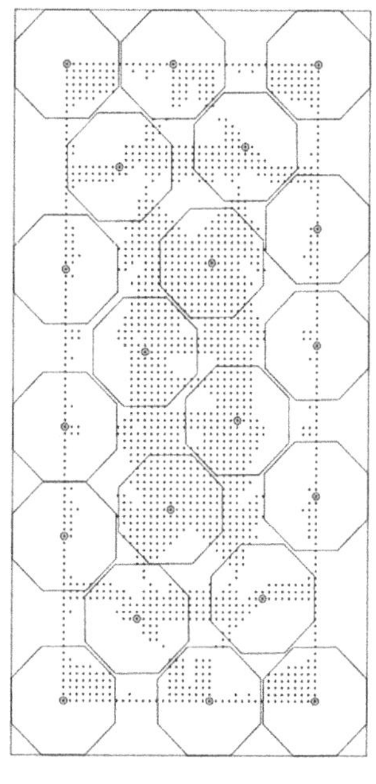

Fig. 5. Octagons, instance 6

As we can see from Tables 1, 3 refining the grid typically (but not always) results in improving the LP-bound. However, solving LP-relaxation for fine grids may be computationally too expensive. Concerning the quality of the integer solution obtained by the heuristic, we may conclude that in most cases (except for the instance 9 for packing circles) the optimal solution was obtained. The use of heuristic reduces CPU time significantly.

5 Conclusions

Integer formulation was proposed for approximated packing circular-like objects in a rectangular container. Different shapes of the objects, such as circles, ellipses, rhombuses, rectangles, octagons can be considered by simply changing the norm used to define the distance. The presented approach can be easily generalized to the three (and more) dimensional case and to different shapes of the container, including irregulars. Valid inequalities are proposed to strengthening the original formulation. A heuristic approach is proposed based on analysis of the reduced costs obtained by LP-relaxation. An interesting direction for the future research is to study the use of Lagrangian relaxation and corresponding heuristics [13] to cope with large dimension of arising problems.

References

1. Akeb, H., Hifi, M.: Solving the circular open dimension problem using separate beams and look-ahead strategies. Comput. Oper. Res. **40**, 1243–1255 (2013)
2. Baltacioglu, E., Moore, J.T., Hill, R.R.: The distributor's three-dimensional pallet-packing problem: a human-based heuristical approach. Int. J. Oper. Res. **1**, 249–266 (2006)
3. Beasley, J.E.: An exact two-dimensional non-guillotine cutting tree search procedure. Oper. Res. **33**, 49–64 (1985)
4. Birgin, E.G., Gentil, J.M.: New and improved results for packing identical unitary radius circles within triangles, rectangles and strips. Comput. Oper. Res. **37**, 1318–1327 (2010)
5. Burtseva, L., Valdez Salas, B., Romero, R., Werner, F.: Recent advances on modeling of structures of multi-component mixtures using a sphere packing approach. Int. J. Nanotechnol. **13**(1–3), 41–56 (2016)
6. Castillo, I., Kampas, F.J., Pinter, J.D.: Solving circle packing problems by global optimization: numerical results and industrial applications. Eur. J. Oper. Res. **191**, 786–802 (2008)
7. Fasano, G.: Solving Non-standard Packing Problems by Global Optimization and Heuristics. Springer, Heidelberg (2014)
8. Fasano, G., Pinter, J.A. (eds.): Optimized Packings with Applications. Springer Optimization and Its Applications, vol. 105. Springer, Switzerland (2015)
9. Frazer, H.J., George, J.A.: Integrated container loading software for pulp and paper industry. Eur. J. Oper. Res. **77**, 466–474 (1994)
10. Galiev, S.I., Lisafina, M.S.: Linear models for the approximate solution of the problem of packing equal circles into a given domain. Eur. J. Oper. Res. **230**, 505–514 (2013)

11. George, J.A.: Multiple container packing: a case study of pipe packing. J. Oper. Res. Soc. **47**, 1098–1109 (1996)
12. Hifi, M., M'Hallah, R.: A literature review on circle and sphere packing problems: models and methodologies, In: Advances in Operations Research, vol. 2009, p. 22 (2009). doi:10.1155/2009/150624
13. Litvinchev, I., Rangel, S., Saucedo, J.: A Lagrangian bound for many-to-many assignment problem. J. Comb. Optim. **19**, 241–257 (2010)
14. Litvinchev, I., Ozuna, L.: Approximate packing circles in a rectangular container: valid inequalities and nesting. J. Appl. Res. Technol. **12**, 716–723 (2014)
15. Litvinchev, I., Infante, L., Ozuna Espinosa, E.L.: Approximate circle packing in a rectangular container: integer programming formulations and valid inequalities. In: González-Ramírez, R.G., Schulte, F., Voß, S., Ceroni Díaz, J.A. (eds.) ICCL 2014. LNCS, vol. 8760, pp. 47–60. Springer, Heidelberg (2014)
16. Litvinchev, I., Infante, L., Ozuna, L.: Approximate packing: integer programming models, valid inequalities and nesting. In: Fasano, G., Pinter, J.A. (eds.) Optimized Packings with Applications. Springer Optimization and Its Applications, vol. 105, pp. 117–135. Springer, Switzerland (2015)
17. Lopez, C.O., Beasley, J.E.: A heuristic for the circle packing problem with a variety of containers. Eur. J. Oper. Res. **214**, 512–525 (2011)
18. Lopez, C.O., Beasley, J.E.: Packing unequal circles using formulation space search. Comput. Oper. Res. **40**, 1276–1288 (2013)
19. Murty, K.G.: Linear Programming. Wiley, New York (1983)
20. Stoyan, Y.G., Yaskov, G.N.: Packing congruent spheres into a multi-connected polyhedral domain. Int. Trans. Oper. Res. **20**, 79–99 (2013)
21. Toledo, F.M.B., Carravilla, M.A., Ribero, C., Oliveira, J.F., Gomes, A.M.: The dotted-board model: a new MIP model for nesting irregular shapes. Int. J. Prod. Econ. **145**, 478–487 (2013)
22. Wang, J.: Packing of unequal spheres and automated radiosurgical treatment planning. J. Comb. Optim. **3**, 453–463 (1999)

Optimisation of Query Processing with Multilevel Storage

Nan N. Noon(✉) and Janusz R. Getta

School of Computer Science and Software Engineering,
University of Wollongong, Wollongong, Australia
nnn326@uowmail.edu.au, jrg@uow.edu.au

Abstract. The typical algorithms for optimization of query processing in database systems do not take under the consideration the availability of different types and sizes of persistent and transient storage resource that can be used to speed up the internal query processing. It is well known that appropriate allocation of storage resources for the internal query processing may significantly improve performance. This paper describes the new algorithms for automatic management of *multilevel transient and persistent storage* resources in order to optimize the performance of query processing in a database system. The algorithms presented in the paper process the concurrently submitted queries and discover the common query processing plans. The algorithms estimate the query processing costs and choose the best allocation of multilevel storage resources to optimise the overall internal query processing costs. The paper presents the outcomes of experiments that confirm the improvements in performance through appropriate allocation of multilevel storage for the internal query processing.

Keywords: Multilevel storage · Automated performance tuning · Internal query processing

1 Introduction

The rapid advances in the technologies of electronic storage make possible construction of new computer systems equipped with many different types and capacities of transient and persistent storage. The availability of different types of storage modules allows for implementation of more sophisticated database query processing systems. Such systems are capable of efficient utilization of different storage resources installed on a single computer system. The main objective of this research is to invent the algorithms for automatic management of *multilevel transient and persistent storage* resources to optimize performance of query processing in a database system.

Multilevel transient and persistent storage is special hardware organization of persistent and transient storage modules that can be visualized as a *storage tiering pyramid* structure [1]. The bottom tier of the *storage pyramid* includes the slowest, the cheapest, and the largest modules of persistent storage, mainly the

N.T. Nguyen et al. (Eds.): ACIIDS 2016, Part II, LNAI 9622, pp. 691–700, 2016.
DOI: 10.1007/978-3-662-49390-8_67

large number of high capacity and slow hard drives. The higher tiers of the *pyramid* are occupied by the smaller numbers of more expensive and faster storage modules that include fast hard drives and solid state drives. The topmost and the smallest tiers of the *storage pyramid* consists of the fastest and the most expensive modules of transient storage.

Due to the large size, the database files are usually kept at the lower levels of slower persistent storage pyramid except for the smaller and more frequently used ones kept at the higher and faster levels. Apart from such simple storage allocation rule it is still not very clear what is the best allocation strategy for a given database workload and given persistent and transient storage resources. An interesting research question is how to organize query processing with multilevel storage devices for a given collection of queries with the best utilisation of all storage levels and the shortest processing time. To solve this problem, we assume availability of n levels $l_1, \ldots, l_n$ of persistent and transient multilevel storage together with information about their speed $s_1, \ldots, s_n$ and capacity $c_1, \ldots c_n$. A solution proposed in the paper finds the best order of processing and the best utilisation of multilevel storage for a given collection of m queries $q_1, \ldots, q_m$. We show how to transform the queries $q_1, \ldots, q_m$ into the syntax trees and later on into a table of individual operations on data containers together with the frequencies of operations and formulas that determine the processing costs for each operation. The new algorithms included in the paper assign the storage resources required by the individual operations to multilevel storage in a way that minimises the total processing costs for a given collection of m queries.

The main contributions of our work are the following.

(1) We show how to discover the common components in the processing of a given collection of queries $q_1, \ldots, q_m$.
(2) We show how to assign the storage resources required by processing component discovered in the previous step to the storage levels of multilevel storage system in a way that minimises the global processing costs of $q_1, \ldots, q_m$.

The paper consists of 6 sections. The next section discusses the previous works in the related research areas. Section 3 introduces the basic concepts needed in the remaining section of the paper. The algorithms are presented and explained in Sect. 4. Section 5 presents and discusses the experimental results. Finally, the paper is concluded in Sect. 6.

2 Related Work

Tuning database system has been motivated area of research in the last two decades [2]. Self-tune database system has no human intervention when tuning database system. Different researchers proposed different ways to tune database system automatically. They are fuzzy-based self-tuning [2], indexing [3,4], materialized view [4], partitioning technique [4], online-indexing [3], cache-based optimization [5], and shared memory multiprocessor database systems [6].

Fuzzy-based [2] self-tuning organized into 5 rules to set the size of new buffer cache size (BCS) according to three inputs, Buffer-Hit-Ration (BHR), Number of Users and Database size (DBSize). The fuzzy-based approach is used buffer cache. Their method is putting the whole table to buffer cache and it might be a problem if buffer cache was not enough to run a large script.

In additional, indexing [3], Materialized view [4] and partitioning [4] are the most common techniques for tuning database system. There are many commercial DBMS tools are being used those techniques to tune database system [4]. On the other hand, on-line index [3] is one of the solutions to tune the unexpected query which means that query is not included in predicted workload.

Furthermore, [5] shows executed the query using cache-base optimization is faster than original hard drive on the computer. The result of shift data storage area from disk to cache memory increases the performance. Their optimization needs no custom-designed hardware support and using TPC-D database. Their experiment achieved 13 % faster than original by using cache-oriented optimization. In addition, shared memory multiprocessor database system also can achieve less execution time [6]. They compute the resources to allocate query by considering disk bandwidth, memory buffers, and general purpose processors.

3 Basic Concepts

In this work, we consider query processing in a typical relational database system where the relational tables are read by the user applications. Let x be a nonempty set of attribute names later on called as a *schema* and let $dom(a)$ denotes a domain of attribute $a \in x$. A *tuple* t defined over a schema x is a full mapping $t : x \rightarrow \cup_{a \in x} dom(a)$ and such that $\forall a \in x$, $t(a) \in dom(a)$. A *relational table* r created on a schema x is a set of tuples over a schema x.

A query processing system transforms SQL statements submitted by the user applications into the query processing plans formulated as the expressions of extended relational algebra. The operations of extended relational algebra include the implementation dependent variants of operations of standard relational algebra such as *selection*, *projection*, *join*, *antijoin*, *set operations*, and operations of *grouping*, *sorting*, and *aggregate functions*. We assume, that query processing is implemented by a collection of the *binary operations* $p_1, \ldots, p_k$ on the relational tables. We also assume that the operations p_i for $i = 1, \ldots, k$ use certain amounts of storage $\langle [left_i], [right_i], [temp_i], [out_i] \rangle$ where $[left_i]$, $[right_i]$ denote the amounts of storage occupied by the left and right arguments of the operation, $[temp_i]$ denotes the amounts of temporary storage required by the operation for its internal processing, and $[out_i]$ denotes the amounts of storage required for the outputs of the operation.

A *multilevel storage* that consists of n levels is a sequence of triples $\langle l_i, c_i, s_i \rangle$ where for $i = 1, \ldots, n$ l_i is a unique identifier of storage level, c_i is a storage capacity at a level l_i, and s_i is a pair $\langle r_i, w_i \rangle$ where r_i and w_i are read and write speeds at a level l_i.

4 Optimisation of Query Processing

The objective of optimisation of query processing with multilevel storage is to minimise query processing time for a given set of queries. To minimise time, we use multilevel storage devices with different speed of read/write block operations at each level. Query compiler transforms the queries into the expressions of extended relational algebra. Next, the query optimizer estimates the storage requirements of each operation included in the expressions and assigns the arguments of the operations to appropriate storage levels. This section presents a sequence of algorithms that optimize query processing with multilevel storage.

4.1 Initial Query Processing

In the preparation stages, the queries $q_1, \ldots, q_n$ are transformed by a database system into the query processing plans formulated as expressions $e_1, \ldots, e_n$ of an extended relational algebra. In a typical relational database system such query processing plans can be obtained through the application of `EXPLAIN PLAN` statement to `SELECT` statements of SQL.

Next, the operations of extended relational algebra included in the expressions $e_1, \ldots, e_n$ are standardised into a set of *binary operations* $p_1, \ldots, p_k$ on the relational tables and/or indexes through elimination of conditions from selections, join and anti-join operations and through incorporation of unary operations into binary operations through pre-processing of the arguments and post-processing of the results.

Next, Algorithm 1 transforms the query processing plans represented by the standarised expressions $e_1, \ldots, e_n$ on operations $p_1, \ldots, p_k$ into a sequence of sets of statements $\mathcal{S} = \langle S_1, \ldots S_p \rangle$ such that each statement in a set S_i, for $i = 1, \ldots, p$ takes a form $t := (r\ p\ s)$ where p is a code operation, r, s are the arguments of the operation and t is a result of operation $(r\ p\ s)$. Both r and s can be either the database relational tables or the results of operations computed earlier.

Algorithm 1. Performs the following actions.

(1) Initially, make a sequence of sets of statements $\mathcal{S}$ empty.
(2) Iterate until all expressions $e_1, \ldots, e_n$ are reduced to a single name of temporary result.
  (2.1) Make the new current set of statements S_i and make it empty.
  (2.2) Find all operations like $(r\ p\ s)$ in the expressions $e_1, \ldots, e_n$ where $p \in \{p_1, \ldots, p_k\}$, r and s are either database relational tables or temporary results of operations.
  (2.3) For each operation found in the previous step include a statement like $t_{ij} := (r\ p\ s)$ into the current set of statements S_i where t_{ij} is a new name of a temporary result of operation $(r\ p\ s)$.
  (2.4) Replace in the expressions $e_1, \ldots, e_n$ all operations like $(r\ p\ s)$ with the respective name of a temporary result t_{ij}.
  (2.5) Append the current set of statements S_i at the end of a sequence of sets of statements $\mathcal{S}$.

As a simple example, consider the queries q_1, q_2, q_3 and their processing plans expressed as the expressions of extended relational algebra, $q_1 = (r\ p_1\ s)$, $q_2 = (v\ p_3\ (s\ p_2\ t))$, and $q_3 = ((v\ p_1\ (s\ p_2\ t))\ p_4\ r)$ where p_1, p_2, and p_3 are the operations and r, s, t, v are the relational tables being the arguments of the operations. Algorithm 1 transforms the expressions into the following sequence of three sets of statements: $\langle\{t_{11} := (r\ p_1\ s),\ t_{12} := (s\ p_2\ t),\ t_{13} := (s\ p_2\ t)\}$, $\{t_{21} := (v\ p_3\ t_{12}),\ t_{22} := (v\ p_1\ t_{13})\}$, $\{t_{31} := (t_{22}\ p_4\ r)\}\rangle$.

Next, Algorithm 2 finds and eliminates the common operations in a sequence of sets of statements

Algorithm 2. Performs the following actions.

(1) Iterate over the sets of statements $\langle S_1, \ldots S_p\rangle$ in a sequence $\mathcal{S}$.
- (2.1) Get the new current set of statements S_i.
  - (2.1.1) Find in S_i all groups of statements such that their right hand side is identical in each group.
  - (2.1.2) Replace each group of statements with the same right hand side like $t_{ij} := (r\ p\ s), \ldots, t_{ik} := (r\ p\ s)$ with a single statement $t_{ij} := (r\ p\ s) : c$ where c is a counter of the total number of statements in a group.
  - (2.1.3) Replace in $\mathcal{S}$ all names of temporary results like t_{ik} with t_{ij}.

When applied to a sequence of sets of statements in the example above Algorithm 2 eliminates a duplicated operations in the statements $t_{12} := (s\ p_2\ t)$, $t_{13} := (s\ p_2\ t)$ and creates a sequence of sets $\langle\{t_{11} := (r\ p_1\ s) : 1,\ t_{12} := (s\ p_2\ t) : 2\}$, $\{t_{21} := (v\ p_3\ t_{12}) : 1,\ t_{22} := (v\ p_1\ t_{12}) : 1\}$, $\{t_{31} := (t_{22}\ p_4\ r) : 1\}\rangle$

Next, Algorithm 3 finds the amounts of storage required for the computations of each statement like $t_{ik} := (r\ p\ s)$ and it appends to the statement the calculated storage requirements tuple $\langle[r], [s], [temp(p)], [t_{ik}]\rangle$ here $[a]$ denotes the total amounts of storage needed for an argument a and $[(temp(p)]$ denotes the total amount of temporary storage required by an operation p.

Algorithm 3. Performs the following actions.

(1) Iterate over the sets of statements $\langle S_1, \ldots S_p\rangle$ in a sequence $\mathcal{S}$.
- (1.1) Get the new current set of statements S_i.
  - (1.1.1) For each statement in S_i like $t_{ik} := (r\ p\ s) : c$ use a specification of operation p to calculate the values of a tuple $\langle[r], [s], [temp(p)], [t_{ik}]\rangle$. If r and s are the relational tables then their storage requirements are collected from data depository of a database system. When a temporary result t_{ik} is used as an argument its storage requirements are collected from a statement that created the temporary result.
  - (1.1.2) Append the storage requirements to each statement in S_i to create a tuple
    $t_{ik} := (r\ p\ s) : c : \langle[r], [s], [temp(p)], [t_{ik}]\rangle$.

4.2 Optimal Allocation of Multilevel Storage

A sequence $\mathcal{S}$ of sets of statements obtained from the Algorithms 1, 2 and 3 is used to allocate multilevel storage needed to compute the statements at the appropriate storage levels. In this procedure, we choose for each statement a storage allocation plan that has the lowest cost computed as $c * cost([r] : l_{[r]}, [s] : l_{[s]}, [temp(p)] : l_{[temp(p)]}, [t_{ik}] : l_{[t_{ik}]})$ where $l_{[r]}, l_{[s]}, l_{[temp(p)]}$, and $l_{[t_{ik}]}$ are devices level of $[r], [s], [temp(p)]$, and $[t_{ik}]$. Algorithm 4 implements a procedure that finds the best multilevel storage allocation plan. The input to the algorithm is a sequence of sets of statements $\mathcal{S} = \langle S_1, \ldots, S_p \rangle$ obtained from Algorithm 3 and the output is a sequence of sets of sequences $OP = \langle \{ \langle t_{ij} := (rps) : c : \langle [r], [s], [temp(p)], [t_{ik}] \rangle, \ldots \}, \ldots \rangle$ where i-th set OP_i in a sequence contains the sequences of all possible processing orders of statements implementing S_i.

Algorithm 4. Performs the following actions.

(1) Initially, make OP empty
(2) Iterate over the sets of statements $\langle S_1, \ldots S_p \rangle$ in a sequence $\mathcal{S}$. Let current set of statements be S_i. Get the new current set of sequences OP_i.
 (2.1) If the total number of statements in S_i is n then generate $n!$ possible sequences of ordered statements and save it in OP_i.
(3) Iteration stops when all sets from $\mathcal{S}$ are used to generate all possible orders.

Next, Algorithm 4.1 calculates the cost for each statement and chooses for each statement a processing plan with the lowest costs. An input to this algorithm is one tuple from OP_i and the output is the best processing plan for each statement.

Algorithm 4.1. Finds the best cost for each OP.

(1) Iterate first to last statements of current tuple from OP_i. Let current operation be p_i and storages needed to compute p_i be $[r]$, $[s]$, $[temp(p_i)]$, and $[t_{ik}]$ and the best cost be $bc_i = 0$.
 (1.1) We get 4!=24 possible distributions for one statements. Next, we iterate over 24 distributions. Let current permutation be $\langle [r], [s], [temp(p_i)], [t_{ik}] \rangle$.
  (1.1.1) We iterate over the storage levels from l_n to l_2. Let current storage level be l_n, temporary storage ts = free size of l_n, and temporary cost $tc = 0$.
   (1.1.1.1) **If** $[r]$ is not fixed in any storage level **and** $[r] \langle = ts$ **and** $l_{[r]}$ is not assigned to any storage level, **then** $l_{[r]} = l_n$ and $ts = ts$ - $[r]$.
   (1.1.1.2) **If** $[s]$ is not fixed in any storage level **and** $[s] \langle = ts$ **and** $l_{[s]}$ is not assigned to any storage level, **then** $l_{[s]} = l_n$ and $ts = ts$ - $[s]$.
   (1.1.1.3) **If** $[temp(p_i)] \langle = ts$ **and** $l_{[temp(p_i)]}$ is not assigned to any storage level, **then** $l_{[temp(p_i)]} = l_n$ and $ts = ts$ - $[temp(p_i)]$.
   (1.1.1.4) **If** $[t_{ik}] \langle = ts$ **and** $l_{[t_{ik}]}$ is not assigned to any storage level, **then** $l_{[t_{ik}]} = l_n$ and $ts = ts$ - $[t_{ik}]$.

(1.1.1.5) **If** all storage requirements are satisfied, **then** calculate cost tc and update $bc_i = max(tc, bc_i)$.

(1.1.2) After (1.1.1), we get the best processing plan for the current permutation.

(1.2) After (1.1), we get 24 processing plans and we choose the best processing plan for the current statement.

(2) After (1), we get the best processing plan for each statement represented by one tuple OP_i. This algorithm can be repeated until all tuples from OP_i gets their own the best processing plans.

We applied Algorithm 4.1 on above example and get best cost for each statement become $OP_i = \{\langle t_{11} := \ldots : bc_{11}, t_{12} := \ldots : bc_{12}\rangle, \langle t_{12} := \ldots : bc_{12}, t_{11} := \ldots : bc_{11}\rangle\}$. Take note that we have to repeat 4.1 only after finish 4.2, because, 4.1 handles one set of statements' possible order. Next, Algorithm 4.2 chooses the best plan for each set of OP and assign that plan according to best allocation. The input of this algorithm are the outcomes of Algorithm 4.1 and the output is the best order of statements for each OP.

Algorithm 4.2. Finds the best plan.

(1) Iterate one to last tuples of OP_i.
  (1.1) Get total cost for current tuple by adding all the costs of operations.

(2) After (1), we get total processing costs of each tuple of OP_i and we choose the best processing plan. Then, according to the best plan, we assign storage requirements of each operation to the different storage levels.

(3) After that, we remove current tuple OP_i from OP and input next tuple OP_{i+1} into an Algorithm 4.1 and we repeat an Algorithm 4.2.

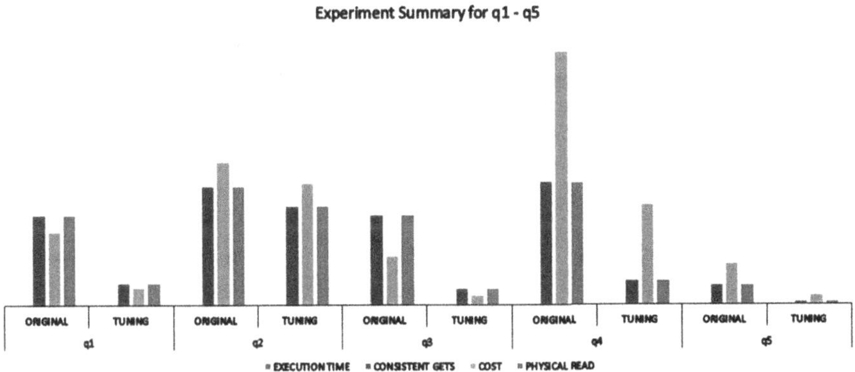

Fig. 1. Experiment summary for q_1-q_5

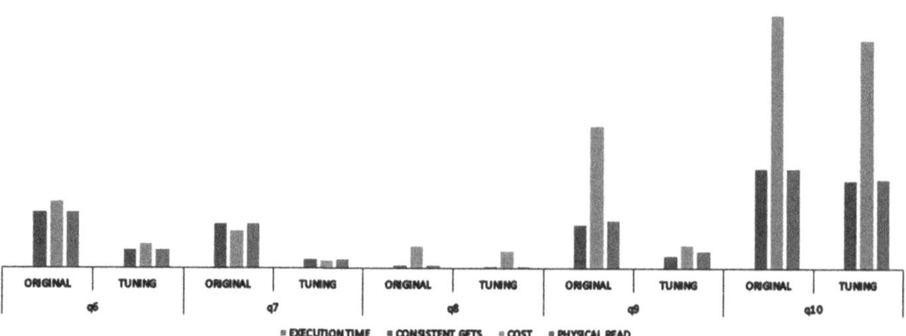

Fig. 2. Experiment summary for q_6-q_{10}

Table 1. Experiment result for q_1 - q_5

Name	Description	Execution time	Consistent gets	I/O cost	Physical read
q_1	Original	00:30.5	304,199.60	248,484.00	303,897.80
	Tuning	00:10.7	73,339.20	61,200.00	73,276.00
	Profit	00:19.7	230,860.40	187,284.00	230,621.80
	Profit %	64.70 %	75.89 %	75.37 %	75.89 %
q_2	Original	00:35.4	400,362.00	483,431.00	400,013.40
	Tuning	00:10.9	334,417.80	413,667.00	334,140.40
	Profit	00:24.4	65,944.20	69,764.00	65,873.00
	Profit %	69.11 %	16.47 %	14.43 %	16.47 %
q_3	Original	00:27.0	304,140.80	165,100.00	303,891.00
	Tuning	00:01.9	57,310.20	31,480.00	57,269.00
	Profit	00:25.2	246,830.60	133,620.00	246,622.00
	Profit %	93.02 %	81.16 %	80.93 %	81.15 %
q_4	Original	00:52.1	414,139	853,804.00	413,711.00
	Tuning	00:08.2	82,887.80	337,397.00	82,819.20
	Profit	00:43.9	331,251.20	516,407.00	330,891.80
	Profit %	84.20 %	79.99 %	60.48 %	79.98 %
q_5	Original	00:06.5	67,045.40	138,481.00	66,814.40
	Tuning	00:00.6	10,536.60	31,692.00	10,463.40
	Profit	00:05.9	56,508.80	106,789.00	56,351.00
	Profit %	90.54 %	84.28 %	77.11 %	84.34 %

5 Experiments

Optimization of query processing with multilevel storage provides a better response time than original processing time. Our experiment used a synthetic TPC-H 4 GB benchmark relational database. [7] implemented on an "off-a-shelf" commercial database software. In this experiment, we used 10 complex queries and 3 storage level device which includes a hard drive (l_1) and solid state drive (l_2) and certain amounts of transient storage in a data buffer cache. The queries are randomly taken from TPC's Template Set. To get the reliable results, the same experiments are repeated for at least 5 times.

First, we transform queries into the sequence of sets of statements. Then, we remove duplicated operations and count frequency for each operation. After that, we calculate the storages requirements for each operation. Next, find the best cost allocation and assign statements into the different storage levels. In our experiment, l_2 level contains enough space for those queries and we assign all statements into this storage level. Then, we iterate above steps for 5 times and record the timings. By executing each test multiple times, we can get the average results for each query and plug into bar chart show in Figs. 1 and 2.

Table 2. Experiment result for q_6 - q_{10}

Name	Description	Execution time	Consistent gets	I/O cost	Physical read
q_6	Original	00:36.0	400,366.80	471,169.00	400,013.00
	Tuning	00:04.1	128,860.40	170,308.00	128,767.00
	Profit	00:31.9	271,506.40	300,861.00	271,246.00
	Profit%	88.57 %	67.81 %	63.85 %	67.81 %
q_7	Original	00:28.5	318,090.80	262,601.00	317,812.40
	Tuning	00:02.4	65,432.60	57,455.00	65,361.00
	Profit	00:26.1	252,658.20	205,146.00	252,451.40
	Profit%	91.72 %	79.43 %	78.12 %	79.43 %
q_8	Original	00:11.1	27,759.20	156,725.00	27,562.80
	Tuning	00:03.2	11,995.60	119,910.00	11,912.00
	Profit	00:08.0	15,763.60	36,815.00	15,650.80
	Profit%	71.56 %	56.79 %	23.49 %	56.78 %
q_9	Original	00:57.6	311,040.00	1,003,688.00	338,363.60
	Tuning	00:19.6	91,827.80	163,122.00	120,595.40
	Profit	00:38.0	219,212.20	840,566.00	217,768.20
	Profit%	65.94 %	70.48 %	83.75 %	64.36 %
q_{10}	Original	01:06.4	705,417.20	1,792,076.00	709,485.00
	Tuning	00:27.8	628,157.40	1,618,736.00	633,049.20
	Profit	00:38.6	77,259.80	173,340.00	76,435.80
	Profit%	58.17 %	10.95 %	9.67 %	10.77 %

To prove that our algorithms are faster than original execution time, consistent gets, physical read and costs, firstly, we execute 5 times for each query (without tuning method) on l_1. After that, record the results and calculate the average results then, record into Tables 1 and 2. We generate Figs. 1 and 2 based on Tables 1 and 2 results. According to Figs. 1 and 2, the performance of our algorithms are better than original script execution time, consistent gets, physical read and cost. In addition, Tables 1 and 2 show that the overall profit percentage for each query is better than original script.

6 Summary and Conclusions

In this paper, we present query optimization technique that uses multilevel storage devices. This method accepts queries from different users and finds expressions of an extended relational algebra that implement the queries. Next, it counts the frequencies of operations involved in the expressions and it calculates the storages requirements of its arguments, internal size and output size needed for each operation. Finally, this method calculates all possible cost by distributing operations into the different layer of devices and gets the best plan to execute the queries. According to our method, we setup experiments with 3 complex queries and two layers of devices. The overall result of those experiment shows that our method performs not only speed up but also less cost execution than the original one.

References

1. HGST: tiered optimization (2015). http://global.hgst.com/science-of-storage/technology-insights/tiered-storage-optimization-data-center-performance-and-tco. Accessed 25 August 2015
2. Rodd, S., Kulkarni, U.: Adaptive self-tuning techniques for performance tuning of database systems: a fuzzy-based approach. In: 2013 2nd International Conference on Advanced Computing, Networking and Security (ADCONS), pp. 124–129. IEEE (2013)
3. Schnaitter, K.: On-line Index Selection for Physical Database Tuning. Ph.D. thesis, Santa Cruz (2010)
4. Surajit, C., Vivek, N.: Self-tuning database systems: a decade of progress. In: Proceedings of the 33rd International Conference on Very Large Data Bases, pp. 3–14. VLDB Endowment Inc. (2007)
5. Trancoso, P., Torrellas, J.: Cache optimization for memory-resident decision support commercial workloads. In: International Conference on Computer Design (ICCD 1999), pp. 546–554 (1999)
6. Murphy, M., Shan, M.C.: Execution plan balancing. In: Proceedings of the Seventh International Conference on Data Engineering, pp. 698–706 (1991)
7. TPC:TPC (2015). http://www.tpc.org/information/benchmarks.asp. Accessed 25 March 2015

Smart Pattern Processing for Sports

A Real-Time Intelligent Biofeedback Gait Patterns Analysis System for Knee Injured Subjects

Putri Wulandari(✉), S.M.N. Arosha Senanayake, and Owais Ahmed Malik

Motion Analysis Lab, Integrated Science, Universiti Brunei Darussalam, Tungku Link, Gadong BE 1410, Brunei Darussalam
{13ml354, arosha.senanayake, 11H1202}@ubd.edu.bn

Abstract. This study presents a real-time visualization system of gait patterns of knee injured subjects for biofeedback monitoring and classification. The developed system includes non-invasive wireless body-mounted motion sensors for kinematics measurements of lower extremities, surface electromyography (EMG) system for relevant specific muscle activity measurements, a motion capture system for recording trial activities and custom-developed intelligent system software implemented using LabVIEW and MATLAB. The real-time biofeedback system provides a visual monitoring of individual and superimposed signals (kinematics, EMG and video data) in order to identify the knee joint abnormality and muscles strength during various ambulation activities performed by the subjects. It can facilitate the clinicians, physiotherapists and physiatrists in determining the impairments in the gait patterns the knee injured based on the data collected and identifying the subjects lacking behind the desired level of recuperation.

Keywords: Real-time system · Biofeedback · Classification · Knee injury · Gait patterns

1 Introduction

Identification of abnormal gait patterns after knee injury or surgery is crucial for avoiding various long term complications. Biofeedback has been found effective in improving the impaired gait patterns and regaining the neuromuscular control after lower limb injuries [1–3]. Since the knee injuries, in particular anterior cruciate ligament (ACL) injury/reconstruction, alter various physiological parameters in the human lower extremity, post-injury or post-surgery rehabilitation requires the assessment of multiple types of bio-signals for comprehensive gait analysis. The analysis of multi-modal bio-signals not only illustrates the correlation among different physiological parameters (qualitative information) but it also can be used to provide a quantitative feedback for human motion.

The monitoring of multiple bio-signals has been used in some of the previous studies to observe the gait patterns alterations in anterior cruciate ligament reconstructed (ACL-R) subjects. In [4] ground reaction force and kinematic data were

N.T. Nguyen et al. (Eds.): ACIIDS 2016, Part II, LNAI 9622, pp. 703–712, 2016.
DOI: 10.1007/978-3-662-49390-8_68

combined with inverse dynamics to predict sagittal plane joint toques and powers from which angular impulse and work were derived for determining the gait variations in ACL-R subjects. In order to test the dynamic knee joint loading, different kinematics and kinetics parameters data were observed for gait analysis of ACL-R subjects during normal walking and stairs use [5]. Similarly, the kinematic and kinetic patterns have also been studied for ACL-R subjects to determine the asymmetries between limbs of the athletes for testing the return-to-sports criteria after ACL rupture [6]. In addition to monitoring the kinematic and kinetic data for gait analysis of ACL-R subject, EMG data from the lower limb muscles (e.g. vastus lateralis and medialis, biceps femoris and adductor longus) have also been recorded and assessed in few studies [7, 8].

However, the integration and superimposition of different bio-signals in real-time has not been much explored in the previous literature for providing comprehensive gait analysis for ACL injured or reconstructed subjects. Such multi-modal data analysis can be used for assessing the role of knee joint movements and relevant muscular activities. This study proposes a real-time visualization system for evaluating and monitoring the gait patterns of knee injured (ACL injured/reconstructed) subjects based on multi-modal data integration and intelligent techniques. The purpose of the system is to provide a complementary decision supporting tool, in conjunction with the existing rehabilitation monitoring mechanisms, to enable the clinicians, trainers and physiotherapists to objectively monitor the rehabilitation progress of athletes and their compliance to the rehabilitation protocol during different convalescence stages. This research is an extension of the off-line system developed in [9, 10].

2 Methodology

2.1 Real-Time System Design

Total of the subjects are four males whom participated in the training session. The mean ages of the participants are 24.75 year-old. The design of real-time system for providing biofeedback and classification of gait patterns consists of two phases:

In the first phase, the trained classification models are generated using multi-modal input data during the learning process for various ambulation activities from healthy/injured subjects [9, 10]. This is an offline process.

In the second phase real-time data processing is performed using Interactive Graphical User Interface (IGUI) which produces visualization of bio-signals and classification of gait patterns for knee injured subjects. For evaluating the current gait patterns of a subject, the transformed pattern set undergoes matching for patterns using trained classifiers. Moreover, individual and overlaid/superimposed kinematics and EMG signals for various ambulation activities of a subject can be displayed in real-time and intra- or inter-subject comparisons can be made using any existing data available in the knowledge base for healthy/injured subjects.

Figure 1 presents the sequence of steps performed in order to generate the real-time biofeedback and classification of gait patterns for knee injured subjects. The system was implemented using LabVIEW 2013 for IGUIs and MATLAB 7.0 software for

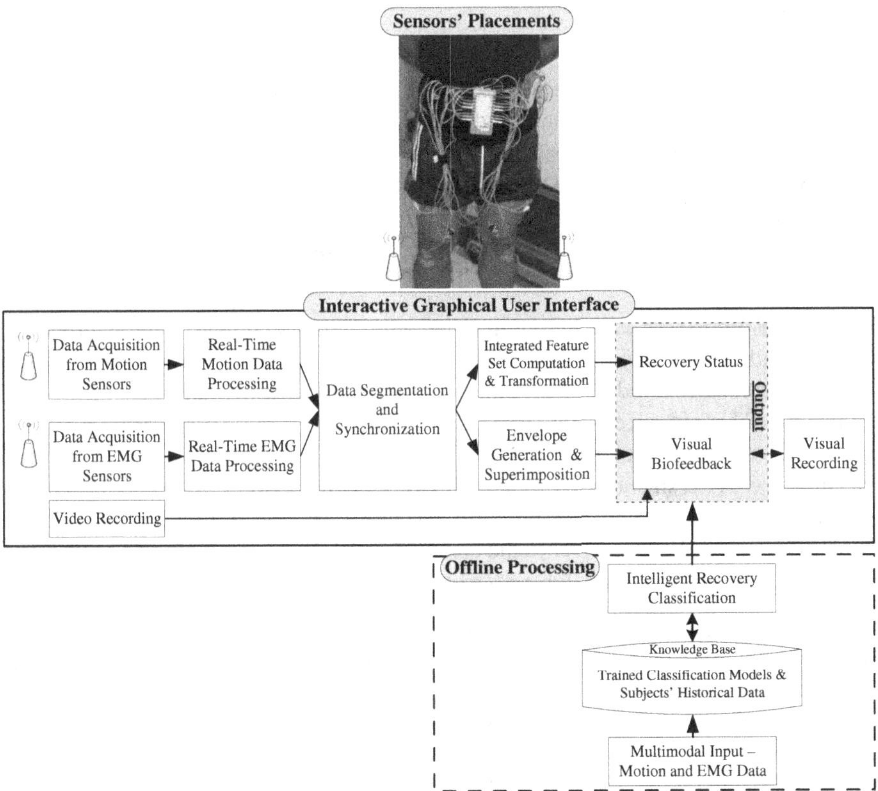

Fig. 1. Real-time system for visual biofeedback and classification of gait patterns

offline processing. The details of different hardware and software components (for real-time system) are described in subsequent sections.

2.2 Hardware Components

The real-time gait patterns monitoring system consisted of three major hardware components: (1) micro-electro-mechanical motion sensor units, (2) surface electromyography monitoring unit with electrodes, and (3) motion capture system.

The wireless micro-electro-mechanical system (MEMS) motion sensors from KinetiSense (ClevMed. Inc) were used in this study to collect the kinematics data. Each sensor unit contains a tri-axial (MEMS) accelerometer and a tri-axial MEMS gyroscope to measure 3-D linear accelerations and 3-D angular velocities, respectively. A BioCapture physiological monitoring system, consisting of BioRadio and USB receiver, was used to record the EMG signals from the relevant muscles around the knee joint. Motion capture system can be used to display reliable data and accurate information about human motion.

2.3 Software Components

Data Acquisition from Motion Sensors. The wireless motion sensors were attached to the identified positions on shanks and thighs (halfway up the surface of the tibia for shank and two thirds up the tensor fascia latae for thigh) for each subject in order to collect 3-D angular rates and 3-D linear acceleration data from knee joint movements. When a testing session was initiated, the command module transferred 3-D angular rates and 3-D linear accelerations from connected motion sensors to the LabVIEW data acquisition module in real time.

Data Acquisition from EMG System. The neuromuscular signals were recorded by placing Kendall disposable pre-gelled snap electrodes on four different knee extensor and flexor muscles including the vastus medialis (VM), vastus lateralis (VL), semitendinosus (ST) and biceps femoris (BF) on both legs of the subjects. For skin preparation and electrodes placement, SENIAM (Surface Electromyography for the Non-Invasive Assessment of Muscles) EMG guidelines were followed [11]. The neuromuscular data, captured using BioRadio and EMG electrodes, were wirelessly transferred through USB receiver to the computer where the LabVIEW data acquisition module recorded the readings for each experiment in real time.

Processing of Kinematics and EMG Signals. The kinematics and EMG data were processed to prepare the input for feature extraction and visual biofeedback generation steps. The angular rate and acceleration measurements obtained from the motion sensors were filtered using 6th order Butterworth filter before computing the orientations. The 3-D orientations of lower extremities were estimated by applying trapezoidal integration method on respective angular rates of both lower limbs [9]. For each trial, the raw EMG data with zero mean were band-pass filtered (20-450 Hz) using 4th order Butterworth filter.

Data Segmentation and Synchronization. Data segmentation of kinematics and EMG signals was performed prior to feature extraction and different features were computed for each segment. Each gait cycle was segmented by detecting the heel strike (HS) event. In order to provide visualization of superimposed signals, both kinematics and EMG signals were synchronized due to different sampling rates and recording delays. The synchronization of both signals was done by detecting a gait cycle using HS event for each subject [12, 13]. For feature extraction, the percentages mentioned in [14] were used to identify each phase from the gait cycle (7 phases of gait cycle: load response, mid-stance, terminal stance, pre-swing, initial swing, mid swing and terminal swing).

Integrated Feature Set Computation and Transformation. Key features were extracted from subjects' motion for collecting relevant data for generating data sets in order to apply classification mechanism. The kinematics feature set for each ambulatory activity included nine parameters from 3-D rotational movements (mean, maximum and std. values for flexion/extension, abduction/adduction and internal/external rotation) for each phase of a gait cycle. Thus, a total of 63 features (3×3 rotational kinematics features $\times$ 7 phases of a gait cycle) were computed from a gait cycle data of an ambulatory activity. For EMG analysis, discrete wavelet transform (DWT) was applied and six wavelet coefficients (cD1…cD5 and cA5) were computed which represent the energy distribution of the EMG signals from four identified muscles. From these coefficients, five statistical features were calculated for each phase of gait cycle. A total of 840 features from EMG data were extracted from each gait cycle of an ambulatory activity (5 EMG features × 6 coefficients × 4 muscles × 7 phases of gait cycle). Thus, a single gait cycle was represented by a feature vector of length 903 (840 EMG features + 63 kinematics features). In order to use the intelligent technique for classification, the stored principal component analysis (PCA) coefficient matrix from knowledge base is applied in real-time to the kinematics and EMG data of current test subject and the result is used for evaluating the class of the subject [9, 10].

Envelope Generation and Signals' Overlaying. In order to generate visual biofeedback, the raw EMG data with zero mean for different muscles were full wave rectified and low pass filtered to generate linear envelopes. The linear envelopes provide useful information for assessing the strength/activation of different muscles for inter- and intra-subjects comparison. For comparing the EMG amplitude, the data were normalized for each subject using mean value of the signal of each stride segment for respective muscles and data were represented as a percentage of mean [15]. The estimated knee orientation (flexion/extension, abduction/adduction and internal/external rotation) in three planes and EMG envelopes from different lower limb muscles were superimposed to observe the changes in both type of signals simultaneously.

Intelligent Recovery Classification. The identification of current class/status of gait patterns of a new ACL-R subject provides useful complementary information in order to make adjustments in his/her rehabilitation process. Based on the patterns of 3-D kinematics and neuromuscular data, an automated real-time identification of class/stage of gait patterns of a subject for an ambulation activity is done by applying the trained intelligent classifier [9, 10]. The classification model was implemented using MATLAB which is called using LabVIEW IGUI. Thus, for each ambulation testing activity, the class of gait patterns of a subject is determined as per his/her performance during the test trials for that particular activity [9, 10]. Four classes/groups (1, 2, 3 and 4) were formed using the previously collected data based on the health condition of the subjects and the real-time system was tested using the same groups [9]. The classes 1 through 4 represent different stages of health/recovery condition of the subjects based on the gait patterns (class 1 represents the initial level of recovery, class 2 represents the next level, class 3 represents the advanced level of recovery and class 4 represents the healthy subjects).

Visual Biofeedback Generation. Visual biofeedback is an effective gait monitoring mechanism which facilitates in identifying intra- and inter-subjects variations of kinematics and neuromuscular signals during different rehabilitation exercises. Additionally, the superimposed/overlapped bio-signals can be used to find the relationship between muscle recruitment and knee joint movements. These relationships assist in detecting the muscles which are causing the changes in the knee 3-D orientation (flexion/extension, internal/external rotation and abduction/adduction). Thus, the physiatrists, physiotherapists and trainers can confirm or disconfirm the contraction of different muscles of a subject with respect to his/her knee angle movements.

3 Results

IGUIs in LabVIEW were designed to display different useful outputs in real-time for individuals for each ambulation monitoring activity (Fig. 2). The IGUI system provides a menu with different option to show the visual biofeedback and classification of gait patterns. The motion and EMG signals can also be stored for offline processing and analysis. Visual biofeedback presents various outputs in real-time e.g. current knee orientation, comparison of the right and left legs, comparison of the strength variations for different flexors/extensors within the muscles of both legs, superimposition of knee orientations and individual muscle plots integrated with full kinematics visualization using motion sensors, wireless EMG sensors and motion capture system. It also presents a comparison of an individual with his/her previous sessions' data (if available) and with subjects from other groups (previously stored in knowledge base) to assess the inter-subject parameters variations during post ACL reconstruction recovery regimen. Some example IGUI screenshots for visual biofeedback are shown in Fig. 2, 3, 4, 5 and 6. Kinematic and neuromuscular data obtained from motion and EMG possesses different signal attributes and they are captured in different sampling rates. To be able to visually monitor the comparison of these two bio-signals; knee flexion/extension and strength of muscle, both signals must be integrated using the superimposition and combined them. This IGUI provides overlapping of data allowing simultaneous monitoring of knee kinematics and muscle movements (variation in muscles' strength, activation timings and durations) for different phases of each gait cycle for ACL-R subjects (Fig. 3). Moreover, motion sensors' signal processing in real time and knee flexion/extension of both legs are shown in Fig. 4. Four motion sensors mounted on both legs of the subject allow displaying motion data from thigh and shank of subject; Sensor 1–2 for the right leg and sensor 3–4 for the left leg. Each sensor can measure 3-D linear accelerations and 3-D angular velocity. Knee flexion/extension data can be obtained through motion sensors which is derived from the angular velocity of the thigh and shank in the sagittal plane. This IGUI can also display a Comparison of strength of the muscle between

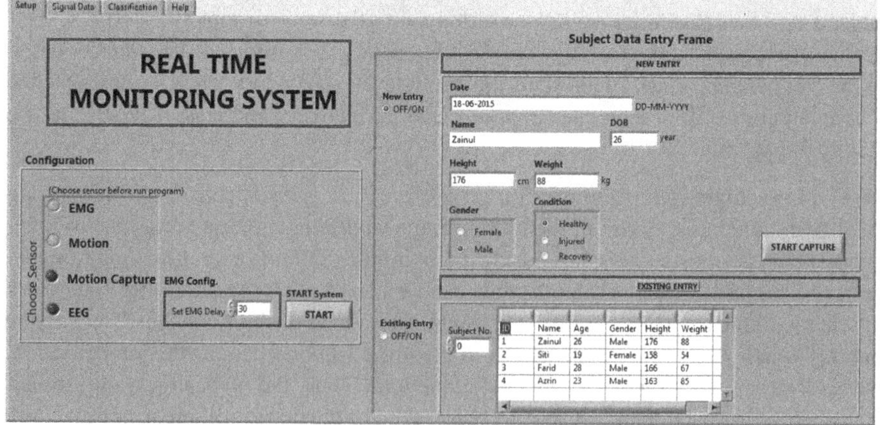

Fig. 2. Setup menu in the IGUI to interact with the real time gait patterns monitoring system

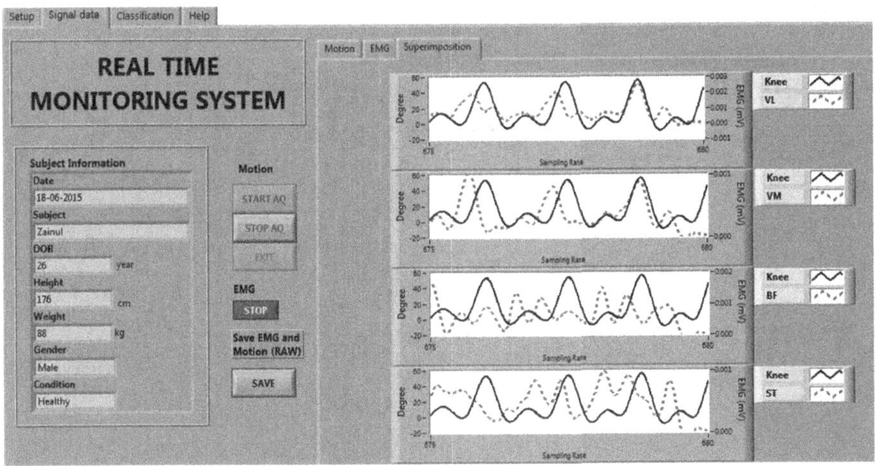

Fig. 3. Superimposition of knee flexion/extension and muscle recruitments (VL, VM, BF and ST) for an ACL-R subject walking at 4 km/h on a treadmill

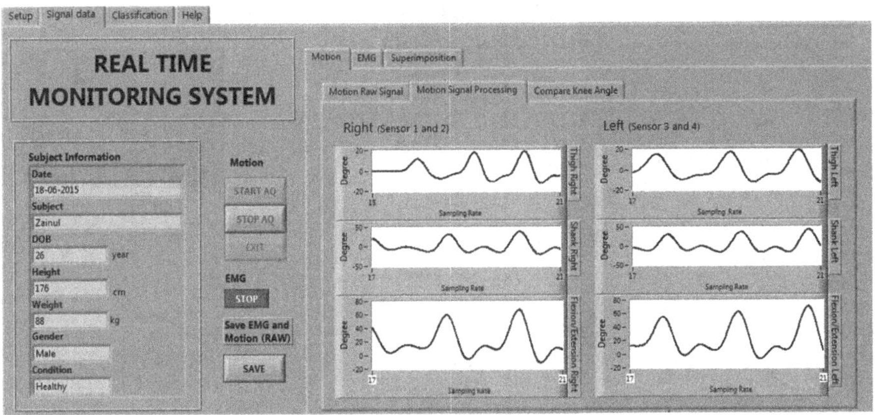

Fig. 4. Motion sensors' signal processing in real time and knee flexion/extension of both legs for an ACL-R subject walking at 4 km/h on a treadmill

muscles in the same leg (Vastus Lateralis – Vastus Medialis Right/Left and Biceps Femoris – Semitendinosus Right/Left) in real time (Fig. 5).

Figure 6 shows an IGUI for real time intelligent biofeedback system with the predicted classification of gait patterns for a subject walking at a speed of 4 km/h. This IGUI will be visualize five gait cycles corresponding to knee flexion/extension signals receiving from motion and EMG sensors for real time monitoring on both legs; right and left. Further, recovery class of the classification of the subject is also displayed depending on the analyzed gait cycles in real time. In order to perform the

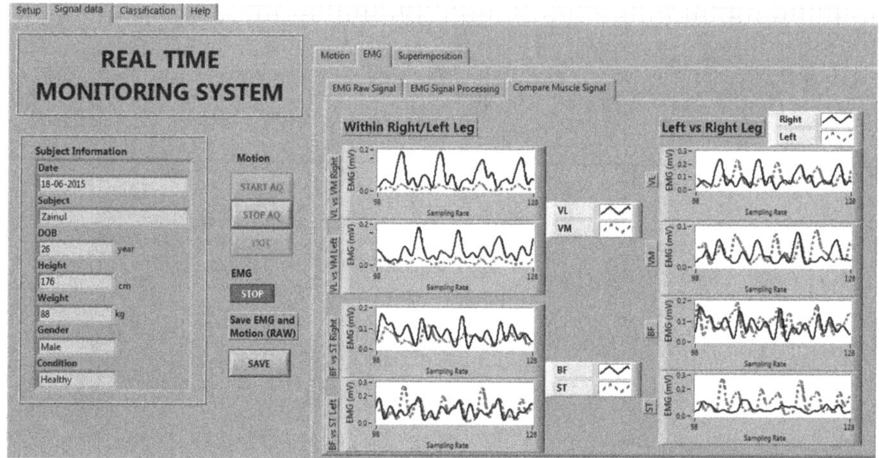

Fig. 5. Comparison of muscle characteristics within same/different leg(s) for an ACL-R subject walking at 4 km/h on a treadmill

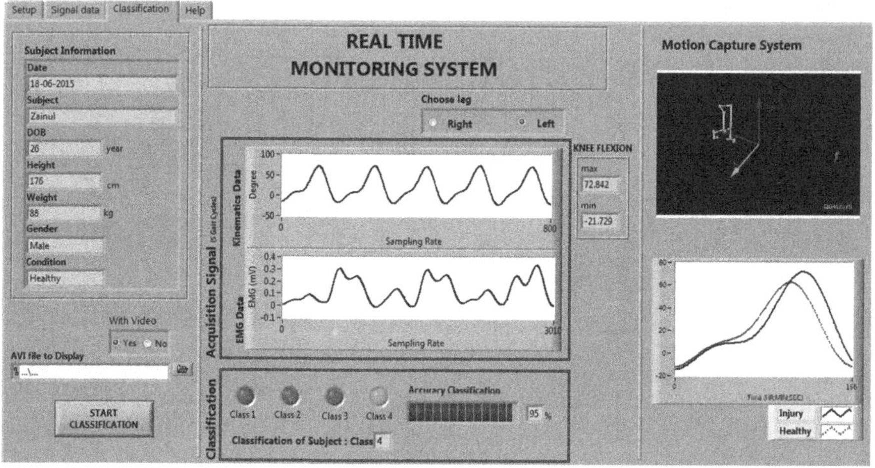

Fig. 6. IGUI designed for real time intelligent biofeedback system with the predicted classification of gait patterns for a subject during walking at 4 km/h on a treadmill

real-time classification of gait patterns for an ambulation activity, respective trained classifier (ANFIS) can be chosen from the user friendly IGUI built in LabVIEW. For a selected ambulation activity (4 km/h walking speed), the IGUI developed in LabVIEW program extracts the kinematics and EMG features from the current completed gait cycle, applies the feature transformation using PCA coefficient matrix and uses the MATLAB classifier to predict the class (1, 2, 3 or 4) of the gait patterns of the subject (Fig. 6). Individual signals for knee joint orientation and muscles movements can be used to monitor the knee dynamics of an ACL-R subject during different gait cycles.

Based on this user friendly visual biofeedback IGUIs developed, the physiatrists, physiotherapists and trainers can apply corrective measures and muscle conditioning in order to improve the recovery condition of the knee. This visual biofeedback can be used as an auxiliary mechanism co-existent with the existing ACL recovery monitoring mechanisms.

4 Conclusion

The IGUI based visual biofeedback and objective rehabilitation progress can be used as a motivational tool for the ACL-R subjects to perform better during the course of rehabilitation. The interactive outputs of IGUIs provide user friendly feedback as supporting information for clinicians, physiatrists, physiotherapists for observing the gait patterns of knee injured and ACL-R subjects. This information can be used for adjusting individual subject's rehabilitation protocol as per the requirements, focusing on specific recovery problem areas, performance comparison during rehabilitation stages for each monitored activity. The hardware and software co-design system has been implemented such a way that additional IGUI tools and routines can be added based on more activities and features identified in the clinical environment.

References

1. Ford, K. R., DiCesare, C. A., Myer, G. D., Hewett, T. E.: Real-time biofeedback to target risk of anterior cruciate ligament injury: A Technical report for injury prevention and rehabilitation, Journal of Sport Rehabilitation (2014) doi: 10.1123/jsr.2013-0138
2. Silkman, C., McKeon, J.: The effectiveness of electromyographic biofeedback supplementation during knee rehabilitation after injury. J. Sport. Rehabil. **19**, 343–351 (2010)
3. Teran-Yengle, P., Birkhofer, R., Weber, M.A., Patton, K., Thatcher, E., Yack, H.J.: Efficacy of gait training with real-time biofeedback in correcting knee hyperextension patterns in young women. J. Orthop. Sports Phys. Ther. **41**, 948–952 (2011)
4. DeVita, P., Hortobagyi, T., Barrier, J.: Gait biomechanics are not normal after anterior cruciate ligament reconstruction and accelerated rehabilitation. Med. Sci. Sports Exerc. **30**, 1481–1488 (1998)
5. Hall, M., Stevermer, C.A., Gillette, J.C.: Gait analysis post anterior cruciate ligament reconstruction: knee osteoarthritis perspective. Gait & Posture **36**, 56–60 (2012)
6. Di Stasi, S.L., Logerstedt, D., Gardinier, E.S., Snyder-Mackler, L.: Gait patterns differ between ACL-reconstructed athletes who pass return-to-sport criteria and those who fail. Am. J. Sports Med. **41**, 1310–1318 (2013)
7. Bulgheroni, P., Bulgheroni, M.V., Andrini, L., Guffanti, P., Giughello, A.: Gait patterns after anterior cruciate ligament reconstruction. Knee Surg. Sports Traumatol. Arthrosc. **5**, 14–21 (1997)
8. Knoll, Z., Kocsis, L., Kiss, R.M.: Gait patterns before and after anterior cruciate ligament reconstruction. Knee Surg. Sports Traumatol. Arthrosc. **12**, 7–14 (2004)

9. Malik, O.A., Senanayake, S.M.N.A., Zaheer, D.: An intelligent recovery progress evaluation system for ACL reconstructed subjects using integrated 3-D kinematics and EMG features. IEEE J. Biomed. Health Inform. **19**, 453–463 (2014). doi:10.1109/JBHI.2014.2320408
10. Senanayake, S.M.N.A., Malik, O.A., Iskandar, P.M., Zaheer, D.: A knowledge-based intelligent framework for anterior cruciate ligament rehabilitation monitoring. App. Soft Comp. **20**, 127–141 (2013)
11. Hermens, H.J., Freriks, B., Disselhorst-Klug, C., Rau, G.: Development of recommendations for semg sensors and sensor placement procedures. J. Electromyogr. Kinesiol. **10**, 361–374 (2000)
12. Jasiewicz, J.M., Allum, J.H.J., Middleton, J.W., Barriskill, A., Condie, P., Purcell, B., Li, R. C.T.: Gait event detection using linear accelerometers or angular velocity transducers in able-bodied and spinal-cord injured individuals. Gait & Posture **24**, 502–509 (2006)
13. Evans, A.L., Duncan, G., Gilchrist, W.: Recording accelerations in body movements. Med. Biol. Eng. Comput. **29**, 102–104 (1991)
14. Perry, J.: Gait Analysis: Normal and Pathological Function. Thorofare, NJ, Slack (1992)
15. Winter, D.A., Yack, H.J.: EMG profiles during normal human walking: stride-to-stride and inter-subject variability. Electroencephalogr. Clin. Neurophysiol. **67**, 402–411 (1987)

Pattern Recognition of Brunei Soldier's Based on 3-Dimensional Kinematics and Spatio-Temporal Parameters

D.N. Filzah P. Damit[1,2(✉)], S.M.N. Arosha Senanayake[1], Owais Ahmed Malik[1], and P.H.N. Jaidi P. Tuah[1]

[1] Faculty of Science, Universiti Brunei Darussalam, Gadong, Brunei Darussalam
filzahdamit@gmail.com, 13h0355@ubd.edu.bn
[2] Performance Optimisation Centre, Ministry of Defence, Bandar Seri Begawan, Brunei Darussalam

Abstract. This paper addresses the designing of a knowledge base system for the Royal Brunei Armed Forces based on the 3-Dimensional gait kinematic patterns of soldiers interfacing with spatio-temporal features extracted; heart rate levels and rating of perceived exertion, during prolonged loaded foot march while on a treadmill under the presence of motion capture system. Kinematic features derived such as angular changes and range of motion in critical joints; ankle, knee, hip, pelvis and trunk from ten healthy soldiers were assessed while walking at 6.4 km/h at 1 % elevation carrying 15 kg loaded military backpack for 30 min. The results achieved have shown significant changes in angular changes of only ankle and knee joint at the end of the trial at frontal plane, during toe-off, and knee joint at transverse plane, during all gait events, respectively; including no significant changes observed in spatio-temporal data. There was a significant increase ($p < 0.05$) in HR and RPE values towards the end of load carriage trial. Based on the results obtained, subjects were able to consistently replicate their kinematic patterns throughout the experimental task leading to establishing a knowledge base system for soldiers' pattern recognition.

Keywords: Load carriage · Gait analysis · Kinematics · Military · Gait pattern

1 Introduction

Walking or running is a periodic movement of the body segments that includes repetitive motions [1]. In order to capture and to measure comprehensive and accurate kinematic, kinetic and temporal gait data motion capture system is used [2]. Military operations or training of soldiers typically involves running and/or marching, with or without load carriage. These foot marches requires movement of troops with their equipment by foot with no or limited support from vehicles [3]. Prolonged duration and heavy loads may induce fatigue which can lead to injuries affecting soldiers' mobility and reduce the effectiveness of the entire military unit [4]. Thus, it is important to investigate the gait patterns of these soldiers in terms of uniformity and detect any anomalies occurring in order to prevent injury and for the success of the mission.

N.T. Nguyen et al. (Eds.): ACIIDS 2016, Part II, LNAI 9622, pp. 713–722, 2016.
DOI: 10.1007/978-3-662-49390-8_69

A comprehensive analysis can be made possible by looking into the gait data that normally appear as a time series pattern [5].

There are existing data patterns published for western military population, where different results were obtained [4, 6–12]. Decreased in step length, greater flexion at the hips and forward inclination of trunk with increase in load weight has also been seen [6–8, 11–13], with results varying for knee and ankle angles, with some reported greater knee flexion and dorsiflexion of the foot and others with no or little changes seen, and this is dependent on the type of test protocols used [6–13]. However, the gait patterns for soldiers in the Asia region, especially for Brunei, may differ from those observed in the western population due to their physiological profile, training regime or exercise protocol, etc. [4] which can contribute to changes in their gait performance.

Hence, the aim of this research is to determine the effects of load carriage on soldiers' gait performance, by addressing the 3-Dimensional kinematics of the trunk and lower limbs including spatio-temporal effects of soldiers in accordance to the standard operating procedure of the Royal Brunei Armed Forces (RBAF); Annual Combat Fitness Test (an annual test to measure aerobic capacity of soldiers while performing loaded march). This research work allows to establish standards of gait patterns for RBAF soldiers in order to build a data base in reference to the 3-D kinematics and spatio-temporal features using motion capture system.

2 Methodology

The overall system consists of data acquisition, 3-D kinematic data processing, 3-D kinematic and spatio-temporal parameter extraction and statistical analysis of all parameters extracted. The design for the load carriage study is illustrated in Fig. 1.

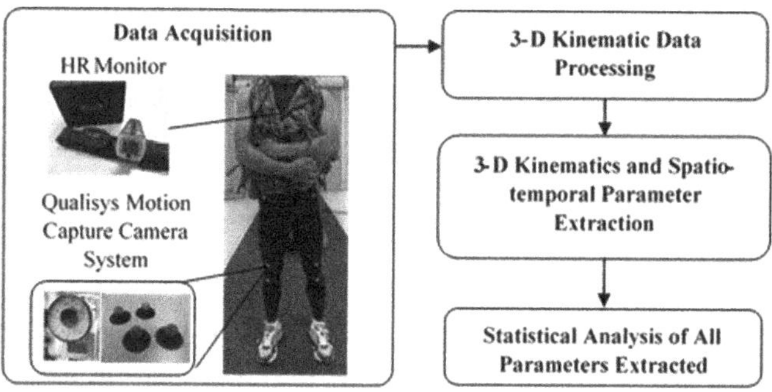

Fig. 1. Overall system overview of the load carriage study

2.1 Participants

Ten healthy RBAF soldiers with age 26.3 ± 5.8 years, height 167.8 ± 5.8 cm, weight 64.9 ± 7.7 kg (mean ± SD), participated in this study. All soldiers have no history of significant musculoskeletal injuries and cardiopulmonary pathology and has experience in carrying military load carriage system. These individuals are also within the normal Body Mass Index range for Brunei soldiers. All participants gave informed consent to participate in the study and investigators adhered to the Universiti Brunei Darussalam's ethical policies for their participants.

2.2 Experimental Setup

For gait analysis, 3-D kinematic data on the trunk and lower extremity, including pelvis, were captured using six Oqus motion capture camera (Qualisys AB, Gothenburg, Sweden) at a sampling rate of 350 Hz. 19 reflective markers were placed at different anatomical parts of the body: superior surface of acromion, anterior superior iliac spine, sacrum, superior of patella, knee joint line, tuberosity of tibia, lateral malleolus, heel, 1st and 5th metatarsal bone (Developed at Lundberg Motion Analysis Laboratory at Sahlgrenska University Hospital for Qualisys). For the load carriage trial, soldiers were instructed to walk on a Biodex treadmill fitted with a 15 kg loaded military backpack (standard issue RBAF webbing and backpack). Heart rate monitor (Polar ST4 watch, Polar, Kempele, Finland) is strapped to their chest in order to monitor changes in their heart rate throughout the trial. Exertion levels were also assessed using Borg's 6–20 Rating of Perceived Exertion (RPE) scale [14]. All participants used individual compressive sports pants, sleeveless shirt and sport shoes.

2.3 Experimental Protocol

Each participant was asked to warm-up prior to testing, which includes stretching and walking on treadmill without load. After warming-up, participants undergo a familiarisation protocol of walking with the prescribed load for five minutes on a self-selected pace. There was a rest period of five to ten minutes before the load carriage trial started to prevent carryover effects and this experiment was conducted in a climate-controlled laboratory. During the load carriage trial, participants were instructed to complete 30 min of marching or walking on the treadmill with a speed of 6.4 km/h and 1 % elevation. Kinematic data were recorded twice for 30 s, during the start, after 5 min of the walking to ensure participants had reached steady state walking, and at the final 30 s of the end of the 30 min trial. HR and RPE were monitored at every 5 min intervals, from zero minutes till the last 30 min of the test.

2.4 3-D Kinematics Data Processing

Qualisys Track Manager software (Qualisys AB, Sweden) was used to track marker data. To calculate the 3-D kinematic data of joint angles on the lower extremity, pelvis and trunk, Visual3D motion analysis software (C-motion, Inc.) was used. The hip joint landmarks where created through estimates of the right and left hip joint centre, where the location of the landmark is defined as: 0.36*ASIS_Distance, − 0.19*ASIS_Distance, − 0.3*ASIS_Distance [15]. Gait events for heel strike and toe-off were computed using only kinematic data based on the heel marker velocity data [16] and mid stance was computed manually. All joint angles were calculated using a Cardan XYZ rotation sequence and Cardan ZYX rotation sequence for pelvis segment [17]. Kinematic data were filtered with low-pass second-order Butterworth filter using a 6 Hz cut-off [10]. Ankle, knee and hip angle was defined as movement of virtual foot, shank and thigh coordinate system relative to the shank, thigh and normalized pelvis coordinate system respectively. Pelvis angle for sagittal and frontal plane was defined as movement of normalized pelvis relative to the global coordinate while pelvis rotation was defined as movement of normalized pelvis relative to trunk coordinate system [8, 13]. Trunk rotation was defined as movement of trunk technical coordinate system relative to the pelvis technical coordinate system. Range of motion (ROM) was computed by calculating the difference between the maximum and minimum joint angles values during one gait cycle [13]. Despite action being taken to minimize missing markers, trials with distorted results due to missing marker or equipment failure where removed. Only six to eight subjects were analyze that had consistent results seen for the different joints at specific planes. The mean data from the 30 s trial of the start and the end for each participant were calculated (consisted of 30 to 40 gait cycles each). The means for each parameter were calculated for all participants in order to obtain a single representation of these data.

2.5 3-D Kinematics and Spatio-Temporal Parameter Extraction

For 3-D kinematic gait analysis, all three planes of sagittal, frontal and transverse plane for ankle, knee, hip, pelvis and trunk joint angles were recorded and analyzed. However after data processing, some results were not used due to inconsistent results or data patterns presented and missing markers data. Spatio-temporal data and ROM values including HR and RPE were also studied. Tables 1 and 2 represents all of the parameters that were extracted. Mean (SD) values for all the parameters were computed.

Table 1. Measured parameters during load carriage trial in a climate-controlled environment

Parameters		Variable	Unit
Heart rate		P_1	Beats/min
Rating of perceived exertion scale		P_2	6–20 scale
Spatio-temporal data	Stride width	S_1	Meters (m)
	Stride length	S_2	
	Step length	S_3	
	Double support time	S_4	Seconds (s)
	Swing time	S_5	
	Stance time	S_6	

2.6 Statistical Analysis of All Parameters Extracted

Paired two –tailed t-test ($\alpha = 0.05$) was performed using Statistical Package for the Social Sciences Version 22 (SPSS, Inc., Chicago, IL) to calculate differences in all the parameters studied between the start and the end of the load carriage trial.

3 Results

3.1 Heart Rate and RPE Values

The mean (SD) values of heart rate and RPE results are shown in Table 3 (P_1, P_2). All physiological data shows an increase of value with increase in time, which were significant ($p <= 0.001$).

3.2 Spatio-Temporal Values

There were no significant differences or changes ($p > 0.05$) seen for stride width, stride length, step length, double support time, swing time and stance time for the subjects (refer to Table 2: S_1 through S_6).

Table 2. Measured kinematic patterns for different joints at all planes during a gait cycle using pre-defined portocols in climate-controlled environment.

Parameters	Joints	Planes	Gait events			ROM	Unit
			Heel strike	Mid stance	Toe-off		
Kinematic data	Ankle	Sagittal (x-axis)	H_1	M_1	T_1	R_1	Degrees (°)
		Frontal (y-axis)	H_2	M_2	T_2	R_2	
		Transverse (z-axis)	H_3	M_3	T_3	R_3	
	Knee	Sagittal	H_4	M_4	T_4	R_4	
		Transverse	H_5	M_5	T_5	R_5	
	Hip	Sagittal	H_6	M_6	T_6	R_6	
		Frontal	H_7	M_7	T_7	R_7	
	Pelvis	Sagittal	H_8	M_8	T_8	R_8	
		Transverse	H_9	M_9	T_9	R_9	
	Trunk	Transverse	H_{10}	M_{10}	T_{10}	R_{10}	

3.3 Kinematic Values

Heel Strike. At heel strike, except for the knee angle at transverse plane, no significant changes or differences were seen in any of the joint angles ($p > 0.05$). The knee showed a significant increase ($p = 0.043$) in external rotation towards the end of the trial at heel strike. Mean (SD) values for the kinematic data of joint angles at heel strike, mid stance and toe-off are shown in Table 4.

Table 3. Changes in mean (SD) in HR, RPE and spatio-temporal data during load carriage trial.

Parameters	Start	End	*p*-Value
P_1	100 (9.79)	132 (12.48)	0.000*
P_2	8.4 (1.51)	11.9 (2.64)	0.001*
S_1	0.0176 (0.0085)	0.0241 (0.0262)	0.405
S_2	1.6096 (0.2947)	1.4082 (0.5110)	0.391
S_3	0.8630 (0.1397)	0.9077 (0.1299)	0.093
S_4	0.2196 (0.0701)	0.2473 (0.0627)	0.089
S_5	0.309 (0.0253)	0.3188 (0.0287)	0.184
S_6	0.5262 (0.0771)	0.5487 (0.0745)	0.126

*Statistical significance (p-value < 0.05)

Mid Stance. At mid stance only the knee joint at transverse plane showed a significant change ($p = 0.029$) while the other parameters did not vary significantly from the start to the end of the load carriage trial. The knee joint showed significant increase in external rotation as time increases at mid stance.

Toe-off. At toe-off, only changes in ankle angle at frontal plane and knee angle at transverse plane were found significant ($p = 0.035$). Other joints have shown no significant angular changes in any plane. The ankle joint showed a significant decrease in inversion with increase in time when the foot push-off from the ground. The knee angle at transverse plane showed a significant decrease in internal rotation with increase in time during toe-off.

Range of Motion (ROM). The mean (SD) values for range of motion for the respective joint angles are shown in Table 5. There were no significant changes seen in all of the range of motion for all of the joint angles at their respective planes seen during the load carriage trial ($p > 0.05$).

4 Discussion

According to the results obtained, prolonged load carriage march did not necessarily have an effect on the gait, i.e. kinematics and spatio-temporal features, although a significant effect on HR and RPE scores were observed. Only kinematic data for the ankle and knee joint at frontal (during toe-off) and transverse plane (all three gait events) respectively showed significant difference between the start and the end of the trial. However, the significant changes proved that these joints are areas of the body that are first and mostly affected by loaded march, which is easily detected by using motion capture system. This study was conducted in a climate-controlled laboratory on a treadmill simulated as a "road" march trial, where the elevation of the treadmill was at 1 % to represent real ground, based on similar studies that have investigated changes in kinematic data of soldiers during load carriage [10, 11, 18].

Table 4. Angular changes in mean (SD) in different planes of different joints at different gait events during load carriage trial

Gait EVENTS	Var	Start	End	*p*-Value
Heel strike	$\mathbf{H_1}$	7.171 (2.443)	5.620 (3.476)	0.184
	$\mathbf{H_2}$	−0.048 (4.109)	−0.879 (3.149)	0.548
	$\mathbf{H_3}$	−5.821 (10.498)	−1757 (8.647)	0.196
	$\mathbf{H_4}$	9.401 (6.756)	7.866 (4.270)	0.346
	$\mathbf{H_5}$	−3.379 (6.720)	−5.894 (5.253)	0.043*
	$\mathbf{H_6}$	28.951 (11.997)	27.885 (9.807)	0.435
	$\mathbf{H_7}$	6.421 (4.253)	5.565 (3.839)	0.109
	$\mathbf{H_8}$	4.855 (9.308)	4.307 (8.091)	0.458
	$\mathbf{H_9}$	−4.068 (2.950)	−7.025 (4.707)	0.117
	$\mathbf{H_{10}}$	4.068 (2.950)	7.025 (4.707)	0.117
Mid stance	$\mathbf{M_1}$	13.198 (3.166)	12.395 (2.087)	0.362
	$\mathbf{M_2}$	−7.600 (6.055)	−7.1702 (4.38)	0.612
	$\mathbf{M_3}$	−3.295 (9.941)	−0.5740 (8.044)	0.286
	$\mathbf{M_4}$	22.265 (9.854)	20.966 (5.915)	0.486
	$\mathbf{M_5}$	−2.095 (6.402)	−4.987 (6.395)	0.029*
	$\mathbf{M_6}$	5.052 (12.163)	3.8412 (9.600)	0.522
	$\mathbf{M_7}$	2.506 (5.045)	1.377 (4.685)	0.534
	$\mathbf{M_8}$	2.358 (10.474)	1.376 (9.753)	0.305
	$\mathbf{M_9}$	−0.956 (1.769)	−2.846 (4.019)	0.214
	$\mathbf{M_{10}}$	0.956 (1.769)	2.846 (4.019)	0.214
Toe-off	$\mathbf{T_1}$	−17.193 (5.289)	−18.915 (4.142)	0.525
	$\mathbf{T_2}$	4.086 (1.291)	1.278 (3.750)	0.05*
	$\mathbf{T_3}$	6.856 (12.906)	10.134 (10.583)	0.207
	$\mathbf{T_4}$	39.582 (4.495)	39.163 (5.152)	0.797
	$\mathbf{T_5}$	1.050 (6.201)	−2.022 (5.525)	0.035*
	$\mathbf{T_6}$	−16.479 (11.729)	−17.650 (12.061)	0.439
	$\mathbf{T_7}$	−9.454 (4.429)	−9.378 (7.404)	0.963
	$\mathbf{T_8}$	1.530 (11.227)	0.8766 (10.978)	0.489
	$\mathbf{T_9}$	4.438 (4.274)	3.687 (4.515)	0.665
	$\mathbf{T_{10}}$	−4.438 (4.274)	−3.687 (4.515)	0.665

*Statistical significance (p-value < 0.05)

Pattern preparation of soldiers in this research consists of comprehensive kinematic and spatio-temporal parameters including extensive data set of the soldiers' gait performance. For this study, specific gait events are of particular interest such as heel strike, mid stance and toe-off compared to the whole gait cycle, as these three gait events are important in enabling the limbs to accomplish its basic task of weight acceptance, single-limb support and limb advancement according to the Rancho Los Amigos system [19]. Thus, in order to build a strong and extensive knowledge base of kinematic patterns, one should consider looking at data more comprehensively and specifically.

Table 5. The range of motion in mean (SD) in different planes of different joints of a gait cycle during load carriage trial.

Parameters	Start	End	p-Value
R_1	33.348 (3.819)	33.634 (3.400)	0.811
R_2	14.645 (3.431)	12.520 (2.619)	0.206
R_3	15.992 (3.397)	15.834 (3.825)	0.861
R_4	67.046 (7.553)	67.264 (7.754)	0.745
R_5	8.662 (1.649)	7.872 (1.257)	0.152
R_6	59.069 (7.575)	60.157 (10.384)	0.519
R_7	16.094 (7.118)	15.389 (6.021)	0.516
R_8	5.816 (1.634)	6.580 (2.496)	0.118
R_9	11.839 (8.174)	11.825 (7.819)	0.986
R_{10}	11.831 (8.245)	11.825 (7.819)	0.994

*Statistical significance (p-value < 0.05)

The significant changes observed in the ankle, at frontal plane during toe-off, shows that this trial has minimal effect on the lower extremities and specifically affects the ankle joint first by altering the degree of inversion, where the medial border of the foot moves towards the medial leg. This was suggested to be due to the military marchers attempt to replicate their kinematic patterns, regardless of backpack loading conditions [20, 21]. This attempt were also observed with other results obtained such as knee angle at sagittal plane for all gait events, hip, pelvis and trunk joint angles in different planes at all three gait events and ROM for all joints. This is also further supported with the lack of changes that were observed in the spatio-temporal features. The knee kinematics result at sagittal plane were consistent with previous studies [10, 18] due to the task performed was of light aerobic intensity and muscular fatigue was not induced. The changes observed in knee joint rotation at transverse plane suggest that as time increases together with addition of load, the knee was observed to rotate more, due to the knee slowly getting tired or fatiguing from the impact forces of the weight carried and thus, affecting the rotation of the knee joint first. This suggest that the first impact of the load carriage has is on the knee joint rotation, and then possibly affecting the flexion and extension of the knee as well, as they work simultaneously with internal and external rotation of the femur.

For other kinematics data, such as the hip, pelvis and trunk joint angles, that has shown lack of significant changes in different planes at all three gait events consistent with [10]. Therefore, subject's stability was maintained throughout the trial leading so neither changes in their gait patterns nor fatigue monitored. In the case of pelvis kinematic data, the lack of significant changes seen due to no changes in pelvic tilt and obliquity due to no accumulative effect of increasing load observed [9]. For trunk kinematics as the study was conducted at a constant speed, there were no changes observed in the pelvis and thus no occurrence of counter balance in the momentum of rotation between the upper and lower body was observed [22].

As mentioned earlier, the spatio-temporal features have shown no significant changes between the start and the end of the load carriage trial due to the climate controlled environment used for gait trials; a treadmill in the motion analysis lab and subjects that were well-trained in carrying loads during their marches consistent with [10, 14]. Thus, subjects were able to maintain their balance and stability during marching, and replicate their kinematic gait patterns at the same time. The research also assess the physiological and subjective consequences of 15 kg load carriage trial for 30 min at 6.4 km/h. The increase observed in the heart rate was consistent with the increase seen in RPE scale during the trial, despite the lack of changes seen in the kinematic and spatio-temporal values which is also reported in [10, 14, 18, 21]. Thus, changes in heart rate and perceived exertion were to be expected in order to account for generic differences in aerobic fitness requirements between soldiers.

5 Conclusions

There were no previous literatures reported a comprehensive study on 3-D kinematic data coupled with spatiotemporal data, for prolonged loaded march trial in the military. There were also no such studies conducted for the Brunei's military population so far that dealt with soldiers' gait and movement performance. Based on the measured parameters, the lack of changes reported in the majority of 3-D kinematics data and spatio-temporal features were due to the experimental protocol used, such as the speed and load magnitude, in accordance to the RBAF guidelines for annual combat fitness test (prolonged loaded march test) and training regime. This study also assessed soldiers carrying backpack only in isolation of other equipment typically carried by soldiers during combat operations, which are established training policies in Brunei. However, the study was able to detect minute changes in specific lower limb joints during the trial. Thus, overall soldiers of RBAF were able to consistently replicate their kinematic patterns throughout the experimental task which allowed the establishment of knowledge base of kinematic patterns of RBAF. The motion capture system introduced in this research is capable to benchmark RBAF soldiers' loaded prolonged march during training regime. In order to expand the knowledge base formed, more parameters to be derived such as; on real ground or looking at whole gait cycle or/and muscles and brain activity during load carriage and analysing the effects of fatigue.

Acknowledgements. Authors would like to thank all participants for their time and effort, as well as Dr Mike Steel of Universiti Brunei Darussalam. At the same time, Authors appreciate Performance Optimization Centre of Ministry of Defense, Brunei and Royal Brunei Armed Forces, Brunei for their contribution and support in the study.

References

1. Tao, W., Liu, T., Zheng, R., Feng, H.: Gait Analysis Using Wearable Sensors. Sensors (Basel) **12**(2), 2255–2283 (2012)
2. DeLisa, J.A.: Gait analysis in the Science of Rehabilitation, 2nd edn. DIANE Publishing, Collingdale (1998). Instrumented Gait Analysis by Ernest Bontrager
3. Van Dijk, J.: Common Military Task: Marching. Optimizing Operational Physical Fitness (2009)

4. Knapik, J., Reynolds, K.L., Harman, E.H.: Soldier Load Carriage: Historical, Physiological. Biomech. Med. Aspects Mil. Med. **169**, 1–45 (2004)
5. Bakar, A.A., Senanayake, S.M.N., Ganeson, R., Wilson, B.D.: Pattern recognition in gait analysis. In: Fuss, F.K., Subic, A., Ujihashi, S. (eds.) The Impact of Technology on Sport II, pp. 169–174. Taylor & Francis, NewYork (2008)
6. Harman, E., et al.: The effects of backpack weight on the biomechanics of load carriage. US Army Research Institute of Environmental Science, Natick, MA (2000)
7. Ling, W., Houston, V., Tsai, Y., Chui, K., Kirk, J.: Women's load carriage performance using modular lightweight load-carrying equipment. Mil. Med. **169**, 11–914 (2004)
8. Attwells, R., Birrell, S., Hooper, R., Mansfield, N.: Influence of carrying heavy loads on soldiers' posture, movements and gait. Ergonomics **49**(14), 1527–1537 (2006)
9. Birrell, S., Haslam, R.: The effect of military load carriage on 3-D lower limb kinematics and spatiotemporal parameters. Ergonomics **52**(10), 1298–1304 (2009)
10. Almosnino, S., Kingston, D.C., Bardana, D.D., Stevenson, J.M., Graham, R.B.: Effects of prolonged load carriage walking on lower extremity and trunk kinematics, heart rate, and subjective responses. In: The American Society of Biomechanics 37th Annual Meeting, 4–7 September 2013, Omaha, Nebraska, USA (2014)
11. Seay, J.F., Fellin, R.E., Sauer, S.G., Frykman, P.E., Bensel, C.K.: Lower Extremity Biomechanical changes associated with symmetrical torso loading during simulated marching. Mil. Med. **179**, 1–85 (2014)
12. Kinoshita, H.: Effects of different loads and carrying systems on selected biomechanical parameters describing walking gait. Ergonomics **28**, 1347–1362 (1985)
13. Qu, X., Yeo, J.C.: Effects of load carriage and fatigue on gait characteristics. J. Biomech. **44**, 1259–1263 (2011)
14. Borg, G.: Psychophysical bases of perceived exertion. Med. Sci. Sports Exerc. **14**(5), 377–381 (1982)
15. Bell, A.L., Pedersen, D.R., Brand, R.A.: A comparison of the accuracy of several hip center location prediction methods. J. Biomech. **23**, 617–621 (1990)
16. Zeni, J.A., Richards, J.G., Higginson, J.S.: Two simple methods for determining gait events during treadmill and overground walking using kinematic data. Gait & Posture **27**, 710–714 (2008)
17. Baker, R.: Pelvic angles: a mathematically rigorous definition which is consistent with a conventional clinical understanding of the terms. Gait & Posture **13**, 1–6 (2001)
18. Beekley, M., Alt, J., Buckley, C., Duffey, M., Crowder, T.: Effects of Heavy load carriage during constant-speed, simulated, road marching. Mil. Med. **172**, 6–592 (2007)
19. Center, Ranchos Los Amigos National Rehabilitation: Observational Gait Analysis. Los Amigos Research and Education Center, Englewood (2001)
20. Hamill, J., Knutzen, K.M.: Biomechanical Basis of Human Movement. Lippincott Williams & Wilkins, Philadelphia (1995)
21. Quesada, P., Mengelkoch, L., Hale, R., Simon, S.: Biomechanical and metabolic effects of varying backpack loading on simulated marching. Ergonomics **43**(3), 293–309 (2000)
22. LaFiandra, M., Wagenaar, R., Holt, K., Obusek, J.: How do load carriage and walking speed influence trunk co-ordination and stride parameters? J. Biomech. **36**, 87–95 (2003)

An Integrated Pattern Recognition System for Knee Flexion Analysis

Joko Triloka(✉), S.M.N. Arosha Senanayake, and Daphne Lai

Motion Analysis Lab, Integrated Science, Universiti Brunei Darussalam, Jalan Tungku Link, Gadong BE 1410, Brunei Darussalam
{12h1052, arosha.senanayake, daphne.lai}@ubd.edu.bn

Abstract. The purpose of this study is to propose an integrated knee-flexion analysis system (IKAS) as a novel tool for recognition pattern of knee muscle for athletes and soldiers based on neuromuscular signals and soft tissue deformation parameter. Different types of parameters from multi-sensors integration are combined to analyze the knee motion. Data fusion of EMG and frames of the video for each knee flexion angle acquired from synchronization of the motion capture system and video cameras interfaced with wireless EMG sensors. Systems are pre-processed in order to prepare the pattern set for a custom-developed artificial neural network and mesh generation technique based intelligent system for classifying the patterns of knee muscle of subjects during walking and squatting activity. Multilayer feed-forward backpropagation networks (FFBPNNs) with different network training algorithm were designed and coefficient correlation (CC) was uses and their classification results were compared. The newly introduced IKAS approach will provides assistance in making an objective and knowledgeable decisions about recognition of patterns from knee muscles.

Keywords: Pattern recognition · Knee flexion · Soft tissue deformation

1 Introduction

The lower-limb motions are very important for the human daily activities [1]. Caused by knee injuries, the actuation patterns of variant muscles over knee are influenced which effect in lower-limb changes in subjects. Recognition of these pattern abnormalities from muscle movements and its features is a demanding research topic as several lower-limb muscles deliver control and strength during activities by contracting at specific intervals in an activity cycle.

A common method to measure the muscle contractions is to use electromyography (EMG) which train of motor unit action potentials discharged by the contraction of striate muscle tissue [2]. In muscle activation patterns, EMG signals are measured during dynamic or isometric muscle contractions. Various types of contraction can cause EMG signals to vary, affecting recognition performance. Furthermore, surface electromyography (EMG) signals was used in the biomechanical lower extremity model to estimate the muscle activation patterns and also be used to design the lower-limb exoskeltal assistive robotic systems for physically challenged persons [6].

N.T. Nguyen et al. (Eds.): ACIIDS 2016, Part II, LNAI 9622, pp. 723–732, 2016.
DOI: 10.1007/978-3-662-49390-8_70

The movement pattern in the knee results from the interaction between external forces and internal forces. Joint reaction typically derived from the knee joint. One of the main movements of the knee joint is flexion. In the flexed position, the collateral ligaments are relaxed while the cruciate ligaments are taut [3]. Because of the oblique position of the cruciate ligaments at least a part of one of them is always tense and these ligaments control the joint as the collateral ligaments are relaxed.

However, it is important to recognize that the entire lower limb moves during exercises. An alteration in any segment of the lower limb can have consequences on the individual's pattern. In this study, alterations during exercises are mainly considered based on musculoskeletal issues. Alterations patterns resulting from musculoskeletal are often caused by strengthen lower-body muscles, related injury resulting from structural abnormalities of a bone, joint, or soft tissue. Other causes of pathologic side include neuromuscular and myopathic conditions [4], soft tissue imbalance, joint alignment or connective tissue after joint [5].

On the other hand, the recognition of soft tissue deformations is also an important aspect of deformability of human body limbs during movement. The consequent inabilities to approximate their dynamics by means of a rigid body model [7] are critical problems in marker less motion analysis. In this context, both for clinical and athletic training purposes appear relevant. While considering a non-pathological trial, during contraction and stretching, muscles change shape and size, and produce a deformation of skin tissues and a modification of the body segment shape. If the subject is affected of skin or muscular pathology, the skin deformation is not natural, presenting non elastic areas: to highlight this phenomenon it is necessary to recognition a pattern of body limbs with non-rigid structures.

Muscular deformations are generally modelled by using triangular meshes. Previous study revealed some techniques and models used to generate realistic warps have been presented [8]. Most of the applications refer to facial models and meshes are typically employed for facial expression animations [9, 10]. Nonetheless, the big disadvantage of a structured mesh is its lack of flexibility in adapting to a domain with a complicated shape [11]. For this reason, given the need to pattern shapes with intricate domains, the less expensive and supportive method are to use the soft tissue deformation (STD) analysis technique [12, 13] to use unstructured meshes, since a structured mesh can arbitrarily adapt to complex domains.

The current activity, muscular and skin deformation methods, however, has some weakness e.g., during isometric muscular contractions exercises [12, 13], or at a static position with a weight backpack load. Moreover, the dynamic muscular activity and soft tissue deformation has not been fully considered in the available literature.

An Integrated Knee-Flexion Analysis System (IKAS) is proposed in this paper as shown in Fig. 1. The novelty of the developed system is mainly on the integration of EMG signals and soft tissue deformation analysis of knee muscles, which is based on the proposed integrated system using motion capture system combined with its video cameras, interfaced with wireless EMG sensors. The input to the IKAS comprises mixed signals (integrated sensors with data fusion) acquired using six Qualisys Oqus 300 infrared optical cameras integrated with three Qualisys Oqus 210 video cameras and EMG sensor data acquired from a Bio-Capture System simultaneously. An IKAS also includes system software which implemented using two main modules; EMG signals processing module and motion and soft tissue deformation processing module. The output produced by the IKAS allows for recognition the pattern of the knee muscle activities of the Vastus Lateralis (VL) and Vastus Medialis (VM) muscles associated with changes in soft tissue deformation during movement.

In order to test the performance of the system, two different exercises i.e. walking and squatting have been taken into account and were administered to ten healthy male soldiers and six male athletes (4 healthy and 2 gaining knee injuries). The means and standard deviations of age, height and weight of the subjects were 26.6 ± 4.05 years, 165.4 ± 3.10 cm and 68.2 ± 11.90 kg, respectively. The participants were recruited from the Performance Optimization Center in the Ministry Of Defense and the Sports Medicine and Research Center in the Ministry of Culture, Youth and Sports, Brunei Darussalam. All subjects read and signed an informed consent form and all procedures were carried out according to the ethics guidelines approved by Universiti Brunei Darussalam's Graduate Research Office and Ethics Committee.

This novel assistive tool was successfully applied in order to recognize the pattern of knee muscles activities and contraction including vastus lateralis and vastus medialis muscles during walking and squatting.

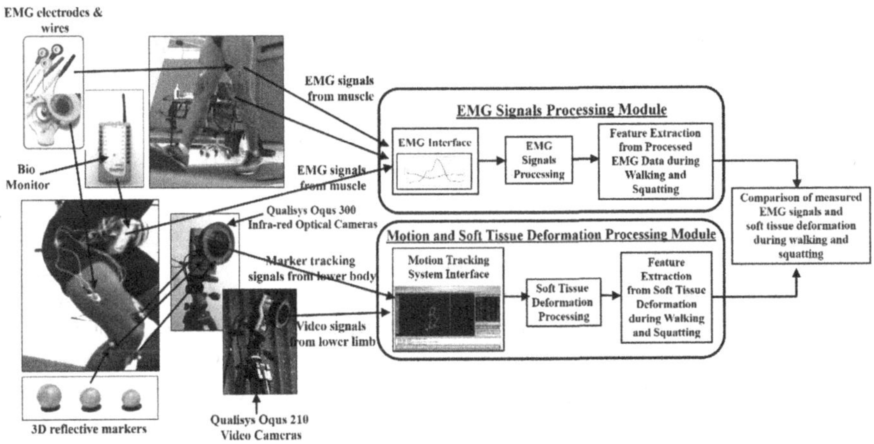

Fig. 1. Integrated knee-flexion analysis system comprises the Qualisys motion capture system and its video cameras interfaced with wireless EMG sensors

2 Walking Pattern Recognition Using IKAS

The parameter was used to recognition the pattern during walking are Root Mean Square (RMS) of EMG signal and Soft Tissue Deformation Parameter (STDP) from frame image that conform to the lower limb segment. RMS is defined as the time windowed RMS value of the raw EMG. RMS is one of a number of methods used to produce waveforms that are more easily analyzable than the noisy raw EMG. RMS value of EMG signal is extracted for different gait phases. The soft tissue deformation is obtained through triangular meshes that automatically conform to the lower limb segment when dynamic muscles are contracted during walking on a treadmill at the constant level. Figure 2 shows a sample of soft tissue deformation processing of VL muscle and the reference points in the Region of Interest (ROI) image has been approximated by motion estimation algorithm.

Raw surface EMG signals and soft tissue deformation from knee muscles were recorded in a population of athletes and soldiers were obtained prior to starting the experiment (actual motion) when the subject were in upright position. These measurements were then extracted from each phase of gait during walking. Pre-processed have been done in order to prepare the pattern set for classification algorithms to be integrated for possible implementation of IKAS. RMS value of EMG signals and STDP were use as the input features for the ANN model. Hence, 461 pattern sets were randomly chosen to train the ANN and the rest of the untrained pattern sets was used to identify walking gait pattern. Multilayer feed-forward backpropagation networks (FFBPNNs) with different network training algorithms were created and their classification results were compared. The pattern set was randomly separated into training (80 % of pattern set for training) and testing (20 % of pattern set as testing pattern set the performance analysis of gait pattern identification) pattern sets. The parameters of each networks are trained by the 12 inputs, 20 neurons in the hidden layer and a single output neuron for identifying healthy/injured (1/0) subject. The performance of the FFBPNNs was determined through a 10-fold cross validation process.

Thus, the pattern recognition for walking were defined consist of RMS and STDP defined as

$$P_{walking} = (IKAS.Walking_{RMS} and_{STDP}, ANN) \tag{1}$$

Analysis was done by correlated two variables defined. The coefficient of correlation is a method of assessing a possible two-way linear association between two continuous variables (EMG Parameter and STDP) divided by the product of their standard deviations. To assuring RMS and STDP value can be used for recognition of gait patterns, the coefficient correlation among two parameters was examined to each muscle with different stage of gait cycle. Based on the mean value from 12 gait cycles of each subject, the mean values of coefficient of correlation for VL muscle were 0.760391 and for VM muscle were found as 0.811655.

Based on the values of coefficient correlation above, RMS and STDP from VL and VM muscles of four subjects were used for design artificial neural network. The FFBPNNs then were created, trained and tested for all subjects while walking on the treadmill at speed 4 km/h. In order to examine a mean squared error, the targeted was setup at 0.001 for higher prediction. Multilayer FFBPNN including four training algorithms were designed for merged RMS and STDP features. The result shows that Levenberg-Marquardt training algorithm can perform well with the maximum training (100.00 %) and testing (98.50 ± 4.11) classification accuracies were achieved.

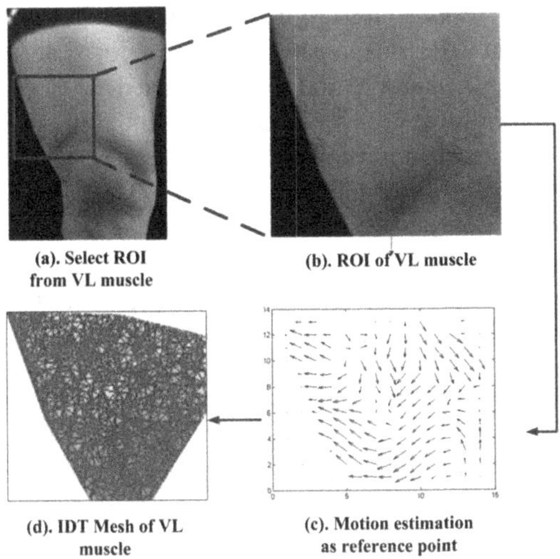

Fig. 2. IDT Mesh processing for walking pattern

Comparison with the rest of training algorithms, Scaled Conjugate Gradient performs next highest classification accuracies with average training (99.68 ± 1.62) and testing (97.50 ± 5.39). The average classification accuracy of the designed FFBPNNs has been compared during training phase and test phase for healthy and injured subjects as presented in Table 1.

Table 1. Classification accuracy comparison between different training algorithms based on combined RMS and STDP feature of healthy and injured subject

Method	Healthy		Injured	
	Training	Test	Training	Test
Gradient descent	69.1 ± 18.34	62.2 ± 24.76	98.7 ± 3.54	98.2 ± 6.42
Levenberg-marquardt	100.0 ± 0	97.0 ± 8.23	100.0 ± 0	100.0 ± 0
Random order weight/bias	100.0 ± 0	95.2 ± 11.05	49.2 ± 18.81	48.6 ± 27.3
Scaled conjugate gradient	99.2 ± 4.18	95.0 ± 10.78	100.0 ± 0	100.0 ± 0

3 Squat Pattern Recognition Using IKAS

The system software records the motion of the subjects during the squat and extracts pertinent features; Soft Tissue Deformation Parameter (STDP) from frames of the video sequences for each knee flexion angle and data fusion of EMG which contain root mean square (RMS), mean absolute value (MAV), and integrated EMG (IEMG) of EMG signals.

Raw surface EMG signals and STDP from knee muscles were recorded in a population of healthy athletes and soldiers were obtained prior to starting the experiment when the subject were in a neutral position (approximately 0° knee flexion) to a depth of approximately 40° knee flexion (half-squat). Pre-processed and experimental setup have been done in order to prepare the pattern for statistical analysis of the muscle activities of the VL and VM muscles associated with changes in soft tissue deformation at four different positions of knee flexion and in two different condition with and without a load. Figure 3 illustrates the video frames and knee flexion positions in the 3D mode of a subject performing half-squats.

The output of the pattern is the coefficient correlation (CC) between two continuous variables (EMG Parameter and STDP) for all knee flexion positions and conditions.

Thus, the pattern recognition for squat were defined consist of RMS, MAV and IEMG of EMG signals and STDP defined as

$$P_{Squat} = (IKAS.Squat_{RMS,MAV,IEMG} and_{STDP}, CC) \tag{2}$$

The coefficient correlation between the parameters of the EMG signals then was calculated in order to perceive the relationship among these features.

To ensure that the IEMG, MAV, and RMS values of the EMG signals and STDP could be used for squat pattern, the relationship among two parameters was tested for every muscle.

The average coefficients correlation for the VL muscle among IEMG, MAV, RMS, and STDP for all subjects was 0.649006, 0.983689, and 0.979508, respectively. For the VM muscle, these values were 0.975926, 0.96621, and 0.962987, respectively.

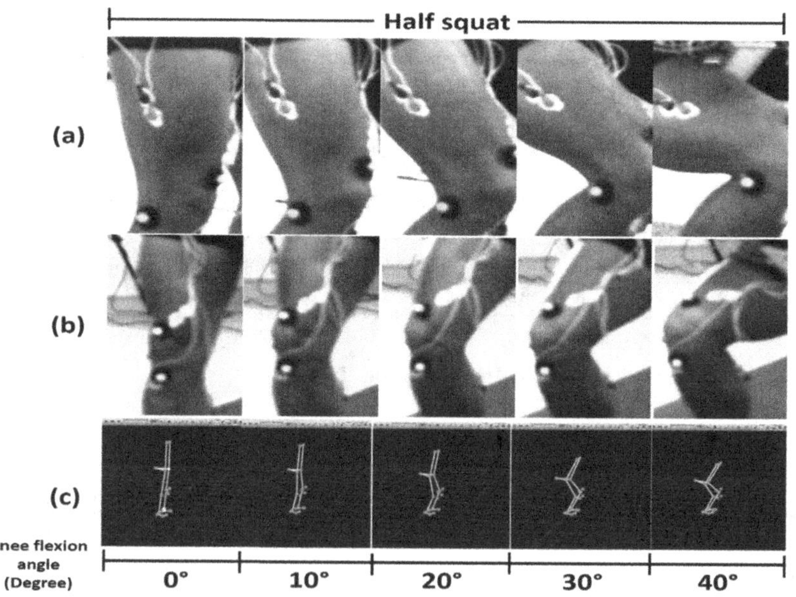

Fig. 3. Video sequence recording for one half repetition squats. (a). Right-side camera. (b). Front camera. (c) Angles of knee flexion in 3D view.

Table 2. Overall Correlations of EMG Parameters and STDP

Parameters	Muscle			
	Vastus lateralis		Vastus medialis	
	Unloaded	Loaded	Unloaded	Loaded
RMS/STDP	0.99	1	0.98	0.99
MAV/STDP	0.98	1	0.97	1
IEMG/STDP	0.98	0.99	0.94	0.97

Based on these coefficients correlation, EMG parameters and STDPs from the VL and VM muscles of all subjects were applied to measure squat exercises with and without load at four angles of knee flexion.

The average normalized EMG amplitude data are presented as the average RMS, MAV, and IEMG using the average of the total volume of muscle activity during each repetition of the squatting exercise for VL and VM muscle. The STDPs are also presented as the averages of the total volume of the body segment contour when the muscle contraction occurs.

The overall correlation among all EMG parameters and the STDP is illustrated in Table 2. The high correlation indicated that the use of EMG parameters and the STDP can provide comprehensive information about muscle contraction during squatting under different conditions and at different positions.

4 Discussion

This paper introduces a method for pattern of knee muscles recognition using EMG and soft tissue deformation parameter that could potentially be used in making an intention and knowledgeable decisions about recognition of patterns from knee muscles and can help to identify important factors for greater muscle activation.

For this purpose, Integrated Knee-Flexion Analysis System (IKAS) were used while recognize EMG activity and soft tissue deformation of Vastus Lateralis (VL) and Vastus Medialis (VM) muscle during walking and squatting.

Recognition the gait pattern developed with Feed-Forward Back Propagation Neural Network (FFBPNN) with several training algorithms showed that FFBPNN demonstrated to be beneficial in objective to recognition of gait patterns from knee muscles. The classification accuracy results were achieved with merged RMS and STDP features were used in the network. It also indicates that RMS and STDP possess a robust correlation relating to acquiring force of a muscle contraction. Furthermore, by inputting large data sets and more features extracted from EMG signals and STDP can improve the overall accuracy of the network. The proposed system therefore provides that a broad range of information about muscle activity during the gait cycle acquires using integration of neuromuscular signals and soft tissue deformation analysis of muscles.

The accuracy between these two features sets during the gait cycle relies on number of factors comprises position of video cameras in different field, placement of surface EMG electrode positions on muscle during movement, different techniques for EMG signal processing and different types of feature extraction from EMG data. A limited number of features have been used for training the network which can further enhance with addition of other features e.g. timing analysis of EMG signals. Moreover, all these data from each phase of gait cycles can be stored and reprocessed in order to compare the result from similar or different subjects

In squatting, the output produced by the IKAS allows for statistical analysis of the muscle pattern of the VL and VM muscles associated with changes in soft tissue deformation. This system can revealed that increasing knee flexion at different angles of knee flexion while squatting under unloaded and loaded conditions has a significant effect on the muscle activity of EMG parameter (RMS) and the STDP of the VL and VM muscles.

However, it is important to recognize that the entire lower limb moves while squatting. An alteration in any segment of the lower limb can have consequences on the individual's squat pattern. In this portion, alterations during squat mainly take into musculoskeletal causes. Squat patterns resulting from musculoskeletal are often caused by strengthen lower-body muscles. Having this consideration, the most important event during squat is to absorb the initial impact from flexion of the knee.

During squat, the system has been tested for a group of healthy athletes and soldiers and shows encouraging results. The high correlation indicated that the use of EMG parameters and the STDP can provide comprehensive information about the patterns of muscle during squatting under different conditions and at different positions. It also suggests that the combinations of EMG parameters and STDP analysis of knee muscle proposed as a supportive device to perceive the changes of neuromuscular patterns in subjects. Furthermore, this result is improving the understanding of knee muscle activity in common with increasing of knee-flexion movement.

As a complex structure results by using IKAS, it is adequate to store the muscles pattern of subjects into the knowledge base as follow:

$$P_i = \sum((IKAS.Walking_{RMS,STDP}, ANN), (IKAS.Squat_{RMS,MAV,IEMG,STDP}, \text{CC})) \quad (3)$$

The knowledge base is used to store the data/signals based on combined parameters at different phase and in different formats for processing and classification. The knowledge base will also include the information from exercise activity which will be updated based on new outcomes produced by the IKAS. This will help the system in learning from previous experiences and reacting properly in the new situations.

The proposed system can be applied as an intelligent mechanism used for classification pattern in other lower extremity muscles. Creating a comprehensive system is still a challenge topic that necessary detail inquiry on features and intelligent systems for classification.

5 Conclusions

This paper introduces a novel method for pattern recognition of knee muscles activities and alteration of the lower limb segment when dynamic muscles are contracted by using combined parameters; EMG signals parameters and soft tissue deformation parameter which are very helpful for recognition of gait patterns between healthy and injured subjects grounded by artificial neural network. Based on the muscles movements acquired using EMG device and captured by video cameras, the trained FFBPNN can classify and categorize the gait patterns of the subjects. Moreover, the results provides the prominence of knee flexion and its contiguous muscles including vastus lateralis and vastus medialis in slightly fraction of knee flexion during squat exercise in two different condition with and without load. This result contributes the detail understanding of the deformation patterns of surface muscle during muscular contraction based on the integrated data from EMG and STDP. Furthermore, development of an IKAS using EMG signals and STDPs can be used as an assistive tool to recognize the patterns of muscle in a variety practice of human motion analysis.

References

1. He, H., Kiguchi, K.: A study on EMG-based control of exoskeleton robots for human lower-limb motion assist. In: 6th International Special Topic Conference on Information Technology Applications in Biomedicine, 2007 (ITAB 2007), pp. 292–295, 8–11 November 2007
2. Fridlund, A.J., Cacioppo, J.T.: Guidelines for human electromyographic research. Psychophysiology **23**, 89–567 (1986)
3. Platzer, W.: Color Atlas of Human Anatomy: Locomotor System, vol. 1, 5th edn, pp. 212–213. Thieme, Stuttgart (2004)
4. Lehmann, J.F., de Lateur, B.J., Price, R.: Biomechanics of abnormal gait. Phys. Med. Rehabil. Clin. North Am. **3**, 125–138 (1992)
5. Dahlkvist, N.J., Mayo, P., Seedhom, B.B.: Forces during squatting and rising from a deep squat. Eng. Med. **11**, 69–76 (1982)
6. Mishra, A.K., Srivastava, A., Tewari, R.P., Mathur, R.: EMG analysis of lower limb muscles for developing robotic exoskeleton orthotic device. In: International Symposium on Robotics and Intelligent Sensors 2012 (IRIS 2012) (2012)
7. Goffredo, M., Schmid, M., Conforto, S., Carli, M., D'Alessio, T.: Posture kinematics reconstruction and body model creation using a marker-free subpixel algorithm. In: IS&T/SPIE's XVI Annual Symposium on Electronic Imaging, San Jose, California, USA (2004)
8. Maa, W., Maa, X., Tsoa, S., Panb, Z.: A direct approach for subdivision surface fitting from a dense triangle mesh. Comput. Aided Des. **36**, 525–536 (2004)
9. Zhanga, Y., Prakash, E.C., Sunga, E.: Face alive. J. Visual Lang. Comput. **15**, 125–160 (2004)
10. Zhanga, Y., Prakash, E.C., Sunga, E.: A Physically-based model with adaptive refinement for facial animation. In: 2001 Proceedings of the Fourteenth Conference on Computer Animation (2001)

11. Bern, M., Eppstein, D.: Mesh generation and optimal triangulation. Computing and Euclidean Geometry, 2nd Ed., pp. 47–123. World Scientific (1995)
12. Carli, M., Goffredo, M., Schmid, M., Neri, A.: Study of muscular deformation based on surface slope estimation. In: Image processing: Algorithms and Systems, Neural Networks, and Machine Learning, vol. 6064, Society of Photo-optical Instrumentation Engineers (SPIE) Conference Series, Dougherty, E.R., Astola, J.T., Egiazarian, K.O., Nasrabadi, N.M., Rizvi, S.A. Eds.: SPIE, 200 (2006)
13. Goffredo, M., Carli, M., Conforto, S., Bibbo, D., Neri, A., D'Alessio, T.: Evaluation of skin and muscular deformations in a nonrigid motion analysis. In: Proceedings of SPIE 5746, Medical Imaging: Physiology, Function, and Structure from Medical Images, p. 535 (2005)

An EMG Knowledge-Based System for Leg Strength Classification and Vertical Jump Height Estimation of Female Netball Players

Umar Yahya(✉), S.M.N. Arosha Senanayake, and Daphne Lai

Motion Analysis Lab, Faculty of Science, Universiti Brunei Darussalam, Jalan Tungku Link, BE1410 Bandar Seri Begawan, Brunei Darussalam
{14h0053,arosha.senanayake,daphne.lai}@ubd.edu.bn

Abstract. This study proposes a framework for the design and implementation of a leg strength Knowledge-Based Classification System based on pattern sets obtained from the statistical analysis of Electromyography (EMG) features (Peak, IEMG, Peak time) extracted from eight different lower limb muscles during a single leg vertical jump test of female netball players. The system employs a novel classification algorithm that takes into account the test subject's anthropometry profile (height, body mass) and age, with respect to pattern sets in the knowledge-base. The algorithm is also able to estimate the vertical jump height with which the test subject attained the corresponding EMG data set. Results indicate that the system is not only able to accurately classify a subject's overall leg strength category (strength category weight ≥ 66.67 %), but also is able to reveal the individual muscle strength classification (strength category weight ≥ 62.5 %). The results offer conditioning coaches and trainers deeper insights into muscle-specific strength variations of players.

Keywords: Classification algorithm · Knowledge-based system · Strength classification · Anthropometry profile · Feature extraction · Pattern set identification · Strength category weight

1 Introduction

Recent years have seen a rising trend in both the study and application of surface Electromyography (EMG) in Sports Biomechanics research for various objectives by sports practitioners and researchers [4–7]. This trend is however increasingly faced with challenges arising from the complex and high variability nature of EMG signals [6, 10] as well as the advancement of EMG signal acquisition technologies which dictates for the need to have smarter processing techniques for smarter acquisition techniques. These two challenges of overlapping nature have led to a multitude of research directions aimed at increasing the reliability, reproducibility, and robustness of EMG signals and EMG-Based systems, respectively. The advancement in Intelligent Information Systems and Knowledge-Based technologies continues to offer extended dimensions through which patterns in EMG Signals can be identified and classified [1, 8].

The purpose of this study was to examine the effectiveness of EMG data when used in identifying a randomly selected subject's leg strength classification from an EMG

N.T. Nguyen et al. (Eds.): ACIIDS 2016, Part II, LNAI 9622, pp. 733–741, 2016.
DOI: 10.1007/978-3-662-49390-8_71

knowledge-base. The knowledge-base's pattern sets were derived from a statistical analysis of EMG features (Peak, IEMG, Peak time) extracted from the EMG activity of eight lower limb muscles recorded during a single leg vertical jump test in female netball players. The muscles from the subjects' preferred (right) leg included muscles from the hamstring group (Biceps Femoris (BF), Semitendinosus (SM)), Quadriceps group (Vastus Lateralis (VL), Vastus Medialis (VM), Rectus Femoris (RF)), and the Calf Muscles (Gastrocnemius Medialis (GM), Gastrocnemius Lateralis (GL), and Peronius Longus (PL)).

This study hypothesized that the formation of such a Knowledge-Base would make it possible to accurately classify a test subject's leg strength, provided her anthropometry profile fits the sample population one's. Unique to this study's hypothesis, unlike previous ones [3, 8], is its consideration of both the EMG features and a combination of more than two other sample-population's characteristics which include; a subject's anthropometry profile (height, body mass), age, and strength classification following a uniform standard leg strength test (wall-squat).

2 Methodology

2.1 Experiment Procedure

Subjects. Twelve healthy professional female netball players of average (Height = 164.94 cm, Age = 22.83 Years, Weight = 62.53 kg) consented to volunteer in this study (Table 1). All subjects were playing for the Brunei National Netball team at the time of experiments. Before the start of the experiment, each subject was allowed sufficient time with all the equipment to be used during the experiment in order to familiarize them with the experience of playing with the wearable apparatus. Written consent to voluntarily participate in the study was obtained from all the 12 subjects in conformity with the ethical policy on experiments involving human test subjects established by the ethics committee of Universiti Brunei Darussalam.

Table 1. All Subjects' descriptive statistics

Measurement	Mean ± SD
Body height (cm)	164.94 ± 5.64
Body mass (kg)	62.53 ± 8.77
Age (years)	22.83 ± 3.66

Activity. Each subject warmed-up from an indoor netball court for 10–15 min as would be the case in preparation for a real netball game [5]. The eight investigated muscle surfaces were prepared for surface Electromyography (EMG) signal detection and 17 Electrodes mounted as per the guidelines [9] of Surface EMG (sEMG) with the help of a physiotherapist and a strength and conditioning coach. The electrode wires were connected to an eight channel wireless BioRadio (Cleveland Medical devices inc. USA) signal transmitter sampling at 960 Hz (Fig. 1).

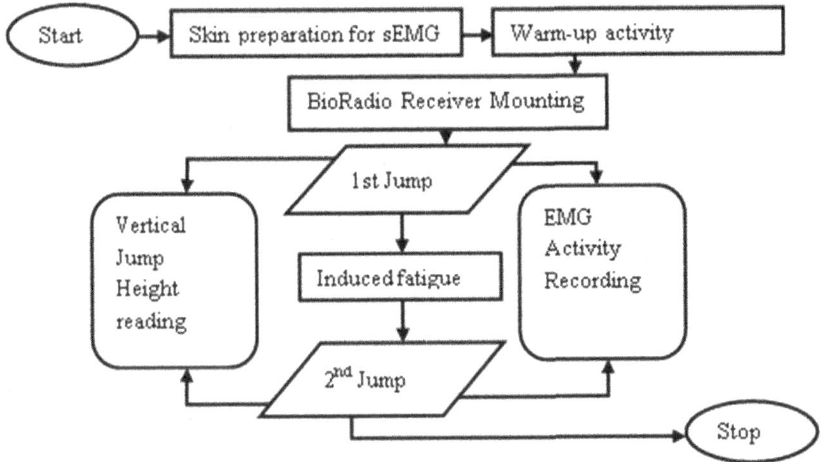

Fig. 1. Experiment protocol for raw EMG signal acquisition

In a standing position, 2 m away from the EMG signal acquisition computer system and adjacent to a vertec device, the subject was asked to jump as high as possible and displace the highest vane they could reach while recording the corresponding EMG activity of the respective muscles in the signal acquisition system. The vertec device's initial reading was adjusted to a 0.00 cm reading corresponding to the subject's fully extended right arm upwards. The reading of the highest displaced vane was recorded as the vertical jump height. The subject performed unloaded step-ups for 2 min on a wooden surface 30 cm above the netball court surface to induce some muscular and cardio fatigue as would be in the case of a real competitive game. Electrode placement was quickly checked for any displacement, and the subject asked to repeat the standing vertical jump while recording both the EMG activity and the vertical jump height reached in the same manner as was in the first attempt.

A day before the experiment, each subject was asked to perform isometric activities (leg curls, calf press, and leg extension) for 3 to 5 trials with increased load variation aiming to achieve a Maximum Voluntary Contraction (MVC) while recording the respective EMG activity for each of the respective muscles. MVC data is collected as a data normalization requirement for all EMG studies involving healthy subjects at the time of the experiment [9, 10]. The EMG reading obtained during the trial with the maximum load was considered as the MVC recording of the corresponding muscle.

The subjects' right leg strength classifications were derived from a standard time-based wall-squat procedure in which each subject stands comfortably with feet approximately shoulder width apart with her back leaning against a firm vertical wall.

The subject then slowly slides her back down the wall to assume a body position with both knees and hips at a 90° angle. The wall-squat timing starts when the left foot is lifted off the ground and stops when the subject cannot maintain the position any longer and the foot is returned to the ground. The Six (n = 6) subjects who are classified as Excellent were able to maintain their single leg squatting positions for a period between 2–3 min, whereas the other six (n = 6) classified as Very Good (Table 2) were able to only maintain their positions for a period between 1–2 min.

Table 2. Subject Right-Leg strength classification using a time-based wall-squat test

Category	Squatting time (min)	Number of subjects
Excellent	2–3	06
Very good	1–2	06

2.2 EMG Signal Processing and Feature Extraction

Signal Processing. The raw EMG signals recorded during two vertical jump trials and the MVC isometric activities were processed before quantifiable features could be extracted for statistical analysis. Using a custom script in MatLab software (MathWorks inc, USA), a fourth order Butterworth high pass filtering design was applied on the raw EMG signals as was applied in studies of similar nature [1, 2]. In-order to generate linear envelope data, full-wave rectification was then applied to the high pass filtered signals before applying a low pass fourth order Butterworth filtering.

Three different MVC Normalized features (Peak, Peak time, Integrated EMG (IEMG)) were extracted for each of the investigated muscles for the two vertical jump trials.

Feature Extractions. A total of 48 MVC Normalized values were extracted for each subject for three EMG features (IEMG, Peak, Peak time) from each of the 8 investigated muscles for the two trials' jump data. Therefore, the total sum of the extracted features' data set 576 EMG values (12 Subjects * 8 Muscles each * 3 features for each muscle *2 Vertical Jump trials by each subject).

2.3 Statistical Analysis for Pattern Set Identification and Classification

Analysis Criterion. The extracted EMG data sets, Anthropometry Data and the Vertical Jump Height data were grouped into two equal groups of Excellent (n = 6) and Very Good (n = 6) categories according to the subjects' corresponding leg strength classification criterion of Excellent and Very Good. Different statistical analyses were performed on the two groups with data sets grouped according to the jump condition (pre or post induced fatigue). This grouping implies a total data set of 288 EMG feature extracts for each of the groups (6 subjects * 8 muscles for each subject * 3 feature extractions for each muscle * 2 trials).

Statistical Analysis Methods Used. *Multivariate Analysis of Variance (MANOVA).* MANOVA was applied for all the two subjects' data as one group to determine whether there is any significant difference ($p < 0.05$) on any of the muscle's EMG activity between the two vertical jump trial data. *T-test.* T-test was then performed on jump height data for the two jump trials for all the 12 subjects to determine if there was a significant difference ($p < 0.05$) between the vertical jump height reached before and after the induced-fatigue task. *Mann-Whitney U Test.* In order to determine whether there was any significant difference on the subjects' muscular activities based on their leg strength classification, a non-parametric statistical analysis was performed on all three EMG feature data grouped according to the subjects' leg strength categories of Excellent ($n = 6$) and Very Good ($n = 6$). *Comparison of Means.* One-way ANOVA was performed on grouped data sets (Excellent and Very Good) for each EMG feature of both trials' jump data as well as on Anthropometry data, and Vertical Jump Height data recorded. This analysis also produced other descriptive statistics (Mean, Standard Deviation, Minimum Value, Maximum Value, Minimum Age, Maximum Age, Minimum Height, Maximum Height) which were used to form the statistical *EMG features*, *Anthropometry Observations*, and *Vertical Jump Observation* entities in the System's Knowledge-Base.

2.4 Formation of the Knowledge-Base

Pattern Set Formation and Classification Criteria. The results obtained from the comparisons of means for grouped data sets (EMG features, Anthropometry data, Vertical Jump Height data) were considered as pattern sets for the respective strength categories and thus were used to populate the corresponding pattern set entities in the Knowledge-Base. The implementation of the Knowledge-Base was done in MySQL Database server and inference rules for the classification algorithm implemented in a custom made script coded in PHP and SQL scripting languages.

Knowledge-Base Entities. The Knowledge-Base design has four entities which store facts, upon which the system's classification algorithm bases to make conclusions. *The first entity is the emgfeatures:* This entity contains 96 possible cases (rows) in which a classification is either excellent or vgood. These values represent all subjects considering all the three EMG features, Jump condition, and their strength classification category by wall-squat test results. That is, ((8 Muscles * 3 EMG features for each muscle * 2 Trials for each feature) * 2 Strength Categories). Each possible case(row) contains 6 values belonging to six different attributes(Strength Category, Jump Condition, EMG feature, Muscle, EMG feature Mean Value, EMG feature SD, Minimum EMG feature value, Maximum EMG feature value). *The second entity is the anthropometrydata:* This entity contains Mean, SD, MinimumVale, and MaximumValue for each of the anthropometric measures (height, body mass, and age) for the entire sample population grouped according to the strength classification categories (excellent and vgood). This entity contains 12 possible cases (3 anthropometric measures * 4 categories of measure). *The third entity is observationjump:* This entity represents the Mean and SD values for each of the two strength categories (excellent and vgood) observed in the vertical jump height data for the two jump trials states (pre and post induced fatigue). It contains 4 possible cases (2 strength categories * for 2 two jump states). *The fourth entity is strengthclass:* This entity represents the criterion used by the Netball Coach to classify leg strength of subjects following a wall-squat time-based test. It has 2 possible cases which are excellent or vgood with time duration of 2–3 min and 1–2 min respectively.

2.5 Classification Algorithm

The algorithm uses the subject's anthropometric measurements (height, body mass), age, vertical jump condition (pre or induced-fatigue), and the extracted EMG-MVC normalized values for the eight investigated muscles (RF, VL, VM, BF, GM, GL, PL), as its classification arguments (Fig. 2). The algorithm implements the classification process as follows:

AN EMG-BASED KNOWLEDGE BASE FOR LEG STRENGTH CLASSIFICATION
AND VERTICAL JUMP HEIGHT ESTIMATION OF FEMALE NETBALL PLAYERS
(EMG KB-LS VJHE SYSTEM)

(provide the system with the following values inorder to classify the subject's leg strength)

EMG Feature:	iemg
Player's Age(Yrs):	23
Player's Height(cm):	168
Player's Body Mass(Kg):	67
Experiment State:	prefatigue
EMG feature values(MVC Normalized) for the following Muscles	
Rectus Femoris(RF):	0.2775
Vastus Lateralis(VL):	0.1538
Vastus Medialis(VM):	0.1746
Biceps Femoris(BF):	0.1257
Semitendinosus(SM):	0.1567
Gastrocnemius Medialis(GM):	0.0666
Gastrocnemius Lateralis(GL):	0.0848
Peroneus Longus(PL):	0.0384
	Classify Player / Reset form

View Classification History | View Knowledge Base | View Classification Rules

Fig. 2. The data entry form for the classification system

Elimination of Outlier Data Sets. The high variability of EMG signals within and between subjects is known to be associated with the differences in their anatomical and physiological characteristics, among other factors [4, 10]. The algorithm fits the user provided values (height, body mass, age) against the established acceptable range in the knowledge base and makes a decision to either proceed with muscle-wise strength classification or notify the user of the out-of-range effect detected on the data sets provided.

Muscle-Wise Strength Classification. Once the data set is found to be in the acceptable range, the classification procedure pings the knowledge base and matches each of the muscle's MVC normalized EMG value against all the possible cases that correspond to the specified EMG feature as well as the jump state condition(pre or post induced-fatigue). This action returns a single classification value for each of the eight muscles which is either excellent or vgood. The procedure also keeps count of each of the two different values in two different classification count variables (excount, and vgcount) which are then used to compute the strength category weight.

Inference and Reporting of Classification Details. The system deduces that a subject falls in a given strength category if its classification count variable value is the larger of the two and reports the percentage ((larger/8) * 100) by which this conclusion was reached. The strength classification's corresponding vertical jump height and wall-squat duration estimations are also reported as well as the individual muscle strength classifications.

2.6 System Testing and Validation

Testing Protocol. Seven (randomly) chosen subjects from the sample population (the 12 female netball players) used to create the knowledge-base were used for system validation. The system was tested on a web-based platform over Apache II web-server. The values for their EMG features (Peak, IEMG, and Peak time) obtained in both jump conditions (pre and post fatigue) were used (one-at-a time) as test data sets to determine the subject's strength category. Each of the six (1 Subject * 2 jump conditions * 3 EMG features) classification runs returned a single strength classification category (excellent or vgood) as well as summations of counts of each category (sumexcellent, sumvgood). The system deduced that a subject is in a given leg strength category if its classification count variable value was the larger of the two and reported the percentage ((larger/6) * 100) by which this conclusion was reached. The System displays results as illustrated in (Table 3).

Table 3. Results of a randomly selected subject's leg strength classification using all three EMG features for both pre and post-fatigue data sets

Subject's overall leg strength is classified as: (75 % excellent)										
EMG feature	Expt. State	Muscle-wise strength classification								Leg strength (based on EMG feature/Expt. state)
		RF	VL	VM	BF	SM	GM	GL	PL	
IEMG	Pre-fatigue	Excellent	Excellent	Vgood	Excellent	Vgood	Excellent	Excellent	Vgood	**(62.5 %) Excellent**
IEMG	Post-fatigue	Vgood	Vgood	Vgood	Excellent	Excellent	Vgood	Vgood	Vgood	**(75 %) Vgood**
Peak	Pre-fatigue	Excellent	Excellent	Vgood	Excellent	Vgood	Excellent	Excellent	Vgood	**(62.5 %) Excellent**
Peak	Post-fatigue	Excellent	Excellent	Vgood	Excellent	Vgood	Excellent	Excellent	Vgood	**(62.5 %) Excellent**
Peak time	Pre-fatigue	Vgood	Vgood	Vgood	Excellent	Excellent	Vgood	Vgood	Vgood	**(75 %) Vgood**
Peak time	Post-fatigue	Excellent	Excellent	Vgood	Excellent	Vgood	Excellent	Excellent	Vgood	**(62.5 %) Excellent**

System Validation. Since the system is able to infer the test subject's leg strength category basing on the strength classification data sets in the knowledge base, the subjects' leg strength categories whose data sets were randomly tested were cross-checked for validation from the independent strength classification data prepared by the netball coach. Where a correct match was observed, the classification result was considered valid and hence 100 % classification accuracy. Otherwise, the result would be considered invalid and hence 0 % classification accuracy.

3 Results

3.1 Statistical Analysis Results

T-test revealed a significant difference ($p < 0.05$) between vertical jump height data obtained in the two trials' data sets. MANOVA revealed no significant difference ($p > 0.05$) in any of the muscle data obtained in both trials for all subjects. Mann-Whitney U test revealed a significant difference ($p < 0.05$) in Peak and Peak time of VM, SM, and BF respectively between the grouped data sets for the second trial data.

3.2 System Testing Results

The system's leg strength classification was validated to be 100 % correct as per the netball coach's classification data while additionally revealing the percentage weightage of the strength category of 66.67 % and above, for each of the seven randomly tested subjects. This overall leg strength considered all the three EMG features for the different Jump conditions (pre or post induced fatigue). Classification also revealed the subject's muscle-specific strength category based on a given EMG feature and jump condition for all the eight muscles. This also subsequently accurately classified her as per coach's classification data with strength category of 62.5 % and above.

4 Discussions and Conclusions

The significant statistical difference ($p < 0.05$) observed between the jump height data recorded before and after-induced fatigue reveals the effectiveness of unloaded step-ups as an enhancement of vertical jump height performance. Previous studies [5, 11] have also reported improved vertical jump performance as a result of different warm-up activities.

Mann-Whitney U Test having revealed a significant difference in Peak and Peak Time of VM, SM and BF respectively for grouped (Excellent and Vgood) data sets and not on the whole data set, is quantitative evidence that the subjects difference in leg strength is more felt when subjected to induced fatigue. Furthermore, the Mean ± SD jump height (cm) data of the Excellent group was 33.5 ± 5.67 and 37 ± 5.65 before and after the induced fatigue, respectively. Whereas, for the Vgood group it was 35.17 ± 1.72 and 38.83 ± 3.13 before and after the induced fatigue, respectively. This difference in jump height data with respect to the subjects' leg strength is an indication of a correlation between leg strength and jump height performance, and thus a justification to derive our knowledge-base's pattern sets on the leg strength basis.

The success of the implementation strategy used in this study also demonstrates the applicability and effectiveness of databases in knowledge representation of EMG data sets alongside other netball player descriptive data. This makes it possible for Netball coaches to be able to precisely and categorically differentiate their players' muscle characteristics by explicitly querying the database (knowledge-base) depending on their need.

To cater for more variability of test data, we propose that more similar experimental data be collected from different populations of varying characteristics in order to enrich the knowledge-base. The knowledge base's EMG features could also be diversified by applying different EMG feature extraction mechanisms thereby providing another classification criterion to the system users. This will make the classification algorithm more inclusive since EMG data sets derived using data processing techniques different from this study's would also be considered.

The results of this study provides strength conditioning and netball coaches with muscle-specific internal differences of players as characterized by the different EMG features and hence allows for appropriate conditioning interventions.

References

1. Arosha, S.M.N., Malik, O.A., Iskandar, P.M., Zaheer, D.: A knowledge-based intelligent framework for anterior cruciate ligament rehabilitation monitoring. Appl. Soft Comput. **20**, 127–141 (2014)
2. Malik, O.A., Arosha, S.M.N.A., Zaheer, D.: An intelligent recovery progress evaluation system for ACL reconstructed subjects using integrated 3-D kinematics and EMG features. IEEE J. Biomed. Health Inform. **19**, 453–463 (2015)
3. Nathaniel, A.B., Gregory, D.M., Timothy, E.H.: Prediction of kinematic and kinetic performance in a drop vertical jump with individual anthropometric factors in adolescent female athletes: implications for cadaveric investigations. Ann. Biomed. Eng. **43**, 926–933 (2015)
4. Mitchell, M.D., Yarossi, M.B., Pierce, D.N., Garbarini, E.L., Forrest, G.F.: Reliability of surface EMG as an assessment tool for trunk activity and potential to determine neurorecovery in SCI. Spinal Cord. **53**, 368–374 (2015)
5. Sotiropoulos, K., Smilios, I., Christou, M., Barzouka, K., Spaias, A., Douda, H., Tokmakidis, S.P.: Effects of warm-up on vertical jump performance and muscle electrical activity using half-squats at low and moderate intensity. J. Sports Sci. Med. **9**, 326–331 (2010)
6. Lucas, A.J., Robert, J.B., Tawnee, L.S., Robin, M.Q.: A single set of biomechanical variables cannot predict jump performance across various jumping tasks. J. Strength Conditioning Res. **29**, 396–407 (2015)
7. Mario, C., Vladimir, M., Stanko, T., Saša, O.: Surface EMG based muscle fatigue evaluation in biomechanics. Clin. Biomech. **24**, 327–340 (2009)
8. Son, J.S., Kim, J.Y., Hwang, S.J., Youngho, K.: The development of an EMG-based upper extremity rehabilitation training system for hemiplegic patients. In: Goh, J.C.H., Lim, C.T. (eds.) ICBME 2008 Proceedings, vol. 23, pp. 1977–1979. Springer, Berlin (2009)
9. Raez, M.B.I., Hussain, M.S., Yasin, F.M.: Techniques of EMG signal analysis: detection, processing, classification and applications. Biol. Proced. **8**, 11–35 (2006)
10. Burden, A.: How should we normalize electromyograms obtained from healthy participants? What we have learned from over 25 years of research. J. Electro. Kinesio. **20**, 1023–1035 (2010)
11. Kirmizigil, B., Ozcaldiran, B., Colakoglu, M.: Effects of three different stretching techniques on vertical jumping performance. J. Strength Conditioning Res. **28**, 1263–1271 (2014)

Intelligent Services for Smart Cities

Approach to Priority-Based Controlling Traffic Lights

Phuoc Vinh Tran[1(✉)], Tha Thi Bui[2], Diem Tran[3], Phuong Quoc Pham[4], and Anh Van Thi Tran[5]

[1] Thudaumot University (TDMU), Binhduong, Vietnam
Phuoc.gis@gmail.com
[2] Hochiminh City Vocational College, Hochiminh City, Vietnam
bththa@gmail.com
[3] University of Information Technology (UIT), Hochiminh City, Vietnam
diemtranbm@gmail.com
[4] Center for Applying GIS of Hochiminh City, Hochiminh City, Vietnam
phuongpq@gmail.com
[5] Hochiminh City Economics College, Hochiminh City, Vietnam
ttvanh26@gmail.com

Abstract. One of causes of repeated traffic congestion at crossroads in Hochiminh City is the unreasonableness of the current system controlling traffic lights that switches lights at fixedly programmed time points. In several cases, vehicles are compelled to stop at stop line while nothing travels on front crossroad, and emergency vehicles can not overpass crowd of vehicles stopping in front of traffic lights. Moreover, the too slow movement of vehicles within crossroad due to conflict of directions is also a cause of congestion and jam. This study proposes an approach, termed the approach to priority-based controlling traffic lights, to serving as a module of traffic light system. The approach eliminates the conflict of vehicles moving within crossroad to decrease the time passing crossroad. Traffic lights are switched according to the priority of directions to reduce the thickness of vehicles in front of traffic lights of directions. As a result, the approach prevents congestion within crossroad and in front of traffic lights at flows entering crossroad.

Keywords: Traffic · Traffic light · Congestion · Traffic light system

1 Introduction

The basic cause of traffic congestion state is the unbalance between supply and demand in communication [1]. In addition, the diversity of types of vehicles and roadway, the behavior of motorists while displacement, especially the unreasonableness of the system of traffic lights in Hochiminh City (HCMC) – Vietnam affect strongly the traffic situation. The current system of traffic lights utilizing fixed-time isolated approach [2] results in several irrational cases in traffic, a lot of vehicles are legally compelled to stop in front of red light while nothing travels on front crossroad (Fig. 1). Moreover, ambulances, fire-engines have also to stop behind crowd of vehicles stopping in front of red light (Fig. 2). These cases not only cause traffic congestions within crossroads

N.T. Nguyen et al. (Eds.): ACIIDS 2016, Part II, LNAI 9622, pp. 745–754, 2016.
DOI: 10.1007/978-3-662-49390-8_72

Fig. 1. Many vehicles are compelled to stop in front of red light while nothing within crossroad (Recorded at 7:30 on September 1st, 2015 at the crossroad of HMG and HH streets, HCMC)

and in front of traffic lights but also result in great losses in socio-economic development [1]. This situation can be improved if the current system of traffic lights is replaced with a more intelligent system, where right-of-way is dedicated to directions going through crossroad based on priority.

This study proposes an approach to priority-based controlling traffic lights at a crossroad. According to the approach, right-of-way is not dedicated to directions based on a fixed program as the current system, on the contrary, it is dedicated to directions of higher priority based on sequence of directions arranged in descended order of priority. The problem to be solved in this study is how to constitute combination of right-of-way directions from the combination of directions going into a crossroad, which are arranged in descended order of priority. The approach supports fast and non-conflict movement within crossroads to prevent congestion.

The remainder of this paper is organized as follows. The next section describes briefly the traffic situation in Hochiminh City, including the specifications of urban road network, the diversity of vehicles, the methods controlling traffic lights, and the possibility of applying an intelligent traffic-light-control system. The third section depicts a system for priority-based controlling traffic lights including a module for detecting traffic priority and a module for priority-based dedicating right-of-way to directions. The fourth section presents the procedure for constituting combination of right-of-way directions, as part of module for priority-based dedicating right-of-way, to switch traffic lights. A case study applying the procedure for a crossroad is illustrated in this fourth section. Finally, the conclusion summarizes the specifications of the proposed procedure and future works.

2 Traffic Situation in Hochiminh City

2.1 Urban Roads in Hochiminh City

Inner Hochiminh City has been built for over three hundred years. It is near 500 km^2 large with over 7 million residents, excluding millions of people coming from other provinces for their education, business, or work, and several tourists. City residents are utilizing over 600,000 cars and over 5 million motorbikes, excluding several vehicles for public transportation and vehicles following people coming from other provinces. The network of urban roads in Hochiminh City may be modeled as a geometric network with several crossroads, each of which is an intersection of many roads, from 3

Fig. 2. An ambulance can not overpass vehicle crowd in front of a red traffic light (Source: http://vietbao.vn/)

to 7 roads, termed n-branch crossroad (n = 3, 4, 5, 6, 7). The network is composed of roads with different widths and geometric shapes. The lengths of road segments between two crossroads are not equal, too short or too long. Some crossroads have roundabouts at centre, others have small islands at corners to form sub-ways for turning right, and others overbridges to form free flows.

According to Vietnamese traffic regulations, all vehicles must move on right-hand side. In organizing roadway, some road segments are used as two-opposite-way roads, others as one-way roads. Opposite ways of two-way roads are separated by way-dividers as painting dual lines, islands, barriers, or lakes. Some ways are divided into flows reserving for four-wheeled vehicles or two-wheeled vehicles, in which four-wheeled vehicles comprise cars, buses, trucks, and two-wheeled vehicles bicycles, motorbikes, tricycles, handcarts, and peddlers. The reserved flows are separated by flow-dividers as painting lines, barriers, or islands. Large four-wheeled flows may be divided into lanes separated by discontinuous painting lines. Traffic lights are installed as vertical sets at road ends joining crossroads. Each set includes three colors, in which the green light at the bottom of the set emits the signal of right-of-way, the yellow at the middle the signal warning a switch from right-of-way to prohibition, and the red at the top the signal prohibiting vehicles' advance.

In traffic, Hochiminh City is now known as a city of too many motorbikes and some of them travel illegally. They attempt to overpass front vehicles or squeeze by themselves to jostle through possible gaps for overtaking advance vehicle crowd, whether those vehicles are two-wheeled or four-wheeled. During rush hours, congestions occur repeatedly at some crossroads out of not only the unbalance between the capacity of transportation infrastructure and the demand on travelling of residents, the occupation of roadway by peddlers and parking cars, but also the unreasonableness of the system for controlling traffic lights with fixed-program.

2.2 Crossroads of Urban-Road Network in Hochiminh City

The development of Hochiminh City through several historical periods results in the diversity of urban-road network, geometric shape, and utilization. Roads connecting to a crossroad may have different widths and be utilized as one-way road entering or leaving crossroad, two-way road entering and leaving crossroad. Each way entering or leaving crossroad may serve as a vehicle-mixed flow or two vehicle-separated flows for two-wheeled and four-wheeled vehicles.

Reaching a three- or four-branch crossroad, a vehicle may go straight, turn right, or turn left; reaching a five-, six-, or seven-branch crossroad, a vehicle may go straight, turn right, turn strongly right, turn left, turn strongly left. In other words, a traffic flow reaching a crossroad is divided into several directions. Most directions need to be coordinated by traffic lights for vehicles going into crossroad; meanwhile, directions going to sub-ways for turning right or strongly right and to overbridges for going straight or turning left need not traffic lights to coordinate vehicles.

2.3 Intelligent Traffic Light Systems

Aiming at reducing waiting time of cars before traffic lights, works on intelligent traffic light control [3] by several authors have approached adaptive control system for multiple close crossroads based on fuzzy logic [4] or expert system with if-then rules for an isolated crossroad [5, 6]. These approaches analyze traffic data on flow rate, density, and velocity of cars on road segments connecting to crossroads [2, 3]. The number of cars entering a road segment is counted in real-time by induction loops, global positioning system (GPS), RFID, or video camera [2–5, 7–9]. Analyzing the data of cars entering a road segment, the approaches predict the number of cars and time when they appear at the next crossroad to switch traffic lights.

The approaches are hard to deploy in Hochiminh City out of the diversity of vehicles, the difference between crossroads, and the complexity in behavior on road. Crossroads in Hochiminh City were built with several branches of different widths and roadway organizations. On many roads, 4-wheeled and 2-wheeled vehicles travel in close proximity. In addition, vehicles from schools, marts, and alleys connecting to roads at any locations between two crossroads make change to the number of vehicles counted by sensors. Moreover, the approaches have not considered the regulated priority of vehicles on flows reaching crossroads as convoy, fire-engine, ambulance, etc. Meanwhile, the system for controlling traffic lights at [10] focuses on clearing roadway for the movement of emergency vehicles and not solve the problem of congestion in front of traffic lights as well as jam within crossroad.

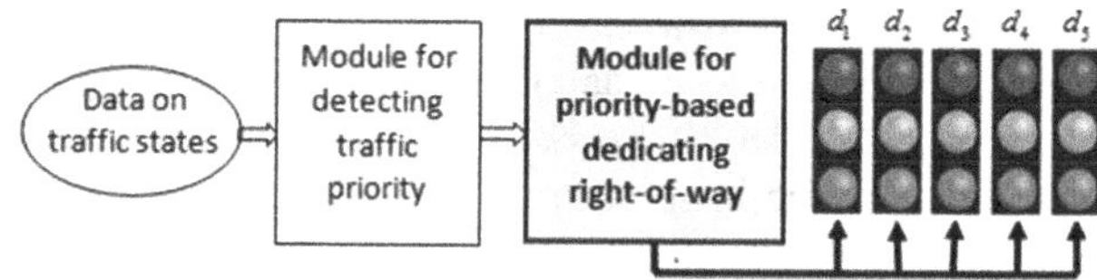

Fig. 3. Diagram of a system for priority-based controlling traffic lights at a crossroad.

3 Approach to Priority-Based Controlling Traffic Lights

3.1 System for Priority-Based Controlling Traffic Lights at a Crossroad

The travel on a flow entering crossroad is separated into directions, each of which is coordinated by a set of traffic lights placed at the end of traffic flow. A system for priority-based controlling traffic lights is composed of two main modules, module for detecting traffic priority and module for priority-based dedicating right-of-way (Fig. 3). The module for detecting traffic priority analyses traffic data in front of traffic lights to arrange directions in descended order of priority. The module for priority-based dedicating right-of-way dedicates right-of-way to higher-priority directions based on the data from module for detecting traffic priority. In the system, a direction of higher priority is denoted as one in which there is the movement of emergency vehicles, maximum-exceeded prohibition time, or higher vehicle thickness. The priority of a direction at a crossroad changes over time.

3.2 Module for Priority-Based Dedicating Right-of-Way

As part of a system for priority-based controlling traffic lights installed at a crossroad, module for priority-based dedicating right-of-way consists of a procedure for constituting right-of-way combination and a matrix for switching traffic lights (Fig. 4). The input data of the procedure are composed of the factors determined by traffic specialists while designing crossroad and the data supplied by the module for detecting traffic priority. The output of the procedure is right-of-way combination corresponding with each priority-descended-order sequence of directions from the module for detecting traffic priority. Finally, the matrix for switching traffic lights decodes the consenting and conflicting combinations for corresponding lights. For a particular crossroad, the input data of the procedure unchanging over time determined by traffic specialists comprise:

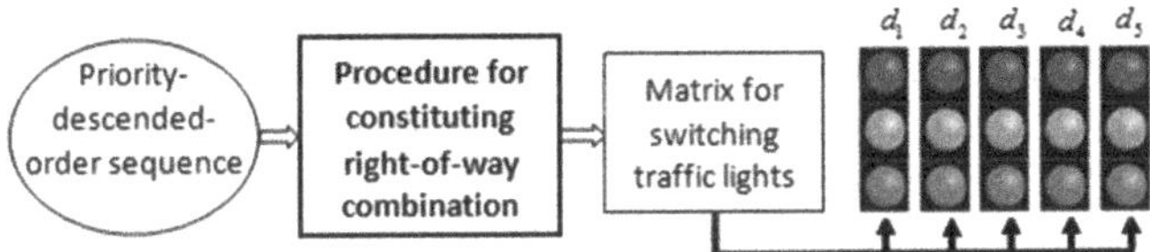

Fig. 4. Diagram of module for priority-based dedicating right-of-way.

- $D = \{d_i, i = 1, 2, \ldots, N\}$: the collection of directions going into a crossroad, excluding the directions consenting to every direction.
- T_d^i: the shortest time necessary for a vehicle on the direction d_i beyond the crossroad, $T_d^i = L_d^i / V_r$, where L_d^i is the length of the direction d_i through the crossroad and V_r is the regulated velocity at the crossroad.
- T_p^i: the maximum time of prohibition for the direction d_i.
- D_{cs}^i, D_{cf}^i: the combinations of directions consenting to and conflicting with $d_i, i = 1, 2, \ldots, N$, respectively; where $D_{cs}^i + D_{cf}^i = D - \{d_i\}$

The input data of the procedure supplied by the module for detecting traffic priority at regular-spaced time points T_s, i.e. at every T_s, or at the moment when an emergency vehicle requests comprise:

- A priority-descended-order sequence of directions $d_i \in D$.
- $D_{row}^{current}$: the current right-of-way combination
- $D_{pro}^{current}$: the current prohibition combination.
- T_g^i: the current duration of green light of the direction i.
- T_r^i: the current duration of red light of the direction i.

The conditions of the problem are analyzed as follows:

- *Condition 1:* $T_g^i \geq T_d^i$ for $\forall i$.
- *Condition 2:* $T_r^i \leq T_p^i$ for $\forall i$.
- *Condition 3:* Vehicles going through crossroad do not conflict with each other.

4 Procedure for Constituting Right-of-Way Combination

4.1 Problem

Input:

- D_{cs}^i for $i = 1, 2, \ldots, N$, structured as a consenting-direction table
- T_d^i and T_p^i for $i = 1, 2, \ldots, N$
- $D = \{d_i, i = 1, 2, \ldots, N\}$, the combination of directions arranged as a priority-descended-order sequence
- $D_{row}^{current}$ and $D_{pro}^{current}$

Output:

- D_{row}^{new}: the new combination of right-of-way directions.
- D_{pro}^{new}: the new combination of prohibition directions.

4.2 Procedure for Constituting Right-of-Way Combination

- **Check condition** $T_g^i \geq T_d^i$. for $i = 1, 2, \ldots, N$, if $\exists d_i : T_g^i < T_d^i$, then preserve the current state of traffic lights.
- **1st right-of-way direction,** d_{pr1}. for $i = 1, 2, \ldots, N$, if $\exists d_i : T_r^i > T_p^i$, then $d_{pr1} \leftarrow d_i$, else $d_{pr1} \leftarrow d^{pr\max}, d^{pr\max} \in D$, where $d^{pr\max}$ is the highest-priority direction of D.
- **2nd right-of-way direction,** d_{pr2}. Consider D_{cs}^{pr1} derived from the consenting-direction table, if $D_{cs}^{pr1} = \emptyset$, then skip to **Result**, else $d_{pr2} \leftarrow d^{pr\max}, d^{pr\max} \in D_{cs}^{pr1}$, where $d^{pr\max}$ is the highest-priority direction of D_{cs}^{pr1}. Note: $D_{cs}^{pr1} \subset D - \{d_{pr1}\}$.
- …

- **i^th^ right-of-way direction,** d_{pri}. Consider $D_{cs}^{pr(i-1)}$ derived from the consenting-direction table, $D_{cs}^{pr(i-1)} \subset D - \{d_{pr(i-1)}\}$. If $\bigcap_i D_{cs}^{pr(i-1)} = \emptyset$, then skip to **Result**, else $d_{pri} \leftarrow d^{pr\max}, d^{pr\max} \in \bigcap_i D_{cs}^{pr(i-1)}$, where $d^{pr\max}$ is the highest-priority direction of $\bigcap_i D_{cs}^{pr(i-1)}$. Note: $d_{pri} \in \bigcap_i D_{cs}^{pr(i-1)}$, where $D_{cs}^{pr0} = D$ and $D_{cs}^{pr0} \cap D_{cs}^{pr1} = D_{cs}^{pr1}$.
- **Result:** $D_{row}^{new} = \{d_{pr1}, d_{pr2}, d_{pr3}, \ldots\}$ and $D_{pro}^{new} = D - D_{row}^{new}$

Application: The current combinations $D_{row}^{current}$ and $D_{pro}^{current}$ are replaced with the new combinations D_{row}^{new} and D_{pro}^{new} as follows.

- The directions belonging to $D_{row}^{current} \cap D_{row}^{new}$ are preserved right-of-way,
- The directions belonging to $D_{pro}^{current} \cap D_{row}^{new}$ are switched to right-of-way,
- The directions belonging to $D_{pro}^{current} \cap D_{pro}^{new}$ are preserved prohibition,
- The directions belonging to $D_{row}^{current} \cap D_{pro}^{new}$ are switched to prohibition. The transition is smoothed by signal of yellow light to slow up vehicles to a stop.

4.3 Properties of the Procedure

The Convergence: The procedure to choose d_{pri} for $D_{row}^{new} = \{d_{pr1}, d_{pr2}, d_{pr3}, \ldots\}$ from $\bigcap_i D_{cs}^{pr(i-1)}$ is finite. Indeed, $(d_{pri} \notin D_{cs}^{pri}) \wedge (\bigcap_i D_{cs}^{pri} = \bigcap_i D_{cs}^{pr(i-1)} \cap D_{cs}^{pri}) \Rightarrow d_{pri} \notin \bigcap_i D_{cs}^{pri}$ for $i = 1, 2, \ldots$. This expression shows that any direction in $\bigcap_i D_{cs}^{pri}$ is taken out for D_{row}^{new} until no direction exists in $\bigcap_i D_{cs}^{pri}$, i.e. until $\bigcap_i D_{cs}^{pri} = \emptyset$.

The Traffic Safety: All the directions belonging to D_{row}^{new} are consenting to each other. The commutative property of "consenting", symbolized "||", results in:

$$d_{pr2} \in D_{cs}^{pr1} \Rightarrow d_{pr2} \parallel d_{pr1} \quad ; \quad d_{pr3} \in D_{cs}^{pr1} \cap D_{cs}^{pr2} \Rightarrow (d_{pr3} \parallel d_{pr1}) \wedge (d_{pr3} \parallel d_{pr2});$$
$$d_{pr4} \in D_{cs}^{pr1} \cap D_{cs}^{pr2} \cap D_{cs}^{pr3} \Rightarrow (d_{pr4} \parallel d_{pr1}) \wedge (d_{pr4} \parallel d_{pr2}) \wedge (d_{pr4} \| d_{pr3}); \ldots$$

4.4 Case Study

Inputs Supplied by Traffic Specialist for the Observation Crossroad. The observation crossroad (Fig. 5) is organized as a four-branch crossroad with five entering flows from four entering ways. The vehicles traveling on the five flows are shifted to 15 directions to go through the crossroad, in which the direction d_3 is indicated as the complete right-of-way direction which consents to every direction. The directions consenting to a chosen direction are designed by traffic specialist.

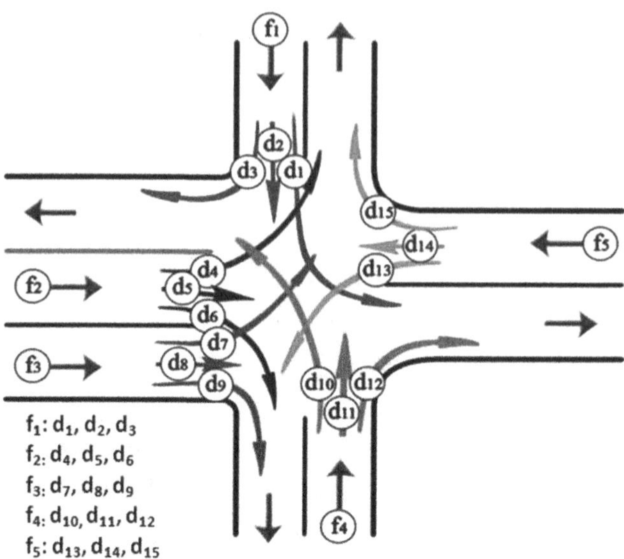

Fig. 5. A 4-branch crossroad with 5 entering flows and 15 directions through crossroad including the direction consenting to all.

Inputs Supplied by the Module for Detecting Traffic Priority: (assumption)

–

$$D = \{d_5, d_2, d_7, d_{15}, d_{12}, d_{11}, d_1, d_4, d_{10}, d_9, d_8, d_{13}, d_6, d_{14}\} \quad (1)$$

–

$$D_{row}^{current} = \{d_1, d_2, d_9, d_{12}, d_{15}\}; D_{pro}^{current} = \{d_4, d_5, d_6, d_7, d_8, d_{10}, d_{11}, d_{13}, d_{14}\} \quad (2)$$

– The current system meets the conditions 1 and 2 for T_g and T_r.

Procedure for Constituting Right-of-way Combination: (from (1) and Table 1)

1st right-of-way direction: $d_{pr1} \leftarrow d_5$, derived from D

2nd right-of-way direction: $D_{cs}^5 = \{d_{15}, d_{12}, d_4, d_9, d_8, d_6, d_{14}\} \neq \emptyset$, then $d_{pr2} \leftarrow d_{15}$

3nd right-of-way direction: $D_{cs}^{15} = \{d_1, d_2, d_4, d_5, d_6, d_7, d_8, d_9, d_{10}, d_{12}, d_{13}, d_{14}\}$, $D_{cs}^5 \cap D_{cs}^{15} = \{d_{12}, d_4, d_9, d_8, d_6, d_{14}\}$, then $d_{pr3} \leftarrow d_{12}$.

4nd right-of-way direction: $D_{cs}^{12} = \{d_1, d_2, d_4, d_6, d_7, d_9, d_{10}, d_{11}, d_{13}, d_{14}, d_{15}\}$, $D_{cs}^5 \cap D_{cs}^{15} \cap D_{cs}^{12} = \{d_4, d_9, d_6, d_{14}\}$, then $d_{pr4} \leftarrow d_4$.

5nd right-of-way direction: $D_{cs}^4 = \{d_5, d_6, d_7, d_8, d_9, d_{11}, d_{12}, d_{13}, d_{15}\}$, $D_{cs}^5 \cap D_{cs}^{15} \cap D_{cs}^{12} \cap D_{cs}^4 = \{d_9, d_6\}$, then $d_{pr5} \leftarrow d_9$.

6nd right-of-way direction: $D_{cs}^9 = \{d_1, d_2, d_4, d_5, d_6, d_7, d_8, d_{10}, d_{11}, d_{12}, d_{14}, d_{15}\}$, $D_{cs}^5 \cap D_{cs}^{15} \cap D_{cs}^{12} \cap D_{cs}^4 \cap D_{cs}^9 = \{d_6\}$, then $d_{pr6} \leftarrow d_6$.

7nd right-of-way direction: $D_{cs}^{6} = \{d_1, d_4, d_5, d_9, d_{10}, d_{11}, d_{12}, d_{14}, d_{15}\}$, $D_{cs}^{5} \cap D_{cs}^{15} \cap D_{cs}^{12} \cap D_{cs}^{4} \cap D_{cs}^{9} \cap D_{cs}^{6} = \emptyset$.

Result: $D_{row}^{new} = \{d_5, d_{15}, d_{12}, d_4, d_9, d_6\}$ and $D_{pro}^{new} = \{d_2, d_7, d_{11}, d_1, d_{10}, d_8, d_{13}, d_{14}\}$
Application: (from (2) and ***Result***)

- The directions preserve green light: $D_{row}^{current} \cap D_{row}^{new} = \{d_9, d_{12}, d_{15}\}$
- The directions preserve red light: $D_{pro}^{current} \cap D_{pro}^{new} = \{d_7, d_8, d_{10}, d_{11}, d_{13}, d_{14}\}$
- The directions switch from red to green light: $D_{pro}^{current} \cap D_{row}^{new} = \{d_4, d_5, d_6\}$
- The directions switch from green light to yellow, then red light: $D_{row}^{current} \cap D_{pro}^{new} = \{d_1, d_2\}$

Table 1. The consenting-direction table, D_{cs}^{i} for $i = 1, 2, \ldots, N$

i	Chosen direction (d_i)	Consenting directions (D_{cs}^{i})
1	d_1	$\{d_2, d_6, d_8, d_9, d_{10}, d_{12}, d_{15}\}$
2	d_2	$\{d_1, d_9, d_{11}, d_{12}, d_{13}, d_{15}\}$
3	d_3	$\{d_1, d_2, d_4, d_5, d_6, d_7, d_8, d_9, d_{10}, d_{11}, d_{12}, d_{13}, d_{14}, d_{15}\}$
4	d_4	$\{d_5, d_6, d_7, d_8, d_9, d_{11}, d_{12}, d_{13}, d_{15}\}$
5	d_5	$\{d_4, d_6, d_8, d_9, d_{12}, d_{14}, d_{15}\}$
6	d_6	$\{d_1, d_4, d_5, d_9, d_{10}, d_{11}, d_{12}, d_{14}, d_{15}\}$
7	d_7	$\{d_4, d_8, d_9, d_{11}, d_{12}, d_{13}, d_{15}\}$
8	d_8	$\{d_1, d_4, d_5, d_7, d_9, d_{12}, d_{14}, d_{15}\}$
9	d_9	$\{d_1, d_2, d_4, d_5, d_6, d_7, d_8, d_{10}, d_{11}, d_{12}, d_{14}, d_{15}\}$
10	d_{10}	$\{d_1, d_6, d_9, d_{11}, d_{12}, d_{15}\}$
11	d_{11}	$\{d_2, d_4, d_6, d_9, d_{10}, d_{12}, d_{15}\}$
12	d_{12}	$\{d_1, d_2, d_4, d_6, d_7, d_9, d_{10}, d_{11}, d_{13}, d_{14}, d_{15}\}$
13	d_{13}	$\{d_4, d_6, d_7, d_9, d_{12}, d_{14}, d_{15}\}$
14	d_{14}	$\{d_5, d_6, d_8, d_9, d_{10}, d_{12}, d_{13}, d_{15}\}$
15	d_{15}	$\{d_1, d_2, d_4, d_5, d_6, d_7, d_8, d_9, d_{10}, d_{12}, d_{13}, d_{14}\}$

5 Conclusion

The approach to priority-based controlling traffic lights starts our works making good the irrationality of the current system controlling traffic lights in Hochiminh City, Vietnam. For preventing congestion risk, the approach not only takes full advantage of roadway but also eliminates factors causing congestion as conflicting movement and traffic confusion. Besides that, it also coordinates reasonably vehicle thickness at directions going into crossroad and clear vehicles waiting for green light in front of unoccupied crossroad. The approach identifies vehicle thickness to control more flexibly traffic lights, not only to reduce the thickness of vehicles in front of traffic lights but also to utilize effectively roadway of crossroad. The approach eliminates conflicting movement within crossroad to increase the speed of vehicles, and clears traffic confusion caused by the bypass of red light of emergency vehicles together with the cling of others to decrease confusion and jam risk.

The proposed approach to priority-based controlling traffic lights offers an integrated solution for clearing roadway for emergency vehicles, preventing congestion in front of traffic light, and avoiding jam out of concurrent movement on conflicting directions through crossroad. The priority defined in the approach is estimated based on requests of emergency vehicles, waiting time of vehicles in front of red light, and vehicle thickness at directions going into and within crossroad. The approaches to detecting vehicle thickness and priority of directions will be discussed in our subsequent studies to apply for the traffic system in Hochiminh City, where innumerable motorbikes mix with 4-wheeled vehicles on urban roads.

References

1. Litman, T.A.: Smart Congestion Relief: Comprehensive Evaluation of Traffic Congestion Costs and Congestion Reduction Strategies. Victoria Transport Policy Institute (2015)
2. Papageorgiou, M., Diakaki, C., Dinopoulou, V., Kotsialos, A., Wang, Y.: Review of road traffic control strategies. Proc. IEEE **91**, 2043–2067 (2003)
3. Wiering, M., van Veenen, J., Vreeken, J., Koopman, A.: Intelligent Traffic Light Control. Institute of Information and Computing Sciences - Utrecht University (2004)
4. Zhou, B., Cao, J., Wu, H.: Adaptive traffic light control of multiple intersections in WSN-based ITS. In: Presented at the IEEE International Conference on Vehicular Technology, Yokohama, Japan (2011)
5. Mehta, S.: Fuzzy control system for controlling traffic lights. In: International MultiConference of Engineers and Computer Scientists IMECS 2008, Hong Kong (2008)
6. Wen, W.: A dynamic and automatic traffic light control expert system for solving the road congestion problem. Expert Syst. Appl. **34**, 2370–2381 (2007)
7. Parmar, R.S., Trivedi, B.: Identification of parameters and sensor technology for vehicular traffic - a survey. Int. J. Traffic Transp. Eng. **3**, 101–106 (2014)
8. Tubaishat, M., Qi, Q., Shang, Y., Shi, H.: Wireless sensor-based traffic light control. IEEE CCNC **2008**, 702–706 (2008)
9. Tubaishat, M., Shang, Y., Shi, H.: Adaptive traffic light control with wireless sensor networks. In: Proceedings of IEEE Consumer Communications and Networking Conference, pp. 187–191 (2007)
10. Mitchell, W.L.: Traffic light control for emergency vehicles. United States Patent 4,443,783, 17 April 1984

ALPR - Extension to Traditional Plate Recognition Methods

Konrad Kluwak[1,2], Jakub Segen[1], Marek Kulbacki[1(✉)], Aldona Drabik[1], and Konrad Wojciechowski[1]

[1] Polish-Japanese Academy of Information Technology, Koszykowa 86, 02-008 Warszawa, Poland
mk@pjwstk.edu.pl

[2] Faculty of Electronics, Wrocław University of Technology, Wybrzeże Wyspiańskiego 27, 50-370 Wrocław, Poland

Abstract. Automatic license plate recognition (ALPR) methods and software are used in toll collection, traffic monitoring and other areas of road transport industry. Majority of ALPR methods and almost all in industrial use, try to recognize a license plate identifier from a single image. However, in a sequence of images, recognition of a license plate in any frame can be improved by considering the information from preceding and succeeding frames, using video object tracking. A new approach is presented, for combining a video tracking and a single frame ALPR method to improve the recognition rate. Unlike earlier techniques which are tied to specific object tracking and identifier recognition methods, the new method can be used with almost any tracking and single frame ALPR methods. Its key part is a method for clustering and alignment of candidate license plate identifiers in a video track. The results from five video sequences taken from a surveillance camera under various weather and light conditions demonstrate the recognition rate improvements.

Keywords: Automatic License Plate Recognition (ALPR) · Automatic Number Plate Recognition (ANPR) · Tracking · Recognition improvement

1 Introduction

Vehicle license plate recognition systems are routinely used in traffic monitoring, traffic speed estimation, access control, automatic toll collection, speed limit control and generally, in situations requiring to identify a vehicle or monitor its location or movement. To recognize a license plate identifier (number), the license plate is first detected and localized in an image, then the plate image is usually processed by a segmentation or feature extraction module, which is followed by character classification and the recognition of the full license plate identifier [1,2,5–9]. The segmentation step can be avoided by cross-correlating the license plate image with each character's image [3]. Also, plate detection and localization can be obtained indirectly, after finding all regions in an image that look like characters [4].

N.T. Nguyen et al. (Eds.): ACIIDS 2016, Part II, LNAI 9622, pp. 755–764, 2016.
DOI: 10.1007/978-3-662-49390-8_73

Fig. 1. Correctly recognized (white square) and incorrectly (black square) license plates for two different captures.

Majority of license plate recognition systems in literature, and almost all in industrial applications recognize the plate characters and its identifier from a single image. To do it reliably for cars in motion, images of adequate resolution and quality are needed, which in turn might require special cameras or imaging conditions, extreme camera settings or added illumination. Such requirements make the cost, per camera, of a license plate recognition system higher than for a typical video surveillance camera, and limit the size of the field of view of a single camera, which increases the number of cameras needed to cover a larger area, such as a parking lot, and it further increases the system cost. To relax these requirements and to lower the system costs, the rate and reliability of recognition of a license plate in an image would need to improve. Such improvements can be obtained by using for the identifier recognition the information from multiple video frames, when a license plate is tracked through a sequence of video frames, using an object tracking method [10–12]. Figure 1 shows two video sequences which are candidates for such an improvement; a license plate is sufficiently distinct to be tracked through both sequences, but it is not recognized in all the frames. The approach of combining video object tracking and a single frame license plate recognition to improve the overall license plate identifier recognition results has been used by Donoser et al. [1]. The authors use Maximally Stable Extremal Region (MSER) based method [13] for license plate detection, localization, tracking and character segmentation, the characters are aligned and recognized using Support Vector Machines (SVM) [14]. The final recognition for each character position, for all aligned plates in a track is obtained by majority voting, significantly improving the recognition rate. Shazad et al. [2] describe another license plate recognition method that combines tracking and single frame recognition, using different steps than [1] for license plate detection, localization, tracking and single frame character recognition, followed by a majority voting as in [1] at each character's position.

This article proposes a new multi frame method that improves license plate recognition through object tracking, markedly different from those described in [1,2]. While each of these two articles presents a specific solutions for detection, localization, tracking and character recognition, our approach permits the use of any of a large number of tracking and recognition (with detection and localization) methods, including closed programs provided only as executable code, and makes it possible to easily combine with the newest and the best of tracking and single frame license plate recognition method. Section 2 describes the proposed method and Sect. 3 shows results on multiple datasets from surveillance cameras.

2 Multi Frame Automatic License Plate Recognition

The proposed method of multi frame license plate recognition consists of three parts, (a) a single frame ALPR, (b) a license plate tracking and (c) In-Track Clustering Correction (ITCC), which correct the identifiers recognized in the first part. Part (a) can be performed by any ALPR method or even a closed program, with output described in Sect. 1.1. For the part (b), any video object tracking method can be used, that can be constrained by the results of license plate detection from the part (a). A simple tracking method described in the Subsect. 2.2 is a placeholder that can be replaced by any tracking method that follows the constraints. The part (c) described in Subsect. 2.3 is the linchpin of the presented method, it aligns and groups candidate identifiers for each track, then finds a composite identifier, which is the corrected result.

2.1 Single Frame ALPR

The input data of application can be taken from any program that recognizes license plates from a single frame, which returns the following information for each processed frame:

- frame number or image file name
- list of License Plate Records (LPRs) one LPR per each identified license plate
  - each LPR should contain
    - license plate identifier as recognized by the program, which we will call a License Plate Candidate Identifier (LPCI). LPCI is a sequence of symbols which may contain gaps (marked with "?") in positions where a symbol is not recognized. The LPCI can be shorter than a true license plate number, and neither its first nor last symbol is expected to correspond to the beginning or the end of the true identifier.
    - image positions of the corners of the license plate

Example of such input data is shown in Fig. 2.

For the purpose of testing, a LPR in the above description can be appended with a manually added annotation that contains a true license plate identifier.

```
192.168.201.1.0.0.jpg;
    SZ2795E, 664, 751, 790, 746, 790, 773, 664, 778;
    SG5292F, 429, 291, 526, 288, 526, 309, 429, 313;
192.168.201.1.0.1.jpg;
    SZ2795E, 717, 846, 853, 842, 853, 871, 717, 877;
    SG5792F, 451, 341, 551, 337, 551, 360, 451, 363;
192.168.201.1.0.2.jpg;
    SZ2795E, 753, 916, 893, 911, 893, 942, 753, 947;
    SE8243F, 1747, 752, 1881, 747, 1881, 777, 1747, 782;
    SG5792F, 465, 371, 568, 366, 568, 389, 465, 393;
```

Fig. 2. Example input data fragment from OpenALPR program.

2.2 Tracking License Plates

Tracking of vehicles is made by assigning the given LPRs to tracks of currently tracked vehicle by criteria written below. The technique described in this subsection presents the methods of tracking in every possible cases like: missing, one or more LPRs in the next frames. The purpose of tracking was shown on Fig. 3. This effect was achieved by analyzing the input set frame by frame and assigning read LPRs to tracks regardless of recognized number of license plate.

Fig. 3. Example of tracking the license plate.

The following parameters are calculated:

1. Perimeter of the LPR. Corners of license plate are denoted as x_l, y_l, where corner index l varies from 1 to 4. Perimeter C of LPR is:

$$C = \sum_{l=1}^{3} \sqrt{(x_{l+1} - x_l)^2 + (y_{l+1} - y_l)^2} + \sqrt{(x_1 - x_4)^2 + (y_1 - y_4)^2} \tag{1}$$

2. Average perimeter. Average perimeter S_j for a frame j that contains one or more LPRs is calculated as an average of the perimeters from all LPRs in frame j, where n is the number of LPR in frame j, and C_{ji} the perimeter for LPR i in frame j.

$$S_j = \frac{1}{n} \sum_{i=0}^{n} C_{ji} \tag{2}$$

3. Normalized distance. Distance between LPR i in current frame and LPR j in previous frame.

$$D(i,j) = \frac{\sum_{l=1}^{4} \sqrt{\left(x_{il} - x'_{jl}\right)^2 + \left(y_{il} - y'_{jl}\right)^2}}{C_i + C'_j} \tag{3}$$

   where x_{il}, y_{il} are corners of current LPR, x'_{jl}, y'_{jl} are corners of previous LPR, C_i is the perimeter for LPR i in a current frame and C'_j is the perimeter for LPR j in previous frame.
4. Match criterion. The match criterion, defined by Eq. 4 determines if LPR i in the current frame and LPR j in the previous frame can belong in the same track.

$$D(i,j) < T \tag{4}$$

   where D is a normalized distance and T is a parameter.
5. Match table. Match table M contains information about matching LPRs in the current frame to those in a previous frame.

$$M = \begin{bmatrix} & P'_1 & P'_2 & \ldots & P'_m \\ P_1 & a_{11} & a_{12} & \ldots & a_{1m} \\ P_2 & a_{21} & a_{22} & \ldots & a_{2m} \\ \ldots & \ldots & \ldots & \ldots & \ldots \\ P_n & a_{n1} & a_{n2} & \ldots & a_{nm} \end{bmatrix} \tag{5}$$

   where $P_{1,2,\ldots}$ are LPRs in current frame, $P'_{1,2,\ldots}$ are LPRs in previous frame, and a_{ij} is the match value: 1 if the match criterion (4) is satisfied, 0 if not.

Tracking. Tracking method builds a list of tracks from the entire sequence of frames. A track is a list of LPRs from consecutive frames, such that there is a single LPR from each frame.

Tracking Method:
1. Start with an empty list of tracks.
2. Make a new track for each LPR in the frame 1.
3. Repeat for each frame, starting with the frame 2:
   - compute the match table M from the current and the previous frames and:
     - compute for each column j the sum, s_j of all values,
     - if s_j is 0 terminate the track in which the last element is P'_j,
     - if s_j is 1 add P_i, where a_{ij} is 1, to the track ending with P'_j,
     - if $s_j > 1$ terminate the track in which the last element is P'_j and start a new track for each P_i, where a_{ij} is 1.

2.3 Correction of License Plate Identifiers Using In-Track Clustering

A license plate cluster C is represented as a sequence of symbol histograms H

$$C = \{H_1, H_2, ..H_k\} \tag{6}$$

where H_k gives the percentage of occurrences of each symbol, including "?" at a position k. We define the function Match(p, C), which computes how well an LPCI p fits a cluster C with the help of a function Match(p, C, s), where is a shift of the sequence p with respect to the sequence of histograms of the cluster C (Fig. 4).

Fig. 4. The rules of comparing one LPR that was assigned to ClusterSet.

Match(p, C, S) is the sum of individual symbol matches when the sequence p and the histogram sequence of C are aligned according to shift s. A symbol match of a symbol x of p, under shift s is the value for x of the histogram H, where H is aligned with x, under shift s. If x = "?" or x has no corresponding histogram, the match value is 0. For each value of shift s, align p with C according to S, calculate Match(p, C, S), and find the shift value S* which maximizes Match(p, H, S). The function [Q, S] = Match(p, C), where p is an LPCI, and C is a cluster returns two values: score Q and shift S, where S:

S = arg maxs Match(p, C, s) and

Q = Match(p, C, S)

Clustering Method:

Start with an empty ClusterSet

For each LPR r of a track:

- Find the cluster C* in the ClusterSet, that gives the highest score value of Match(p, C), where p is the LPCI of r.
- If the ClusterSet is empty or the highest score is less that a threshold T, use p to initiate a new cluster and add it to the ClusterSet (Fig. 5).
- Otherwise, update the histograms of C* according to the p aligned with C* by the shift value of Match(p, C*).

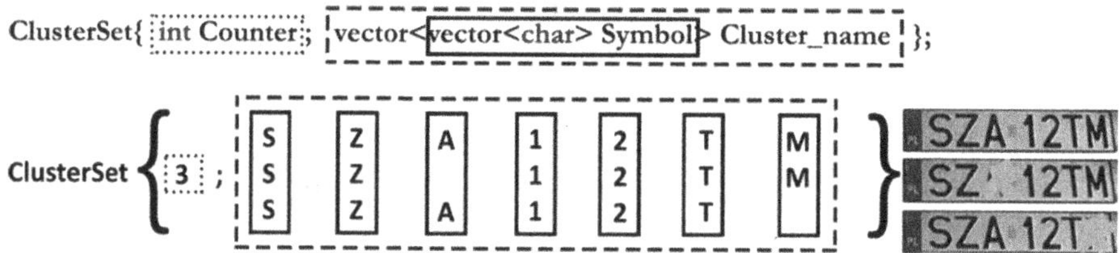

Fig. 5. ClusterSet structure and example of filling it with 3 LPRs.

After the clustering is completed, the corrected identifier is found by selecting the largest cluster in ClusterSet and for each of its histograms selecting the most frequent symbol. The received result is correcting all license plate numbers in the tracks.

3 Results

To run the tests, five video sequences from the supervising camera were obtained. They contained videos recorded in different weather and lighting conditions (Fig. 6). We took care of that sequence contains the license plates partially or fully covered. After checking the recording, hard to see even with human eye license plates were also found on it. The software that was used to recognize the license plates was called OpenALPR [15]. During recognition the default settings recommended by developer were used. The only change was made in main.cpp file to get the information about coordinates of the plate and data that were compatible with required formatting. Each of recorded frames had zero or more license plates. Annotated data of license plates numbers were added to every sequence, to compare how the percentage of correctly recognized license plates was changing before and after correction of license plate identifiers using In-Track Clustering. Every single of them contained 2500 captured frames, which is equal to five minute long recording from the camera. The conditions in which recordings were made are presented below.

1. Dataset 1 (Day) – clear weather, day, afternoon, proper size of license plates.
2. Dataset 2 (Night) – foggy weather, night, proper size of license plates.
3. Dataset 3 (Day) – rainy weather, morning, medium size of license plates.
4. Dataset 4 (Day) – rainy weather, afternoon, small size of license plates.
5. Dataset 5 (Day) – rainy weather, early evening, small size of license plates.

In next step, received in this way DataSets were used in ITCC method which gave the following results in Table 1.

As it is visible in Table 1, the weather and lighting conditions have big influence on the results of license plates recognition. The acquired result confirms the efficiency of described method. Its efficiency is nearly 20 % and it definitely improves recognition of license plates. The next step was to check every Dataset by creating a bar chart of every single Dataset, containing percentage of correctly recognized license plates according to track length before and after using the method. In addition, every Dataset has been watched in terms of plates.

Fig. 6. Every photo shown presents data contained in consecutive DataSets.

Table 1. Results of recognized number plates differing in DataSets.

DataSets	LPRs amount	Recognition rate		Improvement
		Without ITCC	With ITCC	
Dataset 1 (day)	2831	84,10 %	90,96 %	6,85 %
Dataset 2 (night)	2006	54,24 %	86,84 %	32,60 %
Dataset 3 (day)	3272	67,63 %	90,46 %	22,83 %
Dataset 4 (day)	4708	53,12 %	62,83 %	9,71 %
Dataset 5 (day)	3381	64,48 %	75,10 %	10,62 %

As it was expected, results of correctly recognized license plates were lowered because license plates were often lost by ALPR in consecutive tracks, creating tracks of one LPR length. The most interesting bar charts, so those concerning Datasets with the best and the worst improvement of recognition, were included below.

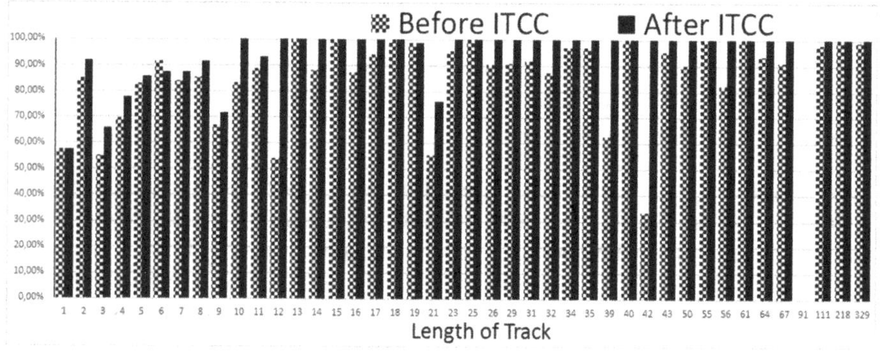

Fig. 7. Percentage result of all correctly and incorrectly recognized number plates differing in length of TRΛCKS from DataSet 1 (day)

Analyzing the shown bar charts, data that was inside sequences were strictly watched to define the reason of big improvements amplitude which had value of 27,75 %. Tracks on Fig. 7 were very well visible, which resulted in good score of correct recognition before using method beyond track of length 91, in which the license plate was not recognized correctly even once. For tracks of length 6 aggravation of recognition was noticed, resulting from 5 incorrectly recognized plates on a sample of length 6. The rest of them were improved, especially those in track of length 42.

In DataSet 2 on Fig. 8 (night) usually the camera capturing sequence is blinded by car headlights. Because of that, the license plate number might be not readable from right or left side. As it is visible on next bar chart which shows night sample, the recognition is very good. Method used was giving good results

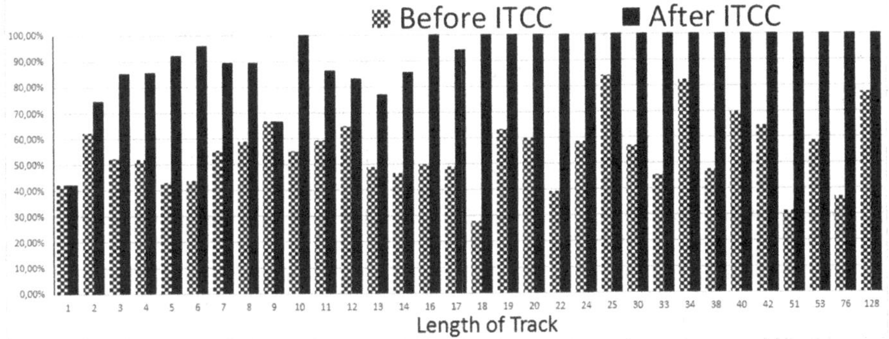

Fig. 8. Percentage result of all correctly and incorrectly recognized number plates differing in length of TRACKS from DataSet 2 (night)

by using tracking. In addition, sometimes license plates were covered by people on pedestrian crossings. To conclude, method described in this article turned out to be successful and it can be further used to increase the successive license plate recognition.

4 Conclusion

A new method of recognizing license plate identifiers from surveillance video has been proposed that includes the information from tracking license plates in a video sequence, which results in a substantial improvement. These improvements are mainly a result of elimination of errors caused by covered or poorly visible license plates. Unlike other methods using a similar approach, the new method can take advantage of most single frame license plate recognition and object tracking methods, by using the presented In-Track Clustering algorithm to combine the recognition with tracking. The results from five video sequences taken by the surveillance cameras in a variety of weather and lighting conditions show improvements in the range of approximately 7 % to 32 %, comparing to the use of a single frame license plate recognition without tracking. Future work will include the most recent, very efficient tracking implementations, and the use of the new method in a pilot parking application.

Acknowledgments. This work has been supported by the National Centre for Research and Development (project UOD-DEM-1-183/001 Intelligent video analysis system for behavior and event recognition in surveillance networks).

References

1. Donoser, M., Arth, C., Bischof, H.: Detecting, tracking and recognizing license plates. In: Yagi, Y., Kang, S.B., Kweon, I.S., Zha, H. (eds.) ACCV 2007, Part II. LNCS, vol. 4844, pp. 447–456. Springer, Heidelberg (2007)

2. Shahzad, A., Fraz, M., Elahi, M.A., Sarfraz, M.S.: Real time localization, tracking and recognition of vehicle license plate. In: VISAPP, pp. 685–688 (2011)
3. Dlagnekov, L., Belongie, S.: Recognizing cars. Technical report CS2005-083, CSE, UCSD (2005)
4. Matas, J., Zimmermann, K.: Unconstrained licence plate and text localization and recognition. In: Proceedings of the IEEE Intelligent Transportation Systems, pp. 225–230 (2005)
5. Mello, C.A.B., Costa, D.C.: A complete system for vehicle license plate recognition. In: WSSIpP, pp. 1–4 (2009)
6. Du, S., Ibrahim, M., Shehata, M., Badawy, W.: Automatic license plate recognition (ALPR): a state-of-the-art review. IEEE Trans. Circ. Syst. Video Technol. **23**, 311–325 (2013)
7. Jia, W., He, X., Piccardi, M.: Automatic license plate recognition: a review. In: Proceedings CISST, pp. 43–49 (2004)
8. Solanki, R., Kapadia, N., Patel, V.: The review on automatic license plate recognition (ALPR). IJRET **02**, 178–183 (2013)
9. Li, X., Hu, W., Shen, C., Zhang, Z., Dick, A., van den Hengel, A.: A survey of appearance models in visual object tracking. In: CoRR (2013). abs/1303.4803
10. Chu, D.M., Cucchiara, R., Calderara, S., Dehghan, A., Shah, M.: Visual tracking: an experimental survey. Pattern Anal. Mach. Intell. **36**, 1442–1468 (2013)
11. Milan, A., Leal-Taix, L., Schindler, K., Reid, I.: Joint Tracking and Segmentation of Multiple Targets CVPR (2015). https://bitbucket.org/amilan/segtracking
12. Staniszewski, M., Kloszczyk, M., Segen, J., Wereszczynski, K., Wojciechowski, K., Kulbacki, M.: Recent developments on tracking based methods (in preparation)
13. Donoser, M., Bischof, H.: Efficient maximally stable extremal region (MSER) tracking. In: Proceedings of Conference on Computer Vision and Pattern Recognition (CVPR) pp. 553–560 (2006)
14. Zheng, L., He, X.: Number plate recognition based on support vector machines. In: Proceedings of the IEEE International Conference on Video and Signal Based Surveillance (AVSS06), p. 13. IEEE Computer Society, Washington (2006)
15. Openalpr.com: OpenALPR – Automatic License Plate Recognition. http://www.openalpr.com/

A New Method for Calibrating Gazis-Herman-Rothery Car-Following Model

Fergyanto E. Gunawan(✉), Elysia Elysia, Benfano Soewito, and Bahtiar S. Abbas

Binus Graduate Programs, Bina Nusantara University, Jakarta 11480, Indonesia
fgunawan@binus.edu, f.e.gunawan@gmail.com

Abstract. Traffic simulation at the microscopic level utilizes car-following model to describe vehicle interactions on a vehicular lane. The most widely used car-following model is the Gazis-Herman-Rothery model, which contains two coefficients: m and l. The coefficients should be determined in calibration tests where the involved vehicles are tracked for their positions, velocities, and accelerations. The existing calibration methods are costly. This study proposes a calibration method using computer vision. Two computer vision algorithms are evaluated, namely, multilayer and Eigen background subtraction. The vehicle movement is tracked on a perspective plane and then is projected to an orthogonal plane. From the verification tests, we determine that the multilayer algorithm has 96.6 % accuracy for the vehicle position and 88.9 % for the velocity. The Eigen algorithm has 92.9 % accuracy for the vehicle position and 84.3 % for the velocity. The estimated model coefficients is 0.4 for m and 1.2 for l. These values are within the range of the most reliable coefficients according to many literatures.

Keywords: Car-following model · Vehicle tracking · Computer vision · Micro-simulation

1 Introduction

Traffic congestion is a huge issue faced by many cities including Jakarta, the capital of Republic of Indonesia [1]. Morichi [2] recommends a number of long-term strategic development to overcome the problem including development of well-structure public transportation systems and decentralized urban form. In 2003, Bus-Rapid-Transit-based public transportation system was initiated in Jakarta [3] and the operational efficiency of the system has been the topics of a number of research works [4–6].

Traffic congestion can also be reduced by traffic management. To yield good traffic management, traffic microsimulation is an indispensable tool. Micro simulation shows the movement of every vehicle that is traced through a road network over time at a small time increment of a fraction of a second [7].

Car-following model is at the heart of the traffic microsimulation. The model governs the vehicle movement along a vehicle lane. Essentially, the model is a

N.T. Nguyen et al. (Eds.): ACIIDS 2016, Part II, LNAI 9622, pp. 765–772, 2016.
DOI: 10.1007/978-3-662-49390-8_74

mathematical description of the stimulus and response interaction [8]. One of the widely used car-following models is Gazis-Herman-Rothery (GHR) model, which has the following mathematical form [9]:

$$a_n(t) = \alpha \frac{v_n^m(t) \cdot \Delta v_n(t - t_d)}{\Delta x_n^l(t - t_d)}, \tag{1}$$

where $a_n(t)$ is the vehicle acceleration, $v_n(t)$ is the vehicle velocity, Δv_n and Δx, respectively, are the relative velocity and the relative position with respect to the leading vehicle. The model has three parameters α, m, and l, and has a delay of t_d. The driver delay or response time t_d is usually about 1 s; α is the driver sensitivity [9].

The central issue to use the micro-simulation model is finding the model parameters for each type of vehicles, road types, traffic conditions, and environment. The parameters are usually determined via a calibration process. There are few existing calibration methods and they are expensive [10–13].

This study intends to establish a significantly low-cost calibration procedure where the vehicle movements will be recorded digitally, tracked with a computer vision technique, and projected with an orthography projection technique.

2 Research Methods

The current proposal is only suitable for vehicle moving along a straight lane. The lane should be marked at four corners. The vehicle position is obtained with the following procedure.

Firstly, a camera is positioned at an elevation and angle from the lane to capture the vehicle movement in the perspective view. The camera is used to produce images of the vehicle movement. Secondly, computer vision is used to obtain the vehicle position in the images [14]. Two algorithms are evaluated in this study: multilayer background subtraction [15] and Eigen background subtraction [16]. Thirdly, the vehicle position is projected from the perspective view to an orthographic view using an orthorectification projection.

The following transformation is used to map the data from the perspective plane to the road/orthographic plane [17]. We consider a unit square S_1 and an arbitrary quadrilateral Q, which is governed by four corner points: x_1', x_2', x_3', and x_4'. These four points have one-to-one relations with those four points in the unit square, see Fig. 1. The following matrix $\mathbf{T}$ projects any point on the unit square to a point on the quadrilateral, or mathematically: $S_1 \xrightarrow{\mathbf{T}} Q$, where the transformation $\mathbf{T}$ is

$$\mathbf{T} = \begin{bmatrix} a_{11} & a_{12} & a_{13} \\ a_{21} & a_{22} & a_{23} \\ a_{31} & a_{32} & a_{33} \end{bmatrix}, \tag{2}$$

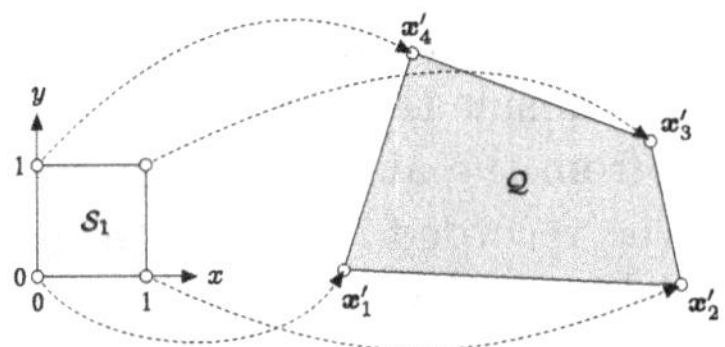

Fig. 1. Projective mapping from the unit square S_1 to an arbitrary quadrilateral Q [17].

where

$$a_{31} = \frac{(x'_1 - x'_2 + x'_3 - x'_4)(y'_4 - y'_3) - (y'_1 - y'_2 + y'_3 - y'_4)(x'_4 - x'_3)}{(x'_2 - x'_3)\cdot(y'_4 - y'_3) - (x'_4 - x'_3)\cdot(y'_2 - y'_3)},$$

$$a_{32} = \frac{(y'_1 - y'_2 + y'_3 - y'_4)(x'_2 - x'_3) - (x'_1 - x'_2 + x'_3 - x'_4)(y'_2 - y'_3)}{(x'_2 - x'_3)\cdot(y'_4 - y'_3) - (x'_4 - x'_3)\cdot(y'_2 - y'_3)},$$

$$a_{11} = x'_2 - x'_1 + a_{31}x'_2 \qquad a_{12} = x'_4 - x'_1 + a_{32}x'_4 \qquad a_{13} = x'_1,$$

$$a_{21} = y'_2 - y'_1 + a_{31}y'_2 \qquad a_{22} = y'_3 - y'_1 + a_{32}y'_4 \qquad a_{23} = y'_1.$$

Inversely, we can project any point on the Q plane to the S_1 plane by the inverse of the transformation matrix $\mathbf{T}^{-1}$.

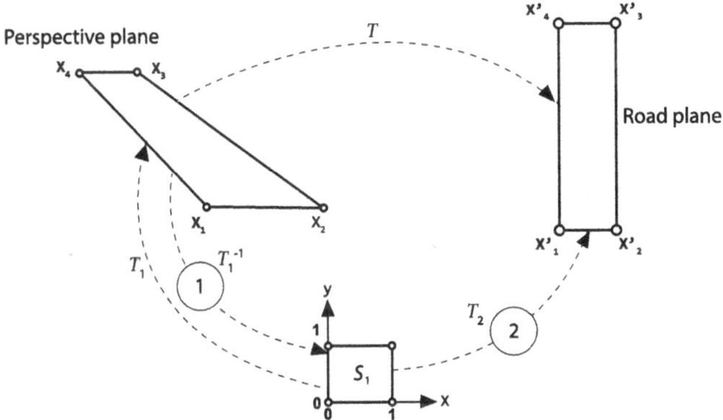

Fig. 2. Two-step projective transformation between perspective plane and road plane. In the first step, the perspective plane Q_1 is transformed to the unit square S_1 by the inverse mapping function $\mathbf{T}_1^{-1}$. In the second step, $\mathbf{T}_2$ transforms the square S_1 to the road plane Q_2. Transformation from the perspective plane Q_1 to the road plane Q_2 can be direct by $\mathbf{T} = \mathbf{T}_1^{-1}\mathbf{T}_2$ [17].

Now, we consider three planes, the perspective plane Q_1, the unit plane S_1, and the road plane Q_2, depicted in Fig. 2. We can map points on the perspective plane Q_1 to points on the road plane Q_2 in two steps: $Q_1 \xrightarrow{\mathbf{T}_1^{-1}} S_1$, and followed by $S_1 \xrightarrow{\mathbf{T}_2} Q_2$. The transformation can be directed from the perspective plane Q_1 to the road plane Q_2 by:

$$\mathbf{T} = \mathbf{T}_2^{-1}\mathbf{T}_1. \tag{3}$$

In the current work, the vehicle movement is also recorded using an accelerometer, which is placed inside the test vehicle. These data are later compared to the data obtained from the above procedure to evaluate the accuracy of the proposed method. The recorded acceleration data are numerically integrated to provide the vehicle velocity and position. The level of accuracy is simply 1 – Relative Error, where the error is defined by

$$\text{Relative Error} = \frac{\|\hat{\mathbf{x}} - \mathbf{x}\|_2}{\|\mathbf{x}\|_2} \times 100\,\%, \tag{4}$$

where $\hat{\mathbf{x}}$ is the vector of the estimated position or velocity and $\mathbf{x}$ is the true value, that is obtained from the accelerometer.

Experimental Procedures

The proposed method was evaluated with a simple experiment described following. The experiment only involved a vehicle. The vehicle was set to travel along a straight trajectory for a distance around 23 m. It was difficult to control the vehicle exact position. Figure 3 respectively show the vehicle initial and final positions. Four cones were placed on the four corners of the vehicle trajectory; see Fig. 3. These four cones were separated by 22.9 m distance longitudinally and 3.5 m distance laterally.

Fig. 3. The test vehicle initial and final positions and the four cones used for orthorectification.

The vehicle movement was recorded by two means: a video camera and an accelerometer. The video camera recorded the vehicle movement at the rate of 25 fps and image size of 720 × 576 pixels. The accelerometer recorded at the sampling time of 0.1 s.

In addition, we also compute the vehicle coefficient m and l by minimizing

$$\text{Error}(t) = a_n(t) - \alpha\,\frac{v_n^m(t)\cdot\Delta v_n(t-t_d)}{\Delta x_n^l(t-t_d)}. \tag{5}$$

All data a_n, Δv_n, Δx_n are from the current proposed method.

3 Results

Two sets of data are necessary for this study. The first set is obtained from the deployed computer-vision-based vehicle tracking methods. The second set is from the accelerometer.

Two different background subtractions in computer vision will be used separately: multilayer and Eigen background subtractions. The computer vision vehicle tracking provides the data of: frame number, blob number, blob area, x blob-centroid, y blob-centroid, and the position of the bottom-right corner of the blob. The frame number and the frame rate data are used to calculate the time associated with the frame number by:

$$\text{Frame Time } t = \frac{\text{Frame Number}}{\text{Frame Rate}}.$$

The frame rate is fixed at 25 fps. Only one vehicle is used in the experiment and it has dimensions of 4.7 m long, 1.8 m wide, and 1.8 m high.

The results of the current study are shown in Fig. 4 for multilayer algorithm and Fig. 5 for Eigen subtraction algorithm. The relative error, absolute error, and accuracy have been computed for every experiments; see Table 1. The results show that the tracking accuracy using multilayer background subtraction is better than that of Eigen background subtraction. Multilayer method has 96.6 % position accuracy and 88.9 % velocity accuracy; meanwhile Eigen method has 92.9 % position accuracy and 84.3 % velocity accuracy.

In Fig. 6, we compare the vehicle acceleration computed the GHR model and the measured one. The estimated GHR car-following parameters are $m = 0.4$ and $l = 1.2$ at 3.2 % of relative error. These estimated parameters are within the range of most reliable estimated parameters according to findings of other researchers for GHR model [18].

Table 1. The relative error of the tracked vehicle position by the computer vision technique with respect to the vehicle position obtained by integrating by the accelerometer data. The relative error is defined by Eq. 4.

Experiment no.	Relative error (%)			
	Position (m)		Velocity (m/s)	
	MultiLayer BGS	Eigen BGS	MultiLayer BGS	Eigen BGS
1	3.41	2.9	13.1	6.9
2	2.49	4.6	6.5	9.1
3	2.40	4.4	8.4	4.0
4	3.84	10.5	12.5	26.3
5	4.37	9.4	10.9	13.1
6	4.30	12.7	12.5	39.3
7	3.07	5.2	13.5	11.2

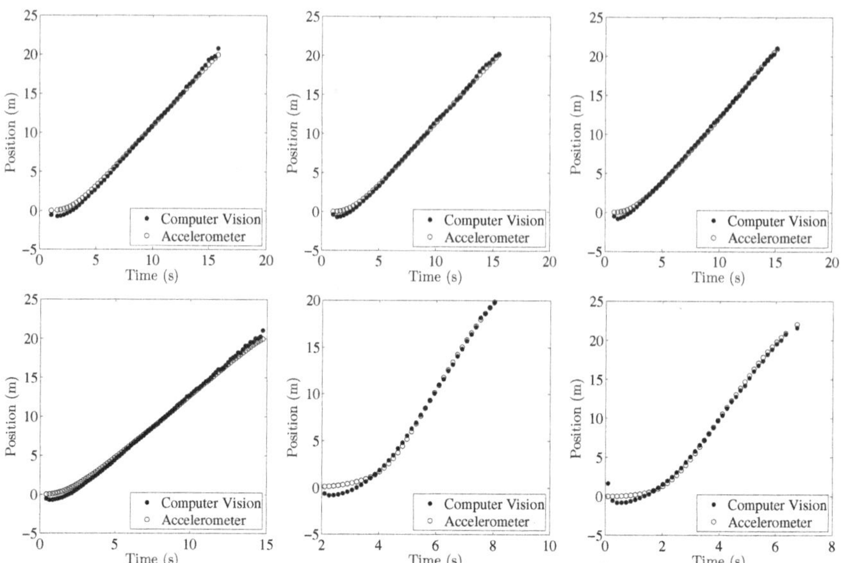

Fig. 4. Comparison of the space-time diagram obtained from computer vision (multi-layer algorithm) and from accelerometer for six experimental replications.

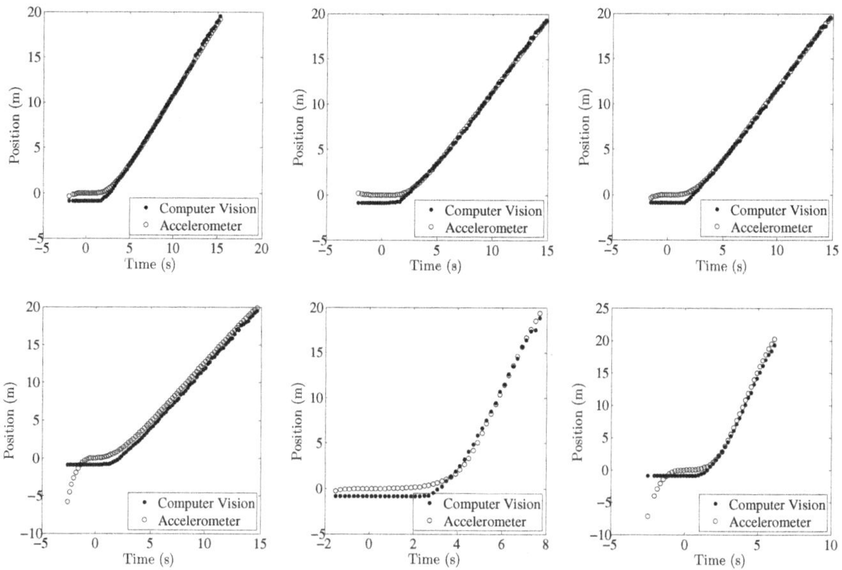

Fig. 5. Comparison of the space-time diagram obtained from computer vision (Eigen background subtraction algorithm) and from accelerometer for six experimental replications.

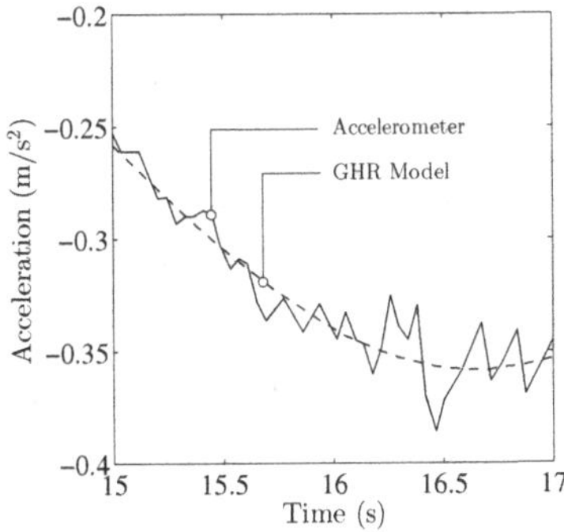

Fig. 6. Comparison of the Vehicle Acceleration between Accelerometer Data and GHR Model Prediction

4 Conclusions

The car-following model has important application in traffic and safety engineering. Unfortunately, finding the model parameters are often costly. For the reason, this research proposed a significantly low-cost method to determine the model coefficients. In the current method, the vehicle movement are recorded digitally and tracked by a computer vision technique. The obtained vehicle position is projected to the road plane with an orthographic projection technique. In the current experiment, to evaluate its accuracy, the method is used to track trajectory of a vehicle moving in a short-straight lane. From comparisons to the data obtained from an accelerometer, it is concluded that the current method is reasonably accurate. The results show that the tracking accuracy of multilayer background subtraction is better than that of Eigen background subtraction. The multilayer method has 96.6 % position accuracy and 88.9 % velocity accuracy; meanwhile, the Eigen method has 92.9 % position accuracy and 84.3 % velocity accuracy. The method estimates the car-following parameters to be $m = 0.4$ and $l = 1.2$ with 3.2 % relative error. These estimated parameters are within the range of the most reliable GHR model parameters according to many references.

References

1. BPS: Perkembangan jumlah kendaraan bermotor menurut jenis tahun 1987–2012 (2013). http://www.bps.go.id/linkTabelStatis/view/id/1413
2. Morichi, S.: Long-term strategy for transport system in Asian megacities. J. East. Asia Soc. Transp. Stud. **6**, 1–22 (2005)
3. Gunawan, F.E.: Empirical assessment on factors affecting travel time of bus rapid transit. Int. J. Eng. Technol. **7**(1), 327–334 (2015). http://www.ijetch.org/show-42-253-1.html
4. Gunawan, F.E.: Design and implementation of discrete-event simulation framework for modeling bus rapid transit system. J. Transp. Syst. Eng. Inf. Technol. **4**, 37–45 (2014). http://www.sciencedirect.com/science/article/pii/S1570667213601397

5. Lumentut, J.S., Gunawan, F.E., Atmadja, W., Abbas, B.S.: A system for real-time passenger monitoring system for bus rapid transit system. In: Nguyen, N.T., Trawiński, B., Kosala, R. (eds.) ACIIDS 2015. LNCS, vol. 9012, pp. 398–407. Springer, Heidelberg (2015). https://dx.doi.org/10.1007/978-3-319-15705-4_39
6. Lumentut, J.S., Gunawan, F.E., Diana, : Evaluation of recursive background subtraction algorithms for real-time passenger counting of bus rapid transit. Procedia Comput. Sci. **59**, 445–453 (2015). http://www.sciencedirect.com/science/article/pii/S1877050915020943
7. Walsh, D., Stewart, D., Luk, J., Tay, J.: The use and application of microsimulation traffic models. Austroads Publications Online (2006). https://www.onlinepublications.austroads.com.au/items/AP-R286-06
8. Ryu, J., Kim, C., Chang, M., Kim, Y., Bae, S.: Simulation and speed classification of car following models. Proc. East. Asia Soc. Transp. Stud. **3**, 123–133 (2001). http://easts.info/on-line/proceedings/vol3no2/320010.pdf
9. Gunawan, F.E.: Two-vehicle dynamics of the car-following models on realistic driving condition. J. Transp. Syst. Eng. Inf. Technol. **12**, 68–75 (2012). http://www.sciencedirect.com/science/article/pii/S1570667211601943
10. Chandler, R.E., Herman, R., Montroll, E.W.: Traffic dynamics: studies in car following. Oper. Res. **6**, 165–184 (1958). http://www.jstor.org/stable/167610
11. Hoogendoorn, S., van Zuylen, H.J., Schreuder, M.: Microscopic traffic data collection by remote sensing. Transp. Res. Rec. **1855**, 121–128 (2007). http://dx.doi.org/10.3141/1855-15
12. Hoogendoorn, S., Schreuder, M.: Toward a robust method for microscopic traffic data collection. In: The 84th Annual Meeting of the Transportation Research Board
13. Ranjitkar, P., Nakatsuji, T., Kawamura, A.: Car-following models: an experiment based benchmarking. J. East. Asia Soc. Transp. Stud. **6**, 1582–1596 (2005). http://www.easts.info/on-line/journal_06/1582.pdf
14. Crist: Blob library for OpenCV, from Google Code (2012). https://code.google.com/p/cvblob/wiki/FAQ. Accessed 11 February 2015
15. Yao, J., Odobez, J.M.: Multi-layer background subtraction based on color and texture. In: IEEE Conference on Computer Vision and Pattern Recognition, June 17–22, pp. 1–8 (2007). http://dx.doi.org/10.1109/CVPR.2007.383497
16. Oliver, N.M., Rosario, B., Pentland, A.P.: A Bayesian computer vision system for modeling human interactions. IEEE Trans. Pattern Anal. Mach. Intell. **22**(8), 831–843 (2000). http://dx.doi.org/10.1109/34.868684
17. Burger, W., Burge, M.: Principles of Digital Image Processing: Core Algorithms. Springer, Heidelberg (2009)
18. Brackstone, M., McDonald, M.: Car-following: a historical review. Transp. Res. Part F **2**, 181–196 (1999). http://www.sciencedirect.com/science/article/pii/S136984780000005X

Control of Smart Environments Using Brain Computer Interface Based on Genetic Algorithm

Guilherme Antonio Camelo[1(✉)], Maria Luiza Menezes[1,2], Anita Pinheiro Sant'Anna[2], Rosa Maria Vicari[1], and Carlos Eduardo Pereira[1]

[1] Electrical Engineering Department and Informatics Department, Federal University of Rio Grande do Sul, Porto Alegre, Brazil
{gacamelo,rosa}@inf.ufrgs.br, cpereira@ece.ufrgs.br
http://www.ufrgs.br

[2] Halmstad University, Halmstad, Sweden
{maria.menezes,anita.santanna}@hh.se
http://www.hh.se/

Abstract. This work deals with the development of an interface to control a smart conference room using passive BCI (Brain Computer Interface). It compares a genetic algorithm developed in a previous project to control the smart conference room with a random control algorithm. The system controls features of the conference room such as air conditioner, lightning systems, electric shutters, entertainment devices, etc. The parameters of the algorithm are extracted from users biosignal using Emotiv Epoc Headset while the user performs an attention test. The tests indicate that the decisions made by the genetic algorithm lead to better results, but in a single execution cannot be considered an effective optimization algorithm.

Keywords: Genetic algorithm · Smart environment · Brain computer interface · Emotive · EEG · Digital signal processing

1 Introduction

Assistive technologies identify resources and services in order to provide or amplify the functional abilities of people both with and without disabilities. Recent works have developed assistive technologies solutions to accompany users into completely supportive environments, pursuing users' autonomy in their daily tasks [13].

In this context, Home and Building Automation and Smart Environments techniques can greatly enhance users autonomy, by provides devices whose behavior is aware of environment changes and are also able to self-manage [18]. Ducatel, in [19], correctly predicted features that smart environment would have, stating that there would be more emphasis on greater user-friendliness, efficient services support, user-empowerment, and support for human interactions. Projects such as this are making possible the idea of ubiquitous computing glimpsed by Mark Weiser [14], in which people interact with devices without perceiving them.

N.T. Nguyen et al. (Eds.): ACIIDS 2016, Part II, LNAI 9622, pp. 773–781, 2016.
DOI: 10.1007/978-3-662-49390-8_75

In order to make the interaction between the system and the user more transparent, one solution is the Brain-Computer Interface (BCI), which tries to incorporate the systems reactions directly to the user's brain activity [9,15,20]. While recent advancements in BCI to control interfaces and Systems are very interesting [16], its use is still challenging as it requires special training and focused attention [8]. By using EEG (Electroencephalography) signals as implicit interaction between human and machine and going beyond direct system control, is possible to make BCI useful for a wider range of scenarios [20]. By implicit interaction, or, in this case, passive BCI [2,3,5], we refer to an action performed by the user that is not primarily aimed to interact with a computerized system but that the system understands as input [4]. George,in [5], defines passive BCI interaction as a situation where the user does not try to control his brain activity.

By reading and analyzing EEG signals from the headset, a representation of the users emotion can be obtained. The proposed goal is to build a smart environment capable of intelligent adaptation in order to increase the engagement level of an individual in a given task. In order to achieve this goal, a genetic algorithm with multiple iterations was used. Each iteration creates a different room scenario that should influence the objective emotion, in order to avoid that the user's engagement level decreases during repetitive tasks. A similar approach was used by Ramirez-Atencia in [21] where a multi-objective genetic algorithm was used to minimize the fuel consumption and makespan of a mission performed by a group of unmanned air vehicles. In [22] Ramirez-Atencia provides analysis for the mission using Temporal Constraint Satisfaction Problem model, and Branch and Bound search algorithm.

This paper is structured as follows: Sect. 2 mentions the equipment and the environment involved, explains the algorithms that will be used, and details the implementation of the genetic algorithm. Section 3 describes the tests and presents the results of each approach. Finally, Sect. 4 explains the relevance of time limitation in the tests performed, analyzes the results, and propose a future work intended to reach better results.

2 Methodology

Our approach comprises the use of a helmet for EEG signal acquisition, capable of interpreting the subjects emotional state. This emotion will be used to passively control a smart environment (a room equipped with a home automation system), adjusting elements such as air conditioning, opening and closing windows, playing media, and other elements available. For signal acquisition, the Emotiv Epoc Headset [12] device was used. The system's quality of biosignal detection is confirmed by Castermans in [17]. It is equipped with 14 electrodes that detect EEG signals from the brain. The Headset is capable of extract data and identify neurological patterns, such as Engagement, which will be referred to as objective emotion or fitness, according the terminology used in the theory of genetic algorithms presented by Sivanandam, in [7].

The environment used in the experiment was an automated conference room placed at the Home Systems company, equipped with six group of lights, five blinds, two air conditioner, a sound system and five monitors. A solution or individual is composed by an array where each position controls one feature of the room, each position assumes values 1 and 0 representing on and off status. An interface was developed to allow the integration of the Emotiv helmet with the home automation system, which made it possible to control the room and make all the decisions based on the information that it gets from the brain. Readers interested on more details on the developed interface should refer to [11].

Considering optimization heuristics, genetic algorithms present some difficulties in terms of parameter calibration. Questions such as mutation rate, crossover strategies, selection of individuals are very sensitive, as stated by Haupt, in [10]. In some cases those questions can only be determined by experimentation.

In order to have a comparison, a random control algorithm was created. The results of the random and genetic approach are compared in order to perform a better analysis.

Genetic algorithms are heuristic optimization and search methods inspired in the evolution of species and populations. According to the natural selection principle and survival of the fittest theories, the rate in which an individual can adapt to its environment is directly related to its survival and reproduction rate [10].

Defining the optimal mutation rate and the crossover process is a challenge, because they are highly dependable on the context and number of iterations. Those concerns are presented by Haupt, in [10]. In this paper, in order to achieve a maximization of the objective function, the generations are sorted by fitness and an elitist selection is used. The genetic operators involved were mutation, one-point crossover, and random mask crossover. The algorithm stops creating new individuals when it reaches a certain fitness threshold, and will stop its execution by manual intervention in some cases and by reaching time limits in others. The threshold can be recalculated along the execution for it may be different for different users. If the threshold criteria of the Engagement is not met in a defined number of generations, the threshold is recalculated in runtime, and the new threshold will be the biggest level of Engagement found in the last n number of iterations. Therefore, when there is an elevation in the engagement, the threshold is recalculated to a new value. With that approach, the threshold is always coherent with the limitation of the users.

2.1 Genetic Algorithm Implementation

A simplified diagram of the genetic algorithm is presented in Fig. 1 with labels identifying each step. Initially we create a set of random solutions that will be the individuals of the first generation (E1). Those solutions still do not have a fitness, each of them has to be evaluated. The conference room will be set in each of the solutions for 3 min, and after that the fitness of the solutions can be measured (E3). Prior to the execution, a fitness threshold is selected, that threshold is an elevated Engagement level that represents the goal. If a fitness of a solution is higher then the goal (D1), the room will maintain the environment configuration (E3). If in any situation the fitness in lower than the threshold (D1), the crossover process starts.

After the evaluation of all the initial solutions, if there is no fitness higher than the goal, parents are selected by elitism (E4) so the elite can be selected(the offspring). A number of individual that are now in the top of the list are considered the elite. In this project we use 3 as the size of the elite, and an elitist selection is applied for the crossover processes.

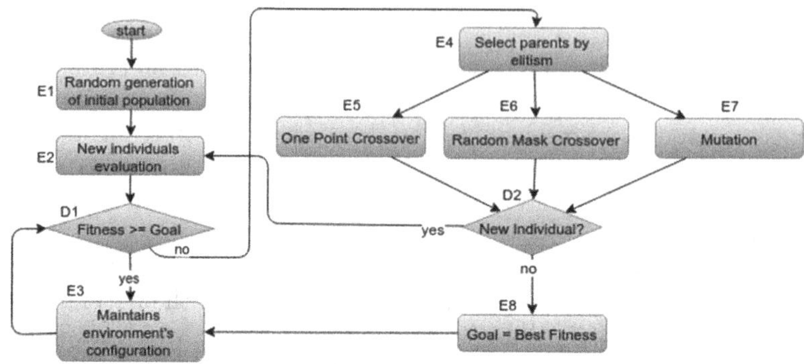

Fig. 1. Genetic algorithm flow diagram

From the elite, the next generation is created using one-point crossover (E5), random mask crossover (E6), and mutation (E7). After that new solutions are created. If the solutions are new (D2) they are evaluated (E2). In order for that to happen, the room will assume the configuration of each of the solutions in order to be able to measure each fitness. If the solutions are not new, the goal will be updated to the highest fitness found (E8), this process is the adaptive threshold.

The adaptive threshold is presented in the researches of Whitley, in [6], and Back, in [1]. If after every setting of the room is evaluated and the threshold have not been reached, it is assumed that the threshold is not adequate for the user, so it will be recalculated. The adaptable objective fitness is needed because each user has a different limitation and different values of Engagement.

After every individual has been evaluated the sort and the selection are made and a new generation is created. The algorithm will stop the creation of new solutions when the threshold is reached, or when all possible scenarios have been evaluated. The process starts to loop until a condition for the execution to stop is met. The stop condition is met when the user finishes the tests.

3 Tests and Results

In order to validate the proposed approach, twelve volunteers aged from 20 to 60 years old and ranging from undergraduate to PhD students were tested. The subjects were asked to perform different attention tasks similar to Brazilian's driving license tests [11] for about 40 min and no pre-test was conducted in order to select initial Engagement states since the goal is to evaluate the gain of the Engagement level.

As shown in [11], the Engagement of users tends to decrease while performing the cited repetitive tasks, as the user tends to get bored. So the purpose of the test is to evaluate if it is possible to maintain or even increase the user's Engagement level during the execution of the repetitive tasks, showing that the system succeeds in adapting the Environment in order to positively affect the user's mental state.

Using the genetic algorithm with such time constrain of 40 min instead of day, weeks or months that the system is intended to be used, is a sensible approach. Usually, the evaluation of solutions in such meta-heuristic problems are given in microseconds or even nanoseconds. Given the tests restrictions, it is not expected to converge to a set of optimal solutions. Nonetheless, the functions responsible for the creation of new solutions are expected to successfully adapt according to the user's Engagement level and to positively affect their mental state.

The initial conditions in the execution of the algorithm were the following: The initial population had 3 individuals created randomly, and each generation that came next had a limit of 4 individuals selected by elitism. Roughly 15 solutions were evaluated after creating 4 generations. The mutation rate was 5 % and two types of crossovers were applied.

In order for the user to be affected by the environment, previous experiments have showed that the state of the room, or scenario, has to stay active for at least three minutes [11]. During all the experiment, the users had the Emotive Epoc Headset in their heads while the Engagement levels were extracted.

The results obtained using the genetic based algorithm were compared with the ones using a random algorithm, when the room is set to a random scenario in every interaction, not taking into account the Engagement levels to make a decision; and also with the results of a baseline algorithm, where no modification on the environment was made whatsoever. During all the tests, the users had the Emotive Epoc Headset while performing the same cited tasks.

The graphs with the highest percentage improvement for the Random, Genetic Based and Baseline algorithms are presented in Figs. 2, 3 and 4 respectively, where the blue line represents the Engagement values, the red dots the mean Engagement for each scenario, the red line the linear regression and the green line the mean Engagement value during the entire test.

In Fig. 2, subject I started with Engagement level of 0.6593, and ended with 0.8643, having an improvement of 0.250, equivalent to 31.0936 %; in Fig. 3, subject A started with Engagement level of 0.6152 and ended with 0.9439, representing an increase of 53.43 % and in Fig. 4 the values for subject D fall from 0.7212 to 0.5854, having a decrease of 18.83 %. In all graphs for the Baseline Test shown in [11] and seen in Fig. 4 we can identify a first excitement or peak of the Engagement level at the beginning of the task, and later a slow decrease of the values, highlighted by the linear regression in red, as the subject becomes bored. More examples of Random and Baseline tests' performance can be seen in [11].

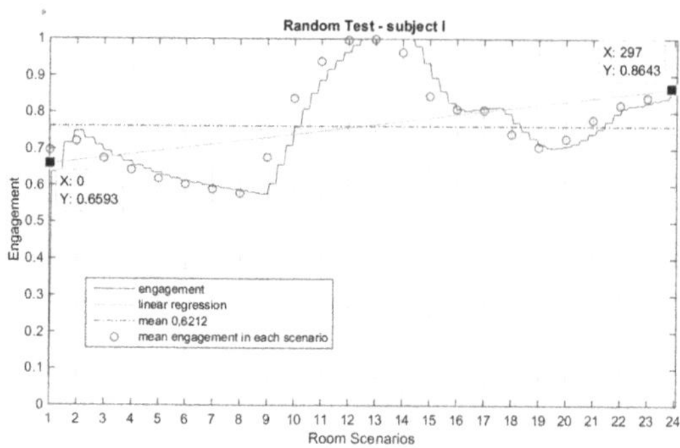

Fig. 2. Graph of subject I, best result of the Random algorithm (Color figure online)

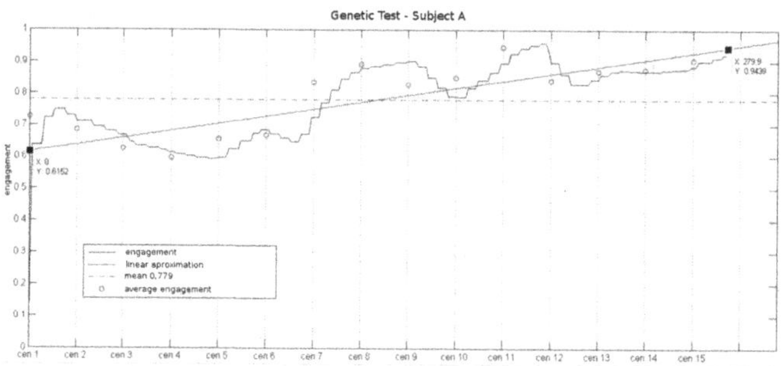

Fig. 3. Graph of Subject A, best result of the Genetic algorithm (Color figure online)

The results of the random algorithm tests are presented in Table 1. At the last column we can see the mean value of 0.723125 for all executions, a positive Engagement difference of 0.042225, meaning a percentage gain of 6.5964 % for all subjects. There were 3 cases with positive results, meaning an increase of the Engagement level, and one negative, representing a decrease of 37.5914 % of the Engagement level for subject F. Three out of four users had higher concentration levels that when they started. The negative value represents that the user lost focus continuously, and the random changes on the Environment's scenario were not able to raise the user Engagement level.

Table 2 presents at the last column the mean of the executions with the genetic approach. A positive Engagement difference of 0.2452 is observed, with a positive percent difference of 38.7766 %. In every execution the users had higher concentration levels that when they started. The smart environment attempted to make changes that successfully improve the concentration levels, achieving better results than the random approach, which changes the scenarios randomly without taking in consideration the user's mental state.

In Table 3 the results for the baseline approach are presented. A negative Engagement difference of −0.156 is observed, with a negative percent difference of −22.0374 %. In every execution the users had lower concentration levels that

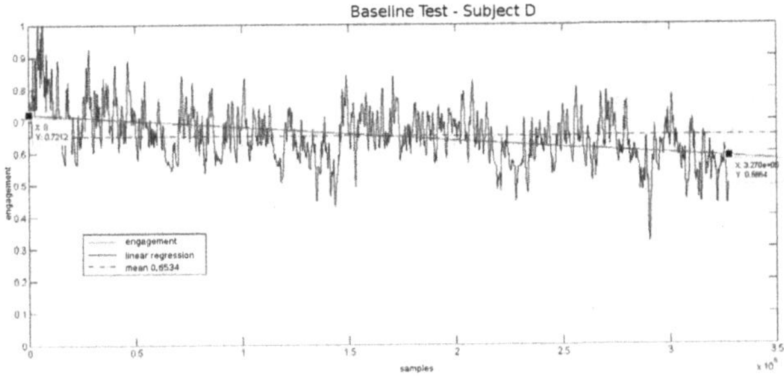

Fig. 4. Graph of Subject D, best result of Baseline algorithm (Color figure online)

Table 1. Random algorithm based test results

	Subject A	Subject F	Subject G	Subject I	Mean
Mean	0.7869	0.6212	0.8632	0.6212	0.723125
Initial engagement	0.7337	0.7656	0.7906	0.6593	0.7373
Final engagement	0.8404	0.4778	0.9356	0.8643	0.779525
Engagement numeric difference	0.1067	−0.2878	0.145	0.2050	0.042225
Engagement % difference	14.5427	−37.5914	18.3405	31.0936	6.5964

Table 2. Genetic algorithm test results

	Subject A	Subject B	Subject C	Subject E	Mean
Mean	0.779	0.6593	0.7883	0.7854	0.753
Initial engagement	0.6152	0.6175	0.6508	0.6392	0.6307
Final engagement	0.9439	0.7011	0.9269	0.9314	0.8758
Engagement numeric diff	0.3287	0.0836	0.2761	0.2922	0.2452
Engagement % diff	53.4298	13.5385	42.4247	45.7134	38.7766

when they started. As stated in [11], with time users tend to get bored because of the execution of the same repetitive task, and the engagement level tends to decrease.

Table 3. Control test results

	Subject A	Subject B	Subject D	Subject F	Mean
Mean	0.636	0.5927	0.6534	0.6465	0.6322
Initial engagement	0.7074	0.6877	0.7212	0.7233	0.7099
Final engagement	0.5642	0.497	0.5854	0.5689	0.5539
Engagement numeric diff	−0.1432	−0.1907	−0.1358	−0.1544	−0.1560
Engagement % diff	−20.2431	−27.7301	−18.8297	−21.3466	−22.0374

4 Conclusion and Future Work

The algorithm was developed for long periods of time, and the genetic algorithm based approach fits well with the real scenario of a daily usage of a Smart Environment. Unfortunately, due to time limitation, the practical tests were not able to simulate the real usage over time. Since the tasks were developed to be repetitive and boring as a way of affecting the user's engagement, longer tests would be discomfort and demand a lot of effort from the user.

Despite the fact that it was not possible to perform tests with the time that the system was intended to be used, the crossover tools, in which new solutions come from previous ones proven to be good (the offspring), we were able to develop an algorithm that in the majority of cases has a positive outcome, raising the Engagement of the user in a repetitive activity.

Since the smart environment was developed to be used in a daily basis, the time restrictions presented in the test will not be an issue and the test gives good indication that the genetic algorithm based approach can be a strong tool to adapt the smart environment in order to increase the user's Engagement levels.

It is also clear that in comparison to random executions, the genetic algorithm approach had a better outcome. This excludes the hypotheses that the gain in the Engagement level is due solely to frequent changes in the scenario, and not due to the system's adaptation according to the user's Engagement.

As future work, we intend to add user profiles, which can add great benefits to the system's usability. With them it would be possible to get closer to the optimal set of solutions. For a following execution with the same user, we would have information about previous states that seem to be preferred by the user, improving the efficacy and efficiency of the system and avoiding unnecessary re-testing.

When we refer to the optimal set of solutions we can only theorize about it. It is not possible to reach an optimal set of solutions in a real execution. As part of the solution we have the attention of the user, which is a volatile parameter, and cannot, and is not treated as a concise and immutable number. Two different executions with the same configurations can vary according to a number of feelings and thoughts that the user may have. Therefore, the observations about the theoretical optimizations are only conjectures.

Acknowledgment. The authors would like to thank the Home Systems Company, as well as the Brazilian research agencies Capes (project PROCAD), FINEP (project CRIAI) and CNPq for their financial support.

References

1. Bäck, T.: Evolutionary Algorithms in Theory and Practice: Evolution Strategies, Evolutionary Programming, Genetic Algorithms. Oxford University Press, Oxford (1996)
2. Girouard, A.: Adaptive brain-computer interface. In: CHI 09 Extended Abstracts on Human Factors in Computing Systems (CHI EA 2009), USA. ACM (2009)

3. Zander, T.O., Kothe, C., Welke, S., Roetting, M.: Utilizing Secondary Input from Passive Brain-Computer Interfaces for Enhancing Human-Machine Interaction. Springer, Heidelberg (2009)
4. Schmidt, A., et al.: Enabling implicit human computer interaction: a wearable RFID-tag reader. In: ISWC, USA. IEEE Computer Society (2000)
5. George, L., Lcuyer, A.: An overview of research on passive brain-computer interfaces for implicit human-computer interaction. In: ICABB 2010-Workshop W1 (2010)
6. Whitley, D.: A Genetic Algorithm Tutorial, Computer Science Department, Colorado State University (1989)
7. Sivanandam, S.N., Deepa, S.N.: Introduction to Genetic Algorithms. Springer, Heidelberg (2008)
8. Cutrell, E., Tan, D.: BCI for Passive Input in HCI. Microsoft Research, USA (2008)
9. Ruscher, G., Kruger, F., Bader, S., Kirste, T.: Controlling smart environments using brain computer interface. In: Proceedings of the 2nd Workshop on Semantic Models for Adaptive Interactive Systems, SEMAIS 2011 (2011)
10. Haupt, L., Haupt, S.E.: Optimum population size and mutation rate for a simple real genetic algorithm that optimizes array factors. Appl. Comput. Electromagn. Soc. J. **15**(2), 92–102 (2000)
11. Menezes, M., Pereira, C.: Proposed Use of Passive Brain-Computer Interface. Universidade Federal do Rio Grande do Sul (UFRGS) Porto Alegre, RS/Brazil (2015)
12. Emotiv Software Research Kit User Manual. Emotiv Systems (2014)
13. Schettini, F., et al.: Assistive device with conventional, alternative, and brain-computer interface inputs to enhance interaction with the environment for people with amyotrophic lateral sclerosis: a feasibility and usability study. Arch. Phys. Med. Rehab. **96**(3), S46–S53 (2014)
14. Neuman, M.R.: Biopotential amplifiers. In: Webster, J.G. (ed.) Medical Instrumentation, pp. 227–288. John Wiley and Sons, New York (1995)
15. Edlinger, G., Holzner, C., Guger, C.: A hybrid brain-computer interface for smart home control. In: Jacko, J.A. (ed.) Human-Computer Interaction, Part II, HCII 2011. LNCS, vol. 6762, pp. 417–426. Springer, Heidelberg (2011)
16. Zich, C., De Vos, M., Kranczioch, C., Debener, S.: Wireless EEG with individualized channel layout enables efficient motor imagery training. Clin. Neurophysiol. **126**(4), 698–710 (2015)
17. Castermans, T.: Detecting Biosignals with Emotiv EPOC headset: a critical review. Universit de Mons, web presentation (2011). http://tinyurl.com/p587qr7
18. Cook, D., Das, S.: Smart Environments. Wiley, New York (2005)
19. Ducatel, K., et al.: Scenarios for ambient intelligence (ISTAG Report). Institute for Prospective Technological Studies (European Commission), Seville (2001)
20. Lin, C.-T., Lin, F.-C., Chen, S.-A., Lu, S.-W., Chen, T.-C., Ko, L.-W.: EEG-based brain-computer interface for smart living environmental auto-adjustment. J. Med. Biol. Eng. **30**, 237–245 (2010)
21. Ramirez-Atencia, C., et al.: A hybrid MOGA-CSP for multi-UAV mission planning. In: GECCO (2015)
22. Ramrez-Atencia, C., Bello-Orgaz, G., R-Moreno, M.D., Camacho, D.: Branching to find feasible solutions in unmanned air vehicle mission planning. In: Corchado, E., Lozano, J.A., Quintián, H., Yin, H. (eds.) IDEAL 2014. LNCS, vol. 8669, pp. 286–294. Springer, Heidelberg (2014)

Comparison of Floor Detection Approaches for Suburban Area

Marcin Luckner(✉) and Rafał Górak

Faculty of Mathematics and Information Science, Warsaw University of Technology, ul. Koszykowa 75, 00-662 Warszawa, Poland
{mluckner,R.Gorak}@mini.pw.edu.pl

Abstract. As a part of smart-buildings, indoor localisation systems – alternative to Global Positioning System localisation – bring constantly improving results. Several localisation methods works with a horizontal localisation error less than few meters. However, for small suburban houses, horizontal localisation is not as important as detection of the current floor, which in is still a challenge in multi-storey buildings. This paper compares several approaches that can be used in fingerprinting-based floor detection systems. The tests include the following fingerprints: pressure measures, Wi-Fi signals, and two generations of cellular networks signals. The tests have been done in the suburban 3-storey building with underdeveloped Wi-Fi and cellular infrastructure. Notwithstanding, the floor detection based on Received Signal Strength from both infrastructures reached from 98 to 100 %. Additionally, we showed that differences in the number of measures and differences in the number of received signals were not a major factor that influenced on accuracy.

1 Introduction

One of aspects of smart buildings is a indoor localisation system that allows the managers to localise people or objects in the building. Most popular localisation methods use Wi-Fi signal strengths [12,13,16,17] and cellular networks strengths [1,3,4,14].

Indoor localisation is mostly associated with horizontal localisation, but floor detection is also important part of issue. When the horizontal localisation estimates x and y coordinates, the floor detection estimates a floor number. The indoor horizontal localisation brings very good results with mean error smaller than 2 m, but the floor detection is still a challenge.

In this work, we want to show how smart buildings localisation solutions dedicated for commercial or academic buildings can be applied for single-family housing. There are several important differences in needs and possibilities when we compare both tasks. First, in a relatively small house horizontal localisation is not as important as floor detection. In most cases it is enough to know a current floor to localise person or object. Second, houses are mostly placed on a suburb. Therefore, both Wi-Fi and cellular network infrastructures – that can

N.T. Nguyen et al. (Eds.): ACIIDS 2016, Part II, LNAI 9622, pp. 782–791, 2016.
DOI: 10.1007/978-3-662-49390-8_76

be used in localisation systems – are underdeveloped. Last, a height of the floors in a house can be lower than in public buildings.

The selection of floor detection methods was based on two different solutions of the floor detection issue. The paper [15] presented a cellular network fingerprinting-based system, which determines the current floor on which a user with a mobile phone is located. The tests were done in 9 to 16-storey buildings. The system classified the floor correctly up to 73 % of cases and nearly 100 % of measures were localised with an error less or equal tree floor. An alternative solution presented in [9] used pressure for floor detection. The proposed system could accurately determine the floor in a 4-storey building. Unfortunately, the methods were not compared on the same testing data.

We tested approaches based on Wi-Fi, cellular network, and pressure. The cellular test have been done using 2nd generation Global System for Mobile Communications (GSM) and 3rd generation Universal Mobile Telecommunications System (UMTS) networks.

Model based on random forest showed that the obtained accuracy could reach from 98 to 100 % for Wi-Fi, GSM, and UMTS. The pressure results are worse, but the error mostly does not exceed one floor.

The remaining part of the paper is organised as follows: Sect. 2 presents basic facts about the analysed data and localisation methods. Section 3 describes measures and collected data, Sect. 4 presents the floor detection results. The work is concluded in Sect. 5.

2 Preliminaries and Notation

This section will briefly describe the fingerprinting method, which is most common in indoor localisation based on Received Signal Strength (RSS).

Data sets consist of vectors of the signal strengths $\mathcal{F} \in \mathbb{R}^n$ where n depends on the kind of measured signals. The vectors are labelled by the floor of measures f. The elements of the data sets are commonly called fingerprints.

The fingerprinting approach for indoor localisation builds a model based on data set of training fingerprints by mapping vectors of signal strengths (fingerprint) to its position. In our case the position is limited to a floor.

The model can be built by several machine learning methods. In work [10] multilayer perceptron was used and in [7] k-Nearest Neighbours was tested, but we considered random forests method [2].

The function $\mathcal{F} \rightarrow f$ that estimates the floor f on the base of a single fingerprint $\mathcal{F}$ is implemented as an ensemble of decision trees. The ensemble contains multiple estimation or prediction trees created on the base of various fragments of the learning set. A linear combination of the results obtained by the trees produces the final, aggregated estimator or prediction. Such ensemble is called a boosting algorithm.

We tested AdaBoostM2 [5] algorithm, where weighted pseudo-loss are calculated for N observations and K classes

(a) measuring area (b) measuring set

Fig. 1. Measuring area and equipment

$$\epsilon_t = \frac{1}{2} \sum_{n=1}^{N} \sum_{k \neq y_n} d_{n,k}^{(t)} (1 - h_t(x_n, y_n) + h_t(x_n, k)), \tag{1}$$

where $h_t(x_n, k)$ is the confidence of prediction, $d_{n,k}^{(t)}$ are observation weights, and y_n is the true class label. Pseudo-loss is a measure of the classification accuracy from any classifier in an ensemble.

We selected the number of trees in the random forest to be 30. We checked that growing more trees does not improve the accuracy of the floor detection.

To evaluate the model we used the following method. By $\hat{f_{\mathcal{F}}}$ we denote the estimator given by the built random forest model for the fingerprint $\mathcal{F}$.

A floor error for the fingerprint $\mathcal{F} \in \mathbb{R}^n$ is defined as

$$fe(\mathcal{F}) = |f - \hat{f_{\mathcal{F}}}| \tag{2}$$

For the test set $\mathcal{T}$ the floor detection is evaluated by accuracy (**ACC**):

$$\mathbf{ACC} = \frac{\sum_{\mathcal{F} \in \mathcal{T}} (1 - \mathrm{sgn}(fe(\mathcal{F})))}{|\mathcal{T}|} \tag{3}$$

In this formula $\mathrm{sgn} : \mathbb{R} \to \{-1, 0, 1\}$, $\mathrm{sgn}(x) = 0 \Leftrightarrow x = 0$.

3 Data

The data were collected in a three floor suburban building (including the ground floor). The building has irregular shape and its outer dimensions are around 15 by 9 m. Figure 1a shows the building and its localisation.

A testing mobile phone was LG Nexus 4 working with Android 4.2 Jelly Bean. This phone was co-developed by Google. Therefore, the operating system was not modified by an additional brand API. The application created for the localisation system [11] collected all data.

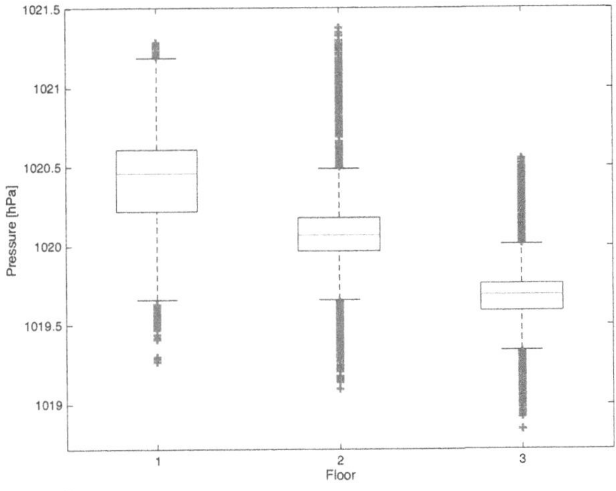

Fig. 2. Measures of pressure on various floors

The mobile phone was attached to a vacuum cleaning robot. The robot started separately on each floor. It was working the whole cleaning cycle by approximately one hour per floor. The measuring set is presented on Fig. 1b.

Data were collected in two days separately from two cellular telecommunication networks: 2nd generation (2G) and 3rd generation (3G). Data sets were labelled as GSM (Global System for Mobile Communications) and UMTS (The Universal Mobile Telecommunications System) respectively. During the measures we collected also data about Wi-Fi signals and pressure. All data were randomly split into two equal parts: the learning set and the testing set.

3.1 Pressure

The pressure was measured 13061 times. Among measures 5457 were taken on the ground floor, on the first floor 4275 measures were taken, and in the attic 3329 measures were taken.

In this case, a fingerprint record contains only one measure. Figure 2 shows distribution of pressure on various floors (1 - the ground, 2 - the first floor, 3 - the attic). Values are given in hectopascals.

On each box, the central mark is the median, the edges of the box are the 25th and 75th percentiles, the whiskers extend to the most extreme data points not considered outliers, and outliers are plotted individually. The 25th and 75th percentiles boxes are well separated, but we can observe many outliers.

3.2 GSM

The cellular signals (GSM) were measured at the same time as the pressure, but not necessary with the same delay between measures. Therefore, 39321 measures

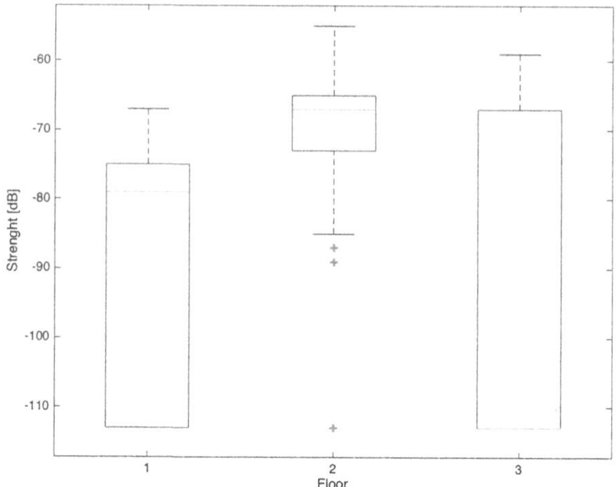

Fig. 3. Measures of a single GSM signal on various floors

were taken. Among measures 16001 were taken on the ground floor, on the first floor 10420 measures were taken, and in the attic 12900 measures were taken.

We observed signals from 12 unique base stations. However – according to Android limitations – each fingerprint contained no more than 7 signals. The obtained values are given in decibel-milliwatts (dBm). According to specification the maximum value is −51 dBm and the minimum value is −113 dBm.

Figure 3 shows the measures of one GSM signal on various floors. The signal was selected on the base of features importance in the floor detection task. In the classification task the Gini coefficient is used to estimate how the data space in the node is divided among classes. The Gini coefficient equals $2(AUC) - 1$. Where AUC is the area underneath the Receiver Operating Characteristic Curve (ROC Curve).

For all floors we can observe a disappearance of signals: the outliers that lie on the level −113 dBm. Signals are strongest on the highest floor, but the variability grows at the same time.

3.3 UMTS

The measures of UMTS signals brought several serious issue. First, the mobile device did not allow the user to choose 3G permanently. The only option is to choose both 2G and 3G. When the coverage of UMTS network is weak the system changes 3G network into 2G network automatically.

Second – according to the Android documentation – for GSM, the measures are returned as an *asu* ranging from 0 to 31. It is used to calculate RSS as $dBm = -113 + 2 * asu$. When *asu* equals 0 then the RSS is −113 dBm or less and 31 means −51 dBm or greater. For UMTS returned value is the absolute power level of the Common Pilot Channel (CPICH) as received by the UE.

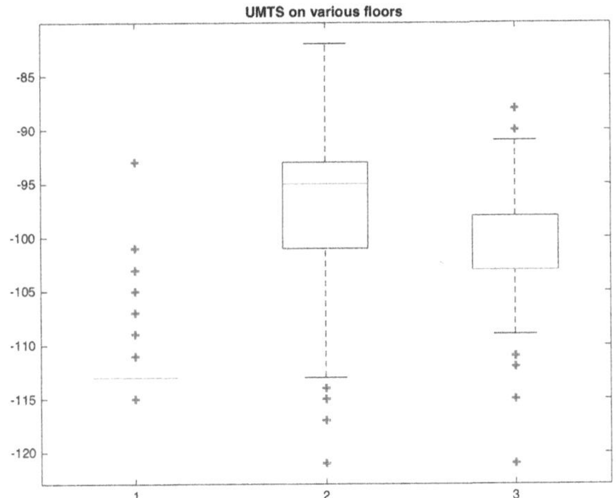

Fig. 4. Measures of a single UMTS signal on various floors

The reporting range for CPICH Received Signal Code Power (RSCP) is from −120 dBm to −25 dBm, but returned values are from the range −5 to 91 and RSS is calculated as $dBm = -115 + value$ with some modifications for border values. As we see the ranges overlaps each other.

For nine unique UMTS signals sources 20311 measures were taken. Among measures 4996 were taken on the ground floor, on the first floor 12381 measures were taken, and in the attic 2934 measures were taken.

Figure 4 shows the measures of the most important UMTS signal on various floors. The values are given in dBm. Several measures exceed the minimum value from the 3GPP specification. Most of the measures on the ground floor return −113 dBm. That means a lack of the signal and can be an important discriminant for elimination of a potential choice in the floor detection task.

3.4 Wi-Fi

The Wi-Fi signals were measured at the same time as the GMS signals and pressure. Among 29459 measures 12074 were taken on the ground floor, on the first floor 8327 measures were taken, and in the attic 9058 measures were taken.

A fingerprint record consists of 13 measures. Only three of the observed access points belonged to the local infrastructure. Two of them were localised in the attic, one on the ground floor. The localisation of the other devices is unknown.

Figure 5 shows the measures of the most important Wi-Fi signal on various floors. Values are given in dBm. The measures are well separated and without far outliers.

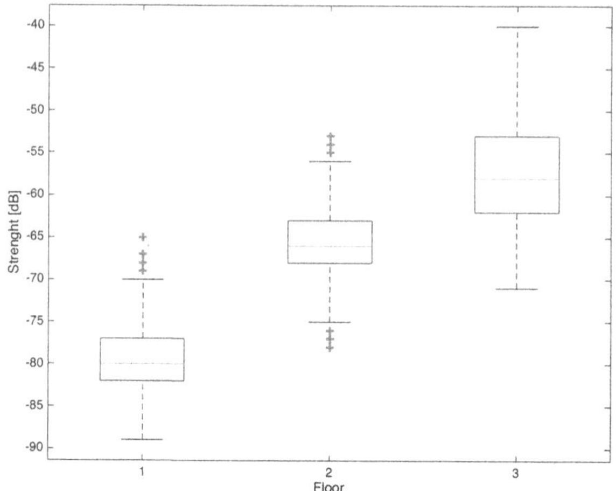

Fig. 5. Measures of a single Wi-Fi signal on various floors

4 Results

The random forest floor detection was tested on 50 % of collected measures. The rest was used to train the model.

Table 1 presents the floor detection accuracy calculated by formula (3). Additionally, the percentage of one-floor errors and two-floor errors are given.

Pressure gave the smallest accuracy and the biggest errors. However, we should remember that the fingerprint in this case contains only one measure.

The detections based on Wi-Fi and GSM have the same accuracy. It is not a typical result. Usually, GSM floor detection is worse than Wi-Fi detection. In case of suburban area the number of the detected sources of signals is relatively small. We observed 13 unique Wi-Fi access points and 12 unique GSM base stations. For comparison in the city center in academic building we observed 46 Wi-Fi access points [6] and 36 GSM base stations [8]. When most of signals have a similar strength, the reduction of the dimension of fingerprints vectors to 7 brings a reduction of accuracy by elimination of important information.

The UMTS based approach reached 100 % accuracy. That was caused by high differences in the receiving signals among the floors.

Table 1. Accuracy and errors for various types of signals

Method	Accuracy [%]	One-floor error [%]	Two-floor error [%]
Pressure	74	24	3
GSM	98	2	0
Wi-Fi	99	1	0
UMTS	100	0	0

Table 2. Accuracy and errors for single signal

Method	Accuracy [%]	One-floor error [%]	Two-floor error [%]
Pressure	74	24	3
GSM	78	20	2
Wi-Fi	84	16	0
UMTS	89	10	1

All described signal based methods gave better results that the detection based on pressure. However, for them the collected fingerprints contained from 9 to 13 values. Therefore, in the next test we compared all approaches using only one – the most important – signal.

Table 2 presents the results of the one signal detection. In all cases the accuracy was better than for pressure. However, the result for GSM are very close to the pressure ones.

During measures various numbers of different signals were measured on the floors. Therefore, to compare all approaches we performed the last test on the equal number of measures for all methods. The number of measures was limited to 10900. Among them 4000 were taken on the ground floor, 4000 were taken on the first floor and 2900 were taken in the attic.

Table 3 shows that the methods based on Wi-Fi and UMTS gave the same results that for the full set of measures. The small reduction is visible for the GSM based method, which had very limited set of measures in comparison to the origin set. The biggest reduction is on the pressure result. However, even the biggest reduction is no greater than 2 %. Therefore, we cannot say that the differences in the number of measures were important.

Table 3. Accuracy and errors for equal number of measures

Method	Accuracy [%]	One-floor error [%]	Two-floor error [%]
Pressure	72	26	2
GSM	97	3	0
Wi-Fi	99	1	0
UMTS	100	0	0

5 Conclusion

We compared several approaches to the floor detection problem. In this task we used the fingerprinting-based method to recognise on which floor a mobile device is. The tests have been done in 3-storey building at suburban area. In all tests the random forest were used as a detection model.

The first approach used measures of pressure to detect the floor. The obtained accuracy was 74 %. The next two approaches used Wi-Fi and GSM signals.

Collected fingerprints contained 13 and 12 measures respectively. The obtained accuracy was 98–99 %.

The results obtained by the GSM based method are very high in comparison to the Wi-Fi based method. Usually, a difference between these approaches is much bigger. We acknowledged that the high GSM result was caused by the small dimension of the collected fingerprints vectors. The reduction of the GSM fingerprints dimension to 7 that was done by Android system was not such important in this case as when the number of observed unique base stations is much bigger.

Even if the UMTS signals are ambiguous, because of automatic change between 2G and 3G cellular network, the obtained accuracy was the best among compared models. This approach recognised all measures without any mistake.

To conclude, the floor detection task examined in this paper can be solved with very high accuracy using GSM, Wi-Fi, or UMTS signals. In future work we want to improve pressure localisation results and check a stability of the system among time.

Acknowledgment. The research is supported by the National Centre for Research and Development, grant No. PBS2/B3/24/2014, application No. 208921.

References

1. Ahriz, I., Oussar, Y., Denby, B., Dreyfus, G.: Carrier relevance study for indoor localization using GSM. In: 7th Workshop on Positioning Navigation and Communication (WPNC 2010), pp. 168–173, March 2010
2. Breiman, L.: Random forests. Mach. Learn. **45**(1), 5–32 (2001)
3. Brida, P., Cepel, P., Duha, J.: The accuracy of RSS based positioning in GSM networks. In: International Conference on Microwaves, Radar Wireless Communications, MIKON 2006, pp. 541–544, May 2006
4. Denby, B., Oussar, Y., Ahriz, I., Dreyfus, G.: High-performance indoor localization with full-band GSM fingerprints. In: IEEE International Conference on Communications Workshops, ICC Workshops 2009, pp. 1–5, June 2009
5. Freund, Y., Schapire, R.E.: A decision-theoretic generalization of on-line learning and an application to boosting. J. Comput. Syst. Sci. **55**(1), 119–139 (1997). http://www.sciencedirect.com/science/article/pii/S002200009791504X
6. Górak, R., Luckner, M.: Malfunction immune Wi–Fi localisation method. In: Núñez, M., Nguyen, N.T., Camacho, D., Trawinski, B. (eds.) ICCCI 2015. LNCS, vol. 9329, pp. 328–337. Springer, Heidelberg (2015). doi:10.1007/978-3-319-24069-5_31
7. Grzenda, M.: On the prediction of floor identification credibility in RSS-based positioning techniques. In: Ali, M., Bosse, T., Hindriks, K.V., Hoogendoorn, M., Jonker, C.M., Treur, J. (eds.) IEA/AIE 2013. LNCS, vol. 7906, pp. 610–619. Springer, Heidelberg (2013). http://dx.doi.org/10.1007/978-3-642-38577-3_63
8. Grzenda, M.: Reduction of signal strength data for fingerprinting-based indoor positioning. In: Jackowski, K., Burduk, R., Walkowiak, K., Wozniak, M., Yin, H. (eds.) IDEAL 2015. LNCS, vol. 9375, pp. 387–394. Springer, Heidelberg (2015). doi:10.1007/978-3-319-24834-9_45

9. He, N., Huo, J., Dong, Y., Li, Y., Yu, Y., Ren, Y.: Atmospheric pressure-aware seamless 3-d localization and navigation for mobile internet devices. Tsinghua Sci. Technol. **17**(2), 172–178 (2012)
10. Karwowski, J., Okulewicz, M., Legierski, J.: Application of particle swarm optimization algorithm to neural network training process in the localization of the mobile terminal. In: Iliadis, L., Papadopoulos, H., Jayne, C. (eds.) EANN 2013, Part I. CCIS, vol. 383, pp. 122–131. Springer, Heidelberg (2013). http://dx.doi.org/10.1007/978-3-642-41013-0_13
11. Korbel, P., Wawrzyniak, P., Grabowski, S., Krasinska, D.: Locfusion API - programming interface for accurate multi-source mobile terminal positioning. In: Federated Conference on Computer Science and Information Systems (FedCSIS), pp. 819–823, September 2013
12. Papapostolou, A., Chaouchi, H.: Scene analysis indoor positioning enhancements. Annales des Télécommunications **66**, 519–533 (2011)
13. Roos, T., Myllymaki, P., Tirri, H., Misikangas, P., Sievanen, J.: A probabilistic approach to WLAN user location estimation. Int. J. Wireless Inf. Netw. **9**(3), 155–164 (2002)
14. Tian, Y., Denby, B., Ahriz, I., Roussel, P., Dreyfus, G.: Hybrid indoor localization using GSM fingerprints, embedded sensors and a particle filter. In: 2014 11th International Symposium on Wireless Communications Systems (ISWCS), pp. 542–547, August 2014
15. Varshavsky, A., LaMarca, A., Hightower, J., de Lara, E.: The skyloc floor localization system. In: Fifth Annual IEEE International Conference on Pervasive Computing and Communications, PerCom 2007, pp. 125–134, March 2007
16. Wang, J., Hu, A., Liu, C., Li, X.: A floor-map-aided wifi/pseudo-odometry integration algorithm for an indoor positioning system. Sensors **15**(4), 7096 (2015). http://www.mdpi.com/1424-8220/15/4/7096
17. Xiang, Z., Song, S., Chen, J., Wang, H., Huang, J., Gao, X.G.: A wireless LAN-based indoor positioning technology. IBM J. Res. Dev. **48**(5–6), 617–626 (2004)

Author Index

Printed in the United States
By Bookmasters